Acta Numerica 2002

Acta Numerica

Volume 11 2002

CAMBRIDGE
UNIVERSITY PRESS

CAMBRIDGE UNIVERSITY PRESS
Cambridge, New York, Melbourne, Madrid, Cape Town,
Singapore, São Paulo, Delhi, Tokyo, Mexico City

Cambridge University Press
The Edinburgh Building, Cambridge CB2 8RU, UK

Published in the United States of America by Cambridge University Press, New York

www.cambridge.org
Information on this title: www.cambridge.org/9780521174305

First published 2002
First paperback edition 2011

A catalogue record for this publication is available from the British Library

ISBN 978-0-521-81876-6 Hardback
ISBN 978-0-521-17430-5 Paperback

Contents

Acta Numerica (2002), pp. 1–71
DOI: 10.1017/S0962492902000014

Structured inverse eigenvalue problems

Moody T. Chu*
Department of Mathematics,
North Carolina State University,
Raleigh, North Carolina, NC 27695-8205, USA
E-mail: chu@math.ncsu.edu

Gene H. Golub†
Department of Computer Science,
Stanford University,
Stanford, California, CA 94305-9025, USA
E-mail: golub@stanford.edu

An inverse eigenvalue problem concerns the reconstruction of a structured matrix from prescribed spectral data. Such an inverse problem arises in many applications where parameters of a certain physical system are to be determined from the knowledge or expectation of its dynamical behaviour. Spectral information is entailed because the dynamical behaviour is often governed by the underlying natural frequencies and normal modes. Structural stipulation is designated because the physical system is often subject to some feasibility constraints. The spectral data involved may consist of complete or only partial information on eigenvalues or eigenvectors. The structure embodied by the matrices can take many forms. The objective of an inverse eigenvalue problem is to construct a matrix that maintains both the specific structure as well as the given spectral property. In this expository paper the emphasis is to provide an overview of the vast scope of this intriguing problem, treating some of its many applications, its mathematical properties, and a variety of numerical techniques.

* This research was supported in part by NSF grants DMS-9803759 and DMS-0073056.
† This work was supported in part by the NSF grant CCR-9971010.

CONTENTS

1. Introduction

In his book *Finite-Dimensional Vector Spaces*, Halmos (1974) wrote:

> Almost every combination of the adjectives proper, latent, characteristic, eigen and secular, with the nouns root, number and value, has been used in the literature for what we call a proper value.

This interesting comment on the nomenclature of eigenvalue echoes the enigmatic yet important role that eigenvalues play in nature. One instance, according to Parlett (1998), is that 'Vibrations are everywhere, and so too are the eigenvalues associated with them.' For that reason, considerable research effort has been expended on eigenvalue computation, especially in the context of matrices. The applications of this research furnish critical insight into the understanding of many vital physical systems.

The process of analysing and deriving the spectral information and, hence, inferring the dynamical behaviour of a system from *a priori* known physical parameters such as mass, length, elasticity, inductance, capacitance, and so on is referred to as a *direct* problem. The *inverse* problem then is to validate, determine, or estimate the parameters of the system according to its observed or expected behaviour. Specifically, in the context of matrices again, an inverse eigenvalue problem (**IEP**) concerns the reconstruction of a matrix from prescribed spectral data.

It is clear that the IEP could trivially be solved if the matrices were subject to no restriction on structure. For the problem to be more significant, either physically or mathematically, it is often necessary to confine the construction to certain special classes of matrices. Matrices with a specified structure, for example, constitute a special class. Thus an IEP is indeed a *structured* IEP (**SIEP**). The solution to an IEP should satisfy two constraints: the *spectral constraint*, referring to the prescribed spectral data, and the *structural constraint*, referring to the desirable structure. The variation of these constraints defines the variety of IEPs, some of which will be surveyed in this paper.

More should be said about these constraints in order to define an IEP. First we recall one condition under which two geometric entities intersect transversally. Loosely speaking, we may assume that the structural constraint and the spectral constraint define, respectively, smooth manifolds in the space of matrices of a fixed size. If the sum of the dimensions of these two manifolds exceeds the dimension of the ambient space, then under some mild conditions one can argue that the two manifolds must intersect and the IEP must have a solution. A more challenging situation is when the sum of dimensions emerging from both structural and spectral constraints does not add up to the transversal property. In that case, it is much harder to tell whether or not an IEP is solvable. Secondly we note that in a complicated physical system it is not always possible to know the entire spectrum. On the other hand, especially in structural design, it is often demanded that certain eigenvectors should also satisfy some specific conditions. The spectral constraints involved in an IEP, therefore, may consist of complete or only partial information on eigenvalues or eigenvectors. We further observe that, in practice, it may occur that one of the two constraints in an IEP should be enforced more critically than the other, due to the physical realizability, say. Without this, the physical system simply cannot be built. There are also situations when one constraint could be more relaxed than the other, due to the physical uncertainty, say. The uncertainty arises when there is simply no accurate way to measure the spectrum, or no reasonable means to obtain all the information. When the two constraints cannot be satisfied simultaneously, the IEP could be formulated in a least squares setting, in which a decision is made as to which constraint could be compromised.

Associated with any IEP are four fundamental questions. These are issues concerning:

- the theory of *solvability*,
- the practice of *computability*,
- the analysis of *sensitivity*, and
- the reality of *applicability*.

A major effort in solvability has been to determine a necessary or a sufficient condition under which an inverse eigenvalue problem has a solution. The main concern in computability, on the other hand, has been to develop a procedure by which, knowing *a priori* that the given spectral data are feasible, a matrix can be constructed in a numerically stable fashion. The discussion on sensitivity concerns how the solution to an IEP is modified by changes in the spectral data. The applicability is a matter of differentiation between whether the given data are exact or approximate, complete or incomplete, and whether an exact value or only an estimate of the parameters of the physical system is needed. Each of these four questions is essential but challenging to the understanding of a given IEP. We are not aware of many IEPs that are comprehensively understood in all these four aspects. Rather, considerably more work remains to be done. For the very same reason, we cannot possibly treat each IEP evenly in this article.

With different emphases and different formulations, studies of IEPs have been intensive and scattered, ranging from acquiring a pragmatic solution to a real-world application to discussing the general theory of an abstract formulation. A timely review that better defines the realm of IEPs as a whole is critical for further research and understanding. Earlier endeavours in this regard include the book by Gladwell (1986*b*), where the emphasis was on applied mechanics, the survey by Boley and Golub (1987), where the emphasis was on numerical computation, the book by Zhou and Dai (1991), which pointed to many publications in Chinese that were perhaps unknown to the West, and the article by Gladwell (1996), which reviewed activities and literature between 1985 and 1995 as a ten-year update of his previous book. In a recent review article, Chu (1998) briefly described a collection of thirty-nine IEPs. These problems were categorized roughly according to their characteristics into three types of IEPs, *i.e.*, *parametrized* (**PIEP**), *structured* (**SIEP**), and *partially described* (**PDIEP**). Since then, many more old results have been unearthed, while new articles have continued to appear, notably the treatise by Ikramov and Chugunov (2000), translated from Russian with the emphasis on finitely solvable IEPs and rational algorithms, and the book by Xu (1998), where many results on the sensitivity issue by Chinese mathematicians are made known in English for the first time. It quickly becomes clear that even for SIEPs alone there is a need to update history and describe recent developments in both theory and application. It is for this purpose that this paper is presented.

Although every IEP should be regarded as an SIEP, that view is certainly too broad to be apprehended by a paper of finite length. Thus, our definition of 'structure' is limited to those structures delineated in this paper. Some of these structures, such as Jacobi or Toeplitz, result in matrices forming linear subspaces; some structures, such as nonnegative or stochastic, limit entries of matrices in a certain range; while others, such as matrices with prescribed entries or with prescribed singular values, lead to some implicitly defined structural constraints. We shall touch upon a variety of SIEPs by describing their formulations, highlighting some theories or numerical procedures, and suggesting some relevant references. Additionally, we shall outline some applications of IEPs from selected areas of disciplines. From time to time, we shall point out some open questions. Let it be noted that, while we sometimes seem to be concentrating on one particular numerical method applied to one particular problem, often the method has enough generality that, with some suitable modifications, it can also be applied to other types of problems. We choose not to encumber readers with the details.

We hope that this presentation, along with previous treatments mentioned above, will help to inspire some additional interest and to stimulate further research that ultimately will lead to a better understanding of this fascinating subject of IEPs.

2. Applications

Inverse eigenvalue problems arise in a remarkable variety of applications. The list includes, but is not limited to, control design, system identification, seismic tomography, principal component analysis, exploration and remote sensing, antenna array processing, geophysics, molecular spectroscopy, particle physics, structural analysis, circuit theory, and mechanical system simulation. In this section we briefly highlight a few applications that, in our judgement, should be of general interest to the readers. So as not to lose sight of the notion of an IEP, it is clear that we have to sacrifice technical details in the description of these applications. We shall divide the discussions into five categories: pole assignment problem, applied mechanics, inverse Sturm–Liouville problem, applied physics, and numerical analysis. Each category covers additional problems.

A common phenomenon that stands out in most of these applications is that the physical parameters of the underlying system are to be reconstructed from knowledge of its dynamical behaviour. The dynamical behaviour is affected by spectral properties in various ways. Vibrations depend on natural frequencies and normal modes, stability controls depend on the location of eigenvalues, and so on. If the physical parameters can be described mathematically in the form of a matrix (as they often are), then we have an IEP. The structure of the matrix is usually inherited from the physical properties of the underlying system.

2.1. Pole assignment problem

Consider first the following dynamic state equation:

$$\dot{\mathbf{x}}(t) = A\mathbf{x}(t) + B\mathbf{u}(t), \tag{2.1}$$

where $\mathbf{x}(t) \in \mathbb{R}^n$ denotes the state of a certain physical system to be controlled by the input $\mathbf{u}(t) \in \mathbb{R}^m$. The two given matrices $A \in \mathbb{R}^{n\times n}$ and $B \in \mathbb{R}^{n\times m}$ are invariant in time. One classical problem in control theory is to select the input $\mathbf{u}(t)$ so that the dynamics of the resulting $\mathbf{x}(t)$ is driven into a certain desired state. Depending on how the input $\mathbf{u}(t)$ is calculated, there are generally two types of controls, both of which have been extensively studied and documented in the literature.

In the state feedback control, the input $\mathbf{u}(t)$ is selected as a linear function of the current state $\mathbf{x}(t)$, that is,

$$\mathbf{u}(t) = F\mathbf{x}(t). \tag{2.2}$$

In this way, the system (2.1) is changed to a closed-loop dynamical system:

$$\dot{\mathbf{x}}(t) = (A + BF)\mathbf{x}(t). \tag{2.3}$$

A general goal in such a control scheme is to choose the *gain matrix* $F \in \mathbb{R}^{m\times n}$ so as to achieve stability and to speed up response. To accomplish this goal, the problem can be translated into choosing F so as to reassign eigenvalues of the matrix $A + BF$. This type of problem is usually referred to in the literature as a (state feedback) pole assignment problem (**PAP**). It should be pointed out that, in contrast to what we described earlier for an IEP, the matrix F in the context of PAPs does not usually carry any further structure at all. A PAP will become a much harder IEP if F needs to satisfy a certain structural constraint.

It is often the case in practice that the state $\mathbf{x}(t)$ is not directly observable. Instead, only the output $\mathbf{y}(t)$ that is related to $\mathbf{x}(t)$ via

$$\mathbf{y}(t) = C\mathbf{x}(t) \tag{2.4}$$

is available. In the above, $C \in \mathbb{R}^{p\times n}$ is a known matrix. The input $\mathbf{u}(t)$ now must be chosen as a linear function of the current output $\mathbf{y}(t)$, that is,

$$\mathbf{u}(t) = K\mathbf{y}(t). \tag{2.5}$$

The closed-loop dynamical system thus becomes

$$\dot{\mathbf{x}}(t) = (A + BKC)\mathbf{x}(t). \tag{2.6}$$

The goal now is to select the *output matrix* $K \in \mathbb{R}^{m\times p}$ so as to reassign the eigenvalues of $A + BKC$. This output feedback PAP once again gives rise to a special type of IEP (with no constraint on the structure of K).

There is a vast literature of research on the subject of PAPs alone. We would suggest the papers by Byrnes (1989) and by Kautsky, Nichols and Van Dooren (1985), which gave an excellent account of activities in this area as a starting point for further exploration. We shall see later that PAPs are a special case of what we call PIEPs in this article.

One important remark should be made at this point. PAPs, as well as many other IEPs, usually have multiple solutions. Among these multiple solutions, the one that is *least* sensitive to perturbations of problem data is perhaps most critical from a practical point of view. Such a solution, termed the *robust solution* in the literature, is usually found by minimizing the condition number associated with the solution. In other words, there are two levels of work when solving an IEP for a robust solution: The first is to develop a means to find a solution, if there is any at all; the second is to use optimization techniques to minimize the condition number associated with the solution. Most of the numerical methods discussed in this paper are for the first task only. Except for PAPs (Kautsky *et al.* 1985), the second task for general IEPs has not been fully explored as yet.

For the state feedback problem, there has also been some interest in the case where K is structured. One such application is the so-called *decentralized dynamic system*, where K is a diagonal matrix. Some background information can be found in a recent paper by Ravi, Rosenthal and Wang (1995). Numerical algorithms are needed for this type of structured problems.

2.2. Applied mechanics

Interpreting the word 'vibration' in a broad sense, we see applied mechanics everywhere. The transverse motion of masses on a string, the buckling of structures, the transient current of electric circuits, and the acoustic sound in a tube are just a few instances of vibration. One of the basic problems in classical vibration analysis is to determine the natural frequencies and normal modes of the vibrating body. But inverse problems are concerned with the construction of a model of a given type, for example, a mass-spring system, a string, an IC circuit, and so on, with prescribed spectral data. Such a reconstruction, if possible, would have practical value to applied mechanics and structure design.

Consider the vibration of beads on a taut string illustrated in Figure 2.1. Assume that the beads, each with mass m_i, are placed along the string with equal horizontal spacing h and are subject to a constant horizontal

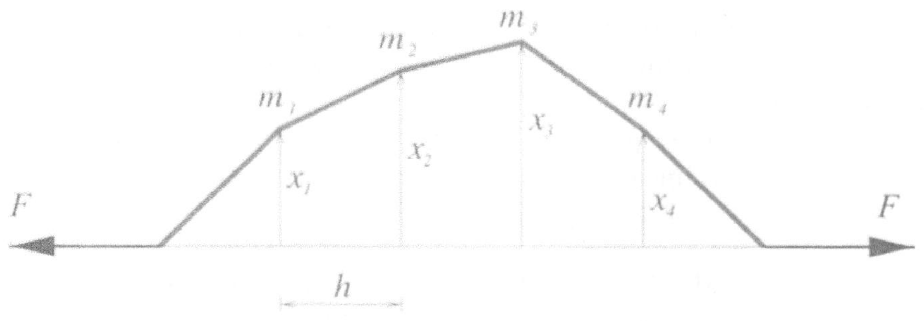

Figure 2.1. Vibration of beads on a string

tension F. Then the equation of motion (for 4 beads) is given by:

$$\begin{aligned}
m_1 \frac{d^2 x_1}{dt^2} &= -F\frac{x_1}{h} + F\frac{x_2 - x_1}{h}, \\
m_2 \frac{d^2 x_2}{dt^2} &= -F\frac{x_2 - x_1}{h} + F\frac{x_3 - x_2}{h}, \\
m_3 \frac{d^2 x_3}{dt^2} &= -F\frac{x_3 - x_2}{h} + F\frac{x_4 - x_3}{h}, \\
m_4 \frac{d^2 x_4}{dt^2} &= -F\frac{x_4 - x_3}{h} - F\frac{x_4}{h}.
\end{aligned}$$

The equation of motion can easily be generalized to the case of n beads, which can conveniently be described in matrix form,

$$\frac{d^2 \mathbf{x}}{dt^2} = -DJ_0\mathbf{x}, \tag{2.7}$$

where $\mathbf{x} = [x_1, x_2, \ldots, x_n]^T$, J_0 is the Jacobi matrix

$$J_0 = \begin{bmatrix} 2 & -1 & 0 & & & \\ -1 & 2 & -1 & & & \\ 0 & -1 & 2 & \ldots & & 0 \\ \vdots & & & \ddots & & \\ 0 & & & & 2 & -1 \\ 0 & & & & -1 & 2 \end{bmatrix}, \tag{2.8}$$

and D is the diagonal matrix $D = \operatorname{diag}(d_1, d_2, \ldots, d_n)$, with $d_i = \frac{F}{m_i h}$. We remark that the system (2.7) may also be thought of the method of lines applied to the one-dimensional wave equation. Eigenvalues of the matrix product DJ_0 are precisely the squares of the so-called *natural frequencies* of the system. An interesting inverse problem that is a special case of the so-called multiplicative IEP (**MIEP**) concerns placing the weights m_i, $i = 1, \ldots, n$ appropriately so that the resulting system has a prescribed set of natural frequencies. An even more fundamental question related to the solvability is whether such a string can have arbitrarily prescribed natural frequencies by adjusting the diagonal matrix D and, if not, what are the reachable frequencies.

More generally, the equation of motion arising in many mechanics applications appears as a linear second-order differential system:

$$M\ddot{\mathbf{x}} + C\dot{\mathbf{x}} + K\mathbf{x} = f(\mathbf{x}), \tag{2.9}$$

where $\mathbf{x} \in \mathbb{R}^n$ and M, C, $K \in \mathbb{R}^{n\times n}$. Usually, the *mass matrix* M is diagonal, and both C and the *stiffness matrix* K are symmetric, positive definite, and tridiagonal. It is known that the general solution to the homogeneous equation is a vital prerequisite for the stability of the subsequent dynamical behaviour. To that end, the fundamental solution can be derived by proposing a solution of the form

$$\mathbf{x}(t) = \mathbf{v}e^{\mu t}.$$

Upon substitution, it turns out that $\mathbf{v}$ and μ are solutions to the quadratic eigenvalue problem

$$(\mu^2 M + \mu C + K)\mathbf{v} = 0. \tag{2.10}$$

Assuming the case that all eigenvalues are distinct, then a general solution to the homogeneous system is given by the superposition principle, that is,

$$\mathbf{x}(t) = \sum_{k=1}^{2n} \alpha_k \mathbf{v}_k e^{\mu_k t},$$

where $(\mu_k, \mathbf{v}_k)$, $k = 1, \ldots, 2n$, are the eigenpair solutions to (2.10).

In the undamped system, where $C = 0$, the quadratic eigenvalue problem is reduced to the generalized eigenvalue problem,

$$(K - \omega^2 M)\mathbf{v} = 0, \tag{2.11}$$

if we write $\lambda = i\omega$. In this case, ω is precisely the natural frequency of the system and $\mathbf{v}$ is the corresponding natural mode. Let $\lambda = \omega^2$, $J := M^{-1/2}KM^{-1/2}$, and $\mathbf{z} = M^{1/2}\mathbf{x}$. The generalized eigenvalue problem can be further reduced to the Jacobi eigenvalue problem

$$J\mathbf{z} = \lambda\mathbf{z}. \tag{2.12}$$

At this point, there are two ways to formulate IEPs in the above context. First, note that the stiffness matrix K is normally more complicated than the mass matrix M. The requirement of maintaining physical feasibility also imposes constraints on the stiffness matrix, making it less flexible and more difficult to construct. Thus, one usual way of forming an IEP is to have the stiffness matrix K determined and fixed from the existing structure, that is, the static constraints, and we want to find the mass matrix M in (2.11) so that some desired natural frequencies are achieved. This inverse problem is equivalent to the MIEP discussed earlier. An alternative formulation is to construct an unreduced, symmetric, and tridiagonal matrix J from its eigenvalues and the eigenvalues of its first leading principal submatrix. This is a special case of the so-called Jacobi IEP (**JIEP**). In Section 4.2, we shall illustrate that such an inverse problem can be identified as configuring a mass-spring system from its spectrum and the spectrum of the same system but with the last mass fixed to have no motion.

The inverse problem for a damped system is considerably more complicated. Assuming that M is normalized to be the identity matrix, the analogous problem to the JIEP for the damped system concerns the reconstruction of matrices C and K from the given spectral information of the damped system. A particular formulation is given as SIEP6b in Section 4.1.

There are many other types of engineering applications for which an IEP formulation could offer useful insight that, in turn, could lead to better control of performance, safety, or effects of the system. A recent paper by Tisseur and Meerbergen (2001) offers an excellent survey of quadratic eigenvalue problems and related applications. Applications of IEPs to structure design problems can be found in Joseph (1992) as well as the conference collection edited by Mottershead and Friswell (2001). By measuring the changes in the natural frequencies, the IEP idea can be employed to detect the size and location of a blockage in a duct or a crack in a beam. See Wu (1990), Gladwell and Morassi (1999) and Gladwell (1996) for additional references. Studies on IEPs with applications to mechanics are especially flourishing. The research began in the former Soviet Union with the work of M. G. Kreĭn (1933). It first became known in the West through the (German) translation of Gantmaher and Kreĭn (1960). The book by Gladwell (1986*b*) and his follow-up review (Gladwell 1996) cover a broad scope of practices and references of IEPs for small mechanical systems. Applications of IEPs to model updating problems and fault detection problems for machine and structure diagnostics are discussed by Starek and Inman (2001). Other individual articles such as Barcilon (1979), Dai (1995), Gladwell (1984), Gladwell and Gbadeyan (1985), Gladwell (1986*a*, 1997, 1999), Ram and Caldwell (1992) and Ram and Gladwell (1994) represent some typical applications to vibrating rods and beams. A more comprehensive bibliography can be found at our web site `http://www4.ncsu.edu/~mtchu`. Discussion for higher-dimensional problems can be found in Barcilon (1990), Gladwell and Zhu (1992), Knobel and McLaughlin (1994), McLaughlin, Polyakov and Sacks (1994), McLaughlin and Hald (1995) and Zayed (1993). An important extension of the Jacobi-type analysis to a tree-like system is given in Duarte (1989).

2.3. Inverse Sturm–Liouville problems

Much of the discussion of IEPs in the literature has been due to an interest in the inverse Sturm–Liouville problem. A classical regular Sturm–Liouville problem concerns a differential equation of the form:

$$-\frac{\mathrm{d}}{\mathrm{d}x}\left(p(x)\frac{\mathrm{d}u(x)}{\mathrm{d}x}\right) + q(x)u(x) = \lambda u(x), \; a < x < b, \tag{2.13}$$

where $p(x)$ and $q(x)$ are piecewise continuous on $[a, b]$ and appropriate boundary conditions are imposed. As a direct problem, it is known that eigenvalues of the system (2.13) are real, simple, countable, and tend to infinity. As an inverse problem, the question is to determine the potential function $q(x)$ from eigenvalues. This inverse problem has generated much interest in the field, for instance, Andrew (1994), Paine (1984) and Zhornitskaya and Serov (1994), and notably the celebrated work by Gel′fand and Levitan (1955) in which the fundamental fact that two data sequences are required to uniquely determine a potential is settled. A quick introduction to this subject can be found in Chadan, Colton, Päivärinta and Rundell (1997, Chapter 3). A more thoroughgoing discussion was done in the translated book by Levitan (1987).

When a numerical solution is sought, the Sturm–Liouville problem is discretized (Pryce 1993). Likewise, the inverse problem leads to a matrix analogue IEP. Assuming that $p(x) \equiv 1$, $[a, b] = [0, 1]$, and mesh size $h = \frac{1}{n+1}$, the differential equation (2.13) is reduced by the central difference scheme, for instance, to the matrix eigenvalue problem

$$\left(-\frac{1}{h^2} J_0 + X\right) \mathbf{u} = \lambda \mathbf{u}, \tag{2.14}$$

where J_0 is given by (2.8) and X is the diagonal matrix representing the discretization of $q(x)$. The inverse problem is to determine a diagonal matrix X so that the matrix on the left side of (2.14) possesses a prescribed spectrum. This is a special case of the so-called additive IEP (**AIEP**). It should be cautioned that there is a significant difference between the behaviour of the discrete problem and that of the continuous case. See the discussion in Hald (1972) and Osborne (1971). The matrix analogue IEP, such as (2.14), however, is of interest in its own right.

We mention one application to geophysics. Assuming that the Earth has spherical symmetry, geophysicists want to infer its internal structure from the frequencies of spheroidal and torsional modes of oscillations. This leads to the generalized Sturm–Liouville problem, that is,

$$u^{(2k)} - (p_1 u^{(k-1)})^{(k-1)} + \cdots + (-1)^k p_k u = \lambda u.$$

Following the work of Gel′fand and Levitan (1955), Barcilon (1974*a*) suggested that $k + 1$ spectra, associated with $k + 1$ distinct sets of boundary conditions, must be present to construct the unknown coefficients $p_1, \ldots, p_k$. See also Barcilon (1974*b*). It is not clear how the matrix analogue for this high-order problem should be formulated.

2.4. Applied physics

The IEP formulation can sometimes be used to explore and alleviate some difficult computational problems in applied physics. We demonstrate two applications in this section.

We first describe an application to quantum mechanics. In computing the electronic structure of an atom, one usually expands the atom's state vector over a convenient basis. The expansion coefficients are determined by solving the eigenvalue problem for a Hamiltonian matrix H. It is known that these expansion coefficients are sensitive to the diagonal elements of H. Yet, in many cases of interest, the diagonal elements of H cannot be determined to sufficient accuracy. On the other hand, eigenvalues of H correspond to energy levels of an atom that can usually be measured to a high degree of accuracy. The idea now is to use these measured energy levels to correct diagonal elements. Furthermore, for practical purpose, all matrices involved are required to be real. Under such a constraint, it is almost always impossible to match the eigenvalues exactly. We therefore formulate a least squares IEP (**LSIEP**) as follows (Deakin and Luke 1992). Given a real symmetric matrix A and a set of real values $\boldsymbol{\omega} = [\omega_1, \ldots, \omega_n]^T$, find a real diagonal matrix D such that

$$\|\sigma(A + D) - \boldsymbol{\omega}\|_2$$

is minimized. Throughout this paper, $\sigma(M)$ denotes *either* the spectrum (set) of the matrix M *or* the column vector formed by these eigenvalues: no ambiguity should arise.

We next describe an application to neuron transport theory. One model for the dynamics in an additive neural network is the differential equation

$$\frac{\mathrm{d}\mathbf{u}}{\mathrm{d}t} = -A\mathbf{u} + \Omega\mathbf{g}(\mathbf{u}) + \mathbf{p}, \tag{2.15}$$

where $A = \operatorname{diag}(a_1, \ldots a_n)$ denotes the decaying factor, $\Omega = [\omega_{ij}]$ denotes connection coefficients between the neurons, $\mathbf{g}(\mathbf{u}) = [g_1(u_1), \ldots, g_n(u_n)]^T$ denotes the *squashing function*, in which each g_i is strictly increasing but bounded in u_i, and $\mathbf{p}$ is a constant input. One of the design problems is, given A, $\mathbf{g}$, and $\mathbf{p}$, to choose the connection matrix Ω so that a predestined point $\mathbf{u}^* \in \mathbb{R}^n$ is a stable equilibrium. This requirement translates into two conditions that must be satisfied simultaneously. First, the linear equation

$$-A\mathbf{u}^* + \Omega\mathbf{g}(\mathbf{u}^*) + \mathbf{p} = 0 \tag{2.16}$$

must hold for Ω. Secondly, all eigenvalues of the Jacobian matrix,

$$\Upsilon = -A + \Omega G(\mathbf{u}^*), \tag{2.17}$$

where $G(\mathbf{u}^*) = \operatorname{diag}(g_1'(\mathbf{u}^*), \ldots g_n'(\mathbf{u}^*))$, must lie in the left half-plane. Upon rearranging the terms, it is easy to see that (2.16) can be rewritten as

$$\Upsilon\mathbf{x} = \mathbf{y}, \tag{2.18}$$

where $\mathbf{x} = G^{-1}(\mathbf{u}^*)\mathbf{g}(\mathbf{u}^*)$ and $\mathbf{y} = A\mathbf{u}^* - AG^{-1}(\mathbf{u}^*)\mathbf{g}(\mathbf{u}^*) - \mathbf{p}$ are known vectors. This is a special case of the equality constrained IEP (**ECIEP**) considered in Li (1997). Given two sets of real vectors $\{\mathbf{x}_i\}_{i=1}^p$ and $\{\mathbf{y}_i\}_{i=1}^p$ with $p \leq n$, and a set of complex numbers $\{\lambda_1, \ldots, \lambda_n\}$, closed under conjugation, find a real matrix A such that $A\mathbf{x}_i = \mathbf{y}_i$, $i = 1, \ldots, p$ and $\sigma(A) = \{\lambda_1, \ldots, \lambda_n\}$. A similar matrix approximation problem with linearly constrained singular values is discussed in Nievergelt (1997).

2.5. Numerical analysis

Finally, we point out that, even within the field of numerical analysis, the notion of IEP helps to shed additional insight on numerical methods and stabilize some numerical algorithms. We comment on four applications: preconditioning, derivation of high-order stable Runge–Kutta schemes, Gaussian quadrature, and low-rank approximations.

Recall first that one of the main ideas in preconditioning a linear equation $Ax = b$ is to transform the original system into an equivalent system that is easier (quicker) to solve with an iterative scheme. The preconditioning of a matrix A can be thought of as implicitly multiplying A by M^{-1}, where M is a matrix for which, hopefully, $Mz = y$ can easily be solved, $M^{-1}A$ is not too far from normal, and $\sigma(M^{-1}A)$ is clustered. This final hope, that the eigenvalues of a preconditioned system $M^{-1}A$ should be clustered, is a loose MIEP criterion. Although, in the context of preconditioning, the locations of eigenvalues need not be exactly specified, the notion of MIEP can certainly help to see what is to be expected of the ideal preconditioner. Many types of unstructured preconditioners have been proposed, including the low-order (coarse-grid) approximation, SOR, incomplete LU factorization, polynomial, and so on. It would be interesting to develop another category of preconditioners where the M is required to possess a certain structure (Forsythe and Straus 1955, Greenbaum and Rodrigue 1989). A related problem that has potential application to optimization is: Given a matrix $C \in \mathbb{R}^{m \times n}$ and a constant vector $\mathbf{b} \in \mathbb{R}^m$, find a vector $\mathbf{x} \in \mathbb{R}^n$ such that the rank-one updated matrix $\mathbf{b}\mathbf{x}^T + C$ has a prescribed set of singular values.

Recall secondly that an s-stage Runge–Kutta method is uniquely determined by the Butcher array

$$\begin{array}{c|cccc}
c_1 & a_{11} & a_{12} & \ldots & a_{1s} \\
c_2 & a_{21} & a_{22} & \ldots & a_{2s} \\
\vdots & \vdots & & & \vdots \\
c_s & a_{s1} & a_{s2} & \ldots & a_{ss} \\
\hline
 & b_1 & b_2 & \ldots & b_s
\end{array}$$

Let $A = [a_{ij}]$, $\mathbf{b} = [b_1, \ldots, b_s]^T$ and $\mathbf{1} = [1, \ldots 1]^T$. It is well established that the stability function for an s-state Runge–Kutta method is given by

$$R(z) = 1 + z\mathbf{b}^T(I - zA)^{-1}\mathbf{1}$$

(see, for example, Lambert (1991)). To attain numerical stability, implicit methods are preferred. However, fully implicit methods are too expensive. Diagonally implicit methods (**DIRK**), *i.e.*, low triangular A with *identical* diagonal entries, are computationally more efficient, but difficult to construct. As an alternative, it is desirable to develop singly implicit methods (**SIRK**) in which the matrix A does not need to be lower-triangular but must have an s-fold eigenvalue. Such a consideration can be approached by an IEP formulation with prescribed entries, as is done by Müller (1992). Given the number s of stages and the desired order p of the method, define $k = \lfloor (p-1)/2 \rfloor$ and constants $\xi_j = 0.5(4j^2-1)^{-1/2}$, $j = 1, \ldots, k$. Find a real number λ and $Q \in \mathbb{R}^{(s-k)\times(s-k)}$ such that $Q + Q^T$ is positive semi-definite and $\sigma(X) = \{\lambda\}$ where $X \in \mathbb{R}^{s\times s}$ is of the form

$$X = \left[\begin{array}{ccccc|c} 1/2 & -\xi_1 & & & & \\ \xi_1 & 0 & & & & \\ 0 & & & & & \\ \vdots & & & \ddots & & \\ & & & & 0 & -\xi_k \\ \hline 0 & & & & \xi_k & Q \end{array}\right],$$

and $q_{11} = 0$ if p is even. Note that, in this formulation, the value of the s-fold eigenvalue λ is one of the unknowns to be determined.

Recall thirdly that orthogonal polynomials play a crucial role in the development of Gaussian quadrature rules. Given a weight function $\omega(x) \geq 0$ on $[a, b]$, an n-point Gauss quadrature rule for the integral

$$\mathcal{I}f = \int_a^b \omega(x) f(x)\, dx \tag{2.19}$$

is a formula of the form

$$\mathcal{G}_n f = \sum_{i=1}^{n} w_i f(\lambda_i), \tag{2.20}$$

with selected nodes $\{\lambda_1, \ldots, \lambda_n\}$ and weights $\{w_1, \ldots, w_n\}$ so that

$$\mathcal{G}_n f = \mathcal{I}f \tag{2.21}$$

for all polynomials $f(x)$ of degree no higher than $2n-1$. With respect to the given $\omega(x)$, a sequence of orthonormal polynomials $\{p_k(x)\}_{k=0}^{\infty}$ satisfying

$$\int_a^b \omega(x) p_i(x) p_j(x)\, dx = \delta_{ij} \tag{2.22}$$

can be defined. It is an established fact that the roots of each $p_k(x)$ are simple, distinct, and lie in the interval $[a, b]$. Indeed, in order that the resulting quadrature should achieve the highest degree of precision $2n-1$, the Gaussian nodes should be the roots $\{\lambda_i\}_{i=1}^n$ of $p_n(x)$. On the other hand, it also known that, with $p_0(x) \equiv 1$ and $p_{-1}(x) \equiv 0$, orthogonal polynomials satisfy a three-term recurrence relationship:

$$p_n(x) = (a_n x + b_n)p_{n-1}(x) - c_n p_{n-2}(x). \tag{2.23}$$

Let $\mathbf{p}(x) = [p_0(x), p_1(x), \ldots, p_{n-1}(x)]^T$. This relationship can be written in matrix form as

$$x\mathbf{p}(x) = \underbrace{\begin{bmatrix} \frac{-b_1}{a_1} & \frac{1}{a_1} & 0 & & 0 \\ \frac{c_2}{a_2} & \frac{-b_2}{a_2} & \frac{1}{a_2} & & \\ 0 & & & & \\ \vdots & & \ddots & \vdots & \\ 0 & & & & \frac{1}{a_{n-1}} \\ 0 & & \cdots & \frac{c_n}{a_n} & \frac{-b_n}{a_n} \end{bmatrix}}_{T} \mathbf{p}(x) + \begin{bmatrix} 0 \\ 0 \\ \\ \vdots \\ 0 \\ \frac{1}{a_n}p_n(x) \end{bmatrix}. \tag{2.24}$$

Observe that $p_n(\lambda_j) = 0$ if and only if

$$\lambda_i \mathbf{p}(\lambda_i) = T\mathbf{p}(\lambda_i).$$

Note that the matrix T can be symmetrized by diagonal similarity transformation into a Jacobi matrix J and that the weight w_j in the quadrature is given by

$$w_i = q_{1i}^2, \quad i = 1, \ldots, n,$$

where $\mathbf{q}_i$ is the ith normalized eigenvector of J. This gives rise to an interesting inverse problem. Given a quadrature with nodes $\{\lambda_1, \ldots, \lambda_n\}$ and weights $\{w_1, \ldots, w_n\}$ satisfying $\sum_{i=1}^n w_i = 1$, determine the corresponding orthogonal polynomials (and the corresponding weight function $\omega(x)$; see Kautsky and Elhay (1984), Ferguson (1980)). We illustrate one interesting application to the derivation of the Gauss–Kronrod quadrature rule. Given a Gaussian quadrature (2.20), the associated Gauss–Kronrod quadrature is a $(2n+1)$-point integral rule

$$\mathcal{K}_{2n+1}f = \sum_{i=1}^{n} \tilde{w}_i f(\lambda_i) + \sum_{j=1}^{n+1} \hat{w}_j f(\hat{\lambda}_j) \tag{2.25}$$

that is exact for all polynomials of degree at most $3n+1$. Note that the original nodes $\{\lambda_1, \ldots, \lambda_n\}$ form a subset of the new nodes in $\mathcal{K}_{2n+1}$. Based on an interesting observation in Laurie (1997), the existence of a Gauss–Kronrod quadrature rule with real distinct nodes and positive weights is equivalent to the existence of a real solution to the following special IEP with prescribed entries (**PEIEP**): determine an $n \times n$ symmetric tridiagonal matrix with prescribed first $n-1$ entries (counting row-wise in the upper-triangular part) and prescribed eigenvalues $\{\lambda_1, \ldots, \lambda_n\}$. More details of the computation can be found in the paper by Calvetti, Golub, Gragg and Reichel (2000).

Finally, we note that the problem of low-rank approximation also belongs to the realm of IEPs, considering that a section of the spectrum for the desirable approximation is preset to zero. Low-rank approximation can be used as a tool for noise removal in signal or image processing where the underling matrix is structured as Toeplitz (Cadzow and Wilkes 1990, Suffridge and Hayden 1993), covariance (Li, Stoica and Li 1999, Williams and Johnson 1993), and so on. The rank to be removed corresponds to the noise level where the signal to noise ratio (SNR) is low (Tufts and Shah 1993). Low-rank approximation can also be used for model reduction problems in speech encoding and filter design with Hankel structure, where the rank to be restored is the number of sinusoidal components in the original signal (Park, Zhang and Rosen 1999). The problem of finding or approximating the greatest common divisor (GCD) of multivariate polynomials can be formulated as a low-rank approximation problem with Sylvester structure whose rank is precisely the degree of the GCD (Corless, Gianni, Trager and Watt 1995, Karmarkar and Lakshman 1998). The molecular structure modelling for protein folding in $\mathbb{R}^3$ involves Euclidean distance matrices whose rank is no more than 5 (Glunt, Hayden, Hong and Wells 1990, Gower 1982). In the factor analysis or latent semantic indexing (LSI) application, the low rank is the number of principal factors capturing the random nature of the indexing matrix (Horst 1965, Zha and Zhang 1999). All of these can be considered as structured IEPs with partial spectrum identically zero.

We have seen from the above illustrations that different applications lead to different IEP formulations. We conclude this section with one additional remark by Gladwell (1996), who suggested that, for application purposes, there should also be a distinction between *determination* and *estimation* in the nature of an inverse problem. When the given data are exact and complete, so that the system can be precisely determined, the IEP is said to be *essentially mathematical*. In contrast, we say that we have an *essentially engineering* IEP when the data are only approximate and often incomplete, in which only an estimate of the parameters of the system is sought and the resulting behaviour is expected to agree only approximately with the prescribed data.

3. Nomenclature

For the ease of identifying the characteristics of various IEPs, we have suggested using a unified name scheme ***IEP#** to categorize an IEP (Chu 1998). When singular values are involved in the spectral constraint, we distinguish ISVPs from IEPs. Letter(s) '$*$' in front of IEP register the type

Table 3.1. Summary of acronyms used in the paper

Acronym	Meaning	Reference
AIEP	Additive IEP	Section 9.3
ECIEP	Equality Constrained IEP	Page 12
ISEP	Inverse Singular/Eigenvalue Problem	Section 11
ISVP	Inverse Singular Value Problem	Section 10
JIEP	Jacobi IEP	Section 4
LSIEP	Least Squares IEP	Page 12
MIEP	Multiplicative IEP	Page 8
MVIEP	Multi-Variate IEP	
NIEP	Nonnegative IEP	Section 6
PAP	Pole Assignment Problem	Page 6
PEIEP	IEP with Prescribed Entries	Section 9
PIEP	Parametrized IEP	Page 51
PDIEP	Partially Described IEP	Page 25
RNIEP	Real-valued Nonnegative IEP	Page 39
SHIEP	Schur–Horn IEP	Page 47
SIEP	Structured IEP	Page 3
SNIEP	Symmetric Nonnegative IEP	Page 39
StIEP	Stochastic IEP	Section 7
STISVP	Sing–Thompson ISVP	Page 49
ToIEP	Toeplitz IEP	Section 5
UHIEP	Unitary Hessenberg IEP	Section 8

of problem. The numeral '#' following IEP, if any, indicates the sequence of variation within type '*IEP'. For convenience of later reference, we summarize the acronyms appearing in this paper in Table 3.1. Also indicated are the page numbers or the section numbers where the problems are first described or where more detailed discussion can be found.

Figure 3.1 depicts a possible inclusion relationship between different problems. In particular, the diagram is intended to imply the following.

- Multivariate IEPs include univariate IEPs as a special case.
- All problems have a natural generalization to a least squares formulation.
- The structural constraints involved in SIEPs can appear in various forms, and hence define different IEPs.
- There is a counterpart ISVP corresponding to any structured IEP, formed by replacing the eigenvalue constraint by a singular value constraint. Very little is known about these structured ISVPs (and hence no diagrams).

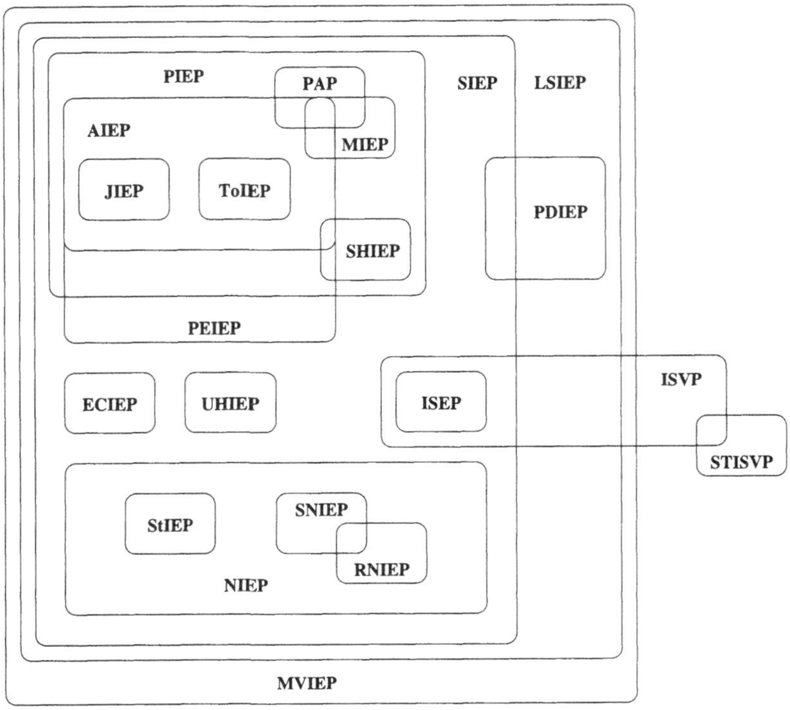

Figure 3.1. Classification of inverse eigenvalue problems

- The class of PIEPs is considered to be a subset of general SIEPs, while many classical IEPs are special cases of PIEPs.
- The relationship depicted in Figure 3.1 is not necessarily definitive because many characteristics may overlap.

This classification along with review articles by Gladwell (1986*c*, 1996), who differentiates problems according to the type of mechanical system, *i.e.*, continuous or discrete, damped or undamped, and the type of prescribed data, *i.e.*, spectral, modal, or nodal, complete or incomplete, should complement each other to offer a fairly broad view of research activities in this area.

This paper concentrates on the SIEP segment only. Even so, the formations and algorithms differ noticeably from problem to problem. Indeed, we pointed out earlier that every IEP should in fact be regarded as an SIEP because of the presence of its structural constraint. That view is too broad to be covered here. Instead, we shall focus on eight selected special structures. These are the IEPs for Jacobi matrices, Toeplitz matrices, nonnegative matrices, stochastic matrices, unitary matrices, matrices with prescribed entries, matrices with prescribed singular values, and matrices with prescribed singular values and eigenvalues. Our criteria of selection are simply that these eight problems are representative of a variety of structural constraints and are slightly better studied in the literature. We choose not to include PAPs because that topic has been well considered in many other places.

We shall consider these eight structured problems in slightly more breadth and depth with regard to the four issues of solvability, computability, sensitivity, and applicability. Some main results, applications, and algorithmic issues will also be presented.

4. Jacobi inverse eigenvalue problems

By a Jacobi structure, we mean a symmetric, tridiagonal matrix of the form

$$J = \begin{bmatrix} a_1 & b_1 & 0 & & & 0 \\ b_1 & a_2 & b_2 & & & 0 \\ 0 & b_2 & a_3 & & & 0 \\ \vdots & & & \ddots & & \\ & & & & a_{n-1} & b_{n-1} \\ 0 & & & & b_{n-1} & a_n \end{bmatrix}, \tag{4.1}$$

with positive subdiagonal elements $b_i > 0$. We have already seen that this structure arises in many important areas of applications, including oscillatory mass-spring systems, composite pendulum, and Sturm–Liouville problems. Eigenvalues of a Jacobi matrix are necessarily real and distinct. Since J is characterized by the $2n-1$ unknown entries, $\{a_i\}_{i=1}^{n}$ and $\{b_j\}_{j=1}^{n-1}$, it is intuitively true that $2n-1$ pieces of information are needed to solve the inverse problems. That is, to fully describe a JIEP we need additional information other than just the spectrum of J. This additional information comes from different sources and defines additional structures for JIEPs. We shall survey a few JIEPs in this section. One unique and important feature for JIEPs is that often the inverse problem can be solved by direct methods in finitely many steps.

Jacobi matrices enjoy many nice properties. These properties make the study of JIEPs more complete and fruitful than other IEPs. For that reason, we shall provide somewhat more details on the theory and development of JIEPs. We shall touch upon all four fundamental questions raised in Section 1 for JIEPs. We hope that this exertion can serve as a study guide for further developments of other IEPs in the future.

Before we move on, we should emphasize that the JIEPs under discussion here are of tridiagonal structure only. The generalization to band matrices is possible. Some initial studies of the IEP for band matrices can be found in the paper by Biegler-König (1981a). Boley and Golub (1987) generalized some of the numerical methods for JIEPs to the banded case. However, be aware that there are some fundamental differences in the generalization. For instance, two sets of eigenvalues generally determine a tridiagonal matrix uniquely (See Theorem 4.1 in Section 4.3), whereas three sets of eigenvalues do not give a pentadiagonal matrix uniquely (and, in fact, sometimes there is a continuum of solutions) (Boley and Golub 1987).

4.1. Variations

There are several variations in formulating a JIEP. Each formulation can be associated with a mass-spring system. In this section, we only describe the setup, a brief history and some relevant references on the original settings for each problem. Topics on physical interpretation, mathematical theory, and computational methods will be discussed in the next few sections.

In the following, J_k denotes the $k \times k$ principal submatrix of J, and J_{n-1} is abbreviated as $\bar{J}$. Whenever possible, we refer to each variation by the identification name used in Chu (1998).

SIEP6a. Given real scalars $\{\lambda_1, \ldots, \lambda_n\}$ and $\{\mu_1, \ldots, \mu_{n-1}\}$ satisfying the interlacing property

$$\lambda_i < \mu_i < \lambda_{i+1}, \quad i = 1, \ldots, n-1, \tag{4.2}$$

find a Jacobi matrix J such that

$$\begin{cases} \sigma(J) = \{\lambda_1, \ldots, \lambda_n\}, \\ \sigma(\bar{J}) = \{\mu_1, \ldots, \mu_{n-1}\}. \end{cases}$$

This problem is perhaps the most fundamental and extensively studied IEP in the literature. It appears that the problem was originally proposed by Hochstadt (1967), although Downing and Householder (1956) had formulated a more general inverse characteristic value problem much earlier. Much of the existence theory and continuous dependence of the solution on data were developed later in Hochstadt (1974), Gray and Wilson (1976), and Hald (1976). Dangerously many numerical methods are available! Some are stable and some are subtly unstable. We shall discuss some of these methods later. For the time being, it suffices to mention some important works in this regard (Boley and Golub 1987, de Boor and Golub 1978, Erra and Philippe 1997, Gragg and Harrod 1984, Hochstadt 1979, Parlett 1998).

SIEP2. Given real scalars $\{\lambda_1, \ldots, \lambda_n\}$, find a Jacobi matrix J such that

$$\begin{cases} \sigma(J) = \{\lambda_1, \ldots, \lambda_n\}, \\ a_i = a_{n+1-i}, \\ b_i = b_{n+2-i}. \end{cases}$$

Let $Xi \in \mathbb{R}^{n\times n}$ denote the unit perdiagonal matrix where

$$\xi_{ij} = \begin{cases} 1, & \text{if } i = n+1-j, \\ 0, & \text{otherwise.} \end{cases} \tag{4.3}$$

A matrix M is said to be persymmetric if and only if $\Xi M \Xi = M^T$. In other words, the entries of M are symmetric with respect to the northeast-to-southwest diagonal. A persymmetric Jacobi matrix involves only n independent entries. The spectral constraint therefore requires spectrum information only. This problem was first considered in Hochstadt (1967) and then in Hald (1976). Numerical methods for SIEP2 usually come along with those for SIEP6a with appropriate modifications (de Boor and Golub 1978, Parlett 1998).

SIEP7. Given real scalars $\{\lambda_1, \ldots, \lambda_n\}$ and $\{\mu_1, \ldots, \mu_{n-1}\}$ satisfying the interlacing property

$$\begin{cases} \lambda_i \leq \mu_i \leq \lambda_{i+1}, \\ \mu_i < \mu_{i+1}, \end{cases} \qquad i = 1, \ldots, n-1, \tag{4.4}$$

and a positive number β, find a periodic Jacobi matrix J of the form

$$J = \begin{bmatrix} a_1 & b_1 & & & & b_n \\ b_1 & a_2 & b_2 & & & 0 \\ 0 & b_2 & a_3 & & & 0 \\ \vdots & & & \ddots & & \\ & & & & a_{n-1} & b_{n-1} \\ b_n & & & & b_{n-1} & a_n \end{bmatrix}$$

such that

$$\begin{cases} \sigma(J) = \{\lambda_1, \ldots, \lambda_n\}, \\ \sigma(\bar{J}) = \{\mu_1, \ldots, \mu_{n-1}\}, \\ \prod_1^n b_i = \beta. \end{cases}$$

A periodic Jacobi matrix differs from a Jacobi matrix in that its eigenvalues need not be strictly separated. The interlacing property (4.4) therefore differs from (4.2) in that equalities are allowed. The notion of periodic Jacobi matrices arise in applications such as periodic Toda lattices or continued fractions (Adler, Haine and van Moerbeke 1993, Andrea and Berry 1992). Spectral properties of the periodic Jacobi matrices were first analysed by Ferguson (1980) using a discrete version of Floquet theory, but numerical methods had been proposed earlier in Boley and Golub (1978). See also discussions in Boley and Golub (1984, 1987).

SIEP8. Given real scalars $\{\lambda_1, \ldots, \lambda_n\}$ and $\{\mu_1, \ldots, \mu_n\}$ satisfying the interlacing property

$$\lambda_i < \mu_i < \lambda_{i+1}, \quad i = 1, \ldots, n,$$

with $\lambda_{n+1} = \infty$, find Jacobi matrices J and $\tilde{J}$ so that

$$\begin{cases} \sigma(J) = \{\lambda_1, \ldots, \lambda_n\}, \\ \sigma(\tilde{J}) = \{\mu_1, \ldots, \mu_n\}, \\ J - \tilde{J} \neq 0, \quad \text{only at the } (n,n) \text{ position.} \end{cases}$$

This problem originally appeared in de Boor and Golub (1978). Note that $\tilde{J}$ is a special rank-one update of J. This problem is closely related to SIEP6a in that the theory and numerical methods for SIEP6a will work almost identically for SIEP8. A similar problem involving the preconditioning of a matrix by a rank-one matrix was mentioned earlier in Section 2.5. An application of rank-one updating involving the inverse quadratic eigenvalue problem was discussed in Datta, Elhay and Ram (1997) and Ram (1995).

SIEP9. Given distinct real scalars $\{\lambda_1, \ldots, \lambda_{2n}\}$ and an $n \times n$ Jacobi matrix $\tilde{J}$, find a $2n \times 2n$ Jacobi matrix J so that

$$\begin{cases} \sigma(J) = \{\lambda_1, \ldots, \lambda_{2n}\}, \\ J_n = \tilde{J}. \end{cases}$$

This problem, first discussed in Hochstadt (1979), corresponds exactly to the problem of computing the Gaussian quadrature of order $2n$ that has degree of precision $4n - 1$, given the Gaussian quadrature of order n that has degree of precision $2n - 1$. Several numerical algorithms are available. See Boley and Golub (1987). An IEP as such is actually a special case of a more general category of IEPs with prescribed entries. The latter, in turn, is a subset of so-called completion problems in the literature. The prescribed entries need not be in a diagonal block as in SIEP9. An interesting question related to the IEP is to find the largest permissible cardinality of the prescribed entries so that the completed matrix has a prescribed spectrum. The first publication devoted to this problem was probably due to London and Minc (1972), followed by the series of work by de Oliveira (1973*a*, 1973*b*, 1975). A most recent and comprehensive survey on this topic was given by Ikramov and Chugunov (2000), who stated that the thesis by Hershkowits (1983) contained the strongest result in this class of problems. Also presented in Ikramov and Chugunov (2000) was a careful treatment showing how the completion problems can be solved by finite rational algorithms. A similar inverse problem for matrices with prescribed entries and characteristic polynomial was considered by Dias da Silva (1974); for matrices with prescribed characteristic polynomial and principal submatrices by Silva (1987*a*); and for matrices with prescribed spectrum and principal submatrices by Silva (1987*b*).

SIEP6b. Given complex scalars $\{\lambda_1, \ldots, \lambda_{2n}\}$ and $\{\mu_1, \ldots, \mu_{2n-2}\}$, distinct and closed under complex conjugation, find tridiagonal symmetric matrices C and K for the λ-matrix $Q(\lambda) = \lambda^2 I + \lambda C + K$ so that

$$\begin{cases} \sigma(Q) = \{\lambda_1, \ldots, \lambda_{2n}\}, \\ \sigma(\bar{Q}) = \{\mu_1, \ldots, \mu_{2n-2}\}. \end{cases}$$

Clearly, SIEP6b is an analogy of SIEP6a applied to a damped system. Strictly speaking, to maintain the physical feasibility a practical solution imposes additional conditions on K and C, that is, both matrices are supposed to have positive diagonal entries, negative off-diagonal entries, and be weakly diagonally dominant. The setup of SIEP6b, where two sets of eigenvalues are given, was considered by Ram and Elhay (1996). Similar inverse problems with prescribed eigenvalues and eigenvectors were studied in a series of works (Starek, Inman and Kress 1992, Starek and Inman 1997, Starek and Inman 2001).

4.2. Physical interpretations

The JIEPs described above can be related to various physical systems, for instance a vibrating beam or rod (Gladwell 1986*b*), a composite pendulum (Hald 1976), or a string with beads (Hochstadt 1967). Correspondingly, the quantities to be determined in a JIEP represent different physical parameters, for instance the stress, the mass, the length, and so on. In this section, we shall use a serially linked, undamped mass-spring system with n particles to demonstrate the physical interpretation of JIEPs. The physical system is depicted in Figure 4.1.

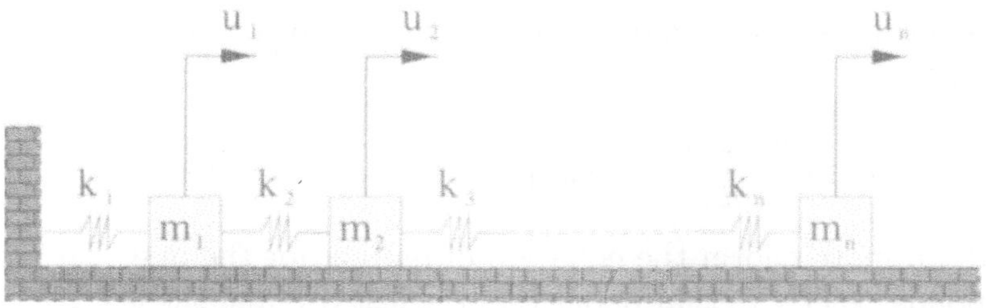

Figure 4.1. Mass-spring system

Suppose that the ith particle has mass m_i, that the springs satisfy Hooke's law, and that the ith spring has spring constant k_i. Let $u_i(t)$ denote the horizontal displacement of the ith particle at time t. Then it is easy to see that the equation of motion is given by

$$\begin{aligned}
m_1 \frac{\mathrm{d}^2 u_1}{\mathrm{d}t} &= -k_1 u_1 + k_2(u_2 - u_1), \\
m_i \frac{\mathrm{d}^2 u_i}{\mathrm{d}t} &= -k_i(u_i - u_{i-1}) + k_{i+1}(u_{i+1} - u_i), \quad i = 2, \ldots, n-1, \\
m_n \frac{\mathrm{d}^2 u_n}{\mathrm{d}t} &= -k_n(u_n - u_{n-1}).
\end{aligned}$$

In matrix form, we have

$$M \frac{\mathrm{d}^2 \mathbf{u}}{\mathrm{d}t} = -K\mathbf{u}, \tag{4.5}$$

where $\mathbf{u} = [u_1, \ldots, u_n]^T$, $M = \mathrm{diag}(m_1, \ldots, m_n)$, and K is the Jacobi matrix given by

$$K = \begin{bmatrix}
k_1 + k_2 & -k_2 & 0 & \ldots & 0 & 0 \\
-k_2 & k_2 + k_3 & -k_3 & & & 0 \\
0 & -k_3 & k_3 + k_4 & & & 0 \\
\vdots & & & \ddots & & \vdots \\
0 & & & & & -k_n \\
0 & & & & -k_n & k_n
\end{bmatrix}.$$

A fundamental solution of the form $\mathbf{u}(t) = e^{i\omega t}\mathbf{v}$ leads to the generalized eigenvalue problem (2.11). A transformation $J = M^{-1/2}KM^{-1}$, $\mathbf{z} = M^{1/2}\mathbf{v}$, and $\lambda = \omega^2$ leads to the Jacobi eigenvalue problem (2.12). The direct problem calculates the natural frequencies and modes of the mass-sprint system from given values of m_i and k_k. The inverse problem requires calculating quantities such as $\frac{k_i + k_{i+1}}{m_i}$ and $\frac{k_{i+1}}{\sqrt{m_i m_{i+1}}}$ from the spectral data. Based on this model, we make the following observations.

If the last mass m_n is fastened to the floor, then the motion of mass m_{n-1} is governed by

$$m_{n-1} \frac{\mathrm{d}^2 u_{n-1}}{\mathrm{d}t} = -k_{n-1}(u_{n-1} - u_{n-2}) + k_n(-u_{n-1}),$$

instead. In matrix form the equation of motion for the first $n-1$ particles corresponds exactly to that of deleting the last row and the last column from (4.5). Thus solving SIEP6a is equivalent to identifying the mass-spring system in Figure 4.1 from its spectrum, and from the spectrum of the reduced system where the last mass is held to have no motion. The recovery of the spring stiffness and the masses from the matrix J is discussed in Gladwell (1986*b*).

Likewise, if another spring from m_n is attached to a wall on the far right side of the system, then the equation of motion for m_n is modified to become

$$m_n \frac{\mathrm{d}^2 u_n}{\mathrm{d}t} = -k_n(u_n - u_{n-1}) + k_{n+1}(-u_n).$$

SIEP2 corresponds to the construction of such a mass-spring system from its spectrum if all parameters m_i and k_i are known *a priori* to be symmetric with respect to the centre of the system.

It is a little bit more complicated to sketch a diagram for the physical layout of SIEP7. Basically, we imagine that masses m_1 and m_n are somehow connected by another spring mechanism so as to form a loop (such as the periodic Toda lattice discussed in the literature). Any displacement in either particle of m_1 or m_n will affect each other via that mechanism, contributing nonzero but equal entries at the $(1, n)$ and $(n, 1)$ positions of K, respectively. Apart from this extra connection, the meaning of SIEP7 is now the same as that of SIEP6a.

We can also identify a mass system from its spectrum and from the spectrum of a new system by replacing the last mass and spring with new parameters $\tilde{m}_n$ and $\tilde{k}_n$ satisfying the relationship

$$\frac{k_n^2}{m_n} = \frac{\tilde{k}_n^2}{\tilde{m}_n}.$$

The resulting inverse problem is precisely SIEP8.

The interpretation of SIEP9 is straightforward. It means completing the construction of a mass-spring system of size $2n$ from its spectrum and from existing physical parameters m_i, k_i of the first half of the particles.

Thus far, we have assumed that the system in Figure 4.1 has no friction. For a damped system, the damping matrix C will be part of the parameters and we shall face a quadratic eigenvalue problem (2.10). Other than this, the physical interpretation for each of the JIEPs described above can be extended to damped systems. For example, SIEP6b is to identify the damped system, including its damper configurations, from its spectrum and from the spectrum of the reduced system where the last mass is held immobile. This problem is still open. The principal difficulty is to find conditions on the (complex) spectra which ensure a realistic solution.

It is important to point out that thus far we have considered using only eigenvalues to construct Jacobi matrices. For large and complex systems, it is often practically impossible to gather the entire spectrum information for reconstruction. Partial information with some from eigenvalues and some from eigenvectors can also be used to determine a Jacobi matrix. This type of problem is referred to as PDIEP in Chu (1998) and is beyond the scope of the present paper.

An interesting question related to PDIEPs is how much eigenvector information is needed to determine such a Jacobi (or any structured) matrix. Gladwell (1996) offered an account from engineering perspectives on why using low-frequency normal modes is important in practice. Some discussion can be found in the books by Zhou and Dai (1991) and Xu (1998). Applications of inverse problems with given spectral and modal data were studied in Gladwell (1986*a*), Starek *et al.* (1992), Starek and Inman (2001), and the many references cited in the review paper by Gladwell (1996). A number of interesting variants of JIEPs may be found in Nylen and Uhlig (1997*a*, 1997*b*), and some corresponding damped problems in Nylen (1999), Gladwell (2001) and Foltete, Gladwell and Lallement (2001).

4.3. *Existence theory*

Among all IEPs, the class of JIEPs probably enjoys the most satisfactory solvability theory. Most of the existence proofs are based on a recurrence relationship among the characteristic polynomials. More precisely, let $p_k(t) = \det(tI - J_k)$ denote the characteristic polynomial of the leading $k \times k$ principal submatrix J_k. Then

$$p_k(t) = (t - a_k)p_{k-1}(t) - b_{k-1}^2 p_{k-2}(t), \quad k = 2, \ldots, n, \tag{4.6}$$

if $p_0 \equiv 1$. Such a recurrence relationship in fact gives rise to a constructive proof that, in turn, can be implemented as a numerical algorithm. Because there is an extensive literature in this regard, and because some of the constructions will be discussed as numerical methods, we shall only state the existence theorems without proof in this section.

Theorem 4.1. Suppose that all the given eigenvalues are distinct. Then:

(1) SIEP6a has a unique solution (Hald 1976);
(2) SIEP2 has a unique solution (Hald 1976);
(3) SIEP8 has a unique solution (de Boor and Golub 1978).

It should be noted that the MIEP (of uniformly spaced beads on a taut string) described on page 8 is very different from the JIEPs described above in several aspects. The former involves only one *single spectrum*; the latter involves *two spectra*. In the former, we have only one set of parameters (the masses) to adjust; in the latter, we have two sets (the m_is and the k_is) to combine. The solution for the latter is often unique while the former is a much harder problem.

Theorem 4.2. (Ram and Elhay 1996) Over the complex field $\mathbb{C}$, suppose that all the given eigenvalues are distinct. Then SIEP6b is solvable and has at most $2^n(2n-3)!/(n-2)!$ different solutions. In the event that there are common eigenvalues, then there are infinitely many solutions for SIEP6b.

We stress that the solvability of SIEP6b established in the above theorem is over the algebraically closed field $\mathbb{C}$. It is not known whether the problem is realistically solvable with positive masses, springs, and dampers (Gladwell 2001).

Theorem 4.3. (Xu 1998) SIEP7 is solvable if and only if

$$\prod_{k=1}^{n} |\mu_j - \lambda_k| \geq 2\beta(1 + (-1)^{n-j+1}),$$

for all $j = 1, \ldots, n-1$. Even in the case of existence, no uniqueness can be ascertained.

It is worth mentioning that Ferguson (1980) characterized periodic Jacobi matrices by a notion of 'compatible' data that can be turned into a numerical algorithm. Each set of compatible data uniquely determines a periodic Jacobi matrix. On the other hand, these sets of eigenvalues $\lambda = \{1, 3, 5\}$ and $\mu = \{2, 4\}$ with $\beta = 1$ server as a counterexample showing that SIEP7 is not solvable (Xu 1998).

Theorem 4.4. (Xu 1998) Assume that all eigenvalues are distinct. Define

$$\Delta_k = \det \begin{bmatrix} 1 & \ldots & 1 & 1 & \ldots & 1 \\ \lambda_1 & \ldots & e_1^T \tilde{J} e_1 & \lambda_{k+1} & \ldots & \lambda_{2n} \\ \vdots & \vdots & \vdots & \vdots & \vdots & \vdots \\ \lambda_1^{2n-1} & \ldots & e_1^T \tilde{J}^{2n-1} e_1 & \lambda_{k+1}^{2n-1} & \ldots & \lambda_{2n}^{2n-1} \end{bmatrix}.$$

Then SIEP9 has a unique solution if and only if

$$\Delta_k > 0$$

for all $k = 1, \ldots, 2n$.

A simple counterexample showing that SIEP9 is not always solvable is as follows. No 2×2 symmetric matrix J can have a fixed $(1,1)$ entry a_1 and eigenvalues satisfying either $a_1 < \lambda_1 < \lambda_2$ or $\lambda_1 < \lambda_2 < a_1$.

4.4. Sensitivity issues

If the numerical computation is to be done using finite precision arithmetic, it is critical to understand the perturbation behaviour of the underlying mathematical problem. The notion of conditioning is normally used as an indication of the sensitivity dependence.

For IEPs, partly because inverse problems are, by nature, harder to analyse than direct problems, and partly because most IEPs have multiple solutions, not many results on sensitivity analysis have been performed. We believe that this is an important yet widely open area for further research.

We mentioned earlier that such a study could have the application of finding a robust solution that is least sensitive to perturbations. Despite the fact that considerable effort has been devoted to the development of numerical algorithms for many of the IEPs discussed in this article, we should make it clear that thus far very little attention has been paid to this direction. The analysis of either the conditioning of IEPs or the stability of the associated numerical methods is lacking. For that reason, we can only partially address the sensitivity issues by demonstrating known results for SIEP6a in this section. Clearly, more work needs to be done.

For the direct problem, it is easy to see that the function $F : \mathbb{R}^n \times \mathbb{R}_+^{n-1} \longrightarrow \mathbb{R}^{2n-1}$ where

$$F(a_1, \ldots, a_n, b_1, \ldots, b_{n-1}) = (\sigma(J), \sigma(\bar{J}))$$

is differentiable. The well-posedness of the inverse problem was initially established by Hochstadt (1974).

Theorem 4.5. (Hochstadt 1974) The unique solution J to SIEP6a depends continuously on the given data $\{\lambda_1, \ldots, \lambda_n\}$ and $\{\mu_1, \ldots, \mu_{n-1}\}$.

Mere continuous dependence is not enough for numerical computation. We need to quantify how the solution is perturbed by the change in problem data. Using the implicit function theorem, Hald (1976) refined this dependence and provided the following sensitivity dependence.

Theorem 4.6. (Hald 1976) Suppose J and $\tilde{J}$ are the solutions to SIEP6a with data

$$\lambda_1 < \mu_1 < \lambda_2 < \cdots < \mu_{n-1} < \lambda_n,$$
$$\tilde{\lambda}_1 < \tilde{\mu}_1 < \tilde{\lambda}_2 < \cdots < \tilde{\mu}_{n-1} < \tilde{\lambda}_n,$$

respectively. Then there exists a constant K such that

$$\|J - \tilde{J}\|_F \leq K \left(\sum_{i=1}^{n} |\lambda_i - \tilde{\lambda}_i|^2 + \sum_{i=1}^{n-1} |\mu_i - \tilde{\mu}_i|^2 \right)^{1/2}, \tag{4.7}$$

where the constant K depends on the quantities

$$d = \max\{\lambda_n, \tilde{\lambda}_n\} - \min\{\lambda_1, \tilde{\lambda}_1\},$$
$$\epsilon_0 = \frac{1}{d} \min_{j,k}\{|\lambda_j - \mu_k|, |\tilde{\lambda}_j - \tilde{\mu}_k|\},$$
$$\delta_0 = \frac{1}{2d} \min_{j \neq k}\{|\lambda_j - \lambda_k|, |\mu_j - \mu_k|, |\tilde{\lambda}_j - \tilde{\lambda}_k|, |\tilde{\mu}_j - \tilde{\mu}_k|, \}.$$

which measure the separation of the given data.

The constant K in (4.7) is significant in that it determines how the perturbation in the given data would be amplified. Its actual quantity, however, remains opaque because the implicit function theorem warrants only its existence but not its content. Xu (1993) introduced a form of condition number that could be explicitly estimated for SIEP6a. As a general rule, the smaller the separation of the given data, the more ill conditioned SIEP6a becomes.

4.5. Numerical methods

We mentioned earlier that numerical algorithms for a JIEP often followed directly from constructive proofs of its existence. Nevertheless, some of the procedures are subtly unstable. To save space, we shall not evaluate each method in this survey. Rather, we shall illustrate the basic ideas of two popular approaches: the Lanczos method (Parlett 1998) and the orthogonal reduction method (Boley and Golub 1987).

We first recall the following theorem, which is the basis of the Lanczos approach.

Theorem 4.7. The orthogonal matrix Q and the upper Hessenberg matrix H with positive subdiagonal entries can be completely determined by a given matrix A and the last (or any) column of Q if the relationship $Q^TAQ = H$ holds.

In our application, we want to construct the symmetric tridiagonal matrix $J = Q^T\Lambda Q$ with $\Lambda = \operatorname{diag}(\lambda_1, \ldots, \lambda_n)$. Thus, if the last column $\mathbf{q}_n$ is known, then the Jacobi matrix J can constructed in finitely many steps:

$$
\begin{aligned}
a_n &:= \mathbf{q}_n^T\Lambda\mathbf{q}_n,\\
b_{n-1} &:= \|\Lambda\mathbf{q}_n - a_n\mathbf{q}_n\|,\\
\mathbf{q}_{n-1} &:= (\Lambda\mathbf{q}_n - a_n\mathbf{q}_n)/b_{n-1},\\
&\text{for } i = 1, \ldots, n-2 \;\{\\
&\qquad a_{n-i} := \mathbf{q}_{n-i}^T\Lambda\mathbf{q}_{n-i},\\
&\qquad b_{n-i-1} := \|\Lambda\mathbf{q}_{n-i} - a_{n-i}\mathbf{q}_{n-i} - b_{n-i}\mathbf{q}_{n-i+1}\|,\\
&\qquad q_{n-i-1} := (\Lambda\mathbf{q}_{n-i} - a_{n-i}\mathbf{q}_{n-i} - b_{n-i}\mathbf{q}_{n-i+1})/b_{n-i-1},\\
&\qquad \}\\
a_1 &:= \mathbf{q}_1^T\Lambda\mathbf{q}_1.
\end{aligned}
$$

It only remains to calculate the column vector $\mathbf{q}_n$. To that end, we recall a classical result by Thompson and McEnteggert (1968).

Theorem 4.8. (Thompson and McEnteggert 1968) Let $(\lambda_i, \mathbf{x}_i)$, $i = 1, \ldots, n$, be orthonormal eigenpairs that form the spectral decomposition of a given symmetric matrix A. Then

$$\operatorname{adj}(\lambda_i I - A) = \prod_{\substack{k=1\\k\neq i}}^{n} (\lambda_i - \lambda_k)\mathbf{x}_i\mathbf{x}_i^T. \tag{4.8}$$

Evaluating both sides of (4.8) at the (n, n) position, we obtain

$$\det(\lambda_i I_{n-1} - A_{n-1}) = x_{ni}^2 \prod_{\substack{k=1\\k\neq i}}^{n} (\lambda_i - \lambda_k)$$

where x_{ni} is the last entry of $\mathbf{x}_i$, and recalling that A_{n-1} denotes the principal submatrix of size $n-1$. In our application, the last column $\mathbf{q}_n$ is precisely $[x_{n1}, \ldots, x_{nn}]^T$, if A is replaced by J. It follows that

$$x_{ni}^2 = \frac{\prod_{k=1}^{n-1}(\lambda_i - \mu_k)}{\prod_{\substack{k=1\\k\neq i}}^{n}(\lambda_i - \lambda_k)}. \tag{4.9}$$

In other words, the last column $\mathbf{q}_n$ for J can be expressed in terms of the spectral data $\{\lambda_1, \ldots, \lambda_n\}$ and $\{\mu_1, \ldots, \mu_{n-1}\}$. The Lanczos algorithm kicks in and SIEP6a is solved in finitely many steps.

We remark that other types of JIEP can be solved in similar ways with appropriate modifications, but we shall not examine them here. Readers are referred to the review paper by Boley and Golub (1987) and the book by Xu (1998) for more details.

We caution that the method by de Boor and Golub (1978) using the orthogonal polynomial approach is entirely equivalent to the above Lanczos approach, but is less stable in the face of roundoff error. We suggest that a reorthogonalization process should take place even along the Lanczos steps to ensure stability.

In the orthogonal reduction method, the given data are used first to construct a bordered diagonal matrix A of the form

$$A = \begin{bmatrix} \alpha & \beta_1 & \ldots & \beta_{n-1} \\ \beta_1 & \mu_1 & & 0 \\ \vdots & & \ddots & \\ \beta_{n-1} & 0 & \ldots & \mu_{n-1} \end{bmatrix}$$

so that $\sigma(A) = \{\lambda_1, \ldots, \lambda_n\}$. Such a construction is entirely possible. First, α is trivially determined as $\alpha = \sum_{i=1}^{n} \lambda_i - \sum_{i=1}^{n-1} \mu_i$. Secondly, the characteristic polynomial is given by

$$\det(\lambda I - A) = (\lambda - \alpha)\prod_{k=1}^{n-1}(\lambda - \mu_k) - \sum_{i=1}^{n-1}\beta_i^2 \left(\prod_{\substack{k=1\\k\neq i}}^{n-1}(\lambda - \mu_k)\right). \tag{4.10}$$

Thus, border elements β_i are given by

$$\beta_i^2 = -\frac{\prod_{k=1}^{n}(\mu_i - \lambda_k)}{\prod_{\substack{k=1\\k\neq i}}^{n-1}(\mu_i - \mu_k)}.$$

Let $\boldsymbol{\beta} = [\beta_1, \ldots, \beta_{n-1}]^T$. The next step is to construct an orthogonal matrix Q efficiently so that

$$\begin{bmatrix} 1 & \mathbf{0}^T \\ \mathbf{0} & Q^T \end{bmatrix} A \begin{bmatrix} 1 & \mathbf{0}^T \\ 0 & Q \end{bmatrix} = \begin{bmatrix} \alpha & b_1\mathbf{e}_1^T \\ b_1\mathbf{e}_1 & \bar{J} \end{bmatrix} \tag{4.11}$$

becomes a Jacobi matrix. For this to happen, we must have $Q^T\boldsymbol{\beta} = b_1\mathbf{e}_1$, where $\mathbf{e}_1$ is the standard first coordinate vector in $\mathbb{R}^{n-1}$. It follows that $b_1 = \|\boldsymbol{\beta}\|$ and that the first column of Q is given by $\boldsymbol{\beta}/b_1$. The Lanczos procedure can now be employed to finish up the construction $Q^T\mathrm{diag}(\mu_1, \ldots, \mu_{n-1})Q = \bar{J}$ and SIEP6a is solved in finite steps.

Finally, we remark that other tridiagonalization process, including Householder transformations, Givens rotations, the Rutishauser method (Gragg and Harrod 1984), and so on, may also be used effectively to explore the bordered diagonal structure (Boley and Golub 1987).

5. Toeplitz inverse eigenvalue problems

Given a column vector $\mathbf{r} = [r_1, \ldots, r_n]^T$, a matrix $T = T(\mathbf{r})$ of the form

$$T := \begin{bmatrix} r_1 & r_2 & \cdots & r_{n-1} & r_n \\ r_2 & r_1 & & r_{n-2} & r_{n-1} \\ \vdots & \ddots & \ddots & & \vdots \\ r_{n-1} & & & r_1 & r_2 \\ r_n & r_{n-1} & & r_2 & r_1 \end{bmatrix}$$

is called a symmetric Toeplitz matrix. An inverse Toeplitz eigenvalue problem (**ToIEP**) concerns finding a vector $\mathbf{r} \in R^n$ so that $T(\mathbf{r})$ has a prescribed set of real numbers $\{\lambda_1, \ldots, \lambda_n\}$ as its spectrum. We mention in passing that a similar IEP could be asked for a Hankel matrix $H(\mathbf{r})$, related to the Toeplitz matrix $T(\mathbf{r})$ via $H(\mathbf{r}) = \Xi T(\mathbf{r})$ and $T(\mathbf{r}) = \Xi H(\mathbf{r})$, where Ξ is defined by (4.3). The set $\mathcal{T}(n)$ of symmetric Toeplitz matrices forms a subset of a larger class

$$\mathcal{C}(n) := \{M \in \mathbb{R}^{n\times n} | M = M^T, M = \Xi M \Xi\}$$

of centrosymmetric matrices, where Ξ is the unit perdiagonal matrix. A vector $\mathbf{v}$ is said to be *even* if $\Xi\mathbf{v} = \mathbf{v}$, and *odd* if $\Xi\mathbf{v} = -\mathbf{v}$. Every eigenspace of a centrosymmetric matrix, and hence a symmetric Toeplitz matrix, has a basis of even and odd vectors. The prescribed eigenvalues $\{\lambda_1, \ldots, \lambda_n\}$ in a ToIEP, therefore, should also carry a corresponding parity assignment, as we shall explore in the next section.

5.1. Symmetry and parity

In Table 5.1 we summarize some characteristics of centrosymmetric matrices (Cantoni and Bulter 1976). Depending on whether n is even or odd, any centrosymmetric matrix M must assume the symmetry as is indicated in the second row of Table 5.1, where $A, C, \Xi \in \mathbb{R}^{\lfloor \frac{n}{2} \rfloor \times \lfloor \frac{n}{2} \rfloor}$, $\mathbf{x} \in \mathbb{R}^{\lfloor \frac{n}{2} \rfloor}$, $q \in \mathbb{R}$, and $A = A^T$. Let K be the orthogonal matrix defined in the table. Then M can be decomposed into 2×2 diagonal blocks via orthogonal similarity transformation by K. The blocks assume the forms shown in the last row of Table 5.1.

Table 5.1. Structure of centrosymmetric matrices

n	even	odd
M	$\begin{bmatrix} A & C^T \\ C & \Xi A \Xi \end{bmatrix}$	$\begin{bmatrix} A & x & C^T \\ x^T & q & x^T\Xi \\ C & \Xi x & \Xi A \Xi \end{bmatrix}$
$\sqrt{2}K$	$\begin{bmatrix} I & -\Xi \\ I & \Xi \end{bmatrix}$	$\begin{bmatrix} I & 0 & -\Xi \\ 0 & \sqrt{2} & 0 \\ I & 0 & \Xi \end{bmatrix}$
KMK^T	$\begin{bmatrix} A - \Xi C & 0 \\ 0 & A + \Xi C \end{bmatrix}$	$\begin{bmatrix} A - \Xi C & 0 & 0 \\ 0 & q & \sqrt{2}x^T \\ 0 & \sqrt{2}x & A + \Xi C \end{bmatrix}$

Effectively, the spectral decomposition of M is reduced to that of two submatrices of about half the size. If Z_1 denotes the $\lfloor \frac{n}{2} \rfloor \times \lfloor \frac{n}{2} \rfloor$ matrix of orthonormal eigenvectors for $A - \Xi C$, then columns from the matrix $K^T \left[\begin{smallmatrix} Z_1 \\ 0 \end{smallmatrix} \right]$ will be eigenvectors of M. These eigenvectors are odd vectors. Similarly, there are $\lceil \frac{n}{2} \rceil$ even eigenvectors of M computable from those of $A + \Xi C$ or

$$\begin{bmatrix} q & \sqrt{2}x^T \\ \sqrt{2}x & A + \Xi C \end{bmatrix}.$$

It is interesting to ask whether a symmetric Toeplitz matrix can have arbitrary spectrum with arbitrary parity. Can the parity be arbitrarily assigned to the prescribed eigenvalues in a ToIEP?

We consider the 3×3 case to further explore this question. Any matrix $M \in \mathcal{C}(3)$ is of the form

$$M = \begin{bmatrix} m_{11} & m_{12} & m_{13} \\ \times & m_{22} & \times \\ \times & \times & \times \end{bmatrix},$$

where quantities denoted by $\times$ can be obtained by symmetry. Without loss of generality, we may assume that the trace of M is zero. In this way, the parameters are reduced to m_{11}, m_{12} and m_{13}. Let $\mathcal{M}_C = \mathcal{M}_C(\lambda_1, \lambda_2, \lambda_3)$ denote the subset of centrosymmetric matrices with eigenvalues $\{\lambda_1, \lambda_2, \lambda_3\}$. Assuming $\sum_{i=1}^3 \lambda_i = 0$, elements in $\mathcal{M}_C$ must satisfy the equations

$$\left(m_{11} - \frac{\lambda_{\varrho 1}}{4}\right)^2 + \frac{1}{2}m_{12}^2 = \frac{(\lambda_{\varrho 2} - \lambda_{\varrho 3})^2}{16},$$
$$m_{13} = m_{11} - \lambda_{\varrho 1},$$

where ϱ denotes any of the six permutations of $\{1, 2, 3\}$. Thus $\mathcal{M}_C$ consists of three ellipses in $\mathbb{R}^3$. It suffices to plot these ellipses in the (m_{11}, m_{12})-plane only, since m_{13} is simply a shift of m_{11}. Several plots with qualitatively different eigenvalues are depicted in Figure 5.1. Observe that it is always the case that one circumscribes the other two.

A 3×3 centrosymmetric matrix is a solution to the ToIEP only if $m_{11} = 0$. By counting the number of m_{12}-intercepts, we should be able to know the number of solutions to the ToIEP. Specifically, we find that there are 4 solutions if eigenvalues are distinct and 2 solutions if one eigenvalue has multiplicity 2. We note further that the parities of the prescribed eigenvalues in the ToIEP cannot be arbitrary. Each of the ellipses corresponds to one particular parity assignment among the eigenvalues. An 'incorrect' parity assignment, such as the two smallest ellipses in the left column of Figure 5.1, implies that there is no m_{12}-intercept and, hence, no isospectral Toeplitz matrix. As far as the ToIEP is concerned, parity assignment is not explicitly given as part of the constraint. As a safeguard for ensuring existence, it has been suggested in the literature that the ordered eigenvalues should have alternating parity.

5.2. Existence

Despite the simplicity of the appearance of a ToIEP, the issue of its solvability has been quite a challenge. Delsarte and Genin (1984) argued that the problem would be analytically intractable if $n \geq 5$. Eventually, using a topological degree argument, Landau (1994) settled the following theorem with a nonconstructive proof.

Theorem 5.1. (Landau 1994) Every set of n real numbers is the spectrum of an $n \times n$ real symmetric Toeplitz matrix.

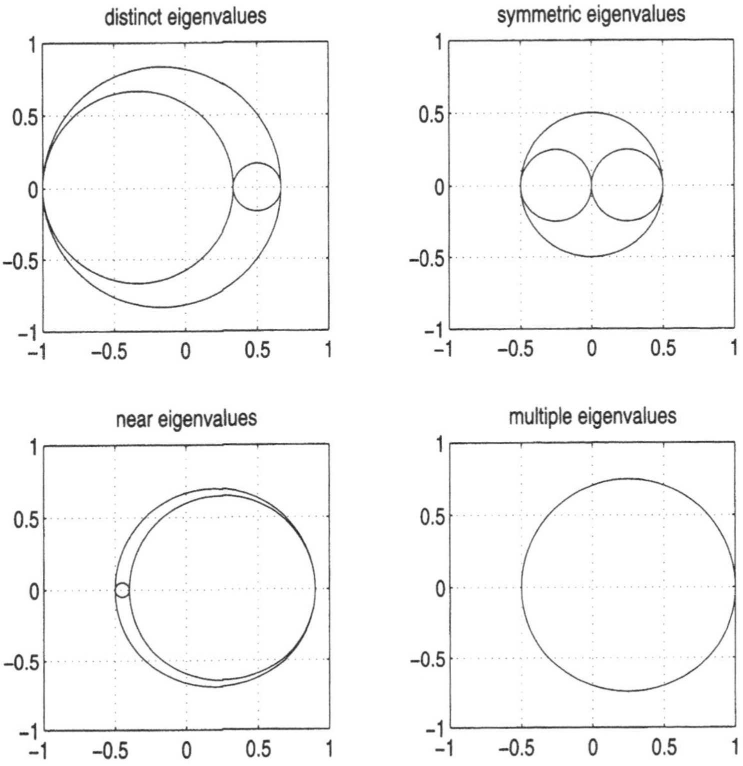

Figure 5.1. Plots of $\mathcal{M}_{\mathcal{C}}$ in the (m_{11}, m_{12})-plane

It might be worthwhile to briefly outline the proof as it shows the existence of a even more restricted class of Toeplitz matrices. A matrix $T(r_1, \ldots, r_n)$ is said to be *regular* if every principal submatrix $T(r_1, \ldots, r_k)$, $1 \le k \le n$, has distinct eigenvalues that, when arranged in ascending order, alternate parity with the largest one even. Consider the map $\varphi : \mathbb{R}^{n-2} \longrightarrow \mathbb{R}^{n-2}$ defined by

$$\varphi(t_3, \ldots, t_n) = (y_2, \ldots, y_{n-1}),$$

where $y_i := -\frac{\mu_i}{\mu_1}$, $i = 2, \ldots, n-1$, if $\mu_1 \le \cdots \le \mu_n$ are eigenvalues of $T(0, 1, t_3, \ldots, t_n)$. Note that since $\sum_{i=1}^{n} \mu_i = 0$, it is necessary that $\mu_1 < 0$. The range of φ is the simplex

$$\Delta := \left\{ (y_2, \ldots, y_{n-1}) \in \mathbb{R}^{n-2} \middle| \begin{array}{c} -1 \le y_2 \le \cdots \le y_{n-1} \\ y_2 + \cdots + y_{n-2} + 2y_{n-1} \le 1 \end{array} \right\}.$$

The key components in the proof by Landau (1994) are as follows. First, the set $\mathcal{F}$ of regular Toeplitz matrices of the form $T(0, 1, t_3, \ldots, t_n)$ is not empty.

Secondly, the map φ restricted to those points $(t_3, \ldots, t_n) \in \mathbb{R}^{n-2}$ such that $T(0, 1, t_3, \ldots, t_n) \in \mathcal{F}$ is a *surjective* map onto the interior of Δ. Finally, any given $\lambda_1 \leq \cdots \leq \lambda_n$ can be *shifted* and *scaled* to a unique point in Δ whose pre-image(s) can then be scaled and shifted backward to a symmetric Toeplitz matrix with eigenvalues $\{\lambda_1, \ldots, \lambda_n\}$.

5.3. Numerical methods

Lack of understanding does not necessarily preclude the development of effective numerical algorithms for the ToIEP. There are two basic approaches to tackle the ToIEP numerically: one by iteration and the other by continuation. We briefly describe the basic ideas for each approach in this section.

Regarding the ToIEP as a nonlinear system of n equations in n unknowns, a natural tactic would be a Newton-type iteration. The schemes in Friedland, Nocedal and Overton (1987), originally proposed for the more general class of PIEPs, are in this class and certainly applicable to the ToIEP. The inverse Rayleigh quotient algorithm in Laurie (1988, 1991) is also equivalent to a Newton-type variation. These methods do not exploit the Toeplitz structure and can suffer from local convergence. The iterative scheme proposed by Trench (1997) seems to have more robust performance, but still no global convergence can be proved. The following discussion is another Newton-type iteration by Chu (1994). The iterations are confined in $\mathcal{C}(n)$. Since the centrosymmetric structure is preserved, the cost is substantially reduced and the case of double eigenvalues can be handled effectively.

Recall that the classical Newton method

$$x^{(\nu+1)} = x^{(\nu)} - (f'(x^{(\nu)}))^{-1} f(x^{(\nu)})$$

for a scalar function $f : \mathbb{R} \longrightarrow \mathbb{R}$ can be thought of as two steps: the tangent step where $x^{(\nu+1)}$ is the x-intercept of the tangent line from $(x^{(\nu)}, f(x^{(\nu)}))$ of the graph of f, and the lift step where the point $(x^{(\nu+1)}, f(x^{(\nu+1)}))$ is a natural lift of the intercept along the y-axis to the graph of f. Let the given eigenvalues $\{\lambda_1, \ldots, \lambda_n\}$ be arranged as $\boldsymbol{\lambda} = [\phi_1, \ldots, \phi_{\lfloor \frac{n}{2} \rfloor}, \psi_1, \ldots, \psi_{\lceil \frac{n}{2} \rceil}]^T$, where ϕ_k and ψ_k are of parity even and odd, respectively. Let $\Lambda = \mathrm{diag}(\boldsymbol{\lambda})$. An analogue of this idea applied to the ToIEP is to think of the isospectral subset $\mathcal{M}_C = \mathcal{M}_C(\Lambda)$ of centrosymmetric matrices as the graph of some unknown f, and the subspace $\mathcal{T}(n)$ of symmetric Toeplitz matrices as the x-axis. We want to do the tangent and lift iterations between these two entities. From Section 5.1, we see that every element $M \in \mathcal{M}_C$ can be characterized by the parameter $Z \in \mathcal{O}(\lfloor \frac{n}{2} \rfloor) \times \mathcal{O}(\lceil \frac{n}{2} \rceil)$ where $M = Q\Lambda Q^T$ and $Z = KQ$. It follows that tangent vectors of $\mathcal{M}_C$ at M are of the form

$$T_M(\mathcal{M}_C) = \tilde{S}M - M\tilde{S}, \tag{5.1}$$

with $\tilde{S} := Q\mathrm{diag}(S_1, S_2)Q^T$ where S_1 and S_2 are arbitrary skew-symmetric matrices in $\mathbb{R}^{\lfloor\frac{n}{2}\rfloor\times\lfloor\frac{n}{2}\rfloor}$ and $\mathbb{R}^{\lceil\frac{n}{2}\rceil\times\lceil\frac{n}{2}\rceil}$, respectively. Thus a tangent step from a given $M^{(\nu)} \in \mathcal{M}_C(\Lambda)$ amounts to finding a skew-symmetric matrix $\tilde{S}^{(\nu)}$ and a vector $\mathbf{r}^{(\nu+1)}$ so that

$$M^{(\nu)} + \tilde{S}^{(\nu)}M^{(\nu)} - M^{(\nu)}\tilde{S}^{(\nu)} = T(\mathbf{r}^{(\nu+1)}). \tag{5.2}$$

Assume the spectral decomposition $M^{(\nu)} = Q^{(\nu)}\Lambda Q^{(\nu)^T}$ and define $Z^{(\nu)} = KQ^{(\nu)}$. The tangent equation is reduced to

$$\Lambda + S^{(\nu)}\Lambda - \Lambda S^{(\nu)} = Z^{(\nu)^T}\left(KT(\mathbf{r}^{(\nu+1)})K^T\right)Z^{(\nu)}, \tag{5.3}$$

where $S^{(\nu)} = Q^{(\nu)^T}\tilde{S}^{(\nu)}Q^{(\nu)}$ remains skew-symmetric. Note that the product $KT(\mathbf{r}^{(\nu+1)})K^T$ is a 2×2 diagonal block matrix, denoted by $\operatorname{diag}(T_1^{(\nu+1)}, T_2^{(\nu+1)})$, because $T(\mathbf{r}^{(\nu+1)})$ is centrosymmetric. We also know from the discussion in Section 5.1 that $Z^{(\nu)} = \operatorname{diag}(Z_1^{(\nu)}, Z_2^{(\nu)})$. Thus the system (5.3) is effectively split in half.

We first retrieve the vector $\mathbf{r}^{(\nu+1)}$. It suffices to compare the diagonal elements on both sides without reference to $S^{(\nu)}$. Note that the right-hand side of (5.3) is linear in $\mathbf{r}^{(\nu+1)}$. This linear relationship can be expressed as

$$\Omega^{(\nu)}\mathbf{r}^{(\nu+1)} = \lambda$$

for $r^{(\nu+1)}$, where the entries in the matrix $\Omega^{(\nu)} = [\Omega_{ij}^{(\nu)}]$ are defined by

$$\Omega_{ij}^{(\nu)} := \begin{cases} (Z_1^{(\nu)})_{*i}^T E_1^{[j]}(Z_1^{(\nu)})_{*i}, & \text{if } 1 \le i \le \lfloor\frac{n}{2}\rfloor, \\ (Z_2^{(\nu)})_{*i}^T E_2^{[j]}(Z_2^{(\nu)})_{*i}, & \text{if } \lfloor\frac{n}{2}\rfloor < i \le n. \end{cases}$$

In the above, $E_1^{[j]}$ and $E_2^{[j]}$ are the diagonal blocks in the 2×2 diagonal block matrix $KT(\mathbf{e}_j)K^T$, $\mathbf{e}_j$ is the jth standard unit vector, and $(Z_k^{(\nu)})_{*i}$ denotes the ith column of the matrix $Z_k^{(\nu)}$. Throughout the calculations, we need only to multiply vectors or matrices of lengths $\lfloor\frac{n}{2}\rfloor$ or $\lceil\frac{n}{2}\rceil$. Once $T(\mathbf{r}^{(\nu+1)})$ is determined, off-diagonal elements in (5.3) determine $S^{(\nu)}$. Specifically, if eigenvalues within each parity group are distinct, then it is easy to see that

$$(S_1^{(\nu)})_{ij} = \frac{(Z_1^{(\nu)})_{*i}^T T_1^{(\nu+1)}(Z_1^{(\nu)})_{*j}}{\phi_i - \phi_j}, \quad 1 \le i < j \le \left\lfloor\frac{n}{2}\right\rfloor,$$

$$(S_2^{(\nu)})_{ij} = \frac{(Z_2^{(\nu)})_{*i}^T T_2^{(\nu+1)}(Z_2^{(\nu)})_{*j}}{\psi_i - \psi_j}, \quad 1 \le i < j \le \left\lceil\frac{n}{2}\right\rceil.$$

This completes the calculation for the tangent step. We remark that the scheme is capable of handling the case of double eigenvalues because such eigenvalues have to be split into one even and one odd (Delsarte and Genin 1984).

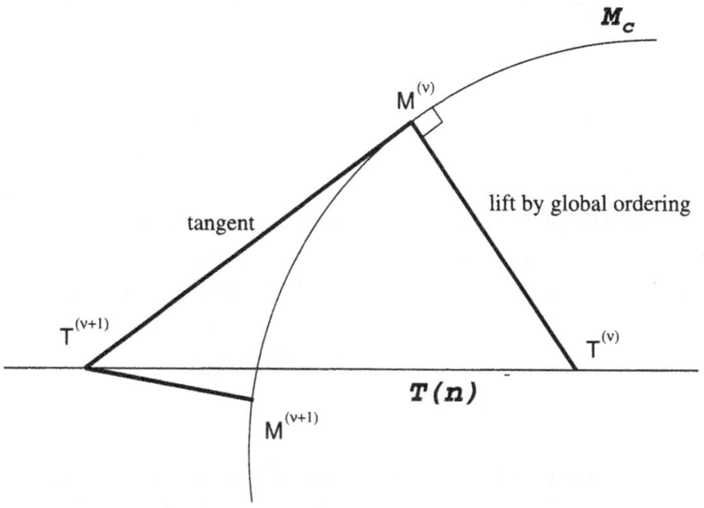

Figure 5.2. Geometry of lift by Wielandt–Hoffman theorem

To obtain a lift from $T(r^{(\nu+1)})$ to $\mathcal{M}_C$, we look for any matrix $M^{(\nu+1)} \in \mathcal{M}_C$ that is nearest to $T(r^{(\nu+1)})$. The idea is sketched in Figure 5.2. Such a nearest approximation can be obtained by the Wielandt–Hoffman theorem. That is, suppose the spectral decomposition of $T(r^{(\nu+1)})$ is given by

$$\overline{Z}^{(\nu+1)^T} KT(r^{(\nu+1)})K^T\overline{Z}^{(\nu+1)} = \begin{bmatrix} \overline{\Lambda}_1^{(\nu+1)} & 0 \\ 0 & \overline{\Lambda}_2^{(\nu+1)} \end{bmatrix}.$$

Rearrange $\{\lambda_1, \ldots, \lambda_n\}$ in the same ordering as in $\overline{\Lambda}_1^{(\nu+1)}$ and $\overline{\Lambda}_2^{(\nu+1)}$ to obtain $\tilde{\Lambda}_1^{(\nu+1)}$ and $\tilde{\Lambda}_2^{(\nu+1)}$. Then the nearest approximation $M^{(\nu+1)}$ is given by

$$M^{(\nu+1)} := K^T\overline{Z}^{(\nu+1)}\mathrm{diag}(\tilde{\Lambda}_1^{(\nu+1)}, \tilde{\Lambda}_2^{(\nu+1)})\overline{Z}^{(\nu+1)^T} K.$$

For computational purposes, this $M^{(\nu+1)}$ never needs to be calculated. We only need to repeat the tangent step with the new parity assignment $\Lambda = \mathrm{diag}(\tilde{\Lambda}_1^{(\nu+1)}, \tilde{\Lambda}_2^{(\nu+1)})$ and the new parameter matrix $Z^{(\nu+1)} := \overline{Z}^{(\nu+1)}$. It can be proved that this method converges quadratically.

An alternative approach, fundamentally different from iterative methods, is to solve the ToIEP by tracing curves defined by differential equations. The idea is to continually transform the matrix Λ until a Toeplitz matrix is found. One such formulation is the initial value problem

$$\begin{cases} \frac{\mathrm{d}X}{\mathrm{d}t} &= [X, k(X)], \\ X(0) &= \Lambda, \end{cases} \tag{5.4}$$

where $[A,B] := AB - BA$ denotes the Lie bracket and $k(X) = [k_{ij}(X)]$ is defined by

$$k_{ij}(X) := \begin{cases} x_{i+1,j} - x_{i,j-1}, & \text{if } 1 \le i < j \le n, \\ 0, & \text{if } 1 \le i = j \le n, \\ x_{i,j-1} - x_{i+1,j}, & \text{if } 1 \le j < i \le n. \end{cases} \tag{5.5}$$

The skew-symmetry of $k(X)$ guarantees that the solution $X(t)$ to (5.4) exists for all t and enjoys a continuous spectral decomposition $X(t) = Q(t)^T \Lambda Q(t)$. The orthogonal transformation $Q(t)$ is determined by the initial value problem

$$\begin{cases} \frac{\mathrm{d}Q}{\mathrm{d}t} &= Qk(Q^T \Lambda Q), \\ Q(0) &= I. \end{cases} \tag{5.6}$$

The limiting behaviour of $Q(t)$ determines the limiting behaviour of $X(t)$ and *vice versa*. The hope of getting a solution to the ToIEP hinges upon that $k(X) = 0$ if and only if X is a Toeplitz matrix. For that reason, $k(X)$ is called a Toeplitz annihilator in Chu (1993). The system (5.6) can be integrated effectively by available geometric integrators (Calvo, Iserles and Zanna 1997, Dieci, Russel and Vleck 1994, Iserles, Munthe-Kaas, Nørsett and Zanna 2000). Numerical experiences seem to suggest that the flows always converge, but a rigorous proof is still missing.

6. Nonnegative inverse eigenvalue problems

The nonnegative inverse eigenvalue problem (**NIEP**) concerns the construction of an entry-wise nonnegative matrix $A \in \mathbb{R}^{n \times n}$ with a prescribed set $\{\lambda_1, \ldots, \lambda_n\}$ of eigenvalues. Partially because of the important Perron–Frobenius theory, this inverse problem has drawn considerable interest in the literature.

The earliest study of this subject was perhaps that of the Russian mathematician Suleĭmanova (1949) on stochastic matrices, followed by Perfect (1953, 1955). The first systematic treatment of eigenvalues of symmetric nonnegative matrices should probably be attributed to Fiedler (1974). A more comprehensive study was conducted by Boyle and Handelman (1991), using the notion of symbolic dynamics to characterize the conditions that a given set is a portion of the spectrum of a nonnegative matrix or primitive matrix. General treatises on nonnegative matrices and their applications include the books by Berman and Plemmons (1979) and Minc (1988). Both books devote extensive discussion to the NIEP as well. Most of the discussions in the literature centre around finding conditions to qualify a given set of values as the spectrum of some nonnegative matrices. A short list of references giving various necessary or sufficient conditions includes Barrett and Johnson (1984), Boyle and Handelman (1991), Friedland (1978), Friedland and Melkman (1979), Loewy and London (1978), de Oliveira (1983) and Reams (1996). The trouble is that the necessary condition is usually too general and the sufficient condition too specific. Under a few special sufficient conditions, the nonnegative matrices can be constructed numerically (Soules 1983). General numerical treatments for NIEPs, even knowing the existence of a solution, are not available at the time of writing.

A further refinement in the posing of NIEPs has also attracted some attention. Suppose the given eigenvalues $\{\lambda_1, \ldots, \lambda_n\}$ are all real. The real-valued nonnegative inverse eigenvalue problem (**RNIEP**) concerns which set of values $\{\lambda_1, \ldots, \lambda_n\}$ occurs as the spectrum of a nonnegative matrix. The symmetric nonnegative inverse eigenvalue problem (**SNIEP**) concerns which set occurs as the spectrum of a symmetric nonnegative matrix. It was proved that there exist real numbers $\{\lambda_1, \ldots, \lambda_n\}$ that solve the RNIEP but *not* the SNIEP (Guo 1996, Johnson, Laffey and Loewy 1996).

6.1. Some existence results

The solvability of the NIEP has been the major issue of discussion in the literature. Existence results, either necessary or sufficient, are too numerous to be listed here. We shall mention only two results that, in some sense, provide the most distinct criteria in this regard.

Given a matrix A, the moments of A are defined to be the sequence of numbers $s_k = \text{trace}(A^k)$. Recall that, if $\sigma(A) = \{\lambda_1, \ldots, \lambda_n\}$, then

$$s_k = \sum_{i=1}^{n} \lambda_i^k.$$

For nonnegative matrices, the moments are always nonnegative. The following necessary condition is due to Loewy and London (1978).

Theorem 6.1. (Loewy and London 1978) Suppose $\{\lambda_1, \ldots, \lambda_n\}$ are eigenvalues of an $n \times n$ nonnegative matrix. Then the inequalities

$$s_k^m \leq n^{m-1} s_{km} \tag{6.1}$$

hold for all $k, m = 1, 2, \ldots$.

Note also that the inequalities in (6.1) are sharp, being equalities for the identity matrix. If we further limit the inverse problem to positive matrices, that is, every entry exceeds zero, it turns out that the eigenvalues can be completely characterized. The following necessary and sufficient condition appeared at the end of the long treatise by Boyle and Handelman (1991, p. 313)

Theorem 6.2. (Boyle and Handelman 1991) The set $\{\lambda_1, \ldots, \lambda_n\} \subset \mathbb{C}$, with $\lambda_1 = \max_{1 \le i \le n} |\lambda_i|$, is the nonzero spectrum of a positive matrix of size $m \ge n$ if and only if:

(1) $\lambda_1 > |\lambda_i|$ for all $i > 1$,
(2) $s_k > 0$ for all $k = 1, 2, \ldots,$ and
(3) $\prod_{i=1}^n (t - \lambda_i)$ has real coefficients in t.

6.2. Symmetric nonnegative inverse eigenvalue problem

We shall discuss a least squares approach for solving the general NIEP in the next section. At present, we touch upon the SNIEP with a few more comments.

First, we remark that solvability of SNIEPs remains open. Some sufficient conditions are listed in Berman and Plemmons (1979, Chapter 4). The set $\lambda = \{\sqrt[3]{51} + \epsilon, 1, 1, 1, -3, -3\}$, with $\epsilon > 0$, however, does not satisfy any of these conditions. In fact, it cannot be the nonzero spectrum of any symmetric nonnegative matrix (Johnson *et al.* 1996).

Friedland and Melkman (1979) limited the consideration of NIEPs to symmetric tridiagonal structure. A simple result can be established.

Theorem 6.3. (Friedland and Melkman 1979) A set of real numbers $\lambda_1 \ge \lambda_2 \ge \cdots \ge \lambda_n$ is the spectrum of an $n \times n$ nonnegative tridiagonal matrix if and only if $\lambda_i + \lambda_{n-i+1} = 0$ for all i. In this case, the matrix is given by $J = \operatorname{diag}(A_1, \cdots, A_{[(n+1)/2]})$, where

$$A_i = \frac{1}{2} \begin{bmatrix} \lambda_i + \lambda_{n-i+1} & \lambda_i - \lambda_{n-i+1} \\ \lambda_i - \lambda_{n-i+1} & \lambda_i + \lambda_{n-i+1} \end{bmatrix}, \quad 1 \le i < (n+1)/2,$$

and $A_i = [\lambda_i]$, if $i = (n+1)/2$ and n is odd.

Note that the matrix J constructed above is reducible when $n > 2$. The problem becomes harder if J is required further to be Jacobi, that is, with positive subdiagonal elements. One sufficient but not necessary condition for this particular JIEP is as follows.

Theorem 6.4. If $\lambda_1 > \lambda_2 > \cdots > \lambda_n$ and if $\lambda_i + \lambda_{n-i+1} > 0$ for all i, then there exists a positive Jacobi matrix with spectrum $\{\lambda_1, \ldots, \lambda_n\}$.

It would be very difficult to achieve a nearly simple characterization of solution to a general SNIEP. On the other hand, we could formulate the SNIEP as a constrained optimization problem of *minimizing* the objective function

$$F(Q, R) := \frac{1}{2} \|Q^T \Lambda Q - R \circ R\|^2,$$

subject to the constraint $(Q, R) \in \mathcal{O}(n) \times \mathcal{S}(n)$, where $\circ$ stands for the

Hadamard product and $\mathcal{S}(n)$ stands for the subspace of $n \times n$ symmetric matrices. The idea is to parametrize any symmetric matrix $X = Q^T \Lambda Q$ that is isospectral to Λ by the orthogonal matrix Q and to parametrize any symmetric nonnegative matrix $Y = R \circ R$ by the symmetric matrix R via entry-wise squares. The SNIEP is solvable if and only if $F(Q, R) = 0$ for some Q and R. Such a formulation offers a handle for numerical computation by optimization techniques. In Chu and Driessel (1991), the dynamical system

$$\begin{cases} \frac{\mathrm{d}X}{\mathrm{d}t} &= [X, [X, Y]], \\ \frac{\mathrm{d}Y}{\mathrm{d}t} &= 4Y \circ (X - Y), \end{cases} \tag{6.2}$$

resulting from projected gradient flow, was studied as a possible numerical means for solving the SNIEP. It is interesting to note that, even if the SNIEP is not solvable, the limit point of the gradient flow gives rise to a least squares solution. We shall discuss an analogous approach of (6.2) for NIEPs in Section 7.2.

7. Stochastic inverse eigenvalue problems

An $n \times n$ nonnegative matrix is a (row) stochastic matrix if all its row sums are 1. The stochastic inverse eigenvalue problem (**StIEP**) concerns the construction of a stochastic matrix with prescribed spectrum. Clearly the StIEP is a special NIEP with the additional row sum structure. It should be noted that, in contrast to the linearly constrained IEPs discussed thus far, the structure involved in the StIEP is nonlinear in the sense that the sum of two (stochastic) structured matrices does not have the same (stochastic) structure.

The Perron–Frobenius theorem asserts that the spectral radius $\rho(A)$ of an irreducible nonnegative matrix A is a positive maximal eigenvalue of A. The corresponding maximal eigenvector can be chosen to have all elements positive. Recall also that the set of reducible matrices forms a subset of measure zero. With this in mind, the spectral properties for stochastic matrices do not differ much from those of other nonnegative matrices, because of the following result (Minc 1988).

Theorem 7.1. Let A be any nonnegative matrix with positive maximal eigenvalue r and a positive maximal eigenvector x. Let $D = \mathrm{diag}(x)$. Then $D^{-1}r^{-1}AD$ is a stochastic matrix.

This motivates the notion that, if we could construct a generic solution for the NIEP, then we would also solve the StIEP by a diagonal similarity transformation. We shall pursue this idea as a numerical method for both the NIEP and the StIEP.

7.1. Existence

The StIEP is a hard problem. First, we note that Minc (1988) called our StIEP an inverse *spectrum* problem to distinguish it carefully from the problem of determining conditions under which one single complex number is an eigenvalue of a stochastic matrix. For the latter, the set Θ_n of points in the complex plane that are eigenvalues of any $n \times n$ stochastic matrices has been completely characterized by Karpelevič (1951). The complete statement of Karpelevič's theorem is rather lengthy (Minc 1988, Theorem 1.8), so we shall only highlight the main points below.

Theorem 7.2. (Karpelevič 1951) The region Θ_n is contained in the unit disk and is symmetric with respect to the real axis. It intersects the unit circle at points $e^{2\pi i a/b}$ where a and b range over all integers such that $0 \le a < b \le n$. The boundary of Θ_n consists of curvilinear arcs connecting these points in circular order. Any λ on these arcs must satisfy one of these equations:

$$\lambda^q(\lambda^p - t)^r = (1-t)^r,$$
$$(\lambda^b - t)^d = (1-t)^d \lambda^q,$$

where $0 \le t \le 1$, and b, d, p, q, r are natural integers determined by certain specific rules (explicitly given in Karpelevič (1951) and Minc (1988)).

An example Θ_4 is sketched in Figure 7.1. It should be stressed that the Karpelevič theorem characterizes only one complex value a time. It does not provide further insights into when two or more points in Θ_n are eigenvalues of the *same* stochastic matrix. It provides only a necessary condition for the StIEP.

We conclude this section with perhaps the first sufficient condition due to Suleĭmanova (1949).

Theorem 7.3. (Suleĭmanova 1949) The n real numbers $1, \lambda_1, \ldots, \lambda_{n-1}$, with $|\lambda_j| < 1$, are the characteristic values of some positive stochastic matrix of order n if $\sum |\lambda_j| < 1$, where the sum is over the js with $\lambda_j < 0$. If the λ_js are all negative the condition is also necessary.

Most of the sufficient conditions for the StIEP are imposed on real eigenvalues (Suleĭmanova 1949, Perfect 1953, Perfect 1955). If the resulting nonnegative matrix is generic, then the sufficient conditions for the NIEP also apply to the StIEP by Theorem 7.1.

7.2. Numerical method

It appears that, except for Soules (1983), none of the proofs for sufficient conditions is constructive, and no numerical algorithms are available even if a sufficient condition is satisfied. Even with Soules (1983), the construction

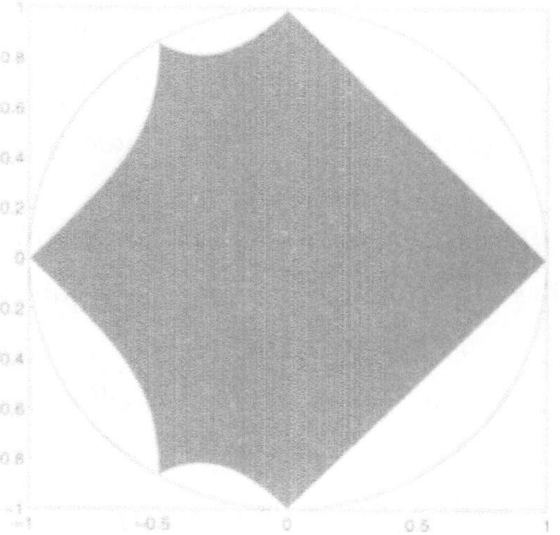

Figure 7.1. Θ_4 by the Karpelevič theorem

is limited in that the components of the Perron eigenvector need to satisfy additional inequalities. Recently, Chu and Guo (1998) proposed the following least squares approach that might be employed to solve the NIEP and the StIEP for generally prescribed eigenvalues. The idea is parallel to that of (6.2).

The diagonal matrix $\Lambda = \text{diag}(\lambda_1, \ldots, \lambda_n)$ can be transformed, if necessary, into a diagonal block matrix with 2×2 real blocks, if some of the given eigenvalues appear in complex conjugate pairs. The set of isospectral matrices

$$\mathcal{M}(\Lambda) = \{P\Lambda P^{-1} \in \mathbb{R}^{n\times n} | P \in \mathbb{R}^{n\times n} \text{ is nonsingular}\}$$

is parametrized by nonsingular matrices P. The cone of nonnegative matrices

$$\pi(R^n_+) := \{B \circ B | B \in \mathbb{R}^{n\times n}\}$$

is characterized by the Hadamard product of general square matrices. A solution to the NIEP must be at the intersection of $\mathcal{M}(\Lambda)$ and $\pi(R^n+)$, if there is any. If such a nonnegative matrix has positive maximal eigenvector, it can be reduced to a stochastic matrix by diagonal similarity transformation. We thus formulate the constrained optimization problem

$$\begin{aligned} &\text{Minimize} \quad F(P,R) := \frac{1}{2}\|P\Lambda P^{-1} - R\circ R\|^2, \\ &\text{Subject to} \quad P \in Gl(n), \ R \in \mathbb{R}^{n\times n}, \end{aligned}$$

where $Gl(n)$ denotes the group of invertible matrices in $\mathbb{R}^{n\times n}$. We use P and R as variables to manoeuvre elements in $\mathcal{M}(\Lambda)$ and $\pi(R^n_+)$ to reduce the objective value. Clearly, the feasible sets are open and a minimum may not exist. With respect to the induced inner product,

$$\langle (X_1, Y_1), (X_2, Y_2) \rangle := \langle X_1, X_2 \rangle + \langle Y_1, Y_2 \rangle,$$

in the product topology of $\mathbb{R}^{n\times n} \times \mathbb{R}^{n\times n}$, the gradient of F can be expressed as the pair

$$\nabla F(P, R) = ([\Delta(P, R), M(P)^T]P^{-T}, -2\Delta(P, R) \circ R), \tag{7.1}$$

where we recall that $[\cdot, \cdot]$ is the Lie bracket, and we abbreviate

$$M(P) := P\Lambda P^{-1},$$
$$\Delta(P, R) := M(P) - R \circ R.$$

The differential system

$$\frac{\mathrm{d}P}{\mathrm{d}t} = [M(P)^T, \Delta(P, R)]P^{-T},$$
$$\frac{\mathrm{d}R}{\mathrm{d}t} = 2\Delta(P, R) \circ R$$

thus provides a steepest descent flow on the feasible set $Gl(n) \times \mathbb{R}^{n\times n}$ for the objective function $F(P, R)$.

There is an unexpected advantage that deserves mentioning. Note that the zero structure in the original matrix $R(0)$ is preserved throughout the integration due to the Hadamard product. This feature may be exploited to construct a Markov chain with prescribed linkages and spectrum. That is, if it is desirable that state i is not allowed to transit into state j, we assign $r_{ij} = 0$ in the initial value $R(0)$. That zero transit status is maintained throughout the evolution.

On the other hand, the solution flow $P(t)$ is susceptible to becoming singular and the involvement of P^{-1} is somewhat worrisome. A remedy is to monitor the analytic singular value decomposition (ASVD) (Bunse-Gerstner, Byers, Mehrmann and Nichols 1991, Wright 1992),

$$P(t) = X(t)S(t)Y(t)^T, \tag{7.2}$$

of the path of matrices $P(t)$. In (7.2), $X(t)$ and $Y(t)$ are orthogonal, $S(t)$ is diagonal, and all are analytic in t. Such an ASVD flow exists because the solution $P(t)$ is analytic, by the Cauchy–Kovalevskaya theorem. The flow $(P(t), R(t))$ is equivalent to the flow of $(X(t), S(t), Y(t), R(t))$, where the differential equations governing $X(t), S(t)$, and $Y(t)$ can be obtained as follows (Wright 1992).

On differentiating (7.2), we have

$$X^T \frac{\mathrm{d}P}{\mathrm{d}t} Y = \underbrace{X^T \frac{\mathrm{d}X}{\mathrm{d}t}}_{Z} S + \frac{\mathrm{d}S}{\mathrm{d}t} + S \underbrace{\frac{\mathrm{d}Y^T}{\mathrm{d}t} Y}_{W}, \tag{7.3}$$

where both Z and W are skew-symmetric matrices. Note that $Q := X^T \frac{dP}{dt} Y$ is known since $\frac{dP}{dt}$ is already specified. Note also that the inverse of $P(t)$ can be calculated from $P^{-1} = YS^{-1}X^T$, whereas the diagonal entries of $S = \text{diag}\{s_1, \ldots, s_n\}$ provide us with information about the proximity of $P(t)$ to singularity. The diagonals on both sides of (7.3) lead to the equation

$$\frac{dS}{dt} = \text{diag}(Q)$$

for $S(t)$. The off-diagonals on both sides of (7.3) give rise to

$$\frac{dX}{dt} = XZ,$$
$$\frac{dy}{dt} = YW,$$

where, if $s_k^2 \neq s_j^2$ (in case of equality, W and Z can still be obtained by other means (Wright 1992)), entries of W and Z are obtained from

$$z_{jk} = \frac{s_k q_{jk} + s_j q_{kj}}{s_k^2 - s_j^2},$$
$$w_{jk} = \frac{s_j q_{jk} + s_k q_{kj}}{s_j^2 - s_k^2}$$

for all $j > k$. The flow is now ready to be integrated by available geometric integrators.

We note that $\frac{dF(P(t),R(t))}{dt} = -\|\nabla F(P(t), R(t))\|^2 \leq 0$. Thus the method fails to solve the NIEP only in two situations: either $P(t)$ becomes singular in finite time or $F(P(t), R(t))$ converges to a least squares local solution. In the former case, a restart might avoid the problem. In the latter case, either the NIEP has no solution at all or the algorithm needs a restart.

8. Unitary Hessenberg inverse eigenvalue problems

Eigenvalues of unitary matrices are on the unit circle. The unitary Hessenberg inverse eigenvalue problem (**UHIEP**) concerns the construction of an unitary Hessenberg matrix with prescribed points on the unit circle. Eigenvalue problems for unitary Hessenberg matrices arise naturally in several signal processing applications, including the frequency estimation procedure and the harmonic retrieval problem for radar or sonar navigation. The characteristic polynomials of unitary Hessenberg matrices are the well-known Szegő polynomials. The Szegő polynomials are orthogonal with respect to a certain measure on the unit circle in just the same way as the characteristic polynomials of the Jacobi matrices are orthogonal with respect to a certain weight on an interval. In many ways, theory and algorithms for the UHIEP are similar to those for the JIEP. Most of the discussion in this section are results adapted from Ammar, Gragg and Reichel (1991) and Ammar and He (1995).

Any upper Hessenberg unitary matrix H with positive subdiagonal entries can uniquely expressed as the product

$$H = G_1(\eta_1) \cdots G_{n-1}(\eta_{n-1})\tilde{G}_n(\eta_n), \tag{8.1}$$

where η_k are complex numbers with $|\eta_k| < 1$ for $1 \leq k < n$ and $|\eta_n| = 1$, each $G_k(\eta_k)$, $k = 1, \ldots, n-1$ is a Givens rotation,

$$G_k(\eta_k) = \begin{bmatrix} I_{k-1} & & & \\ & -\eta_k & \zeta_k & \\ & \zeta_k & \bar{\eta}_k & \\ & & & I_{n-k+1} \end{bmatrix}$$

with $\zeta_k := \sqrt{1 - |\eta_k|^2}$ and $\tilde{G}_n(\eta_n) = \operatorname{diag}[I_{n-1}, -\eta_n]$. In other words, each unitary upper Hessenberg matrix is determined by $2n-1$ *real* parameters, another feature analogous to the Jacobi matrices. For convenience, this dependence is denoted by $H = H(\eta_1, \ldots, \eta_n)$. The decomposition (8.1), referred to as the Schur parametrization, plays a fundamental role in efficient algorithms for upper Hessenberg unitary matrices.

There are considerable similarities between UHIEPs and JIEPs. The following two UHIEPs, for example, are analogous to SIEP8 and SIEP6a, respectively.

Theorem 8.1. (Ammar and He 1995) Given two sets $\{\lambda_1, \ldots, \lambda_n\}$ and $\{\mu_1, \ldots, \mu_n\}$ of strictly interlaced points on the unit circle, there exist a unique unitary upper Hessenberg matrix $H = H(\eta_1, \ldots, \eta_n)$ and a unique complex number α of unit modulus such that $\sigma(H) = \{\lambda_1, \ldots, \lambda_n\}$ and $\sigma(H(\alpha\eta_1, \ldots, \alpha\eta_n)) = \{\mu_1, \ldots, \mu_n\}$.

Note that the matrix $\tilde{H} = H(\alpha\eta_1, \ldots, \alpha\eta_n)$ is a rank-one perturbation of $H(\eta_1, \ldots, \eta_n)$, because $\tilde{H} = (I - (1-\alpha)\mathbf{e}_1\mathbf{e}_1^T)H$. Unlike SIEP8, this perturbation does not just affect the (n, n) entry.

The leading principal submatrix H_{n-1} of a unitary matrix is not unitary, and its eigenvalues do not lie on the unit circle. One way to modify the notion of submatrix is as follows.

Theorem 8.2. (Ammar and He 1995) Given two sets of strictly interlaced points $\{\lambda_1, \ldots, \lambda_n\}$ and $\{\mu_0, \mu_1, \ldots, \mu_{n-1}\}$ on the unit circle, there exist a unique unitary upper Hessenberg matrix $H = H(\eta_1, \ldots, \eta_n)$ such that $\sigma(H) = \{\lambda_1, \ldots, \lambda_n\}$ and $\sigma(H(\eta_1, \ldots, \eta_{n-2}, \rho_{n-1})) = \{\mu_1, \ldots, \mu_{n-1}\}$ with $\rho_{n-1} = (\eta_{n-1} + \bar{\mu}_0\eta_n)/(1 + \bar{\mu}_0\bar{\eta}_{n-1}\eta_n)$.

Just like JIEPs, the proofs of existence for the above results can be turned into numerical methods, such as the Lanczos/Arnoldi algorithm. More details can be found in Ammar *et al.* (1991) and Ammar and He (1995).

9. Inverse eigenvalue problems with prescribed entries

A very large class of SIEPs can be described as inverse eigenvalue problems with prescribed entries (**PEIEP**). The prescribed entries are used to characterize the underlying structure. The most general setting can be delineated as follows (Ikramov and Chugunov 2000). Given a certain subset $\mathcal{L} = \{(i_t, j_t)\}_{t=1}^{\ell}$ of pairs of subscripts, a certain set of values $\{a_1, \ldots, a_\ell\}$ over a field $\mathbb{F}$, and a set of n values $\{\lambda_1, \ldots, \lambda_n\}$, find a matrix $X \in \mathbb{F}^{n\times n}$ such that

$$\begin{cases} \sigma(X) &= \{\lambda_1, \ldots, \lambda_n\}, \\ X_{i_t,j_t} &= a_t \text{ for } t = 1, \ldots, \ell. \end{cases}$$

Let the cardinality ℓ of the index set $\mathcal{L}$ be denoted by $|\mathcal{L}|$. The PEIEP is to determine (complete) values for the $n^2 - |\mathcal{L}|$ positions that do not belong to $\mathcal{L}$ so as to satisfy the spectral constraint. The Jacobi structure can be considered as a special case of PEIEP where, in addition to the desired symmetry of the band, elements outside the tridiagonal band are required to be zero. Another interesting variation of the PEIEP is the completion problem, where only a one-to-one correspondence between the ℓ positions in $\mathcal{L}$ and the ℓ prescribed values $\{a_1, \ldots, a_\ell\}$, but not in any specific order, is required. Two major points that have been the focus of discussion in the literature are to determine the cardinality $|\mathcal{L}|$ so that the problem makes sense, and to study the effect of the locations in $\mathcal{L}$.

9.1. Prescribed entries along the diagonal

Perhaps a natural place to begin the discussion of PEIEPs is the construction of a Hermitian matrix with prescribed diagonal entries and eigenvalues. Recall that a vector $\mathbf{a} \in \mathbb{R}^n$ is said to majorize $\boldsymbol{\lambda} \in \mathbb{R}^n$ if, assuming the orderings $a_{j_1} \leq \cdots \leq a_{j_n}$ and $\lambda_{m_1} \leq \cdots \leq \lambda_{m_n}$ of their elements, the following relationships hold:

$$\begin{cases} \sum_{i=1}^{k} \lambda_{m_i} &\leq \sum_{i=1}^{k} a_{j_i}, \quad k = 1, \ldots, n, \\ \sum_{i=1}^{n} \lambda_{m_i} &= \sum_{i=1}^{n} a_{j_i}. \end{cases} \tag{9.1}$$

The necessary and sufficient relationship between the diagonal entries and the eigenvalues of a Hermitian matrix is completely characterized by the Schur–Horn theorem.

Theorem 9.1. (Horn 1954*a*) A Hermitian matrix H with eigenvalues $\boldsymbol{\lambda}$ and diagonal entries $\mathbf{a}$ exists if and only if $\mathbf{a}$ *majorizes* $\boldsymbol{\lambda}$.

The sufficient condition is the harder part of the proof and that is precisely the heart of the Schur–Horn inverse eigenvalue problem (**SHIEP**): given two vectors $\mathbf{a}$ and $\boldsymbol{\lambda}$ where $\mathbf{a}$ majorizes $\boldsymbol{\lambda}$, construct a Hermitian matrix with diagonals $\mathbf{a}$ and eigenvalues $\boldsymbol{\lambda}$. The original proof was done by mathematical induction. A continuous method was discussed in Chu (1995). A finite iterative method was described by Zha and Zhang (1995).

Without the Hermitian structure, the connection between eigenvalues and diagonal entries of a general matrix is given by the Mirsky theorem.

Theorem 9.2. (Mirsky 1958) A matrix with eigenvalues $\lambda_1, \ldots, \lambda_n$ and main diagonal elements $a_1, \ldots, a_n$ exists if and only if

$$\sum_{i=1}^{n} a_i = \sum_{i=1}^{n} \lambda_i. \tag{9.2}$$

Again, the sufficient condition in the Mirsky theorem constitutes an inverse problem where the prescribed entries are along the diagonal. An inverse problem as such would be of little interest. Later, de Oliveira (1973*a*, 1973*b*, 1975) generalized the Mirsky theorem to the case of non-principal diagonals. Given a permutation ϱ, the positions in a matrix corresponding to the index set $\mathcal{D} = \{(i, \varrho(i))\}_{i=1}^{n}$ is called a ϱ-diagonal of that matrix.

Theorem 9.3. (de Oliveira 1973*b*) Let $\{\lambda_1, \ldots, \lambda_n\}$ and $\{a_1, \ldots, a_n\}$ be two sets of arbitrary numbers over a field $\mathbb{F}$. Suppose that at least one of the disjoint cycles in the product representation $\varrho = \varrho_1 \cdots \varrho_s$ has length > 2. Then there exists a matrix $X \in \mathbb{F}^{n\times n}$ such that $\sigma(X) = \{\lambda_1, \ldots, \lambda_n\}$ and $X_{i,\varrho(i)} = a_i$ for $i = 1, \ldots, n$.

The assumption in the de Oliveira theorem, that at least one cycle has length greater than 2, precludes the case that ϱ is the identity and, hence, the equality (9.2) is not needed. If no cycle is of length > 2, then a similar result holds under some additional restrictions (de Oliveira 1973*b*, Theorem 2). Using the so-called L-transform, the proof of the de Oliveira theorem is constructive, and can be converted into a finite numerical algorithm. We shall comment more on this in the next few sections.

Having described two types of PEIEP arising from the Schur–Horn theorem and the de Oliveira theorem, respectively, we might as well bring forth another class of inverse problem with prescribed entries that involves singular values instead of eigenvalues. The following Sing–Thompson theorem characterizes the relationship between singular values and diagonal entries of a general matrix.

Theorem 9.4. (Sing 1976, Thompson 1977) Assume that elements in two given vectors $\mathbf{d}, \mathbf{s} \in \mathbb{R}^n$ satisfy $s_1 \geq s_2 \geq \cdots \geq s_n$ and $|d_1| \geq |d_2| \geq \cdots \geq |d_n|$. Then a real matrix with singular values $\mathbf{s}$ and main diagonal entries $\mathbf{d}$ (possibly in different order) exists if and only if

$$\begin{cases} \sum_{i=1}^{k} |d_i| & \leq \sum_{i=1}^{k} s_i, \quad \text{for } k = 1, \ldots, n, \\ \left(\sum_{i=1}^{n-1} |d_i|\right) - |d_n| & \leq \left(\sum_{i=1}^{n-1} s_i\right) - s_n. \end{cases} \tag{9.3}$$

The sufficient condition in the Sing–Thompson theorem gives rise to an inverse singular value problem (**STISVP**) of constructing a square matrix with prescribed diagonals and singular values. The original proof was done by mathematical induction. Chu (1999) rewrote it as a divide-and-conquer algorithm that can easily be implemented in any programming language that supports recursion.

9.2. Prescribed entries at arbitrary locations

Note that the PEIEP in the Mirsky theorem really involves only $n-1$ prescribed entries $a_1, \ldots, a_{n-1}$ because a_n is determined from (9.2). London and Minc (1972) showed that the restriction of the $n-1$ prescribed entries to the main diagonal was unnecessary.

Theorem 9.5. (London and Minc 1972) Given two sets $\{\lambda_1, \ldots, \lambda_n\}$ and $\{a_1, \ldots, a_{n-1}\}$ of arbitrary numbers in an algebraically closed field $\mathbb{F}$, suppose $\mathcal{L} = \{(i_t, j_t)\}_{t=1}^{n-1}$ is a set of arbitrary but distinct positions. Then there exists a matrix $X \in \mathbb{F}^{n \times n}$ such that $\sigma(X) = \{\lambda_1, \ldots, \lambda_n\}$ and $X_{i_t, j_t} = a_t$ for $t = 1, \ldots, n-1$.

An alternative proof was given in de Oliveira (1973*b*). Both proofs used mathematical induction. In principle, we think a fast recursive algorithm similar to those for the SHIEP and the STISVP could be devised. We have not yet seen its numerical implementation. Similar inverse problems constructing matrices with arbitrary $n-1$ prescribed entries and prescribed characteristic polynomials are considered in Dias da Silva (1974) and Ikramov and Chugunov (2000).

An interesting follow-up question to the London–Minc theorem is how many more entries of a matrix can be specified while the associated PEIEP is still solvable. Obviously, as we have learned by now, the locations of these prescribed entries also have some effect on the solvability. To help us better grasp the scope of this complicated issue, we first turn our attention to another subclass of PEIEPs before we return to this question in Section 9.4.

9.3. Additive inverse eigenvalue problem

Thus far, we have considered several cases of PEIEPs with small $|\mathcal{L}|$. With $|\mathcal{L}| = n-1$, the London and Minc theorem asserts that the PEIEP is always solvable with no other constraints. With $|\mathcal{L}| = n$, the PEIEP is solvable under some constraints. Indeed, Ikramov and Chugunov (2000,

Section 3b) argued meticulously through various cases to draw the most general conclusion.

Theorem 9.6. (Ikramov and Chugunov 2000) Suppose that the field $\mathbb{F}$ is algebraically closed and that $|\mathcal{L}| = n$. Assume that the following two conditions are met, if they occur:

$$\begin{cases} \text{that (9.2) is satisfied,} & \text{if } \mathcal{L} = \{(i,i)\}_{i=1}^{n}, \quad \text{or} \\ \text{that } a_i = \lambda_j \text{ for some } j, & \text{if } \mathcal{L} = \{(i,j_t)\}_{t=1}^{n} \text{ and } a_t = 0 \text{ for all } j_t \neq i. \end{cases}$$

Then the PEIEP is solvable via rational algorithms in $\mathbb{F}$. (A Maple code that generates a solution in closed form has been implemented by Chugunov.)

In both cases, there is plenty of room, that is, $n^2 - |\mathcal{L}|$ free locations, for constructing such a matrix. In contrast, the classical AIEP (see (2.14)) is another type of PEIEP with much less room for free locations. Recall that an AIEP concerns adding a diagonal matrix D to a given matrix A so that $\sigma(A + D)$ has a prescribed spectrum (note that in a more general context D need not be a diagonal, but can be defined by the *complement* to any index set $\mathcal{L}$). In the AIEP, the prescribed entries consist of all off-diagonal elements, and thus $|\mathcal{L}| = n^2 - n$. In this case, the following brilliant result is due to Friedland (1972). See also Friedland (1977).

Theorem 9.7. (Friedland 1977) The AIEP over any algebraically closed field is always solvable. If n is the order of the problem, then there exist at most $n!$ solutions. For almost all given $\{\lambda_1, \ldots, \lambda_n\}$, there are exactly $n!$ solutions.

Somewhere there is a threshold on the cardinality $|\mathcal{L}|$ of prescribed entries that changes the PEIEP from finitely solvable to finitely unsolvable. It is known that the AIEP in general cannot be solved in finitely many steps. The AIEP in which all off-diagonal entries are 1, for example, is not solvable in radicals for $n \geq 5$. The AIEP for a Jacobi matrix with subdiagonal (and superdiagonal) entries 1 is not solvable in radicals even for $n = 4$ (Ikramov and Chugunov 2000). The AIEP has to be solved by other types of numerical methods (Friedland, Nocedal and Overton 1986).

It is critical to the observer that the solvability assured in both Theorem 9.6 and Theorem 9.7 requires that the underlying field $\mathbb{F}$ is algebraically closed. In Chu (1998), such an AIEP was referred to as AIEP3. The AIEP over the field $\mathbb{R}$ of real numbers was referred to as AIEP1, and AIEP2 if the matrix A is real symmetric. The AIEP is not always solvable over $\mathbb{R}$. It is easy to see, for instance, that a necessary condition for the real solvability of AIEP1 is that $\sum_{i \neq j} (\lambda_i - \lambda_j)^2 \geq 2n \sum_{i \neq j} a_{ij} a_{ji}$. For convenience, define $\pi(M) := \|M - \operatorname{diag}(M)\|_\infty$. The separation of prescribed eigenvalues relative to the size of (prescribed) off-diagonal entries of A renders some sufficient conditions for the real solvability.

Theorem 9.8. Given a set $\boldsymbol{\lambda} = \{\lambda_1, \ldots, \lambda_n\}$ of eigenvalues, define the separation of eigenvalues by

$$d(\boldsymbol{\lambda}) := \min_{i \neq j} |\lambda_i - \lambda_j|. \tag{9.4}$$

Then:

(1) if $d(\lambda) > 2\sqrt{3}(\pi(A \circ A))^{1/2}$, then AIEP2 is solvable (Hadeler 1968);

(2) if $d(\lambda) > 4\pi(A)$, then AIEP1 is solvable (de Oliveira 1970).

The above theorem offers no clue on what will happen when the separation $d(\boldsymbol{\lambda})$ is too small. At the extreme case when two eigenvalues coalesce, we have the following result.

Theorem 9.9. (Shapiro 1983, Sun and Qiang 1986) Both AIEP1 and AIEP2 are unsolvable almost everywhere if there are multiple eigenvalues present in $\boldsymbol{\lambda}$.

Up to this point, we have refrained from venturing into discussion on the class of *parametrized problems* in order to remain focused on the *structured problems*. Nevertheless, as we indicated earlier in Figure 3.1, these problems overlap each other. We have come across the class of PIEPs many times in this paper. It is perhaps fitting to at least describe the PIEP and point to some general results in the context of PIEPs. By a PIEP we mean the IEP of determined parameters $c_1, \ldots, c_\ell$ so that the matrix

$$A(c_1, \ldots, c_\ell) = A_0 + \sum_{t=1}^{\ell} c_t A_t, \tag{9.5}$$

where A_t, $t = 0, \ldots, \ell$, are given matrices, have a prescribed spectrum. It is clear that PEIEPs are a special case of PIEPs by identifying $|\mathcal{L}| = \ell$ and $A_t = \mathbf{e}_{i_t}\mathbf{e}_{j_t}^T$, where $\mathbf{e}_k$ denotes the standard kth coordinate vector in $\mathbb{R}^n$ and (i_t, j_t) is the tth pair of indices in $\mathcal{L}$. Using Brouwer's fixed-point theorem, Biegler-König (1981*b*) derived some sufficient conditions for real solvability of more general PIEPs.

We conclude this section with a sensitivity result for AIEP2 and remarks (Xu 1998, Corollary 4.5.5).

Theorem 9.10. (Xu 1998) Suppose D is a solution to AIEP2 with symmetric matrix A and eigenvalues $\{\lambda_1, \ldots, \lambda_n\}$. Let the spectral decomposition of $A + D$ be written as $A + D = Q(D)^T \text{diag}\{\lambda_1, \ldots, \lambda_n\} Q(D)$, with

$Q(D) = [q_{ij}(D)] = [\mathbf{q}_1, \ldots, \mathbf{q}_n]$. Define

$$\Omega(D) := [q_{ji}^2(D)],$$
$$b(D) := [\mathbf{q}_1(D)^T A \mathbf{q}_1(D), \ldots, \mathbf{q}_n(D)^T A \mathbf{q}_n(D)]^T.$$

Assume that the matrix $\Omega(D)$ is nonsingular and that the perturbation

$$\delta = \|\lambda - \tilde{\lambda}\|_\infty + \|A - \tilde{A}\|_2$$

is sufficiently small. Then the AIEP2 associated with the perturbed data $\tilde{A}$ and $\tilde{\lambda}$ is solvable. Furthermore, for the perturbed problem there is a solution $\tilde{D}$ near to D in the sense that

$$\frac{\|D - \tilde{D}\|_\infty}{\|D\|_\infty} \leq \kappa_\infty(\Omega(D)) \left(\frac{\|\lambda - \tilde{\lambda}\|_\infty + \|A - \tilde{A}\|_2}{\|\lambda - b\|_\infty} \right) + O(\delta^2),$$

where $\kappa_\infty(M)$ stands for the condition number of the matrix M in the infinity norm.

Observe that PEIEPs, including AIEP2, generally have multiple solutions. The above theorem only ensures that, for a given D, there exists in theory a solution $\tilde{D}$ to the perturbed problem. However, the numerical solution $\tilde{\tilde{D}}$ obtained by a computational method, could be very different from D.

9.4. Cardinality and locations

The prescribed entries in the SHIEP and the STISVP are required to be on the diagonal. So certain inequalities (Theorems 9.1 and 9.4) involving the prescribed eigenvalues and entries must be satisfied. The prescribed entries in an AIEP are required to be on the off-diagonal. Complex solvability was addressed in Theorem 9.7, but real solvability is only partially understood. In all these cases, the prescribed entries are located at special positions.

Theorem 9.6 relaxes the specification to arbitrary locations and, under very mild conditions, asserts the existence of a solution to the PEIEP when $|\mathcal{L}| = n$. It is natural to ask what is the interplay between cardinality and locations so that a PEIEP is solvable. To that end, we describe what is possibly the strongest result on $|\mathcal{L}|$ in the class of PEIEPs at arbitrary locations. The original work was presented in the MSc thesis by Hershkowits (1978). We restate the result from Hershkowits (1983).

Theorem 9.11. (Hershkowits 1983) Suppose that the field $\mathbb{F}$ is algebraically closed and that $|\mathcal{L}| = 2n - 3$. Assume that the following two conditions are met, if they occur:

$$\begin{cases} \text{that (9.2) is satisfied,} & \text{if } \mathcal{L} \supseteq \{(i,i)\}_{i=1}^n, \quad \text{or} \\ \text{that } a_i = \lambda_j \text{ for some } j, & \text{if } \mathcal{L} \supseteq \{(i,j_t)\}_{t=1}^n \text{ and } a_t = 0 \text{ for all } j_t \neq i. \end{cases}$$

Then the PEIEP is solvable in $\mathbb{F}$.

Note that the effect of locations of positions in $\mathcal{L}$ is limited to the two necessary conditions stated in the theorem, and are quite general. The proof of the Hershkowits theorem was established by induction. In principle, it was declared in Ikramov and Chugunov (2000) that the construction could be done by a rational algorithm. The seven basic cases plus the many subcases of analysis in the 15-page proof might make a computer implementation quite a challenge. It would be interesting to see if other numerical algorithms could be developed.

10. Inverse singular value problems

The notion of IEPs can naturally be extended to the inverse singular value problems (**ISVP**). An ISVP concerns the construction of a structured matrix with prescribed singular values. Once again, an ISVP should also satisfy a certain structural constraint. To our knowledge, the class of ISVPs is an entirely new territory that has barely been explored in the literature. Adding to the complication is that the underlying matrix should not be symmetric (otherwise, it is reduced to an IEP) and could be rectangular. We have already seen one type of ISVP, that is, the STISVP in Section 9.1, where a matrix is to be constructed with prescribed diagonal entries and singular values. Another type of ISVP was mentioned in Section 2.5, where a given matrix was to be conditioned by a rank-one perturbation. Clearly, every other type of IEP, except for the symmetric problems, has a counterpart under the context of ISVP.

One of the reasons that we include ISVPs in the context of SIEPs is that an ISVP, even without any string of structure, can be converted into an SIEP. Note that eigenvalues of the *structured* symmetric matrix

$$C = \begin{bmatrix} 0 & B \\ B^T & 0 \end{bmatrix} \tag{10.1}$$

are precisely the pluses and minuses of singular values of B. The IEP for C has the fixed structure of zero diagonal blocks plus whatever structure inherited from B. An ISVP for a structured B is solvable if and only if an IEP for C with structure defined in (10.1) is solvable. To establish conditions on the solvability of a structured ISVP should be an interesting question for further research.

To introduce the notion of ISVPs, we shall limit our discussion to a special class of parametrized ISVPs. Given general matrices $B_0, B_1, \ldots, B_n \in \mathbb{R}^{m\times n}$, $m \geq n$ and nonnegative real numbers $s_1 \geq \cdots \geq s_n$, find values of $c := (c_1, \ldots, c_n)^T \in \mathbb{R}^n$ such that singular values of the matrix

$$B(c) := B_0 + \sum_{i=1}^{n} c_i B_i \tag{10.2}$$

are precisely $\{s_1, \ldots, s_n\}$. In analogy to the PIEP (9.5), the matrices B_i can be used to delineate certain quite general structures. We have already discussed a Newton-type iterative procedure for the ToIEP. We now demonstrate how the ISVP can be handled in a similar but more subtle way. The subtlety comes from the fact that, for ISVPs, we have to deal with the left and the right singular vectors at the same time.

For illustration purposes, assume all prescribed singular values $s_1, \ldots, s_n$ are positive and distinct. Let $\Sigma \in \mathbb{R}^{m \times n}$ denote the 'diagonal' matrix with diagonal elements $\{s_1, \ldots, s_n\}$. Define the affine subspace $\mathcal{B} := \{B(c) | c \in \mathbb{R}^n\}$ and the manifold $\mathcal{M}_s(\Sigma) := \{U \Sigma V^T | U \in \mathcal{O}(m), V \in \mathcal{O}(n)\}$ of all matrices with singular values $\{s_1, \ldots, s_n\}$, where $\mathcal{O}(n)$ is the set of all orthogonal matrices in $\mathbb{R}^{n \times n}$. Solving the ISVP is equivalent to finding an intersection of the two sets $\mathcal{M}_s(\Sigma)$ and $\mathcal{B}$. Note that any tangent vector $T(X)$ to $\mathcal{M}_s(\Sigma)$ at $X \in \mathcal{M}_s(\Sigma)$ must be of the form

$$T(X) = XK - HX,$$

for some skew-symmetric matrices $H \in \mathbb{R}^{m \times m}$ and $K \in \mathbb{R}^{n \times n}$. From any given $X^{(\nu)} \in \mathcal{M}_s(\Sigma)$, factorized as

$$U^{(\nu)^T} X^{(\nu)} V^{(\nu)} = \Sigma$$

with $U^{(\nu)} \in \mathcal{O}(m)$ and $V^{(\nu)} \in \mathcal{O}(n)$, our goal is twofold. First, we seek for a $\mathcal{B}$-intercept $B(c^{(\nu+1)})$ from a line that is tangent to the manifold $\mathcal{M}_s(\Sigma)$ at $X^{(\nu)}$. Then we seek for a way to lift the matrix $B(c^{(\nu+1)}) \in \mathcal{B}$ to a point $X^{(\nu+1)} \in \mathcal{M}_s(\Sigma)$.

To determine the intercept, we calculate skew-symmetric matrices $H^{(\nu)} \in \mathbb{R}^{m \times m}$ and $K^{(\nu)} \in \mathbb{R}^{n \times n}$, and a vector $c^{(\nu+1)} \in \mathbb{R}^n$ so that the equation

$$X^{(\nu)} + X^{(\nu)} K^{(\nu)} - H^{(\nu)} X^{(\nu)} = B(c^{(\nu+1)}) \tag{10.3}$$

is satisfied. Equivalently, we calculate skew-symmetric matrices $\tilde{H}^{(\nu)} := U^{(\nu)^T} H^{(\nu)} U^{(\nu)}$ and $\tilde{K}^{(\nu)} := V^{(\nu)^T} K^{(\nu)} V^{(\nu)}$ for the equation

$$\Sigma + \Sigma \tilde{K}^{(\nu)} - \tilde{H}^{(\nu)} \Sigma = \underbrace{U^{(\nu)^T} B(c^{(\nu+1)}) V^{(\nu)}}_{W^{(\nu)}}. \tag{10.4}$$

The values for $c^{(\nu+1)}$, $H^{(\nu)}$, and $K^{(\nu)}$ can be determined separately.

Observe that, in total, there are $m(m-1)/2 + n(n-1)/2 + n$ unknowns and mn equations involved in (10.4). A closer examination of (10.4) shows that the lower-right corner of size $(m-n) \times (m-n)$ in $\tilde{H}^{(\nu)}$ can be arbitrary. For simplicity, we set this part to be identically zero. Then it suffices to consider the mn equations

$$W_{ij}^{(\nu)} = \Sigma_{ij} + \Sigma_{ii} \tilde{K}_{ij}^{(\nu)} - \tilde{H}_{ij}^{(\nu)} \Sigma_{jj}, \quad 1 \le i \le m, \quad 1 \le j \le n, \tag{10.5}$$

where $\tilde{K}_{ij}^{(\nu)}$ is understood to be zero if $i \geq n+1$, for the remaining quantities.

For $1 \leq i = j \leq n$, we obtain

$$\Omega^{(\nu)} c^{(\nu+1)} = \mathbf{s} - \mathbf{b}^{(\nu)}, \tag{10.6}$$

where

$$\begin{aligned} \Omega_{st}^{(\nu)} &:= \mathbf{u}_s^{(\nu)T} B_t \mathbf{v}_s^{(\nu)}, \quad 1 \leq s, t \leq n, \\ \mathbf{s} &:= [s_1, \dots, s_n]^T, \quad \text{and} \\ b_s^{(\nu)} &:= \mathbf{u}_s^{(\nu)T} B_0 \mathbf{v}_s^{(\nu)}, \quad 1 \leq s \leq n, \end{aligned}$$

if $\mathbf{u}_s^{(\nu)}$ and $\mathbf{v}_s^{(\nu)}$ denote column vectors of $U^{(\nu)}$ and $V^{(\nu)}$, respectively. Under mild assumptions, the matrix $\Omega^{(\nu)}$ is nonsingular. The vector $c^{(\nu+1)}$ and, hence, the matrix $W^{(\nu)}$ are thus obtained.

The skew-symmetric matrices $H^{(\nu)}$ and $K^{(\nu)}$ can be obtained by comparing the 'off-diagonal' entries in (10.4) without much trouble. For $n+1 \leq i \leq m$ and $1 \leq j \leq n$, it is clear that

$$\tilde{H}_{ij}^{(\nu)} = -\tilde{H}_{ji}^{(\nu)} = -\frac{W_{ij}^{(\nu)}}{s_j}. \tag{10.7}$$

For $1 \leq i < j \leq n$,

$$\tilde{H}_{ij}^{(\nu)} = -\tilde{H}_{ji}^{(\nu)} = \frac{s_i W_{ji}^{(\nu)} + s_j W_{ij}^{(\nu)}}{s_i^2 - s_j^2}, \tag{10.8}$$

$$\tilde{K}_{ij}^{(\nu)} = -\tilde{K}_{ji}^{(\nu)} = \frac{s_i W_{ij}^{(\nu)} + s_j W_{ji}^{(\nu)}}{s_i^2 - s_j^2}. \tag{10.9}$$

The intercept is now completely determined.

It only remains to lift the intercept $B(c^{(\nu+1)})$ back to $\mathcal{M}_s(\Sigma)$. Towards that end, one possible way is to define the lift as

$$X^{(\nu+1)} := R^T X^{(\nu)} S,$$

where R and S are the Cayley transforms

$$\begin{aligned} R &:= \left(I + \frac{H^{(\nu)}}{2}\right)\left(I - \frac{H^{(\nu)}}{2}\right)^{-1}, \\ S &:= \left(I + \frac{K^{(\nu)}}{2}\right)\left(I - \frac{K^{(\nu)}}{2}\right)^{-1}. \end{aligned}$$

This completes one cycle of the Newton step and the iteration repeats until convergence. Regarding the efficiency of this algorithm, Chu (1992) proved the following result on the rate of convergence.

Theorem 10.1. Suppose that the ISVP (10.2) has an exact solution at c^* and that $B(c^*) = \hat{U}\Sigma\hat{V}^T$ is the corresponding singular value decomposition.

Define the error matrix $E := (E_1, E_2) := (U - \hat{U}, V - \hat{V})$, and suppose that the matrix $\Omega^{(\nu)}$ is nonsingular. Then $\|E^{(\nu+1)}\| = O(\|E^{(\nu)}\|^2)$

We conclude this section with one important remark concerning the case of multiple singular values. Similar remarks might be applicable to IEPs with prescribed multiple eigenvalues as well. It is known that multiple eigenvalues are difficult to compute even in direct problems. Recall earlier in Theorem 9.9 that AIEP1 and AIEP2 are unsolvable almost everywhere in the presence of multiple eigenvalues. Thus, a general rule of thumb is that the ISVP (IEP) may not have a solution if there are repeated singular values (eigenvalues) in the prescribed set. At least, the proximity of multiple spectral data would impose considerable difficulty to the inverse problem. An argument was given in Chu (1992) showing that only a portion of the total set of singular values of $B(c)$ should be specified to give leeway to accommodate the multiplicity.

11. Inverse singular/eigenvalue problems

The structure involved in SIEPs can be quite general. Thus far, we have seen structures including Jacobi, Toeplitz, nonnegative, stochastic, unitary Hessenberg, prescribed entries, and the special form (10.1). Most of these structural constraints are explicitly or, at least, semi-explicitly, given in terms of the appearance of the underlying matrix. It is possible that the structure is described implicitly as the solution set of some nonlinear functions. In this section, we discuss one particular class of SIEPs where the 'structure' is implicitly characterized by the singular values.

Recall that the Schur–Horn theorem identifies the connection between diagonal entries and eigenvalues of a Hermitian matrix. The Mirsky theorem gives the connection between diagonal entries and eigenvalues of a general matrix. The Sing–Thompson theorem characterizes the connection between diagonal entries and singular values of a general matrix. It is natural to ask about the connection between singular values and eigenvalues of a matrix. For Hermitian matrices, the singular values are simply the absolute values of eigenvalues. But for general square matrices, the connection is much more involved, as is given by the Weyl–Horn theorem.

Theorem 11.1. (Weyl 1949, Horn 1954*b*) Given vectors $\boldsymbol{\lambda} \in \mathbb{C}^n$ and $\mathbf{s} \in \mathbb{R}^n$, suppose the entries are arranged in the ordering that $|\lambda_1| \geq \cdots \geq |\lambda_n|$ and $s_1 \geq \cdots \geq s_n$. Then a matrix with eigenvalues $\lambda_1, \ldots, \lambda_n$ and singular values $s_1, \ldots, s_n$ exists if and only if

$$\begin{cases} \prod_{j=1}^{k} |\lambda_j| \leq \prod_{j=1}^{k} s_j, & k = 1, \ldots, n-1, \\ \prod_{j=1}^{n} |\lambda_j| = \prod_{j=1}^{n} s_j. \end{cases} \tag{11.1}$$

If $|\lambda_n| > 0$, then the Weyl–Horn condition is equivalent to the statement that the vector $\log(\mathbf{s})$ majorizes the vector $\log|\boldsymbol{\lambda}|$. The IEP we are concerned with is to construct a matrix with prescribed singular values and eigenvalues (**ISEP**), if the Weyl–Horn condition is met. The original proof was by induction, but Chu (2000) modified the proof to avoid triangularization and derived a divide-and-conquer recursive algorithm, which we outline below.

First, observe that the 2×2 triangular matrix

$$A = \begin{bmatrix} \lambda_1 & \mu \\ 0 & \lambda_2 \end{bmatrix}$$

has singular value $\{s_1, s_2\}$ if and only if

$$\mu = \sqrt{s_1^2 + s_2^2 - |\lambda_1|^2 - |\lambda_2|^2}.$$

The fact that μ is well defined follows from the Weyl–Horn condition that $|\lambda_1| \leq s_1$ and $|\lambda_1|\,|\lambda_2| = s_1 s_2$. It is interesting to note that μ^2 is precisely the so-called departure of A from normality. For the sake of better computational stability, we suggest replacing μ by the definition

$$\mu = \begin{cases} 0, & \text{if } |(s_1 - s_2)^2 - (|\lambda_1| - |\lambda_2|)^2| \leq \epsilon, \\ \sqrt{|(s_1 - s_2)^2 - (|\lambda_1| - |\lambda_2|)^2|}, & \text{otherwise.} \end{cases}$$

The 2×2 matrix serves as the building block in the recursion.

The basic ideas in Weyl–Horn's proof contain three major components:

- the original problem can be reduced to two problems of smaller sizes,
- problems of smaller sizes are guaranteed to be solvable by the induction hypothesis, and
- the subproblems can be affixed together by working on a suitable 2×2 corner that has an explicit solution.

If we repeatedly apply these principles, then the original inverse problem is *divided* into subproblems of size 2×2 or 1×1 that can eventually be *conquered* to build up the original size.

The original idea on how the problems could be divided is quite intriguing. Since this is an entirely new approach different from either the iterative methods or the continuous methods we have discussed thus far for other types of IEPs, we outline the proof as follows. For simplicity, we assume that $s_i > 0$ for all $i = 1, \ldots, n$. It follows that $\lambda_i \neq 0$ for all i. The case of zero singular values can be handled in a similar way. Starting with $\gamma_1 := s_1$, define the sequence

$$\gamma_i := \gamma_{i-1} \frac{s_i}{|\lambda_i|}, \quad i = 2, \ldots, n-1. \tag{11.2}$$

Assume that the maximum $\gamma := \min_{1\leq i\leq n-1} \gamma_i$ is attained at the index j. Define

$$\theta := \frac{|\lambda_1 \lambda_n|}{\gamma}. \tag{11.3}$$

Then the following three sets of inequalities are valid:

$$\begin{cases} |\lambda_1| & \geq & |\lambda_n|, \\ \gamma & \geq & \theta; \end{cases} \tag{11.4}$$

$$\begin{cases} \gamma & \geq & |\lambda_2| & \geq & \cdots & \geq & |\lambda_j|, \\ s_1 & \geq & s_2 & \geq & \cdots & \geq & s_j; \end{cases} \tag{11.5}$$

$$\begin{cases} |\lambda_{j+1}| & \geq & \cdots & \geq & |\lambda_{n-1}| & \geq & \theta, \\ s_{j+1} & \geq & \cdots & \geq & s_{n-1} & \geq & s_n. \end{cases} \tag{11.6}$$

More importantly, the numbers in each of the above sets satisfy the Weyl–Horn condition, respectively, with the first row playing the role of eigenvalues and the second row playing the singular values. Since these are problems of *smaller* sizes, by induction hypothesis, the ISEPs associated with (11.5) and (11.6) are solvable. In particular, there exist unitary matrices U_1, $V_1 \in \mathbb{C}^{j\times j}$ and *triangular* matrices A_1 such that

$$U_1 \begin{bmatrix} s_1 & 0 & \cdots & 0 \\ 0 & s_2 & & 0 \\ \vdots & & \ddots & \\ 0 & 0 & \cdots & s_j \end{bmatrix} V_1^* = A_1 = \begin{bmatrix} \gamma & \times & \times & \cdots & \times \\ 0 & \lambda_2 & & & \times \\ & & & & \\ \vdots & & & \ddots & \\ 0 & 0 & & & \lambda_j \end{bmatrix},$$

and unitary matrices U_2, $V_2 \in \mathbb{C}^{(n-j)\times(n-j)}$, and *triangular* matrix A_2 such that

$$U_2 \begin{bmatrix} s_{j+1} & 0 & \cdots & 0 \\ 0 & s_{j+2} & & 0 \\ \vdots & & \ddots & \\ 0 & 0 & \cdots & s_n \end{bmatrix} V_2^* = A_2 = \begin{bmatrix} \lambda_{j+1} & \times & \cdots & \times & \times \\ 0 & \lambda_{j+2} & & & \times \\ \vdots & & \ddots & & \vdots \\ & & & \lambda_{n-1} & \times \\ 0 & 0 & \cdots & 0 & \theta \end{bmatrix}.$$

Note the positions of γ and θ in the matrices. If we augment A_1 and A_2 to

$$\begin{bmatrix} A_1 & \bigcirc \\ \bigcirc & A_2 \end{bmatrix}, \tag{11.7}$$

then γ and θ reside, respectively, at the $(1,1)$ and the (n,n) positions. In his original proof, Horn claimed that the block matrix could be *permuted* to the triangular matrix

$$\begin{bmatrix} \lambda_2 & \times & \cdots & \times & \times & & & & & \\ 0 & & & & \times & & & & & \\ \vdots & & \ddots & & \vdots & & & \bigcirc & & \\ & & & \lambda_j & \times & & & & & \\ 0 & & \cdots & 0 & \gamma & 0 & & & & \\ 0 & 0 & \cdots & 0 & 0 & \theta & \times & \times & \cdots & \times \\ & & & & & & \lambda_{j+1} & & & \times \\ & & \bigcirc & & & & & & & \\ & & & & & \vdots & & & \ddots & \\ & & \cdots & & & 0 & 0 & & & \lambda_{n-1} \end{bmatrix}.$$

but this is not quite correct. If it were true, it is obvious that the resulting matrix would have singular values $\{s_1, \ldots, s_n\}$ and miss only the eigenvalues $\{\lambda_1, \lambda_n\}$. The next step is to glue the 2×2 corner adjacent to the two blocks together by an equivalence transformation

$$U_0 \begin{bmatrix} \gamma & 0 \\ 0 & \theta \end{bmatrix} V_0^* = A_0 = \begin{bmatrix} \lambda_1 & \mu \\ 0 & \lambda_n \end{bmatrix}$$

that does not affect the eigenvalues $\{\lambda_2, \ldots, \lambda_{n-1}\}$.

In Horn's proof, the ordering of diagonal entries is important and the resulting matrix is upper-triangular. While the final result in the Schur–Horn theorem remains true, it is unfortunate that it takes more than permutations to rearrange the diagonals of a triangular matrix while maintaining the singular values. Such a rearrangement is needed at every conquering step, but it requires new Schur decompositions and is expensive to compute in general.

It was proved in Chu (2000) that the triangular structure was entirely unnecessary, as was the rearrangement of the diagonal entries. It can be shown that modifying the first and the last rows and columns of the block diagonal matrix in (11.7) is sufficient to solve the ISEP, and that the resulting matrix is permutation similar to a triangular matrix. This advance in understanding makes it possible to implement the induction proof as a numerical algorithm.

More precisely, denote the 2×2 orthogonal matrices by $U_0 = [u_{st}^{(0)}]$ and $V_0 = [v_{st}^{(0)}]$. Then the matrix

$$A = \begin{bmatrix} u_{11}^{(0)} & 0 & u_{12}^{(0)} \\ 0 & I_{n-1} & 0 \\ u_{21}^{(0)} & 0 & u_{22}^{(0)} \end{bmatrix} \begin{bmatrix} A_1 & \bigcirc \\ \bigcirc & A_2 \end{bmatrix} \begin{bmatrix} v_{11}^{(0)} & 0 & v_{12}^{(0)} \\ 0 & I_{n-1} & 0 \\ v_{21}^{(0)} & 0 & v_{22}^{(0)} \end{bmatrix}^*$$

has eigenvalues $\{\lambda_1, \ldots, \lambda_n\}$ and singular values $\{s_1, \ldots, s_n\}$. The resulting matrix A has the structure

$$A = \left[\begin{array}{cccccccccc}
\lambda_1 & \otimes & \cdots & \otimes & \otimes & * & * & & & \mu \\
\otimes & \lambda_2 & & & \times & 0 & 0 & & & * \\
\vdots & & \ddots & & \vdots & & & \bigcirc & & \\
 & & & \lambda_{j-1} & \times & & & & & \\
\otimes & \times & \cdots & \times & \lambda_j & & & & & * \\
* & 0 & \cdots & 0 & 0 & \lambda_{j+1} & \times & \times & \cdots & \otimes \\
* & & & & 0 & \times & \lambda_{j+2} & & & \otimes \\
 & & \bigcirc & & & & & & & \\
 & & & & & \vdots & & & \ddots & \\
0 & * & \cdots & * & * & \otimes & \otimes & & & \lambda_n
\end{array}\right],$$

where $\times$ stands for unchanged, original entries from A_1 or A_2, $\otimes$ stands for entries of A_1 or A_2 that are modified by scalar multiplications, and $*$ denotes possible new entries that were originally zero. This pattern repeats itself during the recursion. Note that diagonal entries of A_1 and A_2 are in the fixed orders $\gamma, \lambda_2, \ldots, \lambda_j$ and $\lambda_{j+1}, \ldots, \lambda_{n-1}, \theta$, respectively. Each A_i is similar via permutations, which need not be known, to a lower-triangular matrix whose diagonal entries constitute the same set as the diagonal entries of A_i. Thus the eigenvalues of each A_i are precisely its diagonal entries. The first row and the last row have the same zero pattern except that the lower-left corner is always zero. The first column and the last column have the same zero pattern except that the lower-left corner is always zero. Using graph theory, it can be shown that the affixed matrix A has exactly the same properties.

With this realization, the entire induction process can easily be implemented in any programming language that allows a routine to call itself recursively. The main feature in the routine should be a single divide-and-conquer mechanism as we just described. As the routine is calling itself recursively, the problem is 'divided down' and 'conquered up' accordingly. A sample MATLAB program can be found in Chu (2000).

We illustrate how the divide-and-conquer algorithm works by a 6×6 symbolic example. In the following, the integers j_ℓ, selected randomly only for demonstration, indicate where the problem should be divided.

The dividing process, along with the corresponding eigenvalues and singular values for each subproblems, is depicted in the boxed frames in Figure 11.1. A blank framed box indicates that the division has reached the bottom. In this example, the original 6×6 problem is divided into two 1×1 problems and two 2×2 problems. Each of these small problems can trivially be solved. The pair of numbers beside j_ℓ and in between rows of framed

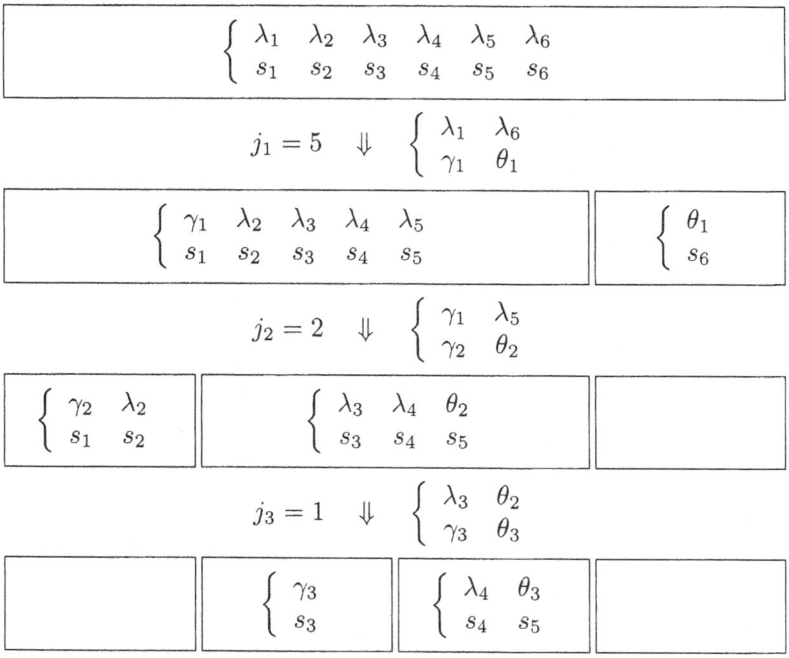

Figure 11.1. An illustration of the dividing process

boxes are the eigenvalues and singular values for the 2×2 matrix used to fasten the smaller matrices together in the conquering process.

The conquering process using the small matrices to build larger matrices is depicted in Figure 11.2. The matrices beside j_ℓ, and in between rows of framed boxes, are the augmented matrices (11.7) with the wrong eigenvalues. After fixing by some appropriated 2×2 matrices, we see in Figure 11.2 that some rows and columns must be modified. The symbols $\times, \otimes$ and $*$, indicating how the values have been changed during the conquering process, have the same meaning as defined before. The final 6×6 matrix with the desirable eigenvalues and singular values has the structure indicated at the top of Figure 11.2.

It is perhaps true that eigenvalues and singular values are two of the most important characteristics of a matrix. Being able to construct a solution for the ISEP might help to create test matrices for numerical linear algebra algorithms.

We should point out promptly that the constructed matrix obtained by the algorithm above is usually complex-valued, if there are complex eigenvalues. It might be desirable to construct a real-valued solution, if all eigenvalues are complete in conjugation. Towards that end, very recently Li and Mathias (2001) extended the Weyl–Horn condition to the case when only $m(\leq n)$ eigenvalues are given. They also proposed a stable algorithm using diagonal

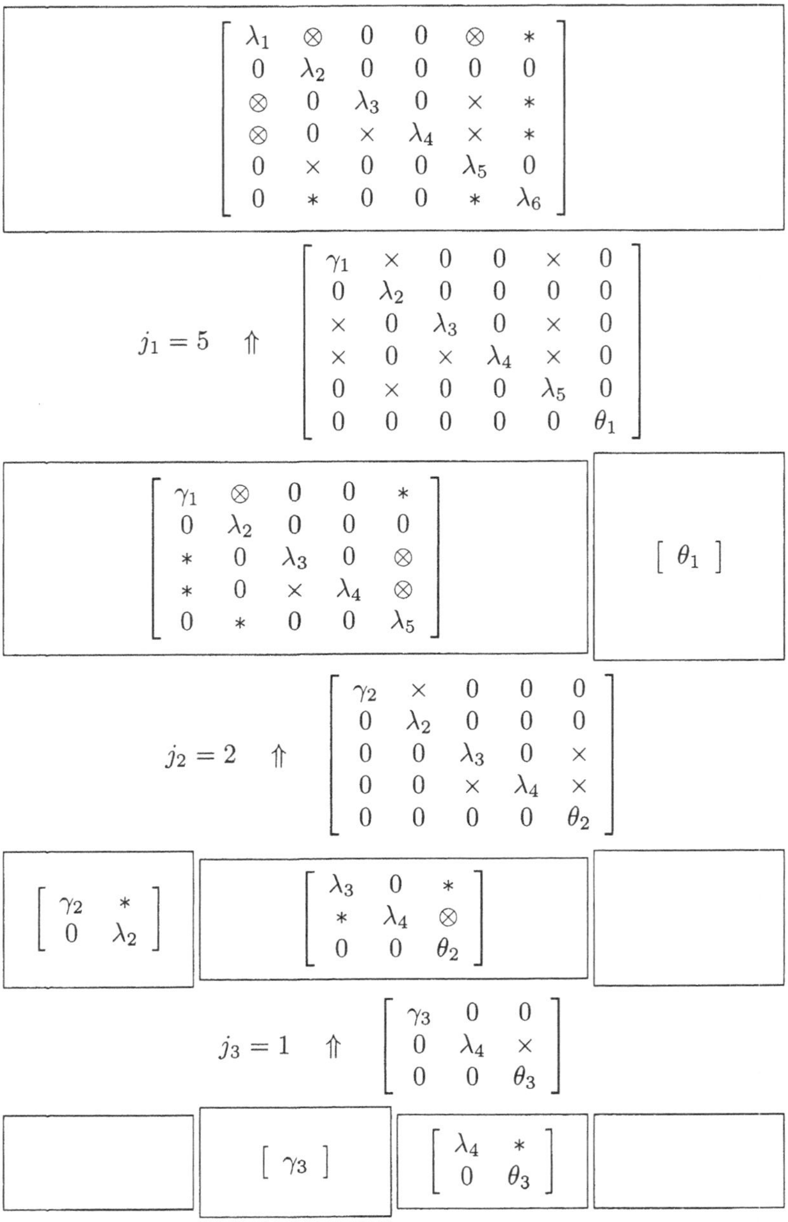

Figure 11.2. An illustration of the conquering process

unitary matrices, permutations, and rotation matrices to construct a real matrix if the specified eigenvalues are closed in complex conjugation. Finally, we remark that the divide-and-conquer feature enables fast computation. Numerical experiments seem to suggest that the overall cost in solving an n-dimensional ISEP is roughly $O(n^2)$.

12. Conclusion

We believe that inverse eigenvalue problems should always be structured problems. In this article, we try to explain, motivate, and review only a small segment of the full scope of structured inverse eigenvalue problems. The structures we selected for study in this presentation are by no means emblematic, but rather reflect our personal preferences. The notions introduced in this paper are by no means conclusive, but rather divulge our limited understanding of this subject. We have collected an extensive bibliography of more than 400 papers on this topic (available at `http://www4.ncsu.edu/~mtchu`). Even with that, the list is far from being comprehensive as we have already overlooked much of the engineering literature. Furthermore, be aware that our consideration has been limited to the setting when the entire spectrum is known and that the structural constraint must be satisfied. We have not discussed the structured problems where only partial eigenvalues and eigenvectors are given. Neither have we examined the case where a least squares solution with approximate spectrum or approximate structure is sufficient for practical purposes.

We hope to have accomplished three goals in this presentation. First, we wanted to demonstrate the breadth of areas where inverse eigenvalue problems can arise. The discipline ranges from practical engineering applications to abstract algebraic theorization. Secondly, we wanted to corroborate the depth of intricacy of inverse eigenvalue problems. While the set-up of an inverse eigenvalue problem seems relatively easy, the solution is not straightforward. The instruments employed to solve such a problem are quite sophisticated, including techniques from orthogonal polynomials, degree theory, optimization, to differential geometry and so on. Finally and most importantly, we wanted to arouse interest and encourage further research into this topic. We have indicated throughout the text that there is much room for further study of the numerical development and theoretical understanding of these fascinating inverse problems.

REFERENCES

M. Adler, L. Haine and P. van Moerbeke (1993), 'Limit matrices for the Toda flow and periodic flags for loop groups', *Math. Ann.* **296**, 1–33.

G. Ammar, W. Gragg and L. Reichel (1991), Constructing a unitary Hessenberg matrix from spectral data, in *Numerical Linear Algebra, Digital Signal Processing and Parallel Algorithms* (Leuven, 1988), Springer, Berlin, pp. 385–395.

G. S. Ammar and C. Y. He (1995), 'On an inverse eigenvalue problem for unitary Hessenberg matrices', *Lin. Alg. Appl.* **218**, 263–271.

S. A. Andrea and T. G. Berry (1992), 'Continued fractions and periodic Jacobi matrices', *Lin. Alg. Appl.* **161**, 117–134.

A. L. Andrew (1994), Some recent developments in inverse eigenvalue problems, in *Computational Techniques and Applications: CTAC93* (D. Stewart, H. Gardner and D. Singleton, eds), World Scientific, Singapore, pp. 94–102.

V. Barcilon (1974*a*), 'On the solution of inverse eigenvalue problems with high orders', *Geophys. J. Royal Astronomical Soc.* **39**, 153–154.

V. Barcilon (1974*b*), 'On the uniqueness of inverse eigenvalue problems', *Geophys. J. Royal Astronomical Soc.* **38**, 287–298.

V. Barcilon (1979), 'On the multiplicity of solutions of the inverse problems for a vibrating beam', *SIAM J. Appl. Math.* **37**, 119–127.

V. Barcilon (1990), 'A two-dimensional inverse eigenvalue problem', *Inverse Problems* **6**, 11–20.

W. W. Barrett and C. R. Johnson (1984), 'Possible spectra of totally positive matrices', *Lin. Alg. Appl.* **62**, 231–233.

A. Berman and R. J. Plemmons (1979), *Nonnegative Matrices in the Mathematical Sciences*, Academic Press, New York. Also Vol. 9 of *Classics in Applied Mathematics*, SIAM, Philadelphia (1994).

F. W. Biegler-König (1981*a*), 'Construction of band matrices from spectral data', *Lin. Alg. Appl.* **40**, 79–87.

F. W. Biegler-König (1981*b*), 'Sufficient conditions for the solubility of inverse eigenvalue problems', *Lin. Alg. Appl.* **40**, 89–100.

D. Boley and G. H. Golub (1984), 'A modified method for reconstructing periodic Jacobi matrices', *Math. Comput.* **24**, 143–150.

D. Boley and G. H. Golub (1987), 'A survey of matrix inverse eigenvalue problems', *Inverse Problems* **3**, 595–622.

D. L. Boley and G. H. Golub (1978), The matrix inverse eigenvalue problem for periodic Jacobi matrices, in *Proc. Fourth Symposium on Basic Problems of Numerical Mathematics* (Plzeň, 1978), Charles University, Prague, pp. 63–76.

C. de Boor and G. H. Golub (1978), 'The numerically stable reconstruction of a Jacobi matrix from spectral data', *Lin. Alg. Appl.* **21**, 245–260.

M. Boyle and D. Handelman (1991), 'The spectra of nonnegative matrices via symbolic dynamics', *Ann. Math.* **133**, 249–316.

A. Bunse-Gerstner, R. Byers, V. Mehrmann and N. K. Nichols (1991), 'Numerical computation of an analytic singular value decomposition of a matrix valued function', *Numer. Math.* **60**, 1–40.

C. I. Byrnes (1989), Pole placement by output feedback, in *Three Decades of Mathematical Systems Theory*, Vol. 135 of *Lecture Notes in Control and Information Sciences*, Springer, pp. 31–78.

J. A. Cadzow and D. M. Wilkes (1990), Signal enhancement and the SVD, in *Proc. 2nd International Workshop on SVD and Signal Processing*, University of Rhode Island, Kingston, RI, pp. 144–151.

D. Calvetti, G. H. Golub, W. B. Gragg and L. Reichel (2000), 'Computation of Gauss–Kronrod quadrature rules', *Math. Comput.* **69**, 1035–1052.

M. P. Calvo, A. Iserles and A. Zanna (1997), 'Numerical solution of isospectral flows', *Math. Comput.* **66**, 1461–1486.

A. Cantoni and F. Bulter (1976), 'Eigenvalues and eigenvectors of symmetric centrosymmetric matrices', *Lin. Alg. Appl.* **13**, 275–288.

K. Chadan, D. Colton, L. Päivärinta and W. Rundell (1997), *An Introduction to Inverse Scattering and Inverse Spectral Problems*, SIAM, Philadelphia, PA.

M. T. Chu (1992), 'Numerical methods for inverse singular value problems', *SIAM J. Numer. Anal.* **29**, 885–903.

M. T. Chu (1993), On a differential equation $\frac{dX}{dt} = [x, k(x)]$ where k is a Toeplitz annihilator. Available at http://www4.ncsu.edu/~mtchu.

M. T. Chu (1994), On the refinement of a Newt

M. T. Chu (1994), On the refinement of a Newton method for the inverse Toeplitz eigenvalue problem. Available at http://www4.ncsu.edu/~mtchu.

M. T. Chu (1995), 'Constructing a Hermitian matrix from its diagonal entries and eigenvalues', *SIAM J. Matrix Anal. Appl.* **16**, 207–217.

M. T. Chu (1998), 'Inverse eigenvalue problems', *SIAM Rev.* **40**, 1–39 (electronic).

M. T. Chu (1999), 'On constructing matrices with prescribed singular values and diagonal elements', *Lin. Alg. Appl.* **288**, 11–22.

M. T. Chu (2000), 'A fast recursive algorithm for constructing matrices with prescribed eigenvalues and singular values', *SIAM J. Numer. Anal.* **37**, 1004–1020 (electronic).

M. T. Chu and K. R. Driessel (1991), 'Constructing symmetric nonnegative matrices with prescribed eigenvalues by differential equations', *SIAM J. Math. Anal.* **22**, 1372–1387.

M. T. Chu and Q. Guo (1998), 'A numerical method for the inverse stochastic spectrum problem', *SIAM J. Matrix Anal. Appl.* **19**, 1027–1039.

R. M. Corless, P. M. Gianni, B. M. Trager and S. M. Watt (1995), The singular value decomposition for polynomial system, in *Proc. Int. Symp. Symbolic and Algebraic Computation*, Montreal, Canada, pp. 195–207.

H. Dai (1995), 'About an inverse eigenvalue problem arising in vibration analysis', *RAIRO Modél. Math. Anal. Numér.* **29**, 421–434.

B. N. Datta, S. Elhay and Y. M. Ram (1997), 'Orthogonality and partial pole assignment for the symmetric definite quadratic pencil', *Lin. Alg. Appl.* **257**, 29–48.

A. S. Deakin and T. M. Luke (1992), 'On the inverse eigenvalue problem for matrices', *J. Phys. A* **25**, 635–648.

P. Delsarte and Y. Genin (1984), Spectral properties of finite Toeplitz matrices, in *Mathematical Theory of Networks and Systems*, Vol. 58 of *Lecture Notes in Control and Information Sciences*, Springer, pp. 194–213.

J. A. Dias da Silva (1974), 'Matrices with prescribed entries and characteristic polynomial', *Proc. Amer. Math. Soc.* **45**, 31–37.

L. Dieci, R. D. Russel and E. S. V. Vleck (1994), 'Unitary integrators and applications to continuous orthonormalization techniques', *SIAM J. Numer. Anal.* **31**, 261–281.

A. C. Downing and A. S. Householder (1956), 'Some inverse characteristic value problems', *J. Assoc. Comput. Mach.* **3**, 203–207.

A. L. Duarte (1989), 'Construction of acyclic matrices from spectral data', *Lin. Alg. Appl.* **113**, 173–182.

R. Erra and B. Philippe (1997), 'On some structured inverse eigenvalue problems', *Numer. Alg.* **15**, 15–35.

W. Ferguson (1980), 'The construction of Jacobi and periodic Jacobi matrices with prescribed spectra', *Math. Comput.* **35**, 79–84.

M. Fiedler (1974), 'Eigenvalues of nonnegative symmetric matrices', *Lin. Alg. Appl.* **9**, 119–142.

E. Foltete, G. M. L. Gladwell and G. Lallement (2001), 'On the reconstruction of a damped vibrating system from two complex spectra, Part II: Experiment', *J. Sound Vib.* **240**, 219–240.

G. E. Forsythe and E. G. Straus (1955), 'On best conditioned matrices', *Proc. Amer. Math. Soc.* **6**, 340–345.

S. Friedland (1972), 'Matrices with prescribed off-diagonal elements', *Israel J. Math.* **11**, 184–189.

S. Friedland (1977), 'Inverse eigenvalue problems', *Lin. Alg. Appl.* **17**, 15–51.

S. Friedland (1978), 'On an inverse problem for nonnegative and eventually nonnegative matrices', *Israel J. Math.* **29**, 43–60.

S. Friedland and A. A. Melkman (1979), 'On the eigenvalues of nonnegative Jacobi matrices', *Lin. Alg. Appl.* **25**, 239–254.

S. Friedland, J. Nocedal and M. L. Overton (1986), Four quadratically convergent methods for solving inverse eigenvalue problems, in *Numerical Analysis* (Dundee, 1985), Longman Sci. Tech., Harlow, pp. 47–65.

S. Friedland, J. Nocedal and M. L. Overton (1987), 'The formulation and analysis of numerical methods for inverse eigenvalue problems', *SIAM J. Numer. Anal.* **24**, 634–667.

F. R. Gantmaher and M. G. Kreĭn (1960), *Oszillationsmatrizen, Oszillationskerne und Kleine Schwingungen Mechanischer Systeme* (*Oscillation Matrices and Kernels and Small Oscillations of Mechanical Systems*), Berlin, Akademie-Verlag, 1960. (In German.)

I. M. Gel′fand and B. M. Levitan (1955), 'On the determination of a differential equation from its spectral function', *Amer. Math. Soc. Transl.* (2) **1**, 253–304.

G. M. L. Gladwell (1984), 'The inverse problem for the vibrating beam', *Proc. R. Soc.* **A 393**, 277–295.

G. M. L. Gladwell (1986*a*), 'The inverse mode problem for lumped-mass systems', *Quart. J. Mech. Appl. Math.* **39**, 297–307.

G. M. L. Gladwell (1986*b*), *Inverse Problems in Vibration*, Martinus Nijhoff, Dordrecht, Netherlands.

G. M. L. Gladwell (1986*c*), 'Inverse problems in vibration', *Appl. Mech. Review* **39**, 1013–1018.

G. M. L. Gladwell (1996), 'Inverse problems in vibration, II', *Appl. Mech. Review* **49**, 2–27.

G. M. L. Gladwell (1997), 'Inverse vibration problems for finite-element models', *Inverse Problems* **13**, 311–322.

G. M. L. Gladwell (1999), 'Inverse finite element vibration problems', *J. Sound Vib.* **211**, 309–324.

G. M. L. Gladwell (2001), 'On the reconstruction of a damped vibrating system from two complex spectra, Part I: Theory', *J. Sound Vib.* **240**, 203–217.

G. M. L. Gladwell and J. A. Gbadeyan (1985), 'On the inverse problem of the vibrating string or rod', *Quart. J. Mech. Appl. Math.* **38**, 169–174.

G. M. L. Gladwell and A. Morassi (1999), 'Estimating damage in a rod from changes in node position', *Inverse Problems in Engineering* **7**, 215–233.

G. M. L. Gladwell and B. R. Zhu (1992), 'Inverse problems for multidimensional vibrating systems', *Proc. Roy. Soc. London Ser. A* **439**, 511–530.

W. Glunt, T. L. Hayden, S. Hong and J. Wells (1990), 'An alternating projection algorithm for computing the nearest Euclidean distance matrix', *SIAM J. Matrix Anal. Appl.* **11**, 589–600.

J. C. Gower (1982), 'Euclidean distance geometry', *Math. Sci.* **7**, 1–14.

W. B. Gragg and W. J. Harrod (1984), 'The numerically stable reconstruction of Jacobi matrices from spectral data', *Numer. Math.* **44**, 317–335.

L. J. Gray and D. G. Wilson (1976), 'Construction of a Jacobi matrix from spectral data', *Lin. Alg. Appl.* **14**, 131–134.

A. Greenbaum and G. H. Rodrigue (1989), 'Optimal preconditioners of a given sparsity pattern', *BIT* **29**, 610–634.

W. Guo (1996), 'An inverse eigenvalue problem for nonnegative matrices', *Lin. Alg. Appl.* **249**, 67–78.

K. P. Hadeler (1968), 'Ein inverse Eigenwertproblem', *Lin. Alg. Appl.* **1**, 83–101.

O. H. Hald (1972), On discrete and numerical inverse Sturm–Liouville problems, PhD thesis, New York University.

O. H. Hald (1976), 'Inverse eigenvalue problems for Jacobi matrices', *Lin. Alg. Appl.* **14**, 63–85.

P. R. Halmos (1974), *Finite-Dimensional Vector Spaces*, 2nd edn, Springer, New York. Reprinting of the 1958 second edition, *Undergraduate Texts in Mathematics.*

D. Hershkowits (1978), Existence of matrices satisfying prescribed conditions, Master's thesis, Technion, Haifa.

D. Hershkowits (1983), 'Existence of matrices with prescribed eigenvalues and entries', *Lin. Multilin. Alg.* **14**, 315–342.

H. Hochstadt (1967), 'On some inverse problems in matrix theory', *Arch. Math.* **18**, 201–207.

H. Hochstadt (1974), 'On construction of a Jacobi matrix from spectral data', *Lin. Alg. Appl.* **8**, 435–446.

H. Hochstadt (1979), 'On the construction of a Jacobi matrix from mixed given data', *Lin. Alg. Appl.* **28**, 113–115.

A. Horn (1954*a*), 'Doubly stochastic matrices and the diagonal of a rotation matrix', *Amer. J. Math.* **76**, 620–630.

A. Horn (1954*b*), 'On the eigenvalues of a matrix with prescribed singular values', *Proc. Amer. Math. Soc.* **5**, 4–7.

P. Horst (1965), *Factor Analysis of Data Matrices*, Holt, Rinehart and Winston, New York.

K. D. Ikramov and V. N. Chugunov (2000), 'Inverse matrix eigenvalue problems', *J. Math. Sci.* (New York) **98**, 51–136.

A. Iserles, H. Munthe-Kaas, S. P. Nørsett and A. Zanna (2000), Lie group methods, in *Acta Numerica*, Vol. 9, Cambridge University Press, pp. 1–151.

C. R. Johnson, T. J. Laffey and R. Loewy (1996), 'The real and the symmetric nonnegative inverse eigenvalue problems are different', *Proc. Amer. Math. Soc.* **124**, 3647–3651.

K. T. Joseph (1992), 'Inverse eigenvalue problem in structural design', *AIAA J.* **30**, 2890–2896.

N. K. Karmarkar and Y. N. Lakshman (1998), 'On approximate GCDs of univariate polynomials', *J. Symbolic Comput.* **26**, 653–666.

F. I. Karpelevič (1951), 'On the characteristic roots of matrices with nonnegative elements', *Izv. Akad. Nauk SSSR Ser. Mat.* **15**, 361–383. (In Russian.)

J. Kautsky and S. Elhay (1984), 'Gauss quadratures and Jacobi matrices for weight functions not of one sign', *Math. Comput.* **43**, 543–550.

J. Kautsky, N. K. Nichols and P. Van Dooren (1985), 'Robust pole assignment in linear state feedback', *Internat. J. Control* **41**, 1129–1155.

R. Knobel and J. R. McLaughlin (1994), 'A reconstruction method for a two-dimensional inverse eigenvalue problem', *Z. Angew. Math. Phys.* **45**, 794–826.

M. G. Kreĭn (1933), 'On the spectrum of a Jacobian matrix, in connection with the torsional oscillations of shafts', *Mat. Sbornik* **40**, 455–478.

J. D. Lambert (1991), *Numerical Methods for Ordinary Differential Systems: The Initial Value Problem*, Wiley, Chichester.

H. J. Landau (1994), 'The inverse eigenvalue problem for real symmetric Toeplitz matrices', *J. Amer. Math. Soc.* **7**, 749–767.

D. P. Laurie (1988), 'A numerical approach to the inverse Toeplitz eigenproblem', *SIAM J. Sci. Statist. Comput.* **8**, 410–405.

D. P. Laurie (1991), 'Solving the inverse eigenvalue problem via the eigenvector matrix', *J. Comput. Appl. Math.* **35**, 277–289.

D. P. Laurie (1997), 'Calculation of Gauss–Kronrod quadrature rules', *Math. Comput.* **66**, 1133–1145.

B. M. Levitan (1987), *Inverse Sturm–Liouville Problems*, VSP, Zeist. Translated from the Russian by O. Efimov.

C. K. Li and R. Mathias (2001), 'Construction of matrices with prescribed singular values and eigenvalues', *BIT* **41**, 115–126.

H. B. Li, P. Stoica and J. Li (1999), 'Computationally efficient maximum likelihood estimation of structured covariance matrices', *IEEE Trans. Signal Processing* **47**, 1314–1323.

N. Li (1997), 'A matrix inverse eigenvalue problem and its application', *Lin. Alg. Appl.* **266**, 143–152.

R. Loewy and D. London (1978), 'A note on an inverse problems for nonnegative matrices', *Lin. Multilin. Alg.* **6**, 83–90.

D. London and H. Minc (1972), 'Eigenvalues of matrices with prescribed entries', *Proc. Amer. Math. Soc.* **34**, 8–14.

J. R. McLaughlin and O. H. Hald (1995), 'A formula for finding a potential from nodal lines', *Bull. Amer. Math. Soc., New Series* **32**, 241–247.

J. R. McLaughlin, P. L. Polyakov and P. E. Sacks (1994), 'Reconstruction of a spherically symmetric speed of sound', *SIAM J. Appl. Math.* **54**, 1203–1223.

H. Minc (1988), *Nonnegative Matrices*, Wiley, New York.

L. Mirsky (1958), 'Matrices with prescribed characteristic roots and diagonal elements', *J. London Math. Soc.* **33**, 14–21.

J. E. Mottershead and M. I. Friswell, eds (2001), *Inverse Problem on Structural Dynamics* (Liverpool, 1999), Academic Press, Berlin. Special Issue of *Mechanical Systems and Signal Processing*.

M. Müller (1992), 'An inverse eigenvalue problem: Computing B-stable Runge–Kutta methods having real poles', *BIT* **32**, 676–688.

Y. Nievergelt (1997), 'Schmidt–Mirsky matrix approximation with linearly constrained singular values', *Lin. Alg. Appl.* **261**, 207–219.

P. Nylen (1999), 'Inverse eigenvalue problem: existence of special mass-damper-spring systems', *Lin. Alg. Appl.* **297**, 107–132.

P. Nylen and F. Uhlig (1997*a*), 'Inverse eigenvalue problem: existence of special spring-mass systems', *Inverse Problems* **13**, 1071–1081.

P. Nylen and F. Uhlig (1997*b*), Inverse eigenvalue problems associated with spring-mass systems, in *Proc. Fifth Conference of the International Linear Algebra Society* (Atlanta, GA, 1995), *Lin. Alg. Appl.* **254**, 409–425.

G. N. de Oliveira (1970), 'Note on an inverse characteristic value problem', *Numer. Math.* **15**, 345–347.

G. N. de Oliveira (1973*a*), 'Matrices with prescribed entries and eigenvalues, I', *Proc. Amer. Math. Soc.* **37**, 380–386.

G. N. de Oliveira (1973*b*), 'Matrices with prescribed entries and eigenvalues, II', *SIAM J. Appl. Math.* **24**, 414–417.

G. N. de Oliveira (1975), 'Matrices with prescribed entries and eigenvalues, III', *Arch. Math.* (Basel) **26**, 57–59.

G. N. de Oliveira (1983), 'Nonnegative matrices with prescribed spectrum', *Lin. Alg. Appl.* **54**, 117–121.

M. R. Osborne (1971), On the inverse eigenvalue problem for matrices and related problems for difference and differential equations, in *Conference on Applications of Numerical Analysis* (Dundee, 1971), Vol. 228 of *Lecture Notes in Mathematics*, Springer, Berlin, pp. 155–168.

J. Paine (1984), 'A numerical method for the inverse Sturm–Liouville problem', *SIAM J. Sci. Statist. Comput.* **5**, 149–156.

H. Park, L. Zhang and J. B. Rosen (1999), 'Low rank approximation of a Hankel matrix by structured total least norm', *BIT* **39**, 757–779.

B. N. Parlett (1998), *The Symmetric Eigenvalue Problem*, SIAM, Philadelphia, PA. Corrected reprint of the 1980 original.

H. Perfect (1953), 'Methods of constructing certain stochastic matrices', *Duke Math. J.* **20**, 395–404.

H. Perfect (1955), 'Methods of constructing certain stochastic matrices, II', *Duke Math. J.* **22**, 305–311.

J. D. Pryce (1993), *Numerical Solution of Sturm–Liouville Problems*, Oxford University Press, New York.

Y. Ram and G. M. L. Gladwell (1994), 'Constructing a finite element model of a vibratory rod from eigendata', *J. Sound Vib.* **169**, 229–237.

Y. M. Ram (1995), Pole-zero assignment of vibratory system by state feedback control, Technical report, University of Adelaide.

Y. M. Ram and J. Caldwell (1992), 'Physical parameters reconstruction of a free-free mass-spring system from its spectra', *SIAM J. Appl. Math.* **52**, 140–152.

Y. M. Ram and S. Elhay (1996), 'An inverse eigenvalue problem for the symmetric tridiagonal quadratic pencil with application to damped oscillatory systems', *SIAM J. Appl. Math.* **56**, 232–244.

M. S. Ravi, J. Rosenthal and X. A. Wang (1995), 'On decentralized dynamic pole placement and feedback stabilization', *IEEE Trans. Automat. Control* **40**, 1603–1614.

R. Reams (1996), 'An inequality for nonnegative matrices and the inverse eigenvalue problem', *Lin. Multilin. Alg.* **41**, 367–375.

A. Shapiro (1983), 'On the unsolvability of inverse eigenvalues problems almost everywhere', *Lin. Alg. Appl.* **49**, 27–31.

F. C. Silva (1987*a*), 'Matrices with prescribed characteristic polynomial and submatrices', *Portugal. Math.* **44**, 261–264.

F. C. Silva (1987*b*), 'Matrices with prescribed eigenvalues and principal submatrices', *Lin. Alg. Appl.* **92**, 241–250.

F. Y. Sing (1976), 'Some results on matrices with prescribed diagonal elements and singular values', *Canad. Math. Bull.* **19**, 89–92.

G. W. Soules (1983), 'Constructing symmetric nonnegative matrices', *Lin. Multilin. Alg.* **13**, 241–251.

L. Starek and D. J. Inman (1997), 'A symmetric inverse vibration problem for nonproportional underdamped systems', *Trans. ASME J. Appl. Mech.* **64**, 601–605.

L. Starek and D. J. Inman (2001), 'Symmetric inverse eigenvalue vibration problem and its applications', *Mechanical Systems and Signal Processing* **15**, 11–29.

L. Starek, D. J. Inman and A. Kress (1992), 'A symmetric inverse vibration problem', *Trans. ASME J. Vibration and Acoustics* **114**, 564–568.

T. J. Suffridge and T. L. Hayden (1993), 'Approximation by a Hermitian positive semidefinite Toeplitz matrix', *SIAM J. Matrix Anal. Appl.* **14**, 721–734.

H. R. Suleĭmanova (1949), 'Stochastic matrices with real characteristic numbers', *Doklady Akad. Nauk SSSR* (NS) **66**, 343–345.

J. G. Sun and Y. Qiang (1986), 'The unsolvability of inverse algebraic eigenvalue problems almost everywhere', *J. Comput. Math.* **4**, 212–226.

R. C. Thompson (1977), 'Singular values, diagonal elements, and convexity', *SIAM J. Appl. Math.* **32**, 39–63.

R. C. Thompson and P. McEnteggert (1968), 'Principal submatrices, II: The upper and lower quadratic inequalities.', *Lin. Alg. Appl.* **1**, 211–243.

F. Tisseur and K. Meerbergen (2001), 'The quadratic eigenvalue problem', *SIAM Review.* See also `http://www.ma.man.ac.uk/~ftisseur`.

W. F. Trench (1997), 'Numerical solution of the inverse eigenvalue problem for real symmetric Toeplitz matrices', *SIAM J. Sci. Comput.* **18**, 1722–1736.

D. W. Tufts and A. A. Shah (1993), 'Estimation of a signal wave-form from noisy data using low rank approximation to a data matrix', *IEEE Trans. Signal Processing* **41**, 1716–1721.

H. Weyl (1949), 'Inequalities between the two kinds of eigenvalues of a linear transformation', *Proc. Nat. Acad. Sci. USA* **35**, 408–411.

D. E. Williams and D. H. Johnson (1993), 'Robust estimation on structured covariance matrices', *IEEE Trans. Signal Processing* **41**, 2891–2906.

K. Wright (1992), 'Differential equations for the analytic singular value decomposition of a matrix', *Numer. Math.* **3**, 283–295.

Q. Wu (1990), Determination of the size of an object and its location in a cavity by eigenfrequency shifts, PhD thesis, University of Sydney.

S. F. Xu (1993), 'A stability analysis of the Jacobi matrix inverse eigenvalue problem', *BIT* **33**, 695–702.

S. F. Xu (1998), *An Introduction to Inverse Algebraic Eigenvalue Problems*, Peking University Press, and Friedr. Vieweg & Sohn, Braunschweig/Wiesbaden, Beijing.

E. M. E. Zayed (1993), 'An inverse eigenvalue problem for an arbitrary multiply connected bounded region: an extension to higher dimensions', *Internat. J. Math. Math. Sci.* **16**, 485–492.

H. Zha and Z. Zhang (1995), 'A note on constructing a symmetric matrix with specified diagonal entries and eigenvalues', *BIT* **35**, 448–452.

H. Zha and Z. Zhang (1999), 'Matrices with low-rank-plus-shift structure: partial SVD and latent semantic indexing', *SIAM J. Matrix Anal. Appl.* **21**, 522–536 (electronic).

L. A. Zhornitskaya and V. S. Serov (1994), 'Inverse eigenvalue problems for a singular Sturm–Liouville operator on $[0, 1]$', *Inverse Problems* **10**, 975–987.

S. Zhou and H. Dai (1991), *Daishu Tezhengzhi Fanwenti* (*The Algebraic Inverse Eigenvalue Problem*), Henan Science and Technology Press, Zhengzhou, China. (In Chinese.)

Acta Numerica (2002), pp. 73–144
DOI: 10.1017/S0962492902000028

Subdivision schemes in geometric modelling

Nira Dyn and David Levin
School of Mathematical Sciences,
Tel Aviv University,
Tel Aviv 69978, Israel
E-mail: niradyn@post.tau.ac.il
levin@post.tau.ac.il

Subdivision schemes are efficient computational methods for the design and representation of 3D surfaces of arbitrary topology. They are also a tool for the generation of refinable functions, which are instrumental in the construction of wavelets. This paper presents various flavours of subdivision, seasoned by the personal viewpoint of the authors, which is mainly concerned with geometric modelling. Our starting point is the general setting of scalar multivariate nonstationary schemes on regular grids. We also briefly review other classes of schemes, such as schemes on general nets, matrix schemes, non-uniform schemes and nonlinear schemes. Different representations of subdivision schemes, and several tools for the analysis of convergence, smoothness and approximation order are discussed, followed by explanatory examples.

CONTENTS

1. Introduction

The first work on a subdivision scheme was by de Rahm (1956). He showed that the scheme he presented produces limit functions with a first derivative everywhere and a second derivative nowhere. The pioneering work of Chaikin (1974) introduced subdivision as a practical algorithm for curve design. His algorithm served as a starting point for extensions into subdivision algorithms generating any spline functions. The importance of subdivision to applications in computer-aided geometric design became clear with the generalizations of the tensor product spline rules to control nets of arbitrary topology. This important step has been introduced in two papers, by Doo and Sabin (1978) and by Catmull and Clark (1978). The surfaces generated by their subdivision schemes are no longer restricted to representing bivariate functions, and they can easily represent surfaces of arbitrary topology.

In recent years the subject of subdivision has gained popularity because of many new applications, such as 3D computer graphics, and its close relationship to wavelet analysis. Subdivision algorithms are ideally suited to computer applications: they are simple to grasp, easy to implement, highly flexible, and very attractive to users and to researchers. In free-form surface design applications, such as in the 3D animation industry, subdivision methods are already in extensive use, and the next venture is to introduce these methods to the more conservative, and more demanding, world of geometric modelling in the industry.

Important steps in subdivision analysis have been made in the last two decades, and the subject has expanded into new directions owing to various generalizations and applications. This review does not claim to cover all aspects of subdivision schemes, their analysis and their applications. It is, rather, a personal view of the authors on the subject. For example, the convergence analysis is not presented in its greatest generality and is restricted to uniform convergence, which is relevant to geometric modelling. On the other hand, the review deals with the analysis and applications of nonstationary subdivision schemes, which the authors view as important for future developments. While most of the analysis presented deals with convergence and regularity, it also relates the results to practical issues such as attaining optimal approximation order and computing limit values.

The presentation starts with the basic notions of nonstationary subdivision: definitions of limit functions and basic limit functions and the refinement relations they satisfy. Different forms of representation of subdivision schemes, and their basic convolution property, are also presented in Section 2. These are later used throughout the review for stating and proving the main results. In the next section we present a gallery of examples of different types of subdivision schemes: interpolatory and non-interpolatory, linear and nonlinear, stationary and nonstationary, matrix subdivision, Hermite-type subdivision, and bivariate subdivision on regular and irregular nets. In the same section we also sketch some extensions of subdivision schemes that are not studied in this review. The material in Sections 2 and 3 is intended to provide a broad map of the subdivision area for tourists and new potential users.

In Section 4, the convergence analysis of univariate and bivariate stationary subdivision schemes, and the smoothness analysis of their limit functions, are presented via the related difference schemes. Analogous analysis is also presented for nonstationary schemes, relating the results to the analysis of stationary subdivision and using smoothing factors and convolutions as main tools. The central results are given, some with full proofs and others with only sketches. The special analysis of convergence and smoothness at extraordinary points, of subdivision schemes on nets of general topology, is reviewed in Section 6. In Section 7 we discuss two issues in the practical application of subdivision schemes. One is the computation of exact limit values of the function (surface), and the limit derivatives, at dyadic points. The other is the approximation order of subdivision schemes, and how to attain it.

For other reviews and tutorials on subdivision schemes and their applications, the reader may turn to Cavaretta, Dahmen and Micchelli (1991), Schröder (2001), Zorin and Schröder (2000) and Warren (1995*b*)

2. Basic notions

This review presents subdivision schemes mainly as a tool for geometric modelling, starting from the general perspective of nonstationary schemes.

2.1. Nonstationary schemes

A subdivision scheme is defined as a set of refinement rules relative to a set of nested meshes of isolated points (nets)

$$N_0 \subseteq N_1 \subseteq N_2 \subseteq \cdots \subseteq \mathbb{R}^s.$$

Each refinement rule maps real values defined on N_k to real values defined on a refined net N_{k+1}. The subdivision scheme is the repeated refinement of initial data defined on N_0 by these rules.

Let us first consider the regular grid case, namely the net $N_0 = \mathbb{Z}^s$ for $s \in \mathbb{Z}_+\backslash 0$ and its binary refinements, namely the refined nets $N_k = 2^{-k}\mathbb{Z}^s$, $k \in \mathbb{Z}_+\backslash 0$. Let $\mathbf{f}^k$ be the values attached to the net $N_k = 2^{-k}\mathbb{Z}^s$,

$$\mathbf{f}^k = \{f^k_\alpha : \alpha \in \mathbb{Z}^s\} \tag{2.1}$$

with f^k_α attached to $2^{-k}\alpha$.

The refinement rule at refinement level k is of the form

$$f_\alpha^{k+1} = \sum_{\beta \in \mathbb{Z}^s} a_{\alpha-2\beta}^k f_\beta^k, \quad \alpha \in \mathbb{Z}^s, \tag{2.2}$$

which we write formally as

$$\mathbf{f}^{k+1} = R_{\mathbf{a}^k} \mathbf{f}^k. \tag{2.3}$$

The set of coefficients $\mathbf{a}^k = \{a_\alpha^k : \alpha \in \mathbb{Z}^s\}$ determines the refinement rule at level k and is termed the kth *level mask*. Let $\sigma(\mathbf{a}^k) = \{\gamma \mid a_\gamma^k \neq 0\}$ be the support of the mask $\mathbf{a}^k$. Here we restrict the discussion to the case that the origin is in the convex hull of $\sigma(\mathbf{a}^k)$, and that $\sigma(\mathbf{a}^k)$ are finite sets, for $k \in \mathbb{Z}_+$. A more general form of refinement, corresponding to a dilation matrix M, is

$$f_\alpha^{k+1} = \sum_{\beta \in \mathbb{Z}^s} a_{\alpha-M\beta}^k f_\beta^k, \tag{2.4}$$

where M is an $s \times s$ matrix of integers with $|\det(M)| > 1$ (see, *e.g.*, Dahmen and Micchelli (1997) and Han and Jia (1998)). In this case the refined nets are $M^{-k}\mathbb{Z}^s$, $k \in \mathbb{Z}_+$. We restrict our discussion to binary refinements corresponding to $M = 2I$, with I the $s \times s$ identity matrix, namely to (2.2).

If the masks $\{\mathbf{a}^k\}$ are independent of the refinement level, namely if $\mathbf{a}^k = \mathbf{a}$, $k \in \mathbb{Z}_+$, the subdivision scheme is termed *stationary*, and is denoted by $S_\mathbf{a}$. In the nonstationary case, the subdivision scheme is determined by $\{\mathbf{a}^k : k \in \mathbb{Z}_+\}$, and is denoted as a collection of refinement rules $\{R_{\mathbf{a}^k}\}$, or by the shortened notation $S_{\{\mathbf{a}^k\}}$.

2.2. Notions of convergence

A continuous function $f \in C(\mathbb{R}^s)$ is termed the *limit function* of the subdivision scheme $S_{\{\mathbf{a}^k\}}$, from the initial data sequence $\mathbf{f}^0$, and is denoted by $S_{\{\mathbf{a}^k\}}^\infty \mathbf{f}^0$, if

$$\lim_{k\to\infty} \max_{\alpha \in \mathbb{Z}^s \cap K} |f_a^k - f(2^{-k}\alpha)| = 0, \tag{2.5}$$

where $\mathbf{f}^k$ is defined recursively by (2.2), and K is any compact set in $\mathbb{R}^s$.

This is equivalent (Cavaretta *et al.* 1991) to f being the uniform limit on compact sets of $\mathbb{R}^s$ of the sequence $\{F_k : k \in \mathbb{Z}_+\}$ of s-linear spline functions interpolating the data at each refinement level, namely

$$F_k(2^{-k}\alpha) = f_\alpha^k, \quad F_k\big|_{2^{-k}(\alpha+[0,1]^s)} \in \pi_1^T, \quad \alpha \in \mathbb{Z}^s, \tag{2.6}$$

where π_1^T is the tensor product space of the spaces of linear polynomials in each of the variables.

From this equivalence we get

$$\lim_{k\to\infty} \|f(t) - F_k(t)\|_{\infty,K} = 0. \tag{2.7}$$

If we do not insist on the continuity of f in (2.5) or on the L_∞-norm in (2.7), we get weaker notions of convergence: for instance, L_p-convergence is defined by requiring the existence of $f \in L_p(\mathbb{R}^s)$ satisfying $\lim_{k\to\infty} \|f(t) - F_k(t)\|_p = 0$ (Villemoes 1994, Jia 1995). The case $p = 2$ is important in the theory of wavelets (Daubechies 1992). In this paper we consider mainly the notion of uniform convergence, corresponding to (2.7), which is relevant to geometric modelling. We also mention here the weakest notion of convergence (Derfel, Dyn and Levin 1995), termed *weak convergence* or *distributional convergence*. A subdivision scheme $S_{\{\mathbf{a}^k\}}$, generating the values $f^{k+1} = R_{a^k} f^k$, for $k \in \mathbb{Z}_+$, converges weakly to an integrable function f if, for any $g \in C_0^\infty$ (infinitely smooth and of compact support),

$$\lim_{k\to\infty} 2^{-k} \sum_{\alpha\in\mathbb{Z}^s} g(2^{-k}\alpha) f_\alpha^k = \int_{\mathbb{R}}^s f(x)g(x)\,\mathrm{d}x.$$

Definition 1. A subdivision scheme is termed *uniformly convergent* if, for any initial data, there exists a limit function in the sense of (2.7) (or equivalently, if, for any initial data, there exists a continuous limit function in the sense of (2.5)) and if the limit function is nontrivial for at least one initial data sequence. A uniformly convergent subdivision scheme is termed C^m, or C^m*-convergent* if, for any initial data, the limit function has continuous derivatives up to order m.

In the following we use the term convergence for uniform convergence, since this notion of convergence is central to the review.

An important initial data sequence is $\mathbf{f}^0 = \boldsymbol{\delta} = \{f_\alpha^0 = \delta_{\alpha,0} : \alpha \in \mathbb{Z}^s\}$. If $S_{\{\mathbf{a}^k\}}$ is convergent, then there exists a nontrivial limit function starting from this initial data sequence:

$$\phi_{\{\mathbf{a}^k\}} = S_{\{\mathbf{a}^k\}}^\infty \boldsymbol{\delta}.$$

By the uniformity of the refinement rules (each refinement rule operates in the same way at all locations), and by their linearity,

$$S_{\{\mathbf{a}^k\}}^\infty \mathbf{f}^0 = \sum_{\alpha\in\mathbb{Z}^s} f_\alpha^0 \phi_{\{\mathbf{a}^k\}}(\cdot - \alpha), \tag{2.8}$$

for any initial $\mathbf{f}^0$. Thus, if $\phi_{\mathbf{a}^k} \in C^m(\mathbb{R}^s)$ for some $m \geq 0$, so is any limit function generated by $S_{\mathbf{a}^k}$, and the scheme is C^m.

When the initial data consist of a sequence of vectors

$$\mathbf{P}^0 = \{P_\alpha^0 \in \mathbb{R}^d : \alpha \in \mathbb{Z}^s\} \in (\ell_\infty(\mathbb{Z}^s))^d,$$

the limit of the subdivision, given by (2.8) with $\mathbf{f}^0$ replaced by $\mathbf{P}^0$, is a parametric representation of a manifold in $\mathbb{R}^d$. In geometric modelling $s = 1$ corresponds to curves in $\mathbb{R}^d$ for $d = 2, 3$ and $s = 2, d = 3$ to surfaces in $\mathbb{R}^3$. The set of refined points $\mathbf{P}^k$, for $k \in \mathbb{Z}_+$, is termed the *control points at level k*.

2.3. The refinement equations

The function $\phi_{\{\mathbf{a}^k\}} = S^{\infty}_{\{\mathbf{a}^k\}}\boldsymbol{\delta}$, termed the *basic limit function* of the subdivision scheme $S_{\{\mathbf{a}^k\}}$, is the first in the family of functions $\{\phi_\ell : \ell \in \mathbb{Z}_+\}$, defined by

$$\phi_\ell = S_\ell^\infty \boldsymbol{\delta}, \tag{2.9}$$

where $S_\ell = \{R_{\mathbf{a}^k} : k \geq \ell,\ k \in \mathbb{Z}_+\}$. Each function in this family is a basic limit function of a subdivision scheme defined in terms of a subset of the masks $\{\mathbf{a}^k\}$. If $S_0 = S_{\{\mathbf{a}^k\}}$ is convergent so is any S_ℓ for $\ell \in \mathbb{Z}_+$ (Dyn and Levin 1995) (see Section 4.1). Thus all the functions $\{\phi_\ell : \ell \in \mathbb{Z}_+\}$ are well defined, if S_0 is convergent. Moreover, by (2.9),

$$S_\ell^\infty \mathbf{f}^0 = \sum_{\alpha \in \mathbb{Z}^s} f_\alpha^0 \phi_\ell(\cdot - \alpha). \tag{2.10}$$

The support of ϕ_ℓ can be determined by the the supports of the masks $\{\mathbf{a}^k\}$. Recalling that $\sigma(\mathbf{a}^k)$ denotes the support of the mask $\mathbf{a}^k$, which is a finite set of points in $\mathbb{Z}^s$, then, by the refinement rules (2.2) and by (2.9), the support $\sigma(\phi_\ell)$ of ϕ_ℓ is given by

$$\sigma(\phi_\ell) = \overline{\sum_{k=\ell}^{\infty} 2^{\ell-k-1}\sigma(\mathbf{a}^k)}, \tag{2.11}$$

where the sum denotes the Minkowski sum of sets (that is, $A + B = \{a + b : a \in A,\ b \in B\}$). In the stationary case and in the univariate case, (2.11) can be further elaborated.

In the univariate case, $s = 1$, let $[\ell^k, u^k] = \langle\sigma(\mathbf{a}^k)\rangle$ be the convex hull of $\sigma(\mathbf{a}^k)$, and let

$$\ell_k = \sum_{j=k}^{\infty} 2^{k-j-1}\ell^j, \qquad u_k = \sum_{j=k}^{\infty} 2^{k-j-1}u^j.$$

Then

$$\sigma(\phi_k) \subseteq [\ell_k, u_k]. \tag{2.12}$$

In the stationary case (Cavaretta *et al.* 1991), (2.11) yields

$$\sigma(\phi_{\mathbf{a}}) \subseteq \langle\sigma(\mathbf{a})\rangle. \tag{2.13}$$

The functions $\{\phi_k : k \in \mathbb{Z}_+\}$ are related by a system of functional equations, termed *refinement equations*. To see this, observe that $(R_{\mathbf{a}^k}\boldsymbol{\delta})_\alpha = a_\alpha^k$,

$\alpha \in \mathbb{Z}^s$ and, by the linearity of the refinement rules,

$$\phi_k = \sum_\alpha a_\alpha^k \phi_{k+1}(2 \cdot -\alpha), \quad k \in \mathbb{Z}_+. \tag{2.14}$$

In the stationary case, namely when $\mathbf{a}^k = \mathbf{a}$, $k \in \mathbb{Z}_+$, this system of equations reduces to a single functional equation

$$\phi_\mathbf{a} = \sum_\alpha a_\alpha \phi_\mathbf{a}(2 \cdot -\alpha), \tag{2.15}$$

with $\mathbf{a} = \{a_\alpha : \alpha \in \mathbb{Z}^s\}$, and $\phi_\mathbf{a} = S_\mathbf{a}^\infty \boldsymbol{\delta}$.

The refinement equation (2.15) is the key to the generation of multiresolution analysis and wavelets (Daubechies 1992, Mallat 1989). When the scheme $S_\mathbf{a}$ converges, the unique compactly supported solution of the refinement equation (2.15) coincides with $S_\mathbf{a}^\infty \boldsymbol{\delta}$. The refinement equation (2.15) suggests another way to compute its unique compactly supported solution. This method is termed the 'cascade algorithm' (see, *e.g.*, Daubechies and Lagarias (1992*a*)). It involves repeated use of the operator

$$T_\mathbf{a} g = \sum_\alpha a_\alpha g(2 \cdot -\alpha).$$

defined on continuous compactly supported functions. The cascade algorithm is as follows.

(1) Choose a continuous compactly supported function, ψ_0, as a 'good' initial guess (*e.g.*, H as in (2.20)).

(2) Iterate: $\psi_{k+1} = T_\mathbf{a} \psi_k$.

It is easy to verify that the operator $T_\mathbf{a}$ is the adjoint of the refinement rule $R_\mathbf{a}$, in the following sense: for any ψ continuous and of compact support, (Cavaretta *et al.* 1991)

$$\sum_\alpha (R_\mathbf{a} f)_\alpha \psi(2 \cdot -\alpha) = \sum_\alpha f_\alpha (T_\mathbf{a} \psi)(\cdot - \alpha). \tag{2.16}$$

Note that, while the refinement rule $R_\mathbf{a}$ is defined on sequences, the operator $T_\mathbf{a}$ is defined on functions. A similar operator to $T_\mathbf{a}$, defined on sequences, is

$$(\widetilde{T}_\mathbf{a} \mathbf{f})_\alpha = \sum_\beta a_\beta f_{2\alpha-\beta} = \sum_\gamma a_{2\alpha-\gamma} f_\gamma. \tag{2.17}$$

This operator is the adjoint of the operator $R_\mathbf{a}$ on the space of sequences defined on $\mathbb{Z}^s$. The operator $\widetilde{T}_\mathbf{a}$ in (2.17) is termed the *transfer operator* (Daubechies 1992), and plays a major role in the analysis of the solutions of refinement equations of the form (2.15) (see, *e.g.*, Jia (1996), Han (2001), Han and Jia (1998) and Jia and Zhang (1999)).

2.4. Representations of subdivision schemes

The notions introduced above regard a subdivision scheme $S_{\{\mathbf{a}^k\}} = \{R_{\mathbf{a}^k}\}$ as a set of operators defined on sequences in $\ell_\infty(\mathbb{Z}^s)$. Each refinement rule can be represented as a bi-infinite matrix with each element indexed by two index vectors from $\mathbb{Z}^s$,

$$f_\alpha^{k+1} = \sum_{\beta \in \mathbb{Z}^s} A_{\alpha,\beta}^k f_\beta^k, \quad \alpha \in \mathbb{Z}^s, \tag{2.18}$$

where the bi-infinite matrix A^k has elements

$$A_{\alpha,\beta}^k = a_{\alpha-2\beta}^k. \tag{2.19}$$

Finite sections of these matrices are used in the analysis of the subdivision scheme $S_{\{\mathbf{a}^k\}}$ (see Section 5).

One may also regard a subdivision scheme as a set of operators $\{R_k : k \in \mathbb{Z}_+\}$ defined on a function space (Dyn and Levin 1995), if one considers the functions $\{F_k\}$ introduced in (2.6). The set of operators $\{R_k\}$ has the property that R_k maps F_k into F_{k+1}. More specifically, let H be defined by

$$H(\alpha) = \delta_{0,\alpha}, \quad H\big|_{(\alpha+[0,1]^s)} \in \pi_1^T, \quad \alpha \in \mathbb{Z}^s. \tag{2.20}$$

Define the operators $\{R_k\}$ on $C(\mathbb{R}^s)$ as

$$R_k g = \sum_{\alpha \in \mathbb{Z}^s} H(2^{k+1} \cdot - \alpha) \sum_{\beta \in \mathbb{Z}^s} a_{\alpha-2\beta}^k g(2^{-k}\beta), \quad k \in \mathbb{Z}_+, \tag{2.21}$$

for any $g \in C(\mathbb{R}^s)$. Then the subdivision scheme $S_{\{\mathbf{a}^k\}}$ is related to the set of operators $\{R_k\}$ in several ways, for example,

$$(R_k g)\big|_{2^{-k-1}\mathbb{Z}^s} = R_{\mathbf{a}^k}(g\big|_{2^{-k}\mathbb{Z}^s}),$$

and the more significant relation

$$S_{\{\mathbf{a}^k\}}^\infty \mathbf{f}^0 = \lim_{k \to \infty} R_k R_{k-1} \cdots R_0 g, \tag{2.22}$$

where $g \in C(\mathbb{R}^s)$ is any interpolant to $\mathbf{f}^0$ on $\mathbb{Z}^s$, namely

$$g(\alpha) = f_\alpha^0, \quad \alpha \in \mathbb{Z}^s.$$

In particular, g can be

$$g = \sum_{\alpha \in \mathbb{Z}^s} H(\cdot - \alpha) f_\alpha^0.$$

Another important relation is

$$\|R_k\| = \|R_{\mathbf{a}^k}\|_\infty = \max_{\alpha \in E_s} \left\{ \sum_{\beta \in \mathbb{Z}^s} |a_{\alpha-2\beta}^k| \right\}, \tag{2.23}$$

where E_s is the set of extreme points of $[0,1]^s$. The representation of subdivision schemes in terms of sequences of operators on $C(\mathbb{R}^s)$ facilitates the application of standard operator-theory tools to the analysis of subdivision schemes, for instance, to deduce convergence properties of nonstationary schemes from those of related stationary ones (Dyn and Levin 1995) (see Section 4.1).

A representation of the refinement rule (2.2), which is a central tool in the convergence and smoothness analysis of stationary schemes, is in terms of z-transforms (Laurent series). Let the symbol of the mask $\mathbf{a}^k$ be defined as the Laurent polynomial

$$a^k(z) = \sum_{\alpha\in\mathbb{Z}^s} a^k_\alpha z^\alpha. \tag{2.24}$$

Here we use the multi-index notation $z^n = z_1^{n_1}\cdots z_s^{n_s}$, for $z\in\mathbb{R}^s$, $n\in\mathbb{Z}^s$, and $z^n = z_1^n\cdots z_s^n$, for $z\in\mathbb{R}^s$, $n\in\mathbb{Z}$. Obviously, a subdivision scheme $S_{\{\mathbf{a}^k\}}$ is identified with the set of its symbols $\{a^k(z)\}$. In our notation we exchange freely between the mask and its symbol, for instance, $\phi_{\{a^k(z)\}}$ denotes the basic limit function of $S_{\{a^k(z)\}} = S_{\{\mathbf{a}^k\}}$.

Let the z-transform of the sequence $\mathbf{f} = \{f_\alpha : \alpha\in\mathbb{Z}^s\}$ be denoted by $L(\mathbf{f};z)$, namely

$$L(\mathbf{f};z) = \sum_{\alpha\in\mathbb{Z}^s} f_\alpha z^\alpha.$$

Then the refinement rule (2.2) can be written in the form

$$L(\mathbf{f}^{k+1};z) = a^k(z)L(\mathbf{f}^k;z^2), \tag{2.25}$$

with the formal meaning of the equality above being that corresponding powers of z on both sides of the equality have equal coefficients. Iterating the relation (2.25), we obtain

$$L(\mathbf{f}^{k+\ell};z) = a^{k+\ell-1}(z)a^{k+\ell-2}(z^2)\cdots a^k(z^{2^{\ell-1}})L(\mathbf{f}^k;z^{2^\ell}). \tag{2.26}$$

Thus, the ℓ-iterated symbol from level k to level $k+\ell$ is

$$a^{[k;\ell]}(z) = \sum_{\alpha\in\mathbb{Z}^s} a^{[k;\ell]}_\alpha z^\alpha = \prod_{j=1}^{\ell} a^{k+\ell-j}(z^{2^{j-1}}). \tag{2.27}$$

In the stationary case we denote the ℓ-iterated symbol by $a^{[\ell]}$:

$$a^{[\ell]}(z) = \prod_{j=1}^{\ell} a(z^{2^{j-1}}). \tag{2.28}$$

2.5. The convolution property

Here we present an important property of schemes, which is easily expressed in terms of the Laurent polynomial representation. This property is presented in three different forms, depending on the notion of convergence used.

(1) Let $S_{\{\mathbf{a}^k\}}$ and $S_{\{\mathbf{b}^k\}}$ each be either (uniformly) convergent or convergent in the sense of (2.5), with corresponding basic limit functions $\phi_{\{\mathbf{a}^k\}}$ and $\phi_{\{\mathbf{b}^k\}}$ continuous in their support. Then the scheme $S_{\{\mathbf{c}^k\}}$ defined by the symbols

$$c^k(z) = 2^{-s}a^k(z)b^k(z) \tag{2.29}$$

is also convergent, and its basic limit function is

$$\phi_{\{\mathbf{c}^k\}} = \phi_{\{\mathbf{a}^k\}} * \phi_{\{\mathbf{b}^k\}}. \tag{2.30}$$

Here the symbol $*$ stands for the s-dimensional convolution (Cavaretta *et al.* 1991, Dyn and Levin 1995).

The convolution property which is repeatedly used in this paper for $s > 1$, is of a different form.

(2) Let $S_{\{\mathbf{b}^k\}}$ be a convergent s-variate subdivision scheme, and let $S_{\{\mathbf{a}^k\}}$ be a univariate scheme, which is convergent in the sense of (2.5) to integrable limit functions. Then the symbols

$$c^k(z) = 2^{-1}a^k(z^\lambda)b^k(z), \tag{2.31}$$

with $\lambda \in \mathbb{Z}^s$, define a convergent scheme $S_{\{\mathbf{c}^k\}}$. Moreover,

$$\phi_{\{\mathbf{c}^k\}}(x) = \phi_{\{\mathbf{a}^k\}} *_\lambda \phi_{\{\mathbf{b}^k\}}(x) \equiv \int_{\mathbb{R}} \phi_{\{\mathbf{a}^k\}}(x - \lambda t)\phi_{\{\mathbf{b}^k\}}(t)\,\mathrm{d}t. \tag{2.32}$$

The convolution property is also valid in the case of weak convergence of $S_{\mathbf{a}}$. This property is used in only one example in the paper.

(3) Let $S_{\{\mathbf{b}^k\}}$ be an s-variate subdivision scheme convergent in the sense of (2.5), with $\phi_{\{\mathbf{b}^k\}}$ continuous in its support, and let $S_{\{\mathbf{a}^k\}}$ be a weakly convergent s-variate scheme, with $\phi_{\{\mathbf{a}^k\}}$ continuous in its support. Then the scheme $S_{\{\mathbf{c}^k\}}$ defined by the symbols in (2.29) is convergent, and $\phi_{\{\mathbf{c}^k\}}$ is given by (2.30).

Here we indicate how to verify convolution property (2) ((2.31) and (2.32)). The verification of the convolution property in its other two forms is based on the same reasoning. Observe that, for $\mathbf{f}^k = R_{\mathbf{a}^{k-1}} \cdots R_{\mathbf{a}^0}\boldsymbol{\delta}$, we have $L(\mathbf{f}^k; z) = a^{[0,\ell]}(z)$, and that in polynomial multiplication the coefficients are computed by convolutions of the coefficients of the factors. Thus, the relations (2.31) and (2.27) yield

$$c^{[0,\ell]}(z) = 2^{-\ell}a^{[0,\ell]}(z^\lambda)b^{[0,\ell]}(z),$$

or equivalently

$$L(\mathbf{g}^\ell; z) = 2^{-\ell} L(\mathbf{f}^\ell; z^\lambda) L(\mathbf{h}^\ell; z), \tag{2.33}$$

with $\mathbf{g}^\ell = R_{\mathbf{c}^{\ell-1}} \cdots R_{\mathbf{c}^0}\boldsymbol{\delta}$, and $\mathbf{h}^\ell = R_{\mathbf{b}^{\ell-1}} \cdots R_{\mathbf{b}^0}\boldsymbol{\delta}$.

Now, (2.32) can be concluded by equating coefficients of equal powers of z on both sides of (2.33), taking into account the convergence of $\{\mathbf{f}^k\}_{k\in\mathbb{Z}_+}$ and of $\{\mathbf{h}^k\}_{k\in\mathbb{Z}_+}$ to the compactly supported limit functions $\phi_{\{\mathbf{a}^k\}}$ and $\phi_{\{\mathbf{b}^k\}}$, respectively.

3. The variety of subdivision schemes

Subdivision schemes were first studied as a tool for generating spline functions (Chaikin 1974, Riesenfeld 1975, Cohen, Lyche and Riesenfeld 1980). The renewed interest in this subject in geometric modelling has evolved as subdivision processes were extended to general topologies (Catmull and Clark 1978, Doo and Sabin 1978). In recent years interesting applications have emerged, such as wavelet theory, and some very challenging theoretical issues have arisen. In the following we discuss the major different types of subdivision schemes, most of them relevant to geometric modelling:

- B-spline and box-spline schemes
- the up-function scheme
- exponential spline and exponential box-spline schemes
- interpolatory schemes
- shape-preserving schemes
- general matrix schemes
- Hermite-type and moment interpolatory schemes
- tensor product schemes
- different topologies for surface subdivision.

While assessing the various types we incorporate the notions of local support and support size, smoothness and approximation order. These issues will be further developed and investigated in the next sections. Here we take the liberty of using these properties in a heuristic manner.

3.1. Elementary schemes and their convolutions

The simplest elementary univariate uniform stationary scheme is the scheme defined by the symbol

$$a^k(z) = a(z) = 1 + z. \tag{3.1}$$

The corresponding basic limit function is the characteristic function of $[0, 1]$, where the convergence is in the sense of (2.5):

$$\phi_{\mathbf{1+z}} = B_0(\cdot) = \chi_{[0,1]}. \tag{3.2}$$

By convolution property (1),

$$\phi_{2^{-m}(1+z)^{m+1}} = B_0(\cdot) * B_0(\cdot) * \cdots * B_0(\cdot) = B_m(\cdot). \tag{3.3}$$

Thus, the scheme with symbol $a(z) = 2^{-m}(1+z)^{m+1}$ has as a basic limit function the mth-degree B-spline function with integer knots, supported in $[0, m+1]$, which is in $C^{m-1}(\mathbb{R})$. As shown in Section 4.2, the symbol of a C^m univariate uniform stationary binary scheme, under an additional mild condition, must contain the factor $(1+z)^{m+1}$. The earliest example of a spline subdivision is Chaikin's algorithm (Chaikin 1974)

$$f_{2i}^{k+1} = \frac{3}{4} f_i^k + \frac{1}{4} f_{i+1}^k, \qquad f_{2i+1}^{k+1} = \frac{1}{4} f_i^k + \frac{3}{4} f_{i+1}^k, \tag{3.4}$$

which converges to the quadratic spline $\sum f_i^0 B_2(\cdot - i)$. Chaikin's algorithm is also the basic example of a 'corner cutting' algorithm, which served as a starting point to various generalizations, for instance in de Boor (1987) and Gregory and Qu (1996). The application of three iterations on a simple control polygon (the polygonal line joining the control points) is presented in Figure 3.1.

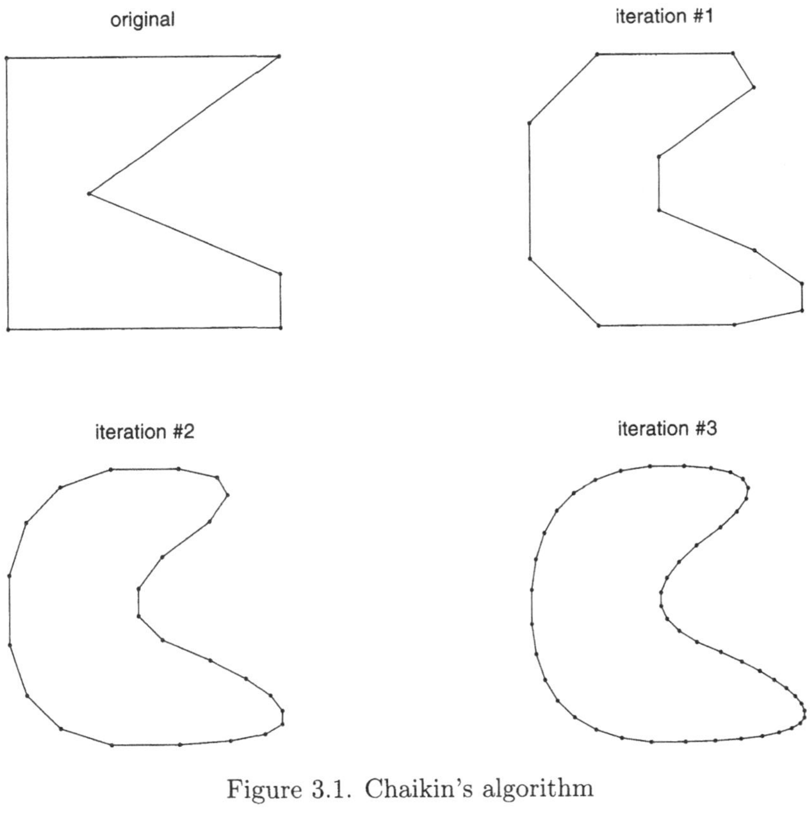

Figure 3.1. Chaikin's algorithm

Another interesting scheme that is constructed by convolutions of the elementary scheme is defined by

$$a^k(z) = 2^{-k+1}(1+z)^k. \tag{3.5}$$

The corresponding basic limit function is Rvachev's up-function (Rvachev 1990, Derfel *et al.* 1995) which is in $C^\infty(\mathbb{R})$ and is supported in $[0,2]$ (see Example 5). The spaces $V_k = \text{span}\{\phi_k(2^k \cdot -\alpha) : \alpha \in \mathbb{Z}^s\}$, $k \in \mathbb{Z}_+$, with $\{\phi_k\}$ defined as (2.9) with respect to the symbols at (3.5), provide spectral approximation order (Dyn and Ron 1995).

Products of the elementary univariate factors in directions in $\mathbb{Z}^s$ generate box-splines in $\mathbb{R}^s$ as basis limit functions. Let $\Lambda = \{\lambda_1, \ldots, \lambda_\ell\} \subset \mathbb{Z}^s$, and define the stationary scheme with the symbol

$$a(z) = 2^{s-\ell} \prod_{j=1}^{\ell} (1 + z^{\lambda_j}). \tag{3.6}$$

This scheme is related to the box-spline when directions Λ (de Boor, Höllig and Riemenschneider 1993, Dahmen and Micchelli 1984). Convergence is guaranteed if there is a subset of s directions $\{\lambda_{i_1}\lambda_{i_2}, \ldots, \lambda_{i_s}\} \in \Lambda$ such that $\det(\lambda_{i_1}\lambda_{i_2} \cdots \lambda_{i_s}) = 1$. Furthermore, if any subset of $\ell - m - 1$ directions spans $\mathbb{R}^s$, then $\phi_{\mathbf{a}}$ is in C^m (de Boor *et al.* 1993).

An important example here is the scheme generating the C^2 quartic 3-directional box-spline, namely, the scheme with the symbol

$$a(z) = 2^{-4}(1 + z^{(1,0)})^2(1 + z^{(0,1)})^2(1 + z^{(1,1)})^2. \tag{3.7}$$

It is easy to check that the above conditions are satisfied with $m = 2$, and thus the basic limit function is a box-spline in C^2.

The uniform nonstationary elementary scheme is again a scheme defined by symbols that are linear polynomials in z, namely

$$a^k(z) = 1 + r_k z, \quad k \in \mathbb{Z}_+. \tag{3.8}$$

The parameters $\{r_k\}_{k\in\mathbb{Z}_+}$ are free parameters which determine the convergence of the subdivision scheme and the regularity of the limit function (Dyn and Levin 1995). To examine these issues we write the scheme explicitly as

$$f_{2i}^{k+1} = f_i^k, \quad f_{2i+1}^{k+1} = r_k f_i^k, \quad i \in \mathbb{Z}. \tag{3.9}$$

Starting the subdivision with initial data sequence $\mathbf{f}^0 = \boldsymbol{\delta}$, the limit at a dyadic point $x = \sum_{i=1}^k d_i 2^{-i} \in [0,1)$, $d_i \in \{0,1\}$, is determined at level k of the subdivision. It is easy to verify that the value of the basic limit function ϕ at a dyadic point is given by

$$\phi(x) = \prod_{i=1}^{k} r_{i-1}^{d_i}. \tag{3.10}$$

Let us define $\phi(x)$ at nondyadic points by

$$\phi(x) = \prod_{i=1}^{\infty} r_{i-1}^{d_i}, \qquad x = \sum_{i=1}^{\infty} d_i 2^{-i} \in [0,1), \tag{3.11}$$

and $\phi(x) = 0$ for all $x \notin [0,1)$. If we assume that the parameters $\{r_k\}$ satisfy $\sum_{k\in\mathbb{Z}_+} |1-r_k| < \infty$, then all the above infinite products converge, and we find out that ϕ is continuous at all nondyadic points. At dyadic points in $[0,1)$ ϕ is right-continuous; hence, it is integrable. As proved in Dyn and Levin (1995), ϕ is also left-continuous at all dyadic points in $(0,1)$ if and only if $r_k = e^{c2^{-k}}$ for some constant c.

Exponential B-splines. The univariate elementary nonstationary scheme defined by

$$a^k(z) = 1 + e^{c2^{-k-1}} z, \quad k \in \mathbb{Z}_+, \tag{3.12}$$

generates the exponential B-spline

$$\phi_{\{\mathbf{a^k}\}}(x) = e^{cx}\chi_{[0,1]}(x). \tag{3.13}$$

Consequently, by convolution property (1), the scheme generating the mth-order exponential B-spline with exponents $c_1, \ldots, c_m$ is

$$a^k(z) = 2^{-m+1} \prod_{j=1}^{m} (1 + e^{c_j 2^{-k-1}} z). \tag{3.14}$$

Similarly, one can derive symbols of schemes generating exponential box-splines and exponential up-functions (Dyn and Levin 1995).

Generating circumscribed circles. A special example of a scheme that is obtained by convolution of elementary schemes is given by the symbol

$$a^k(z) = \frac{1}{2(1+\cos(\alpha_k))}(1+z)(1+e^{i\alpha_k}z)(1+e^{-i\alpha_k}z),$$
$$\alpha_k = 2^{-k-1}\alpha_0, \quad k \in \mathbb{Z}_+. \tag{3.15}$$

This is a C^1 'corner cutting' scheme which reproduces constants and also $\sin(\alpha_0 x)$, $\cos(\alpha_0 x)$. If the initial control polygon is a regular n-gon and $\alpha_0 = 2\pi/n$, then the limit curve is the circle circumscribed in the n-gon (Dyn and Levin 1992). The tensor product of the above scheme with any other stationary scheme generates surfaces of revolution (Morin, Warren and Weimer 2001). It seems that a circle cannot be generated by a linear stationary scheme.

3.2. Interpolatory schemes

A class of subdivision schemes with many specific features is that of 'interpolatory subdivision schemes' (Dyn and Levin 1990). The schemes in this class generate the refined values by retaining the values at the vertices of the current net, and defining new values at the new vertices of the refined net.

Among the B-spline schemes, only those generating B_0 and B_1 are interpolatory schemes, that is, satisfying

$$f_{2j}^{k+1} = f_j^k, \quad j \in \mathbb{Z}, \quad k \in \mathbb{Z}_+, \tag{3.16}$$

together with insertion rules for new points $\{f_{2j+1}^{k+1}\}_{j\in\mathbb{Z}}$. The interpolatory refinement rules on $N_k = 2^{-k}\mathbb{Z}^s$ have the form

$$f_{2\alpha}^{k+1} = f_\alpha^k, \quad f_{\gamma+2\alpha}^{k+1} = \sum_{\beta\in\mathbb{Z}^s} a_{\gamma+2\beta}^k f_{\alpha-\beta}^k, \quad \gamma \in E_s\backslash 0, \quad \alpha \in \mathbb{Z}^s. \tag{3.17}$$

The masks corresponding to an interpolatory subdivision scheme have the feature

$$a_{2\alpha}^k = \delta_{\alpha,0}, \quad \alpha \in \mathbb{Z}^s, \quad k \in \mathbb{Z}_+.$$

It is easy to realize that, in case of a convergent scheme, all the points

$$(2^{-k}\alpha, f_\alpha^k), \quad \alpha \in \mathbb{Z}^s, \quad k \in \mathbb{Z}_+,$$

are on the graph of the limit function. In this setting there is (uniform) convergence if the values generated at the dyadic points $\{f_\alpha^k : \alpha \in \mathbb{Z}^s,\ k \in \mathbb{Z}_+\}$ are continuous.

The basic limit functions $\{\phi_k : k \in \mathbb{Z}_+\}$ satisfy

$$\phi_k(\alpha) = \delta_{\alpha,0}, \quad \alpha \in \mathbb{Z}^s, \quad k \in \mathbb{Z}_+,$$

thus their integer shifts $\{\phi_k(\cdot - \alpha) : \alpha \in \mathbb{Z}^s\}$ are linearly independent for any $k \in \mathbb{Z}_+$.

The following examples are univariate stationary schemes. Nonstationary univariate interpolatory schemes are discussed in Example 2. A bivariate interpolatory scheme is presented in Section 3.5.

The 4-point scheme. The first stationary interpolatory schemes were the 4-point schemes presented in Dubuc (1986) and Dyn, Gregory and Levin (1987). The 4-point scheme is the univariate scheme defined by (3.16) and the insertion rule

$$f_{2j+1}^{k+1} = -wf_{j-1}^k + \left(\frac{1}{2} + w\right) f_j^k + \left(\frac{1}{2} + w\right) f_{j+1}^k - wf_{j+2}^k, \tag{3.18}$$

for $j \in \mathbb{Z}$, and $k \in \mathbb{Z}_+$, where w is a shape parameter of the scheme. The symbol of the 4-point scheme is

$$a_w(z) = \frac{1}{2z}(z+1)^2(1 + wb(z)), \tag{3.19}$$

where

$$b(z) = -2z^{-2}(z-1)^2(z^2+1). \tag{3.20}$$

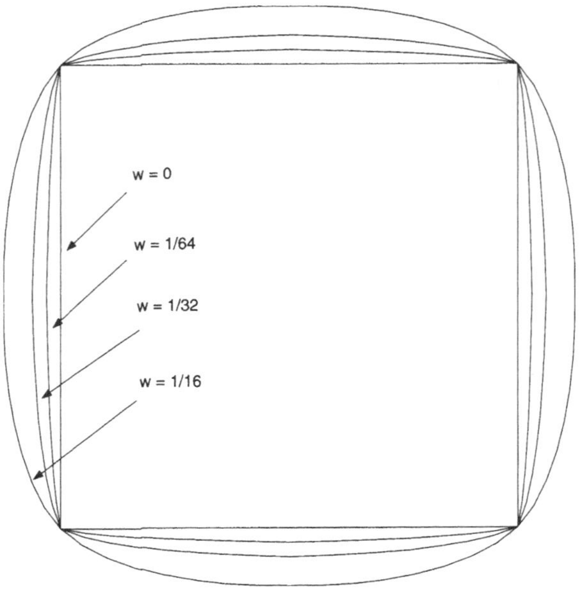

Figure 3.2. Curves generated by the 4-point scheme

For $w = 0$ the limit is the piecewise linear interpolant to the data. As w increases, the limit function is less tight. The symbol contains the elementary factor $(z+1)^2$ necessary for C^1 convergence, and the challenge in Dyn *et al.* (1987) was to determine the range of values of the shape parameter w for which the scheme is C^1. The particular value $w = \frac{1}{16}$ is also analysed in Dubuc (1986). In this case the symbol contains the factor $(z+1)^4$, which means that the scheme reproduces cubic polynomials (see Section 4.2). Yet the limit function is not even C^2. It is shown in Dyn, Gregory and Levin (1991) that the 4-point scheme is C^1 for any $w \in (0, \frac{\sqrt{5}-1}{8})$, and in Deslauriers and Dubuc (1989) that, for $w = \frac{1}{16}$, the first derivative is Hölder-continuous for any Hölder exponent $0 < \nu < 1$, yet the second derivative does not exist at dyadic points (Dyn *et al.* 1987). The application of four iterations of the 4-point scheme, with different shape parameters, on a square control polygon is presented in Figure 3.2.

Dubuc–Deslauriers interpolatory schemes. The 4-point scheme of Dubuc (1986) has been generalized to symmetric $2n$-point interpolatory schemes by Deslauriers and Dubuc (1989). The insertion rule for f_{2j+1}^{k+1} is defined by the value of the interpolation polynomial of degree $2n-1$ at $2^{-k-1}(2j+1)$, interpolating the $2n$ values $f_{j-n+1}^k, \ldots, f_{j+n}^k$. Let us denote the resulting symbol by $d_{(2n)}(z)$. These schemes are studied in Deslauriers and Dubuc (1989) by Fourier analysis, and their convergence is proved. The smoothness of $S_{\mathbf{d}_{(2n)}}$ grows linearly, but slowly, with n (Daubechies 1992).

Generalizations to multidimensional interpolatory schemes are presented in Dyn, Gregory and Levin (1990*a*) and Riemenschneider and Shen (1997).

In analogy to the up-function, it is possible to get C_0^∞ interpolatory basic limit functions using the symbols of Dubuc–Deslauriers interpolatory schemes. This is achieved in Cohen and Dyn (1996) by defining the non-stationary subdivision symbols as $a^k(z) = d_{(2k)}(z)$.

Nonlinear, shape-preserving 4-point schemes. A significant drawback of linear interpolatory schemes is the lack of shape-preservation properties. If one is interested in both interpolation and shape preservation, then linearity has to be given up. A beautiful example of a nonlinear, stationary, shape-preserving interpolatory scheme is the following 4-point C^1 convexity-preserving scheme due to Kuijt and van Damme (1998), where the rule replacing (3.18) is

$$f_{2j+1}^{k+1} = \frac{1}{2}(f_j^k + f_{j+1}^k) - \frac{1}{4\left(\frac{1}{d_j^k} + \frac{1}{d_{j+1}^k}\right)}; \quad d_j^k = f_{j+1}^k - 2f_j^k + f_{j-1}^k. \tag{3.21}$$

Starting with strictly convex initial functional data, it is shown in Kuijt and van Damme (1998) that the limit function is a strictly convex C^1 function. Kuijt and van Damme (1999) have also developed nonlinear schemes preserving monotonicity. It is also possible to use the linear 4-point scheme and to generate a convex limit function from initial strictly convex data, by choosing $w \in (0, w^*)$, where w^* depends on the initial data (Dyn, Kuijt, Levin and van Damme 1999*a*).

3.3. Matrix schemes and Hermite-type schemes

While interpolatory schemes preserve function data at points of the previous level, it is sometimes desirable to preserve other quantities. Two related families of schemes of this kind are Hermite-type schemes and moment-interpolating schemes. We may view interpolatory schemes as schemes generating limit functions with specified values at the integers. Hermite-type schemes generate limit functions with specified function values and certain derivatives' values at the integers. Moment-interpolating schemes produce limit functions with specified moments on the intervals $[i, i+1]$, $i \in \mathbb{Z}$. In both cases, the data attached to the vertices of the nets form a vector of values, and the subdivision operator is defined by a mask with matrix elements.

A univariate uniform stationary matrix subdivision scheme, operating on sequences of vectors in $\mathbb{R}^n$, is defined by a set of real $n \times n$ matrix coefficients $\{A_j : j \in \mathbb{Z}\}$, with a finite number of nonzero A_js, generating sequences of control points in $\mathbb{R}^n$, $\mathbf{v}^k = \{v_j^k \in \mathbb{R}^n : j \in \mathbb{Z}\}$, $k \geq 0$, recursively, by

$$v_i^{k+1} = \sum_{j \in \mathbb{Z}} A_{i-2j} v_j^k, \quad i \in \mathbb{Z}. \tag{3.22}$$

As an example of such a scheme, we consider the scheme generating the double-knot cubic splines. The matrix mask is defined by its matrix symbol,

$$A(z) = \frac{1}{16}\begin{pmatrix} 2+6z+z^2 & 2z+5z^2 \\ 5+2z & 1+6z+2z^2 \end{pmatrix} = \sum_{i\in\mathbb{Z}} A_i z^i. \tag{3.23}$$

Here there are two basic sets of initial data, namely $\mathbf{v}^{1,0} = (1,0)^t\boldsymbol{\delta}$ and $\mathbf{v}^{2,0} = (0,1)^t\boldsymbol{\delta}$. The two basic limit vector functions are

$$S_{\mathbf{A}}^{\infty}\mathbf{v}^{1,0} = (\phi_1,\phi_1)^t, \qquad S_{\mathbf{A}}^{\infty}\mathbf{v}^{2,0} = (\phi_2,\phi_2)^t, \tag{3.24}$$

where ϕ_1 and ϕ_2 are the two different cubic B-splines spanning the space of cubic splines with double knots at the integers (Plonka 1997).

Let us now return to the Hermite-type and moment-interpolating schemes. In the Hermite case we start with Hermite-type data, $\{v_j^0 = (f_j^0, g_j^0)^t\}_{j\in\mathbb{Z}}$ where the values $\{g_j^0\}$ represent derivative data. We now consider the scheme

$$v_{2i}^{k+1} = v_v^k, \quad v_{2i+1}^{k+1} = \sum A_{1-2j}^{(k)} v_{i+j}^k, \quad k \geq 0, \tag{3.25}$$

or, equivalently,

$$v_i^{k+1} = \sum_j A_{i-2j}^{(k)} v_j^k, \quad k \geq 0, \tag{3.26}$$

where $\{A_i^{(k)}\}$ are 2×2 matrices, possibly depending upon the refinement level k, and $A_{2j}^{(k)} = \delta_{j,0} I_{2\times 2}$. The Hermite-type scheme recursively defines values $\{v_j^k = (f_j^k, g_j^k)^t\}_{j\in\mathbb{Z}}$ attached respectively to the dyadic points $\{j2^{-k}\}_{j\in\mathbb{Z}}$. We say that the scheme is C^r if there exists a function $f \in C^r(\mathbb{R})$ such that

$$v_j^k = (f_j^k, g_j^k)^t = (f(j2^{-k}), f'(j2^{-k}))^t, \quad j \in \mathbb{Z}, \quad k \in \mathbb{Z}_+. \tag{3.27}$$

The first interesting example, presented in Merrien (1992), is an extension of the interpolatory Hermite-cubic rule. The nonzero matrices of its mask are

$$A_1^{(k)} = \begin{pmatrix} \frac{1}{2} & \alpha 2^{-k} \\ -\beta 2^k & \frac{1-\beta}{2} \end{pmatrix}, \qquad A_{-1}^{(k)} = \begin{pmatrix} \frac{1}{2} & -\alpha 2^{-k} \\ \beta 2^k & \frac{1-\beta}{2} \end{pmatrix}. \tag{3.28}$$

This scheme with $\alpha = 1/8$ and $\beta = 3/2$ produces the piecewise Hermite-cubic interpolant to the given initial data, and thus it is a C^1-scheme. We note that the matrices depend upon k, and they are even unbounded as $k \to \infty$. However, as shown in Dyn and Levin (1999), if we consider in this case the scheme for transforming the vector of values $u_j^k = (g_j^k, df_j^k)^t$, with $df_j^k = 2^k(f_j^k - f_{j-1}^k)$, this scheme becomes stationary, that is, with a constant matrix mask. Here, if the original scheme is C^1, then both elements of $\{u_j^k\}$ should converge to the same limit function f'.

The moment interpolation problem for m moments is defined as follows. Let $b^\ell(x) = \frac{(m-1)!}{\ell!(m-1-\ell)!}x^l(1-x)^{m-1-l} \cdot \chi_{[0,1]}$ denotes the ℓth Bernstein polynomial of degree $m-1$ for the interval $[0,1]$, truncated to $[0,1]$. Define

$$b_j^\ell(x) = b^\ell(x-j),$$

the translate of b^ℓ that 'lives' on $[j, j+1]$ and has L_1-norm 1.

Given the local moments of a function f,

$$\beta_j^\ell = \langle f, b_j^\ell \rangle, \quad j \in \mathbb{Z}, \quad 0 \le \ell < m, \tag{3.29}$$

the problem is to construct a 'smooth' function $\tilde{f}$ matching those moments. A solution of this problem by a subdivision process is presented in Donoho, Dyn, Levin and Yu (2000). Also shown there is the close relationship between the moment-interpolating subdivision schemes and the Hermite interpolatory subdivision schemes. In the sections on the analysis of subdivision schemes, we consider only schemes with scalar masks. The analysis of schemes with a matrix mask is not reviewed here. The interested reader may consult Plonka (1997), Cohen, Daubechies and Plonka (1997), Cohen, Dyn and Levin (1996), Micchelli and Sauer (1998) and Dyn and Levin (2002).

3.4. Tensor product schemes and related ones

The simplest subdivision schemes on $\mathbb{Z}^2$ are the stationary tensor product schemes, obtained by applying one stationary univariate scheme in the x-direction and then a second (or the same) stationary univariate scheme in the y-direction. Let us denote the symbols of the stationary univariate schemes by $x(z)$ and $y(z)$, respectively; then the symbol of the tensor product scheme S_t is $t(z_1, z_2) = x(z_1)y(z_2)$. Obviously, the tensor product subdivision scheme inherits the convergence and smoothness properties of the univariate schemes. Tensor products of univariate spline schemes are special cases of box-splines, using only two directions in (3.6). For example, the mask generating the biquadratic and the bicubic B-spline functions are, respectively, defined by the symbols

$$a(z_1, z_2) = 2^{-4}(1+z_1)^3(1+z_2)^3, \tag{3.30}$$

$$a(z_1, z_2) = 2^{-6}(1+z_1)^4(1+z_2)^4. \tag{3.31}$$

Yet tensor product schemes have masks of relatively large support for given smoothness. For box-splines, the same smoothness may be achieved by using more directions in (3.6), and fewer linear factors (see Section 4.3).

Considering the case of interpolatory schemes, the tensor product of two 4-point schemes (3.18) has the mask $t_w(z_1.z_2) = a_w(z_1)a_w(z_2)$, with an insertion rule based on 16 points. Yet, as shown in Dyn, Hed and Levin (1993), an interpolatory scheme with insertion rule of smaller support size (12 points) and with the same polynomial precision and smoothness exists.

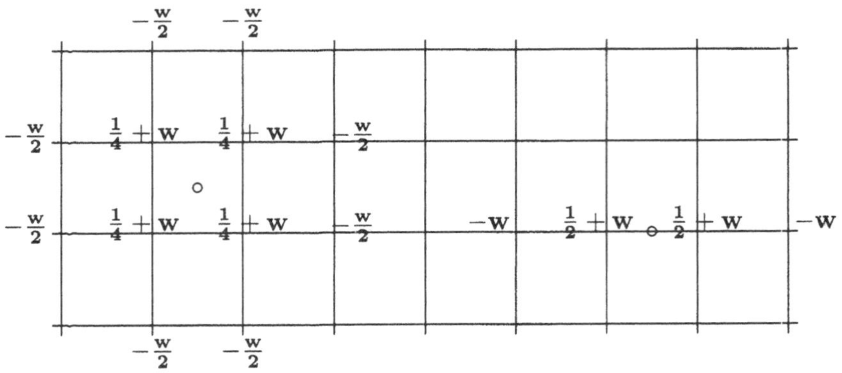

Figure 3.3. The two stencils of the truncated tensor product scheme S_{c_w}

The suggested scheme is obtained by removing all the w^2 terms in t_w. The resulting symbol is

$$c_w(z_1, z_2) = \frac{1}{4}(1+z_1)^2(1+z_2)^2 z_1^{-1} z_2^{-1}(1 - w[b(z_1) + b(z_2)]), \qquad (3.32)$$

where b is given in (3.20).

The stencils (see Section 3.5) of the insertion rule of this truncated tensor product scheme are shown in Figure 3.3.

The scheme S_{c_w} reproduces cubic polynomials for $w = \frac{1}{16}$, and it reduces to the 4-point scheme in one direction, when the data values are constant along the other direction (Dyn *et al.* 1993). An interpolatory subdivision on quadrilateral nets (see Section 3.5), with arbitrary topology based on the 4-point scheme, is proposed by Kobbelt (1996*a*).

3.5. Subdivision on nets

We consider control nets for generating surfaces in $\mathbb{R}^3$, a control net consists of control points in $\mathbb{R}^3$ with topological relations between them. The refinement rules are defined with respect to a control net, and generate a refined control net with new control points. The topological relations in the refined net are determined by the type of net, while the control points are determined by the subdivision scheme as weighted averages of topologically neighbouring control points.

In this section we present subdivision schemes that are defined over nets of arbitrary topology in 3D space. Such nets are valuable for the design of free-form surfaces. The surfaces generated by subdivision schemes on such nets are no longer restricted to bivariate functions, and they can represent surfaces of arbitrary topology. We describe three types of nets: triangular, Catmull–Clark type (primal type) and Doo–Sabin type (dual type), which are the most commonly used.

In addition to the above types of nets, there are hexagonal nets. Very few subdivision schemes with respect to hexagonal nets are available (see, *e.g.*, Dyn, Levin and Liu (1992) and Dyn, Levin and Simoens (2001*b*)), and they are not considered here.

Nets of general topology
A net $N(V,E,F)$, as shown in Figure 3.4, is a configuration of a finite set V of points in $\mathbb{R}^3$ called *vertices*, with two sets of topological relations between them E and F, called *edges* and *faces*. (A similar description of nets can be found in Kobbelt, Hesse, Prautzsch and Schweizerhof (1996).)

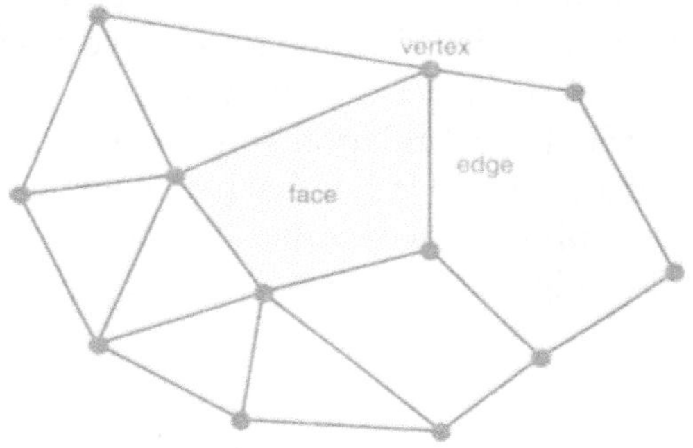

Figure 3.4. A net

An edge denotes a connection between two vertices. A face is a cyclic list of vertices where every pair of consecutive vertices shares an edge. The *valency* of a vertex, or a face, is the number of edges that share that vertex, or that face. While edges can always be represented by straight line segments, the vertices of a face are not necessarily co-planar; therefore a face is not associated with any geometric shape (in contrast to the faces of a polyhedron, which are planar pieces).

An edge e is called a *boundary edge* of $N(V,E,F)$ if it is not shared by two faces. A vertex v is called a *boundary vertex* if it belongs to a boundary edge.

We restrict our attention to nets $N(V,E,F)$ that satisfy the following properties:

(1) any two vertices share at most one edge;
(2) the valency of each vertex is at least 2.;
(3) the valency of each face is at least 3;
(4) every boundary edge belongs to exactly one face;
(5) three boundary edges cannot share a vertex.

$N(V, E, F)$ is said to be *closed* if it has no boundary edges. Otherwise, $N(V, E, F)$ is an *open* net. A triangular net is a net whose faces all have valency 3. A closed triangular net is termed *regular*, or a *regular triangulation*, if the valency of each vertex is 6. A regular triangular net is locally topologically equivalent to a portion of the 3-directional grid, that is, the grid $\mathbb{Z}^2$ with edges connecting (i, j) with $(i+1, j)$, $(i, j+1)$ and $(i+1, j+1)$, for $(i, j) \in \mathbb{Z}^2$. A quad-mesh is a net whose faces all have valency 4. A quad-mesh (quadrilateral net) is termed *regular* if it is topologically equivalent to $\mathbb{Z}^2$, that is, the valency of each vertex is 4.

The subdivision process transforms the net $N(V, E, F)$ into a refined net $N(V', E', F')$, where each new vertex in V' is associated with an element or a configuration c of elements from $N(V, E, F)$. The method for calculating a new vertex $v' \in V'$ can be described as a weighted average (with possibly negative weights) of vertices of V. The weight given to every vertex $v \in V$ depends only on its topological relation to c. The set of weights, together with their topological location in V relative to c, constitute the *stencil* which is determined by the subdivision scheme. There are different stencils for different topological configurations.

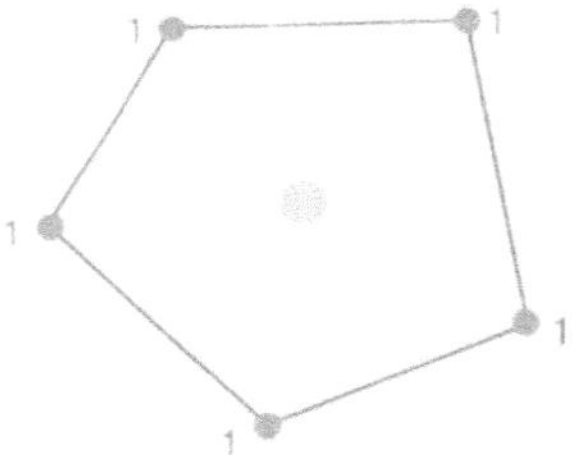

Figure 3.5. A stencil

For example, suppose that a vertex v' is associated with a face $f \in F$ that has valency 5. The stencil in Figure 3.5 represents the rule: v' is the average of the vertices of f. The set of vertices with nonzero weights, the support of the stencil, is topologically related to c, but does not necessarily coincide with c, as occurs in the last example. Together with the definition of V', there is a definition of the new edges E' and faces F', and these are described later for the different types of nets.

Let S denote a subdivision operator for nets. Let $N_0 = N(V, E, F)$ be a given initial net. A sequence of finer nets $N_k = N\left(V^k, E^k, F^k\right)$ is defined by

$$N_{k+1} = SN_k, \quad k = 0, 1, \ldots. \tag{3.33}$$

Ideally, the convergence of the sequence of nets $\{N_k : k \in \mathbb{Z}_+\}$ to a limit surface X should be defined independently of any parametrization of the surface. In the following definition, a surface X is considered as a closed subset of $\mathbb{R}^3$. We say that X is the limit surface of the subdivision scheme (3.33) if

$$\lim_{k\to\infty} \operatorname{dist}\left(V^k, X\right) = 0. \tag{3.34}$$

where $\operatorname{dist}(X, Y) = \max\{\sup_{y\in Y}\inf_{x\in X}\|x-y\|_2, \sup_{x\in X}\inf_{y\in Y}\|x-y\|_2\}$, is the Euclidean Hausdorff distance between two closed subsets $X, Y \subset \mathbb{R}^3$. When a limit surface X exists we denote it by $S^\infty N_0 = X$. In practice, however, the convergence is studied with respect to appropriate local parametrizations of the limit surface.

Triangular subdivision

Triangular subdivision schemes are defined over triangular nets, that is, nets whose faces all have valency 3 and therefore can be regarded as planar triangles. The new vertices are divided to v-vertices, and e-vertices. Each v-vertex in V' is associated with a vertex in V. Every e-vertex in V' is associated with an edge in E. For each type of vertex there is a different stencil. The new edges E' are defined between a new v-vertex and all the e-vertices such that their 'parents' in E share the parent of the v-vertex in V, and between any two e-vertices such that their parent edges share a face in F. Thus every triangle in the original net $N(V, E, F)$ is replaced by four triangles in the new net $N(V', E', F')$. The topology of the new triangular net is shown in Figure 3.6.

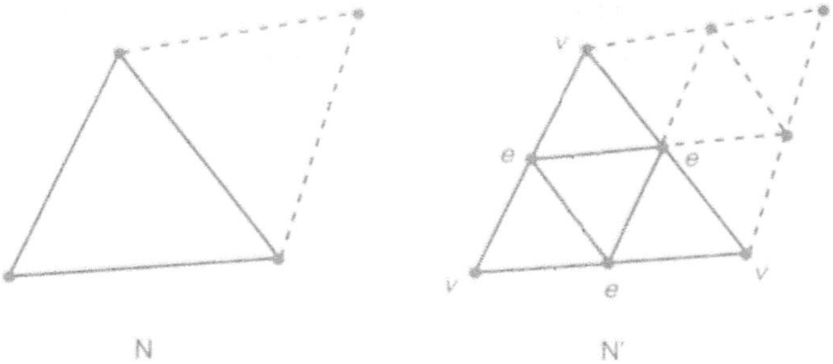

Figure 3.6. Triangular subdivision

A regular vertex in a triangular net is a vertex with valency 6. In a closed net, every new e-vertex has valency 6, and every new v-vertex inherits the valency of its parent vertex. Therefore, the number of irregular vertices in a net remains constant, and most of the net is a regular triangular net.

One of the commonly used triangular subdivision schemes is the Loop subdivision scheme (Loop 1987) defined for closed triangular nets. The stencils for the new e-vertices and v-vertices are depicted in Figure 3.7.

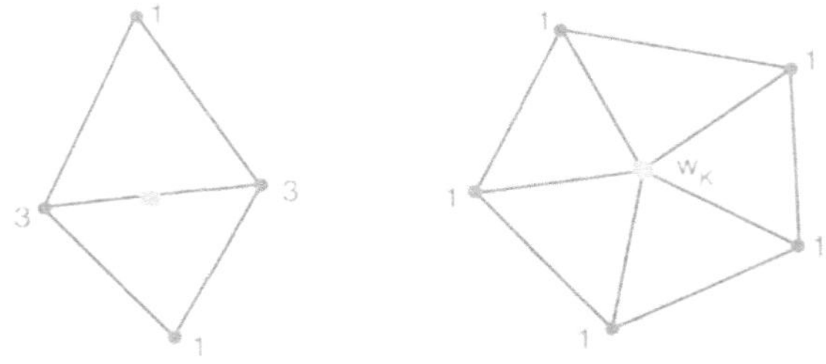

Figure 3.7. Loop scheme: stencils for e-vertex (left) and for v vertex (right)

The weight w_n given to the original vertex, in the stencil for its corresponding new v-vertex, depends on the valency K of that vertex. It is given by the following formula:

$$w_K = \frac{64K}{40 - \left(3 + 2\cos\left(\frac{2\pi}{K}\right)\right)^2} - K, \quad K = 3, 4, \ldots. \tag{3.35}$$

The Loop scheme generalizes the 3-directional box-spline scheme (3.7), in the sense that it coincides with it in the regular parts of the net. This implies that the limit surface is C^2 almost everywhere, and this is achieved with stencils of very small support. Near irregular vertices of the original net, the surface is C^1 (Loop 1987). Another property of this scheme, important for geometric modelling, is shape preservation, which is due to the positivity of the weights in the stencils of the Loop scheme.

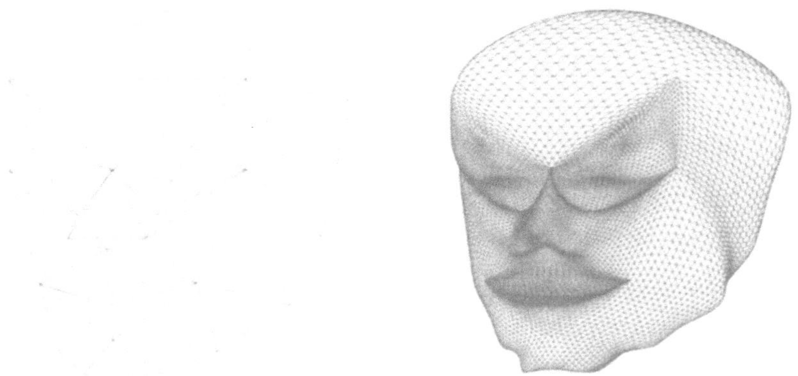

Figure 3.8. Head: initial control net (left), four butterfly iterations (right)

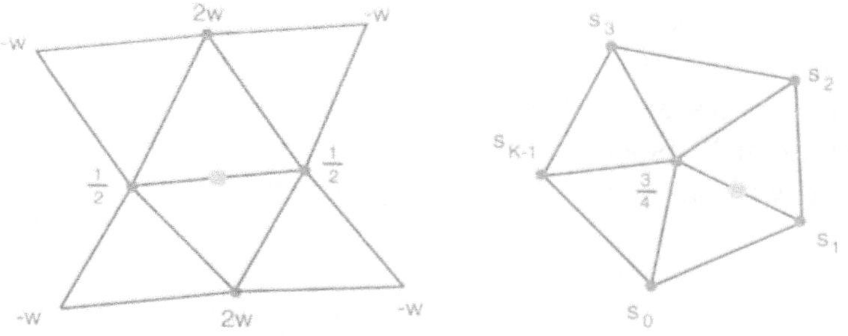

Figure 3.9. Modified Butterfly scheme: stencils corresponding to a 'regular' edge (left) and an 'irregular' edge (right)

An interpolatory triangular subdivision scheme with stencil of small support is the butterfly scheme (Dyn *et al.* 1990*a*). This scheme is defined over closed triangular nets. The application of four iterations of the butterfly insertion rule to an initial closed triangulation is depicted in Figure 3.8. There are modified stencils in the vicinity of irregular vertices (Zorin, Schröder and Sweldens 1996), which produce better-looking and smoother surfaces in the presence of irregular vertices.

As an interpolatory scheme, the new v-vertices inherit their location from their parent vertices. Figure 3.9 shows the stencils for new e-vertices. The butterfly stencil is used to calculate new e-vertices whose parent edge is 'regular', namely, has two regular vertices. A different stencil is used when the parent edge is 'irregular', namely, has one vertex which is regular and one which has valency $K \neq 6$. The weights $\{s_j\}_{j=0,\dots,K-1}$ depend on the valency of the irregular vertex, and are given by

$$s_j = \frac{1}{K}\left(\tfrac{1}{4} + \cos\left(\tfrac{2\pi j}{K}\right) + \tfrac{1}{2}\cos\left(\tfrac{4\pi j}{K}\right)\right), \quad j = 0, \dots, K-1.$$

The case where both of the vertices of the parent edge are irregular can occur only in the initial net. In such a case, in the first refinement step the calculation of the new e-vertex may be done in any reasonable way. The limit surfaces generated by the butterfly scheme are C^1 continuous everywhere, a property valuable for computer graphics applications (Zorin *et al.* 1996). An extended butterfly interpolatory subdivision scheme for the generation of C^2 surfaces on regular grids is presented in Labkovsky (1996).

Subdivision on an arbitrary net

The two types of refinements of nets of arbitrary topological structure are the Catmull–Clark type, also called 'primal', and the Doo–Sabin type, also called 'dual'.

In **primal-type refinement**, every face of valency n in the original net $N(V, E, F)$ is replaced by n quadrilateral faces in the new net $N'(V', E', F')$, as shown in Figure 3.10.

The new vertices are divided into v-vertices, e-vertices and f-vertices. Each v-vertex in V' is associated with a vertex in V. Each e-vertex in V' is associated with an edge in E. Each f-vertex in V' is associated with a face in F.

Figure 3.10 indicates the topological relations in $N(V', E', F')$, with the points v, e, and f indicating v-vertices, e-vertices and f-vertices, respectively. The new edges are marked by line segments and the faces by the quadrilaterals formed.

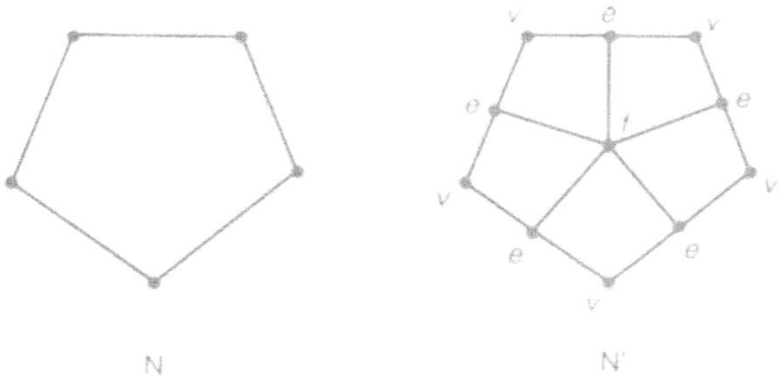

Figure 3.10. Primal-type refinement

A regular vertex in this setting is a vertex with valency 4, and a regular face is also of valency 4, namely, a quadrilateral face. Vertices or faces with valency $\neq 4$ are termed *irregular* or 'extraordinary'. In a closed net, every new e-vertex has valency 4. Every new v-vertex inherits the valency of its parent vertex, and every new f-vertex inherits the valency of its parent face. Therefore, the number of irregularities in a net remains constant throughout the subdivision process. Note that, after one subdivision iteration, all the faces are quadrilateral. The actual locations in $\mathbb{R}^3$ of the vertices V' are determined by the stencils of the subdivision scheme.

Catmull–Clark scheme. The first example of a primal-type scheme is the Catmull–Clark scheme (Catmull and Clark 1978, Doo and Sabin 1978), defined as an extension of the bicubic B-spline scheme (3.31) to closed nets of arbitrary topology. Its stencils are depicted in Figure 3.11.

The stencils for the new e-vertices and v-vertices involve the neighbouring new f-vertices (depicted as empty circles). The weight W_K in the stencil for the new v-vertex depends on the valency K of that vertex. Different formulae for W_K produce different limit surface behaviour near irregular vertices. A commonly used formula for W_K is

$$W_K = K(K-2), \quad K = 3, 4, \ldots.$$

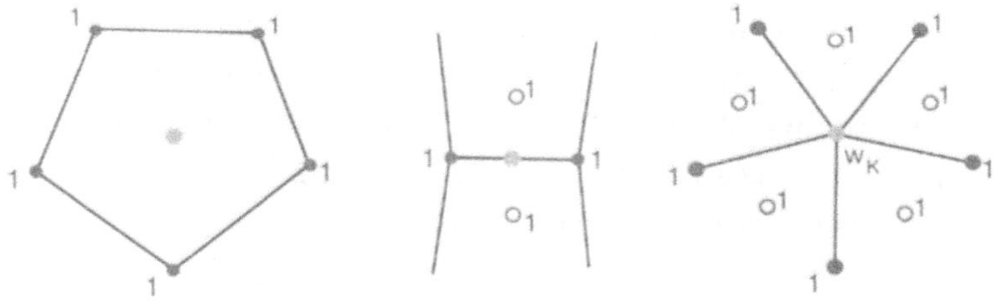

Figure 3.11. Catmull–Clark scheme:
f-stencil (left), e-stencil (middle) and v-stencil (right)

As long as $W_4 = 8$, the limit surfaces of this scheme are C^2 away from irregular points. Different variants of this scheme were investigated by Ball and Storry (1988, 1989). It is observed there that, for every choice of W_K, the surface curvature near an irregular point either tends to zero, or is unbounded. Applications of the Catmull–Clark scheme can be found in DeRose, Kass and Truong (1998) and Halstead, Kass and DeRose (1993).

Here we present an example (see Figure 3.12) of two surfaces generated from an initial triangulation, one by the Loop scheme and the other by the Catmull–Clark scheme, which regards the triangulation as a general net. Note that, in the latter case, most of the initial control points are irregular.

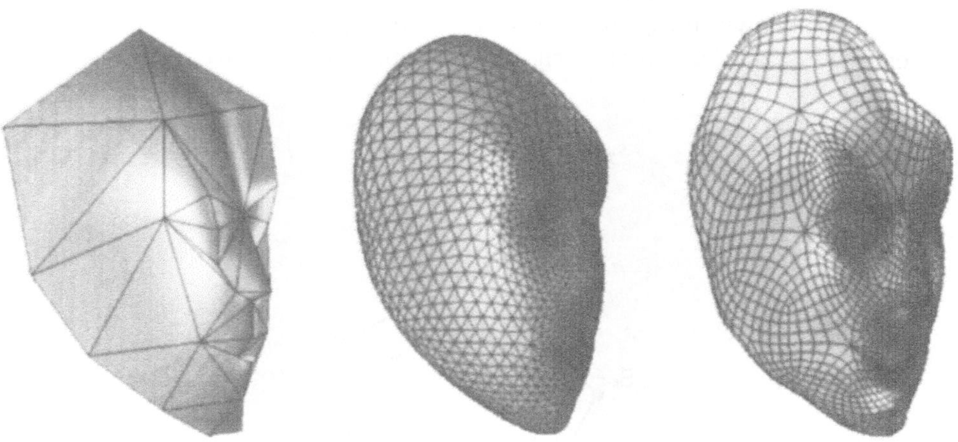

Figure 3.12. Head: initial control net (left), two iterations with the Loop scheme (middle) and with the Catmull–Clark scheme (right)

Dual-type refinement is depicted in Figure 3.13. Every new vertex in $v' \in V'$ corresponds to a pair $(v \in V, f \in F)$ such that v is a vertex of f in the original net N. It is considered a *dual* scheme, since vertices and edges in the original net $N = N(V, E, F)$ correspond to faces in the new net $N' = N(V', E', F')$. A regular vertex in this setting is a vertex with valency 4, and a regular face is a quadrilateral face.

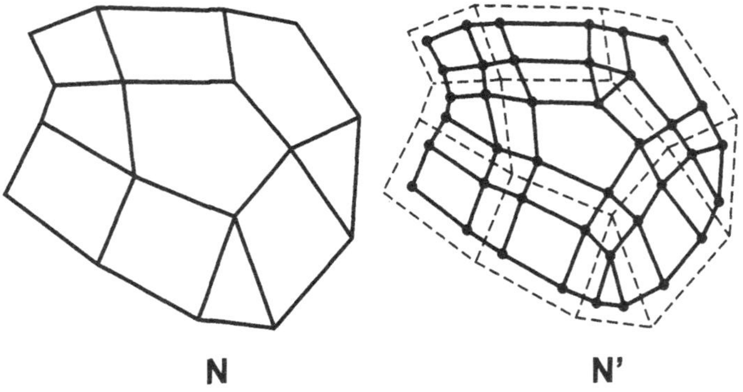

Figure 3.13. Dual-type refinement

Doo–Sabin scheme. The dual scheme due to Doo and Sabin generalizes the biquadratic B-spline scheme to subdivision of closed nets of arbitrary topological type.

The vertex v' is calculated by a weighted average of the vertices of f, with the stencils shown in Figure 3.14. The weights $\{S_j\}_{j=0,\dots,K-1}$ depend on the valency K of the face in the original net, and are given by

$$s_0 = \frac{K+5}{4K}, \qquad s_j = \frac{3 + 2\cos\left(\frac{2\pi j}{K}\right)}{4K}, \qquad j = 1, \dots, K-1.$$

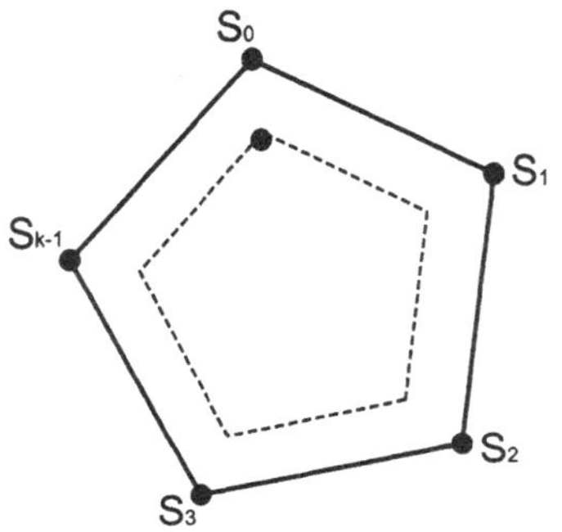

Figure 3.14. Doo–Sabin scheme: stencil for an f vertex

Almost everywhere the new nets are regular quadrilateral nets and the scheme reduces to the scheme defined by the symbol (3.30), giving the C^1 biquadratic spline surface.

In both examples, the Catmull–Clark scheme and the Doo–Sabin scheme, the mask parameters near an extraordinary vertex are chosen to achieve an overall C^1 limit surface. In Section 6 we describe the main results on the analysis of smoothness of stationary subdivision schemes near irregular vertices. Another dual-type subdivision scheme is 'the simplest scheme for smoothing polyhedra' presented in Peters and Reif (1997). In this scheme, given a polyhedron, a new polyhedron is constructed by connecting every edge-midpoint to its four neighbouring edge-midpoints. The limit surface is piecewise quadratic C^1 surface except at some extraordinary vertices. For additional material about subdivision schemes on general nets and their applications in computer graphics see Hoppe, DeRose, Duchamp, Halstead, Jin, McDonald, Schweitzer and Stuetzle (1994), Zorin *et al.* (1996), Zorin, Schröder and Sweldens (1997) and Zorin and Schröder (2000).

3.6. Further extensions

The inspiring iterative refinement idea, which is the basic concept in subdivision and in wavelets, has motivated many new research directions. In this section we briefly mention several extensions and generalizations of the uniform binary subdivision that are not discussed in this review. These include extensions to:

- non-uniform subdivision
- quasi-uniform and combined subdivision
- Lie group valued subdivision
- set-valued subdivision
- polyscale subdivision
- variational subdivision
- quasi-linear subdivision.

Non-uniform schemes. In many applications the data may be given on an irregular mesh and a scheme for iterative refinement of such data should be different from the standard uniform subdivision schemes. Also, convergence and smoothness analysis cannot be performed using the standard tools such as the z-transform or the Fourier transform. The tools that are being used for subdivision schemes over irregular grids are generalizations of the local matrix analysis (Section 5) and of the divided difference schemes (Section 4.2). See, for instance, Warren (1995*a*), Guskov (1998) and Daubechies, Guskov and Sweldens (1999). Another type of non-uniform scheme is still on uniform grids, but the subdivision refinement rules may differ from one point to the other. Here again it seems that the divided difference tools are the only way to analyse convergence and smoothness, as is done by Gregory and Qu (1996) for general corner cutting schemes. A systematic method for deriving the difference schemes, using a variation of the z-transform method, is presented in Levin (1999*e*). A general analysis of shape-preserving schemes for non-uniform data is done in Kuijt and van Damme (2002).

Quasi-uniform and combined subdivision. The analysis presented in this review is restricted to the case of closed nets, that is, there are no boundary edges. In real applications, there are boundaries of surface patches and boundaries may occur inside a patch if the patch should pass through a curve or a system of curves. For a subdivision scheme, a boundary treatment requires the definition of special rules in the vicinity of the boundary, and consequently, a special smoothness analysis. A subdivision scheme, together with special boundary rules, is called a *combined subdivision scheme* in A. Levin (1999*b*, 1999*c*). In these works, analysis tools for combined subdivision schemes are developed, and combined schemes, based on some of the most 'popular' bivariate schemes, are designed. The problem of matching boundary conditions or curve interpolation by subdivision surfaces is also treated in Nasri (1997*a*, 1997*b*) and A. Levin (1999*d*). A boundary may also be the border between two regions, or two patches, where in each patch a different uniform subdivision scheme is applied. This is termed *quasi-uniform* or *piecewise uniform*, and here also a special smoothness analysis is required, as presented in Dyn, Gregory and Levin (1995) for the univariate case and in A. Levin (1999*a*, 1999*c*) and Zorin, Biermann and A. Levin (2000) for surfaces.

Lie group valued subdivision. In some applications the data must lie on a manifold W in $\mathbb{R}^d$, and the limit function is also expected to be a function from $\mathbb{R}^s$ into W. The usual subdivision schemes are defined via linear averaging refinement rules that do not necessarily give points in W. In a recent work (Donoho and Stodden 2001), the general case of Lie group valued data is considered. The main approach is based on the fact that each Lie group has its associated Lie algebra, related through the exponential map, and the subdivision operations are performed in the Lie algebra and mapped back to the group by the exponential map.

Set-valued subdivision. For these schemes the initial data and the refined data generated by the scheme are sequences of sets in $\mathbb{R}^d$, and the limit function is a set-valued function. This is motivated by the problem of the reconstruction of 3D objects from their 2D cross-sections. The given data form a sequence of 2D cross-sections and the set-valued function describes a

3D object. Subdivision schemes for set-valued data require the definition of operations on sets and the study of notions of convergence and smoothness of set-valued functions. These issues, for convex sets using Minkowski averages, and for general compact sets using the 'metric average', are studied in Dyn and Farkhi (2000, 2001*a*, 2001*b*).

Polyscale subdivision. A subdivision scheme is a two-scale process, using data at one refinement level to compute the values at the next refinement level. In Dekel and Dyn (2001), poly-scale subdivision schemes are introduced. Such schemes compute the next refinement level from several previous levels, using several masks. This new idea is also related to the notion of poly-scale refinable functions, and opens up new theoretical convergence and smoothness issues. These issues, several interesting examples, and the relation of poly-scale subdivision schemes to matrix subdivision schemes, are presented in Dekel and Dyn (2001).

Variational subdivision. A variational approach to interpolatory subdivision is presented in Kobbelt (1996*b*). The resulting schemes are global, that is, every new point depends on all the points of the control polygon to be refined. The refinement is defined by minimizing a quadratic 'energy' functional, resulting in a 'fair' limit surface.

Quasi-linear subdivision. Quasi-linear schemes are nonlinear binary interpolatory schemes defined on a regular grid, with linear insertion rules which are data-dependent. In Cohen, Dyn and Matei (2001) a specific class based upon the weighted-ENO interpolation technique is analysed.

4. Convergence and smoothness analysis on regular grids

In this section, analysis of the (uniform) convergence of subdivision schemes on regular grids is presented, together with analysis of the smoothness of the limit functions.

First we present a method which relates the convergence and smoothness of nonstationary schemes to the convergence and smoothness of related stationary schemes (Dyn and Levin 1995); then we present a method for the analysis of stationary schemes, based on difference schemes (see Dyn (1992) and references therein). This method is also applied directly to certain nonstationary schemes.

The other main approaches to the convergence and smoothness analysis are in terms of Fourier transforms, and in terms of the joint spectral radius of a finite set of finite-dimensional matrices. The latter approach is briefly reviewed in Section 5.2. The Fourier analysis approach is not surveyed here: interested readers may consult Cohen and Conze (1992), Deslauriers and Dubuc (1989), Daubechies (1992) and Daubechies and Lagarias (1992*a*).

4.1. Analysis of nonstationary schemes via relations to stationary schemes

The analysis of the convergence of nonstationary schemes presented here, relies on the representation of a subdivision scheme $S_{\{\mathbf{a}^k\}}$ in terms of a sequence of operators $\{R_k : k \in \mathbb{Z}_+\}$ as in (2.22), where each R_k is defined by (2.21).

The main results are based on several properties of sequences of bounded linear operators in a Banach space. From now on all operators considered are bounded and linear. A sequence of operators $\{A_k : k \in \mathbb{Z}_+\}$ in a Banach space $\{X, \|\cdot\|\}$ defines the iterated process $x_{k+1} = A_k x_k$, $k \in \mathbb{Z}_+$, with $x_0 \in X$. Such a sequence is termed *convergent* if, for any $m \in \mathbb{Z}_+$ and any $x \in X$, $\lim_{k\to\infty} x_{m,k}$ exists, where $x_{m,k} = A_{m+k} \cdots A_{m+1} A_m x$. The sequence $\{A_k\}$ is termed *stable* if

$$\|A_{m+k} \cdots A_{m+1} A_m\| \leq M < \infty, \quad \forall m, k \in \mathbb{Z}_+. \tag{4.1}$$

Two sequences of bounded operators $\{A_k\}$ and $\{B_k\}$ are called *asymptotically equivalent* if there exists $L \in \mathbb{Z}$, such that

$$\sum_{k=\max\{0,-L\}}^{\infty} \|A_{k+L} - B_k\| < \infty. \tag{4.2}$$

Proposition 4.1. Let $\{A_k\}$ and $\{B_k\}$ be asymptotically equivalent. Then $\{A_k\}$ is stable if and only if $\{B_k\}$ is stable.

The proof of this proposition (Dyn and Levin 1995) introduces the $\{A_k\}$-norms

$$\|x\|_m = \sup_k \|A_{m+k} \cdots A_m x\|, \quad m \in \mathbb{Z}_+,$$

which are equivalent to the norm of the Banach space when $\{A_k\}$ is stable. It also introduces the Banach spaces $X_m = \{X, \|\cdot\|_m\}$. The key observation is that A_m, as an operator from X_m to X_{m+1}, is bounded in norm by 1. From this observation follows Proposition 4.1. By similar reasoning we get the following.

Proposition 4.2. Let $\{A_k\}$ and $\{B_k\}$ be asymptotically equivalent. Then $\{A_k\}$ is stable and convergent if and only if so is $\{B_k\}$.

This analysis of sequences of operators in a Banach space leads to the important notion of 'asymptotic equivalence' between two subdivision schemes. Here we use the representation of subdivision schemes as operators on $X = C(\mathbb{R}^s)$, with the maximum norm. Two schemes $S_{\{\mathbf{a}^k\}}, S_{\{\mathbf{b}^k\}}$ are defined to be 'asymptotically equivalent' if, for some fixed $L \in \mathbb{Z}$,

$$\sum_{k=\max\{0,-L\}}^{\infty} \|\mathbf{a}^{k+L} - \mathbf{b}^k\|_\infty < \infty, \tag{4.3}$$

where $\|\mathbf{a}^k - \mathbf{b}^j\|_\infty = \max_{\alpha \in E^s} \sum_{\beta \in \mathbb{Z}^s} |a^k_{\alpha-2\beta} - b^j_{\alpha-2\beta}|$.

A scheme $S_{\{\mathbf{a}^k\}}$ is termed *stable* if there exists $M > 0$ such that, for all $k, j \in \mathbb{Z}_+$,

$$\|R_{j+k} \cdots R_{j+1} R_j\|_\infty < M, \tag{4.4}$$

with $\{R_j\}$ the operators corresponding to $S_{\{\mathbf{a}^k\}}$ as in (2.21). It is easy to conclude from (2.10) that a convergent scheme $S_{\{\mathbf{a}^k\}}$ is stable, if and only if the functions $\Phi_k = \sum_{\alpha \in \mathbb{Z}^s} |\phi_k(\cdot - \alpha)|$ are uniformly bounded for $k \in \mathbb{Z}_+$.

Two stable asymptotically equivalent schemes have similar convergence properties. This is easily concluded from Proposition 4.2.

Theorem 4.3. Let $S_{\{a^k\}}$ and $S_{\{b^k\}}$ be asymptotically equivalent. Then $S_{\{a^k\}}$ is stable and convergent if and only if $S_{\{b^k\}}$ is stable and convergent.

If $S_{\{\mathbf{b}^k\}} = S_{\mathbf{b}}$ is stationary, namely $\mathbf{b}^k = \mathbf{b}$ for $k \in \mathbb{Z}_+$, and $S_{\mathbf{b}}$ is convergent, then by (2.8) $S_{\mathbf{b}}$ is stable. Thus we have the following.

Corollary 4.4. Let $S_{\{\mathbf{a}^k\}}$ and $S_{\mathbf{b}}$ be asymptotically equivalent. If $S_{\mathbf{b}}$ is convergent then $S_{\{\mathbf{a}^k\}}$ is stable and convergent.

Example 1. As an example of convergence implied by Corollary 4.4, we consider the nonstationary subdivision scheme given by the symbols

$$a^k(z) = 2\prod_{i=1}^{m} \frac{1}{2}\left(1 + e^{\eta_i 2^{-k}} z\right), \quad k \in \mathbb{Z}_+, \tag{4.5}$$

with $\eta_1, \ldots, \eta_m$ distinct complex constants.

It is easy to verify that $S_{\{\mathbf{a}^k\}}$ is asymptotically equivalent to $S_{\mathbf{b}}$, whose symbol is

$$b(z) = 2^{-m+1}(1+z)^m. \tag{4.6}$$

Thus $S_{\mathbf{b}}$ is a convergent stationary subdivision scheme with basic limit function the polynomial B-spline of order m (degree $m-1$) with integer knots and support $[0, m]$ (see Section 3.1).

Thus the nonstationary scheme (4.5) is convergent. In fact its basic limit function is the exponential B-spline in $\mathrm{span}\{e^{\frac{\eta_i}{2}x} : 1 \le i \le m\}$ with integer knots and support $[0, m]$. (For more about exponential B-splines, see, for example, Schumaker (1980).)

One way to analyse the smoothness of the basic limit function of a nonstationary scheme $S_{\{\mathbf{a}^k\}}$ (and therefore all limit functions generated by $S_{\{a^k\}}$, as implied by (2.8)), is in terms of smoothing factors (Dyn and Levin 1995).

Theorem 4.5. Let the symbols of $S_{\{\mathbf{a}^k\}}$ be of the form

$$a^k(z) = \frac{1}{2}(1 + r_k z^\lambda) b^k(z), \quad k \ge K \in \mathbb{Z}_+, \tag{4.7}$$

with $\lambda \in \mathbb{Z}^s$, where $S_{\{\mathbf{b}^k\}}$ is a stable and convergent subdivision scheme with $\phi_{\{\mathbf{b}^k\}}$ of compact support and in $C^m(\mathbb{R}^s)$. If

$$r_k = e^{\eta 2^{-k}}(1+\varepsilon_k), \qquad \sum_{k=K}^{\infty} |\varepsilon_k| 2^k < \infty, \tag{4.8}$$

then $\phi_{\{\mathbf{a}^k\}}$ and $\partial_\lambda \phi_{\{\mathbf{a}^k\}}$ are in $C^m(\mathbb{R}^s)$

The factors $\frac{1}{2}(1+r_k z^\lambda)$ in (4.7) are termed *smoothing factors* and, for $\varepsilon_k = 0$ in (4.8), are related to the univariate elementary nonstationary scheme of (3.12),

Sketch of proof. The key to the proof is convolution property (2), which in this case has the form

$$\phi_{\{\mathbf{a}^k\}} = \int_{\mathbb{R}} \phi_{\{\mathbf{b}^k\}}(\cdot - \lambda t)\phi_{\{1+r_k z\}}(t)\,dt.$$

Since $\phi_{\{1+r_k z\}}$ is supported on $[0,1]$ and is integrable (as discussed in Section 3.1), $\phi_{\{\mathbf{a}^k\}} \in C^m$. The result $\partial_\lambda \phi_{\{a^k\}} \in C^m$ follows from the general observation that, for a univariate integrable function h with $\sigma(h) = [0,1]$, and for a bounded continuous function $g \in C(\mathbb{R})$,

$$(g * h)(x) = \int_{x-1}^{x} g(t)h(x-t)\,dt \in C^1(\mathbb{R}).$$

For a multivariate function, the conditions $\partial_\lambda f \in C^m$ for $\lambda \in \Lambda$, where Λ is a basis for $\mathbb{R}^s$, imply that $f \in C^{m+1}(\mathbb{R}^s)$. Hence Theorem 4.5 and convolution property (2) give us the following result.

Corollary 4.6. Let

$$a^k(z) = \prod_{i=1}^{s} \frac{1}{2}(1 + r_{i,k} z^{\lambda_i}) b^k(z), \quad k \geq K \in \mathbb{Z}_+,$$

where $S_{\{\mathbf{b}^k\}}$ satisfies the conditions of Theorem 4.5.

If, for $i = 1, \ldots, m$,

$$r_{i,k} = e^{\eta_i 2^{-k}}(1+\varepsilon_{i,k}), \qquad \sum_{k=K}^{\infty} |\varepsilon_{i,k}| 2^k < \infty,$$

and if $\lambda_1, \ldots, \lambda_s \in \mathbb{Z}^s$ are linearly independent, then $\phi_{\{a^k\}} \in C^{m+1}$.

A good example where smoothness is deduced via Theorem 4.5 is provided by the nonstationary, univariate interpolatory schemes that reproduce finite-dimensional spaces of exponential polynomials (Dyn, Levin and Luzzatto 2001*a*).

Example 2. Consider finite-dimensional spaces of univariate exponential polynomials of the form

$$V_{\boldsymbol{\gamma},\boldsymbol{\mu}} = \operatorname{span}\{x^j e^{\gamma_\ell t}, j = 0, \dots, \mu_{\ell-1},\ \ell = 1, \dots, \nu\},$$

where $\boldsymbol{\gamma} = \{\gamma_1, \dots, \gamma_\nu\}$ are the roots with multiplicities $\boldsymbol{\mu} = \{\mu_1, \dots, \mu_\nu\}$ of a real polynomial of degree $N = \sum_{i=1}^{\nu} \mu_i$.

A scheme $S_{\{\mathbf{a}^k\}}$ is termed a *reproducing scheme* of $V_{\boldsymbol{\gamma},\boldsymbol{\mu}}$, if, for any $k \in \mathbb{Z}_+$ and $\mathbf{f}^k = \{f_j^k = f(2^{-k}j) : j \in \mathbb{Z}\}$, with $f \in V_{\boldsymbol{\gamma},\boldsymbol{\mu}}$,

$$S_{\mathbf{a}^k}\mathbf{f}^k = \mathbf{f}^{k+1}.$$

It is proved in Dyn *et al.* (2001*a*) that an interpolatory scheme $S_{\{a^k\}}$ with supports $\sigma(a^k)$ fixed for $k \in \mathbb{Z}_+$, which reproduces $V_{\boldsymbol{\gamma},\boldsymbol{\mu}}$ and does not reproduce any bigger space of exponential polynomials containing $V_{\boldsymbol{\gamma},\boldsymbol{\mu}}$, has the property that its symbols $\{a^k(z) : k \in \mathbb{Z}_+\}$ are Laurent polynomials of degree $2(N-1)$ satisfying

$$\frac{\mathrm{d}^r}{\mathrm{d}z^r} a^k(z_n^k) = 2\delta_{0,r}, \quad \frac{\mathrm{d}^r}{\mathrm{d}z^r} a^k(-z_n^k) = 0,$$
$$r = 0, 1, \dots, \mu_n - 1, \quad n = 1, \dots, \nu, \quad (4.9)$$

where $z_n^k = \exp(2^{-(k+1)}\gamma_n)$, $n = 1, \dots, \nu$, $k \in \mathbb{Z}_+$.

For the case $N = 2n$, it can be concluded from (4.9) that the masks $\{\mathbf{a}^k : k \in \mathbb{Z}_+\}$ with $\sigma(\mathbf{a}^k) = [-n, n]$, tend as $k \to \infty$ to the mask $\mathbf{a}$ with $\sigma(\mathbf{a}) = [-n, n]$ of the interpolatory scheme, introduced in Deslauriers and Dubuc (1989), which reproduces the space π_n of all polynomials of degree not exceeding n (see Section 3.2). More specifically,

$$\|\mathbf{a}^k - \mathbf{a}\|_\infty < 2^{-k}B, \quad 0 < B < \infty,$$

and $a(z)$ is divisible by $(1+z)^n$, as follows from (4.9). Thus $S_{\{\mathbf{a}^k\}}$ is asymptotically equivalent to $S_{\mathbf{a}}$, and since $S_{\mathbf{a}}$ is convergent (Deslauriers and Dubuc 1989) so is $S_{\{\mathbf{a}^k\}}$. To conclude the smoothness of $\phi_0 = \phi_{\{\mathbf{a}^k\}}$ from the smoothness of $\phi_{\mathbf{a}}$, Theorem 4.5 is invoked. Assume $\phi_{\mathbf{a}} \in C^m$. Then, by the theory of smoothness of stationary schemes (see Section 4.2), $m \le n$. Consider, for each $k \in \mathbb{Z}_+$, the m linear factors of $a^k(z)$, $\prod_{i=1}^{m}(1 + (z_{n_i}^k)^{-1}z)$, where $n_1, \dots n_m$ are fixed integers in $\{1, \dots, \nu\}$, such that $\#\{n_i : n_i = j\} \le \mu_j$. The existence of these factors is guaranteed by (4.9). Each of the m factors divided by 2 is a smoothing factor. Now, the symbols $\{c^k(z) : k \in \mathbb{Z}_+\}$, given by

$$c^k(z) = \frac{a^k(z)2^m}{\prod_{i=1}^{m}(1 + (z_{n_i}^k)^{-1}z)}, \quad k \in \mathbb{Z}_+,$$

define a scheme $S_{\{\mathbf{c}^k\}}$ which is asymptotically equivalent to the scheme $S_{\mathbf{c}}$ with symbol

$$c(z) = \frac{a(z)2^m}{(1+z)^m}.$$

Since $\phi_{\mathbf{a}} \in C^m$, it follows from the analysis of stationary schemes (see Section 4.2) that $S_{\mathbf{c}}$ is convergent. Thus $S_{\{\mathbf{c}^k\}}$ is convergent, and by Theorem 4.5 and Corollary 4.6, $\phi_{\{\mathbf{a}^k\}} \in C^m$.

A specific example of this type is the following interpolatory 4-point scheme generating circles. In this scheme the insertion rule is constructed by interpolation with a function from the span of the four functions $H = \mathrm{span}\{1, t, \cos t, \sin t\}$.

The insertion rule turns out to be

$$f_{2j+1}^{k+1} = \frac{-1}{16\cos^2(\theta 2^{-k-2})\cos(\theta 2^{-k-1})}(f_{j-1}^k + f_{j+2}^k)$$
$$+ \frac{(1+2\cos(2\theta 2^{-k}))^2}{16\cos^2(\theta 2^{-k-2})\cos(\theta 2^{-k-1})}(f_j^k + f_{j+1}^k).$$

Note that this insertion rule tends to the 4-point Dubuc–Deslauriers insertion rule as k tends to infinity, at the rate $O(2^{-k})$.

The above insertion rule together with $f_{2j}^{k+1} = f_j^k$, when applied to the equidistributed points on the circle, $\{f_j^0 = R(\cos(j\theta), \sin(j\theta))\}_{j=1}^N$, with $\theta = 2\pi/N$, generates denser sets of points on the circle.

Next we consider a similar example, but in the multivariate setting with general smoothing factors.

Example 3. Let

$$a^k(z) = 2^{s-\ell}\prod_{j=1}^{\ell}\left(1 + r_k^{(j)} z^{\lambda^{(j)}}\right), \quad k \in \mathbb{Z}_+,$$

be symbols with directions $\Lambda = \{\lambda^{(1)}, \ldots, \lambda^{(\ell)}\} \subset \mathbb{Z}^s$. If $r_k^{(1)}, \ldots, r_k^{(\ell)}$ satisfy (4.8), if the set Λ contains a subset of s directions with determinant ± 1, and if any subset of $\ell - m - 1$ directions spans $\mathbb{R}^s$, then $\phi_{\{\mathbf{a}^k\}}$ is in C^m.

To see this, observe that, under the conditions of the example, $S_{\{\mathbf{a}^k\}}$ is asymptotically equivalent to $S_{\mathbf{a}}$ with

$$a(z) = 2^s \prod_{\lambda^{(j)} \in \Lambda}(1 + z^{\lambda^{(j)}})/2.$$

By the conditions on Λ, $S_{\mathbf{a}}$ is convergent and $\phi_{\mathbf{a}}$ is the polynomial box-spline with directions Λ, which is C^m (see Section 3.1). Let $\Lambda_0 \subset \Lambda$ be the smallest subset of Λ for which $S_{\mathbf{b}}$ with $b(z) = 2^s \prod_{\lambda^{(j)} \in \Lambda_0}(1 + z^{\lambda^{(j)}})/2$ is C^0.

The scheme $S_{\{\mathbf{b}^k\}}$ with $b^k(z) = 2^s \prod_{\lambda^{(j)} \in \Lambda_0}(1 + r_k^{(j)} z^{\lambda^{(j)}})/2$ is asymptotically equivalent to $S_{\mathbf{b}}$. Hence, by Corollary 4.4, it follows that $\phi_{\{\mathbf{b}^k\}} \in C(\mathbb{R})$. The maximal m for which $S_{\mathbf{a}}$ is C^m is determined by repeated convolutions with respect to appropriate directions in $\Lambda \setminus \Lambda_0$. The same procedure of adding directions, in view of Theorem 4.5, proves that $S_{\{\mathbf{a}^k\}}$ is also C^m.

4.2. Analysis of univariate schemes via difference schemes

The case $s = 1$ is the simpler to analyse, and the theory for the stationary case is almost complete. This theory provides a method of analysis based on necessary and sufficient conditions for convergence, and in the most interesting cases, also necessary and sufficient conditions for smoothness.

The method presented here is general in the sense that it also applies to nonstationary schemes with symbols that are all divisible by the elementary factor $(1+z)$ and its powers, as in the stationary case. Yet, in the stationary case this divisibility is necessary and sufficient, while in the nonstationary case it is only sufficient.

A necessary condition for convergence (for any $s \in \mathbb{Z}_+ \backslash 0$) (Cavaretta *et al.* 1991, Dyn 1992), which is the key to this analysis in the univariate case, is easily derived from the stationary refinement step

$$f_\alpha^{k+1} = \sum_{\beta \in \mathbb{Z}^s} a_{\alpha - 2\beta} f_\beta^k, \quad \alpha \in \mathbb{Z}^s.$$

Considering large k such that $|f_\alpha^j - (S_{\mathbf{a}}^\infty f^0)(2^{-j}\alpha)| < \varepsilon$, $j = k, k+1$ for sufficiently small ε, and taking into account that $\sigma(\mathbf{a})$ is finite, so that $2^{-k}\beta$ in the above sum is close to $2^{-k-1}\alpha$, we conclude the following.

Theorem 4.7. If $S_{\mathbf{a}}$ is (uniformly) convergent, then

$$\sum_{\beta \in \mathbb{Z}^s} a_{\alpha + 2\beta} = 1, \quad \alpha \in E^s, \tag{4.10}$$

where E^s are the extreme points of $[0,1]^s$.

Analysis of stationary schemes

For a stationary scheme $S_{\mathbf{a}}$ we identify the insertion rule $R_{\mathbf{a}}$ with the scheme. In the univariate case ($s = 1$) conditions (4.10) imply that $a(-1) = 0$, $a(1) = 2$. Thus $a(z)$ is divisible by $(1+z)$, the elementary univariate factor of (3.1). As will become clear hereafter, $(1+z)/2$ is the stationary univariate smoothing factor.

Let the mask $\mathbf{a}$ satisfy (4.10). Then, $a(z) = (1+z)b(z)$, with $S_{\mathbf{b}}$ a scheme related to $S_{\mathbf{a}}$ by

$$S_{\mathbf{b}} \Delta \mathbf{f} = \Delta(S_{\mathbf{a}} \mathbf{f}), \tag{4.11}$$

where $\Delta \mathbf{f} = \{(\Delta \mathbf{f})_j = f_j - f_{j-1} : j \in \mathbb{Z}\}$. The verification of (4.11) is easily done in terms of the z-transform representation of subdivision schemes (2.25). Since

$$L(\Delta \mathbf{f}; z) = \sum_{j\in\mathbb{Z}} (\Delta \mathbf{f})_j \, z^j = (1-z)L(\mathbf{f}; z),$$

it follows from (2.25) and from the factorization of $a(z)$ that

$$\begin{aligned} L(\Delta \mathbf{f}^{k+1}; z) &= (1-z)a(z)L(\mathbf{f}^k; z^2) \\ &= b(z)(1-z^2)L(\mathbf{f}^k; z) \\ &= b(z)L(\Delta \mathbf{f}^k; z^2), \end{aligned}$$

which proves (4.11).

From now on we consider only masks that satisfy (4.10). It is clear that, if $S_{\mathbf{a}}$ is convergent, then $\lim_{k\to\infty} \sup_{j\in\mathbb{Z}} |\Delta f_j^k| = 0$ with $\mathbf{f}^k = S_{\mathbf{a}}^k \mathbf{f}^0$, or $\Delta \mathbf{f}^k = S_{\mathbf{b}}^k \Delta \mathbf{f}^0$. Thus, if $S_{\mathbf{a}}$ is convergent, then $S_{\mathbf{b}}$ maps any initial data to zero; in brief, it is contractive. The converse also holds.

Theorem 4.8. Let $a(z) = (1+z)b(z)$. $S_{\mathbf{a}}$ is convergent if and only if $S_{\mathbf{b}}$ is contractive.

Proof. It remains to prove that if $S_{\mathbf{b}}$ is contractive then $S_{\mathbf{a}}$ is convergent. Consider the sequence $\{F_k(t)\}_{k\in\mathbb{Z}_+}$ defined by (2.6). To show convergence of $S_{\mathbf{a}}$ it is sufficient to show that $\{F_k(t)\}_{k\in\mathbb{Z}_+}$ is a Cauchy sequence with respect to the sup-norm. Now by definition, and by the observation that a piecewise linear function attains its extreme values at its breakpoints

$$\sup_{t\in\mathbb{R}} |F_{k+1}(t) - F_k(t)| = \max\left\{ |\sup_{i\in\mathbb{Z}} |f_{2i}^{k+1} - g_{2i}^{k+1}|, \sup_{i\in\mathbb{Z}} |f_{2i+1}^{k+1} - g_{2i+1}^{k+1}| \right\}, \tag{4.12}$$

where

$$g_{2i}^{k+1} = f_i^k \quad \text{and} \quad g_{2i+1}^{k+1} = \frac{1}{2}(f_i^k + f_{i+1}^k). \tag{4.13}$$

It is easy to verify that (4.13) is represented in terms of the z-transform, by

$$L(\mathbf{g}^{k+1}; z) = \frac{(1+z)^2}{2z} L(\mathbf{f}^k; z^2).$$

Thus

$$\begin{aligned} L(\mathbf{f}^{k+1}; z) - L(\mathbf{g}^{k+1}; z) &= \left(a(z) - \frac{(1+z)^2}{2z}\right) L(\mathbf{f}^k; z^2) \\ &= (1+z)\left(b(z) - \frac{1+z}{2z}\right) L(\mathbf{f}^k; z^2) \\ &= (1+z)d(z)L(\mathbf{f}^k; z^2) \end{aligned}$$

with $d(z) = b(z) - \frac{1+z}{2z}$. Since, by (4.10), $a(1) = 2$, $d(1) = b(1) - 1 = 0$ and hence $d(z) = (1-z)e(z)$. This leads finally to

$$L(\mathbf{f}^{k+1} - \mathbf{g}^{k+1}; z) = e(z)(1-z^2)L(\mathbf{f}^k; z^2) = e(z)L(\Delta\mathbf{f}^k; z^2). \tag{4.14}$$

Recalling that, by (4.12), $\|F_{k+1} - F_k\|_\infty = \sup_{j\in\mathbb{Z}} |f_j^{k+1} - g_j^{k+1}| = \|\mathbf{f}^{k+1} - \mathbf{g}^{k+1}\|_\infty$, and that, by (4.14) and (4.11),

$$\mathbf{f}^{k+1} - \mathbf{g}^{k+1} = S_{\mathbf{e}}\Delta\mathbf{f}^k = S_{\mathbf{e}}S_{\mathbf{b}}^k\Delta\mathbf{f}^0,$$

we finally get

$$\|F_{k+1} - F_k\|_\infty = \|\mathbf{f}^{k+1} - \mathbf{g}^{k+1}\|_\infty \le \|S_{\mathbf{e}}\|_\infty \|S_{\mathbf{b}}^k\Delta\mathbf{f}^0\|_\infty. \tag{4.15}$$

Now, if $S_{\mathbf{b}}$ is contractive, namely if $S_{\mathbf{b}}^k\mathbf{f}$ tends to zero for all $\mathbf{f}$, then there exists $M \in \mathbb{Z}_+\backslash 0$ such that $\|S_b^M\|_\infty = \mu < 1$. Thus (4.15) leads to

$$\|F_{k+1} - F_k\|_\infty = \|\mathbf{f}^{k+1} - \mathbf{g}^{k+1}\|_\infty \le \|S_{\mathbf{e}}\|_\infty \mu^{[\frac{k}{M}]} \max_{0\le j<M} \|\Delta\mathbf{f}^j\| \le C\eta^k, \tag{4.16}$$

where $\eta = (\mu)^{\frac{1}{M}} < 1$ and C is a generic constant. Thus $\{F_k : k \in \mathbb{Z}_+\}$ is uniformly convergent. □

With the analysis presented, we can design an algorithm for checking the convergence of $S_{\mathbf{a}}$ given the mask $\mathbf{a}$. Consider the iterated scheme $S_{\mathbf{b}}^\ell$, transforming data at level k to data at level $\ell + k$. Recall that the symbol of $S_{\mathbf{b}}^\ell$ can be computed by (2.28) as $b^{[\ell]}(z) = \prod_{j=1}^{\ell} b(z^{2^{j-1}})$, and thus, to check the contractivity of $S_{\mathbf{b}}$, the norms of $S_{\mathbf{b}}^\ell$, $\ell = 1, 2, \ldots$, have to be evaluated in terms of $b^{[\ell]}(z) = \sum_{j\in\mathbb{Z}} b_j^{[\ell]} z^k$, according to

$$\|S_b^\ell\|_\infty = \max\left\{ \sum_{j\in\mathbb{Z}} |b_{i-2^\ell j}^{[\ell]}| : 0 \le i < 2^\ell \right\}. \tag{4.17}$$

The norm in (4.17) reflects the fact that there are 2^ℓ different rules in the iterated scheme $S_{\mathbf{b}}^\ell$:

$$\mathbf{g}^{k+\ell} = S_{\mathbf{b}}^\ell \mathbf{g}^k \Leftrightarrow g_i^{k+\ell} = \sum_{j\in\mathbb{Z}} b_{i-2^\ell j}^{[\ell]} g_j^k, \quad i \in \mathbb{Z}.$$

Schemes for which $S_{\mathbf{b}}$ is contractive, but $\|S_{\mathbf{b}}^\ell\|_\infty \ge 1$ for large ℓ ($\ell > 5$), are of no practical value, since a large number of iterations is required to observe convergence (small $\|\Delta\mathbf{f}^k\|_\infty$). Thus the algorithm has an input parameter M_0, such that if $\|S_{\mathbf{b}}^\ell\|_\infty \ge 1$ for $1 \le \ell \le M_0$, the scheme is declared to be practically 'not convergent'. A reasonable choice of M_0 is in the range $5 < M_0 \le 10$.

Algorithm for verifying convergence given the symbol.

Let $a(z)$ be the symbol of the scheme.

If $a(-1) \neq 0$, or $a(1) \neq 2$, then the scheme does not converge. Stop!

Compute $b^{[1]}(z) = a(z)/(1+z) = \sum_j b_j^{[1]} z^j$.

For $\ell = 1, \ldots, M_0$ {

Compute $N_\ell = \max_{0 \leq i < 2^\ell} \sum_{j \in \mathbb{Z}} |b_{i-2^\ell j}^{[\ell]}|$.

If $N_\ell < 1$, the scheme is convergent. Stop!

If $N_\ell \geq 1$, compute $b^{[\ell+1]}(z) = b^{[1]}(z) b^{[\ell]}(z^2) = \sum_{j \in \mathbb{Z}} b_j^{[\ell+1]} z^j$.

} End loop.

$S_{\mathbf{b}}$ is not contractive after M_0 iterations. Stop!

The parameters μ, M from the proof of Theorem 4.8 corresponding to a mask $\mathbf{a}$, also determine the Hölder exponent of $\phi_{\mathbf{a}}$ (or any $S_{\mathbf{a}}^\infty f^0$), and the rate of convergence of the subdivision scheme.

Theorem 4.9. Let $\mathbf{a}, \mu, M, \eta$, be as in the proof of Theorem 4.8, and define $\nu = -(\log_2 \mu)/M$. Then

$$|\phi_{\mathbf{a}}(y) - \phi_{\mathbf{a}}(x)| \leq C|x-y|^\nu.$$

Moreover, the rate of convergence of the sequence $\{F_k(t)\}_{k \in \mathbb{Z}_+}$ defined in (2.6) is

$$\|F_k(t) - S_{\mathbf{a}}^\infty \mathbf{f}^0\|_\infty \leq C\eta^k.$$

Here C is a generic constant.

Proof. Both claims of the theorem follow from (4.16). The second follows directly with the aid of the observation

$$|(S_{\mathbf{a}}^\infty \mathbf{f}^0 - F_k)(x)| = \lim_{\ell \to \infty} |F_\ell(x) - F_k(x)| \leq \sum_{j=k}^{\infty} |F_{j+1}(x) - F_j(x)|.$$

To verify the first claim, we use the second claim in the bound

$$|\phi_{\mathbf{a}}(x) - \phi_{\mathbf{a}}(y)| \leq |\phi_{\mathbf{a}}(x) - F_k(x)| + |\phi_a(y) - F_k(y)| + |F_k(x) - F_k(y)|,$$

and the obvious bound

$$|F_k(x) - F_k(y)| \leq 2\|\Delta \mathbf{f}^k\|_\infty,$$

both holding for any k. The first claim now follows by estimating $\Delta f^k = S_{\mathbf{b}}^k \Delta \delta$ in terms of $\|S_{\mathbf{b}}^M\|_\infty < \mu$, and by the observation that, for $2^{-k} \leq |x-y| \leq 2^{-k+1}$,

$$\mu^{[\frac{k}{M}]} \leq C\mu^{\frac{k}{M}} = C2^{-k\nu} \leq C|x-y|^\nu. \qquad \square$$

The tools for the analysis of smoothness are similar to the tools for the convergence analysis. The analysis of smoothness is based on the observation that in the stationary case $(1+z)/2$ is a smoothing factor.

Theorem 4.10. Let $a(z) = \frac{1+z}{2}q(z)$. If $S_{\mathbf{q}}$ is convergent and C^ℓ, then $S_{\mathbf{a}}$ is convergent and $C^{\ell+1}$.

Sketch of proof. By convolution property (2) and by (3.2), $S_{\mathbf{a}}$ is convergent, and

$$\phi_{\mathbf{a}}(x) = \int_{x-1}^{x} \phi_{\mathbf{q}}(t)\,\mathrm{d}t. \tag{4.18}$$

Thus

$$\phi'_{\mathbf{a}}(x) = \phi_{\mathbf{q}}(x) - \phi_{\mathbf{q}}(x-1). \tag{4.19}$$

Theorem 4.10 supplies a sufficient condition for smoothness. A repeated use of Theorem 4.10 together with Theorem 4.8 leads to

Corollary 4.11. Let $a(z) = \frac{(1+z)^{m+1}}{2^m}b(z)$ with $S_{\mathbf{b}}$ contractive. Then $\phi_{\mathbf{a}} \in C^m(\mathbb{R})$. Moreover

$$\phi_{\mathbf{a}}^{(\ell)} = S^\infty_{a(z)(1+z)^{-\ell}2^\ell}\Delta^\ell\delta, \quad \ell = 0, 1, \ldots, m,$$

where $\Delta^\ell = \Delta\Delta^{\ell-1}$ is defined recursively.

For a scheme with symbol $a(z) = 2^{-m}(1+z)^{m+1}b(z)$, instead of finding the maximal ℓ such that $S_{a(z)2^{\ell-1}(1+z)^{-\ell}}$ is contractive, Rioul (1992) suggests computing the numbers $\|S^\ell_{\mathbf{b}}\|_\infty = \mu_\ell$ and $\nu_\ell = -(\log_2\mu_\ell)/\ell$. If $m - \nu_\ell > 0$, then $\phi_{\mathbf{a}} \in C^{[m-\nu_\ell]}$. Defining $\nu = \sup_{\ell\ge 1}\nu_\ell$, if $m - \nu > 0$, then $\phi_{\mathbf{a}} \in C^{[m-\nu]}$, and $\phi_{\mathbf{a}}^{([m-\nu])}$ has Hölder exponent $n - \epsilon$ for any $\epsilon > 0$ with $n = m - \nu - [m-\nu]$.

Example 4. Consider the stationary interpolatory 4-point scheme with symbol (3.19)

$$a_w(z) = \frac{1}{2z}(1+z)^2[1 - 2wz^{-2}(1-z)^2(z^2+1)].$$

By Theorem 4.8, the range of w for which $S_{\mathbf{a}_w}$ is convergent is the range for which $S_{\mathbf{b}_w}$, with symbol

$$b_w(z) = \frac{1}{2z}(1+z)[1 - 2wz^{-2}(1-z)^2(z^2+1)],$$

is contractive. The condition $\|S_{\mathbf{b}_w}\|_\infty < 1$ yields the range

$$-\frac{3}{8} < w < \frac{-1+\sqrt{13}}{8},$$

while the condition $\|S^2_{\mathbf{b}_w}\|_\infty < 1$ yields the range

$$-\frac{1}{4} < w < \frac{-1+\sqrt{17}}{8}.$$

Thus a range of w for which $S_{\mathbf{a}_w}$ is convergent is (Dyn *et al.* 1991)

$$-\frac{3}{8} < w < \frac{-1+\sqrt{17}}{8} \cong 0.39.$$

By Corollary 4.11, it is sufficient to show that $S_{\mathbf{c}_w}$, with symbol

$$c_w(z) = \frac{1}{z}[1 - 2wz^{-2}(z-1)^2(z^2+1)],$$

is contractive, in order to prove that $S_{\mathbf{a}_w}$ is C^1. Now, $\|R_{\mathbf{c}_w}\|_\infty \geq 1$, while $\|R^2_{\mathbf{c}_w}\|_\infty < 1$ for $0 < w < \frac{\sqrt{5}-1}{8}$, as is shown in Dyn *et al.* (1991).

The fact that $\phi_{\mathbf{a}_w} \notin C^2(\mathbb{R})$ can be deduced from necessary conditions that are violated (see Section 5.2). In Daubechies and Lagarias (1992*a*), it is shown, by methods as in Section 5.2, that $\phi'_{\mathbf{a}_w}$ is differentiable except at all the dyadic points in its support.

After deriving similar results to the above for a class of nonstationary schemes, we return to the stationary case, and show that, in most interesting cases, if $\phi_{\mathbf{a}} \in C^m(\mathbb{R})$ then necessarily the symbol $a(z)$ is divisible by $(1+z)^{m+1}$. In this sense the form of $a(z)$ in Corollary 4.11 is necessary for $S_{\mathbf{a}}$ with C^m limit functions. This result holds if $\phi_{\mathbf{a}}$ is L_∞-stable, namely if for any bounded bi-infinite sequence $\mathbf{f} = \{f_i : i \in \mathbb{Z}\}$,

$$C_2 \sup_{i\in\mathbb{Z}} |f_i| \leq \left\| \sum_{i\in\mathbb{Z}} f_i \phi(x-i) \right\|_\infty \leq C_1 \sup_{i\in\mathbb{Z}} |f_i|, \tag{4.20}$$

with $0 < C_2 \leq C_1 < \infty$. For most interesting schemes the basic limit function is L_∞-stable, for example, for interpolatory schemes and for spline schemes. We also study the related property that, for $S_{\mathbf{a}}$ with $\phi_{\mathbf{a}} \in C^m$ and L_∞-stable, π_m is invariant under $_{\mathbf{a}}$.

Analysis of nonstationary schemes with symbols divisible by stationary smoothing factors

In this section the tools of analysis of Section 4.2 are extended to a class of nonstationary schemes. Theorem 4.10 also holds for a nonstationary scheme with symbols

$$a^k(z) = \frac{(1+z)}{2} q^k(z), \quad k \in \mathbb{Z}_+, \quad k \geq K,$$

with K some positive integer, and such that $S_{\{\mathbf{q}^k\}}$ is convergent. A version of Theorem 4.8 also holds in the nonstationary case. It supplies only a sufficient condition for convergence.

Theorem 4.12. Let a nonstationary scheme be given by the symbols

$$a^k(z) = (1+z)b^k(z), \quad k \in \mathbb{Z}_+, \quad k \geq K \in \mathbb{Z}_+.$$

If $S_{\{\mathbf{b}^k\}}$ is contractive then $S_{\{\mathbf{a}^k\}}$ is convergent.

This theorem holds since $R_{\mathbf{a}^k}$ and $R_{\mathbf{b}^k}$, defined by (2.2) and (2.3), are related by

$$\Delta R_{\mathbf{a}^k}\mathbf{f} = R_{\mathbf{b}^k}\Delta\mathbf{f}, \quad k \in \mathbb{Z}_+, \quad k \geq K, \tag{4.21}$$

and therefore, by the same arguments as in the stationary case, the contractivity of $S_{\{\mathbf{b}^k\}}$ implies the convergence of $S_{\{\mathbf{a}^k\}}$. A simple sufficient condition for the contractivity of $S_{\{\mathbf{b}^k\}}$ is

$$\|R_{\mathbf{b}^k}\|_\infty = \max\left(\sum_{j\in\mathbb{Z}} |b^k_{i-2j}| : i \in \{0,1\}\right) \leq \mu < 1, \quad k \in \mathbb{Z}_+, \quad k \geq K, \tag{4.22}$$

since then, for $\mathbf{g}^k = R_{\mathbf{b}^{k-1}}R_{\mathbf{b}^{k-2}}\cdots R_{\mathbf{b}^0}\mathbf{g}^0$, we have $\|\mathbf{g}^k\|_\infty \leq \mu^k\|\mathbf{g}^0\|_\infty$.

From Theorem 4.12 and the remark above it, we conclude the following.

Corollary 4.13. Let a nonstationary scheme be given by the symbols

$$a^k(z) = \frac{(1+z)^{m+1}}{2^m} b^k(z), \quad k \in \mathbb{Z}_+, \quad k \geq K \in \mathbb{Z}_+.$$

If $S_{\{\mathbf{b}^k\}}$ is contractive then $S_{\{\mathbf{a}^k\}}$ is C^m.

Example 5. In this example we study properties of the up-function introduced in Section 3.1, by applying the analysis tools of this section.

Let a nonstationary scheme be given by the symbols, as in (3.5),

$$a^k(z) = \frac{(1+z)^k}{2^{k-1}}, \quad k \in \mathbb{Z}_+.$$

To show that $\phi_{\{\mathbf{a}^k\}} \in C^\infty(\mathbb{R})$, we show that $\phi_{\{\mathbf{a}^k\}} \in C^m(\mathbb{R})$, for any $m \in \mathbb{Z}_+$.

Now, for $k \geq m+2$,

$$a^k(z) = \frac{(1+z)^{m+1}}{2^m} \cdot \frac{(1+z)^{k-m-1}}{2^{k-m-1}},$$

and, by Corollary 4.13, $\phi_{\{\mathbf{a}^k\}} \in C^m$ if $S_{\{\mathbf{b}^k\}}$ is contractive, with

$$b^k(z) = \frac{(1+z)^{k-m-1}}{2^{k-m-1}}, \quad k \in \mathbb{Z}_+, \quad k \geq m+2.$$

But $\|R_{\mathbf{b}^k}\|_\infty = \frac{1}{2}$ for $k \in \mathbb{Z}_+$, $k \geq m+2$, which proves that $S_{\{\mathbf{b}^k\}}$ is contractive.

Next we show that $\sigma(\phi_{\{\mathbf{a}^k\}}) = [0,2]$. Using (2.11) we get, from (3.5),

$$\sigma(\phi_{\{\mathbf{a}^k\}}) = \sum_{j=0}^{\infty} 2^{-j-1}\sigma(\mathbf{a}^j) = \sum_{j=0}^{\infty} 2^{-j-1}[0, j+1] = [0,2].$$

Polynomials generated by stationary schemes

For stationary interpolatory schemes in $\mathbb{R}^s$ it is easy to show (Dyn and Levin 1990) that $\phi_{\mathbf{a}} \in C^m$ implies that π_m is reproduced by the scheme, namely

$$R_{\mathbf{a}} p|_{\mathbb{Z}^s} = p\left(\frac{\cdot}{2}\right)|_{\mathbb{Z}^s}, \quad \text{and} \quad S_{\mathbf{a}}^{\infty} p|_{\mathbb{Z}^s} = p, \quad \text{for } p \in \pi_m(\mathbb{R}^s). \tag{4.23}$$

For a subdivision scheme with a stable basic limit function, the proof is more involved. It was first proved in Cavaretta *et al.* (1991). Here we present a proof for $s = 1$, which is extendable to univariate matrix subdivison schemes (Dyn and Levin 2002) and to multivariate schemes.

The proof is based on the following important observation in Warren (1995*a*).

Theorem 4.14. Let $S_{\mathbf{a}}$ be a C^m-convergent univariate, stationary subdivision scheme. Let $\mathbb{B}$ denote the set of bi-infinite sequences, and let $\mathbf{v} = \{v_j : j \in \mathbb{Z}\} \in \mathbb{B}$ be an eigenvector of $R_{\mathbf{a}}$ with eigenvalue λ, that is,

$$R_{\mathbf{a}}\mathbf{v} = \lambda\mathbf{v}. \tag{4.24}$$

Then the following hold.

(1) If $|\lambda| \geq 2^{-m}$, either $S_{\mathbf{a}}^{\infty}\mathbf{v} \equiv 0$ or $S_{\mathbf{a}}^{\infty}\mathbf{v} = x^i$ for some $0 \leq i \leq m$, and $\lambda = 2^{-i}$. Also $\lambda = 2^{-i}$, $0 \leq i \leq m$, cannot have a generalized eigenvector $\mathbf{u} \in \mathbb{B}$, satisfying

$$R_{\mathbf{a}}\mathbf{u} = \lambda\mathbf{u} + \mathbf{v}. \tag{4.25}$$

(2) If $|\lambda| < 2^{-m}$ then $(S_{\mathbf{a}}^{\infty}\mathbf{v})^{(\ell)}(0) = 0$, $\ell = 0, \ldots, m$.

(3) If $\lambda \neq 2^{-i}$, $0 \leq i \leq m$, and $\mathbf{u}$ is a corresponding generalized eigenvector satisfying (4.25), then $(S_{\mathbf{a}}^{\infty}\mathbf{u})^{(\ell)}(0) = 0$, $\ell = 0, \ldots, m$.

The proof of Theorem 4.14 is based on the relations

$$(S_{\mathbf{a}}^{\infty}\mathbf{v})(x) = \lambda(S_{\mathbf{a}}^{\infty}\mathbf{v})(2x), \quad (S_{\mathbf{a}}^{\infty}\mathbf{u})(x) = \lambda(S_{\mathbf{a}}^{\infty}\mathbf{u})(2x) + (S_{\mathbf{a}}^{\infty}\mathbf{v})(2x)$$

for $\mathbf{v}, \mathbf{u}$ satisfying (4.24) and (4.25) respectively, and on the continuity at $x = 0$ of the derivatives of order up to m of $S_{\mathbf{a}}^{\infty}\mathbf{u}$, $S_{\mathbf{a}}^{\infty}\mathbf{v}$.

A direct consequence of Theorem 4.14 deals with polynomials generated by a univariate stationary subdivision scheme with smooth limit functions (Dyn *et al.* 1995).

Theorem 4.15. Let $S_{\mathbf{a}}$ be a C^m-subdivision scheme. Then there exist $\mathbf{v}^{[i]} \in \mathbb{B}$, $i = 0, \ldots, m$, such that

$$R_{\mathbf{a}}\mathbf{v}^{[i]} = 2^i \mathbf{v}^{[i]}, \quad S_{\mathbf{a}}^{\infty}\mathbf{v}^{[i]} = x^i, \quad i = 0, \ldots, m. \tag{4.26}$$

The argument leading to (4.26) is that 2^{-i} must be an eigenvalue of $R_{\mathbf{a}}$ for $i = 0, \ldots, m$, otherwise there exists $\ell \in \{0, 1, \ldots, m\}$ such that $2^{-\ell}$ is not an eigenvalue of $R_{\mathbf{a}}$, implying that $\phi_{\mathbf{a}}^{(\ell)} \equiv 0$, in view of Theorem 4.14. But $\phi_{\mathbf{a}}$ is of compact support, $\phi_a \not\equiv 0$, which contradicts $\phi_{\mathbf{a}}^{(\ell)} \equiv 0$. Next we show that $\mathbf{v}^{[i]}$ in Theorem 4.15 is of the form $\mathbf{v}^{[i]} = x^i|_{\mathbb{Z}} + p_i|_{\mathbb{Z}}$ with $p_i \in \pi_{i-1}$, $i = 0, \ldots, m$ (here $p_0 \equiv 0$). For this proof the L_∞-stability of $\phi_{\mathbf{a}}$ is needed. We term a scheme L_∞-stable if its basic limit function is L_∞-stable.

Theorem 4.16. Let $S_{\mathbf{a}}$ be C^m and L_∞-stable. Then there exist polynomials $p_i \in \pi_{i-1}$, $i = 0, \ldots, m$, with $p_0 \equiv 0$, such that

$$S_{\mathbf{a}}^{\infty}(x^i + p_i)|_{\mathbb{Z}} = x^i, \quad i = 0, \ldots, m. \tag{4.27}$$

Sketch of proof. The case $i = 0$ follows directly from (4.10), because $R_{\mathbf{a}}$ maps the constant sequence $\mathbf{1} = \mathbf{u} = \{u_j = 1 : j \in \mathbb{Z}\}$ on itself.

In the following we indicate the proof for $i = 1$. For $i = 2, \ldots, m$, the proof is similar. Let $\mathbf{v} = \mathbf{v}^{[1]}$ satisfy $S_{\mathbf{a}}^{\infty}\mathbf{v} = x$, and for $r \in \mathbb{Z}_+ \backslash 0$ let $\Delta^{(r)}\mathbf{v} = \{v_{j+r} - v_j : j \in \mathbb{Z}\}$. Then the linearity and uniformity of $S_{\mathbf{a}}$ leads to $S_{\mathbf{a}}^{\infty}\Delta^{(1)}\mathbf{v} = x + 1 - x = 1$ or

$$S_{\mathbf{a}}^{\infty}(\Delta^{(1)}\mathbf{v} - \mathbf{1}) \equiv 0. \tag{4.28}$$

If $\Delta^{(1)}\mathbf{v} - \mathbf{1} \in \mathbb{B}$ is bounded, then by the L_∞-stability of $\phi_{\mathbf{a}}$, $\Delta^{(1)}\mathbf{v} = \mathbf{1}$, which is equivalent to $\mathbf{v} = x|_{\mathbb{Z}} + c\mathbf{1}$ for some $c \in \mathbb{R}$. Thus the claim of the theorem for $i = 1$ follows. To show the boundedness of $\Delta^{(1)}\mathbf{v} - \mathbf{1}$ we consider (4.28) at the integers, which in view of (2.8) has the form

$$\sum_{j \in \mathbb{Z}} \left((\Delta^{(1)}\mathbf{v})_j - 1\right)\phi_{\mathbf{a}}(n - j) = 0, \quad n \in \mathbb{Z}. \tag{4.29}$$

Equation (4.29) can be regarded as a finite difference equation for $\Delta^{(1)}\mathbf{v} - \mathbf{1}$, since $\phi_{\mathbf{a}}|_{\mathbb{Z}}$ is finitely supported, and is not identically equal to zero (otherwise $\phi_{\mathbf{a}} \equiv 0$ by (2.15)). As a solution of (4.29), $\Delta^{(1)}\mathbf{v} - \mathbf{1}$ is either bounded or it grows at least polynomially as $j \to \infty$ or $j \to -\infty$. For the latter possibility, $\mathbf{v}$ would have faster than linear growth. This possibility is eliminated, since

$$(R_{\mathbf{a}}\Delta^{(r)}\mathbf{v})_\alpha = \sum_{j \in \mathbb{Z}} a_{\alpha - 2j}(v_{j+r} - v_r) = \frac{1}{2}v_{\alpha+2r} - \frac{1}{2}v_\alpha = \frac{1}{2}(\Delta^{(2r)}\mathbf{v})_\alpha,$$

from which it is concluded, in view of (4.28), that

$$S_{\mathbf{a}}^{\infty}\Delta^{(1)}\mathbf{v} = \lim_{\ell \to \infty} 2^{-\ell}\Delta^{(2^\ell)}\mathbf{v} = 1,$$

or that $v_{\pm 2^\ell} = v_0 \pm 2^\ell + o(1)$, which is in contradiction to faster than linear growth.

As a direct consequence of Theorem 4.16 we get the following result.

Corollary 4.17. Let $S_{\mathbf{a}}$ be C^m and L_∞-stable. Then $\pi_m|_{\mathbb{Z}}$ is invariant under $R_{\mathbf{a}}$ and, for $p \in \pi_i$, $0 \le i \le m$,

$$R_{\mathbf{a}} p|_{\mathbb{Z}} = q\left(\frac{\cdot}{2}\right)|_{\mathbb{Z}},$$

with $q \in \pi_i$ and $p - q \in \pi_{i-1}$, while $p = q$ for $i = 0$.

In the following subsection we derive the factorization of the symbol of a scheme satisfying the requirements of Corollary 4.17.

Factorization of symbols of stationary, smooth, L_∞-stable schemes, and related necessary conditions

First we show that, if $S_{\mathbf{a}}$ is C^m and L_∞-stable, then its symbol has the factor $(1+z)^{m+1}$. Later we show that, necessarily, $S_{2^m a(z)(1+z)^{-m-1}}$ is contractive. A similar result holds for L_p-stability and convergence in the L_p-norm, $1 \le p \le \infty$ (Jia 1995). These results are important in the analysis of smoothness of univariate stationary schemes (see Section 4.2).

Theorem 4.18. Let $S_{\mathbf{a}}$ be C^m and L_∞-stable. Then

$$a(z) = (1+z)^{m+1} b(z) \tag{4.30}$$

with $b(z)$ a Laurent polynomial.

Proof. We use a recursive construction of 'divided difference' schemes with symbols

$$a^{[i]}(z) = 2^i (z+1)^{-i} a(z), \quad i = 1, \dots, m+1.$$

If $a^{[i]}(z)$ is a Laurent polynomial, then, in view of (4.11), $S_{\mathbf{a}^{[i]}}$ is related to $S_{\mathbf{a}}$ by

$$S_{\mathbf{a}^{[i]}} d_k^i \mathbf{f} = d_{k+1}^i S_{\mathbf{a}} \mathbf{f}, \quad \mathbf{f} \in \mathbb{B},$$

where $d_k^i \mathbf{f} = (2^k)^i \Delta^i f$ is the sequence of divided differences of order i on refinement level k. Since, by Corollary 4.17, $R_{\mathbf{a}}$ maps $\mathbf{1} \in \mathbb{B}$ to itself, $\sum_{i\in\mathbb{Z}} a_{2i} = \sum_{i\in\mathbb{Z}} a_{2i+1} = 1$ and $a(z)$ is divisible by $(1+z)$. This guarantees that $a^{[1]}$ exists. Now, $R_{\mathbf{a}}$ maps $\mathbf{v} = x|_{\mathbb{Z}}$, to $R_{\mathbf{a}}\mathbf{v} = \frac{1}{2}x|_{\mathbb{Z}} + c\mathbf{1}$ for some $c \in \mathbb{R}$, so $R_{\mathbf{a}^{[1]}}$ maps $\mathbf{1} \in \mathbb{B}$ into itself. Thus $a^{[1]}(z)$ is divisible by $(1+z)$, and $a^{[2]}$ exists. The general argument is similar.

By applying $(2^k)^i \Delta^i \mathbf{f}$ to $\mathbf{f} = x^i|_{\mathbb{Z}}$ we get a constant sequence. This sequence is mapped by $R_{a^{[i]}}$ to $(2^{k+1})^i \Delta^i R_{\mathbf{a}} \mathbf{f}$, which is the same constant sequence. This is the case since $R_{\mathbf{a}}\mathbf{f} = (\frac{\cdot}{2})^i + q(\frac{\cdot}{2})$ with $q \in \pi_{i-1}$ by Corollary 4.17, and $\Delta^i q(\frac{\cdot}{2}) = 0$. Again, if $R_{a^{[i]}}$ maps the constant sequence on itself, then $a^{[i]}(z)$ is divisible by $(1+z)$.

Using this argument for $i = 0, 1, \ldots, m$ we conclude that $a^{[i]}$ exists for $i = 1, \ldots, m+1$, and thus (4.30) holds. □

Example 6. Consider the symbol

$$a(z) = \frac{1}{4}(1+z)(1+z^2)^2 = \frac{1}{4}(1+z+2z^2+2z^3+z^4+z^5). \tag{4.31}$$

It is easy to see that (4.10) holds, since $a(1) = 2$, $a(-1) = 0$. To verify that $S_{\mathbf{a}}$ is convergent, we show that $S_{\mathbf{b}}$ with $b(z) = \frac{1}{4}(1+z^2)^2$ is contractive. Now, $b(z) = \frac{1}{4}(1+2z^2+z^4)$ and therefore $\|S_{\mathbf{b}}\|_\infty = 1$. Yet from (2.28) and (4.17) we get

$$\begin{aligned} b^{[2]}(z) &= \frac{1}{16}(1+z^4)^2(1+z^2)^2 \\ &= \frac{1}{16}(1+2z^2+3z^4+4z^6+3z^8+2z^{10}+z^{12}), \end{aligned}$$

and therefore $\|S^2_{\mathbf{b}}\|_\infty = \frac{1}{2}$.

Since the symbol $c(z) = 1+z^2$ satisfies $c(1) = 2$, $S_{\mathbf{c}}$ converges weakly (Derfel *et al.* 1995). It is easy to verify that $S^\infty_{\mathbf{c}}\delta = \frac{1}{2}\chi_{[0,1]}$ in the sense of weak convergence. By convolution property (3)

$$\phi_{\mathbf{a}} = \tfrac{1}{4}\chi_{[0,1]} * \chi_{[0,1]} * \chi_{[0,1]}.$$

Thus $\phi_{\mathbf{a}} \in C^1$, while $a(z)$ is not divisible by $(1+z)^2$. This indicates, in view of Theorem 4.18, that $\phi_{\mathbf{a}}$ is not L_∞-stable. Indeed, consider the sequence $\mathbf{u} = \{u_i = (-1)^i : i \in \mathbb{Z}\}$. Clearly $\mathbf{u}$ is bounded. Now in view of (4.31), $R_{\mathbf{a}}\mathbf{u} = \mathbf{0} \in \mathbb{B}$, and therefore $S^\infty_{\mathbf{a}}\mathbf{u} = \sum_{i\in\mathbb{Z}}(-1)^i\phi_{\mathbf{a}}(\cdot - i) \equiv 0$, and $\phi_{\mathbf{a}}$ is not L_∞-stable.

Once we have the factorization of the symbol of a stationary C^m, L_∞-stable scheme,

$$a(z) = (1+z)^{m+1}b(z),$$

we can show that $\frac{2^m a(z)}{(1+z)^{m+1}}$ is the symbol of a contractive scheme. For that we need two results, which are of importance beyond their current use.

Theorem 4.19. Let ϕ be a solution of the functional equation

$$\phi(x) = \sum_{\alpha\in\mathbb{Z}} a_\alpha \phi(2x-\alpha), \tag{4.32}$$

with a mask $\mathbf{a}$ satisfying (4.10). If ϕ is compactly supported, continuous and L_∞-stable, then $S_{\mathbf{a}}$ is convergent.

This theorem was first proved in Cavaretta *et al.* (1991). Here we give a sketch of a different proof (Dyn and Levin 2002).

Sketch of proof. Recalling the relation in (2.16), we observe that, since $\phi = T_{\mathbf{a}}\phi$, and $\mathbf{a} = R_{\mathbf{a}}\boldsymbol{\delta}$,

$$\phi(x) = \sum_{\alpha\in\mathbb{Z}^s} (R_{\mathbf{a}}\delta)_\alpha \phi(2x-\alpha) = \sum_{\alpha\in\mathbb{Z}^s} (R_{\mathbf{a}}^k\delta)_\alpha \phi(2^k x-\alpha), \tag{4.33}$$

and that for all $k \in \mathbb{Z}_+$

$$\sum_{\alpha\in\mathbb{Z}} \phi(x-\alpha) = \sum_{\alpha\in\mathbb{Z}} (R_{\mathbf{a}}^k\mathbf{1})_\alpha \phi(2^k x-\alpha) = \sum_{\alpha\in\mathbb{Z}} \phi(2^k x-\alpha). \tag{4.34}$$

The continuity and L_∞-stability of ϕ together with (4.34) leads, after proper normalization, to

$$\sum_{\alpha\in\mathbb{Z}} \phi(\cdot-\alpha) \equiv 1. \tag{4.35}$$

Combining (4.33) and (4.35) we get

$$0 = \sum_{\alpha\in\mathbb{Z}} \phi(2^k x-\alpha)\big[(R_{\mathbf{a}}^k\delta)_\alpha - \phi(x)\big],$$

which, together with the continuity, compact support and L_∞-stability of ϕ, yields

$$\lim_{k\to\infty} \sup_{\alpha\in\mathbb{Z}\cap K} |(R_{\mathbf{a}}^k\boldsymbol{\delta})_\alpha - \phi(2^{-k}\alpha)| = 0,$$

for any compact set $K \subset \mathbb{R}$. This is the convergence of $S_{\mathbf{a}}$ in the sense of (2.5) to a continuous limit function ϕ, hence uniform convergence.

The second theorem is taken from Dyn and Levin (2002), where it is proved for matrix masks.

Theorem 4.20. Let $a(z) = \frac{1+z}{2}q(z)$, with $S_{\mathbf{a}}$ L_∞-stable and C^1. Then

$$\varphi = \sum_{\alpha\in\mathbb{Z}} \phi_{\mathbf{a}}'(\cdot-\alpha)$$

is a continuous, L_∞-stable solution of

$$\varphi(x) = T_{\mathbf{q}}\varphi(x) = \sum_{\alpha\in\mathbb{Z}} q_\alpha \varphi(2x-\alpha). \tag{4.36}$$

Sketch of proof. The function φ is well defined, continuous and of compact support. It is related to $\phi_{\mathbf{a}}$ by

$$\phi_{\mathbf{a}}(x) = \int_{x-1}^{x} \varphi(t)\,\mathrm{d}t = \varphi * \chi_{[0,1]}. \tag{4.37}$$

Suppose φ is not L_∞-stable; then there exists a bounded nonzero sequence $\mathbf{u} \in \mathbb{B}$ such that

$$\sum_{\alpha \in \mathbb{Z}} u_\alpha \varphi(\cdot - \alpha) \equiv 0.$$

By integrating this relation from $x-1$ to x we obtain

$$\sum_{\alpha \in \mathbb{Z}} \mathbf{u}_\alpha \phi_{\mathbf{a}}(x - \alpha) \equiv 0.$$

This last relation contradicts the L_∞-stability of $\phi_{\mathbf{a}}$. Thus φ is also L_∞-stable. To verify that $\varphi = T_{\mathbf{q}}\varphi$, we observe that $\phi_{\mathbf{a}} = T_{\mathbf{a}}\phi_{\mathbf{a}}$, and after taking the Fourier transform, it is equivalent to

$$\hat{\phi}_{\mathbf{a}}(w) = \frac{1}{2}\hat{a}\left(\frac{w}{2}\right)\hat{\phi}_{\mathbf{a}}\left(\frac{w}{2}\right) \tag{4.38}$$

with $\hat{a}(w) = \sum_{\alpha \in \mathbb{Z}} a_\alpha e^{-iw\alpha}$. Now by (4.37) $\hat{\varphi}(w)\frac{1-e^{-iw}}{w} = \hat{\phi}_{\mathbf{a}}(w)$. Multiplying (4.38) by $\frac{w}{1-e^{-iw}}$, we obtain

$$\hat{\varphi}(w) = \frac{1}{2}\,\frac{2\hat{a}(\frac{w}{2})}{1+e^{-i\frac{w}{2}}}\hat{\varphi}\left(\frac{w}{2}\right) = \frac{1}{2}\hat{q}\left(\frac{w}{2}\right)\hat{\varphi}\left(\frac{w}{2}\right),$$

proving (4.36).

From Theorems 4.19, 4.20, 4.18 and 4.8 we conclude the following.

Corollary 4.21. Let $S_{\mathbf{a}}$ be C^1 and L_∞-stable. Then $b(z) = \frac{2a(z)}{(1+z)^2}$ is a Laurent polynomial and $S_{\mathbf{b}}$ is contractive.

This corollary together with Corollary 4.11 implies the following.

Corollary 4.22. Let $S_{\mathbf{a}}$ be convergent and L_∞-stable. Then the contractivity of $S_{2^m a(z)(1+z)^{-(m+1)}}$ is necessary and sufficient for $S_{\mathbf{a}}$ to be C^m.

4.3. Analysis of bivariate stationary schemes via difference schemes

The analysis of convergence and smoothness of multivariate subdivision schemes defined on regular grids, which is of interest to geometric modelling in $\mathbb{R}^3$, is in the case $s = 2$. Thus, for the sake of simplicity of presentation, we limit the discussion to this case. The results are easily extended to $s > 2$. Here we present similar analysis tools to those in the univariate, stationary case for bivariate, stationary subdivision schemes defined on regular quad-meshes and on regular triangulations. When the symbol factorizes into sufficiently many linear factors (each a univariate smoothing factor in some direction in $\mathbb{Z}^2$), the analysis is almost as simple as in the univariate case (Cavaretta *et al.* 1991, Dyn 1992). This factorization is not the result of (4.10) or of the smoothness of the limit functions, as in the univariate case, but is an additional assumption, which holds for many of the schemes in use.

In fact the same factorization of nonstationary symbols leads to similar results, even for nonstationary schemes. When the symbol is not factorizable to univariate smoothing factors, (4.10) leads to non-unique matrix difference schemes, and the theory of the univariate case can be extended to this case (Cavaretta *et al.* 1991, Dyn 1992, Hed 1990); see Section 4.3.

Analysis of schemes with factorizable symbols

The necessary conditions for convergence of a bivariate scheme $S_{\mathbf{a}}$ defined on $\mathbb{Z}^2$, which are obtained from (4.10), are

$$\sum_{\beta\in\mathbb{Z}^2} a_{\alpha-2\beta} = 1, \quad \alpha \in \{(0,0),(0,1),(1,0),(1,1)\}. \tag{4.39}$$

These conditions imply

$$a(1,1)=4, \qquad a(-1,1)=0, \qquad a(1,-1)=0, \qquad a(-1,-1)=0. \tag{4.40}$$

In contrast to the univariate case ($s=1$), in the bivariate case ($s=2$), the necessary conditions (4.39) and the derived conditions on $a(z)$, (4.40), do not imply a factorization of the mask to linear factors.

If the factorization

$$a(z) = (1+z_1)^m(1+z_2)^m b(z), \quad z=(z_1,z_2), \tag{4.41}$$

is imposed, then, with $m=1$, the convergence can be analysed almost as in the univariate case, and similarly the smoothness if $m>1$.

Theorem 4.23. Let $S_{\mathbf{a}}$ have a symbol of the form (4.41) with $m=1$. If the schemes with the symbols $a_1(z) = \frac{a(z)}{1+z_1} = (1+z_2)b(z)$, $a_2(z) = \frac{a(z)}{1+z_2} = (1+z_1)b(z)$ are both contractive, then $S_{\mathbf{a}}$ is convergent. Conversely, if $S_{\mathbf{a}}$ is convergent then $S_{\mathbf{a}_1}$ and $S_{\mathbf{a}_2}$ are contractive.

The proof of this theorem is similar to the proof of Theorem 4.8, due to the observation that for $\Delta_1\mathbf{f} = \{f_{i,j} - f_{i-1,j} : i,j\in\mathbb{Z}\}$, and $\Delta_2\mathbf{f} = \{f_{i,j} - f_{i,j-1} : i,j\in\mathbb{Z}\}$,

$$S_{\mathbf{a}_\ell}\Delta_\ell\mathbf{f} = \Delta_\ell S_{\mathbf{a}}\mathbf{f}, \quad \ell=1,2.$$

Thus convergence is checked in this case as contractivity of two subdivision schemes $S_{\mathbf{a}_1}, S_{\mathbf{a}_2}$. For schemes having the symmetry of the square grid (topologically equivalent rules for the computation of vertices corresponding to edges), then $a_1(z_1,z_2) = a_2(z_2,z_1)$, and the contractivity of only one scheme has to be checked. Note that the factorization in (4.41) has then the symmetry of $\mathbb{Z}^2$.

For the smoothness result, we introduce the inductive definition of differences: $\Delta^{[i,j]} = \Delta_1\Delta^{[i-1,j]}$, $\Delta^{[i,j]} = \Delta_2\Delta^{[i,j-1]}$, $\Delta^{[1,0]} = \Delta_1$, $\Delta^{[0,1]} = \Delta_2$.

Theorem 4.24. Let $a(z)$ be factorizable as in (4.41). If the schemes with the masks

$$a_{i,j}(z) = \frac{2^{i+j}a(z)}{(1+z_1)^i(1+z_2)^j}, \quad i,j = 0,\dots,m \tag{4.42}$$

are convergent, then

$$\frac{\partial^{i+j}}{\partial t_1^i \partial t_2^j}(S_{\mathbf{a}}^{\infty}\mathbf{f}^0)(t) = (S_{\mathbf{a_{i,j}}}^{\infty}\Delta_1^i\Delta_2^j\mathbf{f}^0)(t), \quad i,j = 0,\dots,m. \tag{4.43}$$

In particular, $S_{\mathbf{a}}$ is C^m.

In geometric modelling the required smoothness of surfaces is at least C^1 and at most C^2. To verify that a scheme $S_{\mathbf{a}}$ generates C^1 limit functions, with the aid of the last two theorems, we have to assume a symbol of the form

$$a(z) = (1+z_1)^2(1+z_2)^2 b(z),$$

and to check the contractivity of the three schemes with symbols

$$2(1+z_1)(1+z_2)b(z), \qquad 2(1+z_2)^2 b(z), \qquad 2(1+z_1)^2 b(z).$$

This analysis applies also to tensor product schemes, but is not needed, since if $a(z) = a_1(z_1)a_2(z_2)$ is the symbol of a tensor product scheme, then $\phi_{\mathbf{a}}(t_1,t_2) = \phi_{\mathbf{a_1}}(t_1)\cdot\phi_{\mathbf{a_2}}(t_2)$, and its smoothness properties are derived from those of $\phi_{\mathbf{a_1}}, \phi_{\mathbf{a_2}}$.

Similar results hold for schemes defined on regular triangulations. For the topology of a regular triangulation, we regard the subdivision scheme as operating on the 3-directional grid. (The vertices of $\mathbb{Z}^2$ with edges in the directions $(1,0),(0,1),(1,1)$.)

Since the 3-directional grid can be regarded also as $\mathbb{Z}^2$, (4.39) and (4.40) hold for convergent schemes on this grid.

A scheme for regular triangulations treats each edge in the 3-directional grid in the same way with respect to the topology of the grid. The symbol of such a scheme, when factorizable, has the form

$$a(z) = (1+z_1)^m(1+z_2)^m(1+z_1z_2)^m b(z). \tag{4.44}$$

Example 7. The symbol of the butterfly scheme on the 3-directional grid has the form (Dyn, Levin and Micchelli 1990*b*)

$$a(z) = \frac{1}{2}(1+z_1)(1+z_2)(1+z_1z_2)(1-wc(z_1,z_2))(z_1z_2)^{-1} \tag{4.45}$$

with

$$\begin{aligned} c(z_1,z_2) &= 2z_1^{-2}z_2^{-1} + 2z_1^{-1}z_2^{-2} - 4z_1^{-1}z_2^{-1} - 4z_1^{-1} - 4z_2^{-1} \\ &\quad + 2z_1^{-1}z_2 + 2z_1z_2^{-1} + 12 - 4z_1 - 4z_2 - 4z_1z_2 + 2z_1^2z_2 + 2z_1z_2^2. \end{aligned} \tag{4.46}$$

Convergence analysis for schemes with factorizable symbols of the form (4.44) is similar to that for schemes with symbols of the form (4.41).

Theorem 4.25. Let $S_{\mathbf{a}}$ have the symbol

$$a(z) = (1+z_1)(1+z_2)(1+z_1z_2)b(z). \tag{4.47}$$

Then $S_{\mathbf{a}}$ is convergent if and only if the schemes with symbols

$$a_1(z) = \frac{a(z)}{1+z_1}, \qquad a_2(z) = \frac{a(z)}{1+z_2}, \qquad a_3(z) = \frac{a(z)}{1+z_1z_2} \tag{4.48}$$

are contractive. If any two of these schemes are contractive, then the third is also contractive.

Note that

$$S_{\mathbf{a_3}}\Delta_3\mathbf{f} = \Delta_3 S_{\mathbf{a}}\mathbf{f},$$

with $(\Delta_3\mathbf{f})_{i,j} = f_{i,j} - f_{i-1,j-1}$. Thus, if two of the schemes $S_{\mathbf{a_i}}$, $i = 1,2,3$ are contractive then the differences in two linearly independent directions tend to zero as $k \to \infty$, which implies, as in the proof of Theorem 4.8, the uniform convergence of the bilinear interpolants to $\{\mathbf{f}^k\}_{k\in\mathbb{Z}_+}$.

The smoothness analysis for a scheme with a symbol (4.47) is different from that for schemes with symbols as in (4.41).

Theorem 4.26. Let $S_{\mathbf{a}}$ have the symbol (4.47), and let $a_i(z)$, $i = 1,2,3$ be as in (4.48). Then $S_{\mathbf{a}}$ generates C^1 limit functions, if the schemes with the symbols $2a_i(z)$, $i = 1,2,3$, are convergent. If any two of these schemes are convergent then the third is also convergent. Moreover,

$$\frac{\partial}{\partial t_i}(S_{\mathbf{a}}^{\infty}\mathbf{f}^0)(t) = (S_{\mathbf{2a_i}}\Delta_i\mathbf{f}^0)(t), \quad i = 1,2,$$

$$\left(\frac{\partial}{\partial t_1} + \frac{\partial}{\partial t_2}\right)(S_{\mathbf{a}}^{\infty}\mathbf{f}^0)(t) = (S_{\mathbf{2a_3}}\Delta_3\mathbf{f}^0)(t).$$

The verification, based on Theorems 4.25 and 4.26, that the scheme $S_{\mathbf{a}}$ with symbol (4.47) is C^1, requires us to check the contractivity of the three schemes with symbols

$$2(1+z_1)b(z), \qquad 2(1+z_2)b(z), \qquad 2(1+z_1z_2)b(z).$$

If these three schemes are contractive, then $S_{\mathbf{a}}$ generates C^1-limit functions. For $a(z)$ with the symmetries of the 3-directional grid, it is sufficient to check the contractivity of only one of the three schemes, as is easily observed in the next example.

Example 8. To verify that the butterfly scheme generates C^1-limit functions, we use the fact that the symbol $a(z)$ of the butterfly scheme, given in (4.45), is of the form (4.47). In view of the observation following

Theorem 4.26, we have to check the contractivity of the three schemes with symbols

$$q_i(z) = (1 + z_i)\big(1 - wc(z_1, z_2)\big)(z_1 z_2)^{-1}, \quad i = 1, 2,$$
$$q_3(z) = (1 + z_1 z_2)\big(1 - wc(z_1, z_2)\big)(z_1 z_2)^{-1}.$$

Noting that

$$c(z_1, z_2) = c(z_2, z_1) = c(z_1 z_2, z_1^{-1}),$$

and that the factor $(z_1 z_2)^{-1}$ in a symbol does not affect the norm of the corresponding subdivision operator, it is sufficient to verify the contractivity of $S_{\mathbf{r}}$, where

$$r(z) = (1 + z_1)\big(1 - wc(z_1, z_2)\big) = \sum_{\alpha \in \mathbb{Z}^2} r_\alpha z^\alpha.$$

Now

$$\|S_{\mathbf{r}}\|_\infty = \max_{\ell, k \in \{0,1\}} \left(\sum_{i,j \in \mathbb{Z}} |r_{k+2i,\ell+2j}| \right),$$

and since

$$\sum_{i,j \in \mathbb{Z}} |r_{2i,2j}| = |1 - 8w| + |8w|,$$

$\|S_r\|_\infty \geq 1$ for all values of w.

Next, we show that for sufficiently small $w > 0$, $\|S_{\mathbf{r}}^2\|_\infty < 1$ (Dyn *et al.* 1990*b*). Ignoring coefficients of $r^{[2]}(z)$ that are not $O(1)$, and computing the others up to order $O(w)$, we get

$$\begin{aligned} r^{[2]}(z) &= r(z) r(z^2) \\ &= (1 + z_1 + z_1^2 + z_1^3)\big(1 - wc(z_1, z_2) - wc(z_1^2, z_2^2) + O(w^2)\big) \\ &= \sum_{i,j \in \mathbb{Z}} r_{ij}^{[2]} z_1^i z_2^j. \end{aligned}$$

Thus, for $j \neq 0$, $r_{i,j}^{[2]} = O(w)$ while $r_{i,0}^{[2]} = 1 + O(w)$, $i = 0, 1, 2, 3$. From this we conclude that it is sufficient to show that, for sufficiently small w,

$$\sum_{i,j \in \mathbb{Z}} |r_{\ell+4i,4j}^{[2]}| < 1, \quad \ell = 0, 1, 2, 3.$$

When $\ell = 0$, all the nonzero coefficients $\{r_{4i,4j}^{[2]}\}$ are

$$\begin{aligned} r_{0.0}^{[2]} &= 1 - 16w + O(w^2), \\ r_{4,0}^{[2]} &= 8w + O(w^2), \\ r_{4,4}^{[2]} &= r_{0,-4}^{[2]} = -2w + O(w^2). \end{aligned}$$

Hence, for sufficiently small $w > 0$,

$$\sum_{i,j\in\mathbb{Z}} \left|r^{[2]}_{4i,4j}\right| = |1 - 16w| + 12|w| + O(w^2) < 1.$$

When $\ell = 1$, the relevant coefficients are

$$r^{[2]}_{1,0} = 1 - 12w + O(w^2),$$
$$r^{[2]}_{5,0} = 4w + O(w^2),$$
$$r^{[2]}_{5,4} = r^{[2]}_{1,-4} = -2w + O(w^2),$$

and, for sufficiently small $w > 0$,

$$\sum_{i,j\in\mathbb{Z}} |r_{1+4i,4j}| = |1 - 12w| + 8|w| + O(w^2) < 1.$$

The cases $\ell = 2$ and $\ell = 3$ are similar to the cases $\ell = 1$ and $\ell = 0$, respectively. Thus, for sufficiently small $w > 0$, the limit surfaces/functions generated by the butterfly scheme on regular triangulations are C^1.

An explicit value of w_0, such that for $w \in (0, w_0)$ the butterfly scheme generates C^1 limit functions on regular triangulations, is computed in Gregory (1991). The computation shows that $w_0 > \frac{1}{16}$. The value $w = \frac{1}{16}$ is of special importance, since for this value the butterfly scheme on $\mathbb{Z}^2$ reproduces cubic polynomials, while for $w \neq \frac{1}{16}$ the scheme reproduces only linear polynomials. These properties are related to the approximation properties of the scheme (see Section 7).

Analysis of general schemes defined on $\mathbb{Z}^2$

The necessary conditions in the bivariate case (4.39) imply four conditions on the symbol (4.40).

These four conditions lead to a subdivision scheme with a matrix mask, for the vector of first differences

$$\boldsymbol{\Delta}\mathbf{f} = \left\{(\boldsymbol{\Delta}\mathbf{f}) = \left(\begin{pmatrix}\Delta_1\\ \Delta_2\end{pmatrix}\mathbf{f}\right)_{ij} = \begin{pmatrix} f_{ij} - f_{i-1,j} \\ f_{ij} - f_{i,j-1}\end{pmatrix} : (i,j) \in \mathbb{Z}^2\right\}. \quad (4.49)$$

Contrary to the univariate case, this matrix mask is *not uniquely* determined. The matrix mask can be derived with the help of the following lemma.

Lemma 4.27. Let $p(z) = p(z_1, z_2)$ be a Laurent polynomial satisfying

$$p(1,1) = p(-1,1) = p(1,-1) = p(-1,-1) = 0. \quad (4.50)$$

Then there exist Laurent polynomials, p_1, p_2, such that

$$p(z) = (1 - z_1^2)p_1(z) + (1 - z_2^2)p_2(z). \quad (4.51)$$

The 'factorization' in (4.51) is not unique, since the term $(1-z_1^2)(1-z_2^2)q(z)$, with q a Laurent polynomial, can be added to the first term on the right-hand side of (4.51) and subtracted from the second.

The proof of the lemma is based on the following two observations.

(a) The Laurent polynomial

$$P(z) = \frac{1}{2}\big[(1+z_2)p(z_1,1) + (1-z_2)p(z_1,-1)\big]$$

coincides with $p(z)$ for $z_2 = 1$ and $z_2 = -1$, and therefore there exists a Laurent polynomial $r(z)$ such that

$$p(z) - P(z) = (1-z_2^2)r(z).$$

(b) $P(z)$ is a Laurent polynomial that is divisible by $(1-z_1^2)$, since, in view of (4.50), $P(\pm 1, z_2) \equiv 0$.

The last lemma guarantees the 'factorization' assumed in (4.52).

Theorem 4.28. Let $a(z) = a(z_1, z_2)$ satisfy (4.40), and let

$$\begin{aligned}(1-z_1)a(z) &= b_{11}(z)(1-z_1^2) + b_{12}(z)(1-z_2^2),\\ (1-z_2)a(z) &= b_{11}(z)(1-z_1^2) + b_{22}(z)(1-z_2^2),\end{aligned} \tag{4.52}$$

where $b_{ij}, i,j = 1,2$, are Laurent polynomials. Then

$$\mathbf{\Delta} R_{\mathbf{a}}\mathbf{f} = R_{\mathbf{B}}\mathbf{\Delta f}, \tag{4.53}$$

where $R_{\mathbf{B}}$ is the refinement rule

$$(R_{\mathbf{B}}\mathbf{v})_\alpha = \sum_{\beta\in\mathbb{Z}^2} B_{\alpha-2\beta}v_\beta, \quad \alpha \in \mathbb{Z}^2, \tag{4.54}$$

with the matrix symbol

$$B(z) = \sum_{\alpha\in\mathbb{Z}^2} B_\alpha z^\alpha = \begin{pmatrix} b_{11}(z) & b_{12}(z)\\ b_{21}(z) & b_{22}(z)\end{pmatrix}, \tag{4.55}$$

and with $\mathbf{v}$ a bi-infinite sequence of vectors in $\mathbb{R}^2$, that is,

$$\mathbf{v} = \{v_\alpha : v_\alpha \in \mathbb{R}^2, \alpha \in \mathbb{Z}^2\}.$$

Sketch of proof. The formalism of the z-transform is the tool for proving the theorem. Observing that

$$L(\mathbf{\Delta f}; z) = \begin{pmatrix} 1-z_1\\ 1-z_2\end{pmatrix} L(\mathbf{f}; z),$$

and recalling the basic relation in (2.25),

$$L(R_{\mathbf{a}}\mathbf{f}; z) = a(z)L(\mathbf{f}; z^2),$$

we obtain from (4.52)

$$\begin{pmatrix} 1-z_1 \\ 1-z_2 \end{pmatrix} L(R_{\mathbf{a}}\mathbf{f}; z) = \begin{pmatrix} b_{11}(z) & b_{12}(z) \\ b_{21}(z) & b_{22}(z) \end{pmatrix} \begin{pmatrix} 1-z_1^2 \\ 1-z_2^2 \end{pmatrix} L(\mathbf{f}; z^2),$$

which is equivalent to (4.53) and (4.54). In the following we let $S_{\mathbf{B}}$ denote the stationary scheme with the refinement rule $R_{\mathbf{B}}$ in (4.54). Theorem 4.28 leads, as in the univariate case, to the following result.

Corollary 4.29. Let $S_{\mathbf{a}}$ be a bivariate subdivision scheme satisfying (4.39). Then $S_{\mathbf{a}}$ is convergent if and only if $S_{\mathbf{B}}$ is contractive for all initial data of the form $\mathbf{\Delta f}$.

A sufficient condition for convergence is thus the contractivity of the scheme $S_{\mathbf{B}}$. This can be verified by considering the numbers $\|S_{\mathbf{B}}^M\|_\infty$ for $M = 1, 2, \ldots$. Here again the formalism of the z-transform leads to the symbol

$$B^{[M]}(z) = B(z)B^{[M-1]}(z^2) = B(z)B(z^2)\cdots B(z^{2^{M-1}})$$

of S_B^M, where the order of the factors in the matrix product is significant. The norm of $S_{\mathbf{B}}^M$ is given by (Hed 1990, Dyn 1992)

$$\|S_{\mathbf{B}}^M\|_\infty = \max_{\alpha \in E_2^M} \left\| \sum_{\beta \in \mathbb{Z}^2} |B^{[M]}_{\alpha - 2^M\beta}| \right\|_\infty$$

where $|A|$ denotes the matrix whose elements are the absolute values of the corresponding elements in the matrix A, $\|A\|_\infty$ denotes the L_∞-norm of the matrix A, and where $E_2^M = \{\alpha = (\alpha_1, \alpha_2) : 0 \le \alpha_1 < 2^M, 0 \le \alpha_2 < 2^M\}$. Thus a similar algorithm to the one given in the univariate case (see Section 4.2), applies also in the bivariate case, although it is based only on a sufficient condition and on a non-unique 'factorization'. It is possible to use optimization techniques to find, among all possible 'factorizations', the one that minimizes $\min\{\|S_{\mathbf{B}}^M\|_\infty : 1 \le M \le 10\}$ (Kasas 1990).

The C^1 analysis is based on the following result.

Theorem 4.30. Let $S_{\mathbf{a}}$ be a convergent subdivision scheme. If $2S_{\mathbf{B}}$ with $\mathbf{B}$ given by (4.55) and (4.52) is convergent for initial data of the form $\mathbf{\Delta f}$, then $S_{\mathbf{a}}$ is C^1.

This result is analogous to Theorem 4.10 in the univariate case. Furthermore,

$$(2S_{\mathbf{B}})^\infty \mathbf{\Delta f}^0 = \begin{pmatrix} \partial_1 \\ \partial_2 \end{pmatrix} S_{\mathbf{a}}^\infty \mathbf{f}^0. \tag{4.56}$$

Equation (4.56) only holds if

$$\sum_{\alpha \in \mathbb{Z}^2} B_{\gamma - 2\alpha} = I_{2\times 2}, \quad \gamma \in \{(0,0), (0,1), (1,0), (1,1)\}, \tag{4.57}$$

which follows from the linear independence of the two components of

$$(2S_{\mathbf{B}})^{\infty}\mathbf{\Delta f} = (\partial_1 S_{\mathbf{a}}^{\infty}\mathbf{f}, \partial_2 S_{\mathbf{a}}^{\infty}\mathbf{f})^T$$

for generic $\mathbf{f}$. From (4.57) and Lemma 4.27 follows the existence of a matrix subdivision scheme $S_{\mathbf{C}}$, for the vectors $2^{-k}\Delta^2\mathbf{f}^k$,

$$\mathbf{\Delta}^2\mathbf{f} = \begin{pmatrix}\Delta_1\\ \Delta_2\end{pmatrix}\mathbf{\Delta f} \in \mathbb{R}^4,$$

with $\mathbf{C}$ a mask of matrices of order 4×4 with symbol

$$\begin{pmatrix} C^{(1,1)}(z) & C^{(1,2)}(z)\\ C^{(2,1)}(z) & C^{(2,2)}(z)\end{pmatrix},$$

where $C^{(i,j)}(z)$ is a matrix of order 2×2 defined by the 'factorization'

$$\begin{pmatrix}1-z_1\\ 1-z_2\end{pmatrix} 2b_{ij}(z) = C^{(i,j)}(z)\begin{pmatrix}1-z_1^2\\ 1-z_2^2\end{pmatrix}.$$

If $S_{\mathbf{C}}$ is contractive then $S_{2\mathbf{B}}$ is convergent and $S_{\mathbf{a}}$ is C^1. The same ideas can be further extended to deal with higher orders of smoothness (Hed 1990, Dyn 1992).

5. Analysis by local matrix operators

Given masks $\{\mathbf{a}^k\}$ of the same finite support, the corresponding refinement rules (2.2) and their representations in matrix form (2.18) are local. For the subdivision scheme $S_{\{\mathbf{a}^k\}}$, this locality is also expressed by the compact supports of the corresponding basic limit functions $\{\phi_k : k \in \mathbb{Z}_+\}$, and the representations (2.10) of the limit functions $S_k^{\infty}\mathbf{f}^0$.

5.1. *The local matrix operators in the univariate setting*

To simplify the presentation we deal here with the case $s = 1$. The results extend to $s > 1$.

The locality of $R_{\mathbf{a}^k}$ can be more emphatically expressed in terms of two finite-dimensional matrices, which are both sections of the bi-infinite matrix A^k in (2.18). First we obtain the two finite-dimensional matrices. Consider

$$S_0^{\infty}\mathbf{f}^0 = S_{\{\mathbf{a}^k\}}^{\infty}\mathbf{f}^0 = \sum_{\alpha\in\mathbb{Z}} f_\alpha^0\phi_0(\cdot - \alpha) = \sum_{\alpha\in\mathbb{Z}} f_\alpha^k\phi_k(2^k\cdot - \alpha), \tag{5.1}$$

and its restriction to a unit interval. Due to the finite support of ϕ_0, there exists a finite set $I \subset \mathbb{Z}$, such that

$$S_{\{\mathbf{a}^k\}}^{\infty}\mathbf{f}^0\Big|_{[j,j+1]} = \sum_{\alpha - j\in I} f_\alpha^0\phi_0(\cdot - \alpha). \tag{5.2}$$

Thus the vector $\{f^0_\alpha : \alpha - j \in I\}$ completely determines the limit function in $[j, j+1]$. By the same reasoning, and since $\sigma(\phi_k) = \sigma(\phi_0)$, $k \in \mathbb{Z}_+$, we deduce, in view of (5.1), that the vector $\{f^k_\alpha : \alpha - j \in I\}$, with $\mathbf{f}^k = R_{\mathbf{a}^{k-1}} \cdots R_{\mathbf{a}^0} f^0$, determines the limit function in $[j, j+1]2^{-k}$. Again, by the linearity of $\{R_{\mathbf{a}^k} : k \in \mathbb{Z}_+\}$, there exists a linear map from $\{f^{k-1}_\alpha : \alpha \in I\}$ to $\{f^k_\alpha : \alpha \in I\}$, which is a square matrix of dimension $|I|$. We denote it by A^k_0. Similarly there is a linear transformation from $\{f^{k-1}_a : \alpha \in I\}$ to $\{f^k_\alpha : \alpha - 1 \in I\}$, denoted by A^k_1. Note that, by the uniformity of $R_{\mathbf{a}^k}$, A^k_ε maps the vector $\{f^{k-1}_\alpha : \alpha - j \in I\}$ to $\{f^k_\alpha : \alpha - j - \varepsilon \in I\}$, $\varepsilon = 0, 1$. It is easy to conclude from the definition of A^k_0, A^k_1 as linear operators, that the matrices A^k_0, A^k_1 are finite sections of the bi-infinite matrix A^k in (2.19), that is,

$$\begin{aligned} (A^k_0)_{\alpha\beta} &= a^k_{\alpha-2\beta}, \quad \alpha, \beta \in I, \\ (A^k_1)_{\alpha\beta} &= a^k_{\alpha+1-2\beta}, \quad \alpha, \beta \in I. \end{aligned} \tag{5.3}$$

In the following we show how to get the value $(S_{\{\mathbf{a}^k\}} f^0)(x)$ for $x \in \mathbb{R}$ in terms of the matrices $\{A^k_0, A^k_1 : k \in \mathbb{Z}_+\}$. It is sufficient to consider the interval $[0, 1)$.

For $x \in [0, 1)$, we use the dyadic representation $x = \sum_{i=1}^\infty d_i 2^{-i}$, $d_i \in \{0, 1\}$, and obtain

$$\left(S^\infty_{\{\mathbf{a}^k\}} f^0\right)(x) = \lim_{k\to\infty} A^k_{d_{k+1}} A^{k-1}_{d_k} \cdots A^0_{d_1} \mathbf{f}^0_{[0,1]} \tag{5.4}$$

where $\mathbf{f}^0_{[0,1)} = \{f^0_\alpha : \alpha \in I\}$. Note that the finite product $A^k_{d_{k+1}} \cdots A^0_{d_1} \mathbf{f}^0_{[0,1)}$ is a vector which determines the limit function in an interval of the form $[j, j+1]2^{-k-1}$ containing x. Thus the convergence and smoothness of the limit function generated by $S_{\{\mathbf{a}^k\}}$ can be deduced from the set of finite matrices

$$\{A^k_0, A^k_1 : k \in \mathbb{Z}_+\} \tag{5.5}$$

and their infinite products of the form appearing in (5.4). In the stationary case there are only two matrices A_0, A_1, and all possible infinite products of them have to be considered (Micchelli and Prautzsch 1989).

5.2. Convergence and smoothness of univariate stationary schemes in terms of finite matrices

In the stationary case the value $(S^\infty_{\mathbf{a}} \mathbf{f}^0)(x)$ for $x = \sum_{j=1}^\infty d_j 2^{-j} \in [0, 1)$, $d_j \in \{0, 1\}$ is given by

$$(S^\infty_{\mathbf{a}} \mathbf{f}^0)(x) = \lim_{k\to\infty} A_{d_k} \cdots A_{d_1} \mathbf{f}^0_{[0,1)} \tag{5.6}$$

with $\mathbf{f}^0_{[0,1)} = \{f^0_\alpha : \alpha \in I\}$.

Note that $S_{\mathbf{a}}$ is contractive if and only if the joint spectral radius of A_0, A_1, $\rho_\infty(A_0, A_1)$, is less than 1, where

$$\rho_\infty(A_0, A_1) = \sup_{k\in\mathbb{Z}_+\backslash 0} \left(\sup\left\{\|A_{\varepsilon_k}A_{\varepsilon_{k-1}}\cdots A_{\varepsilon_1}\|_\infty : \varepsilon_i \in \{0,1\}, i = 1,\ldots,k\right\}\right)^{\frac{1}{k}}. \quad (5.7)$$

Thus the conditions for convergence and smoothness of a stationary scheme given in Section 4.2, which can be expressed as the contractivity of a related scheme, can be formulated in terms of the joint spectral radius of two finite matrices. (See, for instance, Daubechies and Lagarias (1992*b*).) It is easy to conclude that $\rho_\infty(A_0, A_1) \geq \max\{\rho(A_0), \rho(A_1)\}$, where $\rho(A)$ is the spectral radius of the matrix A. From this inequality and from the necessity of the contractivity condition, we obtain necessary conditions for convergence and smoothness (for the latter only in case of L_∞-stability), which are easy to check.

Such necessary conditions are important in the design of new schemes, in the sense that 'bad' schemes can easily be excluded. For example, if $a(z) = \frac{(1+z)^2}{2}b(z)$, and $S_{\mathbf{a}}$ is an interpolatory scheme, then $\rho(B_0) < 1$, and $\rho(B_1) < 1$ (with B_0 and B_1 the local matrix operators corresponding to $S_{\mathbf{b}}$) are necessary for $S_{\mathbf{a}}$ to be C^1.

Here we formulate an open problem: What are the conditions for the contractivity of $S_{\{\mathbf{a}^k\}}$ in terms of the matrices $\{A_0^k, A_1^k : k \in \mathbb{Z}_+\}$?

5.3. L_p-convergence and p-smoothness of univariate stationary schemes in terms of finite matrices

There is a vast literature (see, *e.g.*, Villemoes (1994), Jia (1995, 1999), Ron and Shen (2000), Han (1998), Han and Jia (1998) and Han (2001), and references therein) on the convergence in the L_p-norm of subdivision schemes, and on the p-smoothness of refinable functions. One central method of analysis is in terms of the p-norm joint spectral radius of two operators restricted to a finite-dimensional space.

Let A_0, A_1, be matrices of order $n\times n$. Their p-norm joint spectral radius is

$$\rho_p(A_0, A_1) = \sup_{k\in\mathbb{Z}_+\backslash 0} \left(\left(\sum_{\varepsilon_1,\ldots,\varepsilon_k\in\{0,1\}} \|A_{\varepsilon_k}\cdots A_{\varepsilon_1}\|_p^p\right)^{\frac{1}{p}}\right)^{\frac{1}{k}}, \quad 1 \leq p < \infty.$$

For $\phi \in L_p(\mathbb{R})$ of compact support, the p-smoothness exponent is defined as

$$\nu_p(\phi) = \sup\{v \geq 0 : \|\Delta_h^n\phi\|_p \leq Ch^v\}$$

for some constant $C > 0$ and for sufficiently large n, where $\Delta_h\phi = \phi - \phi(\cdot - h)$ and $\Delta_h^n = \Delta_h\Delta_h^{n-1}$.

Here we bring one result from Jia (1995), which is in some sense an extension of Theorem 4.9 in Section 4.2.

Theorem 5.1. Let $\mathbf{a}$ be a finitely supported mask such that $\sum_{i\in\mathbb{Z}} a_i = 2$. Let $\phi_{\mathbf{a}}$ be a nontrivial solution of the refinement equation

$$\phi_{\mathbf{a}} = \sum_{i\in\mathbb{Z}} a_i \phi_{\mathbf{a}}(2\cdot -i).$$

If there exists $C > 0$ such that $\|\Delta R^n_{\mathbf{a}}\delta\|_p^{1/n} \leq C2^{1/p-\mu}$ for $0 < \mu \leq 1$ and $1 \leq p \leq \infty$, then $v_p(\phi_{\mathbf{a}}) = \mu$.

Since $\|\Delta R^n_{\mathbf{a}}\delta\|_p^{1/n} = \rho_p(A_0|_V, A_1|_V)$ with $V = \{\mathbf{u} \in \mathbb{R}^{|I|} : \sum_{i\in I} u_i = 0\}$ (Jia 1995), the condition of the above theorem can be formulated in terms of two finite-dimensional matrices, which are the restrictions of two operators to a finite-dimensional subspace. In Han (2001), an algorithm is presented for computing $v_2(\phi_{\mathbf{a}})$ efficiently, for $\phi_{\mathbf{a}}$ a multivariate refinable function corresponding to a dilation matrix M and a mask $\mathbf{a}$, both with the same symmetries. For symmetric interpolatory masks there is also an algorithm for the computation of $\nu_\infty(\phi_{\mathbf{a}})$. The situation in the multivariate case is much more complex: there are $|\det M|$ operators, and the finite-dimensional univariate subspace to which these operators are restricted is quite complicated.

6. Extraordinary point analysis

For all the types of subdivision schemes that are defined over nets of arbitrary topology, as described in Section 3.5, the refined nets are regular nets, excluding a fixed number of extraordinary (irregular) points of valency $\neq 6$, in the case of triangular nets, and of valency $\neq 4$, in the case of quadrilateral nets. The smoothness analysis of subdivision schemes over nets of arbitrary topology is thus decomposed into two stages. First, the analysis over the regular part is completed, using the tools described in Sections 4 and 5. After verifying the smoothness over the regular part, we are left with a finite number of isolated points of unknown regularity. The regularity analysis at the extraordinary points has been studied by several authors, starting with the pioneering eigenvalue analysis work by Doo and Sabin (1978), through the works by Ball and Storry (1988, 1989), and completed by Prautzsch (1998), Reif (1995) and Zorin (2000). It is based on the observation that the regularity of the surface is known over a ring of patches Q^k encircling the extraordinary point, and there is a linear transformation T mapping the ring of patches Q^k onto a refined ring of patches, Q^{k+1}. Figure 6.1 displays a graphical description of three rings of patches around a vertex of valency five. The rings, each composed of 15 quadrilaterals, are self-similar, of reducing sizes.

The closure of the union of these rings defines an extraordinary patch covering a 'hole' in the regular part of the surface, and the smoothness of such a patch is completely characterized by the transformation T. In the following we present the key ingredients of the smoothness analysis of such patches and the main results.

Let us denote the basic limit function of the subdivision on a regular net by ϕ. The ring of patches Q^k may be expressed in terms of the control points P^k influencing this ring. Let $P^k = \{P_1^k, P_2^k, \ldots, P_N^k\} \subset \mathbb{R}^3$ be the control points generating Q^k, and let the transformation T be the square matrix such that $P^{k+1} = TP^k$.

Each patch in the ring $Q_\ell^k \in Q^k$ is a parametric patch, triangular or quadrilateral, which is a linear combination of translations of $\phi(2^k\cdot)$ multiplying control points $\{P_r^k\}_{r\in I_\ell} \subset P^k$. In other words,

$$Q^k = \bigcup Q_\ell^k, \tag{6.1}$$

where

$$Q_\ell^k = \left\{ q_\ell^k(u,v) \equiv \sum_{r\in I_\ell} P_r^k \phi(2^k u - i_r, 2^k v - j_r) \mid (u,v) \in \Omega \right\} \tag{6.2}$$

for appropriate $\{i_r, j_r\}_{r\in I_\ell}$. $\Omega = \{(u,v) \mid 0 \le u, v \le 1\}$ for quad-meshes and $\Omega = \{(u,v) \mid 0 \le u, v \ \wedge \ u + v \le 1\}$ for triangular meshes.

Since the regularity of ϕ is assumed to be already known, it is clear that the behaviour at the extraordinary vertex is completely characterized by the matrix T. It is important to note that the conditions for regularity at the extraordinary vertex do not require the knowledge of the explicit formula of ϕ. Using a proper ordering of the points P^k (Doo and Sabin 1978), the matrix T is a block-circulant matrix. The eigenvalue analysis of this matrix

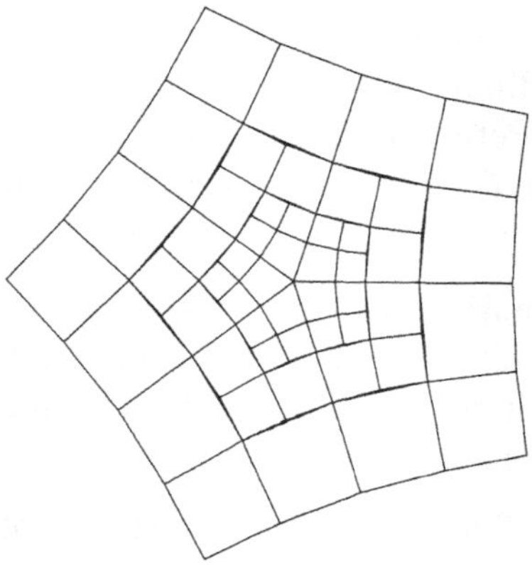

Figure 6.1. Three rings of patches

plays a crucial role in the smoothness analysis, as described in Doo and Sabin (1978), Reif (1995), Zorin (2000) and Prautzsch (1998). The results include necessary and sufficient conditions for geometric continuity, that is, existence of continuous limit normals at the extraordinary vertex and necessary and sufficient conditions for C^m-continuity at an extraordinary vertex – under some assumptions.

Let the eigenvalues $\lambda_0, \ldots, \lambda_{N-1}$ of T be ordered by modulus, that is,

$$|\lambda_0| \geq |\lambda_1| \geq \cdots \geq |\lambda_{N-1}|, \tag{6.3}$$

and let $V_0, V_1, \ldots, V_{N-1} \in \mathbb{R}^N$ denote the corresponding generalized real eigenvectors, assuming they exist.

As first shown in Doo and Sabin (1978), a necessary condition for the continuity of the normal at an extraordinary point is

$$\lambda_0 = 1 > |\lambda_1| = |\lambda_2| > |\lambda_3|, \quad V_0 = \{1, 1, \ldots, 1\}. \tag{6.4}$$

Assuming (6.4) holds, let us consider the particular initial data vector

$$P^0 = \{P_1^0, P_2^0, \ldots, P_N^0\}$$

with

$$P_j^0 = (V_{1,j}, V_{2,j}, 0)^t, \tag{6.5}$$

and let us examine the corresponding rings of patches defined by (6.2).

Injectivity and regularity assumption. We assume that each mapping q_ℓ^0 in (6.2) is regular and injective, and that

$$\bigcap_\ell \operatorname{int}\{Q_\ell^0\} = \emptyset. \tag{6.6}$$

In Reif (1995), the collection of mappings $\{q_\ell^0\}$ is termed the 'characteristic map' and the above assumption is thus referred to as the regularity and injectivity of the characteristic map. The importance of this map is that it defines the natural parametric domain for analysing the smoothness of the surface at the extraordinary vertex. For a discussion and analysis of the characteristic map see Peters and Reif (1998). Under the above assumption sufficient conditions for C^1 regularity are presented in the following result from Reif (1995).

Theorem 6.1. Let (6.3) hold with $\lambda_1 = \lambda_2$ being a real eigenvalue of T with geometric multiplicity 2, and let the characteristic map be regular and injective. Then the limit surface of the subdivision is a regular C^1-manifold in a neighbourhood of the extraordinary vertex for almost any initial data.

The necessary and sufficient conditions for C^m-continuity at an extraordinary vertex were derived independently by Prautzsch (1998) and Zorin (2000). These results are equivalent to the polynomial reproduction result for uniform stationary C^m-schemes on regular meshes.

Theorem 6.2. **(C^m-conditions)** Let the conditions of Theorem 6.1 hold. Then the limit surface of the subdivision is a regular C^m manifold in a neighbourhood of the extraordinary vertex for almost any initial data, if and only if the following condition holds.

For any eigenvalue λ of T satisfying $|\lambda| > \lambda_1^m$:

(a) $|\lambda| = \lambda_1^i$ for some integer $0 \le i \le m$;

(b) for the initial data vector $P^0 = \{P_1^0, P_2^0, \ldots, P_N^0\}$ with

$$P_j^0 = (V_{1,j}, V_{2,j}, V_j)^t \in \mathbb{R}^3, \tag{6.7}$$

and V an eigenvector corresponding to λ, all the patches Q_ℓ^0 lie on a polynomial surface $z = p(x,y)$ in $\mathbb{R}^3$, where p is a homogeneous polynomial of total degree i.

Theorem 6.2 does not give explicit constructive conditions that can help us to build a C^m-scheme. The translation of the conditions in Theorem 6.2 into algebraic conditions on the mask coefficients is rather complicated, and even in the C^2 case is not fully resolved. The partial results in this direction include the construction of schemes with bounded curvatures, in Loop (2001), and the special patch construction by Prautzsch and Umlauf (1998). For some applications it is enough to have curvature integrability of the subdivision surface. Reif and Schröder (2000) show that the Catmull–Clark and Loop schemes (among many others) have square integrable principal curvatures.

7. Limit values and approximation order

In this section we discuss two practical issues in the implementation of subdivision algorithms in geometric modelling. One issue is the computation of limit values and limit derivatives of the subdivision process at the dyadic points of any refinement level. The other important issue, though not yet widely appreciated, is how to actually attain the optimal approximation order for a given scheme: in other words, how to choose the initial control points so that the limit curve/surface will approximate a desired curve/surface with the highest possible approximation power.

7.1. Limit values and derivative values

We consider here only the stationary case, namely when $\mathbf{a}^k = \mathbf{a}$, $k \in \mathbb{Z}_+$, and assume that the basic limit function $\phi \equiv \phi_{\mathbf{a}}$ is C^m. The support of ϕ is contained in the convex hull of the support of the mask, $\sigma(\mathbf{a})$, by (2.13). Furthermore, by (2.10) we can express the limit function of $S_{\mathbf{a}}$ as

$$f \equiv S_{\mathbf{a}}^\infty \mathbf{f}^0 = \sum_{\alpha \in \mathbb{Z}^s} f_\alpha^0 \phi(\cdot - \alpha). \tag{7.1}$$

Thus, the limit values at the integer points $\beta \in \mathbb{Z}^s$ are given by

$$f(\beta) = \sum_{\alpha \in \mathbb{Z}^s} f^0_\alpha \phi(\beta - \alpha). \tag{7.2}$$

By (7.2), knowledge of the values of ϕ at integer points gives one the possibility of computing the limit values of the subdivision process on the integer grid $\mathbb{Z}^s$, using only the initial control points $\mathbf{f}^0$. Similarly, the limit values on the dyadic grid $2^{-k}\mathbb{Z}^s$ are defined by the control points $\mathbf{f}^k$ at level k. In the same way we note that the values of a derivative of f at the integers are linear combinations of the values of the same derivative of ϕ at the integers. The vector of values of ϕ, or of one of its derivatives, at the integer points may each be computed as the eigenvector of a finite matrix.

To see this we recall that ϕ satisfies the refinement equation (2.15), and thus

$$\partial_\lambda \phi = 2^{|\lambda|} \sum_\alpha a_\alpha \partial_\lambda \phi(2 \cdot - \alpha), \tag{7.3}$$

where $\lambda \in \mathbb{Z}^s_+$, $|\lambda| = \sum_{i=1}^s \lambda_i \leq m$. At integer points $\beta \in \mathbb{Z}^s$ we have the linear relations

$$\partial_\lambda \phi(\beta) = 2^{|\lambda|} \sum_\alpha a_\alpha \partial_\lambda \phi(2\beta - \alpha) = 2^{|\lambda|} \sum_\gamma a_{2\beta-\gamma} \partial_\lambda \phi(\gamma). \tag{7.4}$$

Now, since ϕ is of compact support, there is only a finite number N_ϕ of grid points where ϕ is nonzero. Let $\Omega \equiv \mathbb{Z}^s \bigcap \sigma(\phi)$; then $N_\phi = \#\Omega$. The system of equations (7.4), with $\beta \in \Omega$, is a square $N_\phi \times N_\phi$ eigensystem for the values $\{\partial_\lambda \phi(\beta)\}_{\beta \in \Omega}$, and it has a unique solution if we add the side conditions

$$\sum_{-\beta \in \Omega} \beta^\lambda \partial_\lambda \phi(-\beta) = \lambda!, \quad \sum_{-\beta \in \Omega} \beta^\mu \partial_\lambda \phi(-\beta) = 0, \quad \mu \neq \lambda, \quad |\mu| = |\lambda|. \tag{7.5}$$

These side conditions, in view of (7.2), guarantee that the $|\lambda|$ order derivatives of $S^\infty_{\mathbf{a}} x^\mu|_{\mathbb{Z}^s}$ are correctly obtained, for $|\mu| = |\lambda|$. For example, in the univariate case, the vector of values $\{\phi(\beta)\}$ is an eigenvector of the matrix U with elements $U_{i,j} = a_{2i-j}$, corresponding to the eigenvalue 1, and with the normalization $\sum \phi(\beta) = 1$. The vector of values $\{\phi'(\beta)\}$ is an eigenvector of U with eigenvalue 2. Implementing this, the rule for computing the limit derivatives of a curve defined by the 4-point scheme (3.18) turns out to be (Dyn *et al.* 1987):

$$f'(2^{-k}i) = \frac{2^k}{1-4w}\left[\frac{1}{2}(f^k_{i+1} - f^k_{i-1}) - w(f^k_{i+2} - f^k_{i-2})\right]. \tag{7.6}$$

The method for computing limit values is actually applied to non-interpolatory subdivision surfaces, so that at all refinement levels the rendered points are on the limit surface. The shading of the surface at each level is done with normals which are the actual normals of the limit surface. A detailed example of computing limit normals at regular points and at extraordinary points for the case of the butterfly scheme is given in Shenkman (1996) and Dyn, Levin and Shenkman (1999*b*).

7.2. *Attaining the optimal approximation order*

The term *approximation order* of a subdivision scheme $S_{\mathbf{a}}$ refers to the rate by which the limit functions generated by $S_{\mathbf{a}}$, from initial data sampled from a sufficiently smooth function f, get closer to f: in other words, the largest exponent r such that

$$\|f - S_{\mathbf{a}}^{\infty} f|_{h\mathbb{Z}^s}\|_{\infty} \leq ch^r.$$

Yet this order may be improved (for non-interpolatory schemes) by replacing the initial data $f|_{h\mathbb{Z}^s}$ by $Qf|_{h\mathbb{Z}^s}$ with Q a Toeplitz operator of finite support. Our aim is to find the operator Q that yields the largest approximation rate.

Let us start with an example.

Example 9. Let us consider the case of univariate cubic B-splines with integer knots. It is known that the integer shifts of this cubic B-spline, B_3, span π_3, and this implies that the space generated by the integer shifts of the cubic B-spline has potential approximation order 4. If $f \in C^4(\mathbb{R})$, then the use of function values as control points gives a second-order approximation, by the corresponding subdivision scheme

$$\left| f(x) - \sum_{j\in\mathbb{Z}} f(jh) B_3\left(\frac{x}{h} - j\right)\right| \leq c_2 h^2. \tag{7.7}$$

However, using as control points the values

$$\tilde{f}_j = (Q_h f)(jh) \equiv -\frac{1}{6} f((j-1)h) + \frac{4}{3} f(jh) - \frac{1}{6} f((j+1)h), \tag{7.8}$$

we get the optimal fourth-order approximation:

$$\left| f(x) - \sum_{j\in\mathbb{Z}} \tilde{f}_j B_3\left(\frac{x}{h} - j\right)\right| \leq c_4 h^4. \tag{7.9}$$

This special choice of Q_h is made so that the approximation scheme in (7.9) reproduces all polynomials in $\pi_3(\mathbb{R})$, namely, $\sum (Q_h p)(jh) B_3(\frac{x}{h} - j) = p(x)$ for any $p \in \pi_3(\mathbb{R})$. Therefore, to approximate a curve $c(t)$ by a cubic spline subdivision, given a sequence of points $\{P_j\}$ ordered on it, then it is better to start the subdivision process with the control points

$$\widetilde{P}_j = -\frac{1}{6} P_{j-1} + \frac{4}{3} P_j - \frac{1}{6} P_{j+1}. \tag{7.10}$$

The above idea is extended for general subdivision schemes in A. Levin (1999*c*).

For a given uniform stationary scheme $S_{\mathbf{a}}$ we identify the maximal m such that $\pi_m(\mathbb{R}^s)$ is invariant under $S_{\mathbf{a}}$ in the sense that $S_{\mathbf{a}}^{\infty} p|_{\mathbb{Z}^s} \in \pi_m(\mathbb{R}^s)$ for any $p \in \pi_m(\mathbb{R}^s)$. Then, the potential approximation order is $m+1$. To achieve this approximation power we look for a Toeplitz operator Q, of minimal support Σ, of the form

$$(Q\mathbf{f}^0)_\alpha = \sum_{\sigma \in \Sigma} q_\sigma f^0_{\alpha-\sigma}, \tag{7.11}$$

such that

$$S_{\mathbf{a}}^{\infty} Q(p|_{\mathbb{Z}^s}) = p, \quad \forall p \in \pi_m(\mathbb{R}^s). \tag{7.12}$$

In other words, Q is the inverse of $S_{\mathbf{a}}^{\infty}$ on $\pi_m(\mathbb{R}^s)$. If Q exists then it commutes with $S_{\mathbf{a}}^{\infty}$ on π_m. Therefore, we look for Q such that $QS_{\mathbf{a}}^{\infty} p|_{\mathbb{Z}^s} = p, \quad \forall p \in \pi_m(\mathbb{R}^s)$. Using the results of Section 7.1 we can define the polynomials

$$r_\gamma \equiv S_{\mathbf{a}}^{\infty}\{x^\gamma|_{\mathbb{Z}^s}\} = \sum_{\alpha \in \mathbb{Z}^s} \phi(\alpha)(\cdot - \alpha)^\gamma, \quad \gamma \in \mathbb{Z}^s, \quad |\gamma| \le m,$$

which constitute a basis of π_m. Now we look for an operator Q such that on π_m it is the inverse of $S_{\mathbf{a}}^{\infty}$, namely,

$$Qr_\gamma = x^\gamma, \quad \gamma \in \mathbb{Z}^s, \quad |\gamma| \le m. \tag{7.13}$$

This can be formulated as a system of linear equations in the finite-dimensional space π_m,

$$\sum_{\sigma \in \Sigma} q_\sigma r_\gamma(x - \sigma) = x^\gamma, \quad |\gamma| \le m. \tag{7.14}$$

In the above example of the cubic B-splines, the operator Q may also be chosen to be $Qf = f - \frac{1}{6}f''$, or to be the difference operator given in (7.8). The two options act in the same way on π_3, yet, for the purpose of applying Q on the given data points we need the discrete form (7.8). For further examples and applications see A. Levin (1999*c*).

Acknowledgements

The authors wish to thank Nurit Alkalai and Adi Levin for providing the figures, and to Bin Han and Peter Schröder for their help with the references. This work was supported in part by a grant from the Israeli Academy of Sciences (Center of Excellence program), and by the European Union research project 'Multiresolution in Geometric Modelling (MINGLE)' under grant HPRN-CT-1999-00117.

REFERENCES

A. A. Ball and D. J. T. Storry (1988), 'Conditions for tangent plane continuity over recursively generated B-spline surfaces', *ACM Trans. Graphics* **7**, 83–102.

A. A. Ball and D. J. T. Storry (1989), 'Design of an n-sided surface patch', *Computer Aided Geometric Design* **6**, 111–120.

C. de Boor (1987), 'Cutting corners always works', *Computer Aided Geometric Design* **4**, 125–131.

C. de Boor, K. Höllig and S. Riemenschneider (1993), *Box Splines*, Vol. 98 of *Applied Mathematical Sciences*, Springer.

E. Catmull and J. Clark (1978), 'Recursively generated B-spline surfaces on arbitrary topological meshes', *Computer Aided Design* **10**, 350–355.

A. S. Cavaretta, W. Dahmen and C. A. Micchelli (1991), *Stationary Subdivision*, Vol. 453 of *Memoirs of AMS*, American Mathematical Society.

G. M. Chaikin (1974), 'An algorithm for high speed curve generation', *Computer Graphics and Image Processing* **3**, 346–349.

A. Cohen and Conze (1992), 'Regularité des bases d'ondelettes et mesures ergodiques', *Rev. Math. Iberoamer* **8**, 351–366.

A. Cohen and N. Dyn (1996), 'Nonstationary subdivision schemes and multiresolution analysis', *SIAM J. Math. Anal.* **26**, 1745–1769.

A. Cohen, I. Daubechies and G. Plonka (1997), 'Regularity of refinable function vectors', *J. Fourier Anal. Appl.* **3**, 295–324.

A. Cohen, N. Dyn and D. Levin (1996), Stability and inter-dependence of matrix subdivision schemes, in *Advanced Topics in Multivariate Approximation* (F. Fontanella, K. Jetter and P. J. Laurent, eds), World Scientific, pp. 33–45.

A. Cohen, N. Dyn and B. Matei (2001), Quasilinear subdivision schemes with applications to ENO interpolation. Submitted.

E. Cohen, T. Lyche and R. F. Riesenfeld (1980), 'Discrete B-splines and subdivision techniques in computer-aided geometric design and computer graphics', *Computer Graphics and Image Processing* **14**, 87–111.

W. Dahmen and C. A. Micchelli (1984), 'Subdivision algorithms for the generation of box spline surfaces', *Computer Aided Geometric Design* **1**, 115–129.

W. Dahmen and C. A. Micchelli (1997), 'Biorthogonal wavelet expansion', *Constr. Approx.* **13**, 294–328.

I. Daubechies (1992), *Ten Lectures on Wavelets*, SIAM, Philadelphia.

I. Daubechies and J. C. Lagarias (1992*a*), 'Two-scale difference equations, I: Existence and global regularity of solutions', *SIAM J. Math. Anal.* **22**, 1388–1410.

I. Daubechies and J. C. Lagarias (1992*b*), 'Two-scale difference equations, II: Local regularity, infinite products of matrices and fractals', *SIAM J. Math. Anal.* **23**, 1031–1079.

I. Daubechies, I. Guskov and W. Sweldens (1999), 'Regularity of irregular subdivision', *Constr. Approx.* **15**, 381–426.

S. Dekel and N. Dyn (2001), 'Polyscale subdivision schemes and refinability', *Appl. Comput. Harm. Anal.* To appear.

G. Derfel, N. Dyn and D. Levin (1995), 'Generalized refinement equations and subdivision processes', *J. Approx. Theory* **80**, 272–297.

T. DeRose, M. Kass and T. Truong (1998), Subdivision surfaces in character animation, in *Proc. SIGGRAPH 98*, Annual Conference Series, ACM–SIGGRAPH, pp. 85–94.

G. Deslauriers and S. Dubuc (1989), 'Symmetric iterative interpolation', *Constr. Approx.* **5**, 49–68.

D. Donoho and V. Stodden (2001), Multiplicative multiresolution analysis for lie-group valued data indexed by Euclidean parameter. In preparation.

D. Donoho, N. Dyn, D. Levin and T. Yu (2000), 'Smooth multiwavelet duals of Alpert bases by moment-interpolating refinement', *Appl. Comput. Harm. Anal.* **9**, 166–203.

D. Doo and M. Sabin (1978), 'Behaviour of recursive division surface near extraordinary points', *Computer Aided Design* **10**, 356–360.

S. Dubuc (1986), 'Interpolation through an iterative scheme', *J. Math. Anal. Appl.* **114**, 185–204.

N. Dyn (1992), Subdivision schemes in computer aided geometric design, in *Advances in Numerical Analysis II: Subdivision Algorithms and Radial Functions* (W. A. Light, ed.), Oxford University Press, pp. 36–104.

N. Dyn and E. Farkhi (2000), 'Spline subdivision schemes for convex compact sets', *J. Comput. Appl. Math.* **119**, 133–144.

N. Dyn and E. Farkhi (2001*a*), Convexification rates in Minkowski averaging processes. Submitted.

N. Dyn and E. Farkhi (2001*b*), Spline subdivision schemes for compact sets with metric averages, in *Trends in Approximation Theory* (T. L. K. Kopotun and M. Neamtu, eds), Vanderbilt University Press, pp. 93–102.

N. Dyn and D. Levin (1990), Interpolatory subdivision schemes for the generation of curves and surfaces, in *Multivariate Approximation and Interpolation* (W. Haussmann and K. Jetter, eds), Birkhäuser, Basel, pp. 91–106.

N. Dyn and D. Levin (1992), Stationary and non-stationary binary subdivision schemes, in *Mathematical Methods in Computer Aided Geometric Design II* (T. Lyche, and L. L. Schumaker, eds), Academic Press, pp. 209–216.

N. Dyn and D. Levin (1995), 'Analysis of asymptotically equivalent binary subdivision schemes', *J. Math. Anal. Appl.* **193**, 594–621.

N. Dyn and D. Levin (1999), Analysis of Hermite-interpolatory subdivision schemes, in *CRM Proceedings and Lecture Notes*, Vol. 18, Centre de Recherches Mathématiques, pp. 105–113.

N. Dyn and D. Levin (2002), Matrix subdivision: Analysis by factorization, in *Approximation Theory: A Volume Dedicated to Blagovest Sendov* (B. Bojanov, ed.), Darba, Sofia, pp 187–211.

N. Dyn and A. Ron (1995), 'Multiresolution analysis by infinitely differentiable compactly supported functions', *Appl. Comput. Harm. Anal.* **2**, 15–20.

N. Dyn, J. A. Gregory and D. Levin (1987), 'A four-point interpolatory subdivision scheme for curve design', *Computer Aided Geometric Design* **4**, 257–268.

N. Dyn, J. A. Gregory and D. Levin (1990*a*), 'A butterfly subdivision scheme for surface interpolation with tension control', *ACM Trans. Graphics* **9**, 160–169.

N. Dyn, J. A. Gregory and D. Levin (1991), 'Analysis of uniform binary subdivision schemes for curve design', *Constr. Approx.* **7**, 127–147.

N. Dyn, J. A. Gregory and D. Levin (1995), Piecewise uniform subdivision schemes, in *Mathematical Methods for Curves and Surfaces* (M. Dahlen, T. Lyche and L. L. Schumaker, eds), Vanderbilt University Press, Nashville, pp. 111–120.

N. Dyn, S. Hed and D. Levin (1993), Subdivision schemes for surface interpolation, in *Workshop on Computational Geometry* (A. Conte *et al.*, eds), World Scientific, pp. 97–118.

N. Dyn, F. Kuijt, D. Levin and R. van Damme (1999*a*), 'Convexity preservation of the four-point interpolatory subdivision scheme', *Computer Aided Geometric Design* **16**, 789–792.

N. Dyn, D. Levin and D. Liu (1992), 'Interpolatory convexity preserving subdivision schemes for curves and surfaces', *Computer Aided Design* **24**, 211–216.

N. Dyn, D. Levin and A. Luzzatto (2001*a*), Non-stationary interpolatory subdivision schemes reproducing spaces of exponential polynomials. Submitted.

N. Dyn, D. Levin and C. A. Micchelli (1990*b*), 'Using parameters to increase smoothness of curves and surfaces generated by subdivision', *Computer Aided Geometric Design* **7**, 129–140.

N. Dyn, D. Levin and P. Shenkman (1999*b*), 'Normals of the butterfly scheme surfaces and their applications', *J. Comput. Appl. Math.* **102**, 157–180.

N. Dyn, D. Levin and J. Simoens (2001*b*), Face value subdivision schemes on triangulations by repeated averaging. Preprint.

J. Gregory (1991), An introduction to bivariate uniform subdivision, in *Numerical Analysis 1991* (D. Griffiths and G. Watson, eds), Pitman Research Notes in Mathematics, Longman Scientific and Technical, pp. 103–117.

J. A. Gregory and R. Qu (1996), 'Non-uniform corner cutting', *Computer Aided Geometric Design* **13**, 763–772.

I. Guskov (1998), Multivariate subdivision schemes and divided differences. Technical report, Princeton University.

M. Halstead, M. Kass and T. DeRose (1993), Efficient, fair interpolation using Catmull–Clark surfaces, in *Proc. SIGGRAPH 93*, Annual Conference Series, ACM–SIGGRAPH, pp. 35–44.

B. Han (1998), 'Symmetric orthonormal scaling functions and wavelets with dilation factor 4', *Adv. Comput. Math.* **8**, 221–247.

B. Han (2001), Computing the smoothness exponent of a symmetric multivariate refinable function. Preprint.

B. Han and R. Q. Jia (1998), 'Multivariate refinement equations and convergence of subdivision schemes', *SIAM J. Math. Anal.* **29**, 1177–1199.

S. Hed (1990), Analysis of subdivision schemes for surfaces. Master's thesis, Tel Aviv University.

H. Hoppe, T. DeRose, T. Duchamp, M. Halstead, H. Jin, J. McDonald, J. Schweitzer and W. Stuetzle (1994), 'Piecewise smooth surface reconstruction', *Computer Graphics* **28**, 295–302.

R. Q. Jia (1995), 'Subdivision schemes in l_p spaces', *Adv. Comput. Math.* **3**, 309–341.

R. Q. Jia (1996), The subdivision and transition operators associated with a refinement equation, in *Advanced Topics in Multivariate Approximation* (K. J. F. Fontanella and L. Schumaker, eds), World Scientific, pp. 1–13.

R. Q. Jia (1999), 'Characterization of smoothness of multivariate refinable functions in Sobolev spaces', *Trans. Amer. Math. Soc.* **351**, 4089–4112.

R. Q. Jia and S. Zhang (1999), 'Spectral properties of the transition operator associated to a multivariate refinement equation', *Lin. Alg. Appl.* **292**, 155–178.

Y. Kasas (1990), A subdivision-based algorithm for surface/surface intersection. Master's thesis, Tel Aviv University.

L. Kobbelt (1996*a*), 'Interpolatory subdivision on open quadrilateral nets with arbitrary topology', *Computer Graphics Forum* **15**, 409–420.

L. Kobbelt (1996*b*), 'A variational approach to subdivision', *Computer Aided Geometric Design* **13**, 743–761.

L. Kobbelt, T. Hesse, H. Prautzsch and K. Schweizerhof (1996), 'Interpolatory subdivision on open quadrilateral nets with arbitrary topology', *Computer Graphics Forum* **15**, 409–420. *Eurographics '96* issue.

F. Kuijt and R. van Damme (1998), 'Convexity preserving interpolatory subdivision schemes', *Constr. Approx.* **14**, 609–630.

F. Kuijt and R. van Damme (1999), 'Monotonicity preserving interpolatory subdivision schemes', *J. Comput. Appl. Math.* **101**, 203–229.

F. Kuijt and R. van Damme (2002), 'Shape preserving interpolatory subdivision schemes for nonuniform data', *J. Approx. Theory.* To appear.

O. Labkovsky (1996), The extended butterfly interpolatory subdivision scheme for the generation of C^2 surfaces. Master's thesis, Tel Aviv University.

A. Levin (1999*a*), Analysis of quasi-uniform subdivision schemes. In preparation.

A. Levin (1999*b*), 'Combined subdivision schemes for the design of surfaces satisfying boundary conditions', *Computer Aided Geometric Design* **16**, 345–354.

A. Levin (1999*c*), Combined subdivision schemes with applications to surface design. PhD thesis, Tel Aviv University.

A. Levin (1999*d*), Interpolating nets of curves by smooth subdivision surfaces, in *Proc. SIGGRAPH 99*, Annual Conference Series, ACM–SIGGRAPH, pp. 57–64.

D. Levin (1999*e*), 'Using Laurent polynomial representation for the analysis of non-uniform binary subdivision schemes', *Adv. Comput. Math.* **11**, 41–54.

C. Loop (1987), Smooth spline surfaces based on triangles. Master's thesis, University of Utah, Department of Mathematics.

C. Loop (2001), Triangle mesh subdivision with bounded curvature and the convex hull property. Technical Report MSR-TR-2001-24, Microsoft Research.

S. Mallat (1989), 'Theory for multiresolution signal decomposition: The wavelet representation', *IEEE Trans. Pattern Anal. Mach. Intel.* **11**, 674–693.

J. L. Merrien (1992), 'A family of Hermite interpolants by bisection algorithms', *Numer. Alg.* **2**, 187–200.

C. A. Micchelli and H. Prautzsch (1989), 'Uniform refinement of curves', *Lin. Alg. Appl.* **114/115**, 841–870.

C. A. Micchelli and T. Sauer (1998), 'On vector subdivision', *Math. Z.* **229**, 621–674.

G. Morin, J. Warren and H. Weimer (2001), 'A subdivision scheme for surfaces of revolution', *Computer Aided Geometric Design* **18**, 483–502.

A. H. Nasri (1997*a*), 'Curve interpolation in recursively generated B-spline surfaces over arbitrary topology', *Computer Aided Geometric Design* **14**, 13–30.

A. H. Nasri (1997*b*), Interpolation of open curves by recursive subdivision surface, in *The Mathematics of Surfaces VII* (T. Goodman and R. Martin, eds), Information Geometers, pp. 173–188.

J. Peters and U. Reif (1997), 'The simplest subdivision scheme for smoothing polyhedra', *ACM Trans. Graphics* **16**, 420–431.

J. Peters and U. Reif (1998), 'Analysis of algorithms generating B-spline subdivision', *SIAM J. Numer. Anal.* **35**, 728–748.

G. Plonka (1997), Approximation order provided by refinable function vectors, *Constr. Approx.* **13** 221–244.

H. Prautzsch (1998), 'Smoothness of subdivision surfaces at extraordinary points', *Adv. Comput. Math.* **9**, 377–389.

H. Prautzsch and G. Umlauf (1998), Improved triangular subdivision schemes, in *Computer Graphics International 1998* (F. E. Wolter and N. M. Patrikalakis, eds), IEEE Computer Society, pp. 626–632.

G. de Rahm (1956), 'Sur une courbe plane', *J. Math. Pures Appl.* **35**, 25–42.

U. Reif (1995), 'A unified approach to subdivision algorithms near extraordinary points', *Computer Aided Geometric Design* **12**, 153–174.

U. Reif and P. Schröder (2000), 'Curvature integrability of subdivision surfaces', *Adv. Comput. Math.* **12**, 1–18.

D. Riemenschneider and Z. Shen (1997), 'Multidimensional interpolatory subdivision schemes', *SIAM J. Numer. Anal.* **34**, 2357–2381.

R. F. Riesenfeld (1975), 'On Chaikin's algorithm', *Computer Graphics and Image Processing* **4**, 304–310.

O. Rioul (1992), 'Simple regularity criteria for subdivision schemes', *SIAM J. Math. Anal.* **23**, 1544–1576.

A. Ron and Z. Shen (2000), 'The Sobolev regularity of refinable functions', *J. Approx. Theory* **106**, 185–225.

V. A. Rvachev (1990), 'Compactly supported solutions of functional–differential equations and their applications', *Russian Math. Surveys* **45**, 87–120.

P. Schröder (2001), Subdivision, multiresolution and the construction of scalable algorithms in computer graphics, in *Multivariate Approximation and Applications*, Cambridge University Press, pp. 213–251.

L. L. Schumaker (1980), *Spline Functions: Basic Theory*, Wiley-Interscience.

P. Shenkman (1996), Computing normals and offsets of curves and surfaces generated by subdivision schemes. Master's thesis, Tel Aviv university.

L. F. Villemoes (1994), 'Wavelet analysis of refinement equations', *SIAM J. Math. Anal.* **25**, 1433–1460.

J. Warren (1995*a*), Binary subdivision schemes for functions over irregular knot sequences, in *Mathematical Methods in CAGD III* (M. Dahlen, T. Lyche and L. L. Schumaker, eds), Vanderbilt University Press, Nashville.

J. Warren (1995*b*), *Subdivision Methods for Geometric Design*, Rice University.

D. Zorin (2000), 'Smoothness of subdivision on irregular meshes', *Constr. Approx.* **16**, 359–397.

D. Zorin and P. Schröder (2000), *Subdivision for Modeling and Animation*, Course Notes, ACM–SIGGRAPH.

D. Zorin, H. Biermann and A. Levin (2000), Piecewise smooth subdivision surfaces with normal control, in *Proc. SIGGRAPH 2000*, Annual Conference Series, pp. 113–120.

D. Zorin, P. Schröder and W. Sweldens (1996), Interpolating subdivision for meshes with arbitrary topology, in *Proc. SIGGRAPH 96*, Annual Conference Series, ACM–SIGGRAPH, pp. 189–192.

D. Zorin, P. Schröder and W. Sweldens (1997), Interactive multiresolution mesh editing, in *Proc. SIGGRAPH 97*, Annual Conference Series, ACM–SIGGRAPH, pp. 259–268.

Acta Numerica (2002), pp. 145–236
DOI: 10.1017/S096249290200003X

Adjoint methods for PDEs: *a posteriori* error analysis and postprocessing by duality

Michael B. Giles and Endre Süli
University of Oxford, Computing Laboratory,
Wolfson Building, Parks Road,
Oxford OX1 3QD, England

We give an overview of recent developments concerning the use of adjoint methods in two areas: the *a posteriori* error analysis of finite element methods for the numerical solution of partial differential equations where the quantity of interest is a functional of the solution, and superconvergent extraction of integral functionals by postprocessing.

CONTENTS

1. Introduction

Output functionals

In many scientific and engineering applications that lead to the numerical approximation of solutions to partial differential equations, the objective is merely a rough, qualitative assessment of the details of the analytical solution over the computational domain, the quantitative concern being directed towards a few *output functionals*, derived quantities of particular engineering or scientific relevance.

For example, in aeronautical engineering, a CFD calculation of the flow around a transport aircraft at cruise conditions might be performed to investigate whether there are any unexpected shocks on the pylon connecting the engine to the wing, or whether there is an unexpected boundary layer separation caused by the main shock on the wing's suction surface. However, the engineer's overall concern is the impact of such phenomena on the lift and drag on the aircraft, and the quality of the CFD calculation is judged, first and foremost, by the accuracy of the lift and drag predictions. The fine details of the flow field are much less important, and are used only in a qualitative manner to suggest ways in which the design may be modified to improve the lift or drag. This focus on a few output quantities is even clearer in design optimization, when one is trying to maximize or minimize a single objective function, possibly subject to a number of constraints.

Engineering interest in output functionals arises in many applications of CFD. Occasionally, volume integrals are of importance: one example is the infrared signature of a military aircraft, which will depend in part on a volume integral of some function of the temperature in the thermal wake behind the aircraft. However, usually it is surface integrals that are of most concern, as with lift and drag. Other examples in CFD analysis include: the mass flow through a turbomachine; the total heat flux into a turbine blade from the surrounding flow; the total production of nitrous oxides in combustion modelling; the net seepage of a pollutant into an aquifer when modelling soil contamination.

Integral quantities are important in other disciplines as well. In electrochemical simulations of the behaviour of sensors, the quantity of interest is the total current flowing into an electrode (Alden and Compton 1997). In electromagnetics, radar cross-section calculations are concerned with the scattered field emanating from an aircraft. The amplitude of the wave propagating in a particular direction can be evaluated by a convolution integral over a closed surface surrounding the aircraft (Colton and Kress 1991, Monk and Süli 1998). Similar convolution integrals are used in the analysis of multi-port electromagnetic devices, such as microwave ovens and EMR body scanners, to evaluate radiation, transmission and reflection coefficients which characterize the behaviour of the device.

In structural mechanics, one is sometimes concerned with the total force or moment exerted on a surface (Peraire and Patera 1997), but more often the focus of interest is a point functional, such as the maximum stress or temperature. Indeed, as integral quantities can be approximated with much greater accuracy than point functionals, various techniques have been developed to represent point quantities by 'equivalent' integral quantities (see, for example, Babuška and Miller (1984*a*)). In fact, the applicability of these techniques extends beyond structural engineering to other areas where the accurate evaluation of point quantities, rather than integral functionals, is of concern.

The purpose of this article is to explore the question of accurate approximation of output functionals through the use of adjoint problems and duality arguments. As a first step in this direction, we analyse the errors committed in the numerical approximation of linear functionals using an appropriately defined *adjoint* or *dual* problem; hence, we shall quantify the relationship between residual errors in the discretization and the corresponding error in the approximation of the functional. With this information, we then look at ways of improving the accuracy of the computed value of the functional, either through correcting the leading order error in the functional approximation, or through an adaptive mesh refinement algorithm that stems from a residual-based *a posteriori* error bound, aiming to produce the most accurate functional value at a given computational cost.

Partial differential operators and adjoint equations

The application of duality arguments in the theory of differential and integral equations has a long and distinguished history, including the work of Frobenius on two-point boundary value problems, the construction of a Riemann function for a second-order linear hyperbolic partial differential operator, Holmgren's theorem concerning the uniqueness of solutions to parabolic partial differential equations, and Fredholm's theory of integral equations.

The purpose of this section is to highlight, in nonrigorous terms, the intimate connection between adjoint problems and output functionals in the context of partial differential equations. Let us suppose that L is a scalar linear partial differential operator with constant coefficients, and for a sufficiently smooth function f defined over a domain $\Omega \subset \mathbb{R}^n$, let us consider the equation

$$Lu = f \qquad \text{in } \Omega,$$

subject to (unspecified) homogeneous boundary conditions on $\partial\Omega$. Assuming that the Green's function $G : (x, y) \in \Omega \times \Omega \mapsto \mathbb{R}$ of L exists, the solution u can be expressed as

$$u(x) = \int_\Omega G(x, y)\, f(y)\, \mathrm{d}y,$$

where $G(x, y)$ satisfies, for every $x \in \Omega$, the partial differential equation

$$L_y^* G = \delta(x - y), \qquad y \in \Omega,$$

subject to appropriate homogeneous boundary conditions; here L^* denotes the formal adjoint of L, and the subscript y in L_y^* indicates that the partial derivatives are taken with respect to the independent variable $y \in \Omega$.

Now, suppose that one is interested in computing $J(u)$, where $w \mapsto J(w)$ is the linear functional defined by

$$J(w) = \int_\Omega g(x)\, w(x)\, \mathrm{d}x,$$

with g a given, sufficiently smooth, weight function. Clearly, on interchanging the order of integration,

$$J(u) = \int_\Omega \int_\Omega g(x)\, G(x,y)\, f(y)\, \mathrm{d}x\, \mathrm{d}y = \int_\Omega v(y)\, f(y)\, \mathrm{d}y,$$

where we have defined v by

$$v(y) = \int_\Omega g(x)\, G(x,y)\, \mathrm{d}x.$$

Now, v obeys the adjoint partial differential equation

$$L_y^* v = g,$$

subject to appropriate homogeneous boundary conditions.

We see from this discussion that the functional value $J(u)$ may be computed without prior knowledge of u, simply by integrating the 'adjoint/dual solution' v against the forcing function f of the original problem. Indeed, the adjoint solution v may be thought of as a measure of sensitivity of the output functional J to perturbations in the data, f. These simple observations have some far-reaching consequences which will be explored in detail in Section 2.

Adjoint equations in engineering and science

The use of adjoint equations is long-established in optimal control theory (Lions 1971). In the simplest case, one has a control system in which an output $y(t)$ is related to a control input $u(t)$ through a scalar linear ordinary differential equation of the form

$$Ly = u,$$

subject to appropriate boundary conditions. The objective is to choose the control input $u(t)$ to achieve a specified state $y(T)$ at time T, while minimizing the integral of the square of the input.

Using calculus of variations, it can be shown that the optimal input must satisfy the adjoint equation

$$L^* u = 0,$$

subject to appropriate boundary conditions.

The idea of using adjoint equations for design optimization in the context of fluid dynamics was pioneered by Pironneau (1974) but, within the field of aeronautical engineering, the adjoint approach to computing design sensitivities has been primarily developed by Jameson, starting with the potential flow equations (Jameson 1988), and then the Euler equations (Jameson 1995), before proceeding to the Navier–Stokes equations (Jameson 1999, Jameson, Pierce and Martinelli 1998). An overview of recent developments in adjoint design methods for aeronautical applications is provided by Newman, Taylor, Barnwell, Newman and Hou (1999).

In studies of turbulent flow, adjoint equations have been used to investigate the active control of turbulent boundary layers to reduce drag through active re-laminarization (Bewley 2001). They have also been applied to the study of the most unstable modes which lead to the initial onset of turbulence (Airiau 2001).

In weather prediction, adjoint equations are used for a process known as *data assimilation* (Talagrand and Courtier 1997). Due to the chaotic nature of high Reynolds number fluid flow, weather prediction is very sensitive to the initial conditions specified. The idea in data assimilation is to adjust the initial conditions to improve the agreement with a limited number of subsequent measurements. As with engineering design optimization, this is essentially an optimization task, and adjoint solutions are used to find the sensitivity of the objective function, in this case the mismatch between the model and the experimental data, to changes in the initial data.

For further examples of the use of adjoint methods, see the recent special issue of the journal *Flow, Turbulence and Combustion* (Bottaro, Mauss and Henningson, eds 2001) which was devoted to a variety of applications in fluid dynamics.

Adjoint equations in numerical analysis

The use of adjoint equations and duality arguments has also penetrated the field of numerical analysis of partial differential equations. In the subject of *a priori* error estimation, these ideas can be traced back to the work of Aubin (1967), Nitsche (1968) and Oganesjan and Ruhovec (1969).

In the subject of residual-based *a posteriori* error estimation, the application of duality arguments in much more recent. The relevance of duality arguments in *a posteriori* error estimation has been highlighted in the review articles by Eriksson, Estep, Hansbo and Johnson (1996) and Becker and Rannacher (2001) (see also Becker and Rannacher (1996), Hansbo and Johnson (1991), Houston, Rannacher and Süli (2000*a*), Houston and Süli (2001*a*), Larson and Barth (2000), Melenk and Schwab (1999), Oden and Prudhomme (1999), Paraschivoiu, Peraire and Patera (1997), Peraire and Patera (1997), Rannacher (1998), Süli (1998), Süli, Houston and Schwab

(1999) and Houston and Süli (2002*b*)); concerning the use of duality arguments in post-processing and design, we refer to Giles, Larsson, Levenstam and Süli (1997), Giles and Pierce (1997, 1999), Giles (2000) and Pierce and Giles (1998), and references therein. The key ingredient in duality-based error estimation is an auxiliary PDE problem, the dual problem, involving the formal adjoint of the linear partial differential operator under consideration. The data for the dual problem is the quantity of interest: in engineering applications, this is typically an output functional of the analytical solution (Becker and Rannacher 1996, Becker and Rannacher 2001, Giles *et al.* 1997, Giles and Pierce 1997, Giles and Pierce 1999, Giles 2000, Larson and Barth 2000, Oden and Prudhomme 1999, Paraschivoiu *et al.* 1997, Peraire and Patera 1997, Pierce and Giles 1998).

The relevance and generality of duality-based error estimation has been powerfully argued in the work of Johnson and his collaborators; see, for example, Eriksson *et al.* (1996) for an excellent survey. The *a posteriori* error bounds resulting from this analysis involve the finite element residual which is obtained by inserting the computed finite element solution into the partial differential equation; the residual measures the extent to which the finite element approximation to the analytical solution fails to satisfy the PDE. In the framework of Eriksson *et al.* (1996), the error bounds are arrived at by exploiting Galerkin orthogonality (a fundamental property of all finite element methods expressing the fact that the residual is orthogonal to the finite element space), and strong stability (well-posedness/regularity in isotropic Sobolev norms of positive index) of the dual problem.

An overview of the paper

In this article, we are fundamentally interested in the same subject as Becker and Rannacher (2001), the numerical analysis of errors in output functionals. However, while Becker and Rannacher are concerned with Galerkin finite element methods with orthogonality between the residual errors in the primal problem and the trial space for the dual problem, here we shall also discuss discretization methods, such as finite volume methods, which may lack this orthogonality property.

We begin by introducing the notion of linear primal and dual problems in an abstract weak formulation, and prove their equivalence in the Primal–Dual Equivalence Theorem. In Section 3 we show that this equivalence can be maintained in a Galerkin finite element discretization which retains orthogonality between the residual errors of one problem, and the trial space of the other. However, if one uses a Galerkin discretization with entirely different spaces for the primal and dual problems, the equivalence is lost.

Section 4 explores general discretizations, in the absence of Galerkin orthogonality. In this case it is shown how one may evaluate an adjoint error correction which, to leading order, corrects the error in the computed value for the functional. Applying this technique to smoothly reconstructed solutions from finite difference or finite element approximations, one can establish an order of convergence for the corrected functional which is, typically, twice that of the underlying approximate solution.

Section 5 shows that the reconstructed solution can also be used to obtain improved accuracy through a defect correction procedure, but even more accuracy can be achieved by using both defect correction and adjoint error correction. A key to the success of both the adjoint error correction and the defect correction is that the reconstructed solution has an error which is smooth. In Section 6 we present some preliminary analysis of how this may be achieved, given, as a starting point, an initial approximate solution with an error which is pointwise second-order convergent, but whose gradient is first-order convergent.

Section 7 is devoted to the derivation of residual-based *a posteriori* error bounds for h-version finite element approximations of linear output functionals. Specifically, we consider the approximation of the normal flux of the solution to an elliptic boundary value problem through the boundary of the domain. We highlight the significance of Type I *a posteriori* error bounds, where the solution of the adjoint problem appears as local weight to the finite element residual, and we discuss the implementation of Type I *a posteriori* error bounds into h-adaptive finite element algorithms.

In Section 8 we study the effect of mesh-dependent perturbations on the Primal–Dual Equivalence Theorem. We also consider how such perturbations affect the choice of the dual problem. In particular, we show by considering stabilized finite element approximations of a linear hyperbolic problem that if the formal adjoint of the differential operator is used to define the dual problem, then the stabilization term present in the method may lead to an *a posteriori* error bound which exhibits a rate of convergence inferior to that of the error in the output functional. We also show how the bound may be sharpened by using adjoint error correction, and how the problem may be avoided altogether by defining the adjoint problem through the use of the bilinear form of the numerical method.

In Section 9 we develop the *a posteriori* error analysis of hp-version finite element approximation of functionals of solutions to linear and nonlinear hyperbolic problems. Again, we concentrate on Type I *a posteriori* bounds, where the adjoint solution appears in the bound as local weight. We illustrate the ideas through the hp-version of the discontinuous Galerkin finite element method which admits easy and flexible implementation of adaptive local polynomial-degree variation.

We close in Section 10 with some concluding remarks, and discuss areas of further research.

2. The Primal–Dual Equivalence Theorem

Suppose that U and V are two real Hilbert spaces with norms $\|\cdot\|_U$ and $\|\cdot\|_V$, and $U_0 \subseteq U$ and $V_0 \subseteq V$ are either proper real Hilbert subspaces of U and V equipped with norms $\|\cdot\|_U$ and $\|\cdot\|_V$, respectively, or $U_0 = U$, $V_0 = V$. If U_0 is a proper subspace of U and p is a fixed element of U, we define $U_p = p + U_0$; similarly, if V_0 is a proper Hilbert subspace of V and d a fixed element in V, we let $V_d = d + V_0$. Clearly, if $p \in U_0$ then, by linearity, $U_p = U_0$; similarly, if $d \in V_0$ then $V_d = V_0$.

We consider the following variational problem, which we shall henceforth refer to as the *primal problem*.

(P) Suppose that $m : U \to \mathbb{R}$ and $\ell : V \to \mathbb{R}$ are bounded linear functionals, and let $B(\cdot\,,\cdot) : U \times V \to \mathbb{R}$ be a bounded bilinear functional. Find $J_p \in \mathbb{R}$ and $u \in U_p$ such that

$$J_p = m(u) + \ell(v) - B(u,v) \qquad \forall v \in V_d. \tag{2.1}$$

Before we embark on a detailed study of the existence and uniqueness of solutions to (P), let us make some preliminary observations.

Suppose for the moment that there exist $J_p \in \mathbb{R}$ and $u \in U_p$ satisfying (2.1). Then, in particular,

$$J_p = m(u) + \ell(d) - B(u,d). \tag{2.2}$$

On decomposing each $v \in V_d$ in (2.1) as $v = d + v_0$ where $v_0 \in V_0$, and subtracting (2.2) from (2.1), it follows that $u \in U_p$ is a solution of the problem

$$B(u, v_0) = \ell(v_0) \qquad \forall v_0 \in V_0. \tag{2.3}$$

Conversely, suppose that (2.3) has the unique solution $u \in U_p$, and define the real number $J_p \in \mathbb{R}$ by (2.2); it then follows that the pair $(J_p, u) \in \mathbb{R} \times U_p$ is the unique solution to (P). Next, we shall prove that, under suitable assumptions, both (2.1) and (2.3) have unique solutions.

Theorem 2.1. In addition to the hypotheses of (P), suppose that the bilinear form $B(\cdot\,,\cdot)$ is weakly coercive on $U_0 \times V_0$ in the following sense:

(a) there exists a constant $\gamma_0 > 0$ such that

$$\inf_{w_0 \in U_0 \setminus \{0\}} \sup_{v_0 \in V_0 \setminus \{0\}} \frac{|B(w_0, v_0)|}{\|w_0\|_U \|v_0\|_V} \geq \gamma_0;$$

(b) $\qquad \forall v_0 \in V_0 \setminus \{0\} \qquad \sup_{w_0 \in U_0} B(w_0, v_0) > 0.$

Then problem (P) has a unique solution $(J_p, u) \in \mathbb{R} \times U_p$.

Proof. Consider the problem of finding $u_0 \in U_0$ such that

$$B(u_0, v_0) = \ell(v_0) - B(d, v_0) \qquad \forall v_0 \in V_0. \tag{2.4}$$

By virtue of Babuška's generalization of the Lax–Milgram theorem (see, for example, Theorem 6.2 on p. 224 of Oden and Reddy (1983)), under the present hypotheses problem (2.4) has a unique solution $u_0 \in U_0$. Consequently, $u = d + u_0$ is the unique solution to (2.3), and the pair (J_p, u) in $\mathbb{R} \times U_p$, with J_p defined by (2.2), is the unique solution of (2.1). □

Problems of the form (P) will be referred to throughout the text as *measurement problems*, since the process of computing the value

$$J_p = \mathcal{J}_p(u) = m(u) + \ell(d) - B(u, d)$$

of $\mathcal{J}_p(w)$ at $w = u$ has the physical interpretation of sampling the 'output functional' $\mathcal{J}_p(\cdot)$ at u, which can be thought of as making a certain measurement of the solution u to the variational problem (2.3). The relevance of accurate computation of output functionals in engineering applications has been highlighted in the Introduction.

In order to motivate the discussion that will follow, we consider some simple illustrative examples where the quantity of interest is an output functional.

2.1. The elliptic model problem

Suppose that Ω is a bounded open set in $\mathbb{R}^n$, $n \leq 3$, with Lipschitz-continuous boundary $\Gamma = \partial\Omega$. Given that $f \in H^{-1}(\Omega)$ and $g \in H^{1/2}(\Gamma)$ (we refer to Adams (1975) for elements from the theory of Sobolev spaces), consider Poisson's equation subject to a nonhomogeneous Dirichlet boundary condition:

$$\begin{aligned} -\Delta u &= f \qquad \text{in } \Omega, \\ u &= g \qquad \text{on } \Gamma. \end{aligned}$$

In this case, $U = V = H^1(\Omega)$, $U_0 = V_0 = H_0^1(\Omega)$ and

$$U_p = p + H_0^1(\Omega) = \{v \in H^1(\Omega) \,:\, \gamma_{0,\Gamma}(v) = g\},$$

where $\gamma_{0,\Gamma} \,:\, H^1(\Omega) \to H^{1/2}(\Gamma)$ is the classical trace operator and $p \in H^1(\Omega)$ is chosen so that $\gamma_{0,\Gamma}(p) = g$.

The standard weak formulation of the problem is as follows: find $u \in U_p$ such that

$$B(u, v) = \ell(v) \qquad \forall v \in V_0,$$

where

$$B(u, v) = \int_\Omega \nabla u \cdot \nabla v \, \mathrm{d}x, \qquad \ell(v) = \langle f, v\rangle,$$

and $\langle \cdot\,, \cdot\rangle$ denotes the duality pairing between $H^{-1}(\Omega)$ and $H_0^1(\Omega)$.

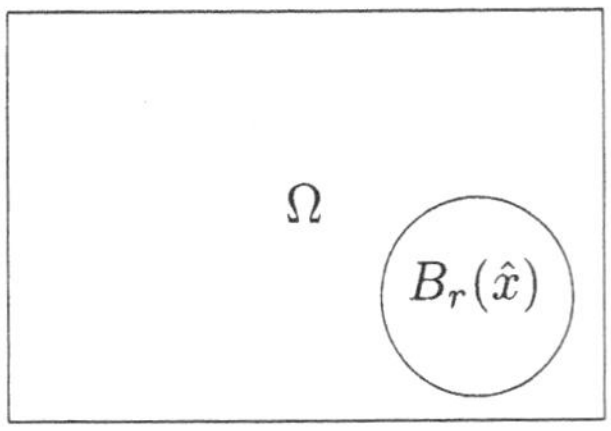

Figure 2.1. Local averaging over $B_r(\hat{x})$: a ball of radius r centred at $\hat{x} \in \Omega$

Local average. Our first example is concerned with calculating the integral average of u over an open ball $B_r(\hat{x}) \subset \Omega$ of radius r centred at $\hat{x} \in \Omega$, as illustrated in Figure 2.1:

$$\mathcal{J}_p(u) = \frac{1}{\operatorname{meas} B_r(\hat{x})} \int_{B_r(\hat{x})} u(x)\,\mathrm{d}x.$$

In this case, we take $d = 0$ and hence $V_d = V_0 = H_0^1(\Omega)$. Consider the problem (2.1) of finding $(J_p, u) \in \mathbb{R} \times U_p$ such that

$$J_p = m(u) + \ell(v) - B(u,v) \qquad \forall v \in V_d,$$

where

$$m(u) = \frac{1}{\operatorname{meas} B_r(\hat{x})} \int_{B_r(\hat{x})} u(x)\,\mathrm{d}x.$$

Hence, recalling that here $V_d = V_0$ and therefore $\ell(v) - B(u,v) = 0$ for all $v \in V_d = V_0$, we find that

$$J_p = \mathcal{J}_p(u) = m(u) = \frac{1}{\operatorname{meas} B_r(\hat{x})} \int_{B_r(\hat{x})} u(x)\,\mathrm{d}x.$$

Point value. If $u \in C(\Omega)$ it is meaningful to consider the point value

$$\mathcal{J}_p(u) = u(\hat{x})$$

of u at $\hat{x} \in \Omega$; to do so, we again take $d = 0$, $V_d = V_0 = H_0^1(\Omega)$, and seek $(J_p, u) \in \mathbb{R} \times U_p$ such that

$$J_p = m(u) + \ell(v) - B(u,v) \qquad \forall v \in V_d,$$

where

$$m(u) = u(\hat{x}).$$

As $V_d = V_0$, it directly follows that $J_p = u(\hat{x})$. Since $m(u) : u \mapsto u(\hat{x})$ is not a bounded linear functional on the space $H_0^1(\Omega)$, $\Omega \subset \mathbb{R}^n$, for $n \geq 2$, this example does not directly fit into the theoretical setting described here in two or more space dimensions, although $m(u)$ can be approximated by

$$m_r(u) = \frac{1}{\operatorname{meas} B_r(\hat{x})} \int_{B_r(\hat{x})} u(x)\,\mathrm{d}x,$$

with $0 < r \ll 1$; alternatively, since $m : u \mapsto u(\hat{x})$ is a bounded linear functional on $W^{1,p}(\Omega)$ for $p > n$, by extending the present Hilbertian theoretical framework to reflexive Banach spaces, the case of $m(u) = u(\hat{x})$ could be covered directly without having to resort to approximations of the type $m(u) \approx m_r(u)$.

Normal flux. Let us assume that $f \in L^2(\Omega)$ and consider the weighted normal flux of u over Γ, with weight function $\psi \in H^{1/2}(\Gamma)$, defined by

$$\mathcal{J}_p(u) = \int_\Gamma \frac{\partial u}{\partial \nu} \psi \,\mathrm{d}s, \tag{2.5}$$

where $\boldsymbol{\nu}$ is the unit outward normal vector to Γ. Strictly speaking, the integral should be thought of as the duality pairing between $H^{-1/2}(\Gamma)$ and $H^{1/2}(\Gamma)$. In this case, let

$$V_d = H^1_{-\psi,\Gamma}(\Omega) = \{v \in H^1(\Omega) : \gamma_{0,\Gamma}(v) = -\psi\},$$

and define $m : U \to \mathbb{R}$ as the trivial linear functional which maps every element of $U = H^1(\Omega)$ into $0 \in \mathbb{R}$. We consider the problem of finding $(J_p, u) \in \mathbb{R} \times U_p$ such that

$$J_p = \ell(v) - B(u, v) \qquad \forall v \in V_d. \tag{2.6}$$

A simple calculation based on Green's second identity shows that

$$J_p = \mathcal{J}_p(u) = \int_\Gamma \frac{\partial u}{\partial \nu} \psi \,\mathrm{d}s. \tag{2.7}$$

The relevance of rewriting the normal flux (2.5) in the form (2.6) will be explained in Section 4. It will be shown that the equality (2.7) is not preserved under discretization; indeed, we shall see that it is *not* the discretization of (2.5) but that of (2.6) that yields the more accurate approximation to (2.5).

2.2. The hyperbolic model problem

Our second model problem is a boundary value problem for a first-order hyperbolic equation. Suppose that $\Omega = (0,1)^n$ and let Γ denote the union of open faces of Ω. Let $\mathbf{b} = (b_1, \ldots, b_n)^{\mathrm{T}}$ belong to $[C^1(\bar{\Omega})]^n$, with each b_i, $i = 1, \ldots, n$, positive on $\bar{\Omega}$; suppose further that $c \in C(\bar{\Omega})$, $f \in L^2(\Omega)$ and $g \in L^2(\Gamma_-)$, where

$$\Gamma_- = \{x \in \Gamma : \mathbf{b}(x) \cdot \boldsymbol{\nu}(x) < 0\}$$

is the *inflow boundary* of Ω and $\boldsymbol{\nu}(x)$ signifies the unit outward normal vector to Γ at $x \in \Gamma$. Consider the transport problem

$$\begin{aligned} \mathbf{b} \cdot \nabla u + cu &= f \qquad \text{in } \Omega, \\ u &= g \qquad \text{on } \Gamma_-. \end{aligned} \tag{2.8}$$

Under our hypotheses, Γ is noncharacteristic (*i.e.*, the vector field $\mathbf{b}$ is, everywhere on Γ, transversal to Γ). We adopt the following (standard) hypothesis: there exists a positive constant γ such that

$$c(x) - \frac{1}{2}\nabla \cdot \mathbf{b}(x) \geq \gamma \qquad \forall x \in \bar{\Omega}. \tag{2.9}$$

In order to deduce the correct weak formulation of (2.8), suppose for the moment that the boundary value problem has a solution u in $H^1(\Omega)$, and let $V_d = V_0 = V = H^1(\Omega)$. On multiplying the partial differential equation in (2.8) by $v \in V$ and integrating by parts, we find that

$$-(u, \nabla \cdot (\mathbf{b}v)) + (cu, v) + \langle u, v\rangle_{\Gamma_+} = (f, v) + \langle g, v\rangle_{\Gamma_-} \qquad \forall v \in V, \tag{2.10}$$

where $(\cdot\,,\cdot)$ denotes the L^2 inner product over Ω,

$$\Gamma_+ = \{x \in \Gamma \,:\, \mathbf{b} \cdot \boldsymbol{\nu} > 0\},$$

and

$$\langle w, v\rangle_{\Gamma_\pm} = \int_{\Gamma_\pm} |\mathbf{b} \cdot \boldsymbol{\nu}| wv \, ds.$$

We consider the inner product $(\cdot\,,\cdot)_U$ defined by

$$(w, v)_U = (w, v) + \langle w, v\rangle_{\Gamma_+},$$

let U denote the closure of V in $L^2(\Omega)$ with respect to the norm $\|\cdot\|_U$ defined by

$$\|w\|_U = (w, w)_U^{1/2},$$

and put $U_p = U_0 = U$. Clearly, U is a Hilbert space. For $w \in U$ and $v \in V$, we now consider the bilinear form $B(\cdot\,,\cdot) \,:\, U \times V \to \mathbb{R}$ defined by

$$B(w, v) = -(w, \nabla \cdot (\mathbf{b}v)) + (cw, v) + \langle w, v\rangle_{\Gamma_+}$$

and for $v \in V$ we introduce the linear functional $\ell \,:\, V \to \mathbb{R}$ by

$$\ell(v) = (f, v) + \langle g, v\rangle_{\Gamma_-}.$$

We shall say that $u \in U$ is a weak solution to the boundary value problem (2.8) if

$$B(u, v) = \ell(v) \qquad \forall v \in V. \tag{2.11}$$

Theorem 2.2. Assuming (2.9), for each $f \in L^2(\Omega)$ and $g \in L^2(\Gamma_-)$ there exists a unique $u \in U$ satisfying (2.11).

For a survey of well-posedness results for linear hyperbolic boundary value problems, we refer to Bardos (1970) and Dautray and Lions (1993).

On choosing $v \in C_0^\infty(\Omega)$ in (2.11), we deduce that $\mathbf{b} \cdot \nabla u + cu = f$ in the sense of distributions on Ω; as $f - cu \in L^2(\Omega)$, it then follows that any weak solution u of (2.11) in U satisfies $\mathbf{b} \cdot \nabla u \in L^2(\Omega)$. Thus, since $\mathbf{b}$ is transversal to Γ at each point of Γ, we conclude that $\gamma_{0,\Gamma_0}(u) \in L^2(\Gamma_0)$ for any open subset $\Gamma_0 \subset \Gamma$. This will be important in our third example below,where we consider the normal flux of u through $\Gamma_0 = \Gamma_+$.

Local average. For this hyperbolic model problem an example of a quantity of physical interest is the integral average of u over an open ball $B_r(\hat{x})$:

$$\mathcal{J}_p(u) = \frac{1}{\operatorname{meas} B_r(\hat{x})} \int_{B_r(\hat{x})} u(x)\,\mathrm{d}x.$$

We set $p = 0$ and $U_p = U_0 = U$ (with U defined above by completion of $H^1(\Omega)$ in the norm $\|\cdot\|_U$), $d = 0$ and $V_d = V_0 = V = H^1(\Omega)$, and seek $(J_p, u) \in \mathbb{R} \times U_p$ such that

$$J_p = m(u) + \ell(v) - B(u, v) \qquad \forall v \in V_d,$$

where

$$m(u) = \frac{1}{\operatorname{meas} B_r(\hat{x})} \int_{B_r(\hat{x})} u(x)\,\mathrm{d}x \qquad \forall v \in V_d.$$

Clearly,

$$J_p = \mathcal{J}_p(u) = m(u) = \frac{1}{\operatorname{meas} B_r(\hat{x})} \int_{B_r(\hat{x})} u(x)\,\mathrm{d}x.$$

Point value. In general, weak solutions to (2.11) exhibit discontinuities across characteristic hypersurfaces; however, if u is continuous in an open neighbourhood of a point $\hat{x} \in \Omega \cup \Gamma_+$, then it is meaningful to consider the point value

$$\mathcal{J}_p(u) = u(\hat{x})$$

of u at $\hat{x}$. Again, we put $d = 0$, let $V_d = V_0 = V = H^1(\Omega)$, and seek $(J_p, u) \in \mathbb{R} \times U_p$ such that

$$J_p = m(u) + \ell(v) - B(u, v) \qquad \forall v \in V_d,$$

where

$$m(u) = u(\hat{x}).$$

Clearly,

$$J_p = \mathcal{J}_p(u) = u(\hat{x}).$$

As $H^1(\Omega)$ is a proper subspace of U and, as we have already seen in the elliptic case, the functional $m : u \mapsto u(\hat{x})$ is not bounded on $H^1(\Omega)$, except

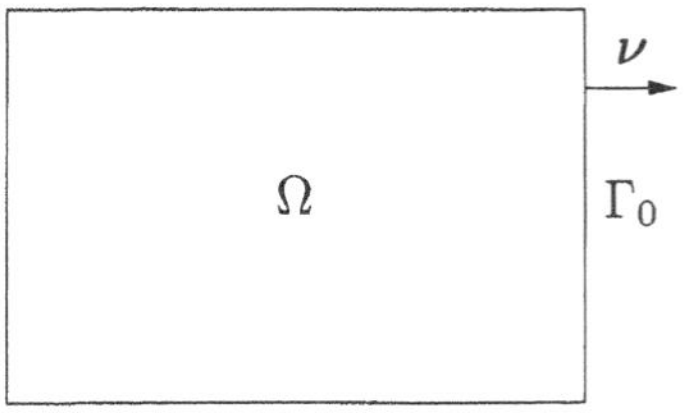

Figure 2.2. Outflow normal flux over a part Γ_0 of the outflow boundary Γ_+ of Ω; $\boldsymbol{\nu}$ denotes the unit outward normal vector to Γ

for $n = 1$, m will not be a bounded linear functional on the space $U \supset H^1(\Omega)$ either, and will need to be approximated through local averaging, as in the elliptic case, to fit into the present theoretical framework.

Outflow normal flux. If the quantity of interest is the weighted normal flux of u over an open subset Γ_0 of Γ_+, as illustrated in Figure 2.2, given by

$$\mathcal{J}_p(u) = \int_{\Gamma_0} |\mathbf{b} \cdot \boldsymbol{\nu}| u\psi \, \mathrm{d}s = \langle u, \psi \rangle_{\Gamma_0},$$

where $\boldsymbol{\nu}$ is the unit outward normal vector to Γ_0 and $\psi \in L^2(\Gamma_0)$ is a fixed weight-function, we let $d = 0$, $V_d = V_0 = V = H^1(\Omega)$ and seek $(J_p, u) \in \mathbb{R} \times U_p$ such that

$$J_p = m(u) + \ell(v) - B(u, v) \qquad \forall v \in V_d,$$

where

$$m(u) = \int_{\Gamma_0} |\mathbf{b} \cdot \boldsymbol{\nu}| u\psi \, \mathrm{d}s \qquad \forall v \in V_d = \langle u, \psi \rangle_{\Gamma_0}.$$

Trivially,

$$J_p = \mathcal{J}_p(u) = \int_{\Gamma_0} |\mathbf{b} \cdot \boldsymbol{\nu}| u\psi \, \mathrm{d}s.$$

After this overview of measurement problems, we introduce the concept of duality, and show that there is an alternative route to obtaining J_p which does not require knowledge of u.

2.3. *The dual problem*

We begin by associating with the measurement problem (P) the following *dual problem*:

(D) Find $J_d \in \mathbb{R}$ and $z \in V_d$ such that

$$J_d = m(w) + \ell(z) - B(w, z) \qquad \forall w \in U_p. \tag{2.12}$$

In order to ascertain the well-posedness of (D), we recall the following result, which is a straightforward consequence of Proposition A.2 in Melenk and Schwab (1999).

Proposition 2.3. The bounded bilinear form $B : U \times V \to \mathbb{R}$ is weakly coercive on $U_0 \times V_0$ in the sense of (a) and (b) of Theorem 2.1 if and only if $B(\cdot\,,\cdot)$ is adjoint-weakly coercive in the following sense:

(a) there exists a constant $\tilde{\gamma}_0 > 0$ such that

$$\inf_{v_0 \in V_0 \setminus \{0\}} \sup_{w_0 \in U_0 \setminus \{0\}} \frac{|B(w_0, v_0)|}{\|w_0\|_U \|v_0\|_V} \geq \tilde{\gamma}_0;$$

(b) $\qquad \forall w_0 \in U_0 \setminus \{0\} \qquad \sup_{v_0 \in V_0} B(w_0, v_0) > 0.$

Now we are ready to make a statement about the well-posedness of (D).

Theorem 2.4. In addition to the hypotheses of (P), suppose that the bilinear form $B(\cdot\,,\cdot)$ is weakly coercive on $U_0 \times V_0$ in the sense of (a) and (b) of Theorem 2.1. Then there exists a unique pair $(J_d, z) \in \mathbb{R} \times V_d$ that satisfies (D).

Proof. As $B(\cdot\,,\cdot) : U_0 \times V_0 \to \mathbb{R}$ is weakly coercive on $U_0 \times V_0$, by Proposition 2.3 it is adjoint-weakly coercive on $U_0 \times V_0$. Therefore the adjoint bilinear form $B' : V_0 \times U_0 \to \mathbb{R}$ defined by

$$B'(v, w) = B(w, v)$$

is weakly coercive on $V_0 \times U_0$. By Theorem 2.1, the following problem has a unique solution $z_0 \in V_0$: find $z_0 \in V_0$ such that

$$B'(z_0, w_0) = m(w_0) - B(w_0, d) \qquad \forall w_0 \in U_0.$$

Equivalently, there exists a unique $z_0 \in V_0$ such that

$$B(w_0, z_0) = m(w_0) - B(w_0, d) \qquad \forall w_0 \in U_0.$$

Consequently, $z = d + z_0$ is the unique solution to the following problem: find $z \in V_d$ such that

$$B(w_0, z) = m(w_0) \qquad \forall w_0 \in U_0. \tag{2.13}$$

Problem (2.13) will be referred to as the *dual* to problem (2.3). Let us define

$$J_d = m(p) + \ell(z) - B(p, z). \tag{2.14}$$

On writing any $w \in U_p$ as $w = p + w_0$ with $w_0 \in U_0$, we deduce from (2.13) and (2.14) that

$$J_d = m(w) + \ell(z) - B(w, z) \qquad \forall w \in U_p.$$

Hence the pair $(J_d, z) \in \mathbb{R} \times V_d$, with J_d defined by (2.14), is the unique solution of problem (D). □

Our next result encapsulates a rather elementary relationship between the primal and dual problems.

Theorem 2.5. (Primal–Dual Equivalence Theorem) Let u and z denote the solutions to the primal problem (P) and the dual problem (D), respectively; then,

$$J_p = \mathcal{J}_p(u) = \mathcal{J}_d(z) = J_d.$$

Proof. On inserting $w = u$ into (D) and $v = z$ into (P) the required identity trivially follows. □

Despite its simplicity, the practical consequences of this result are far-reaching. For, suppose that instead of a single linear functional $\ell : V \to \mathbb{R}$, we have been given N linear functionals $\ell^{(j)} : V \to \mathbb{R}$, $j = 1, \ldots, N$, where $N \gg 1$, and assume that the task is to find $J_p^{(j)} \in \mathbb{R}$ such that $(J_p^{(j)}, u^{(j)}) \in \mathbb{R} \times U_p$ satisfies

$$J_p^{(j)} = m(u^{(j)}) + \ell^{(j)}(v) - B(u^{(j)}, v) \qquad \forall v \in V_d.$$

Note that, in this case, only the numbers $J_p^{(j)}$, $j = 1, \ldots, N$, are required, while $u^{(j)}$, $j = 1, \ldots, N$, are of no interest. The most direct approach to obtaining $J_p^{(j)}$, $j = 1, \ldots, N$, would be to solve each of the following problems:

Find $u^{(j)} \in U_p$ such that

$$B(u^{(j)}, v) = \ell^{(j)}(v) \qquad \forall v \in V_0, \tag{2.15}$$

for $j = 1, \ldots, N$, and then compute $J_p^{(j)}$ via

$$J_p^{(j)} = m(u^{(j)}) + \ell^{(j)}(d) - B(u^{(j)}, d), \qquad j = 1, \ldots, N.$$

Theorem 2.5, however, offers a more attractive alternative. This consists of first solving the following (single) dual problem:

Find $z \in V_d$ such that

$$B(w, z) = m(w) \qquad \forall w \in V_0, \tag{2.16}$$

computing, for each $j \in \{1, \ldots, N\}$,

$$J_d^{(j)} = m(p) + \ell^{(j)}(z) - B(p, z),$$

and exploiting the fact that, by the Primal–Dual Equivalence Theorem,

$$J_p^{(j)} = J_d^{(j)}, \qquad j = 1, \ldots, N.$$

Obviously, the latter approach involves less effort since the complexity of evaluating $\ell^{(j)}(z)$ is typically substantially smaller than that of determining $u^{(j)}$. Of course, in practice neither (P) nor (D) can be solved exactly, and one has to resort to numerical approximations. We shall show, however, in the next section that the Primal–Dual Equivalence Theorem can be preserved under discretization.

3. Galerkin approximations and duality

3.1. The Discrete Primal–Dual Equivalence Theorem

Suppose that $\{U_0^h\}_{h>0}$ and $\{V_0^h\}_{h>0}$ are two families of finite-dimensional subspaces of U_0 and V_0, respectively, parametrized by $h \in (0,1]$. When U_0 is a proper Hilbert subspace of U, we assign to $p \in U$ the affine variety $U_p^h = p + U_0^h \subset U_p \subset U$; similarly, when V_0 is a proper Hilbert subspace of V, we assign to $d \in V$ the affine variety $V_d^h = d + V_0^h \subset V_d$

We consider the following finite-dimensional variational problem, which we shall henceforth refer to as the *discrete primal problem*.

(P^h) Suppose that $m : U \to \mathbb{R}$ and $\ell : V \to \mathbb{R}$ are bounded linear functionals and $B(\cdot\,,\cdot) : U \times V \to \mathbb{R}$ is a bounded bilinear functional. Find $J_p^h \in \mathbb{R}$ and $u^h \in U_p^h$ such that

$$J_p^h = m(u^h) + \ell(v^h) - B(u^h, v^h) \qquad \forall v^h \in V_d^h. \tag{3.1}$$

As in Section 2, it is easily seen that

$$J_p^h = m(u^h) + \ell(d) - B(u^h, d),$$

where $u^h \in U_p^h$ solves

$$B(u^h, v_0^h) = \ell(v_0^h) \qquad \forall v_0^h \in V_0^h. \tag{3.2}$$

Thus the existence of a unique solution to (P^h) is equivalent to the requirement that (3.2) have a unique solution u^h in U_p^h. The latter can be ensured, in a similar manner as for the continuous problem in the previous section, by assuming that the bilinear functional $B(\cdot\,,\cdot)$ is weakly coercive on $U_0^h \times V_0^h$ (with U_0^h and V_0^h equipped with the induced norms $\|\cdot\|_U$, $\|\cdot\|_V$). It is important to note, however, that weak coercivity of $B(\cdot\,,\cdot)$ on $U_0^h \times V_0^h$ is an independent assumption, generally *not* implied by weak coercivity of $B(\cdot\,,\cdot)$ on $U_0 \times V_0$.

Let $\{U_0^H\}_{H>0}$ and $\{V_0^H\}_{H>0}$ be two families of finite-dimensional subspaces of U_0 and V_0, respectively, parametrized by $H \in (0,1]$, typically different from the families $\{U_0^h\}_{h>0}$ and $\{V_0^h\}_{h>0}$. We assign to $p \in U$ the affine variety $U_p^H = p + U_0^H \subset U_p \subset U$; similarly, we assign to $d \in V$ the affine variety $V_d^H = d + V_0^H \subset V_d \subset V$. We now define the *discrete dual problem* as follows:

(D^H) Find $J_d^H \in \mathbb{R}$ and $z^H \in V_d^H$ such that

$$J_d^H = m(w^H) + \ell(z^H) - B(w^H, z^H) \qquad \forall w^H \in U_p^H.$$

In complete analogy with (P^h),

$$J_d^H = m(p) + \ell(z^H) - B(p, z^H),$$

where $z^H \in V_d^H$ solves

$$B(w_0^H, z^H) = m(w_0^H) \qquad \forall w_0^H \in U_0^H. \tag{3.3}$$

Thus, the existence of a unique solution to (D^H) is equivalent to the requirement that (3.3) have a unique solution z^H in V_d^H; the latter can be ensured by requiring that $B(\cdot\,,\cdot)$ is adjoint-weakly coercive on $U_0^H \times V_0^H$, or equivalently (*cf.* Proposition 2.3), that $B(\cdot\,,\cdot)$ is weakly coercive on $U_0^H \times V_0^H$.

Note that there is an interchange in the identity of the test and trial spaces. In the primal problem, U_p^h is the trial space and V_0^h is the test space, whereas in the dual problem it is V_d^H which is the trial space, and U_0^h is the test space.

Next we present representation formulae for the error between J_p, J_d and their approximations J_p^h, J_d^H, respectively.

Theorem 3.1. (Error representation formula) Let $(J_p, u) \in \mathbb{R} \times U_p$ and $(J_d, z) \in \mathbb{R} \times V_d$ denote the solutions to (P) and (D), respectively, and let $(J_p^h, u^h) \in \mathbb{R} \times U_p^h$ and $(J_d^H, z^H) \in \mathbb{R} \times V_d^H$ be the solutions to (P^h) and (D^H), respectively. Then,

$$J_p - J_p^h = B(u - u^h, z - z^h) \qquad \forall z^h \in V_d^h, \tag{3.4}$$

$$J_d - J_d^H = B(u - u^H, z - z^H) \qquad \forall u^H \in U_p^H. \tag{3.5}$$

Proof. Since $V_d^h \subset V_d$, we have from (P) that

$$J_p = m(u) + \ell(v^h) - B(u, v^h) \qquad \forall v^h \in V_d^h.$$

Recalling from (P^h) that

$$J_p^h = m(u^h) + \ell(v^h) - B(u^h, v^h) \qquad \forall v^h \in V_d^h$$

and subtracting, we find that

$$J_p - J_p^h = m(u - u^h) - B(u - u^h, v^h) \qquad \forall v^h \in V_d^h. \tag{3.6}$$

On the other hand, as $u - u^h \in U_0$, we deduce from (2.13) that

$$B(u - u^h, z) = m(u - u^h),$$

which we can use to eliminate $m(u - u^h)$ from (3.6) to deduce (3.4). The proof of (3.5) is analogous. □

The initial hypothesis stated in (P) that $B(\cdot\,,\cdot)$ is a bounded bilinear functional on $U \times V$ implies the existence of a positive constant γ_1 such that

$$|B(w, v)| \leq \gamma_1 \|w\|_U \, \|v\|_V \qquad \forall w \in U \quad \forall v \in V.$$

Thus we deduce the following result.

Corollary 3.2. **(*A priori* error bound)** Let $(J_p, u) \in \mathbb{R}\times U_p$ and $(J_d, z) \in \mathbb{R} \times V_d$ denote the solutions to (P) and (D), respectively, and let $(J_p^h, u^h) \in \mathbb{R} \times U_p^h$ and $(J_d^H, z^H) \in \mathbb{R} \times V_d^H$ be the solutions to (P^h) and (D^H), respectively. Then,

$$|J_p - J_p^h| \le \gamma_1 \|u - u^h\|_U \inf_{v^h \in V_d^h} \|z - v^h\|_V, \tag{3.7}$$

$$|J_d - J_d^H| \le \gamma_1 \|z - z^H\|_V \inf_{w^H \in U_p^H} \|u - w^H\|_U. \tag{3.8}$$

This abstract superconvergence result expresses the fact that the rate of convergence of J_p^h to J_p as $h \to 0$ (respectively, J_d^H to J_d as $H \to 0$) is higher than that of u^h to u in the norm of U as $h \to 0$ (respectively, z^H to z in the norm of V as $H \to 0$).

Our next result is the discrete counterpart of the Primal–Dual Equivalence Theorem.

Theorem 3.3. **(Discrete Primal–Dual Equivalence Theorem)** Let us suppose that $(J_p^h, u^h) \in \mathbb{R}\times U_p^h$ and $(J_d^H, z^H) \in \mathbb{R}\times V_d^H$ denote the solutions to the primal problem (P^h) and the dual problem (D^H), respectively; then,

$$J_d^H = J_p^h + \rho^{hH},$$

where

$$\rho^{hH} = B(u - u^h, z - z^h) - B(u - u^H, z - z^H),$$

for any $u^H \in U_p^H$ and any $z^h \in V_d^h$.

Proof. The result is a direct consequence of the previous theorem, on subtracting (3.4) from (3.5), and recalling from the Primal–Dual Equivalence Theorem that $J_p = J_d$. □

To conclude this section, let us note, in particular, that if

$$\{U_p^h\}_{h>0} = \{U_p^H\}_{H>0} \quad \text{and} \quad \{V_d^h\}_{h>0} = \{V_d^H\}_{H>0},$$

then $\rho^{hH} = 0$ and there is exact equivalence between the primal and dual formulations. When the families are not the same, in general, the error term ρ^{hH} is not equal to 0, but may be made arbitrarily small by sending the discretization parameters h and H to 0. Indeed,

$$|\rho^{hH}| \le \gamma_1 \left\{ \|u - u^h\|_U \inf_{v^h \in V_d^h} \|z - v^h\|_V + \|z - z^H\|_V \inf_{w^H \in U_p^H} \|u - w^H\|_U \right\},$$

so ρ^{hH} will converge to 0, as $h, H \to 0$, whenever the four terms on the right-hand side converge to 0 with h and H. If the bounded bilinear form $B(\cdot\,,\cdot)$

is weakly coercive on the appropriate spaces, then this follows from

$$\lim_{h\to 0}\inf_{w^h\in U_p^h}\|u-w^h\|_U=0,\qquad \lim_{h\to 0}\inf_{v^h\in V_d^h}\|z-v^h\|_V=0,$$

$$\lim_{H\to 0}\inf_{w^H\in U_p^H}\|u-w^H\|_U=0,\qquad \lim_{H\to 0}\inf_{v^H\in V_d^H}\|z-v^H\|_V=0,$$

which, in turn, are the standard approximability hypotheses for abstract Galerkin methods.

3.2. Error representation in terms of residuals

We present a further application of the error representation formulae (3.4) and (3.5), concerned with improving the approximation to the output functional by correcting its value. The discussion in this section is an abstract version of the linear theory presented in Pierce and Giles (2000), Giles (2001), Giles and Pierce (2001, 2002) applied to Galerkin approximations. The extension to non-Galerkin approximations is discussed in the following section.

We begin by noting that, for any $u^h \in U_p^h$, and any $v_0 \in V_0$, it follows from (2.3) that

$$\begin{aligned} B(u-u^h,v_0) &= B(u,v_0)-B(u^h,v_0)\\ &= \ell(v_0)-B(u^h,v_0). \end{aligned}$$

Since $R_p(u^h) : v \mapsto \ell(v) - B(u^h, v)$ is a bounded linear functional on V, we can write

$$B(u-u^h,v_0)=\langle R_p(u^h),v_0\rangle \qquad \forall v_0\in V_0, \tag{3.9}$$

where $\langle\cdot\,,\cdot\rangle$ is the duality pairing between V', the dual space of V, and V.

Recalling from (3.4) that

$$J_p-J_p^h=B(u-u^h,z-z^h)\qquad \forall z^h\in V_d^h,$$

where u^h is the second component of the solution $(J_p^h, u^h) \in \mathbb{R}\times U_p^h$ to the primal problem (P^h), it follows that

$$\begin{aligned} J_p-J_p^h &= \langle R_p(u^h),z-z^h\rangle & (3.10)\\ &= \langle R_p(u^h),z^H-z^h\rangle+\langle R_p(u^h),z-z^H\rangle \qquad \forall z^h\in V_d^h, \end{aligned}$$

where $z^H \in V^H$ is the second component of the solution $(J_d^H, z^H) \in \mathbb{R}\times V_d^H$ to the dual problem (D^H). Taking $z^h = d$, and defining $z_0^H = z^H - d$, we obtain

$$J_p-J_p^h=\langle R_p(u^h),z_0^H\rangle+\langle R_p(u^h),z-z^H\rangle. \tag{3.11}$$

The first term on the right-hand side of (3.11) is computable from u^h, z^H and the data, and can be moved across to the left to yield

$$\begin{aligned} J_p - J_p^{hH} &= \langle R_p(u^h), z - z^H \rangle \\ &= B(u - u^h, z - z^H), \end{aligned} \tag{3.12}$$

where we define J_p^{hH} as

$$\begin{aligned} J_p^{hH} &= J_p^h + \langle R_p(u^h), z_0^H \rangle \\ &= m(u^h) + l(d) - B(u^h, d) + l(z_0^H) - B(u - u^h, z_0^H) \\ &= m(u^h) + l(z^H) - B(u^h, z^H). \end{aligned} \tag{3.13}$$

We shall refer to J_p^{hH} as the corrected functional value.

First, we note that, since

$$\langle R_p(u^h), z_0^h \rangle = \ell(z_0^h) - B(u^h, z_0^h) = 0 \qquad \forall z_0^h \in V_0^h, \tag{3.14}$$

we have that

$$\begin{aligned} |\langle R_p(u^h), z_0 \rangle| &= \inf_{z_0^h \in V_0^h} |\langle R_p(u^h), z_0 - z_0^h \rangle| \\ &= \inf_{z^h \in V_d^h} |\langle R_p(u^h), z - z^h \rangle|. \end{aligned}$$

Hence, on comparing (3.10) with (3.12), we deduce that, if

$$|\langle R_p(u^h), z_0 - z_0^H \rangle| \ll |\langle R_p(u^h), z_0 \rangle|, \tag{3.15}$$

or, equivalently,

$$|\langle R_p(u^h), z - z^H \rangle| \ll \inf_{z^h \in V_d^h} |\langle R_p(u^h), z - z^h \rangle| \tag{3.16}$$

for sufficiently small h and H, then the corrected value J_p^{hH} will represent a more accurate approximation to J_p than J_p^h does.

Clearly, if $z^H \in V_d^h$, then (3.15) does not hold. Thus, to ensure the validity of (3.15) it is necessary to assume that V_d^h and V_d^H differ. Incidentally, this requirement is also reasonable from the computational point of view: since (P) and (D) are driven by different data, their respective solutions u and z will, in general, exhibit different features, and there is no reason to presume that the family of test spaces $\{V_d^h\}_{h>0}$ for the discretization of the primal problem (P) will also be an appropriate choice as a family of trial spaces for the discretization of the dual problem (D).

In the context of h-version finite element methods (Strang and Fix 1973, Ciarlet 1978, Brenner and Scott 1994, Braess 1997) several possible strategies for the selection of V_d^H can be devised. Suppose, for example, that V^h is a finite element space on a subdivision $\mathcal{T}_h$, of granularity h, of the computational domain Ω consisting of (continuous or discontinuous) piecewise polynomials of degree k, and V^H is a finite element space on a possibly different subdivision $\mathcal{T}_H$, of granularity H, of Ω consisting of (continuous or discontinuous) piecewise polynomials of degree K. Below, we list a number of approaches for choosing V_d^H.

(a) Choose $\mathcal{T}_H = \mathcal{T}_h$ and $K > k$, *e.g.*, $K = k + 1$. Thus the numerical solution of the dual problem is performed on the same mesh as for the primal problem, but higher-degree piecewise polynomials are used than for the primal. This approach may be inefficient since the computational cost of solving the dual problem could be considerably larger than that of solving the primal.

(b) Choose $K = k$ and $\mathcal{T}_H = \mathcal{T}_{\lambda h}$ where $\lambda \in (0, 1)$, *e.g.*, $\lambda = 1/2$. Here, the finite element space for the dual is based on a supermesh obtained by global refinement of the primal mesh, and involves piecewise polynomials of the same degree as for the primal. Similarly to (a), the computational cost of solving the dual will be larger than that of solving the primal, although one may benefit from the fact that $K = k$, and therefore the numerical algorithm for the dual is essentially the same as for the primal, albeit on a finer mesh.

(c) Choose $K = k$, and select $\mathcal{T}_H$ adaptively, based on an *a posteriori* error bound for the dual problem. Thereby the mesh $\mathcal{T}_H$ for the dual may be completely different from $\mathcal{T}_h$. In this approach, the dual problem is solved on its own mesh whose choice is governed only by the data for the dual, and not by the choice of the primal mesh. This is perfectly reasonable, since the solutions to the primal and dual problems will in general exhibit completely different behaviour and it is unreasonable to expect that a mesh which is adequate for one will also be appropriate for the other. Of course, a practical drawback is that the adaptive design of the dual mesh $\mathcal{T}_H$ requires additional effort. Further, one needs to transfer information between the different mesh families $\{\mathcal{T}_H\}$ and $\{\mathcal{T}_h\}$ to evaluate the correction term $\langle R(u^h), z_0^H\rangle$.

In the case of *hp*-version finite element methods, which admit variation of both the local mesh size and the local polynomial degree, a further alternative is available:

(d) Choose both $\mathcal{T}_H$ and the local polynomial degree for the dual finite element space adaptively, based on an *a posteriori* error bound. This approach admits even more flexibility in the choice of the dual finite element space V_d^H than (c) – of course, with the added computational cost involved in *hp*-adaptivity in the solution of the dual problem.

Although the discussion in this section was restricted to Galerkin methods, and finite element methods in particular, the idea of error correction is much more general. In order to give a flavour of the scope of the technique, in the next section we consider the question of error correction for a general class of discretization methods for differential equations.

4. Error correction for general discretizations

Now we present a slightly more general version of the argument above, which does not assume that the numerical approximations to u and z stem from a Galerkin type method: it suffices that the approximation u^h to the primal solution $u \in U_p$ is chosen from U_p^h and the approximation z^H to the dual solution $z \in V_d$ is selected from V_d^H; exactly how u^h and z^H are defined is irrelevant. For example, one may suppose that u^h and z^H have been obtained by piecewise polynomial interpolation of finite difference or finite volume approximations to u and z on meshes of size h and H, respectively.

Starting with the equivalence of J_p and J_d, we have from (2.12) that

$$\begin{aligned} J_p &= m(u^h) + \ell(z) - B(u^h, z) \\ &= m(u^h) + \ell(z^H) - B(u^h, z^H) + \ell(z - z^H) - B(u^h, z - z^H). \end{aligned}$$

As

$$\ell(v) = B(u, v) \qquad \forall v \in V_0,$$

on choosing $v = z - z^H \in V_0$ we obtain

$$J_p = m(u^h) + \ell(z^H) - B(u^h, z^H) + B(u - u^h, z - z^H).$$

If we define J_p^h as

$$J_p^h = m(u^h) + l(d) - B(u^h, d),$$

and again define J_p^{hH} as in (3.13) to be

$$J_p^{hH} = J_p^h + \langle R_p(u^h), z_0^H \rangle = m(u^h) + l(z^H) - B(u^h, z^H),$$

we deduce that

$$J_p - J_p^{hH} = B(u - u^h, z - z^H), \tag{4.1}$$

whereas

$$J_p - J_p^h = B(u - u^h, z - z^H) + \langle R_p(u^h), z_0^H \rangle. \tag{4.2}$$

The key point is that if z^H is a very good approximation to z so that

$$|B(u - u^h, z - z^H)| \ll |\langle R_p(u^h), z_0^H \rangle|,$$

then J_p^{hH} will be a much more accurate approximation to J_p than J_p^h.

Next we shall present an experimental illustration of this abstract result. In the example u^h and z^H are computed by means of a finite difference method, in tandem with spline interpolation to construct piecewise polynomial functions from the set of nodal values delivered by the difference scheme.

4.1. Example 1: Elliptic problem in 1D

The example concerns the second-order ordinary differential equation

$$\mathcal{L}u \equiv -u'' = f(x), \qquad x \in (0,1),$$

subject to homogeneous Dirichlet boundary conditions

$$u(0) = 0, \qquad u(1) = 0.$$

Let us suppose that the boundary value problem has been solved on a uniform grid,

$$\{x_j = jh \, : \, j = 0, \ldots, N\},$$

with spacing $h = 1/N$, $N \geq 2$, using the second-order finite difference scheme

$$-\frac{U_{j+1} - 2U_j + U_{j-1}}{h^2} = f(x_j), \qquad j = 1, \ldots, N-1,$$
$$U_0 = 0, \quad U_N = 0.$$

Here U_j denotes the approximation to $u(x_j)$, $j = 0, \ldots, N$. We then define u^h by natural cubic spline interpolation through the values U_j, $j = 0, \ldots, N$, with end conditions $(u^h)''(0) = -f(0)$, $(u^h)''(1) = -f(1)$.

Let us suppose that the quantity of interest is

$$J_p = \mathcal{J}_p(u) = m(u) = \int_0^1 u(x)g(x)\,\mathrm{d}x,$$

where $g \in L^2(0,1)$ is a given weight function. It follows that the corresponding dual problem is then

$$\mathcal{L}^* z \equiv -\, z'' = g(x), \qquad x \in (0,1),$$
$$z(0) = 0, \quad z(1) = 0.$$

The numerical approximation $z^H = z_0^H$ to z is defined analogously to u^h, with the mesh size H for the dual finite difference scheme taken to be *equal* to h, for simplicity.

With $d = 0$, the uncorrected approximation J_p^h is given by

$$J_p^h = m(u^h)$$

whereas the corrected approximation J_p^{hH} is given by

$$\begin{aligned} J_p^{hH} &= m(u^h) + \ell(z^H) - B(u^h, z^H) \\ &= m(u^h) + \langle R(u^h), z^H \rangle, \end{aligned}$$

where

$$B(w,v) = \int_0^1 w'(x)\,v'(x)\,\mathrm{d}x, \qquad \ell(v) = \int_0^1 f(x)\,v(x)\,\mathrm{d}x,$$

for $w \in U_p = U_0 = H_0^1(0,1)$ and $v \in V_0 = V_d = H_0^1(0,1)$. Now, letting $(\cdot,\cdot)$ denote the inner product of $L^2(0,1)$,

$$\begin{aligned}
\langle R(u^h), v\rangle &= \ell(v) - B(u^h, v)\\
&= (f, v) - B(u^h, v)\\
&= \int_0^1 f(x)\,v(x)\,\mathrm{d}x - \int_0^1 (u^h)'(x)\,v'(x)\,\mathrm{d}x\\
&= \int_0^1 \Big\{f(x) + (u^h)''(x)\Big\}\,v(x)\,\mathrm{d}x \qquad \forall v \in H_0^1(0,1),
\end{aligned}$$

where we have made use of the fact that u^h is a cubic spline, and therefore $\mathcal{L}u^h$ is continuous on $[0,1]$. Hence,

$$R(u^h) = f + (u^h)'' = f - \mathcal{L}u^h \in L^2(0,1)$$

and

$$\langle R(u^h), z^H\rangle = (R(u^h), z^H),$$

so that

$$J_p^{hH} = m(u^h) + (R(u^h), z^H).$$

A numerical experiment. The aim of the numerical experiment which we shall now perform is to show that the addition of the 'adjoint correction term' $(R(u^h), z^H)$ to $m(u^h)$ is important, in that J_p^{hH} is a more accurate approximation to $m(u)$ than $J_p^h = m(u^h)$ is.

In the numerical experiment, we took

$$f(x) = -x^3(1-x)^3, \qquad g(x) = -\sin(\pi x).$$

Figure 4.1 depicts the residual

$$R(u^h) = f - \mathcal{L}u^h$$

for $h = \frac{1}{32}$, as well as the values at the three Gaussian quadrature points on each subinterval $[x_{j-1}, x_j]$, $j = 1, \ldots, N$, which have been used in the numerical integration of the inner product $(R(u^h), z_0^H) = (R(u^h), z^H)$, with $H = h$. Since u^h is a cubic spline, $\mathcal{L}u^h$ is continuous and piecewise linear. The best piecewise linear approximation to f has an approximation error

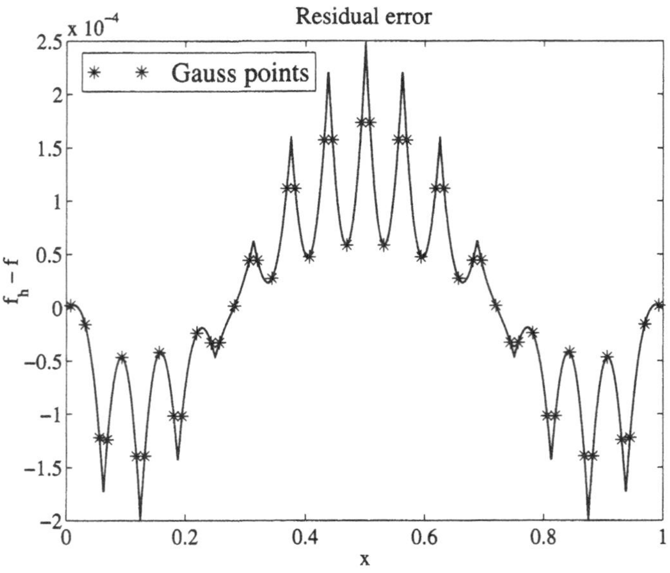

Figure 4.1. Residual $R(u^h)$ for the 1D Poisson equation

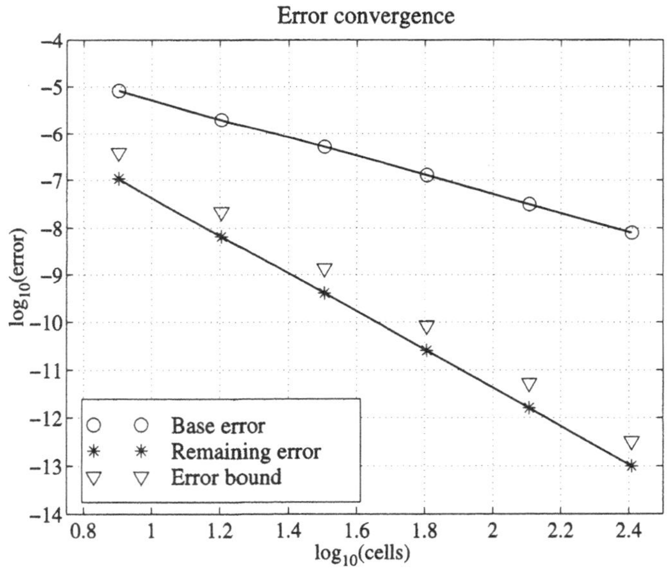

Figure 4.2. Errors in approximating J_p for the 1D Poisson equation

whose dominant term is quadratic on each subinterval; this explains the scalloped shape of $R(u^h)$ in Figure 4.1.

Figure 4.2 is the log-log plot of the three error curves corresponding to:

(a) the 'base error', $|m(u) - m(u^h)|$;

(b) the error $|m(u) - J_p^{hH}|$ resulting after the inclusion of the adjoint correction term $(R(u^h), z^H)$ via $J_p^{hH} = m(u^h) + (R(u^H), z^H)$; and

(c) the error bound $\|\mathcal{L}^{-1}\| \, \|f - \mathcal{L}u^h\| \, \|g - \mathcal{L}z^H\|$ which bounds the magnitude of (b). Here $\mathcal{L}^{-1}$ denotes the inverse of the differential operator $\mathcal{L} : H^2(0,1) \cap H_0^1(0,1) \to L^2(0,1)$.

The superimposed lines have slopes -2 and -4, confirming that the base approximation $m(u^h)$ to $m(u)$ is second-order accurate, while the error in the corrected approximation J_p^{hH} and the error bound are both fourth-order. We note in passing that, on a grid with 16 cells, which might be a reasonable choice for practical computations, the error in the corrected functional value J_p^{hH} is over 200 times smaller than the uncorrected error.

To conclude this experiment, let us explain why the corrected functional value J_p^{hH} converges to the analytical value $J_p = \mathcal{J}_p(u) = m(u)$ as $\mathcal{O}(h^4)$ when $h \to 0$. According to (4.1),

$$J_p - J_p^{hH} = B(u - u^h, z - z^H), \tag{4.3}$$

where, in the present experiment, u^h is the cubic spline interpolant based on the finite difference approximation U_j, $j = 0, \ldots, N$, to the nodal values $u(x_j)$, $j = 0, \ldots, N$, on a uniform mesh of size h; z^H is defined analogously, on a mesh of size $H = h$. Since U approximates u with $\mathcal{O}(h^2)$ error in the discrete L^2-norm based on the internal nodes, and the first-order central difference quotient of U approximates u' with $\mathcal{O}(h^2)$ error in the same norm, it follows that

$$\|u - u^h\|_U = \|u' - (u^h)'\|_{L^2(0,1)} = \mathcal{O}(h^2).$$

Analogously, with $H = h$,

$$\|z - z^H\|_V = \|z' - (z^H)'\|_{L^2(0,1)} = \mathcal{O}(h^2).$$

Thus we deduce from (4.3) that

$$|J_p - J_p^{hH}| = \mathcal{O}(h^4),$$

as required. For further details we refer to the paper of Giles and Pierce (2001).

4.2. Example 2: Elliptic problem in 2D

This example concerns the 2D Laplace equation

$$-\Delta u = 0 \qquad \text{in } \Omega,$$
$$u = g \qquad \text{on } \Gamma = \partial\Omega.$$

The output of interest is a weighted integral of the normal flux

$$\mathcal{J}_p(u) = \int_\Gamma \frac{\partial u}{\partial \nu}\, \psi \, \mathrm{d}s.$$

As described in Section 2.1, the associated primal problem uses

$$B(u,v) = \int_\Omega \nabla u \cdot \nabla v \, \mathrm{d}x, \qquad \ell(u) = 0, \qquad m(v) = 0.$$

The definition of J_p^h required the selection of an appropriate d. If, for example, u^h is twice continuously differentiable, we may integrate by parts to deduce that

$$-B(u^h, d) = (\Delta u^h, d) + \left\langle \frac{\partial u^h}{\partial \nu}, \psi \right\rangle_\Gamma.$$

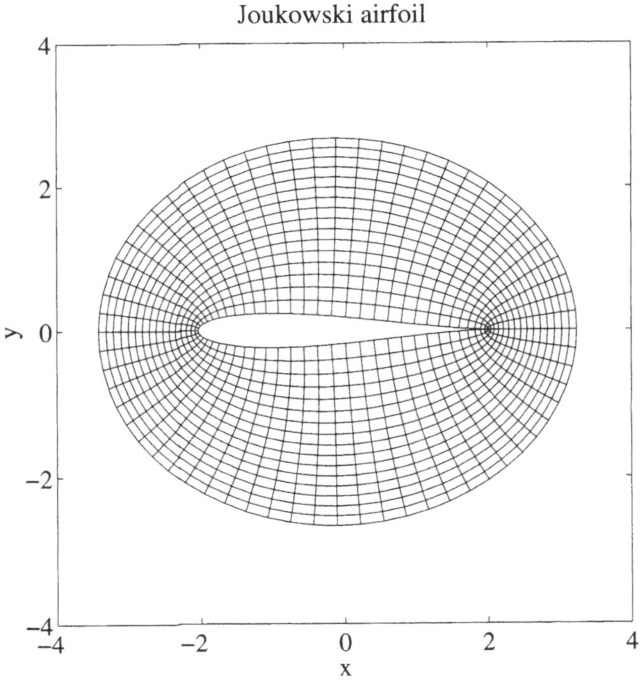

Figure 4.3. The computational grid for a 2D Laplace problem

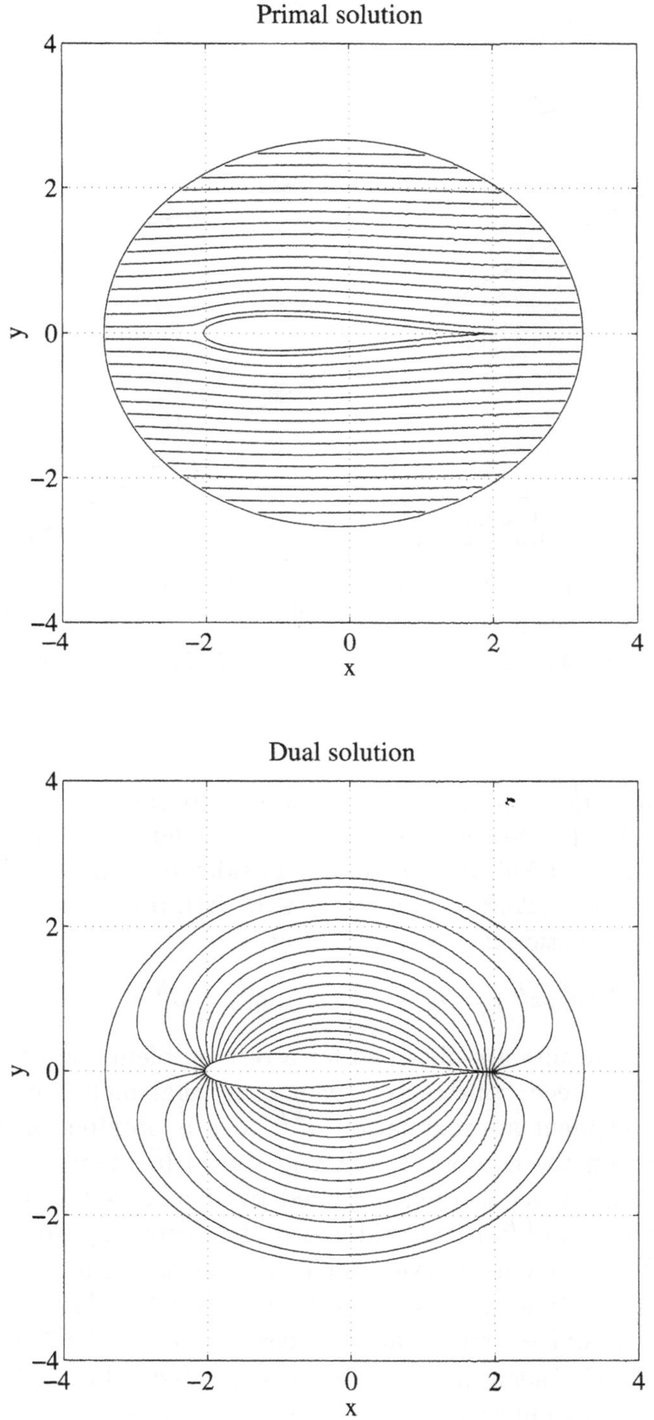

Figure 4.4. The approximate solutions u^h and z^H for a 2D Laplace problem

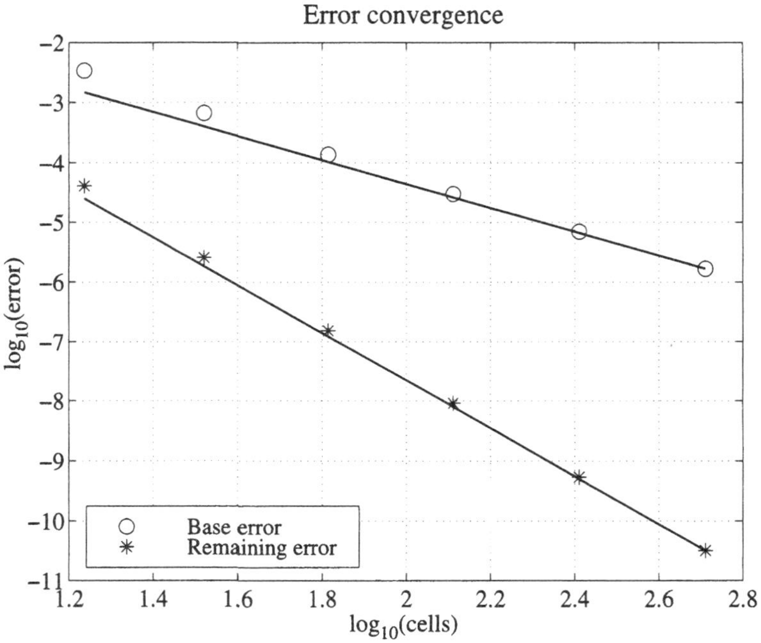

Figure 4.5. Error convergence for a 2D Laplace problem

Thus, we obtain $J_p^h = \mathcal{J}_p(u^h)$ if we choose d to equal 0 on the interior of Ω, and $-\psi$ on Γ. Strictly speaking, this violates the condition that $d \in H^1(\Omega)$, and so $(\Delta u^h, d)$ should be considered to be the limit of an appropriate sequence $(\Delta u^h, d_n)$ where $d_n \in H^1(\Omega)$, $n = 1, 2, \dots$.

The corrected functional is then given by

$$J_p^{hH} = -B(u^h, z^H) = J_p^h + \langle R_p(u^h), z_0^H \rangle = J_p^h + (-\Delta u^h, z^H).$$

The numerical results are obtained for a test problem for which the analytic solution has been constructed by a conformal mapping. An initial Galerkin finite element approximation u^{FE} to u is obtained using bilinear shape functions on the computational grid shown in Figure 4.3. The new approximate solution u^h is then obtained by bicubic spline interpolation through the values of u^{FE} at the grid nodes. The approximate dual solution z^H is obtained similarly using the same computational grid.

The errors in the functional are shown in Figure 4.5. The superimposed lines of slope -2 and -4 show that the base value for the functional, J_p^h, is again second-order accurate whereas the corrected value, J_p^{hH}, is fourth-order accurate. This improvement in the order of accuracy is achieved despite the presence of the singularity in the dual solution.

5. Linear defect error correction

Adjoint error correction is not the only means of improving the accuracy of numerical calculations. In this section, based on Giles (2001), we look at the use of defect correction (Barrett, Moore and Morton 1988, Koren 1988, Skeel 1981, Stetter 1978), and show that it can be extremely effective in reducing the errors in a model 1D Helmholtz problem; the combination of defect and adjoint error correction is even more accurate.

The primary motivation for this investigation is the need for high-order accuracy for aeroacoustic and electromagnetic calculations. In steady CFD calculations, grid adaptation can be used to provide high grid resolution in the limited areas that require it. However, using standard second-order accurate methods, the wave-like nature of aeroacoustic and electromagnetic solutions would lead to grid refinement throughout the computational domain in order to reduce the wave dispersion and dissipation to acceptable levels. The preferable alternative is to use higher-order methods, allowing one to use fewer points per wavelength, which can lead to a very substantial reduction in the total number of grid points for three-dimensional calculations. The difficulty with this is that one often wants to use unstructured grids because of their geometric flexibility, and the construction of higher order approximations on unstructured grids is complicated and computationally expensive.

5.1. General approach for Galerkin approximations

We start with a problem whose weak formulation is to find $u \in U_p$ such that

$$B(u,v) = \ell(v) \qquad \forall v_0 \in V_0,$$

and suppose that we have a Galerkin discretization which defines $u^h \in U_p^h$ to be the solution of

$$B(u^h,v^h) = \ell(v_0^h) \qquad \forall v_0^h \in V_0^h.$$

Next, suppose that we have a method for defining a reconstructed approximation u_R^h from u^h. The purpose of this reconstruction, as with the reconstruction in the last numerical example in the previous section, is to maintain the order of accuracy in the L^2-norm, and improve it in the H^1-norm.

Writing $u = u_R^h + e$ gives

$$B(e,v) = \ell(v) - B(u_R^h, v) \qquad \forall v_0 \in V_0.$$

The error $e \in U_0$ can then be approximated by $e^h \in U_0^h$, which is the solution of

$$B(e^h,v^h) = \ell(v_0^h) - B(u_R^h, v^h) \qquad \forall v_0^h \in V_0^h.$$

An improved value for u_R^h can then be defined by reconstruction from $u^h + e^h$ or, equivalently, by adding the reconstructed error e_R^h to u_R^h. The entire process may then be repeated to further improve the accuracy. This follows the procedure described by Barrett *et al.*, who also showed that it can converge to a solution of an appropriately defined Petrov–Galerkin discretization (Barrett *et al.* 1988).

5.2. 1D Helmholtz problem

The model problem to be solved is the 1D Helmholtz equation

$$-u'' - \pi^2 u = 0, \quad x \in \Omega = (0, 10),$$

subject to the Dirichlet boundary condition $u = 1$ at $x = 0$ and the radiation boundary condition $u' - \mathrm{i}\pi u = 0$ at $x = 10$. The analytic solution is $u = \exp(\mathrm{i}\pi x)$ and the domain contains precisely five wavelengths. The output functional of interest is the value $u(10)$ at the right-hand boundary. This can be viewed as a model of a far-field boundary integral giving the radiated acoustic energy in aeroacoustics, or the radar cross-section in electromagnetics (Monk and Süli 1998).

Multiplying by $\bar{v}$ (the complex conjugate of v) and integrating by parts yields the weak form: find $u \in U_p$ such that

$$B(u, v) = \ell(v) \qquad v \in V_0,$$

where

$$B(u, v) = (u', v') - \pi^2 (u, v) - \mathrm{i}\pi u(10)\bar{v}(10) \qquad \ell(v) = 0,$$

and

$$U_p = \{u \in H^1(\Omega) : u(0) = 1\},$$
$$V_0 = \{v \in H^1(\Omega) : v(0) = 0\}.$$

Note that the inner product $(\cdot\,,\cdot)$ is now a complex inner product

$$(u, v) \equiv \int_\Omega u\bar{v}\, \mathrm{d}x.$$

With this change, the theory presented before for real-valued functions extends naturally to complex-valued functions.

Using a piecewise linear Galerkin discretization, $u^h \in U_p^h$ is defined to be the solution of

$$B(u^h, v^h) = \ell(v_0^h) \qquad \forall v_0^h \in V_0^h.$$

It is well established that this discretization is second-order convergent in $L^2(\Omega)$, producing dispersion but no dissipation on a uniform grid.

The reconstructed solution u_R^h is defined by cubic spline interpolation of the nodal values $u^h(x_j)$. The choice of end conditions for the cubic spline is very important. A natural cubic spline would have $(u_R^h)''=0$ at both ends, but this would introduce small but significant errors at each end since $u''\neq 0$ for the analytic solution. Instead, at $x=10$ we require the splined solution to satisfy the analytic boundary condition by imposing $(u_R^h)'-i\pi u_R^h=0$. At $x=0$, the analytic boundary condition is already imposed through having the correct value for the end-point $U(0)$. Therefore, here we require that $(u_R^h)''+\pi^2 u_R^h=0$, so the splined solution satisfies the original ordinary differential equation at the boundary.

The error e and its Galerkin approximation e^h are defined according to the general approach described above. The reconstruction e_R^h is again obtained by cubic spline interpolation, with the same end conditions.

5.3. *Adjoint error correction*

The output functional of interest is

$$\mathcal{J}(u)=u(10),$$

so we define

$$m(u)=u(10), \qquad \ell(v)=0,$$

to obtain

$$J_p=m(u)+\ell(v)-B(u,v) \qquad \forall v\in V_0.$$

The Galerkin approximation z^h to the dual solution is the piecewise linear solution $z^h\in V_0^h$ for which

$$B(w^h,z^h)=m(w^h) \qquad \forall w^h\in U_0^h.$$

Defect correction can also be applied to the dual solution.

5.4. *Numerical results*

Numerical results have been obtained for grids with 4, 8, 16, 32, 64 and 128 points per wavelength. To test the ability to cope with irregular grids, the coordinates for the grid with N intervals are defined as

$$x_0=0, \qquad x_N=10, \qquad x_j=\frac{10}{N}(j+\sigma_j), \quad 0<j<N,$$

where σ_j is a uniformly distributed random variable in the range $[-0.3,0.3]$.

Figure 5.1 shows the L^2 norm of the error in the reconstructed cubic spline solution before and after defect correction. Without defect correction, the error is second-order, while with defect correction it is fourth-order. Note that a second application of defect correction makes a significant reduction in the error even though it remains fourth-order. This is because one application of the defect correction procedure gives a correction that is second-order in magnitude, with a corresponding error that is second-order in relative

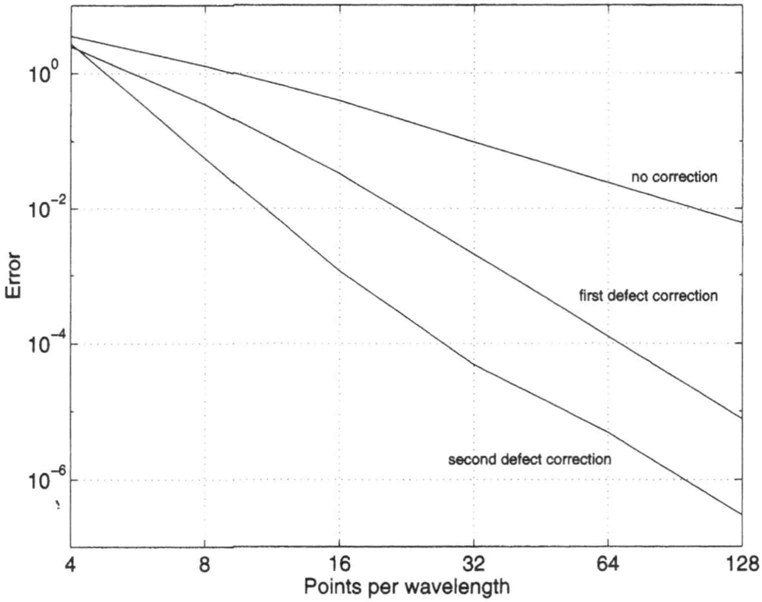

Figure 5.1. L^2 error in the numerical approximation to $u(x)$

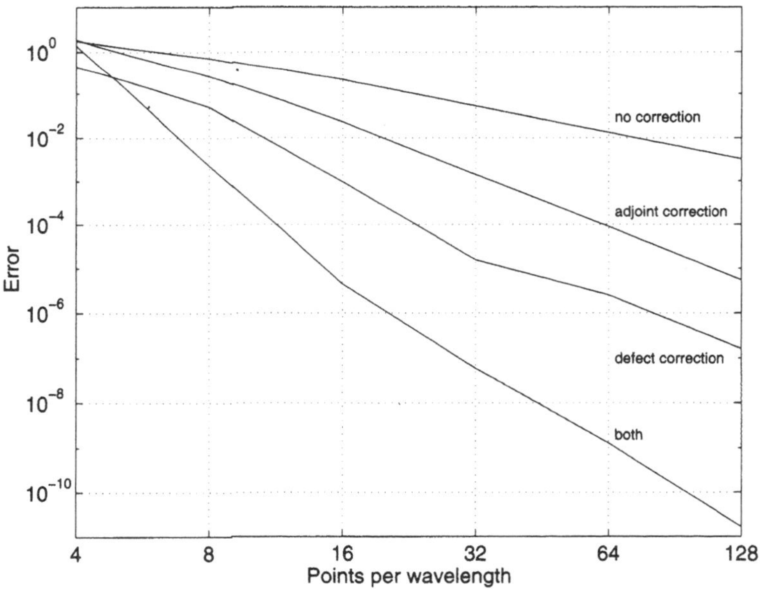

Figure 5.2. Error in the numerical approximation to $u(10)$

magnitude and therefore fourth-order in absolute magnitude. It is this error that is corrected by a second application of the defect correction procedure.

Figure 5.2 shows the error in the numerical value for the output functional $u(10)$. Without any correction, the error is second-order. Using either defect correction or adjoint error correction on their own increases the order of accuracy to fourth-order, but using them both increases the accuracy to sixth-order. Note that the calculation with 8 points per wavelength plus both defect and adjoint error correction gives an error which is approximately 2×10^{-3}. This is more accurate than the calculation with 128 points per wavelength and no corrections, and comparable to the results using 14 points and defect correction, or 30 points with adjoint error correction.

In 3D, the computational cost is proportional to the cube of the number of points per wavelength, so this indicates the potentially huge savings offered by the combination of defect and adjoint error correction. The cost of computing the corrections is five times the cost of the original calculation, due to the additional two calculations for the defect correction, and the one adjoint calculation plus its two defect corrections. In practice, the second defect correction for the primal and adjoint calculations make negligible difference to the value obtained after the adjoint error correction, so these can be omitted, reducing the cost of the corrections to just three times the cost of the original calculation.

6. Reconstruction with biharmonic smoothing

Up to now, we have relied on using cubic spline reconstruction in one space dimension, or its tensor-product version on multidimensional Cartesian product meshes. Here, we consider an alternative reconstruction technique that is applicable on more general nonuniform triangulations.

Suppose that we have a partial differential equation with analytic solution $u \in H^5(\Omega)$, $\Omega = (0,1)^n$, with periodic boundary conditions.

Let u^h be a numerical approximation to u with

$$\|u - u^h\|_{L^2(\Omega)} = \mathcal{O}(h^2).$$

Nothing is assumed about the accuracy of ∇u^h, but in practice, if u^h has come from a piecewise linear finite element approximation, then it will be only a first-order accurate approximation to ∇u in the $L^2(\Omega)$-norm.

We now define a new approximate solution $\tilde{u}$ by

$$h^s \Delta^2 \tilde{u} + \tilde{u} = u^h \tag{6.1}$$

subject to periodic boundary conditions on $\Omega = (0,1)^n$. The purpose of the biharmonic term is to smooth the solution to improve the order of accuracy of the derivative $\nabla \tilde{u}$.

6.1. General analysis

It follows immediately that

$$h^s\,\Delta^2(\tilde{u}-u)+(\tilde{u}-u)=(u^h-u)-h^s\,\Delta^2 u.$$

The analysis proceeds by splitting the error into two components

$$\tilde{u}-u=e+f,$$

with

$$h^s\,\Delta^2 e+e=u^h-u, \tag{6.2}$$

and

$$h^s\,\Delta^2 f+f=-h^s\,\Delta^2 u, \tag{6.3}$$

where e and f satisfy periodic boundary conditions.

Considering equation (6.3) first, multiplying by f and integrating by parts gives

$$h^s\,\|\Delta f\|^2_{L^2(\Omega)}+\|f\|^2_{L^2(\Omega)}=-h^s\,(\Delta^2 u,f)\ \le\ \frac{1}{2}\big(\|h^s\,\Delta^2 u\|^2_{L^2(\Omega)}+\|f\|^2_{L^2(\Omega)}\big),$$

and hence

$$\|f\|_{L^2(\Omega)}\ \le\ h^s\,\|\Delta^2 u\|_{L^2(\Omega)}.$$

Furthermore, taking the gradient of equation (6.3), multiplying both sides by ∇f, and integrating by parts yields similarly that

$$\|\nabla f\|_{L^2(\Omega)}\ \le\ h^s\,\|\Delta^2\nabla u\|_{L^2(\Omega)}.$$

Thus,

$$\|f\|_{H^1(\Omega)}=\mathcal{O}(h^s).$$

Turning now to equation (6.2), multiplying by e and integrating by parts yields

$$h^s\,\|\Delta e\|^2_{L^2(\Omega)}+\|e\|^2_{L^2(\Omega)}\ =\ (u^h-u,e)\ \le\ \frac{1}{2}\big(\|u^h-u\|^2_{L^2(\Omega)}+\|e\|^2_{L^2(\Omega)}\big),$$

and hence

$$\|e\|_{L^2(\Omega)}\ \le\ \|u^h-u\|_{L^2(\Omega)},$$

and

$$h^{s/2}\,\|\Delta e\|_{L^2(\Omega)}\ \le\ \|u^h-u\|_{L^2(\Omega)}.$$

Hence, by the Gagliargo–Nirenberg inequality, we deduce that

$$h^{s/4}\,\|\nabla e\|_{L^2(\Omega)}\ \le\ \|u^h-u\|_{L^2(\Omega)},$$

and thus

$$\|\nabla e\|_{H^1(\Omega)}\ \le\ \mathcal{O}(h^{2-s/4}).$$

Combining the bounds for e and f yields the final result that

$$\|\tilde{u} - u\|_{H^1(\Omega)} \le \mathcal{O}(h^p), \quad p = \min(s, 2 - s/4).$$

The power p is maximized when $s = \frac{8}{5}$, giving $p = \frac{8}{5}$. On the other hand, choosing $s=2$ gives $p=\frac{3}{2}$.

6.2. Fourier analysis

Because we assume periodic boundary conditions, the error component e can be calculated through Fourier analysis. Writing

$$u^h - u = \sum_{\boldsymbol{n}\in\mathbb{Z}^n} a_{\boldsymbol{n}} \exp(2\pi i\, \boldsymbol{n}\cdot\boldsymbol{x}),$$

with $a_{-\boldsymbol{n}} = \bar{a}_{\boldsymbol{n}}$, it follows that

$$e = \sum_{\boldsymbol{n}\in\mathbb{Z}^n} G_{\boldsymbol{n}} a_{\boldsymbol{n}} \exp(2\pi i\, \boldsymbol{n}\cdot\boldsymbol{x}),$$

where

$$G_{\boldsymbol{n}} = \frac{1}{1 + 16\pi^4 h^s |\boldsymbol{n}|^4}.$$

Now,

$$\|u^h - u\|^2_{L^2(\Omega)} = \sum_{\boldsymbol{n}\in\mathbb{Z}^n} |a_{\boldsymbol{n}}|^2 = \mathcal{O}(h^4),$$

and

$$\|e\|^2_{H^1(\Omega)} = \sum_{\boldsymbol{n}\in\mathbb{Z}^n} H_{\boldsymbol{n}} |a_{\boldsymbol{n}}|^2,$$

where

$$H_{\boldsymbol{n}} = \left(1 + 4\pi^2 |\boldsymbol{n}|^2\right) G^2_{\boldsymbol{n}} = \frac{1 + 4\pi^2 |\boldsymbol{n}|^2}{(1 + 16\pi^4 h^s |\boldsymbol{n}|^4)^2}.$$

When $|\boldsymbol{n}| = \mathcal{O}(1)$, $H_{\boldsymbol{n}} = \mathcal{O}(1)$, and when $|\boldsymbol{n}| = \mathcal{O}(h^{-1})$, $H_{\boldsymbol{n}} = \mathcal{O}(h^{6-2s})$. Hence, provided $s \le 3$ and most of the 'energy' of $u^h - u$ is contained in the lowest and highest wave numbers, then

$$\|e\|_{H^1(\Omega)} = \mathcal{O}(h^2).$$

However, $H_{\boldsymbol{n}}$ is a maximum when $|\boldsymbol{n}| = \mathcal{O}(h^{-s/4})$, in which case $H_{\boldsymbol{n}} = \mathcal{O}(h^{-s/2})$. If all of the 'energy' of $u^h - u$ is at these wavelengths, then

$$\|e\|_{H^1(\Omega)} = \mathcal{O}(h^{2-s/4}),$$

which corresponds to the bound obtained from the general analysis.

Thus, the general analysis gives error bounds which are tight with respect to the order of accuracy, but this order is only achieved if most of the initial solution error is in a certain intermediate wave number range. In practice, it seems more likely that the initial solution error will lie in the lowest and highest wave numbers, in which case we will obtain

$$\|\tilde{u} - u\|_{H^1(\Omega)} = \mathcal{O}(h^2),$$

if we use $s=2$.

7. *A posteriori* error estimation by duality

The purpose of this section is to develop another application of duality. Here we shall be concerned with the derivation of *a posteriori* bounds on the error in an output functional of the solution to a differential equation. For recent surveys of the subject of *a posteriori* estimation, see Eriksson, Estep, Hansbo and Johnson (1995), Süli (1998), Becker and Rannacher (2001), and the monographs of Ainsworth and Oden (2000) and Verfürth (1996). In order to motivate the key ideas, it is helpful to begin with a specific example, following Giles *et al.* (1997).

7.1. Elliptic model problem: approximation of the normal flux

Let Ω be a bounded domain in $\mathbb{R}^n$ with Lipschitz-continuous boundary Γ. Given that ψ is an element of $H^{1/2}(\Gamma)$, we let $H^1_{-\psi}(\Omega)$ denote the space of all v in $H^1(\Omega)$ whose trace, $\gamma_{0,\Gamma}(v)$, on Γ is equal to $-\psi$.

We consider the boundary value problem

$$-\nabla \cdot \boldsymbol{\sigma}(u) = f \qquad \text{in } \Omega, \qquad u = 0 \quad \text{on } \Gamma, \tag{7.1}$$

where $f \in L^2(\Omega)$ and $\boldsymbol{\sigma}(u) = A\nabla u$, with A an $n \times n$ matrix-function, uniformly positive definite on $\bar{\Omega}$, with continuous real-valued entries defined on $\bar{\Omega}$. This problem has a unique weak solution $u \in H^1_0(\Omega)$, satisfying

$$B(u, v) = (f, v) \qquad \forall v \in H^1_0(\Omega). \tag{7.2}$$

Here and below $(\cdot\,,\cdot)$ denotes the inner product in $L^2(\Omega)$ and

$$B(v, w) = (\boldsymbol{\sigma}(v), \nabla w) = (\nabla v, A^{\mathrm{T}} \nabla w),$$

for $v, w \in H^1(\Omega)$, where A^{T} is the transpose of A.

Let us suppose that the quantity of interest in the outward normal flux through Γ defined by

$$N(u) = \int_\Gamma \boldsymbol{\nu} \cdot \boldsymbol{\sigma}(u) \, \mathrm{d}s,$$

where $\boldsymbol{\nu}$ denotes the unit outward normal vector to Γ. In order to compute

$N(u)$, for $u \in H_0^1(\Omega)$ denoting the weak solution to problem (7.1) and $\psi \in H^{1/2}(\Gamma)$, we consider the slightly more general problem of computing the weighted normal flux through the boundary, defined by

$$\mathcal{J}_p(u) = N_\psi(u) = \int_\Gamma \boldsymbol{\nu} \cdot \boldsymbol{\sigma}(u)\, \psi \,\mathrm{d}s. \tag{7.3}$$

We note that, since $\boldsymbol{\sigma}(u) \in [L^2(\Omega)]^n$ and $\nabla \cdot \boldsymbol{\sigma}(u) \in L^2(\Omega)$, according to the Trace Theorem for the function space $H(\mathrm{div}, \Omega)$ (see Theorem 2.2 in Girault and Raviart (1986)), the normal stress $\boldsymbol{\nu} \cdot \boldsymbol{\sigma}(u)|_\Gamma$ is correctly defined as an element of $H^{-1/2}(\Gamma)$, and $N_\psi(u)$ is meaningful, provided that the integral over Γ is interpreted as a duality pairing between $H^{-1/2}(\Gamma)$ and $H^{1/2}(\Gamma)$. Moreover, applying a generalization of Green's identity (see Theorem 2.2 in Girault and Raviart (1986)), we deduce that, for any $v \in H^1_{-\psi}(\Omega)$,

$$\begin{aligned} N_\psi(u) &= (f, v) - (\boldsymbol{\sigma}(u), \nabla v) \\ &= (f, v) - B(u, v). \end{aligned} \tag{7.4}$$

Because of (7.2), the value of the expression $(f, v) - B(u, v)$ on the right-hand side of (7.4) is independent of the choice of $v \in H^1_{-\psi}(\Omega)$. Thus, (7.4) can be interpreted as an equivalent (and correct) definition of the weighted normal flux (7.3) of u across Γ.

Next, we construct our finite element approximation to $N_\psi(u)$. We consider a family of finite-dimensional Galerkin trial spaces U^h and test spaces $V^h = U^h$ contained in $H^1(\Omega)$, which consist of continuous piecewise polynomials of degree k defined on a family of regular subdivisions $\mathcal{T}_h$ of Ω into open n-dimensional simplices κ. We denote the diameter of a simplex $\kappa \in \mathcal{T}_h$ by h_κ and assume that the family $\{\mathcal{T}_h\}$ is shape-regular, that is, there exists a positive constant c such that $\mathrm{meas}(\kappa) \geq c\, h_\kappa^n$ for all $\kappa \in \mathcal{T}_h$ and all $\mathcal{T}_h$, where $\mathrm{meas}(\kappa)$ is the n-dimensional volume of κ. Further, for each function $\psi \in H^{1/2}(\Gamma)$ such that $-\psi = v|_\Gamma$ for some $v \in U^h$, we let $U^h_{-\psi} \subset U^h$ be the space of all $w \in U^h$ with $w|_\Gamma = -\psi$. In particular, U_0^h is the space of all $v \in U^h$ which vanish on Γ. Clearly, $U_0^h \subset H_0^1(\Omega)$.

The finite element approximation of (7.1) is defined as follows: find $u^h \in U_0^h$ such that

$$B(u^h, v^h) = (f, v^h) \qquad \text{for all } v^h \in U_0^h. \tag{7.5}$$

Motivated by the identity (7.4), we define the approximation $N_\psi^h(u^h)$ to $N_\psi(u)$ as follows:

$$N_\psi^h(u^h) = (f, v^h) - B(u^h, v^h), \quad v^h \in U^h_{-\psi}. \tag{7.6}$$

We note that, because of (7.5), $N_\psi^h(u^h)$ is independent of the choice of $v^h \in U_{-\psi}^h$. Furthermore, we observe that, in general,

$$N_\psi^h(u^h) \neq \int_\Gamma \boldsymbol{\nu} \cdot \boldsymbol{\sigma}(u^h)\, \psi \,\mathrm{d}s = N_\psi(u^h),$$

in contrast with identity (7.4) satisfied by the analytical solution u. This raises the question as to which of $N_\psi(u^h)$ and $N_\psi^h(u^h)$ is the more accurate approximation to $N_\psi(u)$. As we shall see, the answer to this question is: $N_\psi^h(u^h)$. Indeed, the error estimate in Theorem 7.2 below shows that, for sufficiently smooth data and continuous piecewise polynomial finite elements of degree k, the order of convergence of $N_\psi^h(u^h)$ to $N_\psi(u)$ is $\mathcal{O}(h^{2k})$. In general, this high order of convergence cannot be achieved by using the 'naive' approximation $N_\psi(u^h) = \int_\Gamma \boldsymbol{\nu} \cdot \boldsymbol{\sigma}(u^h)\,\mathrm{d}s$.

In order to derive a representation formula for the error $N_\psi(u) - N_\psi^h(u^h)$ in the boundary flux, we introduce the following dual problem in variational form: find $z \in H_{-\psi}^1(\Omega)$ such that

$$B(v,z) = 0 \qquad \text{for all } v \in H_0^1(\Omega). \tag{7.7}$$

Consider the global error $e = u - u^h$. Upon setting $v = e$ in (7.7) we obtain

$$0 = B(e,z) = B(e, z - \pi^h z) + B(e, \pi^h z), \tag{7.8}$$

where we made use of the fact that the error e is zero on the boundary Γ; here $\pi^h : H_{-\psi}^1(\Omega) \to U_{-\psi}^h$ is a linear operator satisfying the approximation property (7.10) below. Since the definitions of $N_\psi(u)$ and $N_\psi^h(u^h)$ are independent of the choice of $v \in H_{-\psi}^1(\Omega)$ and $v^h \in U_{-\psi}^h$, respectively, we deduce that

$$\begin{aligned} B(e, \pi^h z) &= \left((f, \pi^h z) - B(u^h, \pi^h z)\right) - \left((f, \pi^h z) - B(u, \pi^h z)\right) \\ &= N_\psi^h(u^h) - N_\psi(u). \end{aligned}$$

On substituting this into (7.8) we arrive at the error representation formula

$$\begin{aligned} N_\psi(u) - N_\psi^h(u^h) &= B(e, z - \pi^h z) \\ &= B(u, z - \pi^h z) - B(u^h, z - \pi^h z) \\ &= (f, z - \pi^h z) - B(u^h, z - \pi^h z) \end{aligned} \tag{7.9}$$

where, to obtain the last equality, we made use of the fact that u obeys (7.2) and $z - \pi^h z$ belongs to $H_0^1(\Omega)$.

Our aim is to investigate the problem of approximating $N_\psi(u)$ from two points of view. First, we analyse the convergence rate of the approximation through an *a priori* error analysis following Babuška and Miller (1984*b*) and Barrett and Elliott (1987); we shall then perform an *a posteriori* error analysis and highlight the relevance of the *a posteriori* error bound for adaptive mesh refinement.

***A priori* error analysis.** For the purposes of the *a priori* error analysis we assume that there exists a linear operator $\pi^h : H^1_{-\psi}(\Omega) \to U^h_{-\psi}$ and a positive constant c such that

$$|v - \pi^h v|_{H^1(\Omega)} \leq ch^{s-1}|v|_{H^s(\Omega)}, \quad 1 \leq s \leq k+1, \tag{7.10}$$

for all $v \in H^s(\Omega)$, $1 \leq s \leq k+1$, and all $h = \max_{\kappa \in \mathcal{T}_h} h_\kappa$ (≤ 1, say).

The next theorem is a direct consequence of inequality (3.7) from Corollary 3.2.

Theorem 7.1. Assume that (7.10) holds, $u \in H^s(\Omega) \cap H^1_0(\Omega)$, $s \geq 1$, and $z \in H^t(\Omega) \cap H^1_0(\Omega)$, $t \geq 1$, where u and z are the solutions to (7.2) and (7.7), respectively. Then,

$$|N_\psi(u) - N^h_\psi(u^h)| \leq ch^{\sigma+\tau-2}|u|_{H^\sigma(\Omega)}\, |z|_{H^\tau(\Omega)}, \tag{7.11}$$

where $1 \leq \sigma \leq \min(s, k+1)$, $1 \leq \tau \leq \min(t, k+1)$, c is a constant, and $h = \max_{\kappa \in \mathcal{T}_h} h_\kappa$.

Proof. It follows from the error representation formula (7.9) that

$$|N_\psi(u) - N^h_\psi(u^h)| \leq c|e|_{H^1(\Omega)}\, |z - \pi^h z|_{H^1(\Omega)}.$$

A standard energy-norm error estimate gives

$$|e|_{H^1(\Omega)} \leq ch^{\sigma-1}|u|_{H^\sigma(\Omega)}, \quad 1 \leq \sigma \leq \min(s, k+1).$$

Further, using the approximation property (7.10), we obtain

$$|z - \pi^h z|_{H^1(\Omega)} \leq ch^{\tau-1}|z|_{H^\tau(\Omega)}, \quad 1 \leq \tau \leq \min(t, k+1),$$

Consequently,

$$|N_\psi(u) - N^h_\psi(u^h)| \leq ch^{\sigma+\tau-2}|u|_{H^\sigma(\Omega)}|z|_{H^\tau(\Omega)},$$

for $1 \leq \tau \leq \min(t, k+1)$, $1 \leq \sigma \leq \min(s, k+1)$. □

A numerical experiment. We include a numerical experiment to illustrate the point that $N^h_\psi(u^h)$ is a more accurate approximation to $N_\psi(u)$ than $N(u^h)$ is. Let us consider Laplace's equation in cylindrical polar coordinates given by

$$\frac{\partial^2 u}{\partial r^2} + \frac{1}{r}\frac{\partial u}{\partial r} + \frac{\partial^2 u}{\partial z^2} = 0, \tag{7.12}$$

with boundary conditions

$$\begin{aligned} u &= 0, \quad r \le 1, \; z = 0, \\ \frac{\partial u}{\partial n} &= 0, \quad r > 1, \; z = 0, \\ & \qquad\; r = 0, \; z \ge 0, \\ u &= 1, \quad r, \, z \to \infty, \end{aligned} \tag{7.13}$$

and suppose that the quantity of interest is the weighted normal flux through a part of the boundary so that

$$N_\psi(u) = \int_\Gamma \frac{\partial u}{\partial \nu} \psi r \, \mathrm{d}s, \tag{7.14}$$

where

$$\psi = \begin{cases} -\frac{\pi}{2} & 0 \le r \le 1, \quad z = 0, \\ 0 & \text{elsewhere on } \Gamma. \end{cases} \tag{7.15}$$

It can be shown that the exact solution is $N_\psi(u) = 1$. As described above, we may rewrite $N_\psi(u)$ as

$$N_\psi(u) = -\int_\Omega \nabla u \cdot \nabla v \, r \mathrm{d}r \, \mathrm{d}z \tag{7.16}$$

for any $v \in H^1_{-\psi}(\Omega)$ (where $H^1_{-\psi}(\Omega)$ is defined as before, except that the measure is now $r\mathrm{d}r\,\mathrm{d}z$ rather than $\mathrm{d}x\,\mathrm{d}y$), and we denote the corresponding numerical approximation to (7.16) by $N^h_\psi(u^h)$. To show that $N^h_\psi(u^h)$ is a more accurate approximation to $N_\psi(u)$ than $N_\psi(u^h)$, we consider the convergence of these two quantities to $N_\psi(u) = 1$ on a sequence of regular meshes of mesh size h using a piecewise linear finite element method to compute u^h. The results are shown in Table 7.1, from which it is clear that the approximation $N^h_\psi(u^h)$ converges to $N_\psi(u)$ at over twice the rate at which $N_\psi(u^h)$ converges to $N_\psi(u)$; note that $u \in H^{3/2-\varepsilon}(\Omega)$, $\varepsilon > 0$, thus leading to approximately first-order convergence by virtue of our *a priori* error estimate from the last theorem.

Table 7.1. The numerical approximations $N^h_\psi(u^h)$ and $N_\psi(u^h)$ to $N_\psi(u)$

h	$N_\psi(u^h)$	|error|	order	$N^h_\psi(u^h)$	|error|	order
0.5	0.451	0.549		1.270	0.270	
0.25	0.551	0.449	0.29	1.129	0.129	1.06
0.125	0.644	0.356	0.33	1.064	0.064	1.02
0.0625	0.725	0.275	0.37	1.032	0.032	1.01
0.03125	0.793	0.207	0.41	1.016	0.016	1.00

***A posteriori* error analysis.** We adopt the following local approximation property. There exists a linear operator $\pi^h : H^1_{-\psi}(\Omega) \to U^h_{-\psi}$ and a positive constant c such that

$$\|v - \pi^h v\|_{L^2(\kappa)} + h_\kappa |v - \pi^h v|_{H^1(\kappa)} \leq c h^s_{\hat{\kappa}} |v|_{H^s(\hat{\kappa})}, \quad 1 \leq s \leq k+1, \quad (7.17)$$

for all $v \in H^s(\hat{\kappa})$ and each $\kappa \in \mathcal{T}_h$; here $\hat{\kappa}$ denotes the union of all elements (including κ itself) in the partition $\mathcal{T}_h$ whose closure has non-empty intersection with the closure of κ, and

$$h_{\hat{\kappa}} = \max_{\sigma \in \mathcal{T}_h;\, \sigma \subset \hat{\kappa}} h_\sigma.$$

For the proof of existence of the 'quasi-interpolation' operator π^h we refer to Brenner and Scott (1994), for example.

In addition, we make the following assumption concerning the regularity of the dual problem: there exists a real number $t \geq 1$ such that, for every τ, $1 \leq \tau \leq t$, there is a positive constant C_τ, independent of ψ, such that the solution z to the dual problem (7.7) satisfies the estimate

$$|z|_{H^\tau(\Omega)} \leq C_\tau \|\psi\|_{H^{\tau-1/2}(\Gamma)} \quad (7.18)$$

whenever $\psi \in H^{\tau-1/2}(\Gamma)$. For instance, this bound holds when $\Gamma \in C^{\tau-1,1}$ and the entries of A belong to $C^{[\tau]}(\Omega)$; see Gilbarg and Trudinger (1983).

For each triangle $\kappa \in \mathcal{T}_h$ and u^h denoting the solution to (7.5) in U^h_0, we introduce the *residual term* $\mathcal{R}_\kappa(u^h)$ by

$$\mathcal{R}_\kappa(u^h) = \|\nabla \cdot \boldsymbol{\sigma}(u^h) + f\|_{L^2(\kappa)} + h_{\hat{\kappa}}^{-1/2} \|[\mathbf{n} \cdot \boldsymbol{\sigma}(u^h)]/2\|_{L^2(\partial\kappa \setminus \Gamma)}, \quad (7.19)$$

where $[w]$ is the jump in w across the faces of elements in the partition and $\mathbf{n}$ is the unit outward normal vector to $\partial\kappa$.

Theorem 7.2. Suppose that (7.10) and (7.18) hold and that the weight function ψ belongs to $H^{t-1/2}(\Gamma)$, $t \geq 1$; then we have that

$$|N_\psi(u) - N^h_\psi(u^h)| \leq c \sum_{\kappa \in \mathcal{T}_h} \mathcal{R}_\kappa(u^h) \min_{\{\tau : 1 \leq \tau \leq \min(t,k+1)\}} h^\tau_{\hat{\kappa}} \omega_{\kappa,\tau}, \quad (7.20)$$

where c is a constant, and the local weight $\omega_{\kappa,\tau}$ is defined by

$$\omega_{\kappa,\tau} = |z|_{H^\tau(\hat{\kappa})},$$

and z is the weak solution of (7.7).

Proof. The starting point of the proof is the third line of the error representation formula (7.9); we integrate by parts triangle-by-triangle using

Green's identity to deduce that

$$N_\psi(u) - N_\psi^h(u^h) = (f, z - \pi^h z) - (\boldsymbol{\sigma}(u^h), \nabla(z - \pi^h z)) \tag{7.21}$$
$$= \sum_{\kappa\in\mathcal{T}_h} \int_\kappa (f + \nabla\cdot\boldsymbol{\sigma}(u^h))(z - \pi^h z)\,\mathrm{d}x$$
$$- \sum_{\kappa\in\mathcal{T}_h} \int_{\partial\kappa\setminus\Gamma} ([\mathbf{n}\cdot\boldsymbol{\sigma}(u^h)]/2)(z - \pi^h z)\,\mathrm{d}s$$
$$= \mathrm{I} + \mathrm{II},$$

where we made use of the fact that z, the weak solution of the dual problem (7.7), belongs to $C^{0,\alpha}(\Omega)$, for some α in $(0,1)$ (see Theorem 5.24 in Gilbarg and Trudinger (1983)), so that the jump $[z - \pi^h z](x) = [z](x) = 0$ at any point x of an internal face $\partial\kappa \setminus \Gamma$ for each element κ in the partition.

Next we estimate expressions I and II. In I, we apply the Cauchy–Schwarz inequality and the approximation property (7.17); hence,

$$|\mathrm{I}| \le c \sum_{\kappa\in\mathcal{T}_h} \|f + \nabla\cdot\sigma(u^h)\|_{L^2(\kappa)} \min_{\{\tau\,:\,1\le\tau\le\min(t,k+1)\}} h_{\hat{\kappa}}^{\tau}|z|_{H^\tau(\hat{\kappa})}.$$

Now, we consider II. We begin by recalling that the multiplicative trace inequality

$$\|w\|^2_{L^2(\partial\kappa)} \le c\|w\|_{L^2(\kappa)}\left(h_\kappa^{-1}\|w\|_{L^2(\kappa)} + |w|_{H^1(\kappa)}\right) \qquad \forall w \in H^1(\kappa),\ \kappa\in\mathcal{T}_h \tag{7.22}$$

(see Brenner and Scott (1994)), followed by application of approximation property (7.17), yields

$$\|z - \pi^h z\|_{L^2(\partial\kappa)} \le c h_{\hat{\kappa}}^{\tau-1/2}|z|_{H^\tau(\hat{\kappa})}.$$

Thus,

$$|\mathrm{II}| \le c \sum_{\kappa\in\mathcal{T}_h} \|[\mathbf{n}\cdot\boldsymbol{\sigma}(u^h)]\|_{L^2(\partial\kappa\setminus\Gamma)} \min_{\{\tau\,:\,1\le\tau\le\min(t,k+1)\}} h_{\hat{\kappa}}^{\tau-1/2}|z|_{H^\tau(\hat{\kappa})}.$$

Substituting the bounds on I and II into (7.21) and recalling the definition of the residual term (7.19), we deduce (7.20). □

Since the data for the dual problem (7.7) is generated by a known function, ψ, we may calculate the weight $\omega_{\tau,\alpha}$ by approximating the solution of the dual problem numerically. The right-hand side of the *a posteriori* error estimate can thus be used for quantitative error estimation and local mesh adaptation.

Suppose, for example, that given a positive tolerance TOL, the aim of the computation is to find $N_\psi^h(u^h)$ such that

$$|N_\psi(u) - N_\psi^h(u^h)| \le \mathtt{TOL}. \tag{7.23}$$

In order to achieve (7.23), by virtue of (7.20) it suffices to ensure that

$$c \sum_{\kappa \in \mathcal{T}_h} \mathcal{R}_\kappa(u^h) \min_{\{\tau\,:\,1 \leq \tau \leq \min(t,k+1)\}} h_\kappa^\tau \omega_{\kappa,\tau} \leq \texttt{TOL}. \tag{7.24}$$

This inequality can now be used as a *stopping criterion* in an adaptive mesh refinement algorithm. A second ingredient of an adaptive algorithm is a local *refinement criterion*; assuming that N denotes the number of elements in $\mathcal{T}_h$, a possible refinement criterion might involve checking, on each element $\kappa \in \mathcal{T}_h$, whether

$$c\mathcal{R}_\kappa(u^h) \min_{\{\tau\,:\,1 \leq \tau \leq \min(t,k+1)\}} h_\kappa^\tau \omega_{\kappa,\tau} \leq \frac{\texttt{TOL}}{N}. \tag{7.25}$$

If (7.25) is satisfied on an element κ, then κ is accepted as being of adequate size; if, on the other hand, (7.25) is violated then κ is refined. After (7.25) has been checked on each element κ in $\mathcal{T}^h$ and a new, finer, subdivision $\mathcal{T}_{h'}$ has been generated, a new solution $u^{h'}$ is computed on $\mathcal{T}_{h'}$, thus completing a single step of the adaptive algorithm. Adaptation proceeds until the stopping criterion (7.24) is satisfied. The adaptive algorithm then terminates and delivers $N_\psi^h(u^h)$, accurate to within the specified tolerance TOL, as required by (7.23).

For extensions of the theory discussed in this section to superconvergent lift and drag computations for the Stokes and Navier–Stokes equations, we refer to Giles *et al.* (1997). The technique of postprocessing presented here is based on early ideas of Wheeler (1973), Babuška and Miller (1984*a*, 1984*b*, 1984*c*); see also Barrett and Elliott (1987) for a rigorous error analysis in the presence of variational crimes.

A numerical experiment. The purpose of this experiment is to illustrate the performance of the *a posteriori* error bound (7.20) and to compare it with some heuristic mesh refinement criteria. Let us consider a reaction–diffusion equation in cylindrical polar coordinates,

$$-\nabla^2 u + Ku = 0,$$

in an L-shaped domain with boundary conditions

$$\begin{aligned} u = 1, &\quad r \leq 1,\ z = 0, \\ u \to 0, &\quad r,\ z \to \infty, \\ \frac{\partial u}{\partial \nu} = 0, &\quad r = 0,\ z > 0, \\ &\quad r = 1,\ 0 \leq z \leq 0.5, \\ &\quad r > 1,\ z = 0.5, \end{aligned}$$

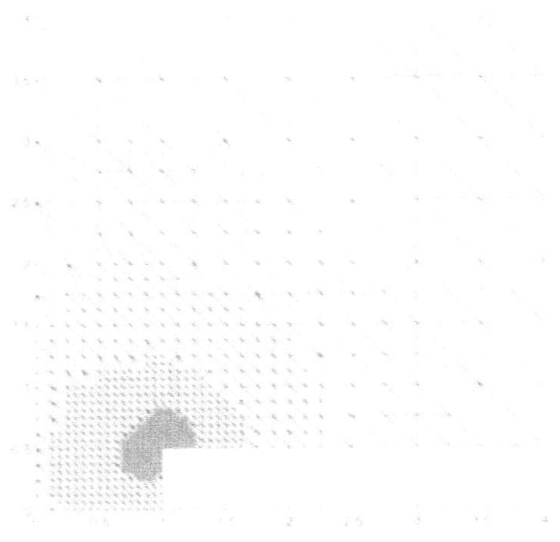

(a) final mesh for $K = 1$, 745 nodes, 1405 triangles

(b) final mesh for $K = 10$, 679 nodes, 1262 triangles

(c) final mesh for $K = 100$, 972 nodes, 1822 triangles

(d) final mesh for $K = 1000$, 3904 nodes, 7494 triangles

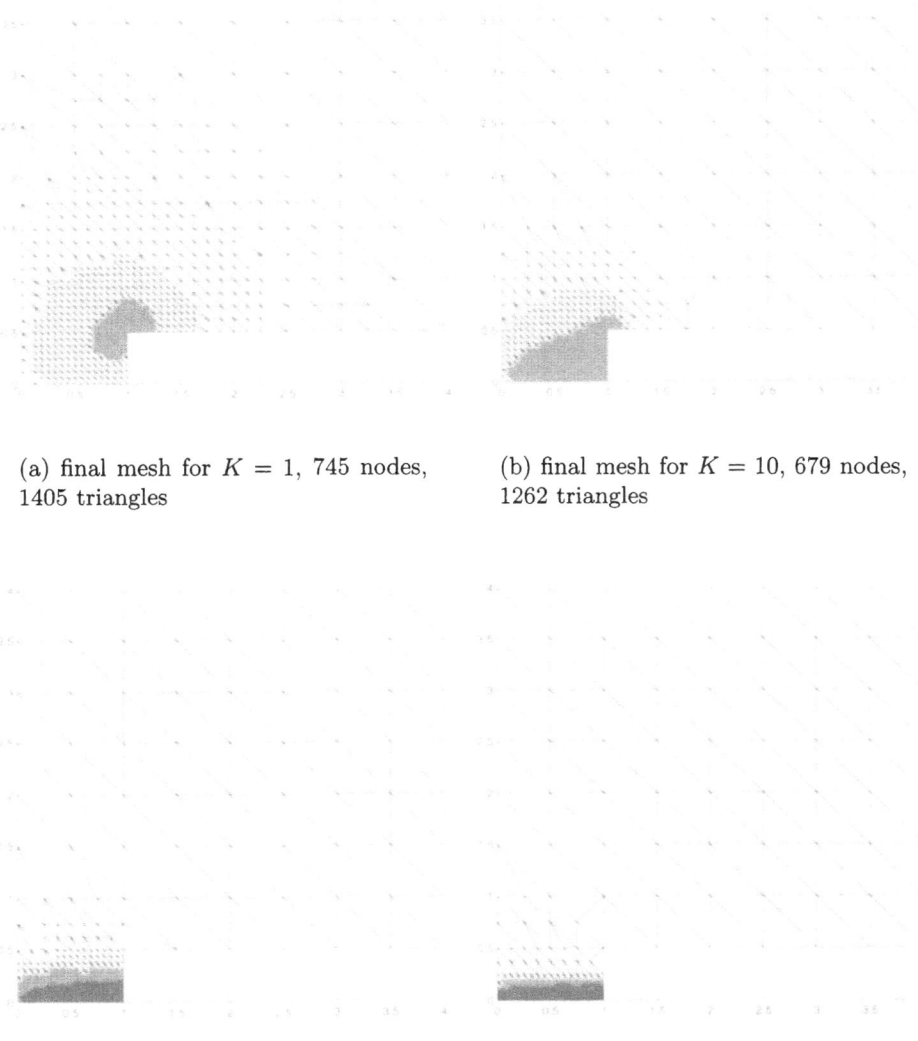

Figure 7.1. The final meshes for calculating the linear functional using a refinement indicator based on (7.24)

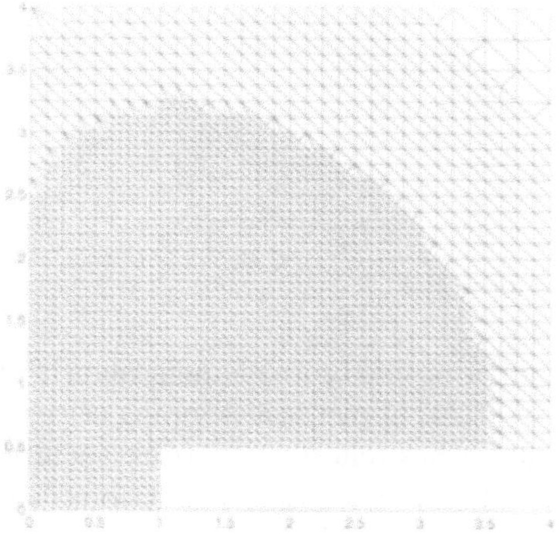

(a) final mesh for $K = 1$, 3311 nodes, 6424 triangles

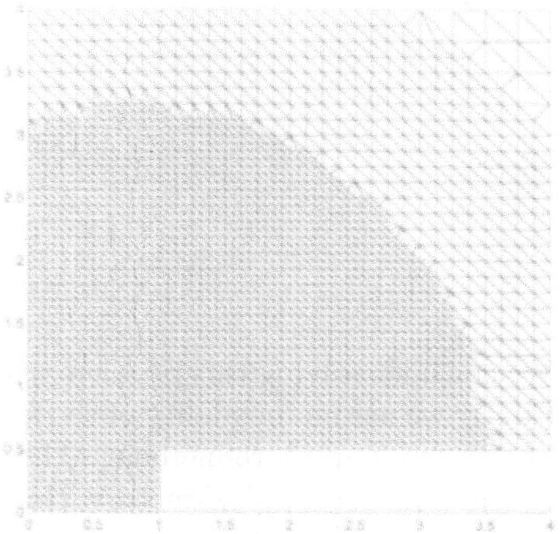

(b) final mesh for $K = 10$, 3277 nodes, 6352 triangles

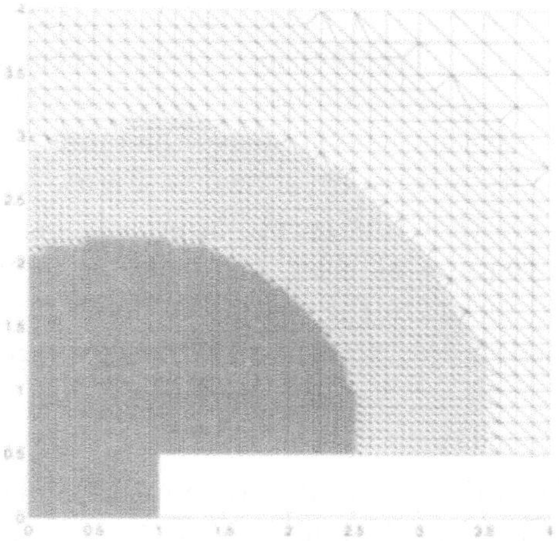

(c) final mesh for $K = 100$, 6463 nodes, 12636 triangles

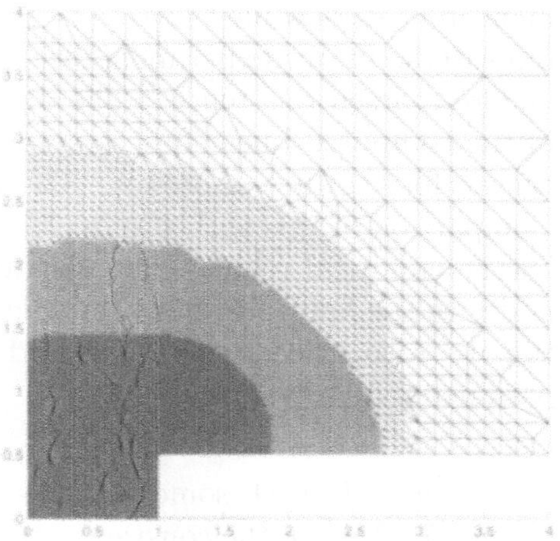

(d) final mesh for $K = 1000$, 24753 nodes, 48941 triangles

Figure 7.2. The final meshes produced using the empirical error indicator $\|\nabla u^h\|_{L^2(\kappa)}$

and suppose we wish to evaluate the linear functional

$$N_\psi(u) = \int_\Gamma \frac{\partial u}{\partial \nu} \psi \, r \, \mathrm{d}s,$$

where

$$\psi = \begin{cases} \frac{\pi}{2} & 0 \le r \le 1, \quad z = 0, \\ 0 & \text{elsewhere on } \Gamma. \end{cases}$$

We consider solving this problem using an adaptive finite element algorithm. Firstly we use an inequality of the form of (7.24) above as a stopping criterion and to guide mesh refinement. The final meshes (chosen to ensure accuracy of the numerical solution to within 1% of the exact linear functional) are shown in Figure 7.1. Secondly, we consider an empirical refinement indicator, namely $\|\nabla u^h\|_{L^2(\kappa)}$. Instead of attempting to equidistribute the error bound we now use the *fixed fraction method* for mesh refinement. In this method we refine and re-coarsen a fixed proportion of the elements at each refinement level. The elements to be refined are those with the largest refinement indicators while those to be re-coarsened have the smallest refinement indicators. In our experiments we chose to refine one-third of the elements and re-coarsen one-tenth. The algorithm was terminated when the error in the computed linear functional was less than the adaptive tolerance TOL chosen for the first approach. The results of this are shown in Figure 7.2. Comparison of this with Figure 7.1 shows that this second approach, based on the empirical refinement indicator, requires at least four times as many nodes as the first to achieve the same accuracy in the computed linear functional. Also, the meshes produced are far less sensitive to the governing parameter K.

We conclude with some remarks. In order to be able to implement the bound (7.20) both the constant c and the dual solution z have to be precomputed. The computation of c in turn involves knowledge of the constants from (7.17) and (7.22); constructive proofs of (7.17) and (7.22) which supply actual values of these constants are available: see, for example, Carstensen (2000). However, it is clear that, while the *a posteriori* error bound (7.20) is reliable (*i.e.*, it consistently overestimates the actual error $|N_\psi(u) - N_\psi^h(u^h)|$), the factor of overestimation may be excessive. Thus, in the next section, we shall develop a minimalistic framework of *Type I a posteriori* error estimation which does not require knowledge of these constants. The guiding principle in the derivation of a Type I error bound is to perform only the absolute minimum in the way of upper bounds on the error representation formula so as to ensure sharpness of the resulting estimate.

7.2. Abstract Type I a posteriori *error estimation*

Let us return to the abstract framework of Section 3.1, and recall that the error between $J_p = \mathcal{J}_p(u)$ and its Galerkin approximation J_p^h is expressed by the error representation formula (3.10), which states that

$$J_p - J_p^h = \langle R_p(u^h), z - z^h \rangle \qquad \forall z^h \in V_d^h, \tag{7.26}$$

where $\langle \cdot\,, \cdot \rangle$ is the duality pairing between V', the dual of the real Hilbert space V, and V; $R_p(u^h) : V \to \mathbb{R}$ denotes the linear functional, called the *residual*, defined by

$$R_p(u^h) \, : \, w \mapsto \ell(w) - B(u^h, w).$$

Now, writing

$$\mathcal{E}_\Omega(u^h; w) \equiv \langle R_p(u^h), w \rangle, \qquad w \in V,$$

and recalling (3.11), we have that

$$\begin{aligned} J_p - J_p^h &= \langle R_p(u^h), z^H - z^h \rangle + \langle R_p(u^h), z - z^H \rangle \\ &\equiv \mathcal{E}_\Omega(u^h; z^H - z^h) + \mathcal{E}_\Omega(u^h; z - z^H) \quad \forall z^h \in V_d^h, \end{aligned} \tag{7.27}$$

where $z^H \in V_d^H$ is the numerical solution of the dual problem (2.13), defined by

$$B(w_0^H, z^H) = m(w_0^H) \qquad \forall w_0^H \in U_0^H.$$

Let us suppose that the aim of the computation is to ensure that, for a given tolerance $\mathtt{TOL} > 0$,

$$|J_p - J_p^h| \le \mathtt{TOL}. \tag{7.28}$$

Clearly, (7.28) is equivalent to requiring that

$$|\mathcal{E}_\Omega(u^h; z - z^h)| \le \mathtt{TOL} \qquad \forall z^h \in V_d^h. \tag{7.29}$$

Hence, a sufficient condition for (7.29) is that

$$|\mathcal{E}_\Omega(u^h; z^H - z^h)| + |\mathcal{E}_\Omega(u^h; z - z^H)| \le \mathtt{TOL} \qquad \forall z^h \in V_d^h. \tag{7.30}$$

In order to ensure that (7.30) holds, we select $\theta \in (0, 1)$, and demand that

$$\mathcal{E}_\mathrm{P} \equiv |\mathcal{E}_\Omega(u^h; z^H - z^h)| \le (1 - \theta)\, \mathtt{TOL} \qquad \forall z^h \in V_d^h, \tag{7.31}$$

$$\mathcal{E}_\mathrm{D} \equiv |\mathcal{E}_\Omega(u^h; z - z^H)| \le \theta\, \mathtt{TOL}. \tag{7.32}$$

Let us discuss each of these two inequalities in detail.

(7.31) The residual $R_p(u^h)$ appearing in $\mathcal{E}_\Omega(u^h; z^H - z^h)$ is computable, since it involves only the numerical solution $u^h \in U_p^h$ and the data (*i.e.*, the functional ℓ, in this case). Further, $z^H \in V_d^H$ is the numerical solution of the dual problem (2.13), defined by

$$B(w_0^H, z^H) = m(w_0^H) \qquad \forall w_0^H \in U_0^H. \tag{7.33}$$

As z^h in (7.31) is an arbitrary element of V_d^h, we need to fix it. It is worth noting at this point that, due to the Galerkin orthogonality property (3.14),

$$\langle R_p(u^h), z^H - z^h\rangle = \langle R_p(u^h), z_0^H - z_0^h\rangle = \langle R_p(u^h), z_0^H\rangle,$$

and therefore the choice of z^h does not influence the value of $\mathcal{E}_\Omega(u^h; z^H - z^h)$. Still, this does not mean that any z^h (such as $z^h = 0$, for example) will be a useful choice. Of course, if $\mathcal{E}_\Omega(u^h; z^H - z^h) = \langle R_p(u^h), z_0^H\rangle$ were all we cared about, the choice of z^h would be immaterial; however, it has to be borne in mind that, in addition to a *stopping criterion* such as (7.31), our adaptive algorithm will also require a *local refinement criterion*. A local refinement criterion can be obtained by localization of the term $\langle R_p(u^h), z^H - z^h\rangle$. By this, we mean the following. Let us suppose that u^h has been computed using a Galerkin finite element method over a subdivision $\mathcal{T}_h$ of the computational domain Ω into finite elements κ. We assume the existence of the following decomposition:

$$\mathcal{E}_\Omega(u^h; w) \equiv \langle R_p(u^h), w\rangle = \sum_{\kappa\in\mathcal{T}_h} \eta_\kappa(u^h|_\kappa, w|_\kappa).$$

Then,

$$\langle R_p(u^h), z^H - z^h\rangle = \sum_{\kappa\subset\mathcal{T}_h} \eta_\kappa(u^h|_\kappa, (z^H - z^h)|_\kappa),$$

and therefore,

$$\begin{aligned} |\mathcal{E}_\Omega(u^h; z^H - z^h)| &\le \sum_{\kappa\in\mathcal{T}_h} |\tilde{\eta}_\kappa| \\ &\equiv \mathcal{E}_{|\Omega|}(u^h; z^H - z^h), \end{aligned} \tag{7.34}$$

where

$$\tilde{\eta}_\kappa = \eta_\kappa(u^h|_\kappa, (z^H - z^h)|_\kappa).$$

Now, $\mathcal{E}_{|\Omega|}(u^h; z^H - z^h)$ is referred to as the *localization* of the expression $|\mathcal{E}_\Omega(u^h; z^H - z^h)|$. While the left-hand side of (7.34) is completely independent of $z^h \in V_d^h$, the right-hand side of this inequality is strongly dependent on z^h because Galerkin orthogonality (*cf.* (3.14)) is a nonlocal property. Ideally, in order to minimize the degree of overestimation in (7.34) which may result from an unfortunate choice of z^h, we would like to choose $z^h \in V_d^h$ so that the right-hand side of (7.34) is as close as possible to the left-hand side. This, of course, is a practically unrealistic demand as it would lead to a complicated optimization problem.

A more reasonable choice from the practical point of view is $z^h = \pi^h z^H$, where $\pi^h : V_d \to V_d^h$ is a finite element interpolation or quasi-interpolation operator; this particular choice of z^h is motivated by the expectation that $\mathcal{E}_{|\Omega|}(u^h; z^H - \pi^h z^H)$ exhibits the same asymptotic behaviour as the expression $\mathcal{E}_\Omega(u^h; z^H - \pi^h z^H) = \mathcal{E}_\Omega(u^h; z_0^H)$ in the limit of $h \to 0$. While this expectation is certainly fulfilled in most situations, it is by no means so in general because, in the presence of global superconvergence effects, $\mathcal{E}_\Omega(u^h; z^H - \pi^h z^H)$ may exhibit a higher rate of convergence than $\mathcal{E}_{|\Omega|}(u^h; z^H - \pi^h z^H)$, as $h \to 0$. However, such global cancellation effects and the related mismatch in the asymptotic behaviour will be largely absent on locally refined unstructured computational meshes, such as those that arise in the course of adaptive mesh refinement.

At any rate, once a choice of z^h has been made, the expression $\mathcal{E}_\text{P}$ is *computable*, and condition (7.31) can be checked. Indeed, a sufficient condition for (7.31), based on localization, is that

$$\mathcal{E}_\text{P}^\text{loc} \equiv \mathcal{E}_{|\Omega|}(u^h; z^H - \pi^h z^H) \leq (1 - \theta)\texttt{TOL}. \tag{7.35}$$

A bound of this kind, which explicitly involves the numerical approximation z^H to the dual solution z, will be referred to as a *Type I a posteriori error bound.*

Next we shall discuss the computation of the dual solution z^H involved in (7.35); we shall see that the correct choice of z^H is closely related to the validity of (7.32).

(7.32) Unlike $\mathcal{E}_\text{P}^\text{loc}$, the term $\mathcal{E}_\text{D}$ involves the (unknown) analytical dual solution z. We note, however, that (7.32) can be restated as a *dual measurement problem* concerned with finding a solution $z^H \in V_d^H$ to (7.33) such that the error in the output functional $\mathcal{E}_\Omega(u^h; \cdot)$ satisfies

$$|\mathcal{E}_\Omega(u^h; z) - \mathcal{E}_\Omega(u^h; z^H)| \leq \theta\,\texttt{TOL}. \tag{7.36}$$

A further important difference between the terms $\mathcal{E}_\text{P}^\text{loc}$ and $\mathcal{E}_\text{D}$ is that (due to the localization) in $\mathcal{E}_\text{P}^\text{loc}$ the absolute value signs appear under the summation over the elements $\kappa \in \mathcal{T}_h$, while in $\mathcal{E}_\text{D}$ the absolute value sign is outside the sum. Thus, we expect $\mathcal{E}_\text{D}$ to be smaller than $\mathcal{E}_\text{P}^\text{loc}$.

Motivated by these observations, we select $\theta \in (0, 1)$ such that $0 < \theta \ll 1$; we then aim to compute $u^h \in U_p^h$ such that

$$\mathcal{E}_\text{P}^\text{loc} \equiv \mathcal{E}_{|\Omega|}(u^h; z^H - \pi^h z^H) \approx (1 - \theta)\,\texttt{TOL}$$

and $z^H \in V_d^H$ such that

$$\mathcal{E}_\text{D} \equiv |\mathcal{E}_\Omega(u^h; z) - \mathcal{E}_\Omega(u^h; z^H)| \leq \theta\,\texttt{TOL}. \tag{7.37}$$

Together, these will imply that $\mathcal{E}_\text{D} \ll \mathcal{E}_\text{P}^\text{loc}$.

The dual measurement problem (7.37) is very similar to the problem (7.28) that we had set out to solve, except that (7.37) concerns the dual solution z while (7.28) involves the primal solution u. To ensure the validity of (7.37) we could derive an *a posteriori* error bound on $|\mathcal{E}_\Omega(u^h; z) - \mathcal{E}_\Omega(u^h; z^H)|$; the corresponding error representation formula would involve the residual associated with the numerical solution of the dual problem (2.13) and the analytical solution of the dual to the dual problem. As the use of a Type I *a posteriori* error bound on $\mathcal{E}_{\mathrm{D}}$ based on such an error representation formula would necessitate the numerical solution of the dual to the dual problem (which is not a particularly appealing prospect), one can instead use a cruder Type II *a posteriori* error bound on $\mathcal{E}_{\mathrm{D}}$ which, in the spirit of Eriksson *et al.* (1995, 1996), eliminates the dual solution from the *a posteriori* error estimate through bounding its norms above by a *stability constant.* This then terminates the potentially infinite sequence of mutually dual problems which would otherwise arise. The crudeness of the Type II bound on $\mathcal{E}_{\mathrm{D}}$ is of no particular concern here: from the practical point of view there appears to be little advantage in performing reliable error control on $\mathcal{E}_{\mathrm{D}}$; our aim, when using the Type II bound on $\mathcal{E}_{\mathrm{D}}$, is merely to generate an adequate sequence of finite element approximations z^H to the dual solution z which we can then use to compute $\mathcal{E}_{\mathrm{P}}^{\mathrm{loc}}$.

Indeed, the numerical experiments in Houston and Süli (2001*a*) indicate (*cf.* also Hartmann (2001) and the remarks in Section 3.2 about the selection of the space V_d^H) that, with a reasonable choice of the dual finite element space V_d^H, $\mathcal{E}_{\mathrm{D}}$ is typically an order of magnitude smaller than $\mathcal{E}_{\mathrm{P}}^{\mathrm{loc}}$. Therefore, as an alternative to the costly exercise of computing z^H through rigorous error control for the dual measurement problem (7.37), we may simply absorb $\mathcal{E}_{\mathrm{D}}$ into $\mathcal{E}_{\mathrm{P}}^{\mathrm{loc}}$, and replace (7.35) by

$$\mathcal{E}_{\mathrm{P}}^{\mathrm{loc}} \leq \mathtt{TOL}, \tag{7.38}$$

without compromising the reliability of the adaptive algorithm.

Of course, for the Type I error bound (7.35) to be an accurate approximation of (7.30) it is essential that $\mathcal{E}_{\mathrm{D}} \ll \mathcal{E}_{\mathrm{P}}^{\mathrm{loc}}$, and for this to be true it is necessary to ensure that the dual finite element space V_d^H is sufficiently different from the primal finite element space V_d^h; for example, if V_d^H is chosen to coincide with V_d^h then $z^H = z^h$ and thereby $\mathcal{E}_{\mathrm{P}}^{\mathrm{loc}} = 0$, so $0 < \mathcal{E}_{\mathrm{D}} \ll \mathcal{E}_{\mathrm{P}}^{\mathrm{loc}}$ cannot hold. We refer to the comments in Section 3.2 for further details on this issue.

In order to construct a working adaptive algorithm, in addition to an *a posteriori* error bound we also need a mesh refinement criterion and a mesh modification strategy. Some of the possible approaches are reviewed in the next section.

7.3. Mesh refinement criteria and mesh modification strategies

Local tolerance criterion. A possible mesh refinement criterion might consist of checking whether on each element κ in the partition $\mathcal{T}_h$ the following inequality holds:

$$|\tilde{\eta}_\kappa| \leq \frac{\mathtt{TOL}}{N}, \tag{7.39}$$

where

$$\tilde{\eta}_\kappa = \eta_\kappa(u^h|_\kappa, (z^H - z^h)|_\kappa),$$

as in (7.34), where N is the number of elements in $\mathcal{T}_h$. If inequality (7.39) is violated on an element $\kappa \in \mathcal{T}_h$ then κ is refined; otherwise κ is accepted as being of adequate size. It is also possible to incorporate derefinement into the algorithm by selecting λ, $0 < \lambda \ll 1$, and marking elements κ with

$$|\tilde{\eta}_\kappa| \leq \lambda \frac{\mathtt{TOL}}{N}$$

for derefinement. It is assumed that the hierarchy of meshes is generated from a coarse *background mesh*, supplied by the user, beyond which no derefinement can occur.

Fixed fraction criterion. Of course, other refinement criteria are also possible. For example, the *fixed fraction strategy* involves choosing two numbers φ_{ref} and φ_{deref} in the interval $(0, 100)$ with $\varphi_{\text{deref}} + \varphi_{\text{ref}} < 100$, ordering the *local refinement indicators* $|\tilde{\eta}_\kappa|$, $\kappa \in \mathcal{T}_h$, according to their size, and then refining those elements κ which correspond to $\varphi_{\text{ref}}\%$ of the largest entries in the ordered sequence (the top 20%, say), and derefining those elements κ which correspond to the $\varphi_{\text{deref}}\%$ of the smallest entries in this ordered sequence (the bottom 10%, say). Further variations on this strategy, with dynamically varying φ_{ref} and φ_{deref}, are also possible.

Optimized mesh criterion. Yet a further technique, called the *optimized mesh strategy* (see, *e.g.*, Giles (1998), Rannacher (1998) and Becker and Rannacher (2001)) aims to design a subdivision $\mathcal{T}_h$ of the computational domain $\Omega \subset \mathbb{R}^n$ (or, equivalently, a mesh function $h(x)$ defined on Ω) for the primal problem so that the number N of elements in the subdivision $\mathcal{T}_h$ is minimized, subject to the constraint that

$$\mathcal{E}_{|\Omega|}(u^h; z - \pi^h z) \approx \mathtt{TOL}.$$

Assuming that the computational domain $\Omega \subset \mathbb{R}^n$ has been subdivided into elements $\kappa \in \mathcal{T}_h$, we can write

$$N = \sum_{\kappa \in \mathcal{T}_h} 1 = \sum_\kappa \int_\kappa \frac{1}{\text{meas}(\kappa)} \, \mathrm{d}x \approx \sum_\kappa \int_\kappa \frac{\mathrm{d}x}{h^n(x)}.$$

On the other hand,

$$\mathcal{E}_{|\Omega|}(u^h; z - \pi^h z) = \sum_{\kappa \in \mathcal{T}_h} |\eta_\kappa|,$$

where

$$\eta_\kappa = \eta_\kappa(u^h; z - \pi^h z).$$

Let us suppose that $|\eta_\kappa|$ can be expressed as

$$|\eta_\kappa| = \int_\kappa A(x)h^k(x)\,\mathrm{d}x \tag{7.40}$$

for some positive real number k, where $A(x) = \mathcal{O}(1)$ as $h \to 0$. Then,

$$\mathcal{E}_{|\Omega|}(u^h; z - \pi^h z) = \int_\Omega A(x)h^k(x)\,\mathrm{d}x.$$

Thus, to find an 'optimal' h, we need to solve the following constrained optimization problem:

$$\int_\Omega \frac{\mathrm{d}x}{h^n(x)} \to \min \quad \text{subject to} \quad \int_\Omega A(x)h^k(x)\,\mathrm{d}x - \mathtt{TOL} = 0.$$

Let us consider the Lagrangian

$$\mathcal{L}(\lambda, h) = \int_\Omega \frac{\mathrm{d}x}{h^n(x)} + \lambda\left(\int_\Omega A(x)h^k(x)\,\mathrm{d}x - \mathtt{TOL}\right),$$

where $\lambda \in \mathbb{R}$ is a Lagrange multiplier. An elementary calculation shows that the Gateaux derivative of $\mathcal{L}$ in the 'direction' $\hat{h}$ is

$$\begin{aligned}\frac{\partial \mathcal{L}}{\partial h}(\lambda, h; \hat{h}) &= \lim_{\epsilon \to O} \frac{\mathcal{L}(\lambda, h + \epsilon\hat{h}) - \mathcal{L}(\lambda, h)}{\epsilon} \\ &= \int_\Omega \left\{k\lambda A(x)h^{k-1}(x) - nh^{-n-1}(x)\right\}\hat{h}(x)\,\mathrm{d}x.\end{aligned}$$

Now, from the requirement that, at a stationary point $(\lambda^{\mathrm{opt}}, h^{\mathrm{opt}})$,

$$\frac{\partial \mathcal{L}}{\partial h}(\lambda^{\mathrm{opt}}, h^{\mathrm{opt}}; \hat{h}) = 0$$

for all $\hat{h}$, we deduce that

$$h^{\mathrm{opt}}(x) = \left(\frac{n}{k\lambda A(x)}\right)^{\frac{1}{k+n}}. \tag{7.41}$$

Substituting this into the constraint

$$\int_\Omega A(x)h^k(x)\,\mathrm{d}x = \mathtt{TOL},$$

we deduce that

$$\left(\frac{n}{k\lambda}\right)^{\frac{k}{k+n}} W = \mathtt{TOL}, \tag{7.42}$$

where

$$W = \int_\Omega A^{\frac{n}{k+n}}(x)\,\mathrm{d}x. \tag{7.43}$$

Eliminating λ from (7.41) using (7.42), we obtain

$$h^{\mathrm{opt}}(x) = \left(\frac{\mathrm{TOL}}{W}\right)^{\frac{1}{k}} A^{-\frac{1}{k+n}}(x), \qquad x \in \kappa, \quad \kappa \in \mathcal{T}_h,$$

where W is defined by (7.43) and A is defined (elementwise) by (7.40); of course, in practice the dual solution z involved in A is replaced by its finite element approximation z^H (*i.e.*, $\tilde{\eta}_\kappa$ is used instead of η_κ). An application of the optimized mesh criterion will be given in the next section.

Any of these criteria can be coupled with a suitable mesh modification algorithm. For example, in two space dimensions a red–green refinement strategy may be used. Here, the user must first specify a coarse *background mesh* upon which any future refinement will be based. Red refinement corresponds to dividing a certain triangle into four similar triangles by connecting the midpoints of the three sides. Since red refinement is performed only locally (rather than in each element in the triangulation), hanging nodes are created in the mesh; green refinement is then used to remove any hanging nodes in the mesh created in the course of red refinement by connecting a hanging node on an edge to the opposite vertex of the triangle. Green refinement is only temporary and is only applied to elements which contain one hanging node; on elements with two or more hanging nodes red refinement is performed. Within this mesh modification algorithm elements may also be removed from the mesh through derefinement provided they do not lie in the original background mesh. It is perhaps worth noting here that the removal of hanging nodes through green refinement is necessary only if U^h is contained in $C(\bar{\Omega})$. In certain nonconforming methods, such as the discontinuous Galerkin finite element method (*cf.* Cockburn, Karniadakis and Shu (2000) and Section 9), it is not assumed that U^h is contained in $C(\bar{\Omega})$, so the existence of hanging nodes in the mesh is perfectly acceptable.

A numerical experiment. The purpose of this numerical experiment is to illustrate the sharpness of a Type I error bound. We consider the reaction–diffusion equation

$$-\nabla^2 u + u = f(x, y) \qquad \text{in } \Omega = (0,1) \times (0,1)$$

with boundary conditions

$$\begin{aligned} u &= 0, & y &= 0, \\ \frac{\partial u}{\partial \nu} &= 0, & x &= 0, \\ & & x &= 1, \\ u &= x^2(1-x)^2, & y &= 1. \end{aligned}$$

Table 7.2. Reliability of a Type I *a posteriori* error bound

h	$N_\psi^h(u^h)$	$\lvert$error$\rvert$	error bound	effectivity index
1/4	-2.082×10^{-2}	5.417×10^{-3}	9.258×10^{-3}	1.709
1/8	-1.693×10^{-2}	1.528×10^{-3}	2.903×10^{-3}	1.900
1/16	-1.579×10^{-2}	3.958×10^{-4}	7.897×10^{-4}	1.995
1/32	-1.550×10^{-2}	9.984×10^{-5}	2.001×10^{-4}	2.005
1/64	-1.542×10^{-2}	2.502×10^{-5}	5.019×10^{-5}	2.006

Here $f(x,y)$ is chosen so that the exact solution to the problem is $u(x,y) = yx^2(1-x)^2$. We consider the numerical approximation of the linear functional

$$N_\psi(u) = \int_\Gamma \psi \frac{\partial u}{\partial \nu} \mathrm{d}s$$

where

$$\psi = \begin{cases} -\cos(2\pi x) & y = 0, \\ 0 & \text{elsewhere on } \Gamma, \end{cases}$$

which has exact value $-3/(2\pi^4)$. As described above we may derive a Type I *a posteriori* error bound:

$$|N_\psi(u) - N_\psi^h(u^h)| \le \sum_\kappa \left| \int_\kappa (u^h - f)(v^h - z)\mathrm{d}x + \frac{1}{2}\int_{\partial\kappa} \left[\frac{\partial u^h}{\partial \nu}\right] |v^h - z|\,\mathrm{d}s \right|.$$

Table 7.2 demonstrates that this really does provide an upper bound on the error in the computed linear functional. Here we have taken a sequence of regular meshes and computed the numerical approximation $N_\psi^h(u^h)$ to the linear functional, the actual error, the error bound and the effectivity index, which is the ratio of the error bound to the actual error and thus measures the extent to which the error bound overestimates the error. In the computations the dual solution z appearing in the inequality above has been replaced by an approximation $\tilde{z}$ computed on the same mesh as the primal approximation u^h, but with a piecewise quadratic finite element space.

8. Mesh-dependent perturbations and duality

In many instances, the bilinear functional $B(\cdot\,,\cdot)$ and the linear functional $\ell(\cdot)$ that appear in the statement of (2.1) have to be replaced by numerical approximations $B_h(\cdot\,,\cdot)$ and $\ell_h(\cdot)$, respectively. For example, numerical quadrature or numerical approximation of a curved computational domain Ω by a polyhedral domain Ω_h may lead to such perturbations of $B(\cdot\,,\cdot)$ and $\ell(\cdot)$. We note in this respect that many finite volume methods can be restated as Petrov–Galerkin finite element methods of the form (3.1) with numerical quadrature.

8.1. *Error correction and primal–dual equivalence*

Let us suppose again that $\{U_0^h\}_{h>0}$ and $\{V_0^h\}_{h>0}$ are two families of finite-dimensional subspaces of U_0 and V_0, respectively, parametrized by $h \in (0,1]$. When U_0 is a proper Hilbert subspace of U, we assign to $p \in U$ the affine variety $U_p^h = p + U_0^h \subset U_p \subset U$; similarly, when V_0 is a proper Hilbert subspace of V, we assign to $d \in V$ the affine variety $V_d^h = d + V_0^h \subset V_d$.

We consider the following *discrete primal problem.*

($\hat{\mathrm{P}}^h$) Suppose that $m : U \to \mathbb{R}$ and $\ell_h : V_d^h \to \mathbb{R}$ are linear functionals and $B_h(\cdot\,,\cdot) : U_p^h \times V_d^h \to \mathbb{R}$ is a bilinear functional. Find $J_p^h \in \mathbb{R}$ and $u^h \in U_p^h$ such that

$$J_p^h = m(u^h) + \ell_h(v^h) - B_h(u^h, v^h) \qquad \forall v^h \in V_d^h. \tag{8.1}$$

It is also possible to include the case when $m(\cdot)$ has been approximated by a linear functional $m_h(\cdot)$, but for the sake of brevity we shall not discuss this here since this extension can be handled similarly.

In analogy with ($\hat{\mathrm{P}}^h$), we define the *discrete dual problem* as follows. Suppose that $\{U_0^H\}_{H>0}$ and $\{V_0^H\}_{H>0}$ are two families of finite-dimensional subspaces of U_0 and V_0, respectively, parametrized by $H \in (0,1]$, typically different from the families $\{U_0^h\}_{h>0}$ and $\{V_0^h\}_{h>0}$. We assign to $p \in U$ the affine variety $U_p^H = p + U_0^H \subset U_p \subset U$; similarly, we assign to $d \in V$ the affine variety $V_d^H = d + V_0^H \subset V_d \subset V$.

($\hat{\mathrm{D}}^H$) Suppose that $m_H : U_p^H \to \mathbb{R}$ and $\ell : V_d \to \mathbb{R}$ are linear functionals, and $B_H(\cdot\,,\cdot) : U_p^H \times V_d^H \to \mathbb{R}$ is a bilinear functional. Find $J_d^H \in \mathbb{R}$ and $z^H \in V_d^H$ such that

$$J_d^H = m_H(w^H) + \ell(z^H) - B_H(w^H, z^H) \qquad \forall w^H \in U_p^H.$$

Again, one may also include the case when $\ell(\cdot)$ has been approximated by a linear functional $\ell_H(\cdot)$; for the sake of brevity, we shall refrain from discussing this.

Next we present representation formulae for the error between J_p, J_d and their respective approximations J_p^h, J_d^H. In particular, we shall see that, when J_p and J_d are appropriately corrected by terms which stem from perturbing the bilinear functional $B(\cdot\,,\cdot)$ and the linear functionals $m(\cdot)$ and $\ell(\cdot)$, we recover error representation formulae analogous to (3.4) and (3.5).

Theorem 8.1. (Error representation formula) Let $(J_p, u) \in \mathbb{R} \times U_p$ and $(J_d, z) \in \mathbb{R} \times V_d$ denote the solutions to (P) and (D), respectively, and let $(J_p^h, u^h) \in \mathbb{R} \times U_p^h$ and $(J_d^H, z^H) \in \mathbb{R} \times V_d^H$ be the solutions to ($\hat{\mathrm{P}}^h$)

and $(\hat{\mathrm{D}}^H)$, respectively. Let us define

$$\hat{J}_p^h = J_p^h + \big[(\ell - \ell_h)(z^h) - (B - B_h)(u^h, z^h)\big], \tag{8.2}$$

$$\hat{J}_d^H = J_d^H + \big[(m - m_H)(u^H) - (B - B_H)(u^H, z^H)\big]. \tag{8.3}$$

Then,

$$J_p - \hat{J}_p^h = B(u - u^h, z - z^h) \qquad \forall z^h \in V_d^h, \tag{8.4}$$

$$J_d - \hat{J}_d^H = B(u - u^H, z - z^H) \qquad \forall u^H \in U_p^H. \tag{8.5}$$

Proof. Since $V_d^h \subset V_d$, we have from (P) that

$$J_p = m(u) + \ell(v^h) - B(u, v^h) \qquad \forall v^h \in V_d^h.$$

Recalling from $(\hat{\mathrm{P}}^h)$ that

$$J_p^h = m(u^h) + \ell_h(v^h) - B_h(u^h, v^h) \qquad \forall v^h \in V_d^h$$

and subtracting, we find that, for any $v^h \in V_d^h$,

$$\begin{aligned} J_p - J_p^h &= \ell(v^h) - \ell_h(v^h) + m(u - u^h) - \big[B(u, v^h) - B_h(u^h, v^h)\big] \\ &= \ell(v^h) - \ell_h(v^h) + m(u - u^h) \\ &\qquad - B(u - u^h, v^h) - \big[B(u^h, v^h) - B_h(u^h, v^h)\big]. \end{aligned} \tag{8.6}$$

On the other hand, as $u - u^h \in U_0$, we deduce from (2.13) that

$$B(u - u^h, z) = m(u - u^h),$$

which we can use to eliminate $m(u - u^h)$ from (8.6) and deduce that

$$\begin{aligned} J_p - &\big\{J_p^h + \big[(\ell(v^h) - B(u^h, v^h)) - (\ell_h(v^h) - B_h(u^h, v^h))\big]\big\} \\ &= B(u - u^h, z - v^h), \end{aligned}$$

for all v^h from V_d^h; hence (8.4). The proof of the identity (3.5) is completely analogous. □

Our next result is a counterpart of the discrete Primal–Dual Equivalence Theorem, Theorem 3.3.

Theorem 8.2. Suppose that $(J_p^h, u^h) \in \mathbb{R} \times U_p^h$ and $(J_d^H, z^H) \in \mathbb{R} \times V_d^H$ denote the solutions to the primal problem $(\hat{\mathrm{P}}^h)$ and the dual problem $(\hat{\mathrm{D}}^H)$, respectively, and define $\hat{J}_p^h$ and $\hat{J}_d^H$ as in (8.2) and (8.3) above; then,

$$\hat{J}_p^h = \hat{J}_d^H + \rho^{hH},$$

where

$$\rho^{hH} = B(u - u^H, z - z^H) - B(u - u^h, z - z^h),$$

for any $u^H \in U_p^H$ and any $z^h \in V_d^h$.

Proof. The result is a direct consequence of the previous theorem, on subtracting (8.4) from (8.5), and recalling from the Primal–Dual Equivalence Theorem that $J_p = J_d$. □

Next, we shall consider an application of these abstract results to a class of stabilized finite element methods that includes the streamline diffusion finite element method (SDFEM), and the least-squares stabilized finite element method for a scalar linear hyperbolic problem. Such stabilized methods arise by perturbing the classical Galerkin finite element method in a consistent manner through the inclusion of a least-squares stabilization term, so as to enhance numerical dissipation in the direction of the characteristic curves of the hyperbolic operator.

8.2. Hyperbolic model problem: the effects of stabilization

Let us consider the transport problem

$$\mathcal{L}u \equiv \mathbf{b} \cdot \nabla u + cu = f, \quad x \in \Omega, \qquad u = g, \quad x \in \Gamma_-, \tag{8.7}$$

where $\Omega = (0,1)^n$, Γ is the union of open faces of Ω, and Γ_- denotes the inflow part of Γ, namely the set of all points $x \in \Gamma$ where the vector $\mathbf{b}(x)$ points into Ω; Γ_+, the outflow part of Γ, is defined analogously.

As before, we assume that the entries $b_1, \ldots, b_n$ of the n-component vector function $\mathbf{b}$ are continuously differentiable and positive on $\bar{\Omega}$; this hypothesis ensures that Γ is noncharacteristic for the operator $\mathcal{L}$ at each point $x \in \Gamma$. Also, we shall suppose that $c \in C(\bar{\Omega})$, $f \in L^2(\Omega)$ and $g \in L^2(\Gamma_-)$. In addition, it will be assumed that there exists $\gamma > 0$ such that

$$c_0^2(x) \equiv c(x) - \frac{1}{2}\nabla \cdot \mathbf{b}(x) \geq \gamma^2 \quad \text{for all } x \in \bar{\Omega}. \tag{8.8}$$

In order to introduce the variational formulation of the boundary value problem (8.7), we associate with $\mathcal{L}$ the graph space

$$H(\mathcal{L}, \Omega) = \{v \in L^2(\Omega) \,:\, \mathcal{L}v \in L^2(\Omega)\}.$$

Let us consider the bilinear form $B(\cdot\,,\cdot) \,:\, H(\mathcal{L},\Omega) \times H(\mathcal{L},\Omega) \to \mathbb{R}$ defined by

$$B(w,v) = (\mathcal{L}w, v) - ((\mathbf{b}\cdot\boldsymbol{\nu})w, v)_{\Gamma_-}$$

and the linear functional $\ell \,:\, H(\mathcal{L},\Omega) \to \mathbb{R}$ given by

$$\ell(v) = (f,v) - ((\mathbf{b}\cdot\boldsymbol{\nu})g, v)_{\Gamma_-}.$$

In these definitions, $(\cdot\,,\cdot)$ denotes the $L^2(\Omega)$ inner product and $(\cdot\,,\cdot)_{\Gamma_-}$ is the $L^2(\Gamma_-)$ inner product with respect to the surface measure ds (with analogous definition of $(\cdot\,,\cdot)_{\Gamma_+}$). In terms of this notation, the boundary value problem (8.7) can be restated in the following variational form: find $u \in H(\mathcal{L},\Omega)$ such that

$$B(u,v) = \ell(v) \qquad \forall v \in H(\mathcal{L},\Omega). \tag{8.9}$$

Suppose that $\mathcal{T}_h$ is a finite element partition of the computational domain Ω into open simplicial element domains κ. It will be assumed that the family $\{\mathcal{T}_h\}_h$ is shape-regular. We then consider on $\mathcal{T}_h$ the finite element trial and test spaces $U^h = V^h \subset H^1(\Omega) \subset H(\mathcal{L},\Omega)$ consisting of continuous piecewise polynomial functions of maximum degree k, $k \geq 1$. The finite element space U^h will be assumed to possess the following approximation property.

(H) Given that $v \in H^{s+1}(\Omega)$ and $v|_{\Gamma_-} \in H^{s+1}(\Gamma_-)$ for some s, $0 \leq s \leq k$, there exists $\pi^h v$ in U^h and a positive constant c_{int}, independent of v and the mesh function h, such that

$$\|v - \pi^h v\|_{L^2(\kappa)} + h_\kappa |v - \pi^h v|_{H^1(\kappa)} \leq c_{\text{int}} h_\kappa^{s+1} |v|_{H^{s+1}(\hat{\kappa})} \quad \forall \kappa \in \mathcal{T}_h,$$

$$\|v - \pi^h v\|_{L^2(\partial\kappa\cap\Gamma_-)} \leq c_{\text{int}} h_\kappa^{s+1} |v|_{H^{s+1}(\partial\hat{\kappa}\cap\Gamma_-)} \; \forall \kappa \in \mathcal{T}_h \, : \, \partial\kappa \cap \Gamma_- \neq \emptyset.$$

In this hypothesis $\hat{\kappa}$ denotes the union of all such elements (including κ itself) whose closure has nonempty intersection with the closure of κ. Hypothesis (H) may be satisfied by taking $\pi^h v$ to be the quasi-interpolant of v based on local averaging that involves the neighbours of κ (see Brenner and Scott (1994), for example). A further possibility is to define $\pi^h v \in U^h$ at the degrees of freedom interior to $\Omega \cup \Gamma_+$ as indicated in the previous sentence, while on Γ_- one can define $\pi^h v|_{\Gamma_-}$ as the orthogonal projection of $v|_{\Gamma_-}$ onto $U^h|_{\Gamma_-}$ with respect to the inner product $\langle \cdot\,,\cdot\rangle_- = ((\mathbf{b}\cdot\boldsymbol{\nu})\cdot\,,\cdot)_{\Gamma_-}$; the inner product $\langle\cdot\,,\cdot\rangle_+$ on $L^2(\Gamma_+)$ is defined analogously.

Next we introduce the stabilized finite element approximation of our model problem. Let δ be a positive function contained in $L^\infty(\Omega)$; δ will be referred to as the *stabilization parameter*. A typical choice of the stabilization parameter, based on *a priori* error analysis, is $\delta = C_\delta h$, where C_δ is a positive constant which should be selected by the user and $x \mapsto h(x)$ is the local mesh size; for instance, $h|_\kappa = h_\kappa$, the diameter of element $\kappa \in \mathcal{T}_h$. The stabilized finite element approximation of (8.7) is then defined as follows:

Find $u^h \in U^h$ such that

$$B_\delta(u^h, v^h) = \ell_\delta(v^h) \quad \forall v^h \in U^h, \tag{8.10}$$

where the bilinear functional $B_\delta \, : \, H(\mathcal{L},\Omega) \times H(\mathcal{L},\Omega) \to \mathbb{R}$ and the linear functional $\ell_\delta \, : \, H(\mathcal{L},\Omega) \to \mathbb{R}$ are given by

$$B_\delta(w,v) = (\mathcal{L}w, v + \delta\hat{\mathcal{L}}v) - \langle w, v\rangle_-, \qquad l_\delta(v) = (f, v + \delta\hat{\mathcal{L}}v) - \langle g, v\rangle_-,$$

with $\mathcal{L}w = \mathbf{b}\cdot\nabla w + cw$ and $\hat{\mathcal{L}}w = \mathbf{b}\cdot\nabla w + \hat{c}w$. Depending on the choice of the coefficient $\hat{c}$, we obtain different stabilization techniques; some typical choices are listed below:

$$\hat{c} = \begin{cases} 0 & \text{SDFEM}, \\ c & \text{least-squares FEM}, \\ \nabla\cdot\mathbf{b} - c & \text{Douglas–Wang stabilization}. \end{cases}$$

Condition (8.8) implies that $B_\delta(v,v) > 0$ for all $v \in U^h \setminus \{0\}$; if the Douglas–Wang stabilization is used, it has to be assumed additionally that $0 < \delta \leq \frac{1}{2}\gamma^2[c^2+(\nabla\cdot\mathbf{b})^2]^{-1}$ on $\bar{\Omega}$ to ensure positivity of $B_\delta(v,v)$ for nontrivial v from U^h. Since (8.10) is a linear problem over a finite-dimensional space U^h, the existence of a unique solution u^h to (8.10) follows from the positivity of $B_\delta(v,v)$, $v \in U^h \setminus \{0\}$.

Let us suppose that we wish to control the discretization error in some linear functional $J(\cdot)$ defined on $H(\mathcal{L},\Omega)+U^h$. To be more precise, suppose that a certain tolerance $\texttt{TOL} > 0$ is given and that the aim of the computation is to find a subdivision $\mathcal{T}_h$ of the computational domain Ω and u^h in the finite element space U^h associated with $\mathcal{T}_h$ such that

$$|J(u) - J(u^h)| < \texttt{TOL}.$$

In order to solve this measurement problem, we consider the *a posteriori* error analysis of the stabilized finite element method (8.10) to derive a 'computable' bound on $|J(u)-J(u^h)|$ and then perform adaptive mesh refinement until the *a posteriori* error bound drops below the specified tolerance. The derivation of the *a posteriori* error bound will be based on a duality argument. The dual problem is defined as follows:

Find $z \in H(\mathcal{L},\Omega)$ such that

$$B(w,z) = J(w) \quad \forall w \in H(\mathcal{L},\Omega). \tag{8.11}$$

Error representation formula and error correction. Our starting point is the following theorem.

Theorem 8.3. The dual problem (8.11) gives rise to the following error representation formula:

$$J(u) - J(u^h) = -\langle r^{h,-}, z - z^h\rangle_- + (r^h, z - z^h) - (r^h, \delta\hat{\mathcal{L}}z^h), \tag{8.12}$$

for all $z^h \in U^h$. Hence,

$$J(u) - \hat{J}_p^h(u^h; z^h) = -\langle r^{h,-}, z - z^h\rangle_- + (r^h, z - z^h) \tag{8.13}$$

for all $z^h \in U^h$, where

$$\hat{J}_p^h(u^h; z^h) = J(u^h) - (r^h, \delta\hat{\mathcal{L}}z^h),$$

$r^h = f - \mathcal{L}u^h$ is the *internal residual*, $r^{h,-} = g - u^h$.

Proof. By virtue of the linearity of J and the definition of the dual problem (8.11), we have that

$$\begin{aligned} J(u) - J(u^h) &= J(u - u^h) \\ &= B(u - u^h, z) \\ &= B(u,z) - B(u^h,z) \end{aligned}$$

$$
\begin{aligned}
&= \ell(z) - B(u^h, z) \\
&= \ell(z - z^h) - B(u^h, z - z^h) + \ell(z^h) - B(u^h, z^h) \\
&= \ell(z - z^h) - B(u^h, z - z^h) \\
&\qquad + \ell(z^h) - \ell_\delta(z^h) - B(u^h, z^h) + B_\delta(u^h, z^h) \\
&= \ell(z - z^h) - B(u^h, z - z^h) \\
&\qquad + \left[(\ell - \ell_\delta)(z^h) - (B - B_\delta)(u^h, z^h)\right],
\end{aligned}
$$

where z_h is any element in U^h. Hence, in agreement with Theorem 8.1, we let

$$\hat{J}_p^h(u^h; z^h) = J(u^h) + \left[(\ell - \ell_\delta)(z^h) - (B - B_\delta)(u^h, z^h)\right]$$

and note that

$$\ell(z - z^h) - B(u^h, z - z^h) = -\langle r^{h,-}, z - z^h\rangle_- + (r^h, z - z^h)$$

and

$$\hat{J}_p^h(u^h; z^h) = J(u^h) - (r^h, \delta\hat{\mathcal{L}}z^h)$$

to complete the proof. □

If we label the three terms on the right-hand side of (8.12) by I_1, II_1 and III_1, then, on general unstructured shape-regular meshes, and continuous piecewise polynomial finite elements of degree $k \geq 1$, hypothesis (H) implies that

$$\mathrm{I}_1 = \mathcal{O}(h^{2k+2}), \qquad \mathrm{II}_1 = \mathcal{O}(h^{2k+1}), \qquad \mathrm{III}_1 = \mathcal{O}(h^{k+1}),$$

and therefore $J(u) - J(u^h) = \mathcal{O}(h^{k+1})$. In fact, we shall see in the next numerical example that, on structured uniform triangular meshes, these rates of convergence may be exceeded, leading to

$$\mathrm{I}_1 = \mathcal{O}(h^{2k+2}), \qquad \mathrm{II}_1 = \mathcal{O}(h^{2k+2}), \qquad \mathrm{III}_1 = \mathcal{O}(h^{k+2}),$$

and hence $J(u) - J(u^h) = \mathcal{O}(h^{k+2})$. One way or the other, the rate of convergence of $J(u) - J(u^h)$ is dominated by III_1 whose convergence order is always inferior to those of I_1 and II_1. This motivates us to move III_1 from the right-hand side of (8.12) to the left-hand side and, as a correction term, combine it with $J(u^h)$. This then leads to the error representation formula (8.13) whose right-hand side is of size $\mathcal{O}(h^{2k+1})$. Hence $\hat{J}_p^h(u^h, z^h)$ will be a better approximation to $J(u)$ than $J(u^h)$ is. Indeed, since the term III_1 is structurally different from terms I_1 and II_1 in that it does not involve the analytical dual solution z, $\hat{J}_p^h(u^h; z^h)$ is a *computable* approximation to $J(u)$. We shall now illustrate these points through a simple numerical example using the streamline diffusion finite element method (SDFEM) with continuous piecewise polynomials of degree $k = 1$ (see Houston *et al.* (2000*a*) for details).

Example 1. Let us take $\Omega = (0,1)^2$, $\mathbf{b} = (1+x, 1+y)$, $c = 0$ and $f = 0$ with boundary condition

$$u(x,y) = \begin{cases} 1 - y^5 & \text{for } x = 0,\ 0 \le y \le 1, \\ \mathrm{e}^{-50x^4} & \text{for } 0 \le x \le 1,\ y = 0. \end{cases}$$

We select $\delta = C_\delta h$ with $C_\delta = 1/4$ and define

$$\psi = \begin{cases} 1 - \sin(\pi(1-y)/2)^2 \cos(\pi y/2) & \text{for } x = 1,\ 0 \le y \le 1, \\ 1 - (1-x)^3 - (1-x)^4/2 & \text{for } 0 \le x \le 1,\ y = 1. \end{cases}$$

We wish to compute the weighted normal flux

$$J(u) = N_\psi(u) = \int_{\Gamma_+} (\mathbf{b} \cdot \boldsymbol{\nu}) u\, \psi \,\mathrm{d}s$$

of the analytical solution u over the outflow boundary Γ_+. For purposes of comparison, the analytical solution u and the dual solution z have been computed to high accuracy using the method of characteristics; in particular, the 'exact' value of the weighted outward normal flux was found to be $N_\psi(u) = 2.4676$.

In Table 8.1 we have displayed the orders of convergence, ρ, of the error in the $L^2(\Omega)$ norm as well as in the functional $N_\psi(\cdot)$, as h tends to zero, on a sequence of uniform triangular meshes obtained from uniform square meshes by cutting each mesh square into two triangles, and U^h consisting of continuous piecewise polynomials of degree 1 ($k = 1$). We observe that $N_\psi(u) - N_\psi(u^h)$ converges like $\mathcal{O}(h^3)$ with $\mathcal{O}(h)$ stabilization, while the $L^2(\Omega)$ norm is of second order.

In Table 8.2 we show the convergence of each of the terms in the error representation formula (8.12). We see, in particular, that the second term in the error representation formula (8.12), *i.e.*, term II_1, is superconvergent; here $\mathrm{II}_1 = \mathcal{O}(h^4)$ as h tends to zero. Term III_1, which arises as the result of the stabilization employed, exhibits $\mathcal{O}(h^3)$ convergence and entirely dominates the error in the weighted outward normal flux. Thus, when term III_1 is interpreted as a computable *correction term* to the functional and is combined with $J(u^h)$, the remaining two terms, I_1 and II_1 exhibit $\mathcal{O}(h^4)$ convergence: hence, by the error representation formula (8.13) for the corrected functional $\hat{J}^h_p(u^h; z^h)$ we see that this approximates $J(u)$ with error $\mathcal{O}(h^4)$. Similar behaviour is observed on unstructured triangular meshes: there, $\mathrm{I}_1 = \mathcal{O}(h^4)$, $\mathrm{II}_1 = \mathcal{O}(h^3)$, so then $\hat{J}^h_p(u^h; z^h)$ approximates $J(u)$ with error $\mathcal{O}(h^3)$.

Table 8.1. Example 1: Convergence of $\|u - u^h\|_{L^2(\Omega)}$ with $\delta = h/4$, and the rate of convergence ρ

Mesh	$\|u - u^h\|_{L^2(\Omega)}$	ρ	$\|N_\psi(u) - N_\psi(u^h)\|$	ρ
17×17	2.927×10^{-3}	–	2.957×10^{-4}	–
33×33	5.195×10^{-4}	2.49	3.860×10^{-5}	2.94
65×65	1.079×10^{-4}	2.27	4.944×10^{-6}	2.96
129×129	2.544×10^{-5}	2.08	6.257×10^{-7}	2.98
257×257	6.260×10^{-6}	2.02	7.874×10^{-8}	2.99

Table 8.2. Example 1: Convergence of the terms in the error representation formula (8.13) with $\delta = h/4$, and the rate of convergence ρ

Mesh	I_1	ρ	II_1	ρ	III_1	ρ
17×17	3.31×10^{-6}	–	3.35×10^{-6}	–	2.96×10^{-4}	–
33×33	1.91×10^{-7}	4.12	2.30×10^{-7}	3.87	3.86×10^{-5}	2.94
65×65	1.17×10^{-8}	4.03	1.52×10^{-8}	3.92	4.95×10^{-6}	2.97
129×129	7.24×10^{-10}	4.01	9.74×10^{-10}	3.96	6.26×10^{-7}	2.98
257×257	4.51×10^{-11}	4.00	6.18×10^{-11}	3.99	7.87×10^{-8}	2.99

Table 8.3. Example 1: Convergence of the terms I_2, II_2 and III_2

Mesh	I_2	ρ	II_2	ρ	III_2	ρ
17×17	1.01×10^{-5}	–	7.43×10^{-5}	–	3.20×10^{-3}	–
33×33	4.94×10^{-7}	4.35	9.12×10^{-6}	3.03	8.16×10^{-4}	1.97
65×65	2.85×10^{-8}	4.12	1.14×10^{-6}	3.00	2.04×10^{-4}	2.00
129×129	1.73×10^{-9}	4.04	1.42×10^{-7}	3.00	5.11×10^{-5}	2.00
257×257	1.08×10^{-10}	4.01	1.78×10^{-8}	3.00	1.28×10^{-5}	2.00

Now let us consider the localized counterparts of the terms I_1, II_1 and III_1, defined by

$$\mathrm{I}_2 = \sum_{\kappa \in \mathcal{T}^h} |\langle r^{h,-}, z - z^h \rangle_{\partial\kappa \cap \Gamma_-}|,$$

$$\mathrm{II}_2 = \sum_{\kappa \in \mathcal{T}^h} |(r^h, z - z^h)_\kappa|, \qquad \mathrm{III}_2 = \sum_{\kappa \in \mathcal{T}^h} |(r^h, \delta \hat{\mathcal{L}} z^h)_\kappa|,$$

respectively.

Table 8.3 demonstrates that localization does not adversely affect the term I_1 but it does slightly affect II_1 whose localized counterpart, II_2, is now only $\mathcal{O}(h^3)$. On unstructured triangular meshes, I_2 and II_2 exhibit the same rates of convergence as I_1 and II_1, namely, $\mathcal{O}(h^4)$ and $\mathcal{O}(h^3)$, respectively. We therefore conclude that on unstructured triangular meshes the convergence rates of I_1 and II_1 are preserved under localization.

Table 8.3 also shows that global superconvergence of the term III_1 is lost under localization: term III_2 is only $\mathcal{O}(h^2)$. However, this is irrelevant, since the error representation formula (8.13) for the corrected functional $\hat{J}_p^h(u^h; z^h)$ only involves the terms I_1 and II_1, while term III_1 has become part of the corrected functional, so its localization is not required.

The insensitivity of the terms I_1 and II_1 in the error representation formula (8.13) to localization on unstructured triangular meshes implies that a Type I error bound on $|J(u) - \hat{J}_p^h(u^h; z^h)|$ will exhibit the same asymptotic rate of convergence as the error itself. Moreover, as any standard mesh refinement criterion will require the localizations I_2 and II_2 to define the local refinement indicators η_κ, the fact that I_2 and II_2 exhibit the same rates of convergence as I_1 and II_1 will be essential for ensuring the optimality of the resulting adaptive meshes.

A mesh-dependent dual problem. Still assuming that the quantity of interest is a certain linear output functional, we now explore an alternative approach to deriving an *a posteriori* error bound where, following Houston *et al.* (2000*a*), instead of $B(\cdot\,,\cdot)$, we use the stabilization-dependent bilinear form $B_\delta(\cdot\,,\cdot)$ to define the dual solution; namely, we now define the dual solution, z_δ, as the solution to the following problem:

$$B_\delta(w, z_\delta) = J(w) \qquad \forall w \in H(\mathcal{L}, \Omega). \tag{8.14}$$

Noting the Galerkin orthogonality property with respect to the bilinear functional

$$B_\delta(u - u^h, v^h) = 0 \qquad \forall v^h \in U^h,$$

we deduce the following error representation formula.

Theorem 8.4. The dual problem (8.14) gives rise to the following error representation formula:

$$J(u) - J(u^h) = -\langle r^{h,-}, z_\delta - z_\delta^h\rangle_- + (r^h, z_\delta - z_\delta^h) + (r^h, \delta\hat{\mathcal{L}}(z_\delta - z_\delta^h)) \quad (8.15)$$

for all $z_\delta^h \in U^h$.

On comparing (8.15) with the error representation formula (8.12) which stems from using the bilinear form $B(\cdot\,,\cdot)$ in the definition of the dual problem, we see that while the first two terms in the two formulae are analogous, the third term in (8.15) has now become more similar to the other terms in the representation formula in that it, too, contains the difference $z_\delta - z_\delta^h$. This is due to the fact that Galerkin orthogonality is with respect to $B_\delta(\cdot\,,\cdot)$ rather than $B(\cdot\,,\cdot)$; indeed, Galerkin orthogonality did not even enter into the derivation of (8.14). This, in turn, has some important consequences.

If we label the three terms on the right-hand side of (8.15) as $\mathrm{I}_{1,\delta}$, $\mathrm{II}_{1,\delta}$ and $\mathrm{III}_{1,\delta}$, and denote their localizations by $\mathrm{I}_{2,\delta}$, $\mathrm{II}_{2,\delta}$ and $\mathrm{III}_{2,\delta}$, repeating the numerical experiment from the previous example, we now find (see Houston *et al.* (2000*a*)) that, both on uniform and on unstructured triangular meshes and continuous piecewise linear basis functions (*i.e.*, $k = 1$),

$$\mathrm{I}_{1,\delta} = \mathcal{O}(h^4), \quad \mathrm{II}_{1,\delta} = \mathcal{O}(h^3), \quad \mathrm{III}_{1,\delta} = \mathcal{O}(h^3).$$

Furthermore,

$$\mathrm{I}_{2,\delta} = \mathcal{O}(h^4), \quad \mathrm{II}_{2,\delta} = \mathcal{O}(h^3), \quad \mathrm{III}_{2,\delta} = \mathcal{O}(h^3).$$

Thus, none of the terms in the error representation formula (8.15) is now sensitive to localization.

In the next example, we show a numerical experiment based on the stabilization-dependent dual problem (8.14). Adaptive mesh refinement is performed based on a Type I *a posteriori* error bound which stems from the error representation formula (8.15), together with the optimized mesh criterion presented in Section 7.3.

More precisely, we define the local refinement indicator

$$\begin{aligned}\eta_\kappa(u^h; z_\delta - z_\delta^h) = -((\mathbf{b}\cdot\boldsymbol{\nu})\, r^{h,-}, z_\delta - z_\delta^h)_{\Gamma_-\cap\partial\kappa} + (r^h, z_\delta - z_\delta^h)_\kappa \\ + (r^h, \delta\hat{\mathcal{L}}(z_\delta - z_\delta^h))_\kappa,\end{aligned}$$

and our Type I *a posteriori* error bound is then

$$\mathcal{E}_{\mathrm{P}}^{\mathrm{loc}} = \sum_{\kappa\in\mathcal{T}_h} |\tilde{\eta}_\kappa| \le \mathtt{TOL},$$

with

$$\tilde{\eta}_\kappa = \eta_\kappa(u^h; z_\delta^H - z_\delta^h),$$

where z_δ^H denotes the numerical solution of the stabilization-dependent dual problem (8.14) and $\mathtt{TOL}$ is the prescribed tolerance.

Let us suppose that the finite element space U^h consists of continuous piecewise polynomials of degree $k = 1$. Guided by the asymptotic behaviour or the terms $\mathrm{I}_{1,\delta}$, $\mathrm{II}_{1,\delta}$ and $\mathrm{III}_{1,\delta}$ and their localizations under mesh refinement, we define

$$\begin{aligned} A(x) &= |r^h(z_\delta - z_\delta^h + \delta r^h \hat{\mathcal{L}}(z_\delta - z_\delta^h)|/h_\Omega^3(x), \\ B(x) &= |(\mathbf{b}\cdot\boldsymbol{\nu})(g-u^h)(z_\delta - z_\delta^h)|/h_{\Gamma_-}^4(x), \end{aligned}$$

where h_Ω and h_{Γ_-} are the mesh functions on $\Omega \cup \Gamma_+$ and Γ_-, respectively. Assuming, for example, that $\Omega \subset \mathbb{R}^2$, *i.e.*, $n = 2$, after an elementary calculation based on the use of Lagrange multipliers, as described in Section 7.3, we arrive at the following optimal mesh functions $h_\Omega^{\mathrm{opt}}(x)$ and $h_{\Gamma_-}^{\mathrm{opt}}(x)$:

$$h_\Omega^{\mathrm{opt}} = \left(\frac{2}{3\lambda A}\right)^{1/5}, \qquad h_{\Gamma_-}^{\mathrm{opt}} = \left(\frac{1}{4\lambda B}\right)^{1/5},$$

where λ is the positive root of

$$\left(\frac{2}{3\lambda}\right)^{3/5}\int_\Omega A^{2/5}\,\mathrm{d}x + \left(\frac{1}{4\lambda}\right)^{4/5}\int_{\Gamma_-} B^{1/5}\,\mathrm{d}\sigma = \mathtt{TOL}.$$

For $\mathtt{TOL} \ll 1$ we expect $\lambda \gg 1$, so that $(1/\lambda)^{4/5} \ll (1/\lambda)^{3/5}$. Thus, for simplicity, we may neglect the boundary integral term in the last equality. We may then explicitly solve for λ in terms of $\mathtt{TOL}$ and the integral of $A^{2/5}$, and substitute the resulting expression into the formula for h_Ω^{opt} to obtain

$$h_\Omega^{\mathrm{opt}}(x) \approx \left(\frac{\mathtt{TOL}}{W}\right)^{1/3}\frac{1}{A^{1/5}(x)}, \qquad \text{where } W = \int_\Omega A^{2/5}(x)\,\mathrm{d}x,$$

with a similar expression for $h_{\Gamma_-}^{\mathrm{opt}}$.

Example 2. Let us again consider the transport equation $\mathbf{b}\cdot\nabla u + cu = f$ in $\Omega = (0,1)^2$, but this time with $\mathbf{b} = (10y^2 - 12x + 1, 1 + y)$, $c = 0$ and $f = 0$. In this problem the characteristics enter Ω through the bottom edge and through the two vertical sides, and exit through the top edge. Thus it is admissible to impose the following inflow boundary condition:

$$u(x,y) = \begin{cases} 0 & \text{for } x = 0,\ \ 0.5 < y \le 1, \\ 1 & \text{for } x = 0,\ \ 0 < y \le 0.5, \\ 1 & \text{for } 0 \le x \le 0.5,\ \ y = 0, \\ 0 & \text{for } 0.5 < x \le 1,\ \ y = 0, \\ \sin^2(\pi y) & \text{for } x = 1,\ \ 0 \le y \le 1. \end{cases}$$

Let us suppose that the objective of the computation is to calculate the weighted normal flux $N_\psi(u)$ of the analytical solution u through the outflow

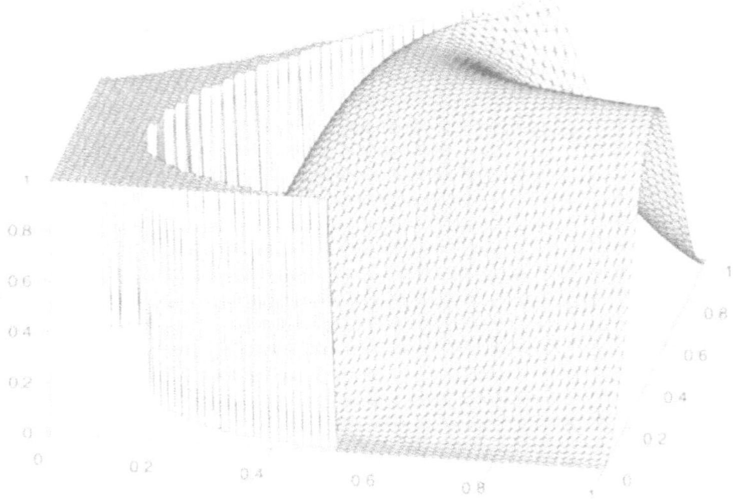

Figure 8.1. Example 2: The analytical solution to the primal problem (from Houston *et al.* (2000*a*))

edge of the square where the weight function ψ is defined by

$$\psi(x) = \sin(\pi x/2) \qquad \text{for } 0 \le x \le 1,\ y = 1.$$

Using the method of characteristics one may compute a highly accurate approximation to u and thereby deduce that $N_\psi(u) = 0.24650$.

The analytical solution to this hyperbolic boundary value problem is shown in Figure 8.1. As the boundary datum is discontinuous along the vertical face $x = 0$ and along the horizontal face $y = 0$, the analytical solution exhibits discontinuities in Ω along the two characteristic curves that stem from the points of discontinuity on the inflow boundary. Nevertheless, numerical experiments analogous to those in Example 1 indicate that the error in the weighted outward normal flux $N_\psi(u)$ is $\mathcal{O}(h^4)$ on uniform triangular meshes and $\mathcal{O}(h^3)$ on unstructured triangular meshes.

We shall aim to compute $N_\psi(u)$ to within a prescribed tolerance $\texttt{TOL}$. Our adaptive mesh refinement is driven by the Type I *a posteriori* error bound which stems from the error representation formula (8.15) for the stabilization-dependent dual problem. The mesh design for the primal problem is based on the *optimal mesh criterion* with $\texttt{TOL} = 5.0 \times 10^{-5}$, and $\delta = h/4$. The background meshes for the primal and dual problems and the adaptively refined meshes which result from them are shown in Figure 8.2. We can see from Figure 8.2(c) that most of the nodes in the adaptively refined mesh for the primal problem are concentrated near the outflow

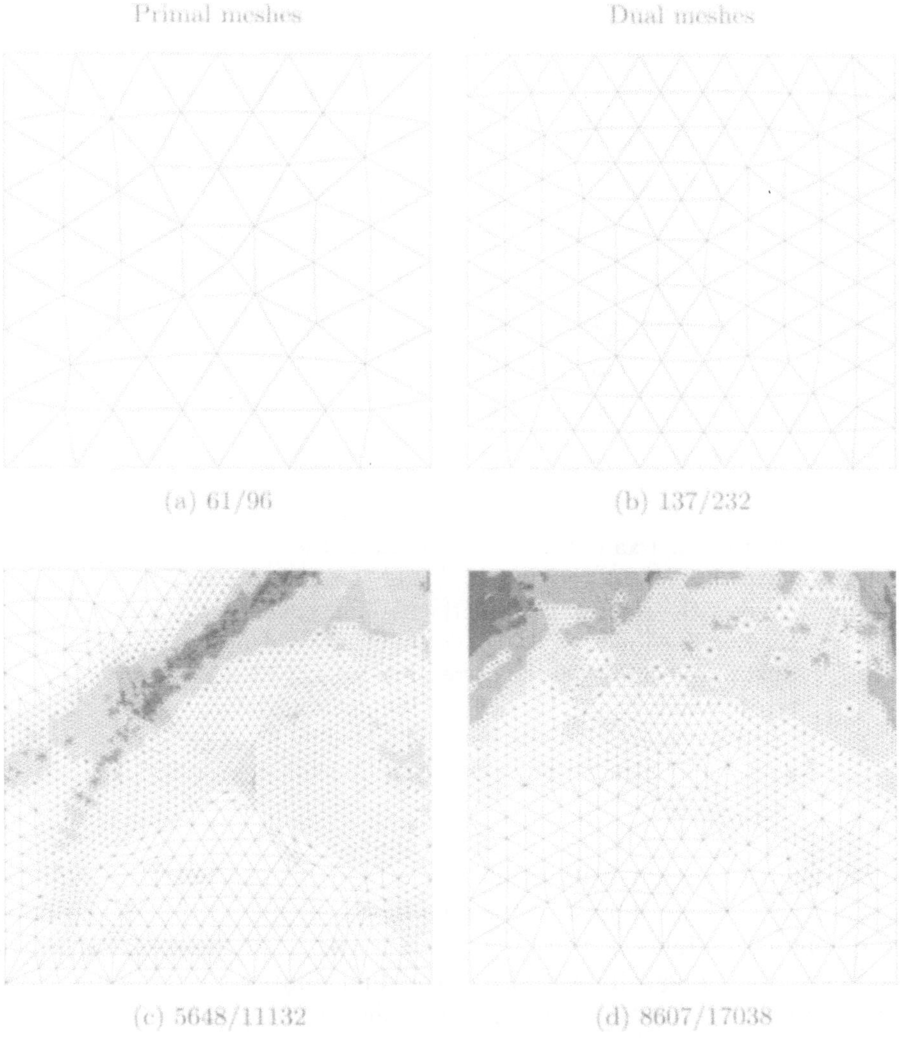

Figure 8.2. Example 2: (a) background mesh for the primal problem with 61 nodes and 96 elements; (b) background mesh for stabilization-dependent dual problem with 137 nodes and 232 elements; (c) mesh for the primal problem based on the use of the optimal mesh criterion ($\mathtt{TOL} = 5.0 \times 10^{-5}$) with 5648 nodes and 11132 elements ($|N_\psi(u) - N_\psi(u^h)| = 6.764 \times 10^{-6}$); (d) adaptively refined mesh for the stabilization-dependent dual problem (*cf.* equation (8.14)) with 7594 nodes and 14199 elements which has been constructed using the fixed fraction criterion (from Houston *et al.* (2000*a*))

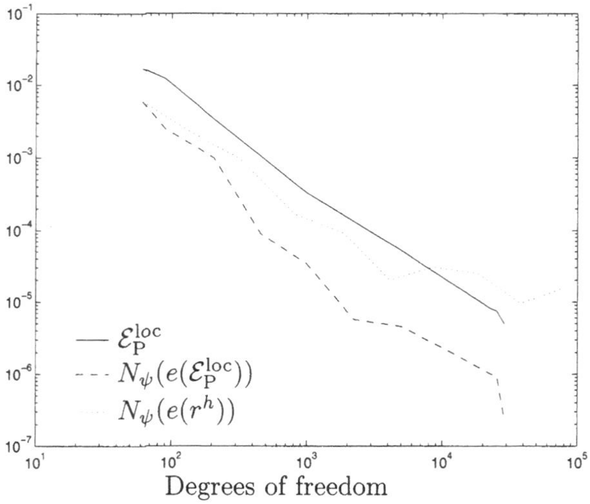

Figure 8.3. Example 2: Performance of the adaptive algorithm for TOL $= 5.0 \times 10^{-6}$ and $\delta = h/4$. $N_\psi(e(\mathcal{E}_{\mathrm{P}}^{\mathrm{loc}}))$ denotes the error in the functional on a sequence of adaptively refined meshes using the stabilization-dependent dual problem; $\mathcal{E}_{\mathrm{P}}^{\mathrm{loc}}$ is the corresponding Type I *a posteriori* error bound; $N_\psi(e(r^h))$ denotes the error in the functional when the adaptive meshes are generated by using $\|r^h\|_{L^2(\kappa)}$ alone as refinement indicator (from Houston *et al.* (2000*a*))

boundary where the quantity of interest, $N_\psi(u)$, is concentrated. We note, in particular, the lack of mesh refinement in the vicinity of the discontinuities as they enter the domain from $y = 0$ and $x = 0$. Had the mesh adaptation been based on disregarding the size of the dual solution and refining according to the local size of the residual alone, the resulting mesh for the primal problem would have contained heavy (and, clearly, unnecessary) refinement along the discontinuities in the primal solution.

This last point is illustrated further by the computations whose results are depicted in Figure 8.3. Here we show the performance of our adaptive algorithm with TOL $= 5.0 \times 10^{-6}$. The initial meshes are as in Figure 8.2. We see that, even though the stabilization-dependent dual problem has been solved numerically, our Type I *a posteriori* error bound $\mathcal{E}_{\mathrm{P}}^{\mathrm{loc}}$, based on the use of the numerically computed dual solution z^H and $\pi^h z^H$ in place of z and $z^h = \pi^h z$, respectively, remains an upper bound on the true error in the approximation of the output functional $N_\psi(u)$. In Figure 8.3, $N_\psi(e(\mathcal{E}_{\mathrm{P}}^{\mathrm{loc}}))$ denotes the true error in the outward normal flux on the sequence of adaptively refined meshes which have been generated using the optimized mesh criterion. Figure 8.3 also shows the true error in the outward normal flux on a sequence of adaptively refined meshes which have been constructed using an empirical refinement indicator based on the local L^2-norm of the residual, $\|r^h\|_{L^2(\kappa)}$ on each element κ in $\mathcal{T}^h$ in conjunction with a fixed fraction strategy. It is clear from Figure 8.3 that the latter approach is inferior: stagnation of the error in the output functional is observed in the course of the adaptive mesh refinement.

9. *hp*-adaptivity by duality

In this section we shall briefly discuss the derivation of Type I *a posteriori* error bounds, based on a duality argument, for *hp*-version finite element methods, and the application of such bounds in *hp*-adaptive finite element algorithms. In addition to local variation of the mesh size h, adaptive algorithms of this kind admit local variation of the degree p of the approximating polynomial in the finite element space, and thereby offer greater flexibility than traditional h-version finite element methods. For the sake of brevity, here we shall focus on one particular algorithm, based on the *hp*-version of the discontinuous Galerkin finite element method (*hp*-DGFEM). The results presented in this section are based on the papers by Süli *et al.* (1999), Houston and Süli (2001*a*), Süli, Houston and Senior (2001). For the *a priori* error analysis of the *hp*-DGFEM and *hp*-SDFEM, we refer to Bey and Oden (1996), Houston, Schwab and Süli (2000*b*), Houston and Süli (2001*b*). The *hp*-version finite element methods were traditionally developed in the context of elliptic boundary value problems. Their use for the numerical solution of first-order hyperbolic systems is more recent and is motivated by the fact that, even though solutions to hyperbolic problems may exhibit local singularities and discontinuities, they are typically piecewise analytic functions. Thus, away from singularities, one may use high-degree piecewise polynomial approximations on course meshes. The relevance of *hp*-version finite element methods for the numerical solution of hyperbolic conservation laws is discussed in Bey and Oden (1996) and Adjerid, Aiffa and Flaherty (1998); see also Flaherty, Loy, Shephard and Teresco (2000) concerning implementational aspects of DGFEM. For a review of recent developments concerning the theory and application of *hp*-version finite element methods, see Ainsworth and Oden (2000), Szabó and Babuška (1991), Schwab (1998).

9.1. The model problem and its hp-DGFEM approximation

Suppose that Ω is a bounded open polyhedral domain in $\mathbb{R}^n$, $n \geq 2$, and let Γ denote the union of open faces of Ω. Let us further assume that $\mathbf{B} = (\mathbf{B}_1, \ldots, \mathbf{B}_n)$ is an n-component matrix function defined on $\bar{\Omega}$ with $\mathbf{B}_i \in [W^1_\infty(\Omega)]^{m\times m}_{\text{symm}}$, $i = 1, \ldots, n$. We shall let $\boldsymbol{\nu} = (\nu_1, \ldots, \nu_n)$ denote the unit outward normal vector to Γ, and we consider the matrix

$$\mathbf{B}(\boldsymbol{\nu}) \equiv \boldsymbol{\nu} \cdot \mathbf{B} = \nu_1 \mathbf{B}_1 + \cdots + \nu_n \mathbf{B}_n.$$

Since $\mathbf{B}(\boldsymbol{\nu})$ is a symmetric matrix, it can be diagonalized; that is, we can write

$$\mathbf{B}(\boldsymbol{\nu}) = X(\boldsymbol{\nu})^{-1} \Lambda(\boldsymbol{\nu}) X(\boldsymbol{\nu}),$$

where $\Lambda(\boldsymbol{\nu})$ is a diagonal matrix, with the (real) eigenvalues of $\mathbf{B}(\boldsymbol{\nu})$ appearing along its diagonal. We shall suppose that Γ is nowhere characteristic, in the sense that none of the diagonal entries of $\Lambda(\boldsymbol{\nu})$ is zero for any choice of the unit outward normal vector $\boldsymbol{\nu}$ on Γ. Let us additively decompose the matrix $\Lambda(\boldsymbol{\nu})$ as

$$\Lambda(\boldsymbol{\nu}) = \Lambda_-(\boldsymbol{\nu}) + \Lambda_+(\boldsymbol{\nu}),$$

where $\Lambda_-(\boldsymbol{\nu})$ is diagonal and negative semidefinite, and $\Lambda_+(\boldsymbol{\nu})$ is diagonal and positive semidefinite. With this notation, we now define the $m \times m$ matrices

$$\mathbf{B}_-(\boldsymbol{\nu}) = X(\boldsymbol{\nu})^{-1} \Lambda_-(\boldsymbol{\nu}) X(\boldsymbol{\nu}) \quad \text{and} \quad \mathbf{B}_+(\boldsymbol{\nu}) = X(\boldsymbol{\nu})^{-1} \Lambda_+(\boldsymbol{\nu}) X(\boldsymbol{\nu}).$$

We then have the following induced decomposition of $\mathbf{B}(\boldsymbol{\nu})$ for each choice of the unit outward normal vector $\boldsymbol{\nu}$ on Γ:

$$\mathbf{B}(\boldsymbol{\nu}) = \mathbf{B}_-(\boldsymbol{\nu}) + \mathbf{B}_+(\boldsymbol{\nu}).$$

Given $\mathbf{C} \in [L^\infty(\Omega)]^{m\times m}$, $\mathbf{f} \in [L^2(\Omega)]^m$ and $\mathbf{g} \in [L^2(\Gamma)]^m$, we consider the following hyperbolic boundary value problem: find $\mathbf{u} \in H(\mathcal{L}, \Omega)$ such that

$$\mathcal{L}\mathbf{u} \equiv \nabla \cdot (\mathbf{B}\mathbf{u}) + \mathbf{C}\mathbf{u} = \mathbf{f} \quad \text{in } \Omega, \qquad \mathbf{B}_-(\boldsymbol{\nu})\mathbf{u} = \mathbf{B}_-(\boldsymbol{\nu})\mathbf{g} \quad \text{on } \Gamma, \tag{9.1}$$

where $H(\mathcal{L}, \Omega) = \{\mathbf{v} \in [L^2(\Omega)]^m : \mathcal{L}\mathbf{v} \in [L^2(\Omega)]^m\}$ denotes the *graph space* of the partial differential operator $\mathcal{L}$ in $L^2(\Omega)$.

Now, let us formulate the *hp*-DGFEM for (9.1). We begin by considering a regular or 1-irregular subdivision $\mathcal{T}_h$ of Ω into disjoint open element domains κ such that $\bar{\Omega} = \cup_{\kappa \in \mathcal{T}_h} \bar{\kappa}$. By *regular or* 1*-irregular* we mean that an $(n-1)$-dimensional face of each element κ in $\mathcal{T}_h$ is allowed to contain at most one hanging (irregular) node – typically, the hanging node is chosen as the barycentre of the face, although this is not essential for what follows. We shall further suppose that the family of subdivisions $\mathcal{T}_h$ is shape-regular and that each $\kappa \in \mathcal{T}_h$ is a bijective affine image of a fixed master element $\hat{\kappa}$; that is, $\kappa = F_\kappa(\hat{\kappa})$ for all $\kappa \in \mathcal{T}_h$, where $\hat{\kappa}$ is either the open unit simplex or the open unit hypercube in $\mathbb{R}^n$.

On the reference element $\hat{\kappa}$, with $(\hat{x}_1, \ldots, \hat{x}_n) \in \hat{\kappa}$ and $(\alpha_1, \ldots, \alpha_n) \in \mathbb{N}_0^n$, we define spaces of polynomials of degree $p \geq 1$ as follows:

$$\mathcal{Q}_p = \operatorname{span}\{\hat{x}_1^{\alpha_1} \cdots \hat{x}_n^{\alpha_n} : 0 \leq \alpha_i \leq p, \ 1 \leq i \leq n\},$$

$$*\mathcal{P}_p = \operatorname{span}\{\hat{x}_1^{\alpha_1} \cdots \hat{x}_n^{\alpha_n} : 0 \leq \alpha_1 + \cdots + \alpha_n \leq p\}.$$

Now, to each $\kappa \in \mathcal{T}_h$ we assign an integer $p_\kappa \geq 1$; collecting the local polynomial degrees p_κ and mappings F_κ in the vectors $\mathbf{p} = \{p_\kappa : \kappa \in \mathcal{T}_h\}$ and $\mathbf{F} = \{F_\kappa : \kappa \in \mathcal{T}_h\}$, respectively, we introduce the finite element space

$$S^{\mathbf{p}}(\Omega, \mathcal{T}_h, \mathbf{F}) = \{\mathbf{v} \in [L^2(\Omega)]^m : \mathbf{v}|_\kappa \circ F_\kappa \in [\mathcal{Q}_{p_\kappa}]^m \text{ if } F_\kappa^{-1}(\kappa) \text{ is the open unit hypercube and } \mathbf{v}|_\kappa \circ F_\kappa \in [\mathcal{P}_{p_\kappa}]^m \text{ if } F_\kappa^{-1}(\kappa) \text{ is the open unit simplex; } \kappa \in \mathcal{T}_h\}.$$

Assuming that $\mathcal{T}_h$ is a subdivision of Ω, we consider the broken Sobolev space $H^{\mathbf{s}}(\Omega, \mathcal{T}_h)$ of composite index $\mathbf{s}$ with nonnegative components s_κ, $\kappa \in \mathcal{T}_h$, defined by

$$[H^{\mathbf{s}}(\Omega, \mathcal{T}_h)]^m = \{\mathbf{v} \in [L^2(\Omega)]^m : \mathbf{v}|_\kappa \in [H^{s_\kappa}(\kappa)]^m \quad \forall \kappa \in \mathcal{T}_h\}.$$

If $s_\kappa = s \geq 0$ for all $\kappa \in \mathcal{T}_h$, we shall simply write $[H^s(\Omega, \mathcal{T}_h)]^m$.

Let us suppose that κ is an element in the subdivision $\mathcal{T}_h$ of the computational domain Ω. We shall let $\partial\kappa$ denote the union of $(n-1)$-dimensional open faces of κ. Let $x \in \partial\kappa$ and suppose that $\mathbf{n}_\kappa(x)$ denotes the unit outward normal vector to $\partial\kappa$ at x. We then define $\mathbf{B}(\mathbf{n}_\kappa)$, $\mathbf{B}_-(\mathbf{n}_\kappa)$ and $\mathbf{B}_+(\mathbf{n}_\kappa)$ analogously to $\mathbf{B}(\boldsymbol{\nu})$, $\mathbf{B}_-(\boldsymbol{\nu})$ and $\mathbf{B}_+(\boldsymbol{\nu})$ above, respectively.

For each $\kappa \in \mathcal{T}_h$ and any $\mathbf{v} \in [H^1(\kappa)]^m$ we let $\mathbf{v}_\kappa^+$ denote the interior trace of $\mathbf{v}$ on $\partial\kappa$ (the trace taken from within κ). Now consider an element κ such that the set $\partial\kappa\backslash\Gamma$ is nonempty; then, for each $x \in \partial\kappa\backslash\Gamma$ (with the exception of a set of $(n-1)$-dimensional measure zero), there exists a unique element κ', depending on the choice of x, such that $x \in \partial\kappa'$. Suppose that $\mathbf{v} \in [H^1(\Omega, \mathcal{T}_h)]^m$. If $\partial\kappa\backslash\Gamma$ is nonempty for some $\kappa \in \mathcal{T}_h$, then we define the outer trace $\mathbf{v}_\kappa^-$ of $\mathbf{v}$ on $\partial\kappa\backslash\Gamma$ relative to κ as the inner trace $\mathbf{v}_{\kappa'}^+$ relative to those elements κ' for which $\partial\kappa'$ has intersection with $\partial\kappa\backslash\Gamma$ of positive $(n-1)$-dimensional measure. The context should always make it clear to which element κ in the subdivision $\mathcal{T}_h$ the quantities $\mathbf{n}_\kappa$, $\mathbf{v}_\kappa^+$ and $\mathbf{v}_\kappa^-$ correspond. Thus, for the sake of simplicity of notation, we shall suppress the letter κ in the subscript and write, respectively, $\mathbf{n}$, $\mathbf{v}^+$, and $\mathbf{v}^-$ instead.

For $\mathbf{v}, \mathbf{w} \in [H^1(\Omega, \mathcal{T}_h)]^m$, we define the bilinear form of the *hp*-DGFEM by

$$B_{\mathrm{DG}}(\mathbf{w}, \mathbf{v}) = \sum_{\kappa \in \mathcal{T}_h} \int_\kappa \mathbf{w} \cdot \mathcal{L}^* \mathbf{v} \, \mathrm{d}x + \sum_{\kappa \in \mathcal{T}_h} \int_{\partial\kappa\backslash\Gamma} \mathcal{H}(\mathbf{w}^+, \mathbf{w}^-, \mathbf{n}) \cdot \mathbf{v}^+ \, \mathrm{d}s + \sum_{\kappa \in \mathcal{T}_h} \int_{\partial\kappa\cap\Gamma} \mathbf{B}_+(\mathbf{n}) \mathbf{w}^+ \cdot \mathbf{v}^+ \, \mathrm{d}s,$$

where $\mathcal{L}^*$ is the formal adjoint of $\mathcal{L}$ defined by $\mathcal{L}^*\mathbf{v} \equiv -(\mathbf{B} \cdot \nabla)\mathbf{v} + \mathbf{C}^T\mathbf{v}$; $\mathcal{H}(\cdot\,,\cdot\,,\cdot)$ is a numerical flux function, assumed to be Lipschitz-continuous, and such that:

(i) $\mathcal{H}$ is consistent, *i.e.*, $\mathcal{H}(\mathbf{u}, \mathbf{u}, \mathbf{n})|_{\partial\kappa\setminus\Gamma} = \mathbf{B}(\mathbf{n})\mathbf{u}|_{\partial\kappa\setminus\Gamma}$ for all κ in $\mathcal{T}_h$;

(ii) $\mathcal{H}(\cdot, \cdot, \cdot)$ is conservative, *i.e.*, $\mathcal{H}(\mathbf{u}^+, \mathbf{u}^-, \mathbf{n})|_{\partial\kappa\setminus\Gamma} = -\mathcal{H}(\mathbf{u}^-, \mathbf{u}^+, -\mathbf{n})|_{\partial\kappa\setminus\Gamma}$.

For example, we may take

$$\begin{aligned}\mathcal{H}(\mathbf{u}^+, \mathbf{u}^-, \mathbf{n}) &= \mathbf{B}_+(\mathbf{n})\mathbf{u}^+ + \mathbf{B}_-(\mathbf{n})\mathbf{u}^- \\ &= \frac{1}{2}\left(\mathbf{B}(\mathbf{n})\mathbf{u}^+ + \mathbf{B}(\mathbf{n})\mathbf{u}^-\right) - \frac{1}{2}|\mathbf{B}(\mathbf{n})|(\mathbf{u}^- - \mathbf{u}^+),\end{aligned}$$

where $|\mathbf{B}(\mathbf{n})| = \mathbf{B}_+(\mathbf{n}) - \mathbf{B}_-(\mathbf{n})$. For $\mathbf{v} \in [H^1(\Omega, \mathcal{T}_h)]^m$, we introduce the linear functional

$$\ell_{\mathrm{DG}}(\mathbf{v}) = \sum_{\kappa\in\mathcal{T}_h} \int_\kappa \mathbf{f} \cdot \mathbf{v} \, \mathrm{d}x - \sum_{\kappa\in\mathcal{T}_h} \int_{\partial\kappa\cap\Gamma} \mathbf{B}_-(\mathbf{n}) \, \mathbf{g} \cdot \mathbf{v}^+ \, \mathrm{d}s.$$

With this notation, the *hp*-DGFEM for (9.1) is defined as follows: find $\mathbf{u}_{\mathrm{DG}} \in S^{\mathbf{p}}(\Omega, \mathcal{T}_h, \mathbf{F})$ such that

$$B_{\mathrm{DG}}(\mathbf{u}_{\mathrm{DG}}, \mathbf{v}) = \ell_{\mathrm{DG}}(\mathbf{v}) \quad \forall \mathbf{v} \in S^{\mathbf{p}}(\Omega, \mathcal{T}_h, \mathbf{F}). \tag{9.2}$$

9.2. A posteriori *error analysis by duality*

Let us suppose that the aim of the computation is to control the error in some linear output functional $J(\cdot)$ defined on a linear space which contains $H(\mathcal{L}, \Omega) + S^{\mathbf{p}}(\Omega, \mathcal{T}_h, \mathbf{F})$. We shall do so by deriving a Type I *a posteriori* bound on the error between $J(\mathbf{u})$ and $J(\mathbf{u}_{\mathrm{DG}})$. For this purpose we introduce the following dual problem: find $\mathbf{z}$ in $H(\mathcal{L}^*, \Omega)$ such that

$$B_{\mathrm{DG}}(\mathbf{w}, \mathbf{z}) = J(\mathbf{w}) \quad \forall \mathbf{w} \in H(\mathcal{L}, \Omega), \tag{9.3}$$

where $H(\mathcal{L}^*, \Omega)$ denotes the graph space of the adjoint operator $\mathcal{L}^*$ in $L^2(\Omega)$. We shall tacitly assume that (9.3) has a unique solution.

Let us define the *internal residual* $\mathbf{r}_{h,\mathbf{p}}$ on $\kappa \in \mathcal{T}_h$ by

$$\mathbf{r}_{h,\mathbf{p}}|_\kappa = (\mathbf{f} - \mathcal{L}\mathbf{u}_{\mathrm{DG}})|_\kappa,$$

which measures the extent to which $\mathbf{u}_{\mathrm{DG}}$ fails to satisfy the differential equation on the union of the elements κ in the mesh $\mathcal{T}_h$; and, for each element κ with $\partial\kappa \cap \Gamma$ nonempty, we define the *boundary residual* $\boldsymbol{\rho}_{h,\mathbf{p}}$ by

$$\boldsymbol{\rho}_{h,\mathbf{p}}|_{\partial\kappa\cap\Gamma} = \mathbf{B}_-(\mathbf{n})(\mathbf{u}_{\mathrm{DG}}^+ - \mathbf{g})|_{\partial\kappa\cap\Gamma}.$$

Analogously, on $\partial\kappa \setminus \Gamma$, we define the *interelement flux residual* $\boldsymbol{\sigma}_{h,\mathbf{p}}$ by

$$\boldsymbol{\sigma}_{h,\mathbf{p}}|_{\partial\kappa\setminus\Gamma} = \left(\mathbf{B}(\mathbf{n})\mathbf{u}_{\mathrm{DG}}^+ - \mathcal{H}(\mathbf{u}_{\mathrm{DG}}^+, \mathbf{u}_{\mathrm{DG}}^-, \mathbf{n})\right)|_{\partial\kappa\setminus\Gamma}.$$

Assuming that $\mathbf{u}$ is sufficiently smooth, it is then a simple matter to verify the following Galerkin orthogonality property:

$$B_{\rm DG}(\mathbf{u}-\mathbf{u}_{\rm DG},\mathbf{v}) = \sum_{\kappa\in\mathcal{T}_h}(\mathbf{r}_{h,\mathbf{p}},\mathbf{v})_\kappa + \sum_{\kappa\in\mathcal{T}_h}(\boldsymbol{\sigma}_{h,\mathbf{p}},\mathbf{v}^+)_{\partial\kappa\backslash\Gamma} + \sum_{\kappa\in\mathcal{T}_h}(\boldsymbol{\rho}_{h,\mathbf{p}},\mathbf{v}^+)_{\partial\kappa\cap\Gamma} = 0 \qquad (9.4)$$

for all $\mathbf{v}$ in $S^{\mathbf{p}}(\Omega,\mathcal{T}_h,\mathbf{F})$. On selecting $\mathbf{w}=\mathbf{u}-\mathbf{u}_{\rm DG}$ in (9.3), the linearity of $J(\cdot)$ and (9.4) yield the following error representation formula:

$$\begin{aligned} J(\mathbf{u})-J(\mathbf{u}_{\rm DG}) &= J(\mathbf{u}-\mathbf{u}_{\rm DG}) \\ &= B_{\rm DG}(\mathbf{u}-\mathbf{u}_{\rm DG},\mathbf{z}) \\ &= B_{\rm DG}(\mathbf{u}-\mathbf{u}_{\rm DG},\mathbf{z}-\mathbf{z}_{h,\mathbf{p}}) \\ &\equiv \mathcal{E}_\Omega(\mathbf{u}_{\rm DG},h,\mathbf{p},\mathbf{z}-\mathbf{z}_{h,\mathbf{p}}) \\ &= \sum_{\kappa\in\mathcal{T}_h}\eta_\kappa, \end{aligned} \qquad (9.5)$$

where

$$\eta_\kappa = (\mathbf{r}_{h,\mathbf{p}},\mathbf{z}-\mathbf{z}_{h,\mathbf{p}})_\kappa + (\boldsymbol{\sigma}_{h,\mathbf{p}},(\mathbf{z}-\mathbf{z}_{h,\mathbf{p}})^+)_{\partial\kappa\backslash\Gamma} + (\boldsymbol{\rho}_{h,\mathbf{p}},(\mathbf{z}-\mathbf{z}_{h,\mathbf{p}})^+)_{\partial\kappa\cap\Gamma}. \qquad (9.6)$$

For a user-defined tolerance TOL, we now consider the problem of designing an *hp*-finite element space $S^{\mathbf{p}}(\Omega,\mathcal{T}_h,\mathbf{F})$ such that

$$|J(\mathbf{u})-J(\mathbf{u}_{\rm DG})| \le \mathtt{TOL}. \qquad (9.7)$$

To do so, we shall use the following Type I *a posteriori* error bound, which stems from (9.5):

$$\mathcal{E}_{\rm P}^{\rm loc} \equiv \sum_{\kappa\in\mathcal{T}_h}|\tilde{\eta}_\kappa| \le \mathtt{TOL}, \qquad (9.8)$$

where $\tilde{\eta}_\kappa$ is defined analogously to η_κ (*cf.* (9.6)), with $\mathbf{z}$ replaced by its numerical approximation $\tilde{\mathbf{z}}_{\rm DG}$ computed by means of the *hp*-DGFEM.

If the stopping criterion (9.8) is violated, then certain elements $\kappa\in\mathcal{T}_h$ will be marked for refinement; in addition to *h*- and *p*-refinement, the adaptive algorithm discussed here will also admit *h*- and *p*-derefinement. Here we shall employ the fixed fraction mesh refinement criterion, based on $\tilde{\eta}_\kappa$, with refinement and derefinement fractions set to 20% and 10%, respectively, to identify elements which will be refined/derefined. In this way, $\mathcal{T}_h$ is partitioned into three disjoint subsets:

$\mathcal{T}_h^{\rm ref}$, consisting of those elements κ in $\mathcal{T}_h$ that are marked for *refinement*;

$\mathcal{T}_h^{\rm deref}$, containing those elements $\kappa\in\mathcal{T}_h$ that are marked for *derefinement*;

$\mathcal{T}_h^{\rm idle}=\mathcal{T}_h\backslash(\mathcal{T}_h^{\rm ref}\cup\mathcal{T}_h^{\rm deref})$, the set of *idle* elements where no action is required.

If $\kappa\in\mathcal{T}_h^{\rm ref}\cup\mathcal{T}_h^{\rm deref}$, then a decision must be made as to whether the local mesh size h_κ or the local degree p_κ of the approximating polynomial should be altered. The choice between h-refinement/derefinement and p-refinement/derefinement is made by assessing the local smoothness of the primal and dual solutions $\mathbf{u}$ and $\mathbf{z}$, respectively. The various possibilities are discussed below in more detail.

`Refinement:` suppose that $\kappa \in \mathcal{T}_h^{\mathrm{ref}}$. If $\mathbf{u}$ or $\mathbf{z}$ are smooth on κ, then p-refinement will be more effective than h-refinement, since the error is then expected to decay quickly within κ as p_κ is increased. If, on the other hand, $\mathbf{u}$ and $\mathbf{z}$ have low regularity within κ, then h-refinement will be performed. In this way, regions in the computational domain where the primal or dual solutions are locally non-smooth are isolated from regions of smoothness; this then reduces the influence of singularities/discontinuities and makes p-refinement more effective. In order to ensure that the desired level of accuracy is achieved efficiently, Houston and Süli (2001*a*) developed an automatic procedure for deciding when to h- or p-refine, based on the Sobolev smoothness estimation strategy proposed in Ainsworth and Senior (1998) in the context of hp-adaptive norm control for second-order elliptic problems; for a review of recent developments on Sobolev regularity estimation, we refer to Houston and Süli (2002*a*).

`Derefinement:` suppose that $\kappa \in \mathcal{T}_h^{\mathrm{deref}}$. The derefinement strategy implemented here is to coarsen the mesh around κ in low error regions where either the primal or dual solutions $\mathbf{u}$ and $\mathbf{z}$, respectively, are smooth, and decrease the degree of the approximating polynomial in low error regions when both $\mathbf{u}$ and $\mathbf{z}$ are insufficiently regular (*cf.* Adjerid *et al.* (1998) and Houston and Süli (2001*a*)).

In Houston and Süli (2001*a*) a fully hp-adaptive algorithm has been developed; hp-adaptivity for the primal problem is controlled by a Type I *a posteriori* error bound, while the hp-adaptive algorithm for the dual is driven by a (cruder) Type II *a posteriori* error bound. Here, for the sake of simplicity, the dual finite element space $\tilde{S}^{\tilde{\mathbf{p}}}(\Omega, \tilde{\mathcal{T}}_h, \tilde{\mathbf{F}})$ that is used to compute the discontinuous Galerkin approximation $\tilde{\mathbf{z}}_{\mathrm{DG}}$ to $\mathbf{z}$ will be constructed using the same mesh as the one employed for $\mathbf{u}_{\mathrm{DG}}$, *i.e.*, $\tilde{\mathcal{T}}_h \equiv \mathcal{T}_h$, with $\tilde{\mathbf{p}} = \mathbf{p} + \mathbf{1}$; this possibility for the numerical approximation of the dual problem was mentioned in Section 3.2. The reader is referred to Houston and Süli (2001*a*) for details concerning the implementation of a more general algorithm where $\tilde{\mathcal{T}}_h \neq \mathcal{T}_h$ and $\tilde{\mathbf{p}}$ is not required to be related to $\mathbf{p}$.

9.3. Numerical experiments

We present a numerical experiment to demonstrate the performance of the hp-adaptive algorithm. The extension to nonlinear problems will be considered in the next section.

Linear advection. In this example, we consider the scalar hyperbolic equation $\nabla \cdot (\mathbf{b}u) + cu = f$ on $\Omega = (0,1)^2$, where $\mathbf{b} = (10y^2 - 12x + 1, 1 + y)$, $c = -\nabla \cdot \mathbf{b}$ and $f = 0$. The characteristics enter the square Ω across three of its sides, *i.e.*, the two vertical faces and the bottom; they exit Ω through the top edge. We prescribe the boundary condition

$$u(x,y) = \begin{cases} 0 & \text{for } x = 0, \quad 0.5 < y \leq 1, \\ 1 & \text{for } x = 0, \quad 0 \leq y \leq 0.5, \\ 1 & \text{for } 0 \leq x \leq 0.75, \quad y = 0, \\ 0 & \text{for } 0.75 < x \leq 1, \quad y = 0, \\ \sin^2(\pi y) & \text{for } x = 1, \quad 0 \leq y \leq 1, \end{cases}$$

on the union Γ_- of the three inflow sides. The objective is to compute the weighted normal flux through the outflow side Γ_+ defined by

$$J(u) = \int_{\Gamma_+} \psi(x) u(x,1)\,\mathrm{d}x,$$

where the weight function ψ is defined by $\psi(x) = \sin(\pi x/2)$ for $0 \leq x \leq 1$; thereby, the true value of the functional is $J(u) = 0.246500283257585$ (*cf.* Houston *et al.* (2000*a*)).

In Table 9.1, we show the performance of the adaptive algorithm: we give the number of nodes (Nds), elements (Els) and degrees of freedom (DOF)

Table 9.1. Adaptive algorithm for the linear advection problem

Nds	Els	DOF	$J(u-u_{\rm DG})$	$\sum_\kappa \tilde{\eta}_\kappa$	θ_1	$\sum_\kappa \lvert\tilde{\eta}_\kappa\rvert$	θ_2
25	16	64	0.1207E-02	0.1023E-02	0.85	0.1938E-01	16.06
30	19	86	-0.8405E-02	-0.8203E-02	0.98	0.1006E-01	1.20
48	31	202	-0.6729E-02	-0.6002E-02	0.89	0.7279E-02	1.08
48	31	244	-0.1611E-02	-0.1623E-02	1.01	0.1927E-02	1.20
57	37	330	-0.9690E-03	-0.9756E-03	1.01	0.1043E-02	1.08
87	61	595	-0.8424E-03	-0.8581E-03	1.02	0.8654E-03	1.03
129	91	1078	-0.1075E-04	-0.4017E-04	3.74	0.4731E-04	4.40
139	100	1439	0.2691E-04	0.2906E-04	1.08	0.3580E-04	1.33
201	148	2490	-0.1456E-05	-0.1290E-05	0.89	0.2808E-05	1.93
263	199	3723	-0.4938E-06	-0.6040E-06	1.22	0.6721E-06	1.36
308	232	4876	-0.1196E-07	-0.1123E-07	0.94	0.4792E-07	4.01
383	292	6793	-0.5294E-08	-0.5296E-08	1.00	0.6621E-08	1.25
429	328	8548	-0.3450E-08	-0.3457E-08	1.00	0.4322E-08	1.25
542	418	12325	-0.1650E-09	-0.1676E-09	1.02	0.2047E-09	1.24

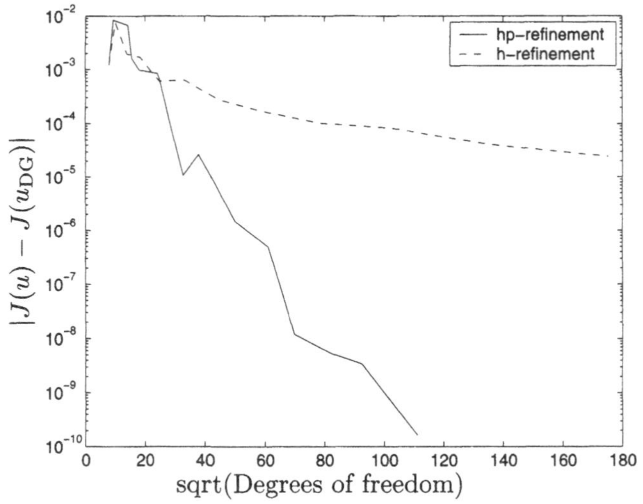

Figure 9.1. Comparison between h- and hp-adaptive mesh refinement for the linear problem

in $S^{\mathbf{p}}(\Omega, \mathcal{T}_h, \mathbf{F})$, the true error in the functional $J(u - u_{\rm DG})$, the computed error representation formula $\sum_{\kappa \in \mathcal{T}_h} \tilde{\eta}_\kappa$, the Type I *a posteriori* error bound $\mathcal{E}_{\rm P}^{\rm loc} = \sum_{\kappa \in \mathcal{T}_h} |\tilde{\eta}_\kappa|$, and their respective effectivity indices θ_1 and θ_2. We see that initially, on very coarse meshes, the quality of the computed error representation formula is quite poor, in the sense that θ_1, the ratio of the error representation formula and the error $J(u - u^h)$, is not close to one; however, as the mesh is refined the effectivity index θ_1 approaches unity. Furthermore, we observe that the Type I *a posteriori* error bound is indeed sharp, in the sense that the second effectivity index $\theta_2 = \mathcal{E}_{\rm P}^{\rm loc}/J(u - u_{\rm DG})$ overestimates the true error in the computed functional by a consistent factor as the finite element space $S^{\mathbf{p}}(\Omega, \mathcal{T}_h, \mathbf{F})$ is enriched.

Figure 9.1 shows $|J(u) - J(u_{\rm DG})|$, using both h- and hp-refinement against the square root of the number of degrees of freedom on a linear-log scale. We see that, after an initial increase, the error in the approximation to the output functional using hp-refinement becomes (on average) a straight line, thereby indicating exponential convergence of $J(u_{\rm DG})$ to $J(u)$, despite the fact that u is only piecewise continuous; this occurs since z is a real analytic function on $\bar{\Omega}$. Figure 9.1 also highlights the superiority of the adaptive hp-refinement strategy over a traditional adaptive h-refinement algorithm. On the final mesh, the true error between $J(u)$ and $J(u_{\rm DG})$ using hp-refinement is almost 6 orders of magnitude smaller than the corresponding quantity when h-refinement is employed alone.

9.4. Adaptivity for nonlinear problems

Multi-dimensional compressible fluid flows are modelled by nonlinear conservation laws whose solutions exhibit a wide range of localized structures, such as shock waves, contact discontinuities and rarefaction waves. The accurate numerical resolution of these features necessitates the use of locally refined, adaptive computational meshes. Here we describe the development of Type I *a posteriori* error bounds for *hp*-version discontinuous Galerkin finite element approximations of nonlinear systems of conservation laws, following Süli *et al.* (2001), which is the extension of the h-version *a posteriori* error analysis in Larson and Barth (2000), Hartmann (2001), Hartmann and Houston (2001) and Süli *et al.* (2001).

Given a bounded open polyhedral domain Ω in $\mathbb{R}^n$, $n \geq 1$, let Γ denote the union of open faces contained in $\partial\Omega$. We consider the following problem: find $\mathbf{u} : \Omega \to \mathbb{R}^m$, $m \geq 1$, such that

$$\operatorname{div}\mathcal{F}(\mathbf{u}) = 0 \qquad \text{in } \Omega, \tag{9.9}$$

where $\mathcal{F} : \mathbb{R}^m \to \mathbb{R}^{m\times n}$ is continuously differentiable. We assume that the system of conservation laws (9.9) may be supplemented by appropriate initial/boundary conditions. In other words, we assume that

$$B(\mathbf{u}, \boldsymbol{\mu}) := \sum_{i=1}^{n} \mu_i \nabla_{\mathbf{u}} \mathcal{F}_i(\mathbf{u})$$

has m real eigenvalues and a complete set of linearly independent eigenvectors for all $\boldsymbol{\mu} = (\mu_1, \dots, \mu_n) \in \mathbb{R}^n$; then at inflow/outflow boundaries, we require that $B_-(\mathbf{u}, \boldsymbol{\nu})(\mathbf{u} - \mathbf{g}) = \mathbf{0}$, where $\boldsymbol{\nu}$ denotes the unit outward normal vector to $\partial\Omega$, $B_-(\mathbf{u}, \boldsymbol{\nu})$ is the negative part of $B(\mathbf{u}, \boldsymbol{\nu})$ and $\mathbf{g}$ is a (given) real-valued function.

The *hp*-DGFEM for (9.9) is defined as follows: find $\mathbf{u}_{\mathrm{DG}} \in S^{\mathbf{p}}(\Omega, \mathcal{T}_h, \mathbf{F})$ such that

$$\sum_{\kappa\in\mathcal{T}_h} \left\{ -\int_\kappa \mathcal{F}(\mathbf{u}_{\mathrm{DG}}) \cdot \nabla \mathbf{v}_{h,\mathbf{p}} \,\mathrm{d}x + \int_{\partial\kappa} \mathcal{H}(\mathbf{u}_{\mathrm{DG}}^+, \mathbf{u}_{\mathrm{DG}}^-, \boldsymbol{\nu}_\kappa)\, \mathbf{v}_{h,\mathbf{p}}^+ \,\mathrm{d}s \right.$$
$$\left. + \int_\kappa \varepsilon \nabla \mathbf{u}_{\mathrm{DG}} \cdot \nabla \mathbf{v}_{h,\mathbf{p}} \,\mathrm{d}x \right\} = 0 \tag{9.10}$$

for all $\mathbf{v}_{h,\mathbf{p}} \in S^{\mathbf{p}}(\Omega, \mathcal{T}_h, \mathbf{F})$ (*cf.* Jaffre, Johnson and Szepessy (1995), Hartmann and Houston (2001), Houston, Hartmann and Süli (2001), Süli *et al.* (2001), for example); here $\boldsymbol{\nu}_\kappa$ denotes the unit outward normal vector to κ, and $\mathcal{H}(\cdot\,,\cdot\,,\cdot)$ is a *numerical flux* function, assumed to be Lipschitz-continuous, consistent and conservative. We emphasize that the choice of the numerical flux function is completely independent of the finite element space employed; in the numerical experiments we use the (local) Lax–Friedrichs flux. Further, the parameter ε denotes the coefficient of artificial viscosity defined, on $\kappa \in \mathcal{T}_h$, by

$$\varepsilon|_\kappa = C_\varepsilon \left(\frac{h_\kappa}{p_\kappa}\right)^{2-\beta} |\mathrm{div}\mathcal{F}(\mathbf{u}_{\mathrm{DG}}|_\kappa)|,$$

where C_ε is a positive constant and $0 < \beta < 1/2$; see Jaffre *et al.* (1995). For elements $\kappa \in \mathcal{T}_h$ whose boundary intersects $\partial\Omega$, $\mathbf{u}_h^-$ is replaced by appropriate boundary/initial conditions on $\partial\kappa \cap \partial\Omega$.

Let us suppose that we are concerned with computing $J(\mathbf{u})$, where $J(\cdot)$ is a given linear output functional. Letting $\mathcal{N}(\mathbf{u}_{\mathrm{DG}}, \mathbf{v}_{h,\mathbf{p}})$ denote the left-hand side of (9.10), we write $\mathcal{M}(\mathbf{u}, \mathbf{u}_{\mathrm{DG}}; \cdot\,, \cdot)$ to denote the mean-value linearization of $\mathcal{N}(\cdot\,, \cdot)$ given by

$$\begin{aligned}\mathcal{M}(\mathbf{u}, \mathbf{u}_{\mathrm{DG}}; \mathbf{u} - \mathbf{u}_{\mathrm{DG}}, \mathbf{v}) &= \mathcal{N}(\mathbf{u}, \mathbf{v}) - \mathcal{N}(\mathbf{u}_{\mathrm{DG}}, \mathbf{v}) \\ &= \int_0^1 \mathcal{N}'_{\mathbf{u}}[\theta\mathbf{u} + (1-\theta)\mathbf{u}_{\mathrm{DG}}](\mathbf{u} - \mathbf{u}_{\mathrm{DG}}, \mathbf{v})\, \mathrm{d}\theta \qquad (9.11)\end{aligned}$$

for all $\mathbf{v}$ in V, where V is a suitable function space such that $S^{\mathbf{p}}(\Omega, \mathcal{T}_h, \mathbf{F}) \subset V$. Here, $\mathcal{N}'_{\mathbf{u}}[\mathbf{w}](\cdot\,, \mathbf{v})$ denotes the Gateaux derivative of $\mathbf{u} \mapsto \mathcal{N}(\mathbf{u}, \mathbf{v})$, for $\mathbf{v} \in V$ fixed, at some $\mathbf{w}$ in V. The linearization introduced in (9.11) is only a *formal* calculation, in the sense that $\mathcal{N}'_{\mathbf{u}}[\mathbf{w}](\cdot\,, \cdot)$ may not in general exist. Instead, a suitable approximation to $\mathcal{N}'_{\mathbf{u}}[\mathbf{w}](\cdot\,, \cdot)$ must be determined, for example, by computing appropriate finite difference quotients of $\mathcal{N}(\cdot\,, \cdot)$ (*cf.* Hartmann and Houston (2001)). Further, we shall suppose that the linearization (9.11) is well defined. Under these hypotheses, we introduce the following *dual* problem: find $\mathbf{z} \in V$ such that

$$\mathcal{M}(\mathbf{u}, \mathbf{u}_{\mathrm{DG}}; \mathbf{w}, \mathbf{z}) = J(\mathbf{w}) \quad \forall \mathbf{w} \in V. \qquad (9.12)$$

As in the linear case considered earlier, we shall tacitly assume that (9.12) possesses a unique solution. We then have the following error representation formula.

Theorem 9.1. Let $\mathbf{u}$ and $\mathbf{u}_{\mathrm{DG}}$ denote the solutions of (9.9) and (9.10), respectively, and suppose that the dual problem (9.12) is well posed. Then,

$$J(\mathbf{u}) - J(\mathbf{u}_{\mathrm{DG}}) = \mathcal{E}_\Omega(\mathbf{u}_{\mathrm{DG}}, h, \mathbf{p}, \mathbf{z} - \mathbf{z}_{h,\mathbf{p}}) \equiv \sum_{\kappa \in \mathcal{T}_h} \eta_\kappa, \qquad (9.13)$$

where

$$\begin{aligned}\eta_\kappa = &\int_\kappa \mathbf{r}_{h,\mathbf{p}}\,(\mathbf{z} - \mathbf{z}_{h,\mathbf{p}})\, \mathrm{d}x + \int_{\partial\kappa} \rho_{h,\mathbf{p}}\,(\mathbf{z} - \mathbf{z}_{h,\mathbf{p}})^+\, \mathrm{d}s \\ &- \int_\kappa \varepsilon \nabla \mathbf{u}_{\mathrm{DG}} \cdot \nabla(\mathbf{z} - \mathbf{z}_{h,\mathbf{p}})\, \mathrm{d}x\end{aligned}$$

for all $\mathbf{z}_{h,\mathbf{p}}$ in $S^{\mathbf{p}}(\Omega, \mathcal{T}_h, \mathbf{F})$. Here,

$$\mathbf{r}_{h,\mathbf{p}}|_{\kappa} = -\mathrm{div}\mathcal{F}(\mathbf{u}_{\mathrm{DG}}) \quad \text{and} \quad \rho_{h,\mathbf{p}}|_{\kappa} = \mathcal{F}(\mathbf{u}_{\mathrm{DG}}) \cdot \boldsymbol{\nu}_{\kappa} - \mathcal{H}(\mathbf{u}_{\mathrm{DG}}^{+}, \mathbf{u}_{\mathrm{DG}}^{-}, \boldsymbol{\nu}_{\kappa})$$

denote internal and boundary finite element residuals, respectively, defined on each $\kappa \in \mathcal{T}_h$.

Proof. The proof is elementary. We choose $\mathbf{w} = \mathbf{u} - \mathbf{u}_{\mathrm{DG}}$ in (9.12), recall the linearity of $J(\cdot)$, and exploit the Galerkin orthogonality property $\mathcal{N}(\mathbf{u}, \mathbf{v}_{h,\mathbf{p}}) - \mathcal{N}(\mathbf{u}_{\mathrm{DG}}, \mathbf{v}_{h,\mathbf{p}}) = 0$ for all $\mathbf{v}_{h,\mathbf{p}}$ in $S^{\mathbf{p}}(\Omega, \mathcal{T}_h, \mathbf{F})$, to deduce that

$$\begin{aligned} J(\mathbf{u}) - J(\mathbf{u}_{\mathrm{DG}}) &= J(\mathbf{u} - \mathbf{u}_{\mathrm{DG}}) \\ &= \mathcal{M}(\mathbf{u}, \mathbf{u}_{\mathrm{DG}}; \mathbf{u} - \mathbf{u}_{\mathrm{DG}}, \mathbf{z}) \\ &= \mathcal{M}(\mathbf{u}, \mathbf{u}_{\mathrm{DG}}; \mathbf{u} - \mathbf{u}_{\mathrm{DG}}, \mathbf{z} - \mathbf{z}_{h,\mathbf{p}}) \\ &= -\mathcal{N}(\mathbf{u}_{\mathrm{DG}}, \mathbf{z} - \mathbf{z}_{h,\mathbf{p}}) \end{aligned}$$

for all $\mathbf{z}_{h,\mathbf{p}}$ in $S^{\mathbf{p}}(\Omega, \mathcal{T}_h, \mathbf{F})$. Equation (9.13) now follows by employing the divergence theorem. □

The error representation formula implies the following Type I *a posteriori* error bound.

Corollary 9.2. Under the assumptions of Theorem 9.1, we have

$$|J(\mathbf{u}) - J(\mathbf{u}_{\mathrm{DG}})| \lessapprox \mathcal{E}_{\mathrm{P}}^{\mathrm{loc}} \equiv \sum_{\kappa \in \mathcal{T}_h} |\tilde{\eta}_{\kappa}|, \tag{9.14}$$

where $\tilde{\eta}_{\kappa}$ is defined in the same way as η_{κ}, except that a numerical approximation to the dual solution is used in $\tilde{\eta}_{\kappa}$, in place of the analytical dual solution $\mathbf{z}$ appearing in η_{κ}.

We see that, in contrast with the linear problems considered earlier on, for nonlinear hyperbolic conservation laws the error representation formula (9.13) depends on the unknown analytical solutions to the primal and dual problems. Thus, to render the Type I *a posteriori* error bound (9.14) computable, now both $\mathbf{u}$ and $\mathbf{z}$ must be replaced by suitable approximations. In particular, the linearization leading to $\mathcal{M}(\mathbf{u}, \mathbf{u}_{\mathrm{DG}}; \cdot, \cdot)$ is performed about $\mathbf{u}_{\mathrm{DG}}$ and the dual solution $\mathbf{z}$ is replaced by a discontinuous Galerkin approximation computed on the same mesh $\mathcal{T}_h$ used for $\mathbf{u}_{\mathrm{DG}}$, but using piecewise polynomials whose local degree is by 1 higher than the local degree of $\mathbf{u}_{\mathrm{DG}}$. Our final example concerns the steady compressible Euler equations of gas dynamics.

Example: Ringleb's flow. We consider the steady compressible Euler equations

$$\sum_{j=1}^{n} \frac{\partial}{\partial x_j} \mathbf{F}_j(\mathbf{U}) = \mathbf{0} \tag{9.15}$$

where

$$\mathbf{U} = [\rho, \rho u_1, \ldots, \rho u_n, \rho E]^{\mathrm{T}}$$

is the vector of *conserved variables*,

$$\mathbf{F}_j = [\rho u_j, \rho u_1 u_j + \delta_{1j} p, \ldots, \rho u_n u_j + \delta_{nj} p, (\rho E + p) u_j]^{\mathrm{T}}, \qquad j = 1, \ldots, n,$$

are the fluxes. For an ideal gas, the density ρ and pressure p are related through the *equation of state*

$$p = (\kappa - 1)\rho\left(E - \frac{1}{2}|\mathbf{u}|^2\right),$$

involving the total energy E and the velocity vector $\mathbf{u} = (u_1, \ldots, u_n)^{\mathrm{T}}$ in Cartesian coordinates. Here, κ is the ratio of specific heats; for dry air, $\kappa = 1.405$.

We consider Ringleb's flow, in two space dimensions, for which an analytical solution may be obtained using the hodograph method. This problem represents a transonic flow which turns around an obstacle; the flow is mostly subsonic, with a small supersonic pocket around the nose of the obstacle (see Barth (1998), Süli *et al.* (2001)).

We take the functional of interest to be the value of the density at the point $(-0.4, 2)$, that is, $J(\mathbf{u}) = \rho(-0.4, 2)$; consequently the true value of the functional is given by $J(\mathbf{u}) = 0.8616065996968034$. Table 9.2 shows the performance of our *hp*-adaptive algorithm; here we see that the quality of the computed error representation formula is extremely good, with $\theta_1 \approx 1$ even on very coarse meshes. Furthermore, the Type I *a posteriori* error bound (9.8) overestimates the true error in the computed functional by

Table 9.2. Adaptive algorithm for Ringleb's flow

Nds	Els	DOF	$J(u-u^h)$	$\sum_\kappa \tilde{\eta}_\kappa$	θ_1	$\sum_\kappa \lvert\tilde{\eta}_\kappa\rvert$	θ_2
91	144	1728	-0.2995E-03	-0.3024E-03	1.01	0.2914E-02	9.73
129	219	3228	0.5143E-04	0.4975E-04	0.97	0.9527E-03	18.52
179	315	5312	-0.1884E-04	-0.1885E-04	1.00	0.2484E-03	13.18
245	445	8560	0.4813E-05	0.4303E-05	0.89	0.1164E-03	24.19
302	554	13480	-0.2541E-05	-0.2662E-05	1.05	0.4230E-04	16.65
352	650	17944	-0.1489E-05	-0.1520E-05	1.02	0.1824E-04	12.25
426	792	26260	-0.5522E-06	-0.5662E-06	1.03	0.5515E-05	9.99
487	912	35280	-0.2602E-07	-0.2615E-07	1.01	0.6618E-06	25.44
622	1171	51544	0.6738E-09	0.6335E-09	0.94	0.1119E-06	166.02

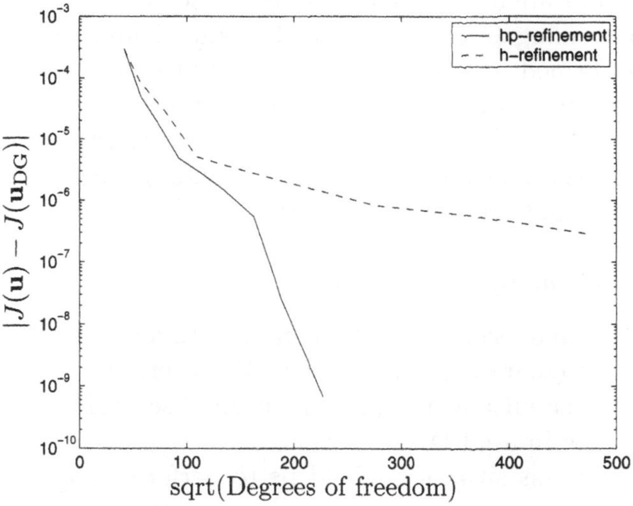

Figure 9.2. Comparison between h- and hp-adaptive mesh refinement for Ringleb's flow

about an order of magnitude, though there is a sharp increase on the last refinement. Figure 9.2 indicates exponential convergence for the error in the computed functional and again highlights the computational advantages of employing hp-mesh refinement when compared with the standard h-method, particularly when the output functional is required with high accuracy.

10. Conclusions and outlook

In this paper we have been concerned with the application of duality arguments to the derivation of *a priori* and *a posteriori* error bounds on the error in output functionals. We also discussed the role of adjoint equations in the process of error correction.

Looking to the future, there are many challenges to be addressed; below we discuss some of these.

Reconstruction on unstructured grids

Section 6 presented some preliminary ideas for reconstruction on unstructured grids. The analysis showed that if the error in the original solution is $\mathcal{O}(h^2)$ in the $L^2(\Omega)$-norm, but $\mathcal{O}(h)$ in $H^1(\Omega)$, then the reconstruction will have an improved accuracy of at least $\mathcal{O}(h^{3/2})$ in $H^1(\Omega)$. However, the ideal would be to achieve an accuracy of $\mathcal{O}(h^2)$ in $H^1(\Omega)$. There was also no discussion of how the analytic reconstruction equation might be approximated numerically.

Clearly an appropriate finite element discretization needs to be formulated using $H^2(\Omega)$-conforming (*e.g.*, C^1) finite elements. Numerical experiments must then be performed for a variety of test cases to establish the accuracy of the reconstruction in practice. If the error of the reconstructed solution is found to be $\mathcal{O}(h^2)$ in the $H^1(\Omega)$-norm, then further research is in order to try to improve the analysis, perhaps by including further assumptions concerning the formulation of the reconstruction algorithm.

Grid adaptation for multiple functionals

The error analysis and grid adaptation in this paper has been driven by concern for one particular output functional. However, in practice one might be interested in the simultaneous approximation of several functionals, such as both lift and drag in a CFD calculation.

One way to treat this situation would be to perform separate error analyses for each functional of interest, and then define a composite grid adaptation criterion. The obvious drawback of this, however, is the increased computational cost.

An alternative approach to grid adaptation might be to use a refinement criterion like (7.25), but instead of the weight $\omega_{\kappa,\mathcal{T}}$ being based on the dual solution for a particular functional, it could instead be constructed to be representative of the dual solutions for a class of functionals. For example, when performing airfoil or aircraft calculations, the functionals of interest are almost always surface integrals. Analysis of the homogeneous adjoint flow equations will reveal the asymptotic behaviour of the dual solution in the far-field, and thus one might construct a weight $\omega_{\kappa,\mathcal{T}}$ which would, at least qualitatively, have approximately the correct magnitude for a range of smoothly weighted boundary integral functionals.

Singularities and discontinuities

A priori error analysis usually leads to results proving that the error in the approximate solution (or the value of an output functional if that is of more interest) is $\mathcal{O}(h^p)$ for some p, provided the analytic solution is sufficiently smooth. Here h is the maximum element size (defined perhaps as the diameter of the smallest enclosing circumsphere) which is proportional to $N^{-1/n}$ for a quasi-uniform n-dimensional grid with N elements.

In practice, the analytic solution often does not satisfy the smoothness conditions required for the maximum value for p. This is frequently due to singularities because of corners in the domain boundary, or due to discontinuities in the boundary data. Under such circumstances, the best that can be hoped for is that the accuracy remains $\mathcal{O}(N^{-p/n})$, with $h \ll N^{-1/n}$ in the neighbourhood of the singularity.

For certain problems, there are indeed *a priori* proofs that this can be achieved with the appropriate local grid resolution. For such cases, to be considered *quasi-optimal*, an adaptive grid strategy should automatically generate such local grid resolution and hence the optimal order of accuracy. However, there has been little work so far on *a priori* proofs of the optimality of grid refinement indicators (see, for example, the papers of Gui and Babuška (1986), Section 3.3.7 of the monograph of Schwab (1998) and the work of Larson (1996)).

Anisotropic adaptation

The adaptive grid strategies discussed in this paper all use grid refinement, adding additional nodes/cells through an isotropic refinement process that improves the grid resolution in each direction. This is appropriate in many applications, but far from ideal in others.

One example is the inviscid flow around a wing. Here the grid resolution normal to the leading edge needs to be much finer than the spanwise resolution. In this case, anisotropic refinement is probably the best solution. This means adding nodes in such a way that the resolution normal to the leading edge is greater than in the spanwise resolution. Another, more extreme, example of the need for anisotropic resolution is a contact discontinuity in the solution to a hyperbolic partial differential equation. In this case, the best solution may well be grid redistribution, moving existing grid nodes to provide the resolution where it is needed.

The questions are how to decide which direction requires additional resolution, and how to move the nodes in grid redistribution? There are existing methods for doing this (Habashi, Fortin, Dompierre, Vallet and Bourgault 1998) but they are somewhat *ad hoc* in nature, although they often work well in practice. The challenge for those developing *a posteriori* adjoint-based refinement indicators is to formulate extensions to address this issue and provide a reliably good adaptive strategy. For recent work in the area of error estimation on anisotropic meshes we refer to Dobrowolski, Gräf and Pflaum (1999), Skalický and Roos (1999), Schötzau, Schwab and Stenberg (1998), Schötzau, Schwab and Stenberg (1999), Dolejši (2001), Apel, Nicaise and Schöberl (2001), Formaggia, Perotto and Zunino (2002).

Shocks

One last challenge we wish to highlight is the problem of shocks. With the quasi-1D Euler equations, it can be proved that, with an appropriate conservative formulation, and a numerical discretization that is second-order accurate when the solution is smooth, the accuracy of output functionals such as the integrated pressure is also second-order (Giles 1996). However, numerical evidence suggests this is not the case in multiple dimensions, and instead there is an error in quantities such as the lift on a transonic airfoil that is proportional to the local grid spacing at the shock. Thus, to get even second-order accuracy in the lift and in the solution on either side of the shock would require anisotropic grid adaptation so that the grid spacing at the shock is $\mathcal{O}(h^2)$, with h here being the average grid spacing in the rest of the grid.

There is another much more fundamental problem in the use of adjoint solutions for error analysis and correction. The approximate primal solution will have an $\mathcal{O}(1)$ error at the shock. This violates the whole basis for the adjoint error analysis since it relies on a linearization of the nonlinear equations that is valid only for small perturbations. The solution to this problem may be to use a regularization in which one numerically approximates a viscous shock with the level of viscosity being $\mathcal{O}(h^2)$. Grid adaptation would be based on the error in approximating the viscous equations, which would automatically lead to termination of the grid refinement in the neighbourhood of the shock once it is sufficiently well resolved. To apply the adjoint error correction to an improved order of accuracy for functionals, one would have to correct for the numerical error in approximating the viscous shock, plus the analytic error in using the viscous shock problem to approximate the inviscid shock problem. Through the use of matched asymptotic expansions, it can be proved that, to leading order, there is a linear dependence of integral functionals on the level of viscosity. Thus the analytic error can be compensated for by using the viscous dual solution to give the sensitivity of the lift to a change in the level of the viscosity.

Acknowledgements

The authors would like to express their sincere gratitude to Rémi Abgrall, Mark Ainsworth, Ivo Babuška, Tim Barth, Bernardo Cockburn, Leszek Demkowicz, Joe Flaherty, Kathryn Harriman, Paul Houston, Claes Johnson, Mats Larson, John Mackenzie, Peter Monk, Niles Pierce, Rolf Rannacher, Christoph Schwab, Thomas Sonar, and Gerald Warnecke for helpful discussions on the subject of this paper. We are particularly indebted to Kathryn Harriman, Paul Houston and Niles Pierce for their kind permission to include several of their computational results.

REFERENCES

R. A. Adams (1975), *Sobolev Spaces*, Academic Press, New York.

S. Adjerid, M. Aiffa and J. E. Flaherty (1998), Computational methods for singularly perturbed systems, in *Singular Perturbation Concepts of Differential Equations* (J. Cronin and R. E. O'Malley, eds), AMS, Providence, RI.

M. Ainsworth and J. T. Oden (2000), *A posteriori Error Estimation in Finite Element Analysis*, Wiley.

M. Ainsworth and B. Senior (1998), 'An adaptive refinement strategy for *hp*-finite element computations', *Appl. Numer. Math.* **26**, 165–178.

C. Airiau (2001), 'Non-parallel acoustic receptivity of a Blasius boundary using an adjoint approach', *Flow, Turbulence and Combustion* **65**, 347–367.

J. A. Alden and R. G. Compton (1997), 'A general method for electrochemical simulations, Part 1: Formulation of the strategy for two-dimensional simulations', *J. Phys. Chem. B* **101**, 8941–8954.

T. Apel, S. Nicaise and J. Schöberl (2001), 'Crouzeix–Raviart type finite elements on anisotropic meshes', *Numer. Math.* **89**, 193–223.

J.-P. Aubin (1967), 'Behavior of the error of the approximate solutions of boundary value problems for linear elliptic operators by Galerkin and finite difference methods', *Ann. Scuola. Norm. Sup. Pisa* **3**, 599–637.

I. Babuška and A. Miller (1984*a*), 'The post processing approach in the finite element method, Part 1: Calculation of displacements, stresses and other higher derivatives of the displacements', *Internat. J. Numer. Methods Engr.* **34**, 1085–1109.

I. Babuška and A. Miller (1984*b*), 'The post processing approach in the finite element method, Part 2: The calculation of stress intensity factors', *Internat. J. Numer. Methods Engr.* **34**, 1111–1129.

I. Babuška and A. Miller (1984*c*), 'The post processing approach in the finite element method, Part 3: *A posteriori* estimates and adaptive mesh selection', *Internat. J. Numer. Methods Engr.* **34**, 1131–1151.

C. Bardos (1970), 'Problèmes aux limites pour les équations aux dérivées partielles du premier ordre à coefficients réels; théorèmes d'approximation; application à l'équation de transport', *Ann. Sci. École Norm. Sup.* **4**, 185–233.

J. W. Barrett, G. Moore and K. W. Morton (1988), 'Optimal recovery in the finite element method, Part 2: Defect correction for ordinary differential equations', *IMA J. Numer. Anal.* **8**, 527–540.

J. W. Barrett and C. M. Elliott (1987), 'Total flux estimates for a finite element approximation of elliptic equations', *IMA J. Numer. Anal.* **7**, 129–148.

T. J. Barth (1998), Numerical methods for gas dynamics systems on unstructured meshes, in *An Introduction to Recent Developments in Theory and Numerics for Conservation Laws* (D. Kröner, M. Ohlberger and C. Rohde, eds), Vol. 5 of *Lecture Notes in Computational Science and Engineering*, Springer, Berlin/ Heidelberg, pp. 195–285.

R. Becker and R. Rannacher (1996), A feed-back approach to error control in finite element methods: Basic analysis and examples, *East–West J. Numer. Math.* **4**, 237–264.

R. Becker and R. Rannacher (2001), An optimal control approach to *a posteriori* error estimation in finite element methods, in *Acta Numerica*, Vol. 10 (A. Iserles, ed.), Cambridge University Press, Cambridge, pp. 1–102.

T. R. Bewley (2001), 'Flow control: New challenges for a new Renaissance', *Progress in Aerospace Sciences* **37**, 21–58.

K. S. Bey and J. T. Oden (1996), '*hp*-version discontinuous Galerkin methods for hyperbolic conservation laws', *Comput. Methods Appl. Mech. Engr.* **133**, 259–286.

A. Bottaro, J. Mauss and D. S. Henningson, eds (2001), *Flow, Turbulence and Combustion*, Special Issue.

D. Braess (1997), *Finite Elements. Theory, Fast Solvers Applications in Solid Mechanics*, Cambridge University Press, Cambridge.

S. C. Brenner and L. R. Scott (1994), *The Mathematical Theory of Finite Element Methods*, Springer, Berlin/Heidelberg.

C. Carstensen and S. A. Funken (2000), 'Constants in Clément-interpolation error and residual based *a posteriori* error estimates in finite element methods', *East–West J. Numer. Math.* **8**, 153–175.

P. G. Ciarlet (1978), *The Finite Element Method for Elliptic Problems*, North-Holland, Amsterdam.

B. Cockburn, G. E. Karniadakis and C.-W. Shu (2000), The development of discontinuous Galerkin methods, in *Discontinuous Galerkin Finite Element Methods* (B. Cockburn, G. E. Karniadakis and C.-W. Shu, eds), Vol. 11 of *Lecture Notes in Computational Science and Engineering*, Springer, Berlin/Heidelberg, pp. 3–50.

D. Colton and R. Kress (1991), *Inverse Acoustic and Electromagnetic Scattering Theory*, Vol. 93 of *Applied Mathematical Sciences*, Springer.

R. Dautray and J.-L. Lions (1993), *Mathematical Analysis and Numerical Methods for Science and Technology*, Vol. 6: *Evolution Problems II*, Springer, Berlin/Heidelberg.

M. Dobrowolski, S. Gräf and C. Pflaum (1999), 'On *a posteriori* error estimators in the finite element method on anisotropic meshes', *Electron. Trans. Numer. Anal.* **8**, 36–45.

V. Dolejši (2001), 'Anisotropic mesh adaptation technique for viscous flow simulation', *East–West J. Numer. Math.* **1**, 1–24.

K. Eriksson, D. Estep, P. Hansbo and C. Johnson (1995), Introduction to adaptive methods for differential equations, in *Acta Numerica*, Vol. 4 (A. Iserles, ed.), Cambridge University Press, Cambridge, pp. 105–158.

K. Eriksson, D. Estep, P. Hansbo and C. Johnson (1996), *Computational Differential Equations*, Cambridge University Press.

J. E. Flaherty, R. M. Loy, M. S. Shephard and J. D. Teresco (2000), Software for the parallel adaptive solution of conservation laws by discontinuous Galerkin methods, in *Discontinuous Galerkin Finite Element Methods* (B. Cockburn, G. E. Karniadakis and C.-W. Shu, eds), Vol. 11 of *Lecture Notes in Computational Science and Engineering*, Springer, Berlin/Heidelberg, pp. 113–124.

L. Formaggia, S. Perotto and P. Zunino (2002), An anisotropic *a-posteriori* error estimate for a convection diffusion equation. EPFL-DMA Analyse et Analyse Numérique Report no. 04. To appear in *Computing and Visualization in Science*.

D. Gilbarg and N. S. Trudinger (1983), *Elliptic Partial Differential Equations of Second Order*, 2nd edn, Springer, Berlin/Heidelberg.

M. B. Giles (1996), 'Analysis of the accuracy of shock-capturing in the steady quasi-1D Euler equations', *Internat. J. Comput. Fluid Dynamics* **5**, 247–258.

M. B. Giles (1998), On adjoint equations for error analysis and optimal grid adaptation in CFD, in *Frontiers of Computational Fluid Dynamics 1998* (D. Caughey and M. Hafez, eds), World Scientific, pp. 155–170.

M. B. Giles (2000), An introduction to the adjoint design approach and analysis, Numerical Analysis Group Research Report NA-00/04, University of Oxford.

M. B. Giles (2001), Defect and adjoint error correction, in *Computational Fluid Dynamics 2000* (N. Satofuka, ed.), Springer.

M. B. Giles and N. A. Pierce (1997), 'Adjoint equations in CFD: Duality, boundary conditions and solution behaviour', AIAA Paper 97–1850.

M. B. Giles and N. A. Pierce (1999), 'Improved lift and drag estimates using adjoint Euler equations', AIAA Paper 99–3293.

M. B. Giles and N. A. Pierce (2001), 'Analysis of adjoint error correction for superconvergent functional estimates'. Submitted to *SIAM J. Numer. Anal.*

M. B. Giles and N. A. Pierce (2002), 'Adjoint error correction for integral outputs', *NASA Ames/VKI Lecture Series on Error Estimation and Solution Adaptive Discretization in CFD* (T. J. Barth and H. Deconinck, eds), *Lecture Notes in Computational Science and Engineering*, Springer. To appear.

M. B. Giles, M. Larsson, M. Levenstam and E. Süli (1997), Adaptive error control for finite element approximations of the lift and drag coefficients in viscous flow, Numerical Analysis Group Research Report NA-97/06, University of Oxford.

V. Girault and P.-A. Raviart (1986), *Finite Element Methods for Navier–Stokes Equations*, Springer, Berlin/Heidelberg.

W. Gui and I. Babuška (1986), 'The *h*, *p* and *h-p* versions of the finite element method in 1 dimension, Part III: The adaptive *h-p* version', *Numer. Math.* **49**, 659–683.

W. G. Habashi, M. Fortin, J. Dompierre, M.-G. Vallet and Y. Bourgault (1998), Anisotropic mesh adaptation: A step towards mesh-independent and user-independent CFD, in *Barriers and Challenges in Fluid Dynamics* (Hampton, VA, 1996), Vol. 6 of *ICASE/LaRC Interdiscip. Ser. Sci. Engr.*, pp. 99–117.

P. Hansbo and C. Johnson (1991), 'Adaptive streamline diffusion finite element methods for compressible flow using conservative variables.', *Comput. Methods Appl. Mech. Engr.* **87**, 267–280.

R. Hartmann (2001), Adaptive FE-methods for conservation equations, in *Eighth International Conference on Hyperbolic Problems: Theory, Numerics, Applications* (HYP2000) (G. Warnecke and H. Freistühler, eds), Birkhäuser, Basel.

R. Hartmann and P. Houston (2001), 'Adaptive discontinuous Galerkin finite element methods for nonlinear hyperbolic conservation laws'. Submitted for publication.

P. Houston and E. Süli (2001*a*), *hp*-adaptive discontinuous Galerkin finite element methods for hyperbolic problems, Numerical Analysis Group Research Report NA-01/05, University of Oxford. *SIAM J. Sci. Comput.* **23**, 1225–1251.

P. Houston and E. Süli (2001*b*), 'Stabilized *hp*-finite element approximation of partial differential equations with non-negative characteristic form', *Computing* **66**, 99–119.

P. Houston and E. Süli (2002*a*), 'Sobolev regularity estimation for *hp*-adaptive finite element methods', in *Proceedings of the ENUMATH 2001 Conference* (F. Brezzi, ed.), Springer. To appear.

P. Houston and E. Süli (2002*b*), 'Adaptive finite element approximation of hyperbolic problems', *NASA Ames/VK1 Lecture Series on Error Estimation and Solution Adaptive Discretization in CFD* (T. J. Barth and H. Deconinck, eds),

Lecture Notes in Computational Science and Engineering, Springer. To appear.

P. Houston, R. Rannacher and E. Süli (2000*a*), '*A posteriori* error analysis for stabilized finite element approximations of transport problems', *Comput. Methods Appl. Mech. Engr.* **190**, 1483–1508.

P. Houston, C. Schwab and E. Süli (2000*b*), Discontinuous *hp*-finite element methods for advection–diffusion problems, Numerical Analysis Group Research Report NA-00/15, University of Oxford. To appear in *SIAM J. Numer. Anal.*

P. Houston, R. Hartmann and E. Süli (2001), Adaptive discontinuous Galerkin finite element methods for nonlinear hyperbolic problems, Numerical Analysis Group Research Report NA-01/06, University of Oxford. In *Numerical Methods for Fluid Dynamics VII* (M. Baines, ed.), ICFD, Oxford, pp. 347–353.

J. Jaffre, C. Johnson and A. Szepessy (1995), 'Convergence of the discontinuous Galerkin finite element method for hyperbolic conservation laws', *Math. Mod. Methods Appl. Sci.* **5**, 367–386.

A. Jameson (1988), 'Aerodynamic design via control theory', *J. Sci. Comput.* **3**, 233–260.

A. Jameson (1995), Optimum aerodynamic design using control theory, in *Computational Fluid Dynamics Review 1995* (M. Hafez and K. Oshima, eds), Wiley, pp. 495–528.

A. Jameson (1999), 'Re-engineering the design process through computation', *J. Aircraft* **36**, 36–50.

A. Jameson, N. A. Pierce and L. Martinelli (1998), 'Optimum aerodynamic design using the Navier–Stokes equations', *J. Theoret. Comput. Fluid Mech.* **10**, 213–237.

B. Koren (1988), 'Defect correction and multigrid for an efficient and accurate computation of airfoil flows', *J. Comput. Phys.* **77**, 183–206.

M. G. Larson (1996), A new error analysis for finite element approximations of indefinite linear elliptic problems, Technical report, Department of Mathematics, Chalmers University of Technology, Sweden.

M. G. Larson and T. J. Barth (2000), *A posteriori* error estimation for adaptive discontinuous Galerkin approximations of hyperbolic systems, in *Discontinuous Galerkin Finite Element Methods* (B. Cockburn, G. E. Karniadakis and C.-W. Shu, eds), Vol. 11 of *Lecture Notes in Computational Science and Engineering*, Springer, Berlin/Heidelberg, pp. 363–368.

J.-L. Lions (1971), *Optimal Control of Systems Governed by Partial Differential Equations*, Springer. Translated by S. K. Mitter.

M. Melenk and C. Schwab (1999), 'An *hp* finite element method for convection-diffusion problems', *IMA J. Numer. Anal.* **19**, 425–453.

P. Monk and E. Süli (1998), 'The adaptive computation of far field patterns by *a posteriori* error estimates of linear functionals', *SIAM J. Numer. Anal.* **36**, 251–274.

J. C. Newman, A. C. Taylor, R. W. Barnwell, P. A. Newman and G. J.-W. Hou (1999), 'Overview of sensitivity analysis and shape optimization for complex aerodynamic configurations', *J. Aircraft* **36**, 87–96.

J. Nitsche (1968), 'Ein Kriterium für die Quasi-Optimalität des ritzschen Verfahrens', *Numer. Math.* **11**, 346–348.

J. T. Oden and S. Prudhomme (1999), 'On goal-oriented error estimation for elliptic problems: Application to control of pointwise errors', *Comput. Methods Appl. Mech. Engr.* **176**, 313–331.

J. T. Oden and J. N. Reddy (1983), *Variational Methods in Theoretical Mechanics*, 2nd edn, Springer, Berlin/Heidelberg.

L. A. Oganesjan and L. A. Ruhovec (1969), 'Investigation of the rate of convergence of variation-difference schemes for second order elliptic equations in two-dimensional region with smooth boundary', *Ž. Vyčisl. Mat. i Mat. Fiz.* **9**, 1102–1120.

M. Paraschivoiu, J. Peraire and A. T. Patera (1997), '*A posteriori* finite element bounds for linear functional outputs of elliptic partial differential equations', *Comput. Methods Appl. Mech. Engr.* **150**, 289–312.

J. Peraire and A. T. Patera (1997), Bounds for linear functional outputs of coercive partial differential equations: Local indicators and adaptive refinement, in *New Advances in Adaptive Computational Methods in Mechanics* (P. Ladeveze and J. T. Oden, eds), Elsevier.

N. A. Pierce and M. B. Giles (1998), Adjoint recovery of superconvergent functionals from approximate solutions of partial differential equations, Numerical Analysis Group Research Report NA-98/18, University of Oxford.

N. A. Pierce and M. B. Giles (2000), 'Adjoint recovery of superconvergent functionals from PDE approximations', *SIAM Review* **42**, 247–264.

O. Pironneau (1974), 'On optimum design in fluid mechanics', *J. Fluid Mech.* **64**, 97–110.

R. Rannacher (1998), Adaptive finite element methods, in *Proc. NATO Summer School on Error Control and Adaptivity in Scientific Computing*, Kluwer Academic, pp. 247–278.

D. Schötzau, C. Schwab and R. Stenberg (1998), 'Mixed *hp*-FEM on anisotropic meshes', *Math. Mod. Methods Appl. Sci.* **8**, 787–820.

D. Schötzau, C. Schwab and R. Stenberg (1999), 'Mixed *hp*-FEM on anisotropic meshes II: Hanging nodes and tensor products of boundary layer meshes', *Math. Mod. Methods Appl. Sci.* **4**, 667–697.

C. Schwab (1998), *p- and hp-Finite Element Methods: Theory and Applications to Solid and Fluid Mechanics*, Oxford University Press, Oxford.

T. Skalický and H. G. Roos (1999), 'Anisotropic mesh refinement for problems with internal and boundary layers', *Internat. J. Numer. Methods Engr.* **11**, 1933–1953.

R. D. Skeel (1981), 'A theoretical framework for proving accuracy results for deferred corrections', *SIAM J. Numer. Anal.* **19**, 171–196.

H. J. Stetter (1978), 'The defect correction principle and discretization methods', *Numer. Math.* **29**, 425–443.

G. Strang and G. J. Fix (1973), *An Analysis of the Finite Element Method*, Prentice-Hall.

E. Süli (1998), *A posteriori* error analysis and adaptivity for finite element approximations of hyperbolic problems, in *An Introduction to Recent Developments in Theory and Numerics for Conservation Laws* (D. Kröner, M. Ohlberger and C. Rohde, eds), Vol. 5 of *Lecture Notes in Computational Science and Engineering*, Springer, Berlin/Heidelberg, pp. 123–194.

E. Süli, P. Houston and C. Schwab (1999), *hp*-finite element methods for hyperbolic problems, in *The Mathematics of Finite Elements and Applications X: MAFELAP 1999* (J. R. Whiteman, ed.), Elsevier, pp. 143–162.

E. Süli, P. Houston and B. Senior (2001), *hp*-Discontinuous Galerkin finite element methods for nonlinear hyperbolic problems, Numerical Analysis Group Research Report NA-01/07, University of Oxford. In *Numerical Methods for Fluid Dynamics VII* (M. Baines, ed.), ICFD, Oxford, pp. 73–86.

B. Szabó and I. Babuška (1991), *Finite Element Analysis*, Wiley, New York.

O. Talagrand and P. Courtier (1997), 'Variational assimilation of meteorological observations with the adjoint vorticity equation, Part 1: Theory', *Quart. J. Royal Met. Soc.* **113**, 1311–1328.

R. Verfürth (1996), *A Review of a posteriori Error Estimation and Adaptive Mesh-Refinement Techniques*, Teubner, Stuttgart.

J. A. Wheeler (1973), 'Simulation of heat transfer from a warm pipeline buried in permafrost', in *Proceedings of the 74th National Meeting of the American Institute of Chemical Engineering.*

Acta Numerica (2002), pp. 237–339
DOI: 10.1017/S0962492902000041

Finite elements in computational electromagnetism

R. Hiptmair
Sonderforschungsbereich 382,
Universität Tübingen,
D-72076 Tübingen, Germany
E-mail: ralf@hiptmair.de

This article discusses finite element Galerkin schemes for a number of linear model problems in electromagnetism. The finite element schemes are introduced as discrete differential forms, matching the coordinate-independent statement of Maxwell's equations in the calculus of differential forms. The asymptotic convergence of discrete solutions is investigated theoretically. As discrete differential forms represent a genuine generalization of conventional Lagrangian finite elements, the analysis is based upon a judicious adaptation of established techniques in the theory of finite elements. Risks and difficulties haunting finite element schemes that do not fit the framework of discrete differential forms are highlighted.

CONTENTS

1. Introduction

Most modern technology is inconceivable without harnessing electromagnetic phenomena. Hence the design and analysis of schemes for the approximate solution of electromagnetic field problems can claim a rightful place as a core discipline of numerical mathematics and scientific computing. However, for a long time it received far less attention among numerical analysts than, for instance, computational fluid dynamics and solid mechanics.

One reason might be that electromagnetism is described by a generically linear theory, in the sense that linear equations arise from basic physical principles. This is in stark contrast to continuum mechanics, where linear models only emerge through linearization of inherently nonlinear governing principles. Being linear, the fundamental laws of electromagnetism might have struck many mathematicians as 'dull'. This view might also have been fostered by the misconception that electromagnetism basically boils down to plain second-order elliptic equations, which have been amply studied and are well understood.

It is one objective of this survey article to refute the idea that one can cope with electromagnetics once one knows how to solve Laplace equations numerically. I aim to convey the richness in subtle mathematical features displayed by apparently 'simple' problems in computational electromagnetism. The problems I have in mind arise from the spatial discretization of electromagnetic fields by means of finite elements. Yet I will not settle for merely specifying and describing the finite element spaces. To gain insight, a comprehensive view is mandatory, encompassing the structural aspects of the physical model, a thorough knowledge of function spaces as well as familiarity with classical finite element techniques. All these issues will be addressed in the paper, and an attempt is made to convince the reader that understanding all of them is necessary for successfully tackling electromagnetic field problems.

Many readers might object to my regular delving into technical details. In my opinion, major breakthroughs in computational electromagnetism have often been brought about by successfully addressing technical issues. This neatly fits my desire to embrace a formal 'rigorous' treatment. Therefore, space permitting, and in order to make the article self-contained, I will not skip proofs. Yet, sometimes I will put forth 'views' – even at the risk of sounding fuzzy and arcane – in a possibly doomed attempt to inspire 'intuitive understanding'.

Plenty of references to original papers and related work will be given. Of course, they can never be exhaustive and will reflect my personal biases and history. In particular, scores of engineering publications that address issues also covered in this article could be cited, but will not be mentioned. My emphasis on theory is reflected by the almost complete absence of numerical results. They can be found in abundance in research papers.

I was pleased to witness a surge in research activities into mathematical aspects of computational electromagnetism in recent years. Now the field is rapidly evolving, which means that this article can hardly be more than a snapshot of the knowledge as of 2001. Many of the results covered are likely to experience significant improvement and extension in years to come. It also means that there is much left to be done. In a sense, I will not balk at stating incomplete results and even conjectures. Maybe this will trigger some fresh research.

Even with a focus on finite element schemes, all that can be covered in a survey article are model problems. Admittedly, they fall way short of matching the complexity of typical engineering applications. For instance, in light of the linear nature of electromagnetism, I will completely restrict my attention to linear problems, that is, only simple 'linear' materials will be considered. In this setting it is possible to skirt any issues of temporal discretization by switching to the *frequency domain*: all quantities are supposed to show a sinusoidal dependence on time with a fixed angular frequency $\omega > 0$. Thus, thanks to linearity, temporal derivation ∂_t can be replaced by the multiplication operator $i\omega\cdot$. This converts all equations into relationships between *complex amplitudes* depending on space only. If $\mathbf{u} = \mathbf{u}(\mathbf{x})$, $\mathbf{x} \in \mathbb{R}^3$ standing for the independent space variable, is such a complex amplitude, the related physical quantity U can be recovered through

$$\mathrm{U}(\mathbf{x}, t) = \mathrm{Re}(\mathbf{u}(\mathbf{x}) \cdot \exp(i\omega t)).$$

The classical notion of finite elements is tied to bounded computational domains. Yet many central problems in computational electromagnetism are posed on unbounded domains. The most prominent example is the scattering of electromagnetic waves. Not all of these problems will be fully treated in this article. Still, when combined with other techniques, for instance boundary element methods, finite elements can play an important role even in these cases. Thus the results reported in this article remain of interest.

Instead of an outline, I am only listing a few points of view that I embrace. They can offer guidance when negotiating through this article.

- In order to discretize the fundamental laws of electromagnetism properly, it is important to appreciate their link with differential geometry and algebraic topology (cohomology theory).
- There is a close relationship between second-order elliptic equations and the governing equations of electromagnetism, but the lack of strong ellipticity introduces subtle new challenges.
- Suitable finite elements for electromagnetic fields should be introduced and understood as discrete differential forms.

- Discrete differential forms are a generalization of $H^1(\Omega)$-conforming Lagrangian finite elements. Their analysis can often use and adapt the tools developed for the latter.
- Finite elements that lack an interpretation as discrete differential forms have to be used with great care.

2. Maxwell's equations

The fundamental governing equations of electromagnetism are Maxwell's equations. Mathematicians usually encounter them in the form of the two first-order partial differential equations

$$\begin{aligned} &\text{Faraday's law:} \quad &\mathbf{curl}\,\mathbf{e} &= -i\omega\,\mathbf{b}, \\ &\text{Ampère's law:} \quad &\mathbf{curl}\,\mathbf{h} &= i\omega\,\mathbf{d} + \mathbf{j}, \end{aligned} \tag{2.1}$$

posed over all of affine space $A(\mathbb{R}^3)$. The equations link (the complex amplitudes of) the electric field $\mathbf{e}$, the magnetic induction $\mathbf{b}$, the magnetic field $\mathbf{h}$, and the displacement current $\mathbf{d}$. Here, $\mathbf{j}$ denotes a (formal) excitation supplied by an imposed current. The equations have to be supplemented by the material laws (also called constitutive laws)

$$\mathbf{d} = \boldsymbol{\epsilon}\mathbf{e}, \quad \mathbf{b} = \boldsymbol{\mu}\mathbf{h}, \tag{2.2}$$

where the dielectric tensor $\boldsymbol{\epsilon}$ and the magnetic permeability tensor $\boldsymbol{\mu}$ are usually introduced as L^∞-functions mapping into the real symmetric, positive definite 3×3 matrices such that $\lambda_{\min}(\boldsymbol{\epsilon}(\mathbf{x})) > \epsilon_0 > 0$ and $\lambda_{\min}(\boldsymbol{\mu}(\mathbf{x})) > \mu_0 > 0$ almost everywhere. Such matrix-valued functions will be referred to as metric tensors in the following. If good conductors are involved, a part of the source current may be given through Ohm's law

$$\mathbf{j} = \boldsymbol{\sigma}\mathbf{e} + \mathbf{j}_0, \tag{2.3}$$

where $\boldsymbol{\sigma}$ stands for the symmetric, positive semi-definite conductivity tensor, yet another metric tensor.

2.1. Fields and forms

Is there more to the unknowns of (2.1) than being plain vector-fields with three components? To answer this question, it is useful to remember the physicists' favourite way of writing Maxwell's equations, namely the *integral form*:

$$\begin{aligned} &\text{Faraday's law:} \quad &\int_{\partial\Sigma} \mathbf{e} \cdot d\mathbf{s} &= -i\omega \int_{\Sigma} \mathbf{b} \cdot \mathbf{n}\, dS, \\ &\text{Ampère's law:} \quad &\int_{\partial\Sigma} \mathbf{h} \cdot d\mathbf{s} &= i\omega \int_{\Sigma} \mathbf{d} \cdot \mathbf{n}\, dS + \int_{\Sigma} \mathbf{j} \cdot \mathbf{n}\, dS. \end{aligned} \tag{2.4}$$

This is to hold for any bounded, two-dimensional, piecewise smooth sub-manifold Σ of $A(\mathbb{R}^3)$, equipped with oriented[1] unit normal vector-field $\mathbf{n}$.

First, the integral form (2.4) reveals that the fields $\mathbf{e}, \mathbf{h}$ and $\mathbf{b}, \mathbf{d}$ have an entirely different nature, as Maxwell remarked in his 'Treatise on Electricity and Magnetism' (Maxwell 1891, Chapter 1):

> Physical vector quantities may be divided into two classes, in one of which the quantity is defined with reference to a line, while in the other the quantity is defined with reference to an area.

Laconically speaking, electromagnetic fields are an abstraction for associating 'voltages' and 'fluxes' to directed paths and oriented surfaces; they are *integral forms* in the sense of the following definition, which is deliberately kept fuzzy because it targets some 'intuitive concepts'.[2]

Definition 1. An integral *form* of degree $l \in \mathbb{N}_0$, $0 \leq l \leq n$, $n \in \mathbb{N}$, on a piecewise smooth n-dimensional manifold $\mathcal{M}$ is a continuous[3] additive mapping from the set $\mathcal{S}_l(\mathcal{M})$ of compact, oriented, piecewise smooth, l-dimensional sub-manifolds of $\mathcal{M}$ into the complex numbers. These so-called integral l-forms on $\mathcal{M}$ form the vector space $\mathcal{F}^l(\mathcal{M})$ (which is to be *trivial* for $l < 0$ or $l > n$).

Here, by 'additive', we mean that the integral form assigns the sum of the respective numbers to the union of disjoint sub-manifolds. Further, flipping the orientation of a sub-manifold should change the sign of the assigned value. This is what we should expect from the integrals occurring in (2.4). Therefore the evaluation $\omega(\Sigma)$, $\omega \in \mathcal{F}^l(\mathcal{M})$, $\Sigma \in \mathcal{S}_l(\mathcal{M})$ is dubbed 'integrating ω over Σ', in symbols $\int_\Sigma \omega$. Now, by merely looking at (2.4), we identify $\mathbf{e}$ and $\mathbf{h}$ as integral 1-forms, whereas $\mathbf{b}$, $\mathbf{d}$, and $\mathbf{j}$ should be regarded as integral 2-forms.

We are accustomed to referring to the 'field at a point in space', that is, a local perspective. Measurement procedures adopt it: measuring an electric field amounts to determining the *virtual work*

$$\delta w = q\,\mathbf{e}(\mathbf{x}) \cdot \delta\mathbf{x} \tag{2.5}$$

needed for the tiny displacement $\delta\mathbf{x}$ of a test charge q at $\mathbf{x}$, with $\cdot$ designating the inner product in Euclidean space $\mathbb{R}^3$. The magnetic induction is

[1] Taking for granted an orientation of the ambient space $A(\mathbb{R}^3)$, we need not distinguish between interior and exterior orientation of manifolds.

[2] It is the subject of geometric measure theory to come up with a more rigorous approach. See Morgan (1995) for an introduction and Federer (1969) for a comprehensive exposition.

[3] Continuity refers to a sort of 'deformation topology' on sets of piecewise smooth manifolds.

measured through the Lorenz force, that is, the work

$$\delta w = q\,(\mathbf{b}(\mathbf{x}) \times \mathbf{v}) \cdot \delta\mathbf{x} \tag{2.6}$$

required for a tiny (transversal) shift of a test charge q at $\mathbf{x}$ moving with velocity $\mathbf{v}$, where $\times$ is the usual cross product of vectors in $\mathbb{R}^3$. From this perspective $\mathbf{e}$ and $\mathbf{b}$ are classical (continuous) *differential forms* of degree 1 and 2, respectively, according to the following definition (*cf.* Lang (1995, Chapter V, Section 3)).

Definition 2. A *differential form* of degree l, $l \in \mathbb{N}_0$, and class C^m, $m \in \mathbb{N}_0$, on a smooth n-manifold $\mathcal{M}$ is an m-times continuously differentiable mapping assigning to each $\mathbf{x} \in \mathcal{M}$ an element of the space $\bigwedge^l(T_{\mathcal{M}}(\mathbf{x}))$ of alternating l-multilinear forms on the tangent space $T_{\mathcal{M}}(\mathbf{x})$. These mappings form the vector space $\mathcal{DF}^{l,m}(\mathcal{M})$.

Any piecewise smooth oriented manifold can be covered and approximated arbitrarily well by tiny flat 'tangential' tiles. Thus, through Riemann summation any differential l-form spawns an integral l-form according to Definition 1 (*cf.* Bossavit (1998*d*, Section 3.2)). This gives us injections $F^l_{\mathcal{M}} : \mathcal{DF}^{l,0}(\mathcal{M}) \mapsto \mathcal{F}^l(\mathcal{M})$, which, of course, are by no means surjective.

A special case is $\mathcal{M} = A(\mathbb{R}^3)$. Then $T_{\mathcal{M}}(\mathbf{x}) = \mathbb{R}^3$ for all $\mathbf{x} \in \mathcal{M}$ and $T_{\mathcal{M}}(\mathbf{x})$ may be endowed with the structure of a Euclidean vector space. Then the identifications of Table 2.1 establish isomorphisms Υ_l between differential l-forms, $0 \leq l \leq 3$, on $A(\mathbb{R}^3)$ and continuous functions/vector-fields, their *vector proxies* (a term coined by A. Bossavit (1998*e*)).

Using the convention put forth in Table 2.1, the integration of forms amounts to the evaluation of the following integrals for vector proxies:

$$\begin{array}{lll}
\omega \in \mathcal{DF}^{0,0}(\Omega): & \int\limits_{\mathbf{x}} \omega = (\Upsilon_0\omega)(\mathbf{x}) & \forall \mathbf{x} \in \Omega, \\
\omega \in \mathcal{DF}^{1,0}(\Omega): & \int\limits_{\gamma} \omega = \int\limits_{\gamma} \Upsilon_1\omega \cdot \mathrm{d}\mathbf{s} & \forall \gamma \in \mathcal{S}_1(\Omega), \\
\omega \in \mathcal{DF}^{2,0}(\Omega): & \int\limits_{\Sigma} \omega = \int\limits_{\Sigma} \Upsilon_2\omega \cdot \mathbf{n}\,\mathrm{d}S & \forall \Sigma \in \mathcal{S}_2(\Omega), \\
\omega \in \mathcal{DF}^{3,0}(\Omega): & \int\limits_{V} \omega = \int\limits_{V} \Upsilon_3\omega\,\mathrm{d}\mathbf{x} & \forall V \in \mathcal{S}_3(\Omega).
\end{array}$$

Here, $\mathbf{n}$ is a unit normal vector-field to Σ whose direction is induced by the orientation of Σ.

It is clear that we have a lot of freedom when defining vector proxies. Choosing an inner product for $\mathbb{R}^3$ gives us other vector proxies for the same differential forms. In short, vector proxies are coordinate-dependent, in contrast to the calculus of differential forms.

More generally, the finite-dimensional spaces $\bigwedge^l(T_{\mathcal{M}}(\mathbf{x}))$ can be equipped with a basis generated by coordinate vectors of charts of $\mathcal{M}$. Thus, any differential l-form on $\mathcal{M}$ can be identified with the $\binom{n}{l}$-tuple of its coefficient

Table 2.1. Relationship between differential forms and vector-fields in three-dimensional Euclidean space ($\mathbf{v}, \mathbf{v}_1, \mathbf{v}_2, \mathbf{v}_3 \in \mathbb{R}^3$).

Differential form	Related function u/vector-field $\mathbf{u}$
$\mathbf{x} \mapsto \omega(\mathbf{x})$	$u(\mathbf{x}) := \omega(\mathbf{x})$
$\mathbf{x} \mapsto \{\mathbf{v} \mapsto \omega(\mathbf{x})(\mathbf{v})\}$	$\langle \mathbf{u}(\mathbf{x}), \mathbf{v}\rangle := \omega(\mathbf{x})(\mathbf{v})$
$\mathbf{x} \mapsto \{(\mathbf{v}_1, \mathbf{v}_2) \mapsto \omega(\mathbf{x})(\mathbf{v}_1, \mathbf{v}_2)\}$	$\langle \mathbf{u}(\mathbf{x}), \mathbf{v}_1 \times \mathbf{v}_2\rangle := \omega(\mathbf{x})(\mathbf{v}_1, \mathbf{v}_2)$
$\mathbf{x} \mapsto \{(\mathbf{v}_1, \mathbf{v}_2, \mathbf{v}_3) \mapsto \omega(\mathbf{x})(\mathbf{v}_1, \mathbf{v}_2, \mathbf{v}_3)\}$	$u(\mathbf{x}) \det(\mathbf{v}_1, \mathbf{v}_2, \mathbf{v}_3) := \omega(\mathbf{x})(\mathbf{v}_1, \mathbf{v}_2, \mathbf{v}_3)$

functions with respect to the bases. Often, calculations with differential forms are greatly facilitated by using a coefficient representation. There is absolutely no objection to using vector proxies as long as their use is consistent with the physical meaning of the fields and confined to legal operations for integral forms. For example, point evaluations of vector proxies of 1-forms should be used with great care, since they fail to make sense for integral 1-forms.

Bibliographical notes

The interpretation of electromagnetic fields as differential forms has a long tradition in mathematical physics and is covered in many textbooks on differential forms. Some references include Grauert and Lieb (1977, Chapter 5), Baldomir and Hammond (1996), and Burke (1985, Chapter VI). This last reference gives lucid explanations for the concept from differential geometry used in this section. Brief presentations of the topic include Bossavit (1998*b*, 1998*c*), Deschamps (1981) and Baldomir (1986). The role of the integral conservation form of Maxwell's equations in space and time is a core theme in Mattiussi (2000).

2.2. Exterior calculus

Prominent in the integral formulation of Maxwell's equations is the evaluation of integral 1-forms over boundaries. This motivates the definition of *exterior derivatives*: these are linear operators $\boldsymbol{d} : \mathcal{F}^l(\mathcal{M}) \mapsto \mathcal{F}^{l+1}(\mathcal{M})$, where $\mathcal{M}$ is a piecewise smooth orientable n-dimensional manifold, $0 \le l < n$, defined by

$$\int_{\Sigma} \boldsymbol{d}\omega := \int_{\partial\Sigma} \omega, \quad \text{for all } \omega \in \mathcal{F}^l(\mathcal{M}),\ \Sigma \in \mathcal{S}_{l+1}(\mathcal{M}). \tag{2.7}$$

The boundary $\partial\Sigma$ bears the induced orientation, and for $\omega \in \mathcal{F}^n(\mathcal{M})$ we set $\boldsymbol{d}\omega = 0$. Forms in the kernel of $\boldsymbol{d}$ are called *closed*. It goes without saying that $\partial\partial\Sigma = \emptyset$, which immediately implies the fundamental relation $\boldsymbol{d} \circ \boldsymbol{d} = 0$. The converse of this statement is the core of the famous Poincaré lemma. There is too little structure in integral forms as we have introduced them to support a proof, but this fundamental lemma expresses that

$$\text{`}\ \omega \in \mathcal{F}^l(\mathcal{M}) : \quad \boldsymbol{d}\omega = 0 \quad \Leftrightarrow \quad \exists \eta \in \mathcal{F}^{l-1}(\mathcal{M}) : \quad \omega = \boldsymbol{d}\eta\ \text{'}, \tag{2.8}$$

if $\mathcal{M}$ is homeomorphic to an n-ball. We will just assume this to hold in the intuitive setting of integral forms.

In Section 2.1 we concluded that the electromagnetic fields can be modelled through integral forms. Hence, the notion of an exterior derivative permits us to recast the integral form (2.4) of Maxwell's equations as

$$\boldsymbol{d}\mathbf{e} = -i\omega\mathbf{b}, \quad \boldsymbol{d}\mathbf{h} = i\omega\mathbf{d} + \mathbf{j}. \tag{2.9}$$

The statement of (2.9) only relies on the topological concepts of orientation and boundaries of manifolds. Therefore, the relationships in (2.9) may be called *topological laws*.

Two conclusions can be drawn from (2.9) and (2.8). First, from $\boldsymbol{d} \circ \boldsymbol{d} = 0$ we get the conservation laws

$$\boldsymbol{d}\mathbf{b} = 0, \quad \boldsymbol{d}(i\omega\mathbf{d} + \mathbf{j}) = 0. \tag{2.10}$$

Second, as (2.9) holds on all of $A(\mathbb{R}^3)$, by (2.8) we can find a magnetic vector potential $\mathbf{a} \in \mathcal{F}^1(A(\mathbb{R}^3))$ and a scalar potential $v \in \mathcal{F}^0(A(\mathbb{R}^3))$ such that

$$\mathbf{b} = \boldsymbol{d}\mathbf{a}, \quad \mathbf{e} = -\boldsymbol{d}v - i\omega\boldsymbol{d}\mathbf{a}. \tag{2.11}$$

A very natural concept is that of the transformation of integral forms under a diffeomorphism $\boldsymbol{\Phi} : \widehat{\mathcal{M}} \mapsto \mathcal{M}$ of manifolds, the so-called *pullback* $\boldsymbol{\Phi}^* : \mathcal{F}^l(\mathcal{M}) \mapsto \mathcal{F}^l(\widehat{\mathcal{M}})$ defined by

$$\boldsymbol{\Phi}^*\omega \in \mathcal{F}^l(\widehat{\mathcal{M}}) : \quad \int_{\widehat{\Sigma}} \boldsymbol{\Phi}^*\omega = \int_{\boldsymbol{\Phi}(\widehat{\Sigma})} \omega \quad \forall \widehat{\Sigma} \in \mathcal{S}_l(\widehat{\mathcal{M}}). \tag{2.12}$$

Straight from the definitions (2.8) and (2.12) we infer that the exterior derivative and pullback commute, that is,

$$\boldsymbol{d} \circ \boldsymbol{\Phi}^* = \boldsymbol{\Phi}^* \circ \boldsymbol{d}. \tag{2.13}$$

This carries the important consequence that, if integral forms satisfy the topological laws (2.9) on some domain $\Omega \subset A(\mathbb{R}^3)$, then their pullbacks will satisfy the same relationships on a transformed domain. In short, the topological laws are invariant under diffeomorphic transformations.

The trace $\mathbf{t}_{\mathcal{N}}\omega$ of a form $\omega \in \mathcal{F}^l(\mathcal{M})$ onto a sub-manifold $\mathcal{N} \in \mathcal{S}_m(\mathcal{M})$, $0 \leq m \leq n$, is straightforward:

$$\int_\Sigma \mathbf{t}_{\mathcal{N}}\omega := \int_\Sigma \omega \quad \forall \Sigma \in \mathcal{S}_l(\mathcal{N}). \tag{2.14}$$

It is clear that the trace commutes with both the exterior derivative and the pullback, that is, $\boldsymbol{d} \circ \mathbf{t}_{\mathcal{N}} = \mathbf{t}_{\mathcal{N}} \circ \boldsymbol{d}$ and $\boldsymbol{\Phi}^* \circ \mathbf{t}_{\mathcal{N}} = \mathbf{t}_{\widehat{\mathcal{N}}} \circ \boldsymbol{\Phi}^*$.

Traces give a meaning to boundary conditions for electromagnetic fields: imposing boundary conditions amounts to fixing the trace of a field on some surfaces. Though a genuine boundary does not exist in electrodynamics, it is often convenient to assume that fields cannot penetrate some surfaces. This is reflected either by the perfect electric conductor (PEC) boundary conditions $\mathbf{t}_\Sigma \mathbf{e} = 0$ on $\Sigma \in \mathcal{S}_2(A(\mathbb{R}^3))$ or magnetic wall boundary conditions (PMC) $\mathbf{t}_\Sigma \mathbf{h} = 0$. Thus, using the trace of 1-forms gives a clear hint about meaningful boundary conditions for electromagnetic fields.

It is the feat of exterior calculus in differential geometry to establish a meaning of the exterior derivative, the pullback, and the trace for differential forms such that the diagrams

$$\begin{array}{ccc} \mathcal{DF}^{l,1}(\mathcal{M}) & \xrightarrow{\ \boldsymbol{d}\ } & \mathcal{DF}^{l+1,0}(\mathcal{M}) \\ {\scriptstyle F^l_{\mathcal{M}}}\big\downarrow & & \big\downarrow{\scriptstyle F^{l+1}_{\mathcal{M}}} \\ \mathcal{F}^l(\mathcal{M}) & \xrightarrow{\ \boldsymbol{d}\ } & \mathcal{F}^{l+1}(\mathcal{M}) \end{array}, \qquad \begin{array}{ccc} \mathcal{DF}^{l,0}(\mathcal{M}) & \xrightarrow{\ \boldsymbol{\Phi}^*\ } & \mathcal{DF}^{l,0}(\widehat{\mathcal{M}}) \\ {\scriptstyle F^l_{\mathcal{M}}}\big\downarrow & & \big\downarrow{\scriptstyle F^l_{\widehat{\mathcal{M}}}} \\ \mathcal{F}^l(\mathcal{M}) & \xrightarrow{\ \boldsymbol{\Phi}^*\ } & \mathcal{F}^l(\widehat{\mathcal{M}}) \end{array}$$

commute. Here, we have used the same symbols for the new operators on differential forms. Given local representations of $\boldsymbol{d}$, $\boldsymbol{\Phi}^*$, and $\mathbf{t}_{\mathcal{N}}$, and the associations of Table 2.1, it merely takes technical manipulations to come up with incarnations for vector proxies in the case of differential forms defined on a domain $\Omega \subset A(\mathbb{R}^3)$. One finds, based on the identifications of Table 2.1,

$$\Upsilon_1 \circ \boldsymbol{d} = \mathbf{grad} \circ \Upsilon_0, \ \Upsilon_2 \circ \boldsymbol{d} = \mathbf{curl} \circ \Upsilon_1, \ \Upsilon_3 \circ \boldsymbol{d} = \operatorname{div} \circ \Upsilon_2. \tag{2.15}$$

This establishes the link between (2.1) and (2.9). The pullbacks, when considered for vector proxies of continuous differential forms, give rise to familiar transformations:

$$\mathfrak{F}^0_{\boldsymbol{\Phi}} := \Upsilon_0 \circ \boldsymbol{\Phi}^* \circ \Upsilon_0^{-1}, \quad (\mathfrak{F}^0_{\boldsymbol{\Phi}} u)(\widehat{\mathbf{x}}) = u(\mathbf{x}), \tag{2.16}$$

$$\mathfrak{F}^1_{\boldsymbol{\Phi}} := \Upsilon_1 \circ \boldsymbol{\Phi}^* \circ \Upsilon_1^{-1}, \quad (\mathfrak{F}^1_{\boldsymbol{\Phi}} \mathbf{u})(\widehat{\mathbf{x}}) = D\boldsymbol{\Phi}(\widehat{\mathbf{x}})^T \mathbf{u}(\mathbf{x}), \tag{2.17}$$

$$\mathfrak{F}^2_{\boldsymbol{\Phi}} := \Upsilon_2 \circ \boldsymbol{\Phi}^* \circ \Upsilon_2^{-1}, \quad (\mathfrak{F}^2_{\boldsymbol{\Phi}} \mathbf{u})(\widehat{\mathbf{x}}) = \det D\boldsymbol{\Phi}(\widehat{\mathbf{x}}) D\boldsymbol{\Phi}(\widehat{\mathbf{x}})^{-1} \mathbf{u}(\mathbf{x}), \tag{2.18}$$

$$\mathfrak{F}^3_{\boldsymbol{\Phi}} := \Upsilon_3 \circ \boldsymbol{\Phi}^* \circ \Upsilon_3^{-1}, \quad (\mathfrak{F}^3_{\boldsymbol{\Phi}} u)(\widehat{\mathbf{x}}) = \det D\boldsymbol{\Phi}(\widehat{\mathbf{x}})\, u(\mathbf{x}). \tag{2.19}$$

Here u stands for a continuous function on Ω, $\mathbf{u}$ for a continuous vector-field with three components, $\boldsymbol{\Phi} : \widehat{\Omega} \mapsto \Omega$ is a diffeomorphism, $D\boldsymbol{\Phi}$ its Jacobian, and $\widehat{\mathbf{x}} \in \widehat{\Omega}$, $\mathbf{x} := \boldsymbol{\Phi}(\widehat{\mathbf{x}})$.

Finally, the trace of continuous vector proxies onto an oriented, piecewise smooth 2-dimensional sub-manifold Γ of Ω generates the following formulae:

$$\begin{aligned}
\gamma &:= \Upsilon_0^\Gamma \circ \mathbf{t}_\Gamma \circ \Upsilon_0^{-1}, & (\gamma u)(\mathbf{x}) &= u(\mathbf{x}),\\
\gamma_\mathbf{t} &:= \Upsilon_1^\Gamma \circ \mathbf{t}_\Gamma \circ \Upsilon_1^{-1}, & (\gamma_\mathbf{t}\mathbf{u})(\mathbf{x}) &= \mathbf{n}(\mathbf{x}) \times (\mathbf{u}(\mathbf{x}) \times \mathbf{n}(\mathbf{x})),\\
\gamma_\mathbf{n} &:= \Upsilon_2^\Gamma \circ \mathbf{t}_\Gamma \circ \Upsilon_2^{-1}, & (\gamma_\mathbf{n}\mathbf{u})(\mathbf{x}) &= \mathbf{u}(\mathbf{x}) \cdot \mathbf{n}(\mathbf{x}).
\end{aligned}$$

for $\mathbf{x}$ in smooth components of Γ. As usual, $\mathbf{n}(\mathbf{x})$ is a unit normal vectorfield whose direction is prescribed by the (external) orientation of Γ. We point out that Υ_0^Γ, Υ_1^Γ, and Υ_2^Γ are isomorphisms asscociating vector proxies to forms on the two-dimensional manifold Γ. These mappings are chosen based on 'projected Euclidean coordinates'. Here, we skip the details.

The traces of differential forms will be important for a particular reason. Consider $\Omega \subset A(\mathbb{R}^3)$ split into two subdomains Ω_1 and Ω_2 separated by a piecewise smooth, oriented interface Γ. When will $\omega \in \mathcal{DF}^{l,0}(\bar{\Omega}_1) \times \mathcal{DF}^{l,0}(\bar{\Omega}_2)$ give rise to a valid integral l-form on all of Ω? The answer is the *patch condition*

$$\omega \in \mathcal{F}^l(\Omega) \quad \Leftrightarrow \quad \mathbf{t}_\Gamma \omega_{|\Omega_1} = \mathbf{t}_\Gamma \omega_{|\Omega_2} \quad \text{on } \Gamma. \tag{2.20}$$

It translates into the requirement of continuity, tangential continuity, and normal continuity as suitable patch conditions for vector proxies of 0-forms, 1-forms, and 2-forms, respectively. From (2.20) it is clear what the *transmission* conditions for electromagnetic fields must look like. These have to make sure that they make sense as global integral forms. Thus, across any piecewise smooth oriented surface Σ, the traces of both $\mathbf{e}$ and $\mathbf{h}$ must be continuous. Denoting by $[\cdot]_\Sigma$ the difference of traces from both sides (the jump), we find, in terms of vector proxies,

$$[\gamma_\mathbf{t}\mathbf{e}]_\Sigma = 0, \quad [\gamma_\mathbf{t}\mathbf{h}]_\Sigma = 0. \tag{2.21}$$

Eventually, in the calculus of differential forms, the Poincaré lemma can be stated as a theorem. In fact, it becomes part of a celebrated, far more general result.

Theorem 2.1. (DeRham theorem for differential forms) There is a finite-dimensional subspace $\mathcal{DH}^{l,0}(\mathcal{M}) \subset \mathcal{DF}^{l,0}(\mathcal{M})$ of closed differential l-forms, whose dimension is equal to the lth Betti number of Ω, such that, for $\omega \in \mathcal{DF}^{l,0}(\mathcal{M})$, we have

$$\boldsymbol{d}\omega = 0 \quad \Leftrightarrow \quad \exists \eta \in \mathcal{DF}^{l-1,1}(\mathcal{M}),\ \tau \in \mathcal{DH}^{l,0}(\mathcal{M}) \quad \text{satisfying } \omega = \boldsymbol{d}\eta + \tau.$$

So far we have not gone much beyond operations already defined for integral forms. Now, we introduce an important device based on a local viewpoint. It is the *exterior product*, a bilinear mapping

$$\wedge : \mathcal{DF}^{l,0}(\mathcal{M}) \times \mathcal{DF}^{m,0}(\mathcal{M}) \mapsto \mathcal{DF}^{l+m,0}(\mathcal{M}) \tag{2.22}$$

pointwise defined via the $\wedge$-product of alternating multilinear forms. It is connected with the other operations through

$$\begin{aligned}
\omega \wedge \eta &= (-1)^{lm}(\eta \wedge \omega) && \forall \omega \in \mathcal{DF}^{l,0}(\mathcal{M}), \eta \in \mathcal{DF}^{m,0}(\mathcal{M}),\\
\boldsymbol{d}(\omega \wedge \eta) &= \boldsymbol{d}\omega \wedge \eta + (-1)^l(\omega \wedge \boldsymbol{d}\eta) && \forall \omega \in \mathcal{DF}^{l,1}(\mathcal{M}), \eta \in \mathcal{DF}^{m,1}(\mathcal{M}),\\
\boldsymbol{\Phi}^*(\omega \wedge \eta) &= \boldsymbol{\Phi}^*\omega \wedge \boldsymbol{\Phi}^*\eta && \forall \omega \in \mathcal{DF}^{l,0}(\mathcal{M}), \eta \in \mathcal{DF}^{m,0}(\mathcal{M}).
\end{aligned}$$

The second equation combined with the definition of the exterior derivative yields the vital *integration by parts formula*

$$\int_{\Sigma} \boldsymbol{d}\omega \wedge \eta + (-1)^l(\omega \wedge \boldsymbol{d}\eta) = \int_{\partial\Sigma} \omega \wedge \eta \tag{2.23}$$

for $\omega \in \mathcal{DF}^{l,0}(\mathcal{M})$, $\eta \in \mathcal{DF}^{m,0}(\mathcal{M})$, $\Sigma \in \mathcal{S}_{l+m+1}(\mathcal{M})$. In terms of Euclidean vector proxies, the exterior product reads

$$\begin{aligned}
\Upsilon_2(\omega \wedge \eta) &= (\Upsilon_1\omega) \times (\Upsilon_1\eta) && \forall \omega, \eta \in \mathcal{DF}^{1,0}(\Omega),\\
\Upsilon_3(\omega \wedge \eta) &= (\Upsilon_2\omega) \cdot (\Upsilon_1\eta) && \forall \omega \in \mathcal{DF}^{2,0}(\Omega), \eta \in \mathcal{DF}^{1,0}(\Omega).
\end{aligned}$$

An exterior product with a 0-form amounts to a pointwise multiplication with its related function. These relationships supply the customary integration by parts formulas (Green's formulae) for functions and vector-fields.

Remark 1. The perspective of differential forms rewards us with insights into hidden relationships. Just tinker with the order of forms in the topological laws (2.9) by viewing $\mathbf{e}$ as a 0-form, $\mathbf{b}$ as a 1-form, $\mathbf{h}$ as a 2-form, and $\mathbf{d}, \mathbf{j}$ as 3-forms. This makes perfect sense and we recover the topological laws underlying the Helmholtz equation of linear acoustics, because in terms of vector proxies we get

$$\mathbf{grad}\,\text{`}\mathbf{e}\text{'} = -i\omega\text{`}\mathbf{b}\text{'}, \quad \operatorname{div}\text{`}\mathbf{h}\text{'} = i\omega\text{`}\mathbf{d}\text{'} + \text{`}\mathbf{j}\text{'}.$$

This looks odd, but now $\mathbf{e}$ must be seen as a scalar potential (pressure) and $\mathbf{h}$ is usually called the flux. Hence Maxwell's equations are a member of a larger family of models, to which the acoustic wave equation belongs as well. Please be aware that, in terms of differential forms, it is related to the scalar wave equation and not to the vectorial one. We also realize a significant difference between the models for acoustics and electromagnetism. In the latter case both $\mathbf{e}$ and $\mathbf{h}$ are 1-forms, which hints at a fundamental symmetry. △

Bibliographical notes

The theory of differential forms is a classical branch of differential geometry covered by many textbooks, of which I would like to mention Cartan (1967). This is the main reference for all results cited above, besides Burke (1985, Chapter IV) and Lang (1995, Chapter V). A lucid presentation is given in Bossavit (1998*d*). There is a close link with algebraic topology, of which Bott and Tu (1982) give a substantial account.

2.3. Variational formulations

So far we have evaded the vexing question of how the material laws fit the framework of forms that we have embraced in the previous sections. Evidently, they link forms of different order, and therefore the multiplications with $\boldsymbol{\epsilon}$, $\boldsymbol{\mu}$, and $\boldsymbol{\sigma}$ must be regarded as special linear operators.

Keep in mind that the material laws arise from averaging microscopic effects. In a sense they are less fundamental than the topological laws (2.9), and make sense only on a macroscopic scale. The material laws introduce the concept of field energy into the model. Let $E_{\mathrm{el}}(\mathbf{e})$ denote the energy contained in the electric field within a bounded control volume $\Omega \subset A(\mathbb{R}^3)$. Since we admit only linear materials, E_{el} is a quadratic form, which arises from a symmetric positive definite sesqui-linear form a_ϵ by $E_{\mathrm{el}}(\mathbf{v}) = \frac{1}{2}a_\epsilon(\mathbf{v},\mathbf{v})$ for all $\mathbf{v} \in \mathcal{F}^1(\Omega)$. Then the displacement current $\mathbf{d}$ has to satisfy

$$\int_\Omega \mathbf{d} \wedge \overline{\mathbf{e}}' = a_\epsilon(\mathbf{e},\mathbf{e}') \quad \forall \mathbf{e}' \in \mathcal{DF}^{1,0}(\Omega), \tag{2.24}$$

where an over-bar indicates the complex conjugate. In fact, $\mathbf{d}$ could be regarded as a linear form on the space of differential 1-forms, that is, a 1-current (Grauert and Lieb 1977, Chapter 5). Yet we will not pursue this, and continue viewing $\mathbf{d}$ as a 2-form.

Similarly, the magnetic induction $\mathbf{b}$ possesses the magnetic energy E_{mag} on Ω. It is related to a symmetric, positive definite sesqui-linear form $a_{1/\mu}$ by $E_{\mathrm{mag}}(\mathbf{v}) = \frac{1}{2}a_{1/\mu}(\mathbf{v},\mathbf{v})$ for all $\mathbf{v} \in \mathcal{F}^2(\Omega)$. Then the magnetic field $\mathbf{h}$ has to fulfil

$$\int_\Omega \mathbf{h} \wedge \overline{\mathbf{b}}' = a_{1/\mu}(\mathbf{b},\mathbf{b}') \quad \forall \mathbf{b}' \in \mathcal{DF}^{2,0}(\Omega). \tag{2.25}$$

As the exterior product introduces a non-degenerate pairing, (2.24) and (2.25) also assign energies to the fields $\mathbf{h}$ and $\mathbf{d}$. Thus we may introduce symmetric, positive definite sesqui-linear forms a_μ and $a_{1/\epsilon}$ and express

$$\int_\Omega \mathbf{b} \wedge \overline{\mathbf{h}}' = a_\mu(\mathbf{h},\mathbf{h}') \quad \forall \mathbf{h}' \in \mathcal{DF}^{1,0}(\Omega), \tag{2.26}$$

$$\int_\Omega \mathbf{e} \wedge \overline{\mathbf{d}}' = a_{1/\epsilon}(\mathbf{d},\mathbf{d}') \quad \forall \mathbf{d}' \in \mathcal{DF}^{2,0}(\Omega). \tag{2.27}$$

The analogous treatment of Ohm's law (2.3) is left to the reader.

The material laws in variational form can be combined with the topological laws and lead to natural weak formulations of Maxwell's equations. We can distinguish between two essentially distinct approaches. They differ in how the two topological laws are taken into account. On the one hand, Ampère's law, when tested with $\mathbf{e}' \in \mathcal{DF}^{1,0}(\bar{\Omega})$ on a bounded control volume Ω, gives rise to

$$\int_\Omega \boldsymbol{d}\mathbf{h} \wedge \overline{\mathbf{e}}' = i\omega \int_\Omega \mathbf{d} \wedge \overline{\mathbf{e}}' + \int_\Omega \mathbf{j} \wedge \overline{\mathbf{e}}'.$$

Now, integration by parts is performed according to (2.23), which means that Ampère's law enters in a weak sense only:

$$\int_\Omega \mathbf{h} \wedge \boldsymbol{d}\overline{\mathbf{e}}' + \int_{\partial\Omega} \mathbf{h} \wedge \overline{\mathbf{e}}' = i\omega \int_\Omega \mathbf{d} \wedge \overline{\mathbf{e}}' + \int_\Omega \mathbf{j} \wedge \overline{\mathbf{e}}'.$$

Two integrals in this equation can be replaced by means of (2.24) and (2.25), which yields

$$a_{1/\mu}(\mathbf{b}, \boldsymbol{d}\mathbf{e}') + \int_{\partial\Omega} \mathbf{h} \wedge \overline{\mathbf{e}}' = i\omega a_\epsilon(\mathbf{e}, \mathbf{e}') + \int_\Omega \mathbf{j} \wedge \overline{\mathbf{e}}' \quad \forall \mathbf{e}' \in \mathcal{DF}^{1,0}(\Omega).$$

Next, Faraday's law is used to express $\mathbf{b}$ through $\boldsymbol{d}\mathbf{e}$, and we end up with the '$\mathbf{e}$-based' *primal variational formulation* of Maxwell's equations: the electric field solution $\mathbf{e}$ satisfies

$$a_{1/\mu}(\boldsymbol{d}\mathbf{e}, \boldsymbol{d}\mathbf{e}') - \omega^2 a_\epsilon(\mathbf{e}, \mathbf{e}') - i\omega \int_{\partial\Omega} \mathbf{h} \wedge \overline{\mathbf{e}}' = -i\omega \int_\Omega \mathbf{j} \wedge \overline{\mathbf{e}}' \tag{2.28}$$

for all $\mathbf{e}' \in \mathcal{DF}^{1,0}(\Omega)$. On the other hand, we may choose to take into account Faraday's law, in weak form, by

$$\int_\Omega \mathbf{e} \wedge \boldsymbol{d}\overline{\mathbf{h}}' - \int_{\partial\Omega} \mathbf{e} \wedge \overline{\mathbf{h}}' = -i\omega \int_\Omega \mathbf{b} \wedge \overline{\mathbf{h}}' \quad \forall \mathbf{h}' \in \mathcal{DF}^{1,0}(\Omega).$$

The alternative variational version of the material laws, namely (2.26) and (2.27), have to be used in this case. In addition, $\mathbf{d}$ can be replaced using the strong form of Ampère's law. We arrive at the '$\mathbf{h}$-based' *dual variational formulation*: for the magnetic field solution $\mathbf{h}$ we have

$$a_{1/\epsilon}(\boldsymbol{d}\mathbf{h}, \boldsymbol{d}\mathbf{h}') - \omega^2 a_\mu(\mathbf{h}, \mathbf{h}') + i\omega \int_{\partial\Omega} \mathbf{e} \wedge \overline{\mathbf{h}}' = a_{1/\epsilon}(\mathbf{j}, \boldsymbol{d}\mathbf{h}') \tag{2.29}$$

for all $\mathbf{h}' \in \mathcal{DF}^{1,0}(\Omega)$. To get valid boundary value problems on bounded domains, the boundary terms in both (2.28) and (2.29) have to be dealt with by imposing suitable boundary conditions. For instance, in the case of (2.28) PEC boundary conditions must be imposed strongly and honoured by demanding that $\mathbf{t}_{\partial\Omega}\mathbf{e}' = 0$. Conversely, PMC boundary conditions $\mathbf{t}_{\partial\Omega}\mathbf{h} = 0$ are taken into account weakly by dropping the boundary term. For the dual variational problem the handling of boundary conditions is reversed.

Remark 2. The typical situation in computational electromagnetism is marked by unbounded domains. In an abstract way the unbounded exterior of Ω can be taken into account by replacing

$$\int_{\partial\Omega} \mathbf{h} \wedge \overline{\mathbf{e}}' \quad \longrightarrow \quad \int_{\partial\Omega} \mathsf{S}_{\mathbf{e}}\mathbf{e} \wedge \overline{\mathbf{e}}'$$

in (2.28), where $\mathsf{S}_{\mathbf{e}}$ is the Poincaré–Steklov operator for the exterior electromagnetic field problem (with radiation conditions at ∞). The Poincaré–Steklov operator can be expressed through boundary integral equations. If Ω is a ball, techniques based on expansions into surface spherical harmonics are available. A survey is given in Nédélec (2001). The numerical treatment of field problems on unbounded exterior domains (scattering problems) is a core area of computational electromagnetism, but will not be covered in this article. △

Inherent in the field model is idealization that the field energy can be strictly localized in terms of an energy density. Assuming some smoothness of the fields and letting the control volume Ω shrink to zero, we finally obtain positive definite quadratic forms on $\bigwedge^l(\mathbb{R}^3)$ ($l = 1$ for $a_\epsilon, a_{1/\mu}$, $l = 2$ for $a_{1/\epsilon}, a_\mu$) for any point $\mathbf{x} \in A(\mathbb{R}^3)$. By simple linear algebra, these define operators $\star : \bigwedge^l(\mathbb{R}^3) \mapsto \bigwedge^{3-l}(\mathbb{R}^3)$, which give rise to *Hodge operators* $\star : \mathcal{DF}^{l,0}(A(\mathbb{R}^3)) \mapsto \mathcal{DF}^{3-l,0}(A(\mathbb{R}^3))$. For vector proxies in Euclidean space $\mathbb{R}^3$ they take the form of the conventional material laws (2.2). This will be enough for our purposes and we are not going to dwell on Hodge operators any further. In short, through the notion of local energy densities we find the conventional expressions

$$a_\epsilon(\mathbf{u},\mathbf{v}) = \int_\Omega \epsilon \Upsilon_1\mathbf{u} \cdot \Upsilon_1\mathbf{v}\,\mathrm{d}\mathbf{x}, \qquad a_{1/\mu}(\mathbf{u},\mathbf{v}) = \int_\Omega \mu^{-1}\Upsilon_2\mathbf{u} \cdot \Upsilon_2\mathbf{v}\,\mathrm{d}\mathbf{x},$$

$$a_{1/\epsilon}(\mathbf{u},\mathbf{v}) = \int_\Omega \epsilon^{-1}\Upsilon_2\mathbf{u} \cdot \Upsilon_2\mathbf{v}\,\mathrm{d}\mathbf{x}, \qquad a_\mu(\mathbf{u},\mathbf{v}) = \int_\Omega \mu\Upsilon_1\mathbf{u} \cdot \Upsilon_1\mathbf{v}\,\mathrm{d}\mathbf{x},$$

where $\mathbf{u}, \mathbf{v}$ are forms of appropriate degree. By their derivation these sesquilinear forms are invariant with respect to the choice of vector proxies, because a change of basis also entails a transformation of the metric tensors.

Remark 3. The terminology 'primal' and 'dual' is borrowed from the study of weak formulations of second-order elliptic problems (*cf.* Brezzi and Fortin (1991, Chapter 1)), that is, the case discussed in Remark 1 where

e is a 0-form. Then $\boldsymbol{d}$ becomes **grad** in the primal variational formulation and the remaining unknown is a plain function. For the dual problem, $\boldsymbol{d}$ is div and the variational formulation is posed for a flux field. In this case the striking difference between the two formulations justifies the labels 'primal' and 'dual' (which stem from convex analysis). In light of the symmetry between **e** and **h** the distinction seems pointless, but it was maintained in order to emphasize the relationships with second-order elliptic problems. These are also reflected by the role reversal of boundary conditions. △

2.4. Function spaces

A Hilbert space framework provides the most powerful tools for the analysis of the linear variational problems (2.28) and (2.29) derived in the previous section. In the remainder of the article $\Omega \subset A(\mathbb{R}^3)$ stands for a bounded Lipschitz polyhedron with plane faces. More generally, in most contexts it could also be a curvilinear Lipschitz polyhedron in the parlance of Costabel and Dauge (1999). This will cover most geometric arrangements that occur in real world simulations. Throughout, $\mathbf{n} \in \boldsymbol{L}^\infty(\Gamma)$ will denote the exterior unit normal vector-field on Γ.

To obtain suitable Hilbert spaces for fields, we follow the usual procedure based on completions of spaces of smooth functions with respect to a norm induced by the total field energy. It is important to be aware that, for both the electric and magnetic field, the total field energy comprises contributions of magnetic and electric field energy. For instance, besides $E_{\mathrm{el}}(\mathbf{e})$ the total energy of an electric field solution of Maxwell's equations also involves the magnetic energy of **curl e**. Appealing to the uniform positivity of the metric tensors, we arrive at the (equivalent) energy norm

$$\|\mathbf{u}\|^2_{\boldsymbol{H}(\mathbf{curl};\Omega)} := \|\mathbf{u}\|^2_{\boldsymbol{L}^2(\Omega)} + \|\mathbf{curl}\,\mathbf{u}\|^2_{\boldsymbol{L}^2(\Omega)}.$$

The space obtained by completion of $\boldsymbol{C}^\infty(\Omega)$ with respect to $\|\cdot\|_{\boldsymbol{H}(\mathbf{curl};\Omega)}$ is customarily denoted by $\boldsymbol{H}(\mathbf{curl};\Omega)$.[4] As the reader might guess, given a background of differential forms, $\boldsymbol{H}(\mathbf{curl};\Omega)$ is only one member of a larger family of Hilbert spaces. We could have introduced an 'energy norm' on smooth differential l-forms on Ω by

$$\|\mathbf{u}\|^2_{\boldsymbol{H}(\boldsymbol{d},\Omega)} := \|\mathbf{u}\|^2_{\boldsymbol{L}^2(\Omega)} + \|\boldsymbol{d}\mathbf{u}\|^2_{\boldsymbol{L}^2(\Omega)},$$

where the $\boldsymbol{L}^2(\Omega)$-norms are computed through some vector proxy. Note that the $\boldsymbol{L}^2(\Omega)$-norm of a vector proxy is, up to equivalence, independent of the Euclidean structure. Then $\boldsymbol{H}(\boldsymbol{d},\Omega)$ can be defined as completion of $\mathcal{DF}^{l,\infty}(\Omega)$ with respect to $\|\cdot\|_{\boldsymbol{H}(\boldsymbol{d},\Omega)}$. Recalling (2.15), this yields the familiar spaces $H^1(\Omega)$ and $\boldsymbol{H}(\mathrm{div};\Omega)$, corresponding to 0-forms and 2-forms,

[4] Bold typeface is meant to distinguish vector-fields and spaces containing them.

respectively. Important closed subspaces will be the kernels of the exterior derivatives, for which we write

$$\boldsymbol{H}(\boldsymbol{d}0,\Omega) := \{\mathbf{u} \in \boldsymbol{H}(\boldsymbol{d},\Omega),\ \boldsymbol{d}\mathbf{u} = 0\},$$

in particular $\boldsymbol{H}(\mathbf{curl}\,0;\Omega)$ and $\boldsymbol{H}(\mathrm{div}\,0;\Omega)$. As it holds for smooth differential forms, the DeRham theorem (Theorem 2.1) can be extended to the function spaces. A particular case is given in the next lemma (*cf.* Girault and Raviart (1986, Theorem 2.9), Amrouche, Bernardi, Dauge and Girault (1998, Proposition 3.14 and Proposition 3.18), Kress (1971)).

Lemma 2.2. There is a finite-dimensional *cohomology space* $\mathcal{H}^1(\Omega) \subset \boldsymbol{H}(\mathbf{curl}\,0;\Omega) \cap \boldsymbol{H}_0(\mathrm{div}\,0;\Omega)$ of *harmonic Neumann vector-fields* whose dimension agrees with the first Betti number of Ω, such that, for any $\mathbf{u} \in \boldsymbol{H}(\mathbf{curl}\,0;\Omega)$ with $\mathbf{curl}\,\mathbf{u} = 0$, we find $\varphi \in H^1(\Omega)$ and $\mathbf{q} \in \mathcal{H}^1(\Omega)$ such that $\mathbf{u} = \mathbf{grad}\,\varphi + \mathbf{q}$.

Every $\mathbf{u} \in \boldsymbol{H}_0(\mathbf{curl}\,0;\Omega)$ has a representation $\mathbf{u} = \mathbf{grad}\,\varphi + \mathbf{q}$, where $\varphi \in H^1_0(\Omega)$ and $\mathbf{q}$ is contained in a *relative cohomology space* $\mathcal{H}^2(\Omega) \subset \boldsymbol{H}_0(\mathbf{curl}\,0;\Omega) \cap \boldsymbol{H}(\mathrm{div}\,0;\Omega)$ of *harmonic Dirichlet vector-fields*, $\dim \mathcal{H}^2(\Omega) =$ 2nd Betti number of Ω.

As is proved in Girault and Raviart (1986, Theorem 2.40, Chapter I), an equivalent definition of the space $\boldsymbol{H}(\mathbf{curl};\Omega)$ is

$$\boldsymbol{H}(\mathbf{curl};\Omega) := \{\mathbf{u} \in \boldsymbol{L}^2(\Omega),\ \mathbf{curl}\,\mathbf{u} \in \boldsymbol{L}^2(\Omega)\},$$

where **curl** has to be understood in the sense of distributions. Therefore, the variational equations (2.28) and (2.29) when considered over $\boldsymbol{H}(\mathbf{curl};\Omega)$ imply the topological laws (2.1) in the sense of distributions. Thus the **e**-based and **h**-based variational formulations are indeed equivalent, and, for example, from (2.28) we can get $\mathbf{h} = \frac{1}{i\omega}\boldsymbol{\mu}^{-1}\,\mathbf{curl}\,\mathbf{e}$. As usual the following result is established (*cf.* Lemmas 6 and 8 in Nédélec (1980)).

Lemma 2.3. If $\Omega \subset \mathbb{R}^3$, $\bar{\Omega} = \bar{\Omega}_1 \cup \bar{\Omega}_2$, $\Omega_1 \cap \Omega_2 = \emptyset$, and $\mathbf{u}_{|\Omega_1} \in \boldsymbol{C}^\infty(\bar{\Omega}_1)$, $\mathbf{u}_{|\Omega_2} \in \boldsymbol{C}^\infty(\bar{\Omega}_2)$, then

$$\begin{aligned} [\gamma_{\mathbf{t}}\mathbf{u}]_{\partial\Omega_1\cap\partial\Omega_2} = 0 \quad &\Leftrightarrow \quad \mathbf{u} \in \boldsymbol{H}(\mathbf{curl};\Omega), \\ [\gamma_{\mathbf{n}}\mathbf{u}]_{\partial\Omega_1\cap\partial\Omega_2} = 0 \quad &\Leftrightarrow \quad \mathbf{u} \in \boldsymbol{H}(\mathrm{div};\Omega). \end{aligned}$$

We can also perform the completion of the space $\boldsymbol{C}_0^\infty(\Omega)$ of smooth vector-fields with compact support in Ω with respect to the energy norm. This results in the space $\boldsymbol{H}_0(\mathbf{curl};\Omega)$, a closed subspace of $\boldsymbol{H}(\mathbf{curl};\Omega)$, which realizes the condition of vanishing trace on Γ. This becomes evident by looking at the integration by parts formula ($\mathbf{u}, \mathbf{v} \in \boldsymbol{C}^\infty(\bar{\Omega})$)

$$\int_\Omega \mathbf{u} \cdot \mathbf{curl}\,\mathbf{v} - \mathbf{curl}\,\mathbf{u} \cdot \mathbf{v}\,\mathrm{d}\mathbf{x} = \int_\Gamma (\gamma_{\mathbf{t}}\mathbf{u} \times \mathbf{n}) \cdot \mathbf{v}_{|\Gamma}\,\mathrm{d}S. \tag{2.30}$$

Appealing to a trace theorem for $\boldsymbol{H}^1(\Omega)$ (Grisvard 1985, Theorem 1.5.1.1), this confirms that the tangential trace $\gamma_t : \boldsymbol{C}^\infty(\bar{\Omega}) \mapsto \boldsymbol{L}^\infty(\Gamma)$ is continuous as a mapping $\boldsymbol{H}(\mathbf{curl};\Omega) \mapsto \boldsymbol{H}^{-\frac{1}{2}}(\Gamma)$. Consequently, it can be extended to $\boldsymbol{H}(\mathbf{curl};\Omega)$. Since, obviously, $\gamma_t\mathbf{u} = 0$ for all $\mathbf{u} \in \boldsymbol{C}_0^\infty(\Omega)$, we get the alternative characterization

$$\boldsymbol{H}_0(\mathbf{curl};\Omega) := \{\mathbf{u} \in \boldsymbol{H}(\mathbf{curl};\Omega),\ \gamma_t\mathbf{u} = 0\}. \tag{2.31}$$

In many respects the space $\boldsymbol{H}(\mathbf{curl};\Omega)$ is rather unwieldy, in contrast to the classical Sobolev space $H^1(\Omega)$. Thus, the next lemma is instrumental in establishing key properties of $\boldsymbol{H}(\mathbf{curl};\Omega)$.

Lemma 2.4. (Regular decomposition lemma) There are continuous maps $\mathsf{R} : \boldsymbol{H}(\mathbf{curl};\Omega) \mapsto \boldsymbol{H}^1(\Omega) \cap \boldsymbol{H}(\operatorname{div} 0;\Omega)$, $\mathsf{N} : \boldsymbol{H}(\mathbf{curl};\Omega) \mapsto H^1(\Omega)$, $\mathsf{N}_{\mathcal{H}} : \boldsymbol{H}(\mathbf{curl};\Omega) \mapsto \boldsymbol{\mathcal{H}}^1(\Omega)$ such that $\mathsf{R} + \mathbf{grad} \circ \mathsf{N} + \mathsf{N}_{\mathcal{H}} = \mathrm{Id}$ on $\boldsymbol{H}(\mathbf{curl};\Omega)$ and $\mathsf{R}_{|\boldsymbol{H}(\mathbf{curl}\,0;\Omega)} = 0$.

In addition, there are continuous maps $\mathsf{R}_0 : \boldsymbol{H}_0(\mathbf{curl};\Omega) \mapsto \boldsymbol{H}^1(\Omega)$, $\mathsf{N}_0 : \boldsymbol{H}_0(\mathbf{curl};\Omega) \mapsto H_0^1(\Omega)$ such that $\mathsf{R}_0 + \mathbf{grad} \circ \mathsf{N}_0 = \mathrm{Id}$ on $\boldsymbol{H}_0(\mathbf{curl};\Omega)$.

The proof will make use of the existence of regular vector potentials.

Lemma 2.5. (Existence of regular vector potentials) For every $r \geq 0$ there is a continuous mapping $\mathsf{L} : \boldsymbol{H}(\operatorname{div} 0;\mathbb{R}^3) \cap \boldsymbol{H}^r(\mathbb{R}^3) \mapsto \boldsymbol{H}^{1+r}_{\mathrm{loc}}(\mathbb{R}^3)$ such that $\mathbf{curl}\,\mathsf{L}\mathbf{v} = \mathbf{v}$ and $\operatorname{div} \mathsf{L}\mathbf{v} = 0$.

The proof boils down to elementary calculations done with the Fourier transforms of the functions. It is given as part of the proof of Lemma 3.5 in Amrouche *et al.* (1998).

Proof of Lemma 2.4. For $\mathbf{u} \in \boldsymbol{H}(\mathbf{curl};\Omega)$ its rotation $\mathbf{curl}\,\mathbf{u}$ belongs to $\boldsymbol{H}(\operatorname{div} 0;\Omega)$. Solving a Neumann problem for Δ outside Ω, a divergence-free extension $\mathbf{v} \in \boldsymbol{H}(\operatorname{div} 0;\mathbb{R}^3)$ of $\mathbf{curl}\,\mathbf{u}$ can be found. Setting $\mathsf{R}\mathbf{u} := \mathsf{L}\mathbf{v}$, we find that $\mathbf{curl}(\mathbf{u} - \mathsf{R}\mathbf{u}) = 0$ in Ω due to the properties of L. Applying Lemma 2.2 for $r = 0$ finishes the first part of the proof.

The second part follows the proof of Proposition 5.1 in Bonnet-BenDhia, Hazard and Lohrengel (1999). A $\mathbf{u} \in \boldsymbol{H}_0(\mathbf{curl};\Omega)$ can be extended by zero to $\widetilde{\mathbf{u}} \in \boldsymbol{H}(\mathbf{curl};\mathbb{R}^3)$. Then $\mathbf{curl}\,\widetilde{\mathbf{u}}$ belongs to the domain of L for $r = 0$. With $\boldsymbol{\Psi} := \mathsf{L}\,\mathbf{curl}\,\widetilde{\mathbf{u}}$ we find that $\widetilde{\mathbf{u}} - \boldsymbol{\Psi}$ is $\mathbf{curl}$-free. Thus there is a scalar potential $\psi \in H^1_{\mathrm{loc}}(\mathbb{R}^3)$ with $\widetilde{\mathbf{u}} - \boldsymbol{\Psi} = \mathbf{grad}\,\psi$. As $\widetilde{\mathbf{u}} = 0$ outside Ω, $\psi \in H^2_{\mathrm{loc}}(\mathbb{R}^3 \setminus \bar{\Omega})$. Write $\phi \in H^2(\Omega)$ for the Sobolev extension of $\psi_{|\mathbb{R}^3\setminus\bar{\Omega}}$ into the interior of Ω. Then

$$\mathbf{u} = (\boldsymbol{\Psi} + \mathbf{grad}\,\phi) + \mathbf{grad}(\psi - \phi)$$

is the desired decomposition and $\mathsf{R}_0\mathbf{u} := \boldsymbol{\Psi} + \mathbf{grad}\,\phi$, $\mathsf{N}_0\mathbf{u} := \psi - \phi$ defines the associated operators. □

Lemma 2.6. (More regular decomposition lemma) We can decompose every $\mathbf{u} \in \boldsymbol{H}(\mathbf{curl};\Omega)$ for which $\mathbf{curl}\,\mathbf{u} \in \boldsymbol{H}^1(\Omega)$, into $\mathbf{u} = \boldsymbol{\Psi} + \mathbf{grad}\,\varphi + \mathbf{h}$, $\boldsymbol{\Psi} \in \boldsymbol{H}^2(\Omega)$, $\varphi \in H^1(\Omega)$, $\mathbf{h} \in \mathcal{H}^1(\Omega)$. Moreover, $\|\boldsymbol{\Psi}\|_{\boldsymbol{H}^2(\Omega)} \leq C\,\|\mathbf{curl}\,\mathbf{u}\|_{\boldsymbol{H}^1(\Omega)}$ for some $C = C(\Omega) > 0$ independent of $\mathbf{u}$.

Proof. The proof follows that of Corollary 3.3 in Girault and Raviart (1986) and, for the sake of simplicity, only tackles the case of a connected boundary $\partial\Omega$. Let $\mathcal{O}$ stand for a large ball containing $\bar{\Omega}$. Then the complement $\Omega' := \mathcal{O} \setminus \bar{\Omega}$ is a connected bounded Lipschitz domain. We exploit the important finding that $\mathbf{grad} : L^2(\Omega')/\mathbb{R} \mapsto \boldsymbol{H}^{-1}(\Omega')$ is injective with closed range (Girault and Raviart 1986, Corollary 2.1). Thus, its $\boldsymbol{L}^2(\Omega')$-adjoint, $\operatorname{div} : \boldsymbol{H}_0^1(\Omega') \mapsto L^2_\bullet(\Omega')$ is surjective, where $L^2_\bullet(\Omega')$ contains all functions in $L^2(\Omega')$ with vanishing mean.

Extend $\mathbf{curl}\,\mathbf{u}$ to $\mathbf{w} \in \boldsymbol{H}_0^1(\mathcal{O})$. Since $\int_{\partial\Omega} \mathbf{curl}\,\mathbf{u} \cdot \mathbf{n}\,\mathrm{d}S = 0$, Gauss's theorem teaches that the mean of $\operatorname{div}\mathbf{w} \in L^2(\Omega')$ vanishes on Ω'. According to the above considerations, we can find $\mathbf{z} \in \boldsymbol{H}_0^1(\Omega')$ such that $\operatorname{div}\mathbf{z} = \operatorname{div}\mathbf{w}$. This means that

$$\mathbf{v} := \begin{cases} 0, & \text{in } \mathbb{R}^3 \setminus \bar{\mathcal{O}}, \\ \mathbf{w} - \mathbf{z}, & \text{in } \Omega', \\ \mathbf{curl}\,\mathbf{u}, & \text{in } \Omega \end{cases}$$

belongs to $\boldsymbol{H}(\operatorname{div} 0;\mathbb{R}^3) \cap \boldsymbol{H}^1(\mathbb{R}^3)$. Using Lemma 2.5 for $r = 1$ the proof can be completed in the same fashion as that of Lemma 2.4. □

Remark 4. In one respect the completion procedure goes too far. Of course, it takes us beyond continuous functions, but we may wonder whether $\boldsymbol{H}(\mathbf{curl};\Omega)$ supplies valid integral 1-forms in the sense that the evaluation of integrals along piecewise smooth curves is a well-defined, that is, continuous, functional on $\boldsymbol{H}(\mathbf{curl};\Omega)$. Unfortunately, the answer is negative, and counterexamples are supplied by gradients of unbounded functions in $H^1(\Omega)$ and paths running through their singularity. This shortcoming of $\boldsymbol{H}(\mathbf{curl};\Omega)$ will be the source of considerable complications in the analysis of numerical methods (*cf.* Section 3.6, in particular Lemma 3.13). △

Bibliographical notes

The theory of the spaces $\boldsymbol{H}(\mathbf{curl};\Omega)$ and $\boldsymbol{H}(\operatorname{div};\Omega)$ is developed, for instance, in Girault and Raviart (1986, Chapter 1), Dautray and Lions (1990, Vol. 3, Chapter IX, Section 1), Fernandes and Gilardi (1997), and Amrouche *et al.* (1998). General information on Sobolev spaces can be found in Adams (1975) and Maz'ya (1985). Those for differential forms are discussed in Iwaniec (1999, Chapter 3) and Schwarz (1995, Chapter 1). The regular decomposition lemma first appeared in Birman and Solomyak (1990) and was implicitly used in the proof of Theorem 2 in Costabel (1990). More sophisticated decomposition theorems can be found in Bonnet-BenDhia *et al.* (1999) and Costabel, Dauge and Nicaise (1999).

3. Discrete differential forms

The perspective of differential forms offers an invaluable guideline for the discretization of Maxwell's equations. It will turn out that discrete differential forms, finite elements for differential forms, are the right tools for this task. Their roots in discrete cohomology theory on cellular complexes guarantee that essential structural properties of the topological laws of electromagnetism are preserved in a discrete setting.

3.1. Cochains

It is the very nature of discrete entities that they can be described by a finite amount of information. Recalling the notion of integral forms from Definition 1, it is natural to demand that a discrete integral l-form on the domain $\Omega \subset A(\mathbb{R}^3)$ is already determined by fixing the numbers it assigns to a *finite* number of compact, oriented, piecewise smooth l-dimensional sub-manifolds of Ω. In order to obtain a meaningful exterior derivative for discrete forms, these sets of special sub-manifolds must support boundary operators ∂. Therefore, one is led to consider triangulations of Ω as the natural device to construct appropriate finite sets of sub-manifolds.

Definition 3. A *triangulation* or *mesh* Ω_h of $\Omega \subset A(\mathbb{R}^3)$ is a finite collection of oriented *cells* (set $\mathcal{S}_3(\Omega_h)$), *faces* (set $\mathcal{S}_2(\Omega_h)$), *edges* (set $\mathcal{S}_1(\Omega_h)$), and *vertices* (set $\mathcal{S}_0(\Omega_h)$) such that:

- all cells are open subsets of $A(\mathbb{R}^3)$, homeomorphic to the unit ball;
- all cells, faces, edges, and vertices form a partition of $\bar{\Omega}$;
- the boundary of each cell is the union of closed faces, the boundary of each face the union of closed edges, and the boundary of each edge consists of vertices;
- each vertex, edge, and face is contained in the boundary of an edge, face, or cell, respectively.

The elements of $\mathcal{S}_l(\Omega_h)$ are called *l-facets*.

These triangulations are special cases of CW-complexes considered in discrete algebraic topology Lundell and Weingram (1969) and Fritsch and Piccini (1990). When augmented by orientation, all finite element meshes without hanging nodes will qualify as valid triangulations in the sense of Definition 3.

It remains to settle the issue of orientation. Let us first consider tetrahedral meshes (*cf.* Bossavit (1998*a*, Section 5.2.1)). In this case any oriented

l-facet is described by an $(l+1)$-tuple of vertices, whose order implies an (internal) orientation of the facet. The orientation of an l-facet induces an orientation of the $(l-1)$-facets contained in its boundary. For a tetrahedron $T = (\mathbf{a}_0, \mathbf{a}_1, \mathbf{a}_2, \mathbf{a}_3) \in \mathcal{S}_3(\Omega_h)$, the boundary faces carrying the induced orientation are given by

$$\partial T \quad \Rightarrow \quad \{(\mathbf{a}_2, \mathbf{a}_1, \mathbf{a}_0), (\mathbf{a}_0, \mathbf{a}_1, \mathbf{a}_3), (\mathbf{a}_3, \mathbf{a}_2, \mathbf{a}_0), (\mathbf{a}_1, \mathbf{a}_2, \mathbf{a}_3)\}.$$

The edges of an oriented face $F = (\mathbf{a}_0, \mathbf{a}_1, \mathbf{a}_2)$ turn out to be

$$\partial F \quad \Rightarrow \quad \{(\mathbf{a}_0, \mathbf{a}_1), (\mathbf{a}_2, \mathbf{a}_0), (\mathbf{a}_1, \mathbf{a}_2)\}.$$

The relative orientation of an $l-1$-facet f in the boundary of an l-facet F is $(-1)^t$, where t is the number of transpositions of vertices it takes to convert the induced ordering of vertices into that fixing the interior orientation of f. The way to define orientation for facets of tetrahedral triangulations is neatly matched by data structures that can be used for their representation: if facet objects possess ordered lists referring to their vertices, an orientation is automatically implied.

Yet, for triangulations containing more general, say pyramidal and hexahedral, cells, it is awkward to rely on interior orientation alone. It is more convenient to rely on the external orientation of faces, which amounts to prescribing a crossing direction. Fixing an orientation of $A(\mathbb{R}^3)$, external and internal orientation are dual to each other (the 'corkscrew rule'). For edges, the usual internal orientation, their direction, is kept. Externally orienting the cells simply means declaring what is 'inside' and 'outside'. Then the induced external orientation of a face, contained in the boundary of a cell, is prescribed by crossing it from the interior to the exterior of the cell. Data structures representing this concept of orientation should supply references from faces to the adjacent cells.

We first look at precursors of discrete differential forms that already display an amazing wealth of structure.

Definition 4. An l-cochain $\vec{\omega}$, $0 \le l \le 3$,[5] on a triangulation Ω_h is a mapping $\mathcal{S}_l(\Omega_h) \mapsto \mathbf{C}$. The vector space of l-cochains on Ω_h will be denoted by $\mathcal{C}^l(\Omega_h)$. The value assigned by $\vec{\omega}$ to a collection of l-faces is computed as the sum of the values assigned to the individual l-faces.

This definition falls short of characterizing 'genuine cochains' studied in discrete cohomology theory on CW complexes. What is missing is the complementary concept of chains. However, Definition 4 suffices to convey the main ideas.

[5] We use the arrow notation to mark cochains, to emphasize that they can be described by 'coefficient vectors' in $\mathbb{C}^N$. No confusion should arise because bold typeface is used for vector-valued quantities.

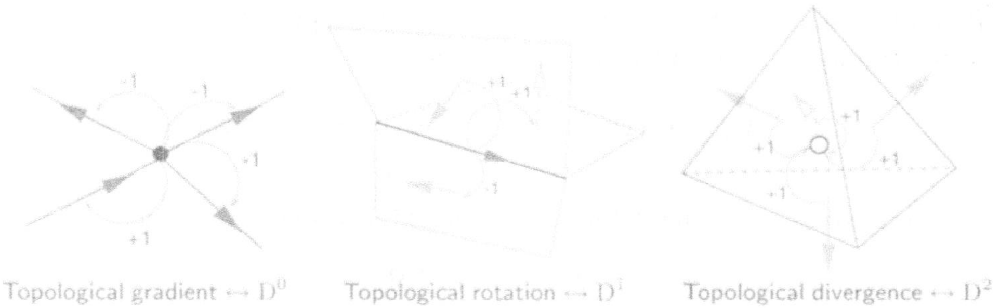

Figure 3.1. Stencils for exterior derivative on cochains

It is clear that $\dim \mathcal{C}^l(\Omega_h) = |\mathcal{S}_l(\Omega_h)|$[6] and that, after ordering the l-faces of Ω_h, we can identify $\mathcal{C}^l(\Omega_h)$ with $\mathbb{C}^{N_l}$, $N_l := |\mathcal{S}_l(\Omega_h)|$. Such an identification will be taken for granted.

It is not difficult to devise cochain counterparts of all operators that we introduced for integral forms. First, a trace operator on l-cochains can be declared, in a natural way, as a mapping isolating the subset of coefficients belonging to l-facets of Ω_h contained in $\partial\Omega$. This gives a meaning to boundary conditions for cochains.

Thanks to the special properties of triangulations, an exterior derivative $\boldsymbol{d}_h : \mathcal{C}^l(\Omega_h) \mapsto \mathcal{C}^{l+1}(\Omega_h)$ can be defined by (2.7): for $\vec{\omega} \in \mathcal{C}^l(\Omega_h)$, its exterior derivative $\boldsymbol{d}_h\vec{\omega}$ assigns to each $F \in \mathcal{S}_{l+1}(\Omega_h)$ the sum of values $\vec{\omega}(f)$, $f \in \mathcal{S}_l(\Omega_h)$, $f \subset \partial F$, weighted with the relative orientations (Gross and Kotiuga 2001, Section 2.1). This exterior derivative of cochains is a linear operator and the associated matrix $\mathrm{D}^l \in \mathbb{C}^{N_{l+1},N_l}$ is the so-called *incidence matrix* of l-facets and $(l+1)$-facets of the triangulation, a sparse matrix with entries $\in \{-1,0,1\}$ determined by adjacency relations and relative orientations (Tonti 2001, Section 4.2). The difference stencils representing D^0, D^1, and D^2 are depicted in Figure 3.1.

I point out that the matrices D^l only depend on the topology of the triangulation. Also, $\mathrm{D}^{l+1}\mathrm{D}^l = 0$ holds as well as a cochain counterpart of the exact sequence property.

Theorem 3.1. (DeRham theorem for cochains) There are finite-dimensional subspaces $\mathcal{HC}^l(\Omega_h) \subset \mathcal{C}^l(\Omega_h)$, $\dim \mathcal{HC}^l(\Omega_h) = l$th Betti number of Ω, such that, for $\vec{\omega} \in \mathcal{C}^l(\Omega_h)$, we have

$$\mathrm{D}^l\vec{\omega} = 0 \quad \Leftrightarrow \quad \exists \vec{\eta} \in \mathcal{C}^{l-1}(\Omega_h),\ \vec{\gamma} \in \mathcal{HC}^l(\Omega_h) \quad \text{satisfying } \vec{\omega} = \mathrm{D}^{l-1}\vec{\eta} + \vec{\gamma}.$$

Formally, we can state the topological laws of electromagnetism in the calculus of cochains as the following systems of linear equations:

$$\mathrm{D}^1\vec{\mathbf{e}} = -i\omega\vec{\mathbf{b}}, \quad \mathrm{D}^1\vec{\mathbf{h}} = i\omega\vec{\mathbf{d}} + \vec{\mathbf{j}}. \tag{3.1}$$

[6] By $|X|$ we denote the cardinality of a finite set X.

It can be connected to the topological Maxwell's equations (2.9) for integral forms through the so-called deRham maps

$$\mathsf{I}_l : \mathcal{F}^l(\Omega) \mapsto \mathcal{C}^l(\Omega_h), \quad \mathsf{I}_l(\omega)(F) = \int_F \omega \quad \forall F \in \mathcal{S}_l(\Omega_h). \tag{3.2}$$

It is immediate from the definition of the exterior derivatives that

$$\mathrm{D}^l \circ \mathsf{I}_l = \mathsf{I}_{l+1} \circ \boldsymbol{d}. \tag{3.3}$$

This has the important consequence that integral forms that are solutions of Maxwell's equations satisfy

$$\mathrm{D}^1 \mathsf{I}_1 \mathbf{e} = -i\omega \mathsf{I}_2 \mathbf{b}, \quad \mathrm{D}^1 \mathsf{I}_1 \mathbf{h} = i\omega \mathsf{I}_2 \mathbf{d} + \mathsf{I}_2 \mathbf{j}. \tag{3.4}$$

Keep in mind that, for $l = 1$, the DeRham map boils down to point evaluation. Also remember the notion of consistency error of finite difference methods. Thus, (3.4) means that the discrete topological laws (3.1) are *consistent* with their continuous counterparts (2.9).

Appealing to Theorem 3.1, we can also introduce cochain potentials $\vec{\mathbf{a}} \in \mathcal{C}^1(\Omega_h)$, $\vec{v} \in \mathcal{C}^0(\Omega_h)$, such that

$$\vec{\mathbf{b}} = \mathrm{D}^1 \vec{\mathbf{a}}, \quad \vec{\mathbf{e}} = -\mathrm{D}^0 \vec{v} - i\omega \vec{\mathbf{a}}. \tag{3.5}$$

Bibliographical notes

Cochains and their application to electromagnetism are discussed, for instance, in Bossavit (1998*a*, Section 5.3), Gross and Kotiuga (2001), Tarhasaari and Kettunen (2001) and Teixeira (2001, Section 2).

3.2. Whitney forms

It was the material laws that forced us to switch from integral forms to (local) differential forms. The material laws also fail to fit into the calculus of cochains. In order to accommodate them we have to use *discrete differential forms*, defined as true differential forms almost everywhere. They will allow the kind of local evaluations required to compute energies. However, we saw that cochains perfectly capture the topological laws. Therefore we opt for discrete differential forms that, sloppily speaking, extend cochains into the interior of cells and provide a model isomorphic to the calculus of cochains in terms of algebraic properties.

Formally, we seek *bijective* linear mappings W^l, the so-called *Whitney maps* (Tarhasaari, Kettunen and Bossavit 1999), from $\mathcal{C}^l(\Omega_h)$ into a space of differential l-forms that are defined almost everywhere on Ω and make sense as integral forms on Ω.

Definition 5. The range space $\mathsf{W}^l(\mathcal{C}^l(\Omega_h))$, $0 \le l \le 3$ is called the space of Whitney l-forms $\mathcal{W}_0^l(\Omega_h)$ on the triangulation Ω_h.

Rather rigorous requirements have to be met by meaningful W^l.

(1) The associated form has to be a true extension of the cochain, in the sense that

$$\mathsf{I}_l \circ \mathsf{W}^l = \mathrm{Id} \quad \Leftrightarrow \quad \int_F \mathsf{W}^l \vec{\omega} = \vec{\omega}(F) \quad \forall F \in \mathcal{S}_l(\Omega_h),\ \vec{\omega} \in \mathcal{C}^l(\Omega_h). \quad (3.6)$$

(2) The exterior derivatives of cochains and related Whitney forms must be linked by the commuting diagram

$$\boldsymbol{d} \circ \mathsf{W}^l = \mathsf{W}^{l+1} \circ \mathrm{D}^l \quad \text{on } \mathcal{C}^l(\Omega_h). \quad (3.7)$$

(3) We demand strict locality: if all cochain coefficients of $\vec{\omega}$ associated with the l-facets belonging to a cell $T \in \mathcal{S}_3(\Omega_h)$ vanish, then $\mathsf{W}^l \vec{\omega}_{|T} = 0$.

(4) The vector proxies of Whitney l-forms should be simple, that is, piecewise polynomial, on Ω_h.

We first study the mappings W^l for tetrahedral triangulations, the most important class of finite element meshes. We start with a 'local extension' on a single tetrahedron $T \in \mathcal{S}_3(\Omega_h)$ with vertices $\mathbf{a}_0, \mathbf{a}_1, \mathbf{a}_2, \mathbf{a}_3$. Cochain coefficients for all its facets are given. We take the cue from the case of 0-forms, for which we know very well how to interpolate point values $\vec{\phi}(\mathbf{a}_i)$, that is, the coefficients of a 0-cochain $\vec{\phi}$, prescribed in the vertices of T. Linear interpolation based on the barycentric coordinate functions $\lambda_0, \ldots, \lambda_3$ gives

$$\phi(\mathbf{x}) = \sum_{i=0}^{3} \vec{\phi}(\mathbf{a}_i)\lambda_i(\mathbf{x}) \quad \forall \mathbf{x} \in \bar{T}.$$

In fact, this definition could be motivated by the identity $\mathbf{x} = \sum_{i=0}^{3} \mathbf{a}_i \lambda_i(\mathbf{x})$, that is, by representing any point in T as a weighted combination of vertices. For 1-forms the role of vertices and points is played by edges and line segments. Any oriented line $(\mathbf{x}, \mathbf{y})$, with $\mathbf{x}, \mathbf{y} \in T$, $\mathbf{x} = \sum_i \lambda_i(\mathbf{x})\mathbf{a}_i$, $\mathbf{y} = \sum_i \lambda_i(\mathbf{y})\mathbf{a}_i$, can be represented as a 'weighted sum of edges of T':

$$\begin{aligned}
(\mathbf{x}, \mathbf{y}) &= \{t\mathbf{x} + (1-t)\mathbf{y}\ ;\ 0 \le t \le 1\} \\
&= \left\{ \sum_i \Big(t\lambda_i(\mathbf{x}) + (1-t)\lambda_i(\mathbf{y}) \Big) \mathbf{a}_i\ ;\ 0 \le t \le 1 \right\} \\
&= \left\{ \sum_i \Big(t \sum_j \lambda_j(\mathbf{y})\lambda_i(\mathbf{x}) + (1-t) \sum_j \lambda_j(\mathbf{x})\lambda_i(\mathbf{y}) \Big) \mathbf{a}_i\ ;\ 0 \le t \le 1 \right\} \\
&= \left\{ \sum_i \sum_j \lambda_i(\mathbf{x})\lambda_j(\mathbf{y})((t\mathbf{a}_i + (1-t)\mathbf{a}_j)\ ;\ 0 \le t \le 1 \right\}.
\end{aligned}$$

Hence, taking into account orientation, we require that the interpolating differential 1-form $\mathsf{W}^1\vec{\omega}$ satisfies

$$\int_{(\mathbf{x},\mathbf{y})} \mathsf{W}^1_{|T}\vec{\omega} := \sum_i \sum_j \lambda_i(\mathbf{x})\lambda_j(\mathbf{y})\vec{\omega}_{(i,j)} = \sum_{i<j} \left(\lambda_i(\mathbf{x})\lambda_j(\mathbf{y}) - \lambda_i(\mathbf{y})\lambda_j(\mathbf{x})\right)\vec{\omega}_{(i,j)}.$$

Here, $\vec{\omega}_{(i,j)}$ is the value that the 1-cochain $\vec{\omega}$ assigns to the oriented edge $(\mathbf{a}_i, \mathbf{a}_j)$. By construction, (3.7) is satisfied. So far, the formula fixes ω as an integral 1-form, but is it even a (local) differential 1-form? It is easy to see that a differential l-form on T can be recovered from integral values through

$$\omega(\mathbf{x})(\mathbf{v}_1, \dots, \mathbf{v}_l) := l! \lim_{t\to 0} \int_{\Sigma_t} \omega, \tag{3.8}$$

where $\Sigma_t \subset T$ is the l-simplex $(\mathbf{x}, \mathbf{x} + t\mathbf{v}_1, \dots, \mathbf{x} + t\mathbf{v}_l)$, $\mathbf{v}_1, \dots, \mathbf{v}_l \in \mathbb{R}^3$. Using the local definition of the exterior derivative, we find

$$\begin{aligned}(\mathsf{W}^1_{|T}\vec{\omega})(\mathbf{x})(\mathbf{v}) &= \lim_{t\to 0} \sum_{i<j} \left(\lambda_i(\mathbf{x})\tfrac{\lambda_j(\mathbf{x}+t\mathbf{v})-\lambda_j(\mathbf{x})}{t} - \lambda_j(\mathbf{x})\tfrac{\lambda_i(\mathbf{x}+t\mathbf{v})-\lambda_i(\mathbf{x})}{t}\right)\vec{\omega}_{(i,j)} \\ &= \sum_{i<j} \left(\lambda_i(\mathbf{x})d\lambda_j(\mathbf{x})(\mathbf{v}) - \lambda_j(\mathbf{x})d\lambda_i(\mathbf{x})(\mathbf{v})\right)\vec{\omega}_{(i,j)}.\end{aligned}$$

Finally, the mapping W^1 is built by combining the local interpolation operators $\mathsf{W}^1_{|T}$ for all cells $T \in \mathcal{S}_3(\Omega_h)$.

Observe that the values that the interpolant $\mathsf{W}^1_{|T}\vec{\omega}$ assigns to line segments contained in a face of T only depend on the cochain coefficients associated with edges of that face. A similar, even simpler statement can be made for edges: $\mathsf{W}^1_{|T}\vec{\omega}$ evaluated for parts of an edge $(\mathbf{a}_i, \mathbf{a}_j)$ only depends on $\vec{\omega}_{(i,j)}$. As a consequence, the above interpolation procedure applied to each tetrahedron of Ω_h will create locally defined smooth differential forms that satisfy the patch condition (2.20) for interelement faces. This confirms that W^1 really maps into $\mathcal{F}^1(\Omega)$.

This procedure can even be generalized to the 'local interpolation formula' on an n-simplex $T \subset A(\mathbb{R}^n)$, $n \in \mathbb{N}$:

$$(\mathsf{W}^l_{|T}\omega)(\mathbf{x}) = \sum_I \sum_{j=0}^{l} (-1)^j \underbrace{\left(\lambda_{i_j} \bigwedge_{k=0,\, k\neq j}^{l} d\lambda_{i_k}\right)}_{\mathbf{b}_I} \cdot \vec{\omega}_I \tag{3.9}$$

where $I = (i_0, \dots, i_l)$, $0 \le l \le n$, runs through all $(l+1)$-subsets of $\{0, \dots, n\}$ and the ordering is induced by the orientation of the corresponding l-facet $([\mathbf{a}_{i_0}, \dots, \mathbf{a}_{i_l})$. The symbol $\vec{\omega}_I$ stands for the cochain coefficient associated with that facet. The $\mathbf{b}_I$ are called local *basis forms*. Their Euclidean vector

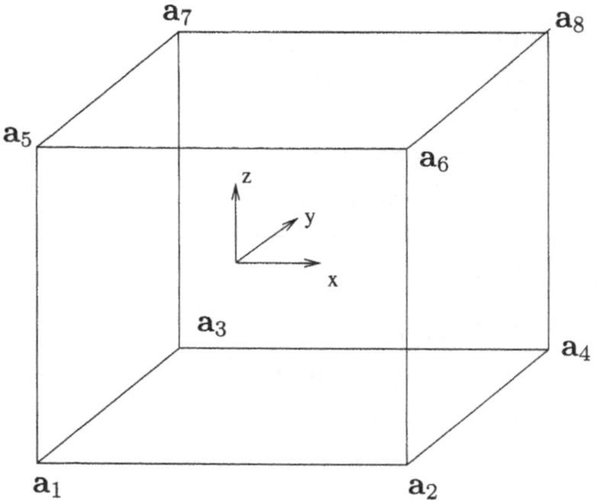

$$\begin{aligned}
\Upsilon_1\mathbf{b}_{(1,2)} &= (1-y)(1-z)\cdot\mathbf{e}_x\\
\Upsilon_1\mathbf{b}_{(3,4)} &= y(1-z)\cdot\mathbf{e}_x\\
\Upsilon_1\mathbf{b}_{(5,6)} &= (1-y)z\cdot\mathbf{e}_x\\
\Upsilon_1\mathbf{b}_{(7,8)} &= yz\cdot\mathbf{e}_x\\
\Upsilon_1\mathbf{b}_{(1,3)} &= (1-x)(1-z)\cdot\mathbf{e}_y\\
\Upsilon_1\mathbf{b}_{(2,4)} &= x(1-z)\cdot\mathbf{e}_y\\
\Upsilon_1\mathbf{b}_{(5,7)} &= (1-x)z\cdot\mathbf{e}_y\\
\Upsilon_1\mathbf{b}_{(6,8)} &= xz\cdot\mathbf{e}_y\\
\Upsilon_1\mathbf{b}_{(1,5)} &= (1-y)(1-x)\cdot\mathbf{e}_x\\
\Upsilon_1\mathbf{b}_{(2,6)} &= (1-y)x\cdot\mathbf{e}_x\\
\Upsilon_1\mathbf{b}_{(3,7)} &= y(1-x)\cdot\mathbf{e}_x\\
\Upsilon_1\mathbf{b}_{(4,8)} &= yx\cdot\mathbf{e}_x
\end{aligned}$$

Figure 3.2. Vector proxies of local basis forms for Whitney 1-forms on a brick, $\mathbf{e}_x = (1,0,0)^T$, $\mathbf{e}_y = (0,1,0)^T$, $\mathbf{e}_z = (0,0,1)^T$

proxies in three dimensions read

$$\begin{aligned}
\Upsilon_0\mathbf{b}_{(i)} &= \lambda_i,\\
\Upsilon_1\mathbf{b}_{(i,j)} &= \lambda_i\,\mathbf{grad}\,\lambda_j - \lambda_j\,\mathbf{grad}\,\lambda_i,\\
\Upsilon_2\mathbf{b}_{(i,j,k)} &= \lambda_i\,\mathbf{grad}\,\lambda_j \times \mathbf{grad}\,\lambda_k\\
&\qquad + \lambda_j\,\mathbf{grad}\,\lambda_k \times \mathbf{grad}\,\lambda_i + \lambda_k\,\mathbf{grad}\,\lambda_i \times \mathbf{grad}\,\lambda_j,\\
\Upsilon_3\mathbf{b}_{(0,1,2,3)} &= 1/\mathrm{Vol}(T).
\end{aligned}$$

If T is a brick, the extension of cochains can be done, too. Then the basis functions arise from a simple tensor-product-based construction. For $l = 1$ they are given in Figure 3.2.

Finally we have come up with objects that fit the conventional notion of a finite element (Ciarlet 1978, Chapter 3) centring on the concept of local spaces and degrees of freedom (d.o.f.). In the case of Whitney forms, the local spaces $\mathcal{W}_0^l(T)$, $T \in \mathcal{S}_3(\Omega_h)$ are supplied by the ranges of the local interpolation operators $\mathsf{W}^l_{|T}$. The integrals over l-facets are linear functionals on $\mathsf{W}^l_{|T}$ and serve as degrees of freedom. They are dual to the basis forms specified above. The expressions for Euclidean vector proxies are given in Table 3.1. By now it should have become clear why the Whitney 1-forms in three dimensions are usually called *edge elements*. This terms alludes to the 'location' of the degrees of freedom.

We stress that Whitney forms are *affine equivalent* in the following sense: If $\mathbf{\Phi} : \widehat{T} \mapsto T$ is the unique affine mapping between two tetrahedra, then

$$\mathcal{W}_0^l(\widehat{T}) = \mathbf{\Phi}^*(\mathcal{W}_0^l(T)), \qquad (3.10)$$

Table 3.1. Local spaces and degrees of freedom for Euclidean vector proxies of tetrahedral Whitney forms on a tetrahedron T with vertices $\mathbf{a}_0, \dots, \mathbf{a}_3$. Vertex indices have to be distinct

Local spaces	Local d.o.f.
$\mathcal{W}_0^0(T) = \{\mathbf{x} \mapsto \mathbf{a}\cdot\mathbf{x} + \beta,\ \mathbf{a} \in \mathbb{R}^3, \beta \in \mathbb{R}\}$	$u \mapsto u(\mathbf{a}_i)$
$\mathcal{W}_0^1(T) = \{\mathbf{x} \mapsto \mathbf{a}\times\mathbf{x} + \mathbf{b},\ \mathbf{a},\mathbf{b} \in \mathbb{R}^3\}$	$\mathbf{u} \mapsto \int_{(\mathbf{a}_i,\mathbf{a}_j)} \mathbf{u}\cdot \mathrm{d}\mathbf{s}$
$\mathcal{W}_0^2(T) = \{\mathbf{x} \mapsto \alpha\mathbf{x} + \mathbf{b},\ \alpha \in \mathbb{R}, \mathbf{b} \in \mathbb{R}^3\}$	$\mathbf{u} \mapsto \int_{(\mathbf{a}_i,\mathbf{a}_j,\mathbf{a}_k)} \mathbf{u}\cdot\mathbf{n}\,\mathrm{d}S$
$\mathcal{W}_0^3(T) = \{\mathbf{x} \mapsto \alpha,\ \alpha \in \mathbb{R}\}$	$u \mapsto \int_T u\,\mathrm{d}\mathbf{x}$

and, as is clear from (2.12), the degrees of freedom retain their value under any pullback. We can even use (3.10) to define the local spaces for any diffeomorphic image of a tetrahedron. Thus we get, for instance, discrete differential forms on cells with curved boundaries, that is, parametric versions of Whitney forms. We could also have obtained them by observing that the interpolation procedure works with any set of functions $\lambda_0, \dots, \lambda_3$ such that $\lambda_i(\mathbf{a}_i) = 1$ and λ_i, $i = 0,1,2,3$, vanishes on the face opposite to $\mathbf{a}_i$.

Combining the local interpolations according to (3.9) for all cells, we get the mappings W^l for $0 \le l \le 3$. This procedure ensures locality. The vector proxies are linear functions/vector-fields in each tetrahedron, as is clear from Table 3.1. It is as obvious that constant differential forms are contained in the local spaces $\mathcal{W}_0^l(T)$. Slightly more technical effort confirms that (3.7) is satisfied. Since the patch condition holds, Lemma 2.3 teaches that Whitney l-forms, $l = 1, 2$, furnish conforming finite elements.

Theorem 3.2. We have $\mathcal{W}_0^1(\Omega_h) \subset \boldsymbol{H}(\mathbf{curl};\Omega)$ and $\mathcal{W}_0^2(\Omega_h) \subset \boldsymbol{H}(\mathrm{div};\Omega)$.

Combining the extension of cochain with the deRham mapping yields the projections

$$\Pi^l := \mathsf{W}^l \circ \mathsf{I}_l : \mathcal{F}^l(\Omega) \mapsto \mathcal{W}_0^l(\Omega_h) \subset \mathcal{F}^l(\Omega), \tag{3.11}$$

called the *local interpolation operators* to hint at their connection with the usual interpolation operators for finite elements that are based on degrees of freedom. This connection is illustrated by

$$\int_F \omega - \Pi^l\omega = 0 \quad \forall \omega \in \mathcal{F}^l(\Omega),\ F \in \mathcal{S}_l(\Omega_h).$$

When researching properties of discrete differential forms, there is a useful guiding principle supported by the fact that conventional Lagrangian finite elements are discrete 0-forms:

Techniques that can be successfully applied to $H^1(\Omega)$-conforming Lagrangian finite elements can usually be adapted for other discrete differential forms, as well. Results obtained for Lagrangian finite elements are often matched by analogous results for other discrete differential forms.

The affine equivalence expressed in (3.10) is just one example of a property bearing out this guideline.

Yet there are aspects of discrete differential forms with no correspondents in the theory of Lagrangian finite elements. They chiefly have to do with relationships of discrete differential forms of different degree: from the properties of W^l we can readily deduce the *commuting diagram property*

$$\boldsymbol{d} \circ \Pi^l = \Pi^{l+1} \circ \boldsymbol{d}. \tag{3.12}$$

Therefore Theorem 3.1 will instantly carry over to Whitney forms.

Corollary 3.3. (DeRham theorem for Whitney forms) For each $1 \leq l \leq n$ there are discrete cohomology spaces $\mathcal{H}^l_h(\Omega_h) \subset \mathcal{W}^l_0(\Omega_h)$, whose dimension agrees with the lth Betti number of Ω, such that, for any $\omega_h \in \mathcal{W}^l_0(\Omega_h)$, we have

$$\boldsymbol{d}\omega_h = 0 \quad \Leftrightarrow \quad \exists \eta_h \in \mathcal{W}^{l-1}_0(\Omega_h),\ \gamma_h \in \mathcal{H}^l_h(\Omega_h) \quad \text{satisfying } \omega_h = \boldsymbol{d}\eta_h + \gamma_h.$$

Provided that Ω is homeomorphic to a ball, we can summarize our findings in the commuting diagram

$$\begin{array}{ccccccccc}
\mathcal{F}^0(\Omega) & \xrightarrow{\boldsymbol{d}} & \mathcal{F}^1(\Omega) & \xrightarrow{\boldsymbol{d}} & \mathcal{F}^2(\Omega) & \xrightarrow{\boldsymbol{d}} & \mathcal{F}^3(\Omega) & \xrightarrow{\mathrm{Id}} & \{0\} \\
\Pi^0 \big\downarrow & & \Pi^1 \big\downarrow & & \Pi^2 \big\downarrow & & \Pi^3 \big\downarrow & & \\
\mathcal{W}^0_0(\Omega_h) & \xrightarrow{\boldsymbol{d}} & \mathcal{W}^1_0(\Omega_h) & \xrightarrow{\boldsymbol{d}} & \mathcal{W}^2_0(\Omega_h) & \xrightarrow{\boldsymbol{d}} & \mathcal{W}^3_0(\Omega_h) & \xrightarrow{\mathrm{Id}} & \{0\},
\end{array}$$

for which all the vertical sequences are exact.

Corollary 3.3 asserts the existence of *discrete potentials* in another space of Whitney forms. This is a really exceptional property. Imagine discrete 1-forms based on a continuous piecewise linear approximation of the components of vector proxies. If those have zero **curl**, there will exist a scalar potential on a domain with trivial topology. This scalar potential can be chosen to be piecewise polynomial of degree two and will feature C^1-continuity. Remember that C^1-finite elements require sophisticated constructions that have to rely on polynomials of degree greater than 2. So we understand that the space of discrete potentials in this case cannot be a finite element space, because it will not possess a localized basis. Consequently, these discrete potentials cannot be handled efficiently in computations. Laconically, Whitney forms provide discrete potentials that are computationally available.

Remark 5. Discrete cohomology spaces can fill the gap between irrotational vector-fields and gradients in the sense that

$$\boldsymbol{H}(\mathbf{curl}\,0;\Omega) = \mathbf{grad}\,H^1(\Omega) \oplus \boldsymbol{\mathcal{H}}_h^1(\Omega_h). \qquad \triangle$$

Remark 6. One finds that the divergence of vector proxies of Whitney 1-forms vanishes locally on each tetrahedron. However, this just happens by accident and is irrelevant, because taking the (strong) divergence of the vector proxy of a discrete 1-form is not quite meaningful. △

Remark 7. In electrodynamics the *Poynting vector* is a 2-form $\mathbf{s}$ defined by $\mathbf{s} := \mathbf{e} \wedge \mathbf{h}$. It describes the flux of electromagnetic energy. The definition of $\mathbf{s}$ hinges on a local interpretation of the forms. In other words, it takes us beyond cochain calculus, and Whitney forms are called for to give it a meaning in a discrete setting. Given discrete fields $\mathbf{e}_h$ and $\mathbf{h}_h$ in $\mathcal{W}_0^1(\Omega_h)$ we can easily compute $\mathbf{e}_h \wedge \mathbf{h}_h$ as a locally polynomial 2-form. However, it is not a Whitney 2-form, because it features quadratic vector proxies.

Therefore, we may define the discrete Poynting vector as $\mathbf{s}_h = \Pi^2(\mathbf{e}_h \wedge \mathbf{h}_h)$. In order to compute the face fluxes of $\mathbf{s}_h$ efficiently, it is essential to recall that the pullback commutes with the $\wedge$-product. We learn that, for any face $F \in \mathcal{S}_2(\Omega_h)$ with edges e_1, e_2, e_3, we can obtain $\int_S \mathbf{s}_h$ as a weighted combination of the d.o.f. $\vec{\mathbf{e}}_h(e_i)$, $\vec{\mathbf{h}}(e_i)$ with weights *independent* of the shape of S. Those can be computed for a single face, and it turns out that

$$\vec{\mathbf{s}}_h(S) = \frac{1}{6}\begin{pmatrix} \pm\vec{\mathbf{h}}_h(e_1) \\ \pm\vec{\mathbf{h}}_h(e_2) \\ \pm\vec{\mathbf{h}}_h(e_3) \end{pmatrix}^H \begin{pmatrix} 0 & 1 & -1 \\ -1 & 0 & 1 \\ 1 & -1 & 0 \end{pmatrix} \begin{pmatrix} \pm\vec{\mathbf{e}}_h(e_1) \\ \pm\vec{\mathbf{e}}_h(e_2) \\ \pm\vec{\mathbf{e}}_h(e_3) \end{pmatrix}.$$

The signs have to be chosen in accordance with the relative orientations of the edges with respect to the face. △

Bibliographical notes

Whitney forms were originally introduced by Whitney (1957) as a tool in algebraic topology. As such they have been used for a long time (Dodziuk 1976, Müller 1978). Their use as finite elements in computational electromagnetism was pioneered by A. Bossavit (1988*a*, 1988*c*, 1992, 1988*b*). A thorough discussion of elementary properties of Whitney forms is given in Bossavit (1998*a*, Chapter 5).

Independently of existing constructions in differential geometry, Whitney forms were rediscovered as mixed finite elements of lowest polynomial order (Raviart and Thomas 1977, Nédélec 1980). Their construction by interpolation is presented in Bossavit (2001, Section 7) and was used in Gradinaru and Hiptmair (1999) to construct Whitney forms on pyramids.

3.3. Discrete Hodge operators

As $\boldsymbol{H}(\mathbf{curl};\Omega)$-conforming finite elements, Whitney 1-forms immediately lend themselves to a straightforward Galerkin discretization of the variational forms (2.28) and (2.29) of Maxwell's equations on bounded domains $\Omega \subset A(\mathbb{R}^3)$. Cochains have just served as a scaffolding to build the Whitney forms. This will be the favoured view in this article.

Yet, seen from a different angle, Whitney forms might be relegated to a mere tool to incorporate the material laws into the cochain framework. This point of view encompasses a much wider range of discretization schemes beyond Galerkin methods, in particular approaches known as finite volume/ finite integration schemes.

For the sake of simplicity we will consider only fields/cochains with zero trace on $\partial\Omega$. To begin with, we can expand the fields into a sum of basis forms, plug this into the variational forms (2.24)–(2.27) of the material laws, also using Whitney forms as test fields. We get the following matrix equations for the expansion coefficients:

$$(2.24) \Rightarrow \widetilde{\mathrm{K}}_2^1\vec{\mathbf{d}} = \mathrm{M}_\epsilon^1\vec{\mathbf{e}}, \qquad (2.26) \Rightarrow \mathrm{K}_2^1\vec{\mathbf{b}} = \widetilde{\mathrm{M}}_\mu^1\vec{\mathbf{h}},$$
$$(2.25) \Rightarrow \widetilde{\mathrm{K}}_1^2\vec{\mathbf{h}} = \mathrm{M}_{1/\mu}^2\vec{\mathbf{b}}, \qquad (2.27) \Rightarrow \mathrm{K}_1^2\vec{\mathbf{e}} = \widetilde{\mathrm{M}}_{1/\epsilon}^2\vec{\mathbf{d}}.$$

The 'mass matrices' M_ϵ^1, $\mathrm{M}_{1/\mu}^2$, $\widetilde{\mathrm{M}}_\mu^1$, $\widetilde{\mathrm{M}}_{1/\epsilon}^2$, being related to energies, have to be real symmetric positive definite. The 'coupling matrices' $\widetilde{\mathrm{K}}_2^1$, $\widetilde{\mathrm{K}}_1^2$, K_2^1, and K_1^2 introduce the exterior products of cochains. They are not regular, nor even square, in general. From the exterior product they will inherit the properties

$$\mathrm{K}_1^2 = (\widetilde{\mathrm{K}}_2^1)^T, \quad \widetilde{\mathrm{K}}_1^2 = (\mathrm{K}_2^1)^T, \tag{3.13}$$

and, more importantly, the 'integration by parts formula'

$$(\mathrm{D}^1)^T\widetilde{\mathrm{K}}_1^2 = \widetilde{\mathrm{K}}_2^1\widetilde{\mathrm{D}}^1 \quad \Leftrightarrow \quad (\widetilde{\mathrm{D}}^1)^T\mathrm{K}_1^2 = \mathrm{K}_2^1\mathrm{D}^1. \tag{3.14}$$

As far as the topological laws are concerned, the fields $\mathbf{e}, \mathbf{b}$ and $\mathbf{h}, \mathbf{d}$ have nothing to do with each other. Recall that it was only the exterior derivative in the topological laws that forced us to use the same triangulation for cochains of different degree. Hence, $\mathbf{e}, \mathbf{b}$ (primal fields) and $\mathbf{h}, \mathbf{d}$ (dual fields) may well be discretized on *unrelated* triangulations Ω_h and $\widetilde{\Omega}_h$, called primal and dual hereafter. No problems are encountered in getting the material laws from their weak version by the Galerkin procedure. This observation explains the ˜ tag for matrices above. It distinguishes matrices acting on cochains defined on the triangulations for dual fields.

The above discrete material laws provide *discrete Hodge operators* for cochains. The idea is that a discretization of Maxwell's equations can be stated in terms of cochains assuming only discrete material laws with positive definite mass matrices and coupling matrices satisfying the algebraic properties (3.13) and (3.14).

Apart from the Galerkin approach there is another avenue to meaningful discrete Hodge operators. It starts with the consideration that the coupling matrices should be regular, reflecting the non-degeneracy of the exterior product. One may even demand that coupling matrices coincide with identity matrices. Then (3.14) involves

$$(\mathrm{D}^1)^T = \widetilde{\mathrm{D}}^1, \quad (\widetilde{\mathrm{D}}^1)^T = \mathrm{D}^1.$$

As the matrices D^1 and $\widetilde{\mathrm{D}}^1$ are the edge-face incidence matrices of the primal and dual triangulations, we can achieve this by choosing $\widetilde{\Omega}_h$ as a triangulation dual to Ω_h (Gross and Kotiuga 2001, Section 5). This means there is a one-to-one correspondence of $\mathcal{S}_l(\Omega_h)$ and $\widetilde{\mathcal{T}}_h^{3-l}$. Prominent examples are geometrically dual triangulations, for which a 2D example is depicted in Figure 3.3. Here, the bijections between primal and dual facets are established by geometric intersection.

Special dual triangulations are orthogonal dual triangulations, for which primal edges are perpendicular to dual faces, as well as dual edges to primal faces Tonti (2001, Section 4), Marrone (2001, Section 3). In this case, provided we have to deal with material tensors that are multiples of the identity matrix, diagonal mass matrices can be chosen. To illustrate this consider the material law $\mathbf{d} = \epsilon\mathbf{e}$ to be translated into $\vec{\mathbf{d}} = \mathrm{M}_\epsilon^1\vec{\mathbf{e}}$. Pick a single primal edge e bearing a coefficient for $\vec{\mathbf{e}}$ and the associated dual face $\widetilde{F}$, for which we seek the coefficient for $\vec{\mathbf{d}}$. The situation is sketched

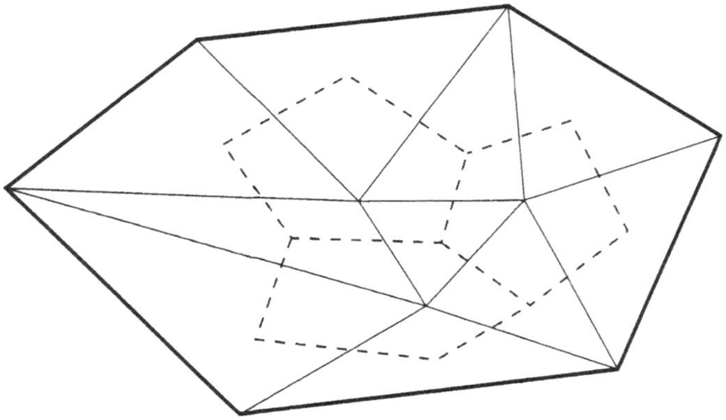

Figure 3.3. Triangulation (solid lines) and dual triangulation (dashed lines) in two dimensions

in Figure 3.4. In terms of vector proxies we have $\mathbf{d} = \boldsymbol{\epsilon}\mathbf{e}$ and, therefore, because the face normal $\mathbf{n}$ runs parallel to e,

$$\vec{\mathbf{d}}(\widetilde{F}) = \int_{\widetilde{F}} \mathbf{d} \cdot \mathbf{n}\, \mathrm{d}S \approx \mathrm{Vol}(\widetilde{F})\mathbf{d}(\mathbf{p}) \cdot \mathbf{n}(\mathbf{p}) = \mathrm{Vol}(\widetilde{F})\boldsymbol{\epsilon}(\mathbf{p})\mathbf{e}(\mathbf{p}) \cdot \mathbf{n}$$

$$= \boldsymbol{\epsilon}(\mathbf{p})\frac{\mathrm{Vol}(\widetilde{F})}{\mathrm{Vol}(e)}\mathbf{e}(\mathbf{p}) \cdot e \approx \boldsymbol{\epsilon}(\mathbf{p})\frac{\mathrm{Vol}(\widetilde{F})}{\mathrm{Vol}(e)} \int_e \mathbf{e} \cdot \mathrm{d}\mathbf{s} = \boldsymbol{\epsilon}(\mathbf{p})\frac{\mathrm{Vol}(\widetilde{F})}{\mathrm{Vol}(e)}\vec{\mathbf{e}}(e).$$

Thus, the entry of M^1_ϵ associated with e is equal to $\boldsymbol{\epsilon}\mathrm{Vol}(\widetilde{F})/\mathrm{Vol}(e)$. Yet the existence of orthogonal dual triangulations hinges on the Delaunay property of Ω_h, which might be difficult to ensure for tetrahedral triangulations. If Ω_h consists of bricks, the same will be true of the dual grid and the construction is straightforward. The discrete Hodge operators obtained thus characterize the venerable Yee scheme (Yee 1966) and its more general version, the finite integration technique (Weiland 1996, Clemens and Weiland 2001).

Using the topological laws (3.1) for cochains, the discrete material laws, and the integration by parts formulas (3.14), we can perform the complete elimination of either the primal and dual quantities. Thus we arrive at the 'primal' discrete equations

$$(\mathrm{D}^1)^T\mathrm{M}^2_{1/\mu}\mathrm{D}^1\vec{\mathbf{e}} - \omega^2\mathrm{M}^1_\epsilon\vec{\mathbf{e}} = \widetilde{\mathrm{K}}^1_2\vec{\mathbf{j}},$$

and their 'dual' versions

$$(\widetilde{\mathrm{D}}^2)^T\mathrm{M}^2_{1/\epsilon}\widetilde{\mathrm{D}}^1\vec{\mathbf{h}} - \omega^2\mathrm{M}^1_\mu\vec{\mathbf{h}} = (\widetilde{\mathrm{D}}^2)^T\mathrm{M}^2_{1/\epsilon}\vec{\mathbf{j}}.$$

Surprisingly, the coupling matrices have largely disappeared, and we end up with equations that differ from the Galerkin equations only by a more general

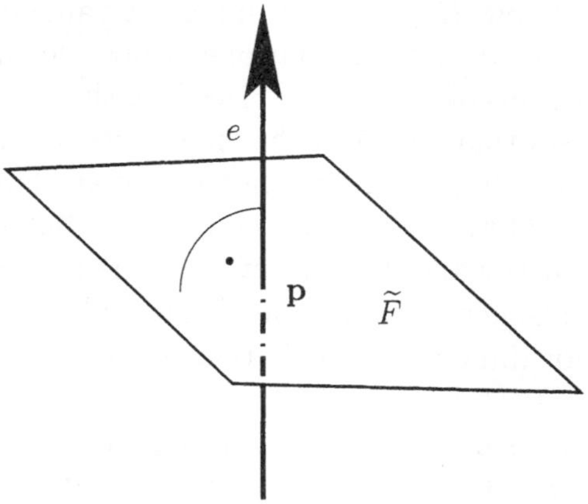

Figure 3.4. Primal edge and orthogonal dual face

choice of mass matrices. The conclusion is that discretization schemes that can be put into the framework of discrete Hodge operators can be analysed as Galerkin schemes with perturbed mass matrices. This is why we mainly look at the Galerkin approach as the generic case.

Bibliographical notes

Recently, the theoretical foundations of discrete models for the laws of electromagnetism have seen a surge in interest (Bossavit and Kettunen 1999*a*, Bossavit and Kettunen 1999*b*, Tarhasaari *et al.* 1999, Teixeira and Chew 1999, Teixeira 2001, Tonti 2001, Tonti 1996, Mattiussi 2000). The rationale behind the concept of discrete Hodge operators is elucidated in Tarhasaari *et al.* (1999) and Mattiussi (2000). A comprehensive theory of discrete Hodge operators including convergence estimates has been developed in Hiptmair (2001*b*) and (2001*a*), extending ideas of Nicolaides and Wang (1998) and Nicolaides and Wu (1997). Many apparently different schemes, known as finite volume methods (Yee 1966), mimetic finite differences (Hyman and Shashkov 2001, 1999), cell methods (Marrone 2001), and finite integration techniques (Clemens and Weiland 2001, Weiland 1996, Schuhmann and Weiland 2001), emerge from particular realizations of discrete Hodge operators. We remind readers that the discrete formulation of electrodynamics, based on cochains and discrete Hodge operators, is only one example of a discrete 'lattice model', which plays an increasingly important role in physics (Dezin 1995, Mercat 2001).

3.4. Higher-order Whitney forms

The Whitney forms constructed in Section 3.2 are piecewise linear. Yet, for any $p \in \mathbb{N}_0$ we know $H^1(\Omega)$-conforming Lagrangian finite elements on tetrahedral and hexahedral triangulations, whose local spaces contain all multivariate polynomials of degree $\leq p$. Heeding the guideline of Section 3.2, this strongly suggests that, for all other members of the family of discrete differential forms, such higher-order schemes are available.

First, only simplicial meshes Ω_h are considered, but in dimension $n \in \mathbb{N}$. This generality is worthwhile, because we will make heavy use of induction and recursion with respect to dimension. Thus, in fact, it is easier to present the developments for dimension $n \in \mathbb{N}$ than to discuss the cases $n = 2$ and $n = 3$ separately.

Initially, we only consider a single generic simplex $T \subset A(\mathbb{R}^n)$ with vertices $\mathbf{a}_0, \ldots, \mathbf{a}_n$. Let λ_i, $i = 0, \ldots, n$, denote the related barycentric coordinate functions in T.

The local spaces $\mathcal{W}_p^l(T)$ should contain polynomials. To state this, we introduce $\mathcal{P}_p(T)$ as the space of n-variate polynomials of total degree $\leq p$

on T. Then we define the space of polynomial differential l-forms on T, $0 \leq l \leq n$,

$$\mathcal{DP}_p^l(T) := \left\{ \omega = \sum_{I \subset \{1,\dots,n\}, |I|=l} \alpha_I \boldsymbol{d}\lambda_I,\ \alpha_I \in \mathcal{P}_p(T) \right\}, \tag{3.15}$$

where $I = \{i_1, \dots, i_l\}$, $\boldsymbol{d}\lambda_I := \boldsymbol{d}\lambda_{i_1} \wedge \cdots \wedge \boldsymbol{d}\lambda_{i_l}$, $i_1 < i_2 < \cdots < i_l$. Stated differently, a polynomial differential form possesses (component-wise) polynomials as vector proxies, regardless of the choice of basis.

The local spaces $\mathcal{W}_p^l(T)$ should contain all of $\mathcal{DP}_p^l(T)$, and should themselves be polynomial. This inspires the requirement

$$\mathcal{DP}_p^l(T) \subset \mathcal{W}_p^l(T) \subset \mathcal{DP}_{p'}^l(T), \quad \text{for some } p' \geq p \geq 0. \tag{3.16}$$

We also argue that the exact sequence property of Whitney forms must remain true for higher-order discrete differential forms. A necessary condition is

$$\{\omega \in \mathcal{W}_p^l(T),\ \boldsymbol{d}\omega = 0\} = \boldsymbol{d}\mathcal{W}_p^{l-1}(T) \quad \forall 1 \leq l \leq n,\ p \in \mathbb{N}_0. \tag{3.17}$$

The idea of the construction of the local spaces $\mathcal{W}_p^l(T)$ takes the cue from the p-version for $H^1(\Omega)$-conforming finite elements (Schwab 1998, Babuška and Suri 1994). There the usual approach is to associate with each m-facet $S \in \mathcal{S}_m(T)$ spaces of polynomials of degree p_S, whose traces on ∂S vanish (Khoromskij and Melenk 2001, Section 3.1). These polynomials are extended into the interior of T such that they have zero trace on any other m-facet of T. The span of the extended polynomials constitutes the desired local space. This simple procedure for 0-forms is beset with tremendous technical difficulties for $l > 0$. The culprits are the constraint (3.17), which is no longer immaterial for $l > 0$, and the lack of a straightforward 'trace compliant' extension into T.

To begin with, suitable polynomial spaces $\mathcal{X}_p^l(S)$ associated with an m-facet $S \in \mathcal{S}_m(T)$ of T, $l \leq m \leq n$, have to be found. Analogously to the case of 0-forms they must satisfy (for some $p' \geq p$)

$$\mathcal{X}_p^l(S) \subset \mathcal{Z}_{p'}^l(S) := \begin{cases} \{\omega \in \mathcal{DP}_{p'}^l(S),\ \mathbf{t}_{\partial S}\omega = 0\}, & \text{for } l < m, \\ \{\omega \in \mathcal{DP}_{p'}^m(S),\ \int_S \omega = 0\}, & \text{for } l = m. \end{cases} \tag{3.18}$$

In addition they have to accommodate the restriction of (3.16) and (3.17) to S, that is, $\mathcal{Z}_p^l(S) \subset \mathcal{X}_p^l(S)$ and

$$\{\omega \in \mathcal{X}_p^l(S),\ \boldsymbol{d}\omega = 0\} = \boldsymbol{d}\mathcal{X}_p^{l-1}(S) \quad \forall 1 \leq l \leq m,\ p \in \mathbb{N}_0. \tag{3.19}$$

As a next step, functions from the facet-based spaces $\mathcal{X}_p^l(S)$ have to be extended into the interior of T, without 'polluting' spaces on other facets of the same dimension. For polynomial 0-forms this can easily be achieved.

If $S = (\mathbf{a}_0, \ldots, \mathbf{a}_m)$, every $\omega \in \mathcal{X}_p^0(S)$, $p > m$, has a representation

$$\omega = \alpha(\lambda_0, \ldots, \lambda_m)\, \lambda_0 \cdots \lambda_m,$$

where α is a homogeneous polynomial of degree $p - (m+1)$. This representation instantly provides the desired formula for a polynomial extension of ω that is zero on all other m-facets of T. For $l > 0$ the extension to the interior of T will rely on a generalization of this procedure and uses a symmetric barycentric representation formula. Writing $\lambda_I := \lambda_{i_1} \cdots \lambda_{i_l}$ for $I = \{i_1, \ldots, i_l\}$, it can be stated as follows.

Lemma 3.4. (Symmetric barycentric form) If $S = (\mathbf{a}_{s_0}, \ldots, \mathbf{a}_{s_m})$, $\{s_0, \ldots, s_m\} \subset \{0, \ldots, n\}$, then any $\omega \in \mathcal{Z}_p^l(S)$, $l \leq m$, $p > m - l$, can be written as

$$\omega = \sum_{I \subset \{s_0, \ldots, s_m\},\, |I| = l} \alpha_I(\lambda_{s_0}, \ldots, \lambda_{s_m}) \lambda_{I'} \boldsymbol{d}\lambda_I,$$

with α_I belonging to the space $\widetilde{\mathcal{P}}_q(\mathbb{R}^{m+1})$ of $(m+1)$-variate homogeneous polynomials of degree q, $q := p - (m + 1 - l)$, and $I \cup I' = \{s_0, \ldots, s_m\}$, $I \cap I' = \emptyset$.

Proof. Starting from a true basis representation of ω in terms of $\boldsymbol{d}\lambda_I$, $I \subset \{s_1, \ldots, s_m\}$, the zero trace conditions on sub-facets of S give additional relationships between the coefficients. The symmetric barycentric representation formula emerges after tedious technical manipulations. □

At first, this formula involves the restrictions of the barycentric coordinates to S, but it also makes sense for barycentric coordinates on T. In this reading it defines a *barycentric extension* $\mathsf{E}_{p,S}^l \omega$ of $\omega \in \mathcal{Z}_p^l(S)$ to $\mathcal{DP}_p^l(T)$. The representation according to Lemma 3.4 is by no means unique, nor is the barycentric extension, but all choices have the elementary property

$$\omega \in \mathcal{Z}_p^l(S): \quad \mathbf{t}_{S'}(\mathsf{E}_{p,S}^l \omega) = 0 \quad \forall S' \in \mathcal{S}_m(T) \setminus \{S\}. \tag{3.20}$$

In plain English, the barycentric extension features zero trace on all other m-facets of T. Of course, any barycentric extension also preserves the polynomial degree of ω.

The construction of facet-based spaces $\mathcal{X}_p^l(S)$ is possible thanks to the following key result.

Lemma 3.5. (Polynomial potentials with boundary conditions) For any m-facet $S \in \mathcal{S}_m(T)$, $1 \leq m \leq n$, we find, for all $m < l \leq n$,

$$\mathcal{N}_p^l(S) := \{\omega \in \mathcal{Z}_p^l(S),\, \boldsymbol{d}\omega = 0 \text{ on } S\} = \boldsymbol{d}\mathcal{Z}_{p+1}^{l-1}(S).$$

Proof. We may choose a coordinate system with origin in vertex $\mathbf{a}_0$ of T. The crucial device is the Poincaré mapping $\mathfrak{K}^l : \mathcal{DF}^{l,0}(T) \mapsto \mathcal{DF}^{l-1,1}(T)$, defined by

$$(\mathfrak{K}^l\omega)(\mathbf{x})(\mathbf{v}_1,\ldots,\mathbf{v}_{l-1}) := \int_0^1 t^{l-1}\omega(t\mathbf{x})(\mathbf{x},\mathbf{v}_1,\ldots,\mathbf{v}_{l-1})\,\mathrm{d}t, \tag{3.21}$$

for all $\mathbf{v}_1,\ldots.\mathbf{v}_{l-1} \in \mathbb{R}^n$, $\mathbf{x} \in T$, $\omega \in \mathcal{DF}^{l,0}(T)$. According to Cartan (1967), formula (2.13.2), it satisfies

$$\boldsymbol{d}(\mathfrak{K}^l\omega) + \mathfrak{K}^{l+1}(\boldsymbol{d}\omega) = \omega. \tag{3.22}$$

The remainder of the proof relies on induction with respect to m. For $m = 1$, plain integration along the edge S of a polynomial 1-form $\omega \in \mathcal{Z}^1_p(S)$ with zero average will provide a 0-form π, $\boldsymbol{d}\pi = \omega$, of polynomial degree $p+1$ vanishing in the endpoints of S.

For $m > 1$ pick some $\omega \in \mathcal{Z}^l_p(S)$, $\boldsymbol{d}\omega = 0$, and set $\eta := \mathfrak{K}^l\omega$. The identity (3.22) guarantees $\boldsymbol{d}\eta = \omega$, and by a simple inspection of (3.21) we see that $\eta \in \mathcal{DP}^{l-1}_{p+1}(S)$. A closer scrutiny of (3.21) also shows that $\mathbf{t}_F\eta = 0$ for all faces $F \in \mathcal{S}_{m-1}(S)$ that have $\mathbf{a}_0$ as vertex. A single $(m-1)$-facet $\widetilde{F}$ of S remains, on which the trace of η may not be zero. However, if $l < m$, then $\mathbf{t}_{\partial\widetilde{F}}\eta = 0$ so that $\mathbf{t}_{\widetilde{F}}\eta \in \mathcal{Z}^{l-1}_{p+1}(\widetilde{F})$. To see $\mathbf{t}_{\widetilde{F}}\eta \in \mathcal{Z}^{l-1}_{p+1}(\widetilde{F})$ for $l = m$, observe that

$$\int_{\widetilde{F}} \eta = \int_{\partial S} \eta = \int_S \boldsymbol{d}\eta = \int_S \omega = 0$$

by definition of $\mathcal{Z}^m_p(S)$. Thus, the induction hypothesis for $m-1$ applied to $\widetilde{F}$ gives $\widetilde{\tau} \in \mathcal{Z}^{l-2}_{p+2}(\widetilde{F})$ with $\boldsymbol{d}\widetilde{\tau} = \mathbf{t}_{\widetilde{F}}\eta$. A barycentric extension gives $\tau \in \mathcal{DP}^{l-2}_{p+2}(S)$, whose trace on all other $(m-1)$-facets except $\widetilde{F}$ vanishes. Then $\pi := \eta - \boldsymbol{d}\tau$ yields the desired potential, because differentiation invariably reduces the polynomial degree by at least one. □

A simple construction of $\mathcal{X}^l_p(S)$ is immediately on hand, based on the choice

$$\mathcal{X}^l_p(S) = \mathcal{Z}^l_{p+1}(S), \quad \mathcal{X}^{l-1}_p(S) = \mathcal{Z}^{l-1}_{p+2}(S). \tag{3.23}$$

Lemma 3.5 will make (3.19) hold, but we have to put up with non-uniform polynomial degrees in the exact sequence $\mathcal{X}^0_{p+m}(S) \overset{\boldsymbol{d}}{\mapsto} \mathcal{X}^1_{p+m-1}(S) \overset{\boldsymbol{d}}{\mapsto} \ldots \overset{\boldsymbol{d}}{\mapsto} \mathcal{X}^m_p(S)$. However, usually the focus will be on discrete differential forms of a particular degree. Then (3.23) supplies perfect discrete differential forms of polynomial degree p. Using the barycentric extension procedure, this results in the so-called *second family* of discrete differential forms. Hardly surprisingly, the local spaces turn out to be $\check{\mathcal{W}}^l_p(T) = \mathcal{DP}^l_{p+1}(T)$. Symbols related to this second family will be tagged with $\check{}$, *e.g.*, $\check{\mathcal{W}}^l_p(\Omega_h)$.

Yet a true generalization of Whitney forms should avoid the increase in polynomial degree accompanying a decrease of l. Witness the case of Whitney forms, whose local spaces are contained in $\mathcal{DP}_1^l(T)$ for all l. Therefore we tighten the requirement (3.19) by demanding $p' = p+1$, which leads to the *first family* of discrete differential form, which includes the Whitney forms as lowest-order representatives.

The key to the construction is the use of *some* direct decomposition

$$\mathcal{Z}_p^l(S) = \mathcal{Y}_p^l(S) \oplus \mathcal{N}_p^l(S). \tag{3.24}$$

We find that the choice

$$\mathcal{X}_p^l(S) := \mathcal{Z}_p^l(S) + \mathcal{Y}_{p+1}^l(S) = \boldsymbol{d}\mathcal{Y}_{p+1}^{l-1}(S) \oplus \mathcal{Y}_{p+1}^l(S) \tag{3.25}$$

meets all the requirements put forth for $\mathcal{X}_p^l(S)$. Despite the considerable freedom involved in choosing $\mathcal{X}_p^l(T)$, the dimensions of these spaces are already fixed.

Theorem 3.6. (Dimensions of $\mathcal{X}_p^l$) If the facet-associated spaces $\mathcal{X}_p^l(S)$ satisfy (3.19) and $\mathcal{Z}_p^l(S) \subset \mathcal{X}_p^l(S) \subset \mathcal{Z}_{p+1}^l(S)$, they fulfil $\dim \mathcal{X}_p^l(S) = \binom{m}{l}\binom{p+l}{m}$ if $l < n$, and $\dim \mathcal{X}_p^l(S) = \binom{p+m}{m} - 1$ if $l = m$. In particular, $\mathcal{X}_p^l(S) = \{0\}$ if $m > l + p$.

Proof. Compare the considerations in Section 3 of Hiptmair (2001*c*). □

Now let us fix the barycentric extension operators $\mathsf{E}_{p,S}^l : \mathcal{X}_p^l(S) \mapsto \mathcal{DP}_p^l(T)$, $0 \le l \le n$, $p \in \mathbb{N}$, $S \in \mathcal{S}_m(T)$, $l \le m \le n$. Using these, define $\mathsf{F}_{p,S}^l : \mathcal{X}_p^l(S) \mapsto \mathcal{DP}_p^l(T)$ by

$$\mathsf{F}_{p,S}^l\omega := \mathsf{E}_{p+1,S}^l\omega^\perp + \boldsymbol{d}\mathsf{E}_{p+1,S}^{l-1}\eta, \quad \omega = \omega^\perp + \boldsymbol{d}\eta, \tag{3.26}$$

where $\omega^\perp \in \mathcal{Y}_{p+1}^l(S)$ and $\eta \in \mathcal{Y}_{p+1}^{l-1}(S)$ are unique by (3.25). Now, we are in a position to build the local spaces as a direct sum of facet contributions

$$\mathcal{W}_p^l(T) := \mathcal{W}_0^l(T) \oplus \left(\bigoplus_{m=l}^{n} \bigoplus_{S \in \mathcal{S}_m(T)} \mathsf{F}_{p,S}^l(\mathcal{X}_p^l(S)) \right). \tag{3.27}$$

The facet-based construction paves the way for introducing the global space $\mathcal{W}_p^l(\Omega_h)$ of discrete differential l-forms on a simplicial triangulation Ω_h. It can be achieved by using extensions according to (3.26) from $\mathcal{X}_p^l(S)$ into adjacent simplices for all S, $S \in \mathcal{S}_m(\Omega_h)$, $l \le m \le n$. This yields a space of Ω_h-piecewise polynomial l-forms, for which the patch condition (2.20) is automatically satisfied, that is, $\mathcal{W}_p^l(\Omega_h) \in \mathcal{F}^l(\Omega)$.

There is no reason why the same polynomial degree p should be assigned to each m-facet S. Rather, we could use a vector $\underline{\mathbf{p}} := (p_S)_S$, $S \in \mathcal{S}_m(\Omega_h)$, $l \leq m \leq n$, of different polynomial degrees $p_S \in \mathbb{N}_0$. The construction of the resulting space $\mathcal{W}^l_{\underline{\mathbf{p}}}(\Omega_h) \subset \mathcal{F}^l(\Omega)$ is carried out as above. This leads to the so-called variable degree p-version for discrete differential forms. As in the case of Whitney forms, discrete potentials exist.

Theorem 3.7. (DeRham theorem for higher-order discrete forms) Using the discrete cohomology spaces $\mathcal{H}^l_h(\Omega_h)$ introduced in Corollary 3.3, for all $\omega_h \in \mathcal{W}^l_{\underline{\mathbf{p}}}(\Omega_h)$, we have

$$\boldsymbol{d}\omega_h = 0 \quad \Leftrightarrow \quad \exists \eta_h \in \mathcal{W}^{l-1}_{\underline{\mathbf{p}}}(\Omega_h),\, \gamma_h \in \mathcal{H}^l_h(\Omega_h) \quad \text{satisfying } \omega_h = \boldsymbol{d}\eta_h + \gamma_h.$$

Proof. By the commuting diagram property (3.12) we know $\boldsymbol{d}\Pi^l\omega_h = 0$, which, thanks to the discrete exact sequence property, implies the existence of $\eta^0_h \in \mathcal{W}^l_0(\Omega_h)$ with $\boldsymbol{d}\eta^0_h = \Pi^l\omega_h$.

For $\omega'_h := (\mathrm{Id} - \Pi^l)\omega_h$ we can find purely *local* discrete potentials by a recursive procedure. As first step pick any $S \in \mathcal{S}_l(\Omega_h)$ and note that both $\mathbf{t}_S\omega'_h \in \mathcal{X}^l_{p_S}(S)$ and $\boldsymbol{d}\mathbf{t}_S\omega'_h = 0$. This means that $\mathbf{t}_S\omega_h = \boldsymbol{d}\eta_S$ for some $\eta_S \in \mathcal{Y}^{l-1}_{p_S+1}(S)$. Adding the extensions according to (3.26) of all η_S provides η_l.

At the beginning of the second step we replace $\omega'_h \leftarrow \omega'_h - \boldsymbol{d}\eta_l$. Then consider $S \in \mathcal{S}_{l+1}(\Omega_h)$ and find a potential $\eta_S \in \mathcal{X}^{l-1}_{p_S}(S)$. The sum of the extended η_S will be η_{l+1}. This $(l-1)$-form will have zero trace on all l-facets on Ω_h.

Carrying on with $(l+2)$-facets and $\omega'_h \leftarrow \omega'_h - \boldsymbol{d}\eta_{l+1}$, we finally reach the n-facets and $\eta_h := \sum_{k=l}^{n} \eta_k$ will provide the desired potential $\in \mathcal{W}^{l-1}_{\underline{\mathbf{p}}}(\Omega_h)$. □

Remark 8. The discrete potentials of closed forms in the p-hierarchical surplus space $(\mathrm{Id} - \Pi^l)\mathcal{W}^l_{\underline{\mathbf{p}}}(\Omega_h)$ are *local* in the sense that, if the support of the form is confined to the elements adjacent to some facet, the same will hold true for some discrete potential. The theorem also illustrates that, in terms of cohomology, the higher-order discrete differential forms do not convey the slightest additional information. △

From (3.25) we can conclude that we have $\mathcal{Z}^l_{p+1}(S) = \mathcal{X}^l_p(S) + \boldsymbol{d}\mathcal{X}^{l-1}_{p+1}(S)$. This carries over to the global spaces and means that the second family of discrete differential forms can be obtained as

$$\check{\mathcal{W}}^l_p(\Omega_h) = \mathcal{W}^l_p(\Omega_h) + \boldsymbol{d}\,\mathcal{W}^{l-1}_{p+1}(\Omega_h). \tag{3.28}$$

The spaces of the second family simply emerge from those of the first family by adding some closed forms of the next-higher polynomial degree. This implies $\boldsymbol{d}\check{\mathcal{W}}^l_p(\Omega_h) = \boldsymbol{d}\mathcal{W}^l_p(\Omega_h)$.

One issue still looms large, namely how to construct concrete basis functions (shape functions) for higher-order discrete differential forms. In general, one starts with building bases for $\mathcal{X}^l_p(S)$ on a generic m-facet

S, $l \leq m \leq 3$, the reference facet. This can be done based on the symmetric barycentric representation from Lemma 3.4 (aided by some symbolic expression manipulation software). Decisions on the choice of the algebraic complement $\mathcal{Y}_p^l(S)$ and the special properties of the basis functions (orthogonality, hierarchical arrangement) translate into linear constraints on the coefficients of the homogeneous polynomials. Ultimately it will be the variational problem to be discretized that determines what are good sets of basis functions. The objective may be to achieve the best possible condition for element matrices. As no variational problem has been specified, no recommendations can be given at this stage.

Once basis functions for reference facets are available in symmetric barycentric representation, the same formulas can be used for all other facets. This is the gist of an *affine invariant* construction of discrete differential forms (*cf.* Section 3.2).

We want to mention special bases of $\mathcal{X}_p^l(S)$, the *p-hierarchical* bases $\mathsf{HB}(\mathcal{X}_p^l(S))$, for $p \geq 1$ distinguished by the properties

- $\mathsf{HB}(\mathcal{X}_p^l(S))$ is a basis of $\mathcal{X}_p^l(S)$ for all p,
- $\mathsf{HB}(\mathcal{X}_p^l(S)) = \mathsf{HB}(\mathcal{X}_{p-1}^l(S)) \cup \widetilde{\mathsf{HB}}(\mathcal{X}_p^l(S))$,

where $\widetilde{\mathsf{HB}}(\mathcal{X}_p^l(S))$ is a suitable subset of $\mathcal{X}_p^l(S)$. This induces a partitioning

$$\mathsf{HB}(\mathcal{X}_p^l(S)) = \mathsf{HB}(\mathcal{X}_0^l(S)) \cup \widetilde{\mathsf{HB}}(\mathcal{X}_1^l(S)) \cup \cdots \cup \widetilde{\mathsf{HB}}(\mathcal{X}_p^l(S)).$$

Note that $|\widetilde{\mathsf{HB}}(\mathcal{X}_p^l(S))| := \binom{k+l-1}{n-1}$. We can even select basis functions reflecting the splitting (3.24). Thus we get a hierarchical basis of the form

$$\widetilde{\mathsf{HB}}(\mathcal{X}_p^l(S)) = \{\beta_{p,1}^l, \ldots, \beta_{p,M_p^l}^l, \boldsymbol{d}\beta_{1,p}^{l-1}, \ldots, \boldsymbol{d}\beta_{p,M_p^{l-1}}^{l-1}\}, \qquad (3.29)$$

Table 3.2. Numbers of basis functions associated with sub-simplices for tetrahedral discrete differential 1- and 2-forms and uniform polynomial degree p

p		0	1	2	3	4	5	6
$l = 1$	edges	1	2	3	4	5	6	7
	faces	0	2	6	12	20	30	42
	cell	0	0	3	12	30	60	105
$l = 2$	faces	1	3	6	10	15	21	28
	cell	0	3	12	30	60	105	168

where $M_p^0 = \binom{k-1}{n-1}$ and $M_p^l + M_p^{l-1} = \binom{m}{l}\binom{p+l-1}{m-1}$. In words, we can find a p-hierarchical basis for which all closed discrete l-forms in p-hierarchical surpluses are linear combinations of closed basis forms. The p-hierarchical basis of $\mathcal{W}^l_{\underline{\mathbf{p}}}(\Omega_h)$ is straightforward by extension and combination.

From (3.28) it is immediate that we get a basis for $\breve{\mathcal{W}}^l_p(\Omega_h)$ for uniform polynomial degree $p \in \mathbb{N}_0$ by augmenting the basis of $\mathcal{W}^l_p(\Omega_h)$ by (linearly independent) exterior derivatives of basis functions of $\mathcal{W}^{l-1}_{p+1}(\Omega_h)$.

Returning to tetrahedral triangulations, that is, $n = 3$, the numbers of basis functions associated with different facets are listed in Table 3.2. The numbers of extra non-closed basis functions needed to build a hierarchical basis according to (3.29) are recorded in Table 3.3. Note that all higher-order basis functions on l-facets are closed.

In this fashion we can find the hierarchical basis for first-order 1-forms of the first family. According to Table 3.2 we expect two basis functions associated with each edge, and two more with each face. Their vector proxies in symmetric barycentric representation read

$$\left.\begin{array}{l}\lambda_i\,\mathbf{grad}\,\lambda_j - \lambda_j\,\mathbf{grad}\,\lambda_i\\ \lambda_i\,\mathbf{grad}\,\lambda_j + \lambda_j\,\mathbf{grad}\,\lambda_i\end{array}\right\}\quad\text{on edge } [\mathbf{a}_i, \mathbf{a}_j],$$

$$\left.\begin{array}{l}\lambda_j\lambda_i\,\mathbf{grad}\,\lambda_k - \lambda_j\lambda_k\,\mathbf{grad}\,\lambda_i\\ \lambda_i\lambda_j\,\mathbf{grad}\,\lambda_k - \lambda_i\lambda_k\,\mathbf{grad}\,\lambda_j\end{array}\right\}\quad\text{on face } [\mathbf{a}_i, \mathbf{a}_j, \mathbf{a}_k].$$

The basis functions belonging to the edges constitute a basis for the lowest-order discrete 1-forms of the second family.

Table 3.3. Numbers M_p^l for a tetrahedron (uniform polynomial degree)

p		[0]	1	2	3	4	5	6
$l = 0$	vertices	[1]	0	0	0	0	0	0
	edges	[0]	1	1	1	1	1	1
	faces	[0]	0	1	2	3	4	5
	cell	[0]	0	0	1	3	6	10
$l = 1$	edges	[1]	0	0	0	0	0	0
	faces	[0]	2	3	4	5	6	7
	cell	[0]	0	3	8	15	24	35
$l = 2$	faces	[1]	0	0	0	0	0	0
	cell	[0]	3	6	10	15	21	28

For hexahedral triangulations similar strategies lead to higher-order discrete differential forms. We skip the discussion of the details and just present the results. For the first family of discrete differential forms the local spaces on a generic brick T read

$$\mathcal{W}_p^1(T) = \mathcal{Q}_{p,p+1,p+1}(T) \times \mathcal{Q}_{p+1,p,p+1}(T) \times \mathcal{Q}_{p+1,p+1,p}(T),$$
$$\mathcal{W}_p^2(T) = \mathcal{Q}_{p+1,p,p}(T) \times \mathcal{Q}_{p,p+1,p}(T) \times \mathcal{Q}_{p,p,p+1}(T).$$

Here $\mathcal{Q}_{p_1,p_2,p_3}(T)$ denotes the space of 3-variate polynomials of degree $\leq p_i$ in the ith variable, $i = 1, 2, 3$. The number of basis functions associated with edges, faces and cells is given in Table 3.4.

As for the case of tetrahedra, a second family of discrete differential forms on hexahedral features complete spaces of 3-variate polynomials of degree $\leq p + 1$ in each independent variable as local spaces.

Bibliographical notes

Higher-order discrete 1-forms in two dimensions are known as $\boldsymbol{H}(\text{div}; \Omega)$-conforming Raviart–Thomas finite elements, and were introduced in Raviart and Thomas (1977). The first family of discrete 1- and 2-forms in 3D was first presented in Nédélec (1980), and the second family in Nédélec (1986). The second family for $\boldsymbol{H}(\text{div}; \Omega)$-conforming finite elements is also known as BDM-elements (Brezzi, Douglas and Marini 1985). A complete survey of various discrete $(n-1)$-forms, $n = 2, 3$, is given in Brezzi and Fortin (1991, Chapter 3). The treatment in this section was first adopted in Hiptmair (1999*a*, 2001*c*) and is also sketched in Demkowicz (2001).

The choice of p-hierarchical higher-order shape functions is dealt with in Sun, Lee and Cendes (2001) for tetrahedra, and Monk (1994) and Ainsworth

Table 3.4. Numbers of basis functions associated with edges, faces and cells for hexahedral discrete differential 1- and 2-forms of the first family and uniform polynomial degree p

p		0	1	2	3	4	5	6
$l = 1$	edges	1	2	3	4	5	6	7
	faces	0	4	12	24	40	60	84
	cell	0	6	36	108	240	450	756
$l = 2$	faces	1	4	9	16	25	36	49
	cell	0	12	54	144	300	540	882

and Coyle (2001*a*, 2001*b*) for the use of Legendre polynomials on hexahedra/quadrilaterals. Shape functions for the second family of discrete differential forms on tetrahedra with variable polynomial degree are given in Demkowicz (2001). This survey article also discusses criteria for choosing basis functions. Issues of implementation are addressed in Vardapetyan and Demkowicz (1999) and Rachowicz and Demkowicz (2002).

3.5. Interpolation operators

With the spaces $\mathcal{W}^l_{\underline{\mathbf{p}}}(\Omega_h)$ being completely defined, we may consider analogues of the local interpolation operators Π^l, which satisfy a commuting diagram property according to (3.12). The reader should be aware that these interpolation operators are never needed in an implementation of a finite element code based on discrete differential forms. In this respect only the shape functions are relevant. The interpolation operators discussed in this section are merely theoretical tools. For the same reason as in the previous section, we take a look at the general situation in dimension $n \in \mathbb{N}$.

In the case of interpolation operators and local bases alike we find that their construction is canonical only for Whitney forms, whereas for higher-order discrete differential forms we have numerous sensible options. First, we examine *moment-based* interpolation operators for the first family of simplicial discrete differential forms. Following a classical idea in the field of finite elements, they are based on *degrees of freedom* (Ciarlet 1978, Section 2.3), that is, a set $\Theta^l_{\underline{\mathbf{p}}}(\Omega_h)$ of linear forms $\mathcal{W}^l_{\underline{\mathbf{p}}}(\Omega_h) \mapsto \mathbb{R}$, such that:

(i) $\Theta^l_{\underline{\mathbf{p}}}(\Omega_h)$ is a dual basis of $\mathcal{W}^l_{\underline{\mathbf{p}}}(\Omega_h)$ (unisolvence);

(ii) to each facet $S \in \mathcal{S}_m(\Omega_h)$, $l \le m \le n$, we can associate a set $\Theta^l_{p_S}(S) \subset \Theta^l_{\underline{\mathbf{p}}}(\Omega_h)$. Each $\kappa \in \Theta^l_{\underline{\mathbf{p}}}(\Omega_h)$ belongs to exactly one of these sets, and

$$\kappa(\omega_h) = 0 \quad \forall \kappa \in \Theta^l_{p_R}(R),\ R \in \mathcal{S}_k(S),\ l \le k \le m \quad \Leftrightarrow \quad \mathbf{t}_S \omega_h = 0$$

for every $\omega_h \in \mathcal{W}^l_{\underline{\mathbf{p}}}(m)$. Thus the degrees of freedom have to be local.

Locality suggests a facet-based approach. For each $S \in \mathcal{S}_m(\Omega_h)$, $l \le m \le n$, we find linear forms $\zeta_i : \mathcal{X}^l_{p_S}(S) \mapsto \mathbb{R}$, $i = 1, \ldots, \dim \mathcal{X}^l_{p_S}(S)$, that form a dual basis for $\mathcal{X}^l_{p_S}(S)$. Then set $\zeta^S_i := \zeta_i \circ \mathbf{t}_S$ and observe that

$$\Theta^l_{\underline{\mathbf{p}}}(\Omega_h) := \{\zeta^S_i,\ S \in \mathcal{S}_m(S),\ l \le m \le n,\ i = 1, \ldots, \dim \mathcal{X}^l_{p_S}(S)\}$$

constitutes a set of degrees of freedom meeting both of the above requirements. Just note that $\{\zeta^S_i\}_i$ is simply the subset of $\Theta^l_{\underline{\mathbf{p}}}(\Omega_h)$ associated with S. It remains to find degrees of freedom for all $\mathcal{X}^l_{p_S}(S)$.

Theorem 3.8. For every $\omega \in \mathcal{X}_p^l(S)$, $S \in \mathcal{S}_m(\Omega_h)$, $l \leq m \leq n$, $p \geq m-l$, we have

$$\int_S \omega \wedge \mu = 0 \quad \forall \mu \in \mathcal{DP}_{p-m+l}^{m-l}(S) \quad \Leftrightarrow \quad \omega = 0.$$

Proof. (See the proof of Lemma 10 in Hiptmair (1999*a*).) The proof employs (descending) induction with respect to l. For $l = m$ we saw that $\mathcal{X}_p^m(S) + \mathcal{DP}_0^m(S) = \mathcal{DP}_p^m(S)$ and the assertion is evident, because for both $\mathcal{DP}_p^m(S)$ and $\mathcal{DP}_p^0(S)$ the vector proxies belong to the same space $\mathcal{P}_p(S)$.

In the case $l < m$, use $\mathbf{t}_{\partial S}\omega = 0$ and integration by parts to obtain

$$\int_S \boldsymbol{d}\omega \wedge \mu = \pm \int_S \omega \wedge \boldsymbol{d}\mu = 0$$

for all $\mu \in \mathcal{DP}_{p-m+l+1}^{m-(l+1)}(S)$. As $\boldsymbol{d}\omega \in \mathcal{X}_p^{l+1}(S)$, the induction hypothesis for $l+1$ involves $\boldsymbol{d}\omega = 0$. By definition of $\mathcal{X}_p^l(S)$ in formula (3.25), we can conclude that $\omega \in \mathcal{DP}_p^l(S)$. In particular, we can find a representation

$$\omega = \sum_{I \subset \{1,\dots,m\},\, |I|=l} \beta_I(\lambda_1, \dots, \lambda_m)\lambda_{I'} \boldsymbol{d}\lambda_I, \tag{3.30}$$

$I' \cup I = \{1, \dots, m\}$, $I \cap I' = \emptyset$, $\beta_I \in \mathcal{P}_{p-(m-l)}(\mathbb{R}^m)$. Here λ_i, $i = 1, \dots, m$, are local barycentric coordinates of S. Pick $\mu = \lambda_J \boldsymbol{d}\lambda_J$, $J \subset \{1, \dots, m\}$, $|J| = m - l$, and verify

$$\int_S \omega \wedge \mu = \pm \int \beta_{J'} \lambda_J^2 \, \boldsymbol{d}\lambda_1 \wedge \cdots \wedge \boldsymbol{d}\lambda_m.$$

As $\mu \in \mathcal{DP}_{m-l}^{m-l}(S)$ this has evaluate to zero, which immediately implies $\beta_{J'} = 0$, $J \cup J' = \{1, \dots, m\}$. By varying J, all β_I in (3.30) are seen to vanish. □

Using well-known formulas for dimensions of spaces of polynomials, it is immediate that, for $m > l$,

$$\dim \mathcal{DP}_{p-m+l}^{m-l}(S) = \binom{m}{l}\binom{p+l}{m} = \dim \mathcal{X}_p^l(S).$$

Moreover, for $l = m$ we calculate $\dim \mathcal{DP}_p^0(S) = \dim \mathcal{X}_p^m(S) + \dim \mathcal{W}_0^m(S)$. The agreement of dimensions has an important consequence.

Corollary 3.9. (Unisolvence of higher moments) Let $N = \dim \mathcal{X}_p^l(S)$ and $\{\beta_1, \dots, \beta_N\}$ be any bases of $\mathcal{DP}_{p-m+l}^{m-l}(S)$. The functionals

$$\eta_i : \mathcal{X}_p^l(S) \mapsto \mathbb{R}, \quad \omega \mapsto \int_S \omega \wedge \beta_i, \quad i = 1, \dots, N,$$

provide a dual basis of $\mathcal{X}_p^l(S)$ for $l < m$, or $\mathcal{X}_p^m(S) + \mathcal{W}_0^m(S)$ for $l = m$. Conversely, any set of degrees of freedom for $\mathcal{X}_p^l(S)$, $l < m$, or $\mathcal{X}_p^m(S) + \mathcal{W}_0^m(S)$, respectively, has the above representation.

The degrees of freedom are integrals of an l-form weighted with polynomials. This is why they are called moment-based.

Given degrees of freedom on $\mathcal{W}_{\underline{\mathbf{p}}}^l(\Omega_h)$ that satisfy (i) and (ii), we can define the moment-based interpolation operators $\Pi_{\underline{\mathbf{p}}}^l : \mathcal{F}^l(\Omega) \mapsto \mathcal{W}_{\underline{\mathbf{p}}}^l(\Omega_h)$ by

$$\kappa(\omega - \Pi_{\underline{\mathbf{p}}}^l \omega) = 0 \quad \forall \kappa \in \Theta_{\underline{\mathbf{p}}}^l(\Omega_h). \tag{3.31}$$

An important property of the interpolation operators Π^l for Whitney forms is commuting diagram property (3.12). If we want this to carry over to the variable p-version of discrete differential forms, we have to impose the following minimum degree rule on the local polynomial degrees p_S:

$$\forall S \in \mathcal{S}_m(\Omega_h),\, l \le m \le n : \; p_S \le \min\{p_R,\, R \in \mathcal{S}_{m-1}(\Omega_h),\, R \subset \partial S\}. \tag{3.32}$$

Theorem 3.10. (Commuting diagram property) If $\underline{\mathbf{p}}$ complies with (3.32), any moment-based local interpolation operator according to (3.31) satisfies the *commuting diagram property*

$$\boldsymbol{d} \circ \Pi_{\underline{\mathbf{p}}}^l = \Pi_{\underline{\mathbf{p}}}^{l+1} \circ \boldsymbol{d}.$$

Proof. For $\omega \in \mathcal{DF}^{l,0}(\Omega)$, integration by parts yields

$$\int_S \boldsymbol{d}(\omega - \Pi_{\underline{\mathbf{p}}}^l \omega) \wedge \mu = \pm \int_S (\mathrm{Id} - \Pi_{\underline{\mathbf{p}}}^l)\omega \wedge \boldsymbol{d}\mu + \int_{\partial S} (\mathrm{Id} - \Pi_{\underline{\mathbf{p}}}^l)\omega \wedge \mu.$$

If $\mu \in \mathcal{DP}_{p+1-m+l}^{m-1-l}(S)$, the integrals on the right-hand side represent valid functionals (in ω) for the definition of $\Pi_{\underline{\mathbf{p}}}^l$. Hence, by (3.31) the entire right-hand side evaluates to zero. Observing that $\boldsymbol{d}\mathcal{W}_{\underline{\mathbf{p}}}^l(\Omega_h) \in \mathcal{W}_{\underline{\mathbf{p}}}^{l+1}(\Omega_h)$ gives the assertion. □

The moment-based local interpolation operators are perfectly suitable for discrete differential forms with uniform polynomial degree. Yet the rule (3.32) is just the opposite of what is usually demanded for variable p-version finite element schemes, namely

$$\forall S \in \mathcal{S}_m(\Omega_h),\, l \le m \le n : \quad p_S \le \min\{p_T,\, T \in \mathcal{S}_n(\Omega_h),\, S \subset \partial T\}. \tag{3.33}$$

Hence, we will need another class of local *projection-based* interpolation operators, also denoted by $\Pi_{\underline{\mathbf{p}}}^l$, for which the commuting diagram property remains true, even if the local polynomial degrees are chosen according to (3.33). We give their construction for the first family on simplices and assume non-uniform polynomial degrees $\underline{\mathbf{p}}$. Thanks to locality, it is sufficient to specify $\Pi^l_{\underline{\mathbf{p}}}$ on a single element T. The following tools are required for all $S \in \mathcal{S}_m(T)$, $m \geq l$.

- Liftings $\mathsf{PL}^l_S : \{\omega \in \mathcal{X}^l_{p_S}(S),\ \boldsymbol{d}\omega = 0\} \mapsto \mathcal{X}^{l-1}_{p_S}(S)$ such that $\boldsymbol{d} \circ \mathsf{PL}^l_S = \mathrm{Id}$. An example of such a lifting was explicitly constructed in the proof of Lemma 3.5. It should be remarked that the liftings PL^l_S have to be compatible with the choice of spaces $\mathcal{Y}^l_{p_S+1}(S)$. Conversely, their ranges can even be used to fix these spaces. This approach was pursued in Hiptmair (1999*a*) and (2001*c*).
- Extension operators $\mathsf{PE}^l_S : \mathcal{X}^l_{p_S}(S) \mapsto \mathcal{W}^l_{p_T}(T)$ with $\mathbf{t}_{S'} \circ \mathsf{PE}^l_S = 0$ for all $S' \in \mathcal{S}_m(S) \setminus \{S\}$. Moreover, the choice of extension operators on different cells has to take into account the patch condition (2.20). Examples include barycentric extensions, but there are many choices.
- Projections PQ^l_S from l-forms on S onto $\{\omega \in \mathcal{X}^l_{p_S}(S),\ \boldsymbol{d}\omega = 0\}$.

Then, given $\omega \in \mathcal{DF}^{l,0}(T)$ its interpolant $\tau_h := \Pi^l_{\underline{\mathbf{p}}}\omega$ on T is defined by the algorithm of Figure 3.5. Observe that, for $l = 0$, the extensions are the only relevant mappings and Figure 3.5 describes the common interpolation procedure for p-version Lagrangian finite elements.

Lemma 3.11. The projection-based interpolation operators

$$\Pi^l_{\underline{\mathbf{p}}} : \mathcal{DF}^{l,0}(T) \mapsto \mathcal{W}^l_{\underline{\mathbf{p}}}(T)$$

are linear projections and possess the commuting diagram property.

Proof. Both assertions of the theorem can be established by simple induction arguments inspired by the recursive definition of $\Pi^l_{\underline{\mathbf{p}}}$ in Figure 3.5. □

$\omega \leftarrow \omega - \Pi^l\omega;\quad \tau_h := \Pi^l\omega;$
for$(m = l; m \leq n; m++)$
{
  foreach $(S \in \mathcal{S}_m(T))$
  {
    $\xi_h := \mathsf{PL}^{l+1}_S(\mathsf{PQ}^{l+1}_S \boldsymbol{d}(\mathbf{t}_S\omega));$
    $\eta_h := \mathsf{PL}^l_S(\mathsf{PQ}^l_S(\mathbf{t}_S\omega - \xi_h));$
    $\zeta_h := \mathsf{PE}^l_S\xi_h + \boldsymbol{d}(\mathsf{PE}^{l-1}_S\eta_h);$
    $\omega \leftarrow \omega - \zeta_h;\quad \tau_h \leftarrow \tau_h + \zeta_h;$
  }
}

Figure 3.5. Algorithm defining the local projection-based interpolation operator $\Pi^l_{\underline{\mathbf{p}}}$ through the computation of $\tau_h := \Pi^l_{\underline{\mathbf{p}}}\omega$

From now on we will always take for granted that the families of spaces of discrete differential forms are equipped with local interpolation operators Π_p^l, $\Pi_{\underline{\mathbf{p}}}^l$, for which the commuting diagram property holds.

Bibliographical notes

The moment-based degrees of freedom are the classical choice for spaces of discrete differential forms, when introduced as 'mixed finite elements' (Nédélec 1980, Nédélec 1986, Brezzi and Fortin 1991). The projection-based construction is due to Demkowicz *et al.* (Demkowicz, Monk, Schwab and Vardapetyan 1999, Monk and Demkowicz 2001, Demkowicz and Babuška 2001).

3.6. Interpolation estimates

Eventually discrete differential forms are meant to approximate integral forms representing physical fields. To gauge their efficacy, we study how well differential forms can be approximated by their discrete counterparts in the natural metric provided by the energy norms. Often, sufficient information can be obtained from examining the interpolation error for the local projections. By and large, this can be accomplished drawing on techniques developed for Lagrangian finite elements.

Due to their relevance for electromagnetism, we will largely restrict our attention to 1-forms in three dimensions. In addition, we will chiefly consider discrete forms on tetrahedral triangulations. Approximation by discrete differential forms can be enhanced in two ways:

(1) by using finer meshes (*h-version* of discrete differential forms), and
(2) by raising the polynomial degree p on a fixed mesh (p-version of discrete differential forms).

The h-version of discrete differential forms, that is, the case of fixed uniform polynomial degree $p \in \mathbb{N}_0$ will be investigated first. In this case moment-based local interpolation operators are natural. Following the classical approach (Ciarlet 1978, Section 3.1), for a mesh Ω_h we introduce its shape regularity measure by

$$\rho(\Omega_h) := \max_{T \in \mathcal{S}_3(\Omega_h)} \rho_T, \quad \rho_T := h_T / r_T,$$

where $h_T := \max\{|\mathbf{x} - \mathbf{y}|, \mathbf{x}, \mathbf{y} \in T\}$ is the diameter of a cell T, and $r_T := \max\{\exists \mathbf{x} \in T : |\mathbf{x} - \mathbf{y}| \leq \rho \Rightarrow \mathbf{y} \in T\}$ is the radius of the largest inscribed ball. We remark that $\rho(\Omega_h)$ can be computed from bounds for the largest and smallest angles enclosed by faces of the tetrahedra. From now on the index h of Ω_h should be read as meshwidth

$$h := \max\{h_T, \ T \in \mathcal{S}_3(\Omega_h)\}. \tag{3.34}$$

Techniques based on the pullback of differential forms are instrumental in getting interpolation estimates. They are available if the spaces of discrete differential forms in Ω_h are affine-equivalent in the following sense. There is a *reference simplex* $\widehat{T}$ such that the local space on any $T \in \mathcal{S}_3(\Omega_h)$ can be obtained from that on $\widehat{T}$ via affine pullback (*cf.* (2.16)–(2.19))

$$\mathcal{W}_p^l(T) = (\mathbf{\Phi}_T^{-1})^* \mathcal{W}_p^l(\widehat{T}), \tag{3.35}$$

where $\mathbf{\Phi}_T : \widehat{T} \mapsto T$, $\mathbf{\Phi}(\widehat{\mathbf{x}}) := \mathrm{B}_T \widehat{\mathbf{x}} + \mathbf{t}_T$, $\mathrm{B}_T \in \mathbb{R}^{3,3}$ regular, $\mathbf{t}_T \in \mathbb{R}^3$, is the unique affine mapping taking $\widehat{T}$ to T. Moreover, the local projections on $\widehat{T}$ and affine pullbacks must commute:

$$\Pi_{p|T}^l \circ (\mathbf{\Phi}_T^{-1})^* = (\mathbf{\Phi}_T^{-1})^* \circ \Pi_{p|\widehat{T}}^l. \tag{3.36}$$

We saw in Section 3.2 that Whitney forms are affine-equivalent. For higher-order discrete differential forms this can be ensured by first constructing the local spaces on a reference element. Afterwards (3.35) is used to get all local spaces. The relationship (3.36) is enforced by using affine pullbacks, too, in order to assemble the local interpolation operators from those on $\widehat{T}$ (*cf.* the discussion in Section 4 of Hiptmair (2001*c*)).

For the rest of the article we fix $\widehat{T}$ to be the customary reference tetrahedron spanned by the canonical basis vectors in Euclidean space $\mathbb{R}^3$. Using this, we find, for every $T \in \mathcal{S}_3(\Omega_h)$ (Ciarlet 1978, Theorem 3.1.3),

$$\begin{aligned} \|\mathrm{B}_T\| &\le 4\rho_T h_T, & \|\mathrm{B}_T^{-1}\| &\le \rho_T h_T^{-1}, \\ |\det \mathrm{B}_T| &\le 6h_T^3, & |\det \mathrm{B}_T|^{-1} &\le \tfrac{1}{8\pi}\rho_T h_T^{-3}, \end{aligned}$$

with $\|\mathrm{B}_T\|$ standing for the Euclidean matrix norm. The gist of affine equivalence techniques is to use these estimates in combination with the pullback formulas (2.16)–(2.19) for vector proxies. Thanks to (3.36) any local interpolation error can then be bounded in terms of interpolation errors on $\widehat{T}$.

The estimates of interpolation errors on $\widehat{T}$ rely on extra smoothness of the fields, which is measured by certain Sobolev (semi-)norms of the (components of the) vector proxies. For a vector proxy $\mathbf{u} = (u_1, \ldots, u_K)$, $K \in \mathbb{N}$, they read

$$|\mathbf{u}|_{H^m(\Omega)}^2 := \sum_{k=1}^{K} \int_{\Omega} |D^m u_k(\mathbf{x})|^2 \, \mathrm{d}\mathbf{x}, \quad m \in \mathbb{N}_0,$$

$$|\mathbf{u}|_{H^s(\Omega)}^2 := \sum_{k=1}^{K} \int_{\Omega} \int_{\Omega} \frac{|u_k(\mathbf{x}) - u_k(\mathbf{y})|^2}{|\mathbf{x} - \mathbf{y}|^{3+2s}} \, \mathrm{d}\mathbf{x}\, \mathrm{d}\mathbf{y}, \quad 0 \le s < 1.$$

The associated Sobolev spaces of vector proxies are $\boldsymbol{H}^m(\Omega)$ and $\boldsymbol{H}^s(\Omega)$. The behaviour of these norms under the pullbacks given by (2.16)–(2.18) has to be examined. We include the well-known estimates for 0-forms in order to highlight the pattern.

Lemma 3.12. (Transformation of norms under pullbacks) Let $T = \mathbf{\Phi}_T(\widehat{T})$ be the image of the reference tetrahedron under the bijective affine mapping $\mathbf{\Phi}_T$, $\mathbf{\Phi}_T(\widehat{\mathbf{x}}) := \mathrm{B}_T\widehat{\mathbf{x}} + \mathbf{t}_T$ as above. Then, for all $m \in \mathbb{N}_0$ and all vector-fields/functions in $\boldsymbol{H}^m(T)/H^m(T)$,

$$\begin{aligned}
\left|\mathfrak{F}^0_{\mathbf{\Phi}_T} v\right|^2_{H^m(\widehat{T})} &\leq \|\mathrm{B}_T\|^{2m} |\det \mathrm{B}_T|^{-1} |v|^2_{H^m(T)} &&\leq C\rho_T h_T^{2m-3} |v|^2_{H^m(T)}, \\
\left|\mathfrak{F}^1_{\mathbf{\Phi}_T} \mathbf{u}\right|^2_{\boldsymbol{H}^m(\widehat{T})} &\leq \|\mathrm{B}_T\|^{2+2m} |\det \mathrm{B}_T|^{-1} |\mathbf{u}|^2_{\boldsymbol{H}^m(T)} &&\leq C\rho_T h_T^{2m-1} |\mathbf{u}|^2_{\boldsymbol{H}^m(T)}, \\
\left|\mathfrak{F}^2_{\mathbf{\Phi}_T} \mathbf{u}\right|^2_{\boldsymbol{H}^m(\widehat{T})} &\leq \|\mathrm{B}_T\|^{2m} \left\|\mathrm{B}_T^{-1}\right\|^2 |\det \mathrm{B}_T| |\mathbf{u}|^2_{\boldsymbol{H}^m(T)} &&\leq C\rho_T h_T^{2m+1} |\mathbf{u}|^2_{\boldsymbol{H}^m(T)},
\end{aligned}$$

with universal constants $C > 0$. For any $0 \leq s < 1$, we have

$$\begin{aligned}
\left|\mathfrak{F}^0_{\mathbf{\Phi}_T} v\right|^2_{H^s(\widehat{T})} &\leq \|\mathrm{B}_T\|^{2s+3} |\det \mathrm{B}_T|^{-2} |v|^2_{H^s(T)} &&\leq C\rho_T h_T^{2s-3} |v|^2_{H^s(T)}, \\
\left|\mathfrak{F}^1_{\mathbf{\Phi}_T} \mathbf{u}\right|^2_{\boldsymbol{H}^s(\widehat{T})} &\leq \|\mathrm{B}_T\|^{2s+5} |\det \mathrm{B}_T|^{-2} |\mathbf{u}|^2_{\boldsymbol{H}^s(T)} &&\leq C\rho_T h_T^{2s-1} |\mathbf{u}|^2_{\boldsymbol{H}^s(T)}, \\
\left|\mathfrak{F}^2_{\mathbf{\Phi}_T} \mathbf{u}\right|^2_{\boldsymbol{H}^s(\widehat{T})} &\leq \|\mathrm{B}_T\|^{2s+3} \left\|\mathrm{B}_T^{-1}\right\|^2 |\mathbf{u}|^2_{\boldsymbol{H}^s(T)} &&\leq C\rho_T h_T^{2s+1} |\mathbf{u}|^2_{\boldsymbol{H}^s(T)},
\end{aligned}$$

for all functions/vector-fields in $\boldsymbol{H}^s(T)$, where $C > 0$ are universal constants.[7]

Proof. Using the pullback formulas (2.16)–(2.18), the definitions of the norms, and the transformation formulas for integrals, everything reduces to elementary calculations. For the fractional Sobolev norms the trick is to use

$$|\mathbf{x} - \mathbf{y}| = |\mathrm{B}_T\mathrm{B}_T^{-1}(\mathbf{x} - \mathbf{y})| \leq \|\mathrm{B}_T\| \, |\mathrm{B}_T^{-1}(\mathbf{x} - \mathbf{y})|.$$

The details can be looked up in Ciarlet (1978, Theorem 3.1.2), Alonso and Valli (1999, Section 5), and Ciarlet, Jr, and Zou (1999, Section 3). □

A first application of Lemma 3.12 confirms the h-uniform L^2-*stability* of local bases of $\mathcal{W}^l_p(\Omega_h)$. If the local shape functions on any $T \in \mathcal{S}_3(\Omega_h)$ are obtained by means of affine pullback from those on $\widehat{T}$, we find, for all $\alpha_{\mathbf{b}} \in \mathbb{C}$,

$$C \sum_{\mathbf{b}} \alpha_{\mathbf{b}}^2 \|\mathbf{b}\|^2_{\boldsymbol{L}^2(\Omega)} \leq \|\sum_{\mathbf{b}} \alpha_{\mathbf{b}}\mathbf{b}\|^2_{\boldsymbol{L}^2(\Omega)} \leq C \sum_{\mathbf{b}} \alpha_{\mathbf{b}}^2 \|\mathbf{b}\|^2_{\boldsymbol{L}^2(\Omega)}, \tag{3.37}$$

[7] Most of the estimates will be asymptotic in nature, featuring 'generic constants', for which the symbol C, sometimes tagged with a subscript, is used throughout. The value of these generic constants may vary between different occurrences, but it will always be made clear what they depend upon. When the h-version of discrete differential forms is considered, the constants may not depend on meshwidth.

where $\mathbf{b}$ runs through the set of basis functions of $\mathcal{W}_p^l(\Omega_h)$ and the constants depend on p and $\rho(\Omega_h)$ only. For 0-form the constants do not even depend on $\rho(\Omega_h)$, but this can *not* be expected for $l > 0$. The transformation rules of Lemma 3.12 also establish a Bernstein-type *inverse estimate* for discrete 1-forms: for all $\mathbf{u}_h \in \mathcal{W}_p^1(\Omega_h)$ we find

$$\|\mathbf{curl}\,\mathbf{u}_h\|_{L^2(\Omega)} \leq Ch^{-1}\,\|\mathbf{u}_h\|_{L^2(\Omega)},$$

with $C = C(\rho(\Omega_h)) > 0$.

Affine equivalence techniques owe their power to Bramble–Hilbert arguments on the reference element (Ciarlet 1978, Section 3.1). These are available if, firstly, the interpolation operators $\widehat{\Pi}_p^l$ on $\widehat{T}$ preserve spaces of polynomials. Secondly, they have to be continuous on spaces of sufficiently smooth functions. The latter requirement is addressed by the following lemma (*cf.* Lemma 4.7 in Amrouche *et al.* (1998)).

Lemma 3.13. (Continuity of edge moments) Let $\hat{e}$ be an edge of the reference element $\widehat{T}$ and $\hat{F}$ be a face adjacent to $\hat{e}$. If $\varphi \in W_q^{1-1/q}(\hat{e})$ with $1 < q < 2$, then, for $p^{-1} + q^{-1} = 1$, any smooth vector-field $\mathbf{u}$ on $\widehat{T}$, and $C = C(p) > 0$,

$$\left|\int_{\hat{e}} \varphi\mathbf{u}\cdot \mathrm{d}\mathbf{s}\right| \leq C\big(\,\|\mathbf{curl}\,\mathbf{u}\|_{L^p(\widehat{T})} + \|\gamma_{\mathbf{t}}\mathbf{u}\|_{L^p(\hat{F})}\big)\,\|\varphi\|_{W_q^{1-1/q}(\hat{e})}.$$

Proof. Denote by $\bar{\varphi}$ the extension by zero of φ to $\partial\hat{F}$. As smooth functions with compact support are dense in $W_q^{1-1/q}(\hat{e})$, we also have $\bar{\varphi} \in W_q^{1-1/q}(\partial\hat{F})$. By Theorem 1.5.1.3 in Grisvard (1985) we can extend $\bar{\varphi}$ to a function $\hat{\varphi} \in W_q^1(\hat{F})$ in a stable fashion. Again, $\hat{\varphi}$ can be extended by zero onto $\partial\widehat{T}$, which yields a function $\tilde{\varphi} \in W_q^{1-1/q}(\partial\widehat{T})$. Once more appealing to Theorem 1.5.1.3 in Grisvard (1985), we extend $\tilde{\varphi}$ to $\check{\varphi} \in W_q^1(\widehat{T})$. The continuity of these extensions is reflected by the norm estimates

$$\begin{aligned}\|\check{\varphi}\|_{W_q^1(\widehat{T})} &\leq C\,\|\tilde{\varphi}\|_{W_q^{1-1/q}(\partial\widehat{T})} \leq C\,\|\hat{\varphi}\|_{W_q^1(\hat{F})}\\ &\leq C\,\|\bar{\varphi}\|_{W_q^{1-1/q}(\partial\hat{F})} \leq C\,\|\varphi\|_{W_q^{1-1/q}(\hat{e})},\end{aligned}$$

where C stands for generic positive constants depending only on p. Next, we use integration by parts, more precisely Stokes' theorem on $\hat{F}$ combined with Green's formula in $\widehat{T}$:

$$\begin{aligned}\int_{\hat{e}} \varphi\mathbf{u}\cdot\mathrm{d}\mathbf{s} &= \int_{\partial\hat{F}} \bar{\varphi}\mathbf{u}\cdot\mathrm{d}\mathbf{s} = \int_{\hat{F}} \mathbf{curl}(\mathbf{u}\bar{\varphi})\cdot\mathbf{n}\,\mathrm{d}S\\ &= \int_{\hat{F}} \hat{\varphi}\,\mathbf{curl}\,\mathbf{u}\cdot\mathbf{n}\,\mathrm{d}S - \int_{\hat{F}} (\mathbf{u}\times\mathbf{grad}\,\hat{\varphi})\cdot\mathbf{n}\,\mathrm{d}S\end{aligned}$$

$$= \int_{\partial\widehat{T}} \tilde{\varphi}\,\mathbf{curl}\,\mathbf{u}\cdot\mathbf{n}\,\mathrm{d}S + \int_{\widehat{F}} \mathbf{grad}\,\hat{\varphi}\cdot(\mathbf{u}\times\mathbf{n})\,\mathrm{d}S$$

$$= \int_{\widehat{T}} \mathbf{grad}\,\check{\varphi}\cdot\mathbf{curl}\,\mathbf{u}\,\mathrm{d}\mathbf{x} + \int_{\widehat{F}} \mathbf{grad}\,\hat{\varphi}\cdot(\mathbf{u}\times\mathbf{n})\,\mathrm{d}S.$$

Thus, by Hölder's inequality,

$$\left|\int_{\hat{e}} \varphi\mathbf{u}\cdot\mathrm{d}\mathbf{s}\right| \leq \|\check{\varphi}\|_{W^1_q(\widehat{T})}\,\|\mathbf{curl}\,\mathbf{u}\|_{L^p(\widehat{T})} + \|\hat{\varphi}\|_{W^1_q(\widehat{F})}\,\|\gamma_{\mathbf{t}}\mathbf{u}\|_{L^p(\widehat{F})}.$$

Along with the stability of the extensions this gives the result. □

Transformations of L^p-norms are awkward to handle. Thus it is desirable to switch back to Sobolev spaces $\boldsymbol{H}^s(\widehat{T})$, for which Lemma 3.12 provides transformation rules. This can be achieved by means of the continuous embedding (Adams 1975, Section 5.4)

$$H^s(\widehat{T}) \hookrightarrow L^{\frac{6}{3-2s}}(\widehat{T}), \quad 0 \leq s < \frac{3}{2},$$

and the continuity of the tangential trace mappings

$$(\gamma_{\mathbf{t}})_{|\widehat{F}} : \boldsymbol{H}^{\frac{1}{2}+s}(\widehat{T}) \mapsto \boldsymbol{H}^s(\widehat{F}), \quad s > 0,$$

for each face $\widehat{F}$ of $\widehat{T}$. Recall that the moment-based interpolation operators for discrete 1-forms rely on weighted path integrals along edges, moments of tangential traces on faces and weighted integrals over cells. The weights are fixed polynomials. For $s > \frac{1}{2}$ the continuity on $\boldsymbol{H}^s(\widehat{T})$ of d.o.f. associated with faces and cells is straightforward. This gives continuity

$$\left\|\widehat{\Pi}^1_p\mathbf{u}\right\|_{\boldsymbol{L}^2(\widehat{T})} \leq C\Big(\|\mathbf{u}\|_{\boldsymbol{H}^s(\widehat{T})} + \|\mathbf{curl}\,\mathbf{u}\|_{\boldsymbol{H}^{s-\frac{1}{2}}(\widehat{T})}\Big) \tag{3.38}$$

for all sufficiently smooth $\mathbf{u}$, $s > \frac{1}{2}$, with $C > 0$ depending only on s and p.

Theorem 3.14. (Moment-based interpolation estimates for discrete 1-forms) For any tetrahedral mesh Ω_h with meshwidth $h > 0$, $s \in]\frac{1}{2};1[\cup\mathbb{N}$, and affine equivalent discrete 1-forms $\mathcal{W}^1_p(\Omega_h)$ of uniform polynomial degree $p \in \mathbb{N}_0$, we have the interpolation error estimate

$$\left\|\mathbf{u} - \Pi^1_p\mathbf{u}\right\|_{\boldsymbol{L}^2(\Omega)} \leq Ch^{\min\{s,p+1\}}\big(\|\mathbf{u}\|_{\boldsymbol{H}^s(\Omega)} + \|\mathbf{curl}\,\mathbf{u}\|_{\boldsymbol{H}^s(\Omega)}\big)$$

for all sufficiently smooth vector-fields $\mathbf{u}$, with $C > 0$ depending only on the shape regularity measure of the mesh Ω_h, s, and the specification of $\widehat{\Pi}^1_p$.

Proof. First, we treat the case $s = m \in \mathbb{N}$. Pick any $T \in \mathcal{S}_3(\Omega_h)$. Then use (3.36) and the transformation rule for the $\boldsymbol{L}^2(T)$-norm of 1-forms from

Lemma 3.12 and note that $\mathcal{W}_p^1(\widehat{T})$ contains all polynomials of degree $\leq p$. This gives

$$\begin{aligned}
&\|\mathbf{u} - \Pi^1_{p|T}\mathbf{u}\|^2_{\boldsymbol{L}^2(T)} \\
&\quad\leq C\rho(\Omega_h)h_T\|\mathfrak{F}^1_{\Phi_T}(\mathbf{u} - \widehat{\Pi}^1_p\mathbf{u})\|^2_{\boldsymbol{L}^2(\widehat{T})} \\
&\quad\leq C\rho(\Omega_h)h_T\|(\mathrm{Id} - \widehat{\Pi}^1_p)\widehat{\mathbf{u}}\|^2_{\boldsymbol{L}^2(\widehat{T})} \\
&\quad\leq C\rho(\Omega_h)h_T \inf_{\mathbf{p}\in(\mathcal{P}_p(\widehat{T}))^3}\Big(\|\widehat{\mathbf{u}} - \mathbf{p}\|^2_{\boldsymbol{H}^{m'}(\widehat{T})} + \|\mathbf{curl}(\widehat{\mathbf{u}} - \mathbf{p})\|^2_{\boldsymbol{H}^{r-\frac{1}{2}}(\widehat{T})}\Big) \\
&\quad\leq C\rho(\Omega_h)h_T\Big(|\widehat{\mathbf{u}}|^2_{\boldsymbol{H}^{m'}(\widehat{T})} + |\mathbf{curl}\,\widehat{\mathbf{u}}|^2_{\boldsymbol{H}^{m'}(\widehat{T})}\Big)
\end{aligned}$$

for $\widehat{\mathbf{u}} := \mathfrak{F}^1_{\Phi_T}\mathbf{u}$, $m' = \min\{m, p+1\}$, and constants depending only on p and m. The final step rests on the Peetre–Tartar lemma (Girault and Raviart 1986, Theorem 2.1), which shows the existence of $C = C(m') > 0$, such that, for all smooth $\widehat{\mathbf{u}}$,

$$\begin{aligned}
\inf_{\mathbf{p}\in(\mathcal{P}_p(\widehat{T}))^3}&\Big(\|\widehat{\mathbf{u}} - \mathbf{p}\|_{\boldsymbol{H}^{m'}(\widehat{T})} + \|\mathbf{curl}(\widehat{\mathbf{u}} - \mathbf{p})\|_{\boldsymbol{H}^{m'}(\widehat{T})}\Big) \\
&\qquad\leq C\Big(|\widehat{\mathbf{u}}|_{\boldsymbol{H}^{m'}(\widehat{T})} + |\mathbf{curl}\,\widehat{\mathbf{u}}|_{\boldsymbol{H}^{m'}(\widehat{T})}\Big).
\end{aligned}$$

The remaining semi-norms have to be taken back to T using the transformation rules of 1-forms for $\mathbf{u}$ and those for 2-forms for $\mathbf{curl}\,\mathbf{u}$. This gives

$$\|\mathbf{u} - \Pi^1_{p|T}\mathbf{u}\|^2_{\boldsymbol{L}^2(T)} \leq C\rho(\Omega_h)h_T^{2m'}\Big(|\mathbf{u}|^2_{\boldsymbol{H}^{m'}(T)} + |\mathbf{curl}\,\mathbf{u}|^2_{\boldsymbol{H}^{m'}(T)}\Big),$$

and the assertion for integer s follows by adding these estimates for all cells.

If $\frac{1}{2} < s < 1$, we adapt the above arguments and find

$$\begin{aligned}
&\|\mathbf{u} - \Pi^1_{p|T}\mathbf{u}\|^2_{\boldsymbol{L}^2(T)} \\
&\quad\leq C \inf_{\mathbf{p}\in(\mathcal{P}_0(\widehat{T}))^3}\Big(\|\widehat{\mathbf{u}} - \mathbf{p}\|^2_{\boldsymbol{H}^{s}(\widehat{T})} + \|\mathbf{curl}\,\widehat{\mathbf{u}}\|^2_{\boldsymbol{H}^{s-\frac{1}{2}}(\widehat{T})}\Big) \\
&\quad\leq C\rho(\Omega_h)h_T\Big(|\widehat{\mathbf{u}}|^2_{\boldsymbol{H}^{s}(\widehat{T})} + \|\mathbf{curl}\,\widehat{\mathbf{u}}\|^2_{\boldsymbol{L}^2(\widehat{T})} + |\mathbf{curl}\,\widehat{\mathbf{u}}|^2_{\boldsymbol{H}^{s-\frac{1}{2}}(\widehat{T})}\Big),
\end{aligned}$$

where a Bramble–Hilbert argument in fractional Sobolev spaces was invoked (Dupont and Scott 1980, Proposition 6.1). Then an application of Lemma 3.12 finishes the proof. □

In a similar fashion to Lemma 3.13, the following result can be established.

Lemma 3.15. (Continuity of face moments) For a face $\hat{F}$ of the reference element $\widehat{T}$ and $\varphi \in W_q^{1-1/q}(\hat{F})$ with $1 < q < 2$, we have

$$\left|\int_{\hat{F}} \varphi\mathbf{u}\cdot\mathbf{n}\,\mathrm{d}S\right| \leq C\Big(\|\mathbf{u}\|_{\boldsymbol{L}^p(\widehat{T})} + \|\mathrm{div}\,\mathbf{u}\|_{L^2(\widehat{T})}\Big)\|\varphi\|_{W_q^{1-1/q}(\hat{F})},$$

for any smooth vector-field $\mathbf{u}$ on $\widehat{T}$, and $C = C(p) > 0$.

From this lemma we conclude the continuity

$$\left\|\widehat{\Pi}_p^2 \mathbf{u}\right\|_{L^2(\widehat{T})} \leq C\Big(\|\mathbf{u}\|_{H^s(\widehat{T})} + \|\operatorname{div} \mathbf{u}\|_{L^2(\widehat{T})} \Big) \tag{3.39}$$

for all $s > 0$ and $C = C(s,p) > 0$. In the same fashion as Theorem 3.14, we can establish the following interpolation error estimate.

Theorem 3.16. (Moment-based interpolation estimate for discrete 2-forms) Let Ω_h be a tetrahedral mesh with meshwidth h, and Π_p^2 the local projection onto a space $\mathcal{W}_p^2(\Omega_h)$ belonging to the first family of discrete 2-forms on Ω_h. Then

$$\left\|\mathbf{u} - \Pi_p^2 \mathbf{u}\right\|_{L^2(\Omega)} \leq C h^{\min\{s,p+1\}} \left(\|\mathbf{u}\|_{H^s(\Omega)} + \|\operatorname{div} \mathbf{u}\|_{H^{\max\{0,s\}}(\Omega)} \right)$$

for all sufficiently smooth vector-fields $\mathbf{u}$, $s \in]0,1[\cup \mathbb{N}$, and with a constant $C = C(s, p, \rho(\Omega_h)) > 0$.

This result has been included because, thanks to the commuting diagram property $\Pi_p^2 \circ \mathbf{curl} = \mathbf{curl} \circ \Pi_p^1$, we can instantly infer interpolation estimates for 1-forms in the $\|\mathbf{curl} \cdot\|_{L^2(\Omega)}$-seminorm.

Corollary 3.17. If $\mathbf{u}$ belongs to the domain of Π_p^1 and $\mathbf{curl}\,\mathbf{u}$ is sufficiently smooth, we have, for all $s > 0$,

$$\left\|\mathbf{curl}(\mathbf{u} - \Pi_p^1 \mathbf{u})\right\|_{L^2(\Omega)} \leq C h^{\min\{s,p+1\}} \left\|\mathbf{curl}\,\mathbf{u}\right\|_{H^s(\Omega)}$$

with a constant $C = C(s,p) > 0$ independent of $\mathbf{u}$.

Bibliographical notes

Some more technical details on moment-based interpolation estimates can be found in Girault and Raviart (1986, Chapter 3, Section 5), Alonso and Valli (1999, Section 5), and Ciarlet, Jr, and Zou (1999, Section 3).

Remark 9. All the above estimates hinge on the shape regularity measure, which deteriorates even for tetrahedra like

$$T = [(0,0,0)^T, (h_1,0,0)^T, (0,h_2,0)^T, (0,0,h_3)^T]$$

when h_1, h_2, h_3 differ strongly. We may recall the advice from Section 3.2: it is known that the usual interpolation estimates for the standard local interpolation operator for Whitney 0-forms still hold on such elements with the local interpolation error depending on $\max\{h_1, h_2, h_3\}$ only (Apel and Dobrowolski 1992, Apel 1999). A corresponding result was shown for Whitney 2-forms in Nicaise (2001) and Apel, Nicaise and Schöberl (2001). We may conjecture that Whitney 1-forms behave identically, but proof is at present elusive (Some preliminary investigations for edge elements have been conducted by Nicaise (2001)). △

Remark 10. For the first family of discrete differential forms we get the same order of the interpolation error in the $\boldsymbol{L}^2(\Omega)$-norm and the $\|\mathbf{curl}\cdot\|_{L^2(\Omega)}$-seminorm. This is surprising because one would expect the application of **curl** to reduce one power of h in the error estimate. △

Remark 11. The spaces $\breve{\mathcal{W}}_p^l(\Omega_h)$ of the second family of discrete differential forms locally contain all polynomials up to a total degree of $p+1$. Therefore, in Theorem 3.14 the order of the interpolation error can be raised to $p+2$, smoothness of $\mathbf{u}$ permitting. No improvement in the estimate of Theorem 3.16 can be achieved. Thus, the second family really displays the typical poorer decrease of the interpolation error in norms involving a differential operator. From the perspective of theoretical results it is hard to assess the gain from using the second family in Galerkin schemes: the simultaneous approximation of the vector-field and its **curl** will always be needed. Therefore the **curl** of the interpolation error $\mathbf{u}-\Pi_p^1\mathbf{u}$ will dominate in *a priori* error estimates (*cf.* Sections 4 and 5). △

Remark 12. For parametric elements, that is, the case of a general diffeomorphism $\mathbf{\Phi}_T : \widehat{T} \mapsto T$, apart from the infimum/supremum on $\widehat{T}$ of the expressions $\|D\mathbf{\Phi}_T(\widehat{\mathbf{x}})\|$, $|\det D\mathbf{\Phi}_T(\widehat{\mathbf{x}})|$, norms of higher derivatives of $\mathbf{\Phi}_T$ enter the bounds of Lemma 3.12. This renders the shape regularity measure an inadequate tool for the analysis of interpolation errors (Girault and Raviart (1986, Chapter I, A.2) and Ciarlet (1978, Chapter 4, Section 4.3)). We also have to impose bounds for the deviation of $\mathbf{\Phi}_T$ from an affine mapping. No investigations have yet been carried out for discrete 1-forms, but the results for Lagrangian finite elements should carry over. △

For the p-version of discrete differential forms, the projection-based interpolation operators have to be considered. Their continuity properties are crucial. These hinge on the continuity of the mappings PL_S^l, PE_S^l, and PQ_S^l used for the definition of the interpolation operators in Section 3.5. We examine these separately for tetrahedral discrete 1-forms, starting with lifting mappings. As Ω_h is fixed and the interpolation is local, we need only scrutinize a single element T. All argument functions are supposed to be sufficiently smooth and the rule (3.33) is to be in effect.

- For any edge e the lifting PL_e^1 amounts to integration along e. We conclude that, uniformly in the polynomial degree p,

$$\left\|\mathsf{PL}_e^1(u)\right\|_{L^2(e)} \leq C\left(\|u\|_{H^{-1}(e)} + \left|\int_e l(s)u(s)\,\mathrm{d}s\right|\right),$$

  where $l(s)$ is a linear function on e vanishing at one endpoint and assuming the value 1 at the other.

- For a triangular face F and a tetrahedral cell T, the liftings PL_F^1, PL_T^1 boil down to finding scalar potentials, that is, $\mathbf{grad}\,\mathsf{PL}_S^1\mathbf{u} = \mathbf{u}$, $S \in \{F, T\}$, in terms of vector proxies. Imposing vanishing averages on the scalar potentials, it is elementary that, uniformly in p and $\mathbf{u}$,

$$\left\|\mathsf{PL}_S^1(\mathbf{u})\right\|_{H^1(S)} \leq C \left\|\mathbf{u}\right\|_{\boldsymbol{L}^2(S)}, \quad S \in \{F, T\}.$$

- The lifting mappings PL_F^2, F face, and PL_T^2, T cell, for 2-forms are to yield vector potentials with zero boundary conditions. Suitable mappings can be constructed as in the proof of Lemma 3.5 based on the Poincaré mapping. The L^2-continuity of the Poincaré mapping (confirmed by applying the Cauchy–Schwarz inequality to (3.21)) combined with extension theorems for polynomials, known in the theory of p-version Lagrangian finite elements (Babuška, Craig, Mandel and Pitkäranta 1991, Section 7), gives

$$\left\|\mathsf{PL}_S^2(\mathbf{u})\right\|_{\boldsymbol{L}^2(S)} \leq C \left\|\mathbf{u}\right\|_{\boldsymbol{L}^2(S)}, \quad S \in \{F, T\},$$

  with $C > 0$ independent of $\mathbf{u}$ and p. Details in the case of faces can be found in Demkowicz and Babuška (2001, Section 3).

As mappings PQ_e^1 we choose $H^{-1}(e)$-orthogonal projections. All the other required projections PQ_F^1, PQ_F^2, F face, and PQ_T^1, PQ_T^2, T cell, may be $\boldsymbol{L}^2(S)$-orthogonal, $S \in \{F, T\}$.

Finally, the extensions PE_e^0, e edge, and PE_F^0, F face, are extensions of polynomials with zero traces on ∂e and ∂F, respectively. We demand that they satisfy

$$\left\|\mathsf{PE}_e^0(q)\right\|_{H^1(T)} \leq C \left\|q\right\|_{L^2(e)},$$
$$\left\|\mathsf{PE}_F^0(q)\right\|_{H^1(T)} \leq C \left\|q\right\|_{H^1(F)},$$

for all admissible polynomials q and uniformly in p. For the extension PE_e^0 a related result can be found in Pavarino and Widlund (1996, Section 4.4). The existence of a stable extension from the faces can be concluded from Munoz-Solar (1997, Theorem 1). One more extension PE_F^1 of polynomial vector-fields on faces is needed. Unfortunately, no results about possible p-uniformly continuous constructions are available. Therefore we make the *assumption* that, for all p and $\mathbf{u} \in \mathcal{X}_p^1(F)$,

$$\left\|\mathsf{PE}_F^1(\mathbf{u})\right\|_{\boldsymbol{L}^2(T)} \leq C \left\|\mathbf{u}\right\|_{\boldsymbol{L}^2(F)}. \tag{3.40}$$

From Adams (1975, Theorem 7.58) we learn the continuous embedding of $H^1(I)$ into $W_q^{1-1/q}(I)$, $1 \leq q \leq 2$, for a bounded interval $I \subset \mathbb{R}$. Recalling Lemma 3.13, we conclude, for any edge e of T, sufficiently smooth vector-fields $\mathbf{u}$, and $q > 2$, that

$$\|\mathbf{u}\cdot\mathbf{t}_e\|_{H^{-1}(e)} + \left|\int_e l(s)u(s)\,\mathrm{d}s\right| \leq C\left(\|\mathbf{curl}\,\mathbf{u}\|_{L^q(T)} + \|\gamma_{\mathbf{t}}\mathbf{u}\|_{L^q(F)}\right)$$

with $C = C(q,T) > 0$. As in Lemma 3.13, F has to be a face of T adjacent to e, and $\mathbf{t}_e$ stands for a unit vector in the direction of e. A straightforward inspection of the construction given in Figure 3.5, the above stabilities of liftings and projections, and embeddings and standard trace theorems for Sobolev spaces thus reward us with the continuity

$$\left\|\Pi^1_{p|T}\mathbf{u}\right\|_{L^2(T)} \leq C\Big(\|\mathbf{u}\|_{H^{\frac{1}{2}+s}(T)} + \|\mathbf{curl}\,\mathbf{u}\|_{H^s(T)}\Big), \tag{3.41}$$

with $C = C(s,T) > 0$ independent of p and for any $s > 0$.

Theorem 3.18. (p-interpolation for discrete 1-forms) Under the assumptions made above, the projection-based interpolation operators Π^1_p, $p \in \mathbb{N}_0$, satisfy

$$\left\|\mathbf{u} - \Pi^1_p\mathbf{u}\right\|_{L^2(\Omega)} \leq Cp^{\frac{1}{2}-\epsilon}\Big(\|\mathbf{u}\|_{H^1(\Omega)} + \|\mathbf{curl}\,\mathbf{u}\|_{H^1(\Omega)}\Big),$$

for any $\epsilon > 0$, and for all sufficiently smooth vector-fields $\mathbf{u}$ satisfying $C = C(\rho(\Omega_h),\epsilon) > 0$ independent of p. For all $r > 1$ and $\epsilon > 0$ we have, with $C = C(r,\epsilon,\rho(\Omega_h)) > 0$,

$$\left\|\mathbf{u} - \Pi^1_p\mathbf{u}\right\|_{L^2(\Omega)} \leq Cp^{r-1-\epsilon}\,\|\mathbf{u}\|_{H^r(\Omega)}.$$

Proof. Thanks to scaling arguments it suffices to consider a single element T. Pick $\mathbf{u} \in H^1(T)$, $\mathbf{curl}\,\mathbf{u} \in H^1(T)$, and use the decomposition of Lemma 2.6

$$\mathbf{u} = \boldsymbol{\Psi} + \mathbf{grad}\,\varphi, \quad \boldsymbol{\Psi} \in H^2(T),\ \varphi \in H^2(T).$$

Since φ is continuous on T, $\Pi^0_p\varphi$ is well defined. The same holds true for $\Pi^1_p\boldsymbol{\Psi}$ and we can use the commuting diagram property in order to get

$$\left\|\mathbf{u} - \Pi^1_p\mathbf{u}\right\|_{L^2(T)} \leq \left\|\boldsymbol{\Psi} - \Pi^1_p\boldsymbol{\Psi}\right\|_{L^2(T)} + \left|\varphi - \Pi^0_p\varphi\right|_{H^1(T)}.$$

The interpolation operators Π^1_p and Π^0_p locally preserve polynomials of degree p and $p+1$, respectively. Thus,

$$\begin{aligned}\left\|\mathbf{u} - \Pi^1_p\mathbf{u}\right\|_{L^2(T)} \leq &\inf_{\mathbf{p}\in(\mathcal{P}_p(T))^3}\left\|(\mathrm{Id}-\Pi^1_p)(\boldsymbol{\Psi}-\mathbf{p})\right\|_{L^2(T)} \\ &+ \inf_{p\in\mathcal{P}_{p+1}(T)}\left|(\mathrm{Id}-\Pi^0_p)(\varphi-p)\right|_{H^1(T)}.\end{aligned}$$

The p-uniform continuity of the interpolation operators leads to

$$l\left\|\mathbf{u}-\Pi_p^1\mathbf{u}\right\|_{L^2(T)} \leq$$
$$C\left(\inf_{\mathbf{p}\in(\mathcal{P}_p(T))^3}\|\mathbf{\Psi}-\mathbf{p}\|_{H^{1+\epsilon}(T)} + \inf_{p\in\mathcal{P}_{p+1}(T)}\|\varphi-p\|_{H^{\frac{3}{2}+\epsilon}(T)}\right),$$

for any $\epsilon > 0$ and with $C = C(\epsilon, T) > 0$. Standard estimates for best approximation by polynomials (Braess and Schwab 2000, Theorem 3.3) and the stability of the splitting yield the first assertion. Using the continuity

$$\left\|\Pi_{p|T}^1\mathbf{u}\right\|_{L^2(T)} \leq C\left\|\mathbf{u}\right\|_{H^{1+s}(T)} \quad \forall\mathbf{u}\in \boldsymbol{H}^{1+s}(T),$$

the proof of the second only involves the last two steps in the above considerations. □

Bibliographical notes

Estimates for a projection-based interpolation operator for discrete 1-forms in 2D are given in Demkowicz and Babuška (2001). Approximation estimates for higher-order discrete 1-forms on quadrilaterals and hexahedra are covered in Monk (1994) and Ainsworth and Pinchedez (2001).

4. Maxwell eigenvalue problem

In our first model problem $\Omega \subset A(\mathbb{R}^3)$ plays the role of a bounded cavity filled with non-conducting dielectric material and lined by ideally conducting walls. The material parameters $\boldsymbol{\epsilon}, \boldsymbol{\mu}$ are supposed to be piecewise constant. We aim to compute *resonant frequencies* and related eigenmodes of Ω, that is, both $\omega \neq 0$ and electromagnetic fields have to be determined, such that Maxwell's equations (2.1), (2.2) and $\gamma_t\mathbf{e} = 0$ on $\partial\Omega$ are satisfied. This task is called the (electric) Maxwell eigenvalue problem.

Following the reasoning in Section 2.3, primal and dual variational formulations of the Maxwell eigenvalue problem can be derived. In terms of vector proxies the primal, $\mathbf{e}$-based formulation reads: Find $\mathbf{e} \in \boldsymbol{H}_0(\mathbf{curl};\Omega)$, $\omega \neq 0$ such that

$$\left(\boldsymbol{\mu}^{-1}\,\mathbf{curl}\,\mathbf{e}, \mathbf{curl}\,\mathbf{e}'\right)_{L^2(\Omega)} = \omega^2\left(\boldsymbol{\epsilon}\mathbf{e}, \mathbf{e}'\right)_{L^2(\Omega)} \quad \forall\mathbf{e}' \in \boldsymbol{H}_0(\mathbf{curl};\Omega). \tag{4.1}$$

The dual, $\mathbf{h}$-based formulation seeks $\mathbf{h} \in \boldsymbol{H}(\mathbf{curl};\Omega)$, $\omega \neq 0$, such that

$$\left(\boldsymbol{\epsilon}^{-1}\,\mathbf{curl}\,\mathbf{h}, \mathbf{curl}\,\mathbf{h}'\right)_{L^2(\Omega)} = \omega^2\left(\boldsymbol{\mu}\mathbf{h}, \mathbf{h}'\right)_{L^2(\Omega)} \quad \forall\mathbf{h}' \in \boldsymbol{H}(\mathbf{curl};\Omega). \tag{4.2}$$

The treatment of (4.1) and (4.2) is very similar. In the following we restrict our attention to (4.1).

4.1. Embeddings

As far as structure is concerned, the eigenvalue problem (4.1) much resembles the Dirichlet eigenvalue problem for the Laplacian. However, the infinite-dimensional kernel of **curl** introduces a pronounced difference: In contrast to Δ the operator **curl curl** fails to have a compact resolvent in $\boldsymbol{L}^2(\Omega)$. A fundamental prerequisite for the application of the powerful Riesz–Schauder spectral theory for compact operators is missing.

Yet, as can be seen by testing with irrotational functions, due to $\omega \neq 0$, (4.1) can be equivalently stated on the space

$$\boldsymbol{Z}_0(\boldsymbol{\epsilon},\Omega) := \{\mathbf{u} \in \boldsymbol{H}_0(\mathbf{curl};\Omega),\ (\boldsymbol{\epsilon}\mathbf{u},\mathbf{z})_{\boldsymbol{L}^2(\Omega)} = 0\ \forall \mathbf{z} \in \boldsymbol{H}_0(\mathbf{curl}\,0;\Omega)\}. \quad (4.3)$$

Seek $\mathbf{e} \in \boldsymbol{Z}_0(\boldsymbol{\epsilon},\Omega)$, $\omega \neq 0$ such that

$$\left(\mu^{-1}\,\mathbf{curl}\,\mathbf{e}, \mathbf{curl}\,\mathbf{e}'\right)_{\boldsymbol{L}^2(\Omega)} = \omega^2\left(\boldsymbol{\epsilon}\mathbf{e}, \mathbf{e}'\right)_{\boldsymbol{L}^2(\Omega)} \quad \forall \mathbf{e}' \in \boldsymbol{Z}_0(\boldsymbol{\epsilon},\Omega). \quad (4.4)$$

The advantage of the formulation (4.4) is clear from the following fundamental embedding result. To state it, we let $\boldsymbol{\alpha}$ denote a generic piecewise smooth metric tensor and introduce the Hilbert space

$$\boldsymbol{X}_0(\boldsymbol{\alpha},\Omega) := \{\mathbf{u} \in \boldsymbol{H}_0(\mathbf{curl};\Omega),\ \mathrm{div}(\boldsymbol{\alpha}\mathbf{u}) \in L^2(\Omega)\},$$

equipped with the natural graph norm. It is essential that the extra constraint on the divergence involved in the definition of $\boldsymbol{X}_0(\boldsymbol{\alpha},\Omega)$ enforces a slightly enhanced regularity of the vector-fields (*cf.* Amrouche *et al.* (1998, Proposition 3.7)).

Theorem 4.1. There exists $s_0 > 0$ such that the space $\boldsymbol{X}_0(\boldsymbol{\alpha},\Omega)$ is continuously embedded in $\boldsymbol{H}^s(\Omega)$ for all $s < s_0$. If $\boldsymbol{\alpha}$ is uniformly Lipschitz-continuous, we can choose $s_0 > \frac{1}{2}$.

Proof. Pick any $\mathbf{u} \in \boldsymbol{X}_0(\boldsymbol{\alpha},\Omega)$ and use the stable regular decomposition from Lemma 2.4:

$$\mathbf{u} = \boldsymbol{\Psi} + \mathbf{grad}\,\varphi, \quad \boldsymbol{\Psi} \in \boldsymbol{H}^1(\Omega) \cap \boldsymbol{H}_0(\mathbf{curl};\Omega),\ \varphi \in H_0^1(\Omega). \quad (4.5)$$

Formally, we find that φ satisfies

$$\varphi \in H_0^1(\Omega): \quad -\mathrm{div}(\boldsymbol{\alpha}\,\mathbf{grad}\,\varphi) = \mathrm{div}(\boldsymbol{\alpha}\boldsymbol{\Psi}) - \mathrm{div}(\boldsymbol{\alpha}\mathbf{u}). \quad (4.6)$$

Since $\boldsymbol{\alpha}$ is piecewise smooth and $\boldsymbol{\Psi} \in \boldsymbol{H}^1(\Omega)$, we conclude that $\mathrm{div}(\boldsymbol{\alpha}\boldsymbol{\Psi}) \in H^{-\frac{1}{2}-\epsilon}(\Omega)$ for any $\epsilon > 0$.

Now we invoke lifting theorems for the Dirichlet problem for second-order elliptic operators on curvilinear Lipschitz polyhedra, and in the case of piecewise smooth coefficients. These can be found in Dauge (1988) and Costabel *et al.* (1999, Section 4). They guarantee the existence of $s^*_{\mathrm{Dir}}(\boldsymbol{\alpha}) > 1$ such that for all $1 \le t < s^*_{\mathrm{Dir}}(\boldsymbol{\alpha})$, the solution of

$$\phi \in H_0^1(\Omega): \quad \mathrm{div}(\boldsymbol{\alpha}\,\mathbf{grad}\,\phi) = f \in H^{t-2}(\Omega) \quad \text{in } \Omega$$

satisfies $\varphi \in H^t(\Omega)$ and $\|\varphi\|_{H^t(\Omega)} \le C\,\|f\|_{H^{t-2}(\Omega)}$. This sophisticated result is obtained by studying corner, edge and interface singularities of solutions of the Dirichlet problem for the Laplacian. Thus, we learn that $\varphi \in H^t(\Omega)$ for all $t < \min\{s^*_{\mathrm{Dir}}(\boldsymbol{\alpha}), \frac{3}{2}\}$, which involves the first assertion of the theorem with $s_0 := \min\{s^*_{\mathrm{Dir}}(\boldsymbol{\alpha}) - 1, \frac{1}{2}\}$.

To get the second, remember that it is known from the work of Dauge (1988, Chapter 6) (see also Costabel and Dauge (1998)) that $s^*_{\mathrm{Dir}}(\mathbf{I}) > \frac{3}{2}$. Further, note that for uniformly Lipschitz-continuous $\boldsymbol{\alpha}$ the right-hand side in (4.6) belongs to $L^2(\Omega)$. □

The same technique enables us to establish another embedding theorem (*cf.* Amrouche *et al.* (1998, Proposition 3.7) and Hazard and Lenoir (1996, Lemma B8)).

Lemma 4.2. If $\mathbf{u} \in \boldsymbol{H}(\mathbf{curl};\Omega)$, $\operatorname{div}(\boldsymbol{\alpha}\mathbf{u}) = 0$, and $\gamma_{\mathbf{n}}(\boldsymbol{\alpha}\mathbf{u}) = 0$ on $\partial\Omega$, then there exists $r_0 > 0$ such that $\mathbf{u} \in \boldsymbol{H}^r(\Omega)$ for all $0 < r < r_0$ and

$$\|\mathbf{u}\|_{\boldsymbol{H}^r(\Omega)} \le C\Big(\|\mathbf{u}\|_{\boldsymbol{L}^2(\Omega)} + \|\mathbf{curl}\,\mathbf{u}\|_{\boldsymbol{L}^2(\Omega)} + \|\operatorname{div}(\boldsymbol{\alpha}\mathbf{u})\|_{L^2(\Omega)}\Big),$$

with $C = C(\Omega, r) > 0$. If $\boldsymbol{\alpha}$ is uniformly Lipschitz-continuous, then we can choose $r_0 > \frac{1}{2}$.

Recalling Rellich's theorem for scales of classical Sobolev space we can infer a compact embedding (*cf.* Amrouche *et al.* (1998, Theorem 2.8) and Jochmann (1997)).

Corollary 4.3. The embedding $\boldsymbol{X}_0(\boldsymbol{\epsilon}, \Omega) \hookrightarrow \boldsymbol{L}^2(\Omega)$ is compact.

This compact embedding immediately implies a Poincaré–Friedrichs-type inequality, because $\operatorname{div}(\boldsymbol{\epsilon}\mathbf{u}) = 0$ for all $\mathbf{u} \in \boldsymbol{Z}_0(\boldsymbol{\epsilon}, \Omega)$.

Corollary 4.4. (Poincaré–Friedrichs-type inequality) There is a constant $C > 0$ depending on Ω only, such that

$$\|\mathbf{u}\|_{\boldsymbol{L}^2(\Omega)} \le C\,\|\mathbf{curl}\,\mathbf{u}\|_{\boldsymbol{L}^2(\Omega)} \quad \forall \mathbf{u} \in \boldsymbol{Z}_0(\boldsymbol{\epsilon}, \Omega).$$

Therefore, by the Lax–Milgram lemma, the variational equation

$$\left(\boldsymbol{\mu}^{-1}\,\mathbf{curl}\,\mathsf{T}\mathbf{u}, \mathbf{curl}\,\mathbf{u}'\right)_{\boldsymbol{L}^2(\Omega)} = \left(\boldsymbol{\epsilon}\mathbf{u}, \mathbf{u}'\right)_{\boldsymbol{L}^2(\Omega)} \quad \forall \mathbf{u}' \in \boldsymbol{Z}_0(\boldsymbol{\epsilon}, \Omega),$$

defines a continuous operator $\mathsf{T} : \boldsymbol{L}^2(\Omega) \mapsto \boldsymbol{Z}_0(\boldsymbol{\epsilon}, \Omega)$, which is compact as an operator $\mathsf{T} : \boldsymbol{L}^2(\Omega) \mapsto \boldsymbol{L}^2(\Omega)$. As such, T is also self-adjoint by virtue of the symmetry of the bilinear forms involved in its definition.

By means of T the eigenvalue problem (4.4) can be recast as

$$\left(\boldsymbol{\mu}^{-1}\,\mathbf{curl}\,\mathbf{e}, \mathbf{curl}\,\mathbf{e}'\right)_{\boldsymbol{L}^2(\Omega)} = \omega^2\left(\boldsymbol{\mu}^{-1}\,\mathbf{curl}\,\mathsf{T}\mathbf{e}, \mathbf{curl}\,\mathbf{e}'\right)_{\boldsymbol{L}^2(\Omega)} \quad \forall \mathbf{e}' \in \boldsymbol{Z}_0(\boldsymbol{\epsilon}, \Omega),$$

and converted into the operator eigenvalue problem

$$\mathsf{T}\mathbf{e} = \omega^{-2}\mathbf{e}.$$

Thus, the Riesz–Schauder theory for the spectrum of compact, self-adjoint operators in Hilbert space implies that (4.1) has an increasing sequence $0 < \omega_1 < \omega_2 < \cdots$ of nonzero real 'Maxwell eigenvalues' tending to ∞. They all have finite multiplicity and the corresponding eigenspaces are mutually $\boldsymbol{L}^2(\Omega)$-orthogonal. Hence, by switching to the complement $\boldsymbol{Z}_0(\boldsymbol{\epsilon}, \Omega)$ of Ker(**curl**) we have recovered a situation typical of second-order elliptic eigenproblems. We point out that $\omega_k \in \mathbb{R}$ makes it possible to work with *real* field amplitudes only, when solving a Maxwell eigenvalue problem.

Bibliographical notes
The main reference for this section is Amrouche *et al.* (1998, Sections 2,3). The statement of Corollary 4.3 was first proved in Weber (1980) and can also be found in Picard (1984), and an extension to mixed boundary conditions can be found in Jochmann (1997).

4.2. Fortin projectors

Using the space $\mathcal{W}^1_{p,0}(\Omega_h) := \boldsymbol{H}_0(\mathbf{curl}; \Omega) \cap \mathcal{W}^1_p(\Omega_h)$ of discrete 1-forms on a triangulation Ω_h of Ω a Galerkin discretization of (4.1) is straightforward. Seek $\mathbf{e}_h \in \mathcal{W}^1_{p,0}(\Omega_h)$, $\omega_h \neq 0$ such that

$$\left(\boldsymbol{\mu}^{-1} \mathbf{curl}\, \mathbf{e}_h, \mathbf{curl}\, \mathbf{e}'_h\right)_{\boldsymbol{L}^2(\Omega)} = \omega_h^2 \left(\boldsymbol{\epsilon} \mathbf{e}_h, \mathbf{e}'_h\right)_{\boldsymbol{L}^2(\Omega)} \quad \forall \mathbf{e}'_h \in \mathcal{W}^1_{p,0}(\Omega_h). \tag{4.7}$$

Below we examine the convergence of ω_h and $\mathbf{u}_h$ for the h-version of discrete differential forms. Throughout we take for granted affine equivalence of the spaces $\mathcal{W}^1_{p,0}(\Omega_h)$ and $\mathcal{W}^2_{p,0}(\Omega_h)$ of discrete forms as well as the commuting diagram property for the moment-based local projectors Π^1_p and Π^2_p.

Parallel to the continuous case, a variational problem equivalent to (4.7) can be posed on the space

$$\boldsymbol{Z}_{h,0}(\boldsymbol{\epsilon}, \Omega_h) = \{\mathbf{u}_h \in \mathcal{W}^1_{p,0}(\Omega_h),\ (\boldsymbol{\epsilon}\mathbf{u}_h, \mathbf{z}_h)_{\boldsymbol{L}^2(\Omega)} = 0\ \forall \mathbf{z}_h \in \mathcal{W}^1_{p,0}(\Omega_h) \cap \mathrm{Ker}(\mathbf{curl})\}.$$

Does this pave the way for studying the approximation of Maxwell eigenvalues along the same lines as for the discrete Laplacian, namely by appealing to the theory of eigenvalue approximation for self-adjoint positive definite operators with compact resolvent? Unfortunately, this hope is dashed by the observation that, in general,

$$\boldsymbol{Z}_{h,0}(\boldsymbol{\epsilon}, \Omega_h) \not\subset \boldsymbol{Z}_0(\boldsymbol{\epsilon}, \Omega).$$

Bluntly speaking, in terms of (4.4) the variational problem (4.7) when restricted to $\boldsymbol{Z}_{h,0}(\boldsymbol{\epsilon}, \Omega_h)$ is a *non-conforming* discretization. What is available in this case is the theory of spectral approximation of compact operators from Babuška and Osborn (1991). It requires us to introduce a discrete operator $\mathsf{T}_h : \boldsymbol{L}^2(\Omega) \mapsto \boldsymbol{Z}_{h,0}(\boldsymbol{\epsilon}, \Omega_h)$ that approximates T. To begin with, we

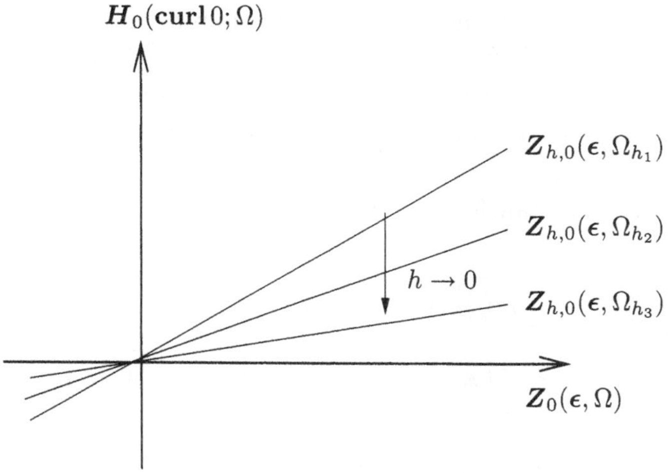

Figure 4.1. Graphical illustration of the message of Lemma 4.5. The plane of the paper represents $\boldsymbol{H}(\mathbf{curl};\Omega)$ with its natural geometry

badly need a link between $\boldsymbol{Z}_0(\epsilon,\Omega)$ and its non-conforming approximating space $\boldsymbol{Z}_{h,0}(\epsilon,\Omega_h)$. It is provided by the *Hodge mapping* $\mathsf{H}_\epsilon : \boldsymbol{H}_0(\mathbf{curl};\Omega) \mapsto \boldsymbol{Z}_0(\epsilon,\Omega)$, an $(\epsilon\cdot,\cdot)_{\boldsymbol{L}^2(\Omega)}$-orthogonal projection, defined by

$$\mathsf{H}_\epsilon\mathbf{u} \in \boldsymbol{Z}_0(\epsilon,\Omega): \quad \mathbf{curl}\,\mathsf{H}_\epsilon\mathbf{u} = \mathbf{curl}\,\mathbf{u}, \quad \mathbf{u} \in \boldsymbol{H}_0(\mathbf{curl};\Omega). \tag{4.8}$$

Thanks to Corollary 4.4 this is a valid definition, because an element of $\boldsymbol{Z}_0(\epsilon,\Omega)$ is already uniquely determined by its **curl**. By the following approximation result we see that $\boldsymbol{Z}_0(\epsilon,\Omega)$ and $\boldsymbol{Z}_{h,0}(\epsilon,\Omega_h)$ are close on fine meshes, provided that ϵ is regular enough: the gap between them will shrink to zero as $h \to 0$, as illustrated in Figure 4.1.

Lemma 4.5. (Approximation estimate for the Hodge map on $\mathcal{Z}_{h,0}$) If ϵ is uniformly Lipschitz-continuous, we have

$$\|\mathbf{u}_h - \mathsf{H}_\epsilon\mathbf{u}_h\|_{\boldsymbol{L}^2(\Omega)} \leq Ch^s \|\mathbf{curl}\,\mathbf{u}_h\|_{\boldsymbol{L}^2(\Omega)} \quad \forall \mathbf{u}_h \in \boldsymbol{Z}_{h,0}(\epsilon,\Omega_h),$$

for $\frac{1}{2} < s < s_0$, s_0 from Theorem 4.1, and a constant $C > 0$ depending only on s, ϵ, Ω, and the shape regularity of Ω_h.

The proof will make use of the following technical result.

Lemma 4.6. Let $\frac{1}{2} < s \leq 1$, and $\mathbf{u} \in \boldsymbol{H}^s(\Omega) \cap \boldsymbol{H}_0(\mathbf{curl};\Omega)$. If $\mathbf{curl}\,\mathbf{u}$ belongs to the space $\mathcal{W}^2_{p,0}(\Omega_h)$ of discrete 2-forms on Ω_h, then

$$\left\|\mathbf{u} - \Pi^1_p\mathbf{u}\right\|_{\boldsymbol{L}^2(\Omega)} \leq Ch^s \, |\mathbf{u}|_{\boldsymbol{H}^s(\Omega)},$$

with $C = C(\Omega, s, p, \rho(\Omega_h)) > 0$ independent of $\mathbf{u}$ and h.

Proof. The idea of the proof can be found in Amrouche *et al.* (1998, Section 4). Consider a single element $T \in \mathcal{S}_3(\Omega_h)$. We conclude from Lemma 3.13 and trace/embedding theorems for Sobolev spaces that the moment-based local interpolant $\Pi^l_{p|T}\mathbf{u}$ is well defined, since $\gamma_{\mathbf{t}}\mathbf{u} \in \boldsymbol{H}^{s-\frac{1}{2}}(F)$ for each face F of T and $\mathbf{curl}\,\mathbf{u}_{|T}$ is smooth.

Then use the transformation formulae of Lemma 3.12 to get, with $\widehat{\mathbf{u}} := \mathfrak{F}^1_{\Phi_T}\mathbf{u}$,

$$\begin{aligned}\left\|\mathbf{u} - \Pi^1_p\mathbf{u}\right\|_{\boldsymbol{L}^2(T)} &\leq C h_T^{\frac{1}{2}} \|(\mathrm{Id} - \widehat{\Pi}^1_p)\widehat{\mathbf{u}}\|_{\boldsymbol{L}^2(\widehat{T})} \\ &\leq C h_T^{\frac{1}{2}} \inf_{\mathbf{p} \in (\mathcal{P}_p(\widehat{T}))^3} \Big(\|\widehat{\mathbf{u}} - \mathbf{p}\|_{\boldsymbol{H}^s(\widehat{T})} + \|\mathbf{curl}(\widehat{\mathbf{u}} - \mathbf{p})\|_{\boldsymbol{H}^{s-\frac{1}{2}}(\widehat{T})} \Big).\end{aligned}$$

Observe that $\mathbf{curl}(\widehat{\mathbf{u}} - \mathbf{p})$ belongs to $\mathcal{W}^2_p(\widehat{T})$. All norms on the finite-dimensional space $\mathbf{curl}\,\mathcal{W}^1_p(\widehat{T}) \subset \mathcal{W}^2_p(\widehat{T})$ are equivalent, whence

$$\|\mathbf{curl}(\widehat{\mathbf{u}} - \mathbf{p})\|_{\boldsymbol{H}^{s-\frac{1}{2}}(\widehat{T})} \leq C \, \|\widehat{\mathbf{u}} - \mathbf{p}\|_{\boldsymbol{H}^s(\widehat{T})}.$$

This permits us to continue the estimate by

$$\begin{aligned}&\left\|\mathbf{u} - \Pi^1_p\mathbf{u}\right\|_{\boldsymbol{L}^2(T)} \leq \\ &\quad C h_T^{\frac{1}{2}} \inf_{\mathbf{p} \in (\mathcal{P}_p(\widehat{T}))^3} \|\widehat{\mathbf{u}} - \mathbf{p}\|_{\boldsymbol{H}^s(\widehat{T})} \leq C h_T^{\frac{1}{2}} \, |\widehat{\mathbf{u}}|_{\boldsymbol{H}^s(\widehat{T})} \leq C h_T^{s} \, \|\mathbf{u}\|_{\boldsymbol{H}^s(T)},\end{aligned}$$

where we used the Bramble–Hilbert lemma for fractional Sobolev norms on $\widehat{T}$ (Dupont and Scott 1980, Proposition 6.1) and Lemma 3.12. The generic constants $C > 0$ only depend on s, p, and the shape regularity measure $\rho(\Omega_h)$. Summing this estimate over all elements yields the assertion. □

Proof of Lemma 4.5. Pick any $\mathbf{u}_h \in \boldsymbol{Z}_{h,0}(\boldsymbol{\epsilon}, \Omega_h)$. As $\mathsf{H}_{\boldsymbol{\epsilon}}\mathbf{u}_h \in \boldsymbol{Z}_0(\boldsymbol{\epsilon}, \Omega)$, we know from Theorem 4.1 ($\boldsymbol{\epsilon}$ uniformly Lipschitz-continuous) that

$$\|\mathsf{H}_{\boldsymbol{\epsilon}}\mathbf{u}_h\|_{\boldsymbol{H}^s(\Omega)} \leq C\Big(\|\mathsf{H}_{\boldsymbol{\epsilon}}\mathbf{u}_h\|_{\boldsymbol{L}^2(\Omega)} + \|\mathbf{curl}\,\mathsf{H}_{\boldsymbol{\epsilon}}\mathbf{u}_h\|_{\boldsymbol{L}^2(\Omega)} + \|\mathrm{div}(\boldsymbol{\epsilon}\mathsf{H}_{\boldsymbol{\epsilon}}\mathbf{u}_h)\|_{L^2(\Omega)} \Big), \tag{4.9}$$

with $\frac{1}{2} < s < s_0$. By definition of $\mathsf{H}_{\boldsymbol{\epsilon}}$, Theorem 4.1, and the previous lemma we learn that $\Pi^1_p\mathsf{H}_{\boldsymbol{\epsilon}}\mathbf{u}_h$ is well defined. This permits us to apply Nédélec's trick, from Nédélec (1980),

$$\begin{aligned}&\|\boldsymbol{\epsilon}^{\frac{1}{2}}(\mathbf{u}_h - \mathsf{H}_{\boldsymbol{\epsilon}}\mathbf{u}_h)\|^2_{\boldsymbol{L}^2(\Omega)} \\ &\quad = \big(\boldsymbol{\epsilon}(\mathbf{u}_h - \mathsf{H}_{\boldsymbol{\epsilon}}\mathbf{u}_h), \mathbf{u}_h - \Pi^1_p\mathsf{H}_{\boldsymbol{\epsilon}}\mathbf{u}_h + \Pi^1_p\mathsf{H}_{\boldsymbol{\epsilon}}\mathbf{u}_h - \mathsf{H}_{\boldsymbol{\epsilon}}\mathbf{u}_h\big)_{\boldsymbol{L}^2(\Omega)} \\ &\quad = \big(\boldsymbol{\epsilon}(\mathbf{u}_h - \mathsf{H}_{\boldsymbol{\epsilon}}\mathbf{u}_h), \Pi^1_p\mathsf{H}_{\boldsymbol{\epsilon}}\mathbf{u}_h - \mathsf{H}_{\boldsymbol{\epsilon}}\mathbf{u}_h\big)_{\boldsymbol{L}^2(\Omega)} \\ &\quad \leq \|\boldsymbol{\epsilon}^{\frac{1}{2}}(\mathbf{u}_h - \mathsf{H}_{\boldsymbol{\epsilon}}\mathbf{u}_h)\|_{\boldsymbol{L}^2(\Omega)} \|\boldsymbol{\epsilon}^{\frac{1}{2}}(\Pi^1_p - \mathrm{Id})\mathsf{H}_{\boldsymbol{\epsilon}}\mathbf{u}_h\|_{\boldsymbol{L}^2(\Omega)}.\end{aligned}$$

The commuting diagram property is pivotal here. It guarantees that

$$\mathbf{curl}(\mathbf{u}_h - \Pi_p^1 \mathsf{H}_\epsilon \mathbf{u}_h) = \mathbf{curl}(\Pi_p^1(\mathbf{u}_h - \mathsf{H}_\epsilon \mathbf{u}_h)) = \Pi_p^2(\mathbf{curl}(\mathbf{u}_h - \mathsf{H}_\epsilon \mathbf{u}_h)) = 0,$$

by definition of H_ϵ. Since both $\boldsymbol{Z}_0(\epsilon, \Omega)$ and $\boldsymbol{Z}_{h,0}(\epsilon, \Omega_h)$ are $(\epsilon\cdot, \cdot)_{\boldsymbol{L}^2(\Omega)}$-orthogonal to Ker($\mathbf{curl}$) $\cap\, \mathcal{W}_{p,0}^1(\Omega_h)$ the above manipulations are justified. Then we can use the interpolation estimate of Lemma 4.5, $\operatorname{div}(\epsilon \mathsf{H}_\epsilon \mathbf{u}_h) = 0$, and the fact that ϵ is uniformly bounded from above and below, and get from (4.9)

$$\|\mathbf{u}_h - \mathsf{H}_\epsilon \mathbf{u}_h\|_{\boldsymbol{L}^2(\Omega)} \leq C h^s \|\mathsf{H}_\epsilon \mathbf{u}_h\|_{\boldsymbol{H}^s(\Omega)} \leq C h^s \|\mathbf{curl}\, \mathsf{H}_\epsilon \mathbf{u}_h\|_{\boldsymbol{L}^2(\Omega)}.$$

This amounts to the contention of the theorem. □

Now we are in a position to establish a discrete version of the Poincaré–Friedrichs inequality from Corollary 4.4.

Theorem 4.7. (Discrete Poincaré–Friedrichs inequality) There is a positive constant C depending only on Ω, ϵ, p and the shape regularity of the mesh, such that

$$\|\mathbf{u}_h\|_{\boldsymbol{L}^2(\Omega)} \leq C \|\mathbf{curl}\, \mathbf{u}_h\|_{\boldsymbol{L}^2(\Omega)} \quad \forall \mathbf{u}_h \in \boldsymbol{Z}_{h,0}(\epsilon, \Omega_h).$$

Proof. Let $\mathbf{I}$ stand for the 3×3 identity matrix. Then, pick some $\mathbf{u}_h \in \boldsymbol{Z}_{h,0}(\epsilon, \Omega_h)$ and use its *Helmholtz decompositions*

$$\mathbf{u}_h = \begin{cases} \mathbf{z} + \mathbf{q}, & \mathbf{z} \in \boldsymbol{Z}_0(\mathbf{I}, \Omega),\ \mathbf{curl}\, \mathbf{q} = 0, \\ \mathbf{z}_h + \mathbf{q}_h, & \mathbf{z}_h \in \boldsymbol{Z}_{h,0}(\mathbf{I}, \Omega_h),\ \mathbf{curl}\, \mathbf{q}_h = 0. \end{cases}$$

From the definition of $\mathsf{H}_\mathbf{I}$ it is clear that $\mathbf{z} = \mathsf{H}_\mathbf{I} \mathbf{z}_h$. Hence, by Lemma 4.5 we find $s > \frac{1}{2}$ and $C = C(\Omega, \rho(\Omega_h)) > 0$ such that

$$\|\mathbf{z} - \mathbf{z}_h\|_{\boldsymbol{L}^2(\Omega)} \leq C h^s \|\mathbf{curl}\, \mathbf{u}_h\|_{\boldsymbol{L}^2(\Omega)}.$$

We can exploit the orthogonality in the definition of $\boldsymbol{Z}_{h,0}(\epsilon, \Omega_h)$ and get

$$\begin{aligned} (\epsilon \mathbf{u}_h, \mathbf{u}_h)_{\boldsymbol{L}^2(\Omega)} &= (\epsilon \mathbf{u}_h, \mathbf{z}_h + \mathbf{q}_h)_{\boldsymbol{L}^2(\Omega)} = (\epsilon \mathbf{u}_h, \mathbf{z}_h)_{\boldsymbol{L}^2(\Omega)} \\ &= (\epsilon \mathbf{u}_h, \mathbf{z}_h - \mathbf{z} + \mathbf{z})_{\boldsymbol{L}^2(\Omega)} \\ &\leq C\, (\epsilon \mathbf{u}_h, \mathbf{u}_h)_{\boldsymbol{L}^2(\Omega)}^{\frac{1}{2}} \left(\|\mathbf{z} - \mathbf{z}_h\|_{\boldsymbol{L}^2(\Omega)} + \|\mathbf{z}\|_{\boldsymbol{L}^2(\Omega)} \right), \end{aligned}$$

with $C > 0$ depending on ϵ. Assembling the estimates and using Corollary 4.4 yields the assertion. □

The Hodge mapping takes us from $\boldsymbol{Z}_{h,0}(\epsilon, \Omega_h)$ to $\boldsymbol{Z}_0(\epsilon, \Omega)$. The opposite direction is covered by the *Fortin projector* $\mathsf{F}_h : \boldsymbol{H}_0(\mathbf{curl}; \Omega) \mapsto \boldsymbol{Z}_{h,0}(\epsilon, \Omega_h)$, a term that originated in the theory of mixed finite elements (Brezzi and

Fortin 1991, Section II.2.3). Due to the discrete Poincaré–Friedrichs inequality, $\mathsf{F}_h\mathbf{u} \in \boldsymbol{Z}_{h,0}(\boldsymbol{\epsilon}, \Omega_h)$ can be defined for all $\mathbf{u} \in \boldsymbol{H}_0(\mathbf{curl}; \Omega)$ by

$$\left(\boldsymbol{\mu}^{-1}\,\mathbf{curl}\,\mathsf{F}_h\mathbf{u}, \mathbf{curl}\,\mathbf{u}'_h\right)_{\boldsymbol{L}^2(\Omega)} = \left(\boldsymbol{\mu}^{-1}\,\mathbf{curl}\,\mathbf{u}, \mathbf{curl}\,\mathbf{u}'_h\right)_{\boldsymbol{L}^2(\Omega)} \qquad \forall \mathbf{u}'_h \in \boldsymbol{Z}_{h,0}(\boldsymbol{\epsilon}, \Omega_h). \tag{4.10}$$

Therefore $\mathbf{curl}\,\mathsf{F}_h\mathbf{u} = \mathsf{Q}_h^{1/\boldsymbol{\mu}}\,\mathbf{curl}\,\mathbf{u}$, where $\mathsf{Q}_h^{1/\boldsymbol{\mu}}$ denotes the $\left(\boldsymbol{\mu}^{-1}\cdot, \cdot\right)_{\boldsymbol{L}^2(\Omega)}$-orthogonal projection onto $\mathbf{curl}\,\mathcal{W}^1_{p,0}(\Omega_h)$. The following theorem mirrors the approximation estimate for the Hodge mapping.

Theorem 4.8. (Fortin projector approximation estimate) If $\boldsymbol{\epsilon}$ is uniformly Lipschitz-continuous, then there exist $r = r(\Omega, \boldsymbol{\mu}, \boldsymbol{\epsilon}) > 0$ and $C = C(r, \Omega, \boldsymbol{\epsilon}, \boldsymbol{\mu}) > 0$ such that

$$\left\|\mathsf{F}_h\mathbf{u} - \mathbf{u}\right\|_{\boldsymbol{L}^2(\Omega)} \leq Ch^r \left\|\mathbf{curl}\,\mathbf{u}\right\|_{\boldsymbol{L}^2(\Omega)} \quad \forall \mathbf{u} \in \boldsymbol{Z}_0(\boldsymbol{\epsilon}, \Omega).$$

Proof. We employ a duality technique invented by D. Boffi (2000). Pick $\mathbf{u} \in \boldsymbol{Z}_0(\boldsymbol{\epsilon}, \Omega)$ and fix $\mathbf{u}^* \in \boldsymbol{Z}_0(\boldsymbol{\epsilon}, \Omega)$ by demanding $\mathbf{curl}\,\mathbf{u}^* = \mathsf{Q}_h^{1/\boldsymbol{\mu}}\,\mathbf{curl}\,\mathbf{u}$. This is meaningful because of the Poincaré–Friedrichs inequality of Corollary 4.4. First, observe that the approximation estimate of Lemma 4.5 for the Hodge mapping readily yields (for a suitable $s > 0$)

$$\left\|\mathsf{F}_h\mathbf{u} - \mathbf{u}^*\right\|_{\boldsymbol{L}^2(\Omega)} = \left\|(\mathrm{Id} - \mathsf{H}_{\boldsymbol{\epsilon}})\mathsf{F}_h\mathbf{u}\right\|_{\boldsymbol{L}^2(\Omega)} \leq Ch^s \left\|\mathbf{curl}\,\mathsf{F}_h\mathbf{u}\right\|_{\boldsymbol{L}^2(\Omega)}. \tag{4.11}$$

We proceed with a duality estimate. For $\mathbf{g} \in \boldsymbol{L}^2(\Omega)$ consider the saddle point problem that seeks $\mathbf{w} = \mathbf{w}(\mathbf{g}) \in \boldsymbol{H}_0(\mathbf{curl}; \Omega)$, $\mathbf{p} = \mathbf{p}(\mathbf{g}) \in \mathbf{curl}\,\boldsymbol{H}_0(\mathbf{curl}; \Omega)$ such that

$$\begin{aligned} \left(\boldsymbol{\epsilon}\mathbf{w}, \mathbf{w}'\right)_{\boldsymbol{L}^2(\Omega)} \quad + \left(\mathbf{curl}\,\mathbf{w}', \boldsymbol{\mu}^{-1}\mathbf{p}\right)_{\boldsymbol{L}^2(\Omega)} &= 0, \\ \left(\boldsymbol{\mu}^{-1}\,\mathbf{curl}\,\mathbf{w}, \mathbf{p}'\right)_{\boldsymbol{L}^2(\Omega)} \qquad\qquad &= \left(\boldsymbol{\mu}^{-1}\mathbf{g}, \mathbf{p}'\right)_{\boldsymbol{L}^2(\Omega)}, \end{aligned}$$

for all $\mathbf{w}' \in \boldsymbol{H}_0(\mathbf{curl}; \Omega)$, $\mathbf{p}' \in \mathbf{curl}\,\boldsymbol{H}_0(\mathbf{curl}; \Omega)$. As $\mathbf{curl}\,\boldsymbol{H}_0(\mathbf{curl}; \Omega)$ is a closed subspace of $\boldsymbol{L}^2(\Omega)$ and the inf-sup conditions (Brezzi and Fortin 1991, Chapter 2) are trivially satisfied, we get a unique solution $(\mathbf{w}(\mathbf{g}), \mathbf{p}(\mathbf{g}))$.

Obviously, $\mathbf{w}(\mathbf{g}) \in \boldsymbol{Z}_0(\boldsymbol{\epsilon}, \Omega)$, and for $\mathbf{v} := \boldsymbol{\mu}^{-1}\mathbf{p}$ we deduce from the variational equations

$$\mathbf{curl}\,\mathbf{v} = -\boldsymbol{\epsilon}\mathbf{w} \in \boldsymbol{L}^2(\Omega), \quad \mathrm{div}(\boldsymbol{\mu}\mathbf{v}) = 0 \quad \text{in } \Omega, \quad \gamma_{\mathbf{n}}(\boldsymbol{\mu}\mathbf{v}) = 0 \quad \text{on } \partial\Omega.$$

Moreover, since $\mathbf{p} \in \mathbf{curl}\,\boldsymbol{H}_0(\mathbf{curl}; \Omega)$, $\mathbf{v}$ is orthogonal to the space of $\boldsymbol{\mu}$-harmonic Neumann vector-fields. Since those provide all functions $\mathbf{h}$ satisfying $\mathbf{curl}\,\mathbf{h} = 0$, $\mathrm{div}(\boldsymbol{\mu}\mathbf{h}) = 0$, $\gamma_{\mathbf{n}}(\boldsymbol{\mu}\mathbf{h}) = 0$, we conclude from Lemma 4.2 that there is some $0 < r < \frac{1}{2}$, $r < r_0$, r_0 from Theorem 4.2, such that

$$\left\|\mathbf{v}\right\|_{\boldsymbol{H}^r(\Omega)} \leq C\Big(\left\|\mathbf{curl}\,\mathbf{v}\right\|_{\boldsymbol{L}^2(\Omega)} + \left\|\mathrm{div}(\boldsymbol{\mu}\mathbf{v})\right\|_{L^2(\Omega)}\Big).$$

The piecewise smooth matrix function μ is a multiplier in $H^r(\Omega)$. Thus, we end up with an estimate for $\mathbf{p}$:

$$\begin{aligned}\|\mathbf{p}(\mathbf{g})\|_{H^r(\Omega)} &\le C\,\|\mathbf{v}\|_{H^r(\Omega)} \le C\,\|\mathbf{curl}(\mu^{-1}\mathbf{p})\|_{L^2(\Omega)}\\ &\le C\,\|\epsilon\mathbf{w}\|_{L^2(\Omega)} \le C\,\|\mathbf{curl}\,\mathbf{w}\|_{L^2(\Omega)} \le C\,\|\mathbf{g}\|_{L^2(\Omega)},\end{aligned}$$

where we used Corollary 4.4 in the penultimate step. From here on, the generic constants are allowed to depend on the material parameters ϵ and μ.

In the spirit of classical duality techniques, we find $\mathbf{u}-\mathbf{u}^* = \mathbf{w}(\mathbf{g})$, if $\mathbf{g} = (\mathrm{Id}-\mathsf{Q}_h^{1/\mu})\,\mathbf{curl}\,\mathbf{u}$. From the saddle point problem we can extract

$$\begin{aligned}\|\mathbf{u}-\mathbf{u}^*\|_{L^2(\Omega)} &\le C\,(\epsilon(\mathbf{u}-\mathbf{u}^*),\mathbf{u}-\mathbf{u}^*)_{L^2(\Omega)}\\ &\le C\,\left(\mu^{-1}\mathbf{p},\mathbf{curl}(\mathbf{u}-\mathbf{u}^*)\right)_{L^2(\Omega)}\\ &\le C\,\left(\mu^{-1}(\mathrm{Id}-\Pi_p^2)\mathbf{p},(\mathrm{Id}-\mathsf{Q}_h^{1/\mu})\,\mathbf{curl}\,\mathbf{u}\right)_{L^2(\Omega)}.\end{aligned}$$

This is legal, because by (3.39) the interpolant $\Pi_p^2\mathbf{p}$ is well defined. As, by Lemma 2.4, there exists $\mathbf{q}\in H^1(\Omega)\cap H_0(\mathbf{curl};\Omega)$ such that $\mathbf{p}=\mathbf{curl}\,\mathbf{q}$, the commuting diagram property gives

$$\Pi_p^2\mathbf{p} = \mathbf{curl}\,\Pi_p^1\mathbf{q} \in \mathbf{curl}\,\mathcal{W}_{p,0}^1(\Omega_h).$$

Finally, the properties of $\mathsf{Q}_h^{1/\mu}$ justify the above manipulations. The proof is finished by using Theorem 3.16,

$$\begin{aligned}\|\mathbf{u}-\mathbf{u}^*\|_{L^2(\Omega)} &\le C\,\left\|(\mathrm{Id}-\Pi_p^2)\mathbf{p}\right\|_{L^2(\Omega)}\left\|\mu^{-\frac{1}{2}}(\mathrm{Id}-\mathsf{Q}_h^{1/\mu})\,\mathbf{curl}\,\mathbf{u}\right\|_{L^2(\Omega)}\\ &\le Ch^r\,\|\mathbf{p}\|_{H^r(\Omega)}\left\|\mu^{-\frac{1}{2}}\,\mathbf{curl}\,\mathbf{u}\right\|_{L^2(\Omega)}\\ &\le Ch^r\,\|\mathbf{curl}\,\mathbf{u}\|_{L^2(\Omega)},\end{aligned}$$

and taking into account (4.11). □

Compared to the local interpolation operator Π_p^1, the big advantage of the Fortin projector F_h is that it is well defined on all of $H_0(\mathbf{curl};\Omega)$. For both projectors Ker($\mathbf{curl}$) (restricted to their domains) is an invariant subspace. Yet properties of F_h are much more difficult to establish because of its non-local nature.

Remark 13. The assumption that ϵ be uniformly Lipschitz-continuous is very restrictive, because the common case of Ω being filled with different dielectric materials invariably leads to discontinuous ϵ. At present we need this assumption, in order to be able to apply Lemma 4.6. It seems likely that the statement of this lemma can be extended to less regular vector-fields by following the lines of the proofs of Theorem 4.1 and of Theorem 3.5

in Costabel *et al.* (1999), and using information about interface singularities of solutions of the Dirichlet boundary value problem for $\operatorname{div}(\boldsymbol{\epsilon}\,\mathbf{grad}\,\cdot)$. △

4.3. Error estimates

First, owing to the discrete Poincaré–Friedrichs inequality of Theorem 4.7, we can define a meaningful operator $\mathsf{T}_h : \boldsymbol{L}^2(\Omega) \mapsto \boldsymbol{Z}_{h,0}(\boldsymbol{\epsilon}, \Omega_h)$ by

$$\left(\boldsymbol{\mu}^{-1}\,\mathbf{curl}\,\mathsf{T}_h\mathbf{u}, \mathbf{curl}\,\mathbf{u}_h'\right)_{\boldsymbol{L}^2(\Omega)} = \left(\boldsymbol{\epsilon}\mathbf{u}, \mathbf{u}_h'\right)_{\boldsymbol{L}^2(\Omega)} \quad \forall \mathbf{u}_h' \in \mathcal{W}^1_{p,0}(\Omega_h).$$

In particular, this implies

$$\left(\boldsymbol{\mu}^{-1}\,\mathbf{curl}(\mathsf{T}_h - \mathsf{T})\mathbf{u}, \mathbf{curl}\,\mathbf{u}_h'\right)_{\boldsymbol{L}^2(\Omega)} = 0 \quad \forall \mathbf{u}_h' \in \mathcal{W}^1_{p,0}(\Omega_h),$$

which can be expressed through the Fortin projector as

$$\mathsf{T}_h = \mathsf{F}_h \circ \mathsf{T}.$$

Thus, Theorem 4.8 and the continuity of T lead to

$$\begin{aligned}\|(\mathsf{T} - \mathsf{T}_h)\mathbf{u}\|_{\boldsymbol{L}^2(\Omega)} &\le \|(\mathrm{Id} - \mathsf{F}_h)\mathsf{T}\mathbf{u}\|_{\boldsymbol{L}^2(\Omega)} \\ &\le Ch^r\,\|\mathbf{curl}\,\mathsf{T}\mathbf{u}\|_{\boldsymbol{L}^2(\Omega)} \le Ch^r\,\|\mathbf{u}\|_{\boldsymbol{L}^2(\Omega)}\end{aligned}$$

for all $\mathbf{u} \in \boldsymbol{L}^2(\Omega)$ and some $r > 0$, provided that $\boldsymbol{\epsilon}$ is uniformly Lipschitz-continuous. Under this assumption we conclude *uniform* convergence $\mathsf{T}_h \overset{h\to 0}{\longrightarrow} \mathsf{T}$ in $\boldsymbol{L}^2(\Omega)$, that is,

$$\|\mathsf{T} - \mathsf{T}_h\|_{\boldsymbol{L}^2(\Omega)\to\boldsymbol{L}^2(\Omega)} \le Ch^r \to 0 \quad \text{for } h \to 0. \tag{4.12}$$

This paves the way for applying the powerful results on spectral approximation put forth in Babuška and Osborn (1991, Section 7), because a key assumption of this theory is that of uniform convergence $\mathsf{T}_h \overset{h\to 0}{\longrightarrow} \mathsf{T}$. Writing $\mathbf{v}_1, \ldots, \mathbf{v}_m$ for the orthonormalized eigenfunctions of T belonging to an eigenvalue $\omega > 0$ of multiplicity $m \in \mathbb{N}$, we conclude from Babuška and Osborn (1991, Theorem 7.3) that *on sufficiently fine meshes* we will find m discrete eigenvalues $\omega_1, \ldots, \omega_m$ such that

$$\begin{aligned}|\omega^{-2} - \omega_k^{-2}| \le C\Bigg(&\sum_{n,l=1}^{m} ((\mathsf{T} - \mathsf{T}_h)\mathbf{v}_n, \mathbf{v}_l)_{\boldsymbol{L}^2(\Omega)} \\ &+ \left\|(\mathsf{T} - \mathsf{T}_h)_{|\operatorname{Span}\{\mathbf{v}_1,\ldots,\mathbf{v}_m\}}\right\|^2_{\boldsymbol{L}^2(\Omega)\to\boldsymbol{L}^2(\Omega)}\Bigg).\end{aligned} \tag{4.13}$$

Moreover, for each $\mathbf{v}_k$ we find a discrete eigenfunction $\mathbf{v}_h$ such that

$$\|\mathbf{v}_k - \mathbf{v}_h\|_{\boldsymbol{L}^2(\Omega)} \le C\left\|(\mathsf{T} - \mathsf{T}_h)_{|\operatorname{Span}\{\mathbf{v}_1,\ldots,\mathbf{v}_m\}}\right\|_{\boldsymbol{L}^2(\Omega)\to\boldsymbol{L}^2(\Omega)}. \tag{4.14}$$

In both cases the constants are independent of the choice of the finite element spaces.

Hence, information about the convergence of eigenvalues can be obtained from studying the *pointwise convergence* $\mathsf{T}_h \to \mathsf{T}$. The latter can be derived from the following saddle point problem. Seek $\mathsf{T}\mathbf{u} \in \boldsymbol{H}_0(\mathbf{curl};\Omega)$, $\mathbf{p} \in \boldsymbol{H}_0(\mathbf{curl}\,0;\Omega)$ such that

$$\begin{aligned} \left(\mu^{-1}\,\mathbf{curl}\,\mathsf{T}\mathbf{u}, \mathbf{curl}\,\mathbf{v}'\right)_{\boldsymbol{L}^2(\Omega)} + \left(\mathbf{v}', \mathbf{p}\right)_{\boldsymbol{L}^2(\Omega)} &= \left(\epsilon\mathbf{u}, \mathbf{v}'\right)_{\boldsymbol{L}^2(\Omega)}, \\ \left(\mathsf{T}\mathbf{u}, \mathbf{p}'\right)_{\boldsymbol{L}^2(\Omega)} &= 0, \end{aligned} \tag{4.15}$$

for all $\mathbf{v}' \in \boldsymbol{H}_0(\mathbf{curl};\Omega)$, $\mathbf{p}' \in \boldsymbol{H}_0(\mathbf{curl}\,0;\Omega)$. The discrete Poincaré–Friedrichs inequality guarantees that the Galerkin discretization of (4.15) based on $\mathcal{W}_p^1(\Omega_h)$ and $\mathcal{W}_p^1(\Omega_h)\cap\mathrm{Ker}(\mathbf{curl})$ will satisfy 'ellipticity on the kernel', the critical one of the LBB-conditions for (4.15) (Brezzi and Fortin 1991, Chapter 2). Please be aware that in (4.13) and (4.14) only eigenvectors occur as arguments to T and T_h. These belong to $\boldsymbol{Z}_0(\epsilon,\Omega)$ and this makes the solution for $\mathbf{p}$ and its discrete approximations vanish. Summing up, for $\mathbf{v} \in \boldsymbol{Z}_0(\epsilon,\Omega)$ we get the asymptotic *a priori* error estimate

$$\|(\mathsf{T}-\mathsf{T}_h)\mathbf{v}\|_{\boldsymbol{H}(\mathbf{curl};\Omega)} \le C \inf_{\mathbf{u}_h\in\mathcal{W}_p^1(\Omega_h)} \|\mathsf{T}\mathbf{v}-\mathbf{u}_h\|_{\boldsymbol{H}(\mathbf{curl};\Omega)}, \tag{4.16}$$

where $C > 0$ depends on the constant in Corollary 4.7 and the material parameters. Ultimately the rate of convergence of eigenvalues and eigenvectors will be governed by the smoothness of the eigenvectors. Its discussion will be postponed until the end of Section 5.2. One might object that using (4.16) gives sub-optimal estimates, since the $\boldsymbol{L}^2(\Omega)$-norm should be targeted. However, it seems that better estimates for the $\boldsymbol{L}^2(\Omega)$-norm are hard to get.

Bibliographical notes

The above plan for analysing the Maxwell eigenvalue problem was first applied in Boffi, Fernandes, Gastaldi and Perugia (1999), based on techniques of Boffi, Brezzi and Gastaldi (2000). The Fortin projector F_h was introduced in Boffi (2000).

4.4. Discrete compactness

The use of the Fortin projector to tackle the Maxwell eigenvalue problem is fairly recent. The first successful convergence analysis of Kikuchi (1989) pursued a different course and employed the notions of discrete compactness and collective compactness (Anselone 1971). They permit us to reduce convergence of eigenvalues/eigenfunctions to the pointwise convergence $\mathsf{T}_h \to \mathsf{T}$ without a detour via uniform convergence. It will turn out that thus we can get rid of the restrictive assumptions on ϵ that were necessary in the previous section. It should be stressed that discrete compactness is a very interesting property of discrete 1-forms in its own right, important beyond the analysis of eigenvalue problems.

However, the arguments are based on sequential compactness. Therefore, they always target a family $(\Omega_h)_{h\in\mathbb{H}}$ of triangulations, for which the sequence $\mathbb{H}$ of meshwidths is decreasing and converging to 0. Moreover, $\rho(\Omega_h)$, $h \in \mathbb{H}$, is to be bounded uniformly in h. In the terminology of Ciarlet (1978, Section 3.2) this characterizes a shape-regular family of meshes.

Theorem 4.9. (Discrete compactness property) Let $(\Omega_h)_{h\in\mathbb{H}}$ be a uniformly shape-regular family of triangulations of Ω with meshwidths tending to 0. Any sequence $(\mathbf{u}_h)_{h\in\mathbb{H}}$ with $\mathbf{u}_h \in \boldsymbol{Z}_{h,0}(\epsilon, \Omega_h)$ that is uniformly bounded in $\boldsymbol{H}(\mathbf{curl}; \Omega)$ contains a subsequence that converges strongly in $\boldsymbol{L}^2(\Omega)$.

Proof. (See the proof of Theorem 4.1 in Kirsch and Monk (2000), and Caorsi, Fernandes and Raffetto (2000).) As in the proof of Theorem 4.7 we first look at the $\boldsymbol{L}^2(\Omega)$-orthogonal Helmholtz decompositions

$$\mathbf{u}_h = \begin{cases} \mathbf{z}^h + \mathbf{q}^h, & \mathbf{z}^h \in \boldsymbol{Z}_0(\mathbf{I}, \Omega),\ \mathbf{curl}\,\mathbf{q}^h = 0, \\ \mathbf{z}_h^h + \mathbf{q}_h^h, & \mathbf{z}_h^h \in \boldsymbol{Z}_{h,0}(\mathbf{I}, \Omega_h),\ \mathbf{curl}\,\mathbf{q}_h^h = 0. \end{cases}$$

As both Helmholtz decompositions are uniformly stable in the $\boldsymbol{H}(\mathbf{curl}; \Omega)$-norm, $(\mathbf{z}^h)_{h\in\mathbb{H}}$ is a bounded sequence in $\boldsymbol{Z}_0(\mathbf{I}, \Omega)$. Hence, by Corollary 4.3, it possesses a subsequence, still denoted by $(\mathbf{z}^h)_{h\in\mathbb{H}}$, that converges in $\boldsymbol{L}^2(\Omega)$ to some $\mathbf{z} \in \boldsymbol{L}^2(\Omega)$ as $h \to 0$. Appealing to Lemma 4.5 we conclude that $\|\mathbf{z}_h - \mathbf{z}_h^h\|_{\boldsymbol{L}^2(\Omega)} \to 0$. Hence, $(\mathbf{z}_h^h)_{h\in\mathbb{H}}$ must converge in $\boldsymbol{L}^2(\Omega)$ to the same limit $\mathbf{z}$ as $(\mathbf{z}^h)_{h\in\mathbb{H}}$.

By definition of $\boldsymbol{Z}_{h,0}(\epsilon, \Omega_h)$,

$$\left(\epsilon \mathbf{q}_h^h, \mathbf{p}_h\right)_{\boldsymbol{L}^2(\Omega)} = -\left(\epsilon \mathbf{z}_h^h, \mathbf{p}_h\right)_{\boldsymbol{L}^2(\Omega)} \quad \forall \mathbf{p}_h \in \mathrm{Ker}(\mathbf{curl}) \cap \mathcal{W}^1_{p,0}(\Omega_h).$$

This can be regarded as a perturbed Galerkin approximation of the following continuous variational problem. Seek $\mathbf{q} \in \boldsymbol{H}_0(\mathbf{curl}\,0; \Omega)$ such that

$$(\epsilon \mathbf{q}, \mathbf{p})_{\boldsymbol{L}^2(\Omega)} = -(\epsilon \mathbf{z}, \mathbf{p})_{\boldsymbol{L}^2(\Omega)} \quad \forall \mathbf{p} \in \boldsymbol{H}_0(\mathbf{curl}\,0; \Omega).$$

Strang's lemma (Ciarlet 1978, Theorem 4.4.1) teaches us that

$$\|\mathbf{q} - \mathbf{q}_h^h\|_{\boldsymbol{L}^2(\Omega)} \le C\left(\inf_{\mathbf{v}_h \in \mathcal{W}^1_{p,0}(\Omega_h) \cap \mathrm{Ker}(\mathbf{curl})} \|\mathbf{q} - \mathbf{v}_h\|_{L^2(\Omega)} + \|\mathbf{z}_h^h - \mathbf{z}\|_{L^2(\Omega)}\right), \tag{4.17}$$

where $C > 0$ depends only on ϵ. Next, recall that there is a representation

$$\mathbf{q} = \mathbf{grad}\,\phi + \mathbf{h}, \quad \phi \in H_0^1(\Omega),\ \mathbf{h} \in \boldsymbol{\mathcal{H}}^2(\Omega).$$

The cohomology space, by orthogonalization with respect to all gradients, can always be chosen to be a subspace of $\boldsymbol{X}_0(\mathbf{I},\Omega) \cap \boldsymbol{H}_0(\mathbf{curl}\,0;\Omega)$. Therefore, $\Pi_p^1\mathbf{h}$ makes sense. From Theorem 4.1 in combination with Lemma 4.6 we learn that $\Pi_p^1\mathbf{h}$ converges to $\mathbf{h}$ in $\boldsymbol{L}^2(\Omega)$.

It is also known that $\bigcup_{h\in\mathbb{H}} \mathcal{W}_{p,0}^0(\Omega_h)$ is dense in $H_0^1(\Omega)$ (Schatz and Wang 1996). Hence, $\mathbf{v}_h$ in (4.17) can be built from the gradient of the best approximation of ϕ in $\mathcal{W}_{p,0}^0(\Omega_h)$ and the interpolant $\Pi_p^1\mathbf{h}$. This will achieve $\|\mathbf{q}-\mathbf{v}_h\|_{\boldsymbol{L}^2(\Omega)} \to 0$ as $h \to 0$. As a consequence, we also have $\mathbf{q}_h^h \to \mathbf{q}$ in $\boldsymbol{L}^2(\Omega)$. Summing up, $\mathbf{u}_h \to \mathbf{z}+\mathbf{q}$ in $\boldsymbol{L}^2(\Omega)$ for $h \to 0$. □

Remark 14. It is hardly surprising that the discrete Friedrich's inequality is implied by the discrete compactness property (Monk and Demkowicz 2001, Caorsi *et al.* 2000). △

The discrete compactness property enables us to use the theory of collectively compact operators (Kress 1989, Section 10.3). We recall the principal definition.

Definition 6. A family $\mathcal{A}$ of linear operators mapping a normed space $\mathcal{X}$ into a normed space $\mathcal{Y}$ is called *collectively compact*, if, for each bounded set $U \subset \mathcal{X}$, the image $\mathcal{A}(U) := \{Ax,\, x \in U,\, A \in \mathcal{A}\}$ is relatively compact in $\mathcal{Y}$.

Theorem 4.10. (Collective compactness of T_h) On a shape-regular family of meshes with meshwidth $h \in \mathbb{H}$ tending to zero, the family $\{\mathsf{T}_h\}_{h\in\mathbb{H}}$ of operators is collectively compact.

Proof. The proof is adapted from Monk and Demkowicz (2001). We first take a look at a sequence $(\mathbf{w}_n)_{n\in\mathbb{N}}$ in the space

$$\mathcal{W} := \bigcup_h \boldsymbol{Z}_{h,0}(\epsilon,\Omega_h) \subset \boldsymbol{H}(\mathbf{curl};\Omega)$$

that is bounded in $\boldsymbol{H}_0(\mathbf{curl};\Omega)$. Write h_n for the largest meshwidth $\in \mathbb{H}$ for which $\mathbf{w}_n$ belongs to $\boldsymbol{Z}_{h,0}(\epsilon,\Omega_h)$. If the set $\{h_n\}_{n\in\mathbb{N}}$ is finite, the sequence $(\mathbf{w}_n)_n$ is contained in a space of finite dimension and must have a convergent subsequence in $\boldsymbol{L}^2(\Omega)$.

Otherwise, we can assume $h_n \to 0$ as $n \to \infty$. By the discrete compactness property we may extract a subsequence that converges in $\boldsymbol{L}^2(\Omega)$. Thus, it is immediate that $\mathcal{W}$ is compactly embedded in $\boldsymbol{L}^2(\Omega)$.

Next, let U be a bounded set in $\boldsymbol{L}^2(\Omega)$. As

$$\|\mathbf{curl}\,\mathsf{T}_h\mathbf{f}\|_{\boldsymbol{L}^2(\Omega)} \le C\,\|\mathbf{f}\|_{\boldsymbol{L}^2(\Omega)},$$

with $C > 0$ depending only on $\boldsymbol{\epsilon}$ and $\boldsymbol{\mu}$, the set

$$\{\mathsf{T}_h\mathbf{f},\, h \in \mathbb{H},\, \mathbf{f} \in U\}$$

is bounded in $\mathcal{W}$. Therefore, it must be relatively compact in $\boldsymbol{L}^2(\Omega)$. □

Citing Theorems 4 and 5 from Osborn (1975), collective compactness of $(\mathsf{T}_h)_{h\in\mathbb{H}}$ along with the pointwise convergence $\mathsf{T}_h \stackrel{h\to 0}{\longrightarrow} \mathsf{T}$, established at the end of the previous section, gives us that the estimates (4.13) and (4.14) still hold for an arbitrary metric tensor $\boldsymbol{\epsilon}$.

Remark 15. *A priori convergence estimates for the p-version of discrete 1-forms applied to the variational eigenvalue problem (4.1) could not be obtained up to now. Apart from gaps in the theory of p-version interpolation error estimates (see Section 3.6), the main obstacle is that Lemma 4.6 cannot be directly adapted to a p-version setting. In particular, writing T for a fixed tetrahedron, one would need there to exist $r > 0$ and a positive constant $C = C(T)$, such that*

$$\left\|\mathbf{u} - \Pi_p^1 \mathbf{u}\right\|_{\boldsymbol{L}^2(T)} \leq C p^{-r} \left\|\mathbf{u}\right\|_{\boldsymbol{H}^s(T)}$$

for all $\mathbf{u} \in \boldsymbol{H}^s(T) \cap \boldsymbol{H}(\mathbf{curl}; T)$, $s > \frac{1}{2}$, for which $\mathbf{curl}\,\mathbf{u}$ is contained in $\mathcal{W}_p^2(T)$. △

Bibliographical notes

A profound analysis of the Maxwell eigenvalue problem, giving sufficient and necessary conditions for viable $\boldsymbol{H}(\mathbf{curl}; \Omega)$-conforming finite element spaces, is presented in Caorsi, Fernandes and Raffetto (1999, 2000). A less technical discussion is given in Fernandes and Raffetto (2000). Discrete compactness for edge elements is also a core subject in Monk and Demkowicz (2001) and Demkowicz *et al.* (1999).

4.5. Lagrangian finite elements

There is a big temptation raised by looking at the variational formulation (4.1) only, oblivious of the subtle algebraic relationships between the electromagnetic fields. Then one might argue that any $\boldsymbol{H}(\mathbf{curl}; \Omega)$-conforming finite element space on Ω_h can be used for the sake of Galerkin approximation, provided that any $\mathbf{u} \in \boldsymbol{H}_0(\mathbf{curl}; \Omega)$ can be approximated arbitrarily well for $h \to 0$. In particular, piecewise linear, globally continuous finite elements might be used as trial spaces for the Cartesian components of $\mathbf{u}$. However, this is strongly deprecated for two reasons.

(1) Thanks to the discrete Poincaré–Friedrichs inequality, the nonzero discrete eigenvalues are bounded away from zero, if discrete differential forms are used. This cannot be guaranteed for componentwise discretization with Lagrangian finite elements. In general, the kernel of **curl** will give rise to numerous nonzero discrete eigenvalues swamping the discrete spectrum.

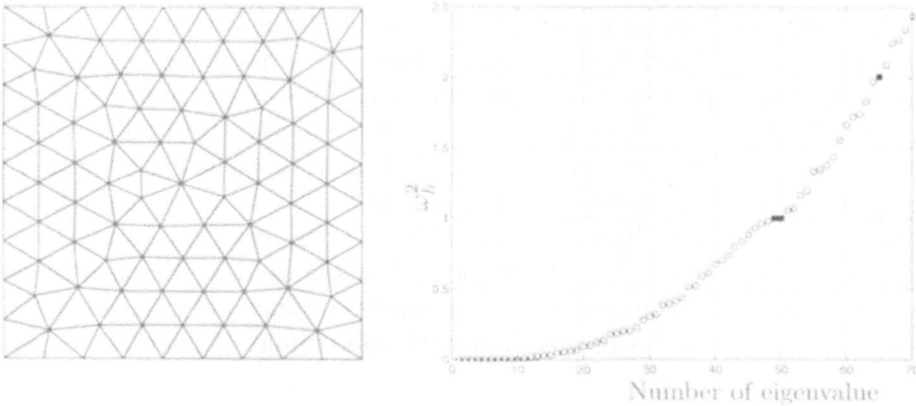

Figure 4.2. Numerical experiment for the 2D Maxwell eigenvalue problem on the unit square. Componentwise discretization of the field by means of Lagrangian finite elements on an unstructured mesh (left) was employed. (Figure 8 in Boffi *et al.* (1999), courtesy of D. Boffi, F. Fernandes, L. Gastaldi, and I. Perugia)

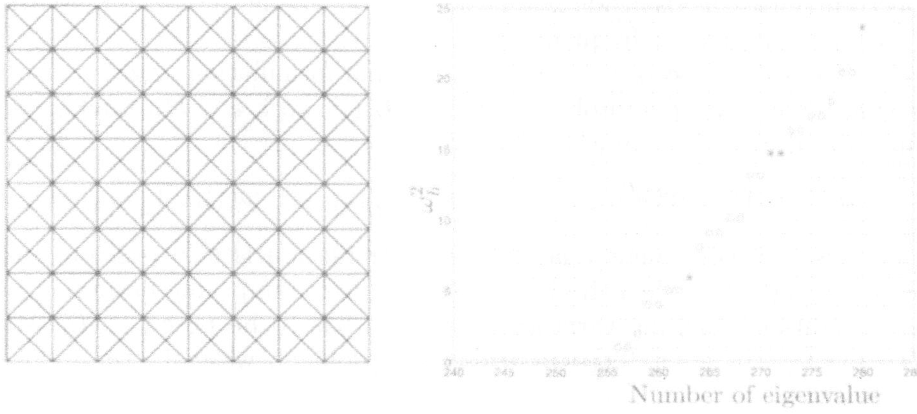

Figure 4.3. Discrete eigenvalues of the 2D Maxwell eigenvalue problem on the unit square discretized by means of Lagrangian finite elements on a 'criss-cross' mesh (left) for the components of $\mathbf{e}$. (Figure 10 in Boffi *et al.* (1999), courtesy of D. Boffi, F. Fernandes, L. Gastaldi, and I. Perugia)

(2) On special meshes it might be possible to capture the kernel of **curl** by continuous finite elements. This will keep the smallest nonzero eigenvalue uniformly bounded away from zero. Yet in this case perilous *spurious* eigenvalues might still arise. They are dotted among correct eigenvalues and it seems impossible to single them out.

Numerical computations confirming these statements are reported in Boffi *et al.* (1999) and here we restate the striking findings. For the sake of simplicity a two-dimensional situation was considered by assuming translation invariance with respect to one spatial direction. This means that $\mathbf{e} = (E_1, E_2)$ and $\mathbf{curl}\,\mathbf{e} = \partial_1 E_2 - \partial_2 E_1$ in (4.1). The discrete spectrum for a truly unstructured mesh is given in Figure 4.2. A glaring pollution effect occurs: most of the small eigenvalues (circles in Figure 4.2) should actually be zero. The computed eigenvalues for a special 'criss-cross' grid that allows the resolution of the discrete kernel are plotted in Figure 4.3. Eigenvalues that have no continuous counterpart ('spurious eigenvalues') are marked by solid squares, whereas circles represent correct approximations of eigenvalues.

5. Maxwell source problem

As in the case of the Maxwell eigenvalue problem, we consider a bounded cavity $\Omega \subset A(\mathbb{R}^3)$ with piecewise smooth Lipschitz boundary, on which PEC boundary conditions are imposed. The cavity is filled with a non-conducting dielectric medium described by piecewise smooth material parameters $\boldsymbol{\epsilon}$ and $\boldsymbol{\mu}$. We rely on the framework developed in Section 2.3 and assume that electromagnetic fields inside Ω are excited by a time-harmonic source current $\mathbf{j}$ with a fixed angular frequency $\omega > 0$.

We focus on the $\mathbf{e}$-based primal variational formulation (2.28) of the following boundary value problem. For $\mathbf{j} \in \boldsymbol{L}^2(\Omega)$ seek $\mathbf{e} \in \boldsymbol{H}_0(\mathbf{curl};\Omega)$ such that, for all $\mathbf{e}' \in \boldsymbol{H}_0(\mathbf{curl};\Omega)$,

$$\left(\boldsymbol{\mu}^{-1}\,\mathbf{curl}\,\mathbf{e}, \mathbf{curl}\,\mathbf{e}'\right)_{\boldsymbol{L}^2(\Omega)} - \omega^2 \left(\boldsymbol{\epsilon}\mathbf{e}, \mathbf{e}'\right)_{\boldsymbol{L}^2(\Omega)} = -i\omega \left(\mathbf{j}, \mathbf{e}'\right)_{\boldsymbol{L}^2(\Omega)}. \tag{5.1}$$

Of course, an equivalent dual $\mathbf{h}$-based formulation is also possible and it arises from (2.29). Then the PEC case amounts to imposing natural boundary conditions and the variational problem has to be posed on the space $\boldsymbol{H}(\mathbf{curl};\Omega)$. In order to get a well-posed problem, we make the following assumption.

Assumption 1. The angular frequency ω is distinct from any resonant frequency (Maxwell eigenvalue) of the dielectric cavity Ω.

This assumption is equivalent to demanding that the variational problem (5.1) has at most one solution. The Galerkin discretization of (5.1) by means of discrete 1-forms is straightforward. Seek $\mathbf{e}_h \in \mathcal{W}^1_{p,0}(\Omega_h)$ such that, for all $\mathbf{e}'_h \in \mathcal{W}^1_{p,0}(\Omega_h)$,

$$\left(\boldsymbol{\mu}^{-1}\,\mathbf{curl}\,\mathbf{e}_h, \mathbf{curl}\,\mathbf{e}'_h\right)_{\boldsymbol{L}^2(\Omega)} - \omega^2 \left(\boldsymbol{\epsilon}\mathbf{e}_h, \mathbf{e}'_h\right)_{\boldsymbol{L}^2(\Omega)} = -i\omega \left(\mathbf{j}, \mathbf{e}'_h\right)_{\boldsymbol{L}^2(\Omega)}. \tag{5.2}$$

As already indicated by the notation, the focus will be on the h-version of discrete 1-forms with uniform polynomial degree $p \in \mathbb{N}_0$. For the sake of brevity, the sesqui-linear form on the left-hand side of (5.1)/(5.2) will be denoted by $a(\cdot\,,\cdot)$.

We aim at showing asymptotically optimal convergence of the energy norm of the discretization error $\mathbf{e}-\mathbf{e}_h$ in the sense that, on sufficiently fine meshes,

$$\|\mathbf{e}-\mathbf{e}_h\|_{\boldsymbol{H}(\mathbf{curl};\Omega)} \leq C \inf_{\mathbf{v}_h \in \mathcal{W}_p^1(\Omega_h)} \|\mathbf{e}-\mathbf{v}_h\|_{\boldsymbol{H}(\mathbf{curl};\Omega)} \tag{5.3}$$

with $C > 0$ only depending on parameters of the continuous problem and the shape regularity measure of Ω_h. Note that $a(\cdot\,,\cdot)$ is Hermitian but obviously indefinite, which thwarts very simple convergence estimates based on $\boldsymbol{H}(\mathbf{curl};\Omega)$-ellipticity and Cea's lemma.

5.1. *Coercivity*

Let us take a brief look at the closely related Dirichlet problem for the Helmholtz equation. For $f \in L^2(\Omega)$ seek $\varphi \in H_0^1(\Omega)$ such that, for all $\varphi' \in H_0^1(\Omega)$,

$$\left(\mathbf{grad}\,\varphi, \mathbf{grad}\,\varphi'\right)_{\boldsymbol{L}^2(\Omega)} - \omega^2 \left(\varphi, \varphi'\right)_{\boldsymbol{L}^2(\Omega)} = \left(f, \varphi'\right)_{\boldsymbol{L}^2(\Omega)}. \tag{5.4}$$

A priori error estimates for finite element schemes for this problem usually employ (i) the coercivity of the underlying sesqui-linear form $a_H(\cdot\,,\cdot)$, that is, the fact that the zero-order term is a *compact perturbation* of the second-order term, the principal part, and (ii) a Gårding inequality of the form

$$|a_H(\varphi,\varphi)| \geq C\,\|\varphi\|^2_{H^1(\Omega)} - C'\,\|\varphi\|^2_{L^2(\Omega)} \quad \forall \varphi \in H_0^1(\Omega). \tag{5.5}$$

This is definitely not the case for $a(\cdot\,,\cdot)$ from (5.1), which has to be blamed on the infinite-dimensional kernel of the **curl**-operator. We can view the situation from a different angle. In electromagnetism the electric and magnetic energies of a field are perfectly symmetric. None can be relegated to a compact perturbation of the other. On the contrary, in acoustics, which is described by the Helmholtz equation, the two energies involved are potential energy and kinetic energy and one of them can be identified as the principal energy.

In the case of the eigenvalue problem of Section 4, we found a remedy by restricting the variational problem to $\boldsymbol{Z}_0(\boldsymbol{\epsilon},\Omega)$. Thus $(\boldsymbol{\mu}^{-1}\mathbf{curl}\,\cdot, \mathbf{curl}\,\cdot)_{\boldsymbol{L}^2(\Omega)}$ could be promoted to a principal part, whereas $(\boldsymbol{\epsilon}\cdot,\cdot)_{\boldsymbol{L}^2(\Omega)}$ played the role of a compact perturbation. For the source problem the solution $\mathbf{e}$ does not necessarily belong to $\boldsymbol{Z}_0(\boldsymbol{\epsilon},\Omega)$. Yet the idea is still fruitful, when refined into the *splitting principle*:

Consider the Maxwell source problem on fields decomposed into components on which $\left(\boldsymbol{\mu}^{-1}\,\mathbf{curl}\,\cdot, \mathbf{curl}\,\cdot\right)_{\boldsymbol{L}^2(\Omega)}$ *becomes a principal part, and other components, usually* Ker(**curl**)*, on which the zero-order terms in* $a(\cdot\,,\cdot)$ *are 'principal'.*

It is natural to retain a close relative of $\boldsymbol{Z}_0(\boldsymbol{\epsilon},\Omega)$ as the first component in such a decomposition. Therefore, we consider the classical $\boldsymbol{L}^2(\Omega)$-orthogonal *Helmholtz decomposition*

$$\boldsymbol{H}_0(\mathbf{curl};\Omega) = \boldsymbol{Z}_0(\mathbf{I},\Omega) \oplus \boldsymbol{\mathcal{N}}(\Omega), \tag{5.6}$$

where we abbreviated $\boldsymbol{\mathcal{N}}(\Omega) := \boldsymbol{H}_0(\mathbf{curl}\,0;\Omega)$. Merging the variational problem (5.1) and (5.6) we arrive at the following. Seek $\mathbf{e}_\perp \in \boldsymbol{Z}_0(\mathbf{I},\Omega)$, $\mathbf{e}_0 \in \boldsymbol{\mathcal{N}}(\Omega)$ such that

$$\begin{aligned}
&\left(\boldsymbol{\mu}^{-1}\,\mathbf{curl}\,\mathbf{e}_\perp, \mathbf{curl}\,\mathbf{e}'_\perp\right)_{\boldsymbol{L}^2(\Omega)} \\
&\quad -\omega^2\left(\boldsymbol{\epsilon}\mathbf{e}_\perp, \mathbf{e}'_\perp\right)_{\boldsymbol{L}^2(\Omega)} - \omega^2\left(\boldsymbol{\epsilon}\mathbf{e}_0, \mathbf{e}'_\perp\right)_{\boldsymbol{L}^2(\Omega)} = -i\omega\left(\mathbf{j}, \mathbf{e}'_\perp\right)_{\boldsymbol{L}^2(\Omega)}, \\
&\quad \omega^2\left(\mathbf{e}_\perp, \mathbf{e}'_0\right)_{\boldsymbol{L}^2(\Omega)} + \omega^2\left(\mathbf{e}_0, \mathbf{e}'_0\right)_{\boldsymbol{L}^2(\Omega)} = i\omega\left(\mathbf{j}, \mathbf{e}'_0\right)_{\boldsymbol{L}^2(\Omega)},
\end{aligned}$$

for all $\mathbf{e}'_\perp \in \boldsymbol{Z}_0(\boldsymbol{\epsilon},\Omega)$, $\mathbf{e}_0 \in \boldsymbol{\mathcal{N}}(\Omega)$. It is important to notice that the sign of the second equation has been flipped. Therefore, in the sesqui-linear form corresponding to the above split problem, the 'principal' diagonal parts have the same (positive) sign. That they are really principal is rigorously confirmed by the following lemma. To state it, we introduce the $\boldsymbol{L}^2(\Omega)$-orthogonal projections $\mathsf{P}^\perp : \boldsymbol{H}_0(\mathbf{curl};\Omega) \mapsto \boldsymbol{Z}_0(\mathbf{I},\Omega)$, $\mathsf{P}^0 : \boldsymbol{H}_0(\mathbf{curl};\Omega) \mapsto \boldsymbol{\mathcal{N}}(\Omega)$, $\mathsf{P}^\perp + \mathsf{P}^0 = \mathrm{Id}$, associated with (5.6).

Lemma 5.1. The sesqui-linear form $k : \boldsymbol{H}_0(\mathbf{curl};\Omega) \times \boldsymbol{H}_0(\mathbf{curl};\Omega) \mapsto \mathbb{C}$,

$$k(\mathbf{u},\mathbf{v}) := \left(\boldsymbol{\epsilon}\mathsf{P}^\perp\mathbf{u}, \mathsf{P}^\perp\mathbf{v}\right)_{\boldsymbol{L}^2(\Omega)} - \left(\boldsymbol{\epsilon}\mathsf{P}^\perp\mathbf{u}, \mathsf{P}^0\mathbf{v}\right)_{\boldsymbol{L}^2(\Omega)} + \left(\boldsymbol{\epsilon}\mathsf{P}^0\mathbf{u}, \mathsf{P}^\perp\mathbf{v}\right)_{\boldsymbol{L}^2(\Omega)},$$

is compact.

Proof. We can split $k = k_1 - k_2 + k_3$,

$$\begin{aligned}
k_1(\mathbf{u},\mathbf{v}) &:= \left(\boldsymbol{\epsilon}\mathsf{P}^\perp\mathbf{u}, \mathsf{P}^\perp\mathbf{v}\right)_{\boldsymbol{L}^2(\Omega)}, \\
k_2(\mathbf{u},\mathbf{v}) &:= \left(\boldsymbol{\epsilon}\mathsf{P}^\perp\mathbf{u}, \mathsf{P}^0\mathbf{v}\right)_{\boldsymbol{L}^2(\Omega)}, \\
k_3(\mathbf{u},\mathbf{v}) &:= \left(\boldsymbol{\epsilon}\mathsf{P}^0\mathbf{u}, \mathsf{P}^\perp\mathbf{v}\right)_{\boldsymbol{L}^2(\Omega)},
\end{aligned}$$

and deal with the individual terms separately. First, note that $k_3(\mathbf{u},\mathbf{v}) = \overline{k_2(\mathbf{v},\mathbf{u})}$. Therefore, compactness of k_3 will be implied by that of k_2. Recall that a sesqui-linear form is defined to be compact if this holds for the associated operator.

Write $K_1 : \boldsymbol{H}_0(\mathbf{curl};\Omega) \mapsto \boldsymbol{H}_0(\mathbf{curl};\Omega)'$ for the bounded linear operator spawned by k_1. We find a $C > 0$ only depending on $\boldsymbol{\epsilon}$ such that

$$\|K_1\mathbf{u}\|_{\boldsymbol{H}_0(\mathbf{curl};\Omega)'} = \sup_{\mathbf{v}\in\boldsymbol{H}_0(\mathbf{curl};\Omega)} \frac{|k_1(\mathbf{u},\mathbf{v})|}{\|\mathbf{v}\|_{\boldsymbol{H}(\mathbf{curl};\Omega)}} \leq C\left\|\mathsf{P}^\perp\mathbf{u}\right\|_{\boldsymbol{L}^2(\Omega)}.$$

Then we can appeal to the compact embedding $\boldsymbol{Z}_0(\mathbf{I},\Omega) \hookrightarrow \boldsymbol{L}^2(\Omega)$ stated in Corollary 4.3 to verify the compactness of K_1. Exactly the same argument settles the compactness of K_2. □

However, what we discretize is (5.1) and not a split problem. In order to express the splitting idea in terms of $a(\cdot,\cdot)$, we resort to the isometric involution $\mathsf{X} : \boldsymbol{H}_0(\mathbf{curl};\Omega) \mapsto \boldsymbol{H}_0(\mathbf{curl};\Omega)$, $\mathsf{X} := \mathsf{P}^\perp - \mathsf{P}^0$. It naturally emerges from the flipping of signs that accompanied the formulation of the split variational problem. Thus we get

$$a(\mathbf{u},\mathsf{X}\mathbf{v}) = \left(\boldsymbol{\mu}^{-1}\,\mathbf{curl}\,\mathsf{P}^\perp\mathbf{u},\mathbf{curl}\,\mathsf{P}^\perp\mathbf{v}\right)_{\boldsymbol{L}^2(\Omega)} + \omega^2\left(\boldsymbol{\epsilon}\mathsf{P}^0\mathbf{u},\mathsf{P}^0\mathbf{v}\right)_{\boldsymbol{L}^2(\Omega)} - k(\mathbf{u},\mathbf{v}). \quad (5.7)$$

In view of Lemma 5.1, the properties of metric tensors, and Corollary 4.4, this implies a *generalized Gårding inequality*, in the sense that, for all $\mathbf{u} \in \boldsymbol{H}_0(\mathbf{curl};\Omega)$ and $C_G = C_G(\Omega,\boldsymbol{\epsilon},\boldsymbol{\mu}) > 0$,

$$|a(\mathbf{u},\mathsf{X}\mathbf{u}) + k(\mathbf{u},\mathbf{u})| \geq C_G\,\|\mathbf{u}\|^2_{\boldsymbol{H}(\mathbf{curl};\Omega)}. \quad (5.8)$$

Standard theory based on the Fredholm alternative gives existence of continuous solutions.

Theorem 5.2. (Existence and uniqueness of continuous solution) Under Assumption 1 a solution of the variational problem (5.1) exists for every admissible right-hand side.

Existence and uniqueness of solutions of linear variational problems guarantee the following inf-sup condition for the associated bilinear form. There exists $C_S > 0$ such that

$$\sup_{\mathbf{v}\in\boldsymbol{H}_0(\mathbf{curl};\Omega)} \frac{|a(\mathbf{u},\mathbf{v})|}{\|\mathbf{v}\|_{\boldsymbol{H}(\mathbf{curl};\Omega)}} \geq C_S\,\|\mathbf{u}\|_{\boldsymbol{H}(\mathbf{curl};\Omega)} \quad \forall \mathbf{u} \in \boldsymbol{H}_0(\mathbf{curl};\Omega). \quad (5.9)$$

5.2. A priori *error estimates*

We know from Babuška's theory (Babuška 1971) that we have to establish a discrete analogue of (5.9), that is,

$$\sup_{\mathbf{v}_h\in\mathcal{W}^1_{p,0}(\Omega_h)} \frac{|a(\mathbf{u}_h,\mathbf{v}_h)|}{\|\mathbf{v}_h\|_{\boldsymbol{H}(\mathbf{curl};\Omega)}} \geq C_D\,\|\mathbf{u}_h\|_{\boldsymbol{H}(\mathbf{curl};\Omega)} \quad \forall \mathbf{u}_h \in \mathcal{W}^1_{p,0}(\Omega_h), \quad (5.10)$$

because it permits us to conclude, for any $\mathbf{w}_h \in \mathcal{W}^1_p(\Omega_h)$,

$$\begin{aligned}\|\mathbf{e}-\mathbf{e}_h\|_{\boldsymbol{H}(\mathbf{curl};\Omega)} &\leq \|\mathbf{e}-\mathbf{w}_h\|_{\boldsymbol{H}(\mathbf{curl};\Omega)} + \|\mathbf{w}_h-\mathbf{e}_h\|_{\boldsymbol{H}(\mathbf{curl};\Omega)} \\ &\leq \|\mathbf{e}-\mathbf{w}_h\|_{\boldsymbol{H}(\mathbf{curl};\Omega)} + C_D^{-1} \sup_{\mathbf{v}_h\in\mathcal{W}^1_{p,0}(\Omega_h)} \frac{|a(\mathbf{e}_h-\mathbf{w}_h,\mathbf{v}_h)|}{\|\mathbf{v}_h\|_{\boldsymbol{H}(\mathbf{curl};\Omega)}}\end{aligned}$$

$$\leq \|\mathbf{e} - \mathbf{w}_h\|_{\boldsymbol{H}(\mathbf{curl};\Omega)} + C_D^{-1} \sup_{\mathbf{v}_h \in \mathcal{W}^1_{p,0}(\Omega_h)} \frac{|a(\mathbf{e} - \mathbf{w}_h, \mathbf{v}_h)|}{\|\mathbf{v}_h\|_{\boldsymbol{H}(\mathbf{curl};\Omega)}}$$

$$\leq (1 + C_D^{-1} C_A) \|\mathbf{e} - \mathbf{w}_h\|_{\boldsymbol{H}(\mathbf{curl};\Omega)},$$

as we have Galerkin orthogonality $a(\mathbf{e} - \mathbf{e}_h, \mathbf{v}_h) = 0$ for all $\mathbf{v}_h \in \mathcal{W}^1_{p,0}(\Omega_h)$. Here,

$$C_A := \max\{\|\boldsymbol{\mu}^{-1}\|_{L^\infty(\Omega)}, \|\boldsymbol{\epsilon}\|_{L^\infty(\Omega)}\}$$

is a bound for the norm of the bilinear form $a(\cdot\,,\cdot)$.

As a first step towards a discrete inf-sup condition, we have to find a suitable candidate for $\mathbf{v}$ in (5.9). To that end, for $\mathbf{u} \in \boldsymbol{H}_0(\mathbf{curl};\Omega)$ fix $\mathsf{S}\mathbf{u} \in \boldsymbol{H}_0(\mathbf{curl};\Omega)$ by

$$a(\mathbf{u}', \mathsf{S}\mathbf{u}) = k(\mathbf{u}, \mathbf{u}') \quad \forall \mathbf{u}' \in \boldsymbol{H}_0(\mathbf{curl};\Omega).$$

This defines a bounded linear operator, $\mathsf{S} : \boldsymbol{H}_0(\mathbf{curl};\Omega) \mapsto \boldsymbol{H}_0(\mathbf{curl};\Omega)$, thanks to Theorem 5.2.

Lemma 5.3. The operator $\mathsf{S} : \boldsymbol{H}_0(\mathbf{curl};\Omega) \mapsto \boldsymbol{H}_0(\mathbf{curl};\Omega)$ is compact.

Proof. By (5.9) we find

$$\|\mathsf{S}\mathbf{u}\|_{\boldsymbol{H}(\mathbf{curl};\Omega)} \leq C \sup_{\mathbf{v} \in \boldsymbol{H}_0(\mathbf{curl};\Omega)} \frac{|a(\mathsf{S}\mathbf{u}, \mathbf{v})|}{\|\mathbf{v}\|_{\boldsymbol{H}(\mathbf{curl};\Omega)}} \leq C \sup_{\mathbf{v} \in \boldsymbol{H}_0(\mathbf{curl};\Omega)} \frac{|k(\mathbf{u}, \mathbf{v})|}{\|\mathbf{v}\|_{\boldsymbol{H}(\mathbf{curl};\Omega)}},$$

and the compactness of S follows from that of k. □

Using the definition of S it is immediate that, for all $\mathbf{u} \in \boldsymbol{H}_0(\mathbf{curl};\Omega)$,

$$a(\mathbf{u}, (\mathsf{X} + \mathsf{S})\mathbf{u}) = a(\mathbf{u}, \mathsf{X}\mathbf{u}) + k(\mathbf{u}, \mathbf{u}) \geq C_G \|\mathbf{u}\|^2_{\boldsymbol{H}(\mathbf{curl};\Omega)}. \tag{5.11}$$

The choice $\mathbf{v} := (\mathsf{X}+\mathsf{S})\mathbf{u}$ will make (5.9) hold with $C_S = C_G$. The challenge is that $(\mathsf{X}+\mathsf{S})\mathbf{u}_h$ will not be a finite element function even for $\mathbf{u}_h \in \mathcal{W}^1_{p,0}(\Omega_h)$. It will be necessary to project it back to the finite element space by applying a suitable continuous projection operator $\widetilde{\mathsf{F}}_h : \boldsymbol{H}_0(\mathbf{curl};\Omega) \mapsto \mathcal{W}^1_{p,0}(\Omega_h)$. Then, for an arbitrary $\mathbf{u}_h \in \mathcal{W}^1_{p,0}(\Omega_h)$, we can hope that $\mathbf{v}_h := \widetilde{\mathsf{F}}_h(\mathsf{X}+\mathsf{S})\mathbf{u}_h$ is an appropriate choice for $\mathbf{v}_h$ in (5.10). Making use of (5.11), we see that

$$\begin{aligned}
&|a(\mathbf{u}_h, \mathbf{v}_h)| \\
&\quad = |a(\mathbf{u}_h, (\mathsf{X}+\mathsf{S})\mathbf{u}_h) - a(\mathbf{u}_h, (\mathrm{Id} - \widetilde{\mathsf{F}}_h)(\mathsf{X}+\mathsf{S})\mathbf{u}_h)| \\
&\quad \geq C_G \|\mathbf{u}_h\|^2_{\boldsymbol{H}(\mathbf{curl};\Omega)} - C_A \|\mathbf{u}_h\|_{\boldsymbol{H}(\mathbf{curl};\Omega)} \left\|(\mathrm{Id} - \widetilde{\mathsf{F}}_h)(\mathsf{X}+\mathsf{S})\mathbf{u}_h\right\|_{\boldsymbol{H}(\mathbf{curl};\Omega)}.
\end{aligned}$$

Obviously, the projector $\widetilde{\mathsf{F}}_h$ has to effect *uniform* convergence $(\mathrm{Id} - \widetilde{\mathsf{F}}_h)(\mathsf{X}+\mathsf{S}) \to 0$ in $\boldsymbol{H}_0(\mathbf{curl};\Omega)$. Again, let us compare the situation with the boundary value problem for the Helmholtz equation. There the choice $\mathsf{X} = \mathrm{Id}$ is possible, but we also have to choose a projector. This is easy, because the compactness of S converts pointwise convergence into uniform convergence.

Lemma 5.4. If $\widetilde{\mathsf{F}}_h \to \mathrm{Id}$ pointwise in $\boldsymbol{H}_0(\mathbf{curl};\Omega)$, then there is a function $\epsilon :]0;\infty[\to \mathbb{R}^+$ with $\lim_{h\to 0}\epsilon(h) = 0$ such that

$$\|(\mathrm{Id} - \widetilde{\mathsf{F}}_h)\mathsf{S}\mathbf{u}\|_{\boldsymbol{H}(\mathbf{curl};\Omega)} \leq \epsilon(h)\, \|\mathbf{u}\|_{\boldsymbol{H}(\mathbf{curl};\Omega)} \quad \forall \mathbf{u} \in \boldsymbol{H}_0(\mathbf{curl};\Omega).$$

Proof. A similar, even more general, statement is made in Kress (1989, Corollary 10.4). □

This settles everything for the Helmholtz problem. However, the components of the Helmholtz decomposition of a discrete 1-form $\in \mathcal{W}_p^1(\Omega_h)$ do not necessarily belong to finite element spaces. Therefore, the projector $\widetilde{\mathsf{F}}_h$ will not necessarily act as identity on $\mathsf{X}(\mathcal{W}_p^1(\Omega_h))$; we cannot escape estimates of the operator norm of $(\mathrm{Id} - \widetilde{\mathsf{F}}_h)\mathsf{X}$ on $\boldsymbol{H}_0(\mathbf{curl};\Omega)$ and have to show that it tends to zero as $h \to 0$.

Pondering these requirements, the Fortin projector (for $\boldsymbol{\epsilon} = \mathbf{I}$) from Section 4.2 looks promising. It can be used as the key ingredient of the projector $\widetilde{\mathsf{F}}_h$, which is chosen as

$$\widetilde{\mathsf{F}}_h : \boldsymbol{H}_0(\mathbf{curl};\Omega) \mapsto \mathcal{W}_{p,0}^1(\Omega_h), \quad \widetilde{\mathsf{F}}_h := \mathsf{F}_h\mathsf{P}^\perp + \mathsf{Q}_h^{\mathcal{N}}\mathsf{P}^0,$$

where $\mathsf{Q}_h^{\mathcal{N}}$ is the $\boldsymbol{L}^2(\Omega)$-orthogonal projection $\boldsymbol{\mathcal{N}}(\Omega) \mapsto \boldsymbol{\mathcal{N}}(\Omega_h)$, $\boldsymbol{\mathcal{N}}(\Omega_h) := \mathcal{W}_{p,0}^1(\Omega_h) \cap \mathrm{Ker}(\mathbf{curl})$.

Lemma 5.5. We have pointwise convergence $\lim_{h\to 0}\|\mathbf{u} - \widetilde{\mathsf{F}}_h\mathbf{u}\|_{\boldsymbol{H}(\mathbf{curl};\Omega)} = 0$ for any $\mathbf{u} \in \boldsymbol{H}_0(\mathbf{curl};\Omega)$.

Proof. From Theorem 4.8, for some $s > 0$ and $C = C(\Omega, p, s, \rho(\Omega_h))$,

$$\|(\mathrm{Id} - \mathsf{F}_h)\mathsf{P}^\perp\mathbf{u}\|_{\boldsymbol{L}^2(\Omega)} \leq Ch^s\, \|\mathbf{curl}\,\mathbf{u}\|_{\boldsymbol{L}^2(\Omega)} \to 0 \quad \text{for } h \to 0.$$

Moreover, as $h \to 0$ any $\mathbf{v} \in \boldsymbol{\mathcal{N}}(\Omega)$ can be arbitrarily well approximated by a $\mathbf{v}_h \in \boldsymbol{\mathcal{N}}(\Omega_h)$, which means $\|\mathbf{v} - \mathsf{Q}_h^{\mathcal{N}}\mathbf{v}\|_{\boldsymbol{L}^2(\Omega)} \to 0$ for $h \to 0$. □

Combined with Lemma 5.5 this shows uniform convergence $(\mathrm{Id} - \widetilde{\mathsf{F}}_h)\mathsf{S} \to 0$ in $\boldsymbol{H}_0(\mathbf{curl};\Omega)$. It turns out that the essential 'uniform convergence' $(\mathrm{Id} - \widetilde{\mathsf{F}}_h)\mathsf{X}_{|\mathcal{W}_p^1(\Omega_h)} \to 0$ also holds.

Lemma 5.6. There is an $s > 0$ and $C_* = C_*(\Omega, p, s, \rho(\Omega_h)) > 0$ such that

$$\|(\mathrm{Id} - \widetilde{\mathsf{F}}_h)\mathsf{X}\mathbf{u}_h\|_{\boldsymbol{H}(\mathbf{curl};\Omega)} \leq C_* h^s\, \|\mathbf{curl}\,\mathbf{u}_h\|_{\boldsymbol{L}^2(\Omega)} \quad \forall \mathbf{u}_h \in \mathcal{W}_{p,0}^1(\Omega_h).$$

Proof. The key to the proof is to employ the $\boldsymbol{L}^2(\Omega)$-orthogonal discrete Helmholtz decomposition

$$\mathcal{W}_{p,0}^1(\Omega_h) = \boldsymbol{Z}_{h,0}(\mathbf{I}, \Omega_h) \oplus \boldsymbol{\mathcal{N}}(\Omega_h),$$

and the associated orthogonal projections $\mathsf{P}_h^\perp : \mathcal{W}_{p,0}^1(\Omega_h) \mapsto \boldsymbol{Z}_{h,0}(\mathbf{I}, \Omega_h)$, $\mathsf{P}_h^0 : \mathcal{W}_{p,0}^1(\Omega_h) \mapsto \boldsymbol{\mathcal{N}}(\Omega_h)$, $\mathsf{P}_h^\perp + \mathsf{P}_h^0 = \mathrm{Id}$. As $\boldsymbol{\mathcal{N}}(\Omega_h) \subset \boldsymbol{\mathcal{N}}(\Omega)$ it is immediate that

$$\mathsf{P}^0 \circ \mathsf{P}_h^0 = \mathsf{P}_h^0 \quad \text{and} \quad \mathsf{P}^\perp \circ \mathsf{P}_h^0 = 0.$$

Alert readers will have noticed that $\mathsf{Q}_h^{\mathcal{N}} = \mathsf{P}_h^0$ and $\mathsf{P}^\perp = \mathsf{H}_{\mathbf{I}}$ with the Hodge mapping $\mathsf{H}_{\boldsymbol{\epsilon}}$ from Section 4.2 (for $\boldsymbol{\epsilon} = \mathbf{I}$). Thus, using the projection properties, in particular $\widetilde{\mathsf{F}}_h \circ \mathsf{P}_h^0 = \mathrm{Id}$, we can rewrite

$$\begin{aligned}
(\mathrm{Id} - \widetilde{\mathsf{F}}_h)\mathsf{X} &= (\mathsf{P}^\perp - \mathsf{F}_h\mathsf{P}^\perp + \mathsf{P}^0 - \mathsf{Q}_h^{\mathcal{N}}\mathsf{P}^0)(\mathsf{P}^\perp - \mathsf{P}^0)(\mathsf{P}_h^\perp + \mathsf{P}_h^0) \\
&= (\mathrm{Id} - \mathsf{F}_h)\mathsf{P}^\perp\mathsf{P}_h^\perp - (\mathrm{Id} - \mathsf{P}_h^0)\mathsf{P}^0\mathsf{P}_h^\perp \\
&= (\mathrm{Id} - \mathsf{F}_h)\mathsf{P}^\perp\mathsf{P}_h^\perp - (\mathrm{Id} - \mathsf{P}_h^0)\mathsf{P}^0(\mathsf{P}_h^\perp - \mathsf{P}^\perp)\mathsf{P}_h^\perp \\
&= (\mathsf{H}_{\mathbf{I}} - \mathrm{Id})\mathsf{P}_h^\perp - (\mathrm{Id} - \mathsf{P}_h^0)\mathsf{P}^0(\mathrm{Id} - \mathsf{H}_{\mathbf{I}})\mathsf{P}_h^\perp \\
&= (\mathrm{Id} + (\mathrm{Id} - \mathsf{P}_h^0)\mathsf{P}^0)(\mathsf{H}_{\mathbf{I}} - \mathrm{Id})\mathsf{P}_h^\perp.
\end{aligned}$$

Now it remains to use Lemma 4.5. □

Summing up, the two previous lemmas give us

$$\|(\mathrm{Id} - \widetilde{\mathsf{F}}_h)(\mathsf{X} + \mathsf{S})\mathbf{u}_h\|_{\boldsymbol{H}(\mathbf{curl};\Omega)} \leq \epsilon(h) + C_* h^s \|\mathbf{u}_h\|_{\boldsymbol{H}(\mathbf{curl};\Omega)}.$$

Thus, for h sufficiently small to ensure $1 - C_A(\epsilon(h) + C_* h^s)/C_G > \frac{1}{2}$, we have the discrete inf-sup condition (5.10). This yields the main result.

Theorem 5.7. Provided that Assumption 1 holds, there exists $h^* > 0$ depending on the parameters of the continuous problem and the shape regularity measure of the triangulation, such that a unique solution $\mathbf{e}_h$ of the discrete problem (5.2) exists, provided that $h < h_*$. It provides an asymptotically optimal approximation to the continuous solution $\mathbf{e}$ of (5.1) in the sense of (5.3).

It must be emphasized that the result is asymptotic in nature. Computational resources hardly ever permit us to use very fine meshes on three-dimensional domains. Hence, the relevance of the statement of Theorem 5.7 for a concrete problem and mesh is not clear.

Remark 16. Even if we accept asymptotic optimality as an essential property, actual rates of convergence will still hinge on the smoothness of the solution $\mathbf{e}$. The comprehensive investigations in Costabel *et al.* (1999) and Costabel and Dauge (1998) send a daunting message: even for smooth $\boldsymbol{\epsilon}$, $\boldsymbol{\mu}$, and $\mathbf{j}$, the solution for $\mathbf{e}$ might not belong to $\boldsymbol{H}^1(\Omega)$ if the boundary possesses re-entrant edges/corners. Only a regularity of the form $\mathbf{e} \in \boldsymbol{H}^{\frac{1}{2}+\epsilon}(\Omega)$, $\epsilon > 0$, can be expected. If there are jumps in $\boldsymbol{\epsilon}$ the situation can be even worse, reducing the regularity of $\mathbf{e}$ down to a mere $\mathbf{e} \in \boldsymbol{H}^{\epsilon}(\Omega)$, $\epsilon > 0$. Therefore, in Section 3.6 we have taken great pains to get interpolation estimates in rather weak fractional Sobolev norms. Despite the effort, the (theoretical) rates of convergence of the h-version of discrete differential forms can be abysmally poor.

Yet the results of Costabel *et al.* (1999) and Costabel and Dauge (1998) also imply (*cf.* Costabel and Dauge (2002, Section 6)) that, for analytic $\epsilon, \boldsymbol{\mu}, \mathbf{j}$, the solution $\mathbf{e}$ can be decomposed according to $\mathbf{e} = \boldsymbol{\Psi} + \mathbf{grad}\,\varphi$, where both $\boldsymbol{\Psi}$ and φ belong to countably normed spaces defined by the intersection of weighted Sobolev spaces. The weight functions are powers of distances from the geometric singularities of the boundary.

Again, it is wise to remember the guideline of Section 3.2. Barring highly irregular (in practical terms) coefficients and data, the so-called hp-version of finite elements (Babuška and Suri 1994, Schwab 1998, Khoromskij and Melenk 2001) can achieve exponential order of convergence of approximate solutions of second-order elliptic problems by judiciously combining local mesh refinement and local adjustments of polynomials degrees in $\underline{\mathbf{p}}$. It is prerequisite that the solution belongs to those special countably normed spaces mentioned above. Thus we realize that conditions for the exponential convergence of a hp-version of discrete differentials forms are also met in the cases of the Maxwell source problem and the Maxwell eigenvalue problem. This justifies the following prediction, though a complete theory is still missing: *The (adaptive) hp-version of discrete differential forms will provide the most efficient discretization for the Maxwell source problems/Maxwell eigenvalue problem, provided that the coefficients $\boldsymbol{\mu}, \epsilon$ and excitation $\mathbf{j}$ are (piecewise) analytic.* △

Remark 17. Of course, the constant in (5.3) will blow up as ω approaches a Maxwell eigenvalue. We may hope to avoid this in the case of complex ϵ, *i.e.*, for conducting media or by imposing absorbing boundary conditions (*cf.* Remark 2) instead of PEC. This seems to cure the problem. However, investigations of the behaviour of Lagrangian finite elements for the Helmholtz problem (5.4) disclosed that, for h-version finite elements, a blow-up of C still occurs as ω increases (Ihlenburg 1998). This is the so-called *pollution effect* notorious for fixed degree FEM in acoustics. There is empirical evidence that discrete 1-forms are also vulnerable to pollution. A partial remedy is the use of higher-order schemes. This provides another rationale for studying these carefully in Section 3.4. △

Bibliographical notes

The first convergence analysis of the time-harmonic Maxwell's equations, when discretized by means of edge elements, was given by P. Monk (1992*a*). The above proof follows the lines of Buffa, Hiptmair, von Petersdorff and Schwab (2002, Section 4.1). In Monk and Demkowicz (2001) and Kirsch and Monk (2000, Section 4) it was shown that the analysis can also rely on discrete compactness (*cf.* Section 4.4). The application of abstract theory, for nonlinear problems (Brezzi, Rappaz and Raviart 1980) to the Maxwell source problem, is studied in Boffi and Gastaldi (2001). Matching their potential, much effort has recently been put into the analysis and implementation of *hp*-FEM for discrete 1-forms (Ainsworth and Pinchedez 2001, Rachowicz and Demkowicz 2002, Vardapetyan and Demkowicz 1999).

5.3. *Duality estimates*

For the Maxwell source problem, we expect the error norm $\|\mathbf{e}-\mathbf{e}_h\|_{\boldsymbol{L}^2(\Omega)}$ to decrease faster than $\|\mathbf{e}-\mathbf{e}_h\|_{\boldsymbol{H}(\mathbf{curl};\Omega)}$ as $h \to 0$. In finite element theory, duality estimates, often referred to as Aubin–Nitsche tricks (Ciarlet 1978, Section 3.2), are an effective technique with which such questions are tackled. In light of the guideline stated in Section 3.2 and Remark 1, it is not surprising that they can be adapted to the Maxwell source problem.

Elliptic regularity plays a pivotal role in duality estimates. Therefore, we have to investigate the regularity of solutions of the Maxwell source problem. With reasonable effort this is only possible in the case $\boldsymbol{\mu}, \boldsymbol{\epsilon} = \mathbf{I}$ (for more general results see Costabel *et al.* (1999)), which will be the only one considered in this subsection.

Theorem 5.8. (*A priori* **error estimate in $L^2(\Omega)$-norm)** If $\boldsymbol{\mu}, \boldsymbol{\epsilon} = \mathbf{I}$, $\mathbf{j} \in \boldsymbol{H}(\mathrm{div};\Omega)$, and Assumption 1 holds, there exists $s > 1/2$ such that the solutions $\mathbf{e}$ and $\mathbf{e}_h$ of (5.1) and (5.2) satisfy

$$\|\mathbf{e}-\mathbf{e}_h\|_{\boldsymbol{L}^2(\Omega)} \leq Ch^s \|\mathbf{e}-\mathbf{e}_h\|_{\boldsymbol{H}(\mathbf{curl};\Omega)},$$

with $C > 0$ independent of the meshwidth h.

Proof. Small wonder that the splitting idea is successful again. We start with

$$\mathbf{e}-\mathbf{e}_h = \delta\mathbf{e}^{\perp} + \delta\mathbf{e}^0, \quad \delta\mathbf{e}^{\perp} \in \boldsymbol{Z}_0(\mathbf{I},\Omega),\ \delta\mathbf{e}^0 \in \boldsymbol{\mathcal{N}}(\Omega).$$

Both components will be estimated separately. A genuine duality technique is required for the first: define $\mathbf{g} \in \boldsymbol{Z}_0(\mathbf{I},\Omega)$ as the solution of

$$a(\mathbf{g},\mathbf{v}) = \left(\delta\mathbf{e}^{\perp},\mathbf{v}\right)_{\boldsymbol{L}^2(\Omega)} \quad \forall \mathbf{v} \in \boldsymbol{Z}_0(\mathbf{I},\Omega). \tag{5.12}$$

This is legal due to Theorem 5.2. From Theorem 4.1 and $\mathrm{div}\,\mathbf{g} = 0$ we learn that $\mathbf{g} \in \boldsymbol{H}^s(\Omega)$ for some $\frac{1}{2} < s \leq 1$, and that

$$\|\mathbf{g}\|_{\boldsymbol{H}^s(\Omega)} \leq C\,\|\mathbf{g}\|_{\boldsymbol{H}(\mathbf{curl};\Omega)}. \tag{5.13}$$

Moreover, (5.12) means that $\mathbf{g}$ satisfies, in the sense of distributions,

$$\mathbf{curl}\,\mathbf{curl}\,\mathbf{g} - \omega^2\mathbf{g} = \delta\mathbf{e}^{\perp} \quad \text{in } \Omega.$$

Thus, $\mathbf{w} := \mathbf{curl}\,\mathbf{g}$ fulfils $\mathbf{curl}\,\mathbf{w} = \delta\mathbf{e}^{\perp} + \omega^2\mathbf{g} \in \boldsymbol{L}^2(\Omega)$, $\mathrm{div}\,\mathbf{w} = 0$, and

$\gamma_{\mathbf{n}}\mathbf{w} = 0$. Using Lemma 4.2, we infer that $\mathbf{w} \in \boldsymbol{H}^s(\Omega)$ for some $\frac{1}{2} < s \leq 1$ and that

$$\|\mathbf{w}\|_{\boldsymbol{H}^s(\Omega)} \leq C \|\mathbf{w}\|_{\boldsymbol{H}(\mathbf{curl};\Omega)} \leq C\Big(\|\mathbf{g}\|_{\boldsymbol{H}(\mathbf{curl};\Omega)} + \|\delta\mathbf{e}^{\perp}\|_{\boldsymbol{L}^2(\Omega)}\Big).$$

This s may be different from the one above. In what follows we keep the symbol s to designate whichever is the smaller. Using (5.9) and (5.12) we obtain

$$\|\mathbf{g}\|_{\boldsymbol{H}(\mathbf{curl};\Omega)} \leq C_S^{-1} \|\delta\mathbf{e}^{\perp}\|_{\boldsymbol{L}^2(\Omega)}.$$

Combined with the previous estimates, this yields, with $C = C(\Omega) > 0$,

$$\|\mathbf{g}\|_{\boldsymbol{H}^s(\Omega)} + \|\mathbf{curl}\,\mathbf{g}\|_{\boldsymbol{H}^s(\Omega)} \leq C \|\delta\mathbf{e}^{\perp}\|_{\boldsymbol{L}^2(\Omega)}. \tag{5.14}$$

Armed with these estimates, and Galerkin orthogonality $a(\mathbf{e} - \mathbf{e}_h, \mathbf{v}_h) = 0$ for all $\mathbf{v}_h \in \mathcal{W}^1_{p,0}(\Omega_h)$, we obtain

$$\begin{aligned}\|\delta\mathbf{e}^{\perp}\|^2_{\boldsymbol{L}^2(\Omega)} &= a(\mathbf{g}, \delta\mathbf{e}^{\perp}) = a(\mathbf{g}, \mathbf{e} - \mathbf{e}_h) = a(\mathbf{g} - \Pi^1_p\mathbf{g}, \delta\mathbf{e}) \\ &\leq C_A \|\mathbf{g} - \Pi^1_p\mathbf{g}\|_{\boldsymbol{H}(\mathbf{curl};\Omega)} \|\mathbf{e} - \mathbf{e}_h\|_{\boldsymbol{H}(\mathbf{curl};\Omega)}.\end{aligned}$$

We also used the fact that $(\mathbf{g}, \delta\mathbf{e}^0)_{\boldsymbol{L}^2(\Omega)} = 0$ and hence, by Lemma 3.13 and (5.13), $\mathbf{g}$ is sufficiently smooth to ensure that moment-based local interpolation operators be well defined. Next, we rely on (5.14), Theorem 3.14 and Corollary 3.17 and get, with a positive constant $C = C(\Omega, p, s, \rho(\Omega_h))$,

$$\|\mathbf{g} - \Pi^1_p\mathbf{g}\|_{\boldsymbol{H}(\mathbf{curl};\Omega)} \leq Ch^s \|\delta\mathbf{e}^{\perp}\|_{\boldsymbol{L}^2(\Omega)}.$$

The last two inequalities give the assertion for $\delta\mathbf{e}^{\perp}$.

In order to tackle $\delta\mathbf{e}^0$, we use

$$-\omega^2 \left(\delta\mathbf{e}^0, \mathbf{z}_h\right)_{\boldsymbol{L}^2(\Omega)} = a(\delta\mathbf{e}^0, \mathbf{z}_h) = 0 \quad \forall \mathbf{z}_h \in \boldsymbol{\mathcal{N}}(\Omega_h),$$

which is a consequence of Galerkin orthogonality. For the time being, let us assume that $\Pi^1_p)\delta\mathbf{e}^0$ is well defined. Then, by virtue of the commuting diagram property (3.12) this involves

$$\|\delta\mathbf{e}^0\|_{\boldsymbol{L}^2(\Omega)} \leq \|(\mathrm{Id} - \Pi^1_p)\delta\mathbf{e}^0\|_{\boldsymbol{L}^2(\Omega)}. \tag{5.15}$$

The observation $\delta\mathbf{e}^{\perp} = \mathsf{P}^{\perp}(\mathbf{e} - \mathbf{e}_h)$ confirms the identity

$$\begin{aligned}(\mathrm{Id} - \Pi^1_p)\delta\mathbf{e}^0 &= (\mathrm{Id} - \Pi^1_p)(\mathsf{P}^{\perp} + \mathsf{P}^0)\mathbf{e} - (\mathrm{Id} - \Pi^1_p)\delta\mathbf{e}^{\perp} \\ &= (\mathrm{Id} - \Pi^1_p)\mathsf{P}^0\mathbf{e} + (\mathrm{Id} - \Pi^1_p)\mathsf{P}^{\perp}\mathbf{e}_h.\end{aligned}$$

The second term is amenable to the policy pursued in the proof of Lemma 4.5 and we get

$$\|(\mathrm{Id} - \Pi^1_p)\mathsf{P}^{\perp}\mathbf{e}_h\|_{\boldsymbol{L}^2(\Omega)} \leq Ch^s \|\mathbf{curl}\,\mathbf{e}_h\|_{\boldsymbol{L}^2(\Omega)}.$$

To deal with the first term, we resort to Lemma 2.2, which gives a splitting $\mathsf{P}^0\mathbf{e} = \mathbf{grad}\,\varphi + \mathbf{q}$, $\varphi \in H_0^1(\Omega)$, $\mathbf{q} \in \boldsymbol{\mathcal{H}}^2(\Omega) \subset \boldsymbol{H}^s(\Omega)$. Note that φ satisfies

$$\omega\,(\mathbf{grad}\,\varphi, \mathbf{grad}\,\psi)_{\boldsymbol{L}^2(\Omega)} = i\,(\operatorname{div}\mathbf{j}, \psi)_{\boldsymbol{L}^2(\Omega)} \quad \forall \psi \in H_0^1(\Omega).$$

Remembering the connection between s and the optimal lifting exponent for the Dirichlet problem for the Laplacian on Ω (*cf.* Theorem 4.1), we see that $\mathbf{grad}\,\varphi \in \boldsymbol{H}^s(\Omega)$, too. In particular, now we know that $\Pi_p^1 \delta \mathbf{e}^0$ really makes sense. Moreover, we can invoke Theorem 3.14, which gives

$$\left\|(\mathrm{Id} - \Pi_p^1)\mathsf{P}^0\mathbf{e}\right\|_{\boldsymbol{L}^2(\Omega)} \leq C h^s \Big(\left\|\mathsf{P}^0\mathbf{e}\right\|_{\boldsymbol{L}^2(\Omega)} + \|\operatorname{div}\mathbf{j}\|_{L^2(\Omega)} \Big).$$

Combining all estimates, the proof is finished. □

Remark 18. If Ω has a smooth boundary or is convex, which involves 2-regularity of the Dirichlet and Neumann problem for the Laplacian on Ω, we can pick $s = 1$ in the above theorem. △

Bibliographical notes

Duality estimates for Maxwell source problems were pioneered by P. Monk, in Monk (1992*b*), and later refined in Monk (2001). They have been used in various contexts, for instance in Ciarlet, Jr, and Zou (1999, Section 4).

5.4. Zero frequency limit

Apart from problems faced for resonant ω the formulation (5.1) is also prone to instability when $\omega \to 0$. In the limit case $\omega = 0$ a solution does not exist if $\mathbf{j} \notin \boldsymbol{H}(\operatorname{div} 0; \Omega)$. Even if this is satisfied, we can only expect uniqueness of $\mathbf{e}$ up to a contribution from $\boldsymbol{H}_0(\mathbf{curl}\,0; \Omega)$.

Such stability problems are inherent in the derivation of (5.1) through the elimination of the magnetic field. Therefore, $\mathbf{e}$ is of a twin magnetic and electric nature. However, for $\omega = 0$ the magnetic and electric field are completely decoupled: information on the magnetic field can no longer be conveyed through $\mathbf{e}$.

We cannot help separating magnetic and electric quantities in the variational problem (5.1). A way to achieve this is the introduction of potentials. To avoid difficulties caused by the artificial electric walls, we will assume trivial topology of Ω in the remainder of this section.

Following (2.11), we write

$$\mathbf{e} = -\,\mathbf{grad}\,v - i\omega\,\mathbf{curl}\,\mathbf{a},$$

supplemented by a *gauge condition* to fix the potentials. For low frequencies the Coulomb gauge $\operatorname{div}\mathbf{a} = 0$ is most natural. Imposing the gauge constraint in weak form the variational problem is thus converted into the following saddle point problem. Seek $\mathbf{a} \in \boldsymbol{H}_0(\mathbf{curl};\Omega)$, $v \in H_0^1(\Omega)$ such that

$$\begin{aligned}(\mu^{-1}\,\mathbf{curl}\,\mathbf{a}, \mathbf{curl}\,\mathbf{a}')_{\boldsymbol{L}^2(\Omega)} \\ -\omega^2\,(\epsilon\mathbf{a},\mathbf{a}')_{\boldsymbol{L}^2(\Omega)} + i\omega\,(\epsilon\mathbf{a}', \mathbf{grad}\,v)_{\boldsymbol{L}^2(\Omega)} &= (\mathbf{j},\mathbf{a}')_{\boldsymbol{L}^2(\Omega)}, \\ i\omega\,(\epsilon\mathbf{a}, \mathbf{grad}\,v')_{\boldsymbol{L}^2(\Omega)} &= 0,\end{aligned}$$

for all $\mathbf{a}' \in \boldsymbol{H}_0(\mathbf{curl};\Omega)$, $v' \in H_0^1(\Omega)$. Here, v can be regarded as a Lagrangian multiplier enforcing the divergence constraint on $\mathbf{a}$ (Kikuch 1987).

Theorem 5.9. If $\omega \geq 0$ complies with Assumption 1, then the above saddle point problem has a unique solution $(\mathbf{a}, v)$.

Proof. As, evidently, $\mathbf{a} \in \boldsymbol{Z}_0(\epsilon,\Omega)$, first consider the following restriction of the saddle point problem to $\boldsymbol{Z}_0(\epsilon,\Omega)$. Seek $\mathbf{a} \in \boldsymbol{Z}_0(\epsilon,\Omega)$ such that

$$(\mu^{-1}\,\mathbf{curl}\,\mathbf{a}, \mathbf{curl}\,\mathbf{a}')_{\boldsymbol{L}^2(\Omega)} - \omega^2\,(\epsilon\mathbf{a},\mathbf{a}')_{\boldsymbol{L}^2(\Omega)} = (\mathbf{j},\mathbf{a}')_{\boldsymbol{L}^2(\Omega)} \qquad (5.16)$$

for all $\mathbf{a}' \in \boldsymbol{Z}_0(\epsilon,\Omega)$. By Corollary 4.3 we have coercivity, and Assumption 1 ensures uniqueness, which implies existence by the Fredholm alternative.

Secondly, observe that, thanks to $\mathbf{grad}\,H_0^1(\Omega) \subset \boldsymbol{H}_0(\mathbf{curl};\Omega)$,

$$\sup_{\mathbf{v}\in\boldsymbol{H}_0(\mathbf{curl};\Omega)} \frac{(\epsilon\mathbf{v}, \mathbf{grad}\,v)_{\boldsymbol{L}^2(\Omega)}}{\|\mathbf{v}\|_{\boldsymbol{H}(\mathbf{curl};\Omega)}} \geq \frac{(\epsilon\,\mathbf{grad}\,v, \mathbf{grad}\,v)_{\boldsymbol{L}^2(\Omega)}}{\|\mathbf{grad}\,v\|_{\boldsymbol{H}(\mathbf{curl};\Omega)}} \geq C\,|v|_{H^1(\Omega)}. \qquad (5.17)$$

From this and the Poincaré–Friedrichs inequality in $H_0^1(\Omega)$, we conclude the uniqueness of v. □

In a discrete setting based on discrete differential forms the nice properties of the continuous problem are preserved.

Theorem 5.10. For any $\omega \geq 0$ complying with Assumption 1 the saddle point problem discretized over $\mathcal{W}^1_{p,0}(\Omega_h) \times \mathcal{W}^0_{p,0}(\Omega_h)$ has a unique solution that converges quasioptimally provided that the meshwidth is sufficiently small.

Proof. Following the proof of the previous theorem, we first consider the following variational problem on $\boldsymbol{Z}_{h,0}(\epsilon,\Omega_h)$. Find $\mathbf{a}_h \in \boldsymbol{Z}_{h,0}(\epsilon,\Omega_h)$ with

$$(\mu^{-1}\,\mathbf{curl}\,\mathbf{a}_h, \mathbf{curl}\,\mathbf{a}'_h)_{\boldsymbol{L}^2(\Omega)} - \omega^2\,(\epsilon\mathbf{a}_h,\mathbf{a}'_h)_{\boldsymbol{L}^2(\Omega)} = (\mathbf{j},\mathbf{a}'_h)_{\boldsymbol{L}^2(\Omega)}$$

for all $\mathbf{a}'_h \in \boldsymbol{Z}_{h,0}(\epsilon,\Omega_h)$. Using the techniques of Section 5.2 we get the assertion for $\mathbf{a}_h$. The details are left to the reader.

Then, thanks to $\mathbf{grad}\,\mathcal{W}^0_{p,0}(\Omega_h) \subset \mathcal{W}^1_{p,0}(\Omega_h)$ we can compute a unique $v_h \in \mathcal{W}^0_{p,0}(\Omega_h)$ in a second step. The argument relies on the discrete version of (5.17). □

6. Regularized formulations

Using the discretization of the Maxwell eigenvalue problem studied in Section 4 we end up with a generalized eigenvalue problem for the coefficient vector $\vec{\mathbf{e}}$ describing the eigenmode. It has the form

$$\mathrm{A}_{\mu}\vec{\mathbf{e}}_h = \omega^2 \mathrm{M}^1_{\epsilon}\vec{\mathbf{e}}_h,$$

where both matrices are real, symmetric, M^1_{ϵ} is a positive definite mass matrix, but A_{μ} is only positive semidefinite, beset with a large kernel. Usually, a few of the smallest nonzero eigenvalues and eigenfunctions of (4.7) are of interest. The vast numbers of degrees of freedom – inevitable for problems in three dimensions – force us to use iterative methods of inverse iteration type (Adam, Arbenz and Geus 1997, Section 7), most of which are crippled by the presence of a kernel. Thus, it becomes desirable to get rid of the kernel, of course, without affecting the approximation of eigenvectors and eigenvalues.

The kernel of $\mathrm{A}_{\mu}\vec{\mathbf{e}}_h$ might also hamper the solution of the discretized Maxwell source problem. For instance, when applying Krylov subspace methods, *e.g.*, MINRES (Hackbusch 1993), for its iterative solution, convergence will deteriorate, if there are large numbers of both negative and positive eigenvalues of the system matrix corresponding to (5.2). The presence of many negative eigenvalues may also hurt preconditioning techniques.

6.1. Discrete regularization

Any solution $\mathbf{e}_h$ of the Maxwell eigenvalue problem (4.7) will satisfy $\mathbf{e}_h \in \boldsymbol{Z}_{h,0}(\epsilon, \Omega_h)$. Therefore all discrete eigenvalues and eigenfunctions of (4.7) can also be obtained from the following problem. Find $\mathbf{e} \in \mathcal{W}^1_{p,0}(\Omega_h)$, $\varphi_h \in \mathcal{W}^0_{p,0}(\Omega_h)$, $\omega \neq 0$ such that, for some regularization parameter $s > 0$,

$$\begin{aligned}
(\boldsymbol{\mu}^{-1}\,\mathbf{curl}\,\mathbf{e}_h, \mathbf{curl}\,\mathbf{e}'_h)_{\boldsymbol{L}^2(\Omega)} &+ (\boldsymbol{\epsilon}\mathbf{e}'_h, \mathbf{grad}\,\varphi_h)_{\boldsymbol{L}^2(\Omega)} &&= \omega^2\,(\boldsymbol{\epsilon}\mathbf{e}_h, \mathbf{e}'_h)_{\boldsymbol{L}^2(\Omega)},\\
(\boldsymbol{\epsilon}\mathbf{e}_h, \mathbf{grad}\,\varphi'_h)_{\boldsymbol{L}^2(\Omega)} &- \tfrac{1}{s}d(\varphi_h, \varphi'_h) &&= 0,
\end{aligned}$$

for all $\mathbf{e}'_h \in \mathcal{W}^1_p(\Omega_h)$, $\varphi'_h \in \mathcal{W}^0_{p,0}(\Omega_h)$. Note that, if $\mathbf{e}_h$ is a discrete eigenvalue of the original problem (4.7), the associated φ_h turns out to be zero. Thus, $d(\cdot\,,\cdot)$ can be *any* $H^1(\Omega)$-continuous bilinear form on $\mathcal{W}^0_{p,0}(\Omega_h)$. In matrix notation the above eigenvalue problem reads:

$$\begin{aligned}
\mathrm{A}_{\mu}\vec{\mathbf{e}}_h &+ \mathrm{G}\vec{\varphi}_h &&= \omega^2\mathrm{M}^1_{\epsilon}\vec{\mathbf{e}}_h,\\
\mathrm{G}^T\vec{\mathbf{e}}_h &- \tfrac{1}{s}\mathrm{D}\vec{\varphi}_h &&= 0.
\end{aligned}$$

The case $d = 0$ was already treated in Section 5.4. It leads to a saddle point problem, which is not the desirable outcome of regularization. Thus, the only sensible choices for $d(\cdot\,,\cdot)$ are symmetric, positive definite bilinear forms.[8]

[8] Remember that the eigenvalue problem can be tackled in an entirely real setting.

Then D is invertible, and we can eliminate $\vec{\varphi}_h$, providing the symmetric generalized eigenvalue problem

$$(\mathrm{A}_\mu + s\mathrm{G}\mathrm{D}^{-1}\mathrm{G}^T)\vec{\mathbf{e}}_h = \omega^2 \mathrm{M}_\epsilon^1 \vec{\mathbf{e}}_h. \tag{6.1}$$

Lemma 6.1. If Ω has trivial topology, that is, its first and second Betti number vanish, then $\mathrm{A}_\mu + s\mathrm{G}\mathrm{D}^{-1}\mathrm{G}^T$ is positive definite. Its eigenvalues are bounded from below by $\min\{C_1^{1/2}, sC_0\|d\|^{-1/2}\}$, where C_1 is the constant of the discrete Poincaré–Friedrichs inequality from Theorem 4.7, and C_0 depends on ϵ and the constant of the Poincaré–Friedrichs inequality of $H_0^1(\Omega)$. Further,

$$\|d\| := \sup\{d(\varphi_h, \psi_h),\ \varphi_h, \psi_h \in \mathcal{W}_p^0(\Omega_h),\ \|\varphi_h\|_{H^1(\Omega)} \leq 1, \|\psi_h\|_{H^1(\Omega)} \leq 1\}.$$

Proof. We can analyse the generalized saddle point problem behind the regularized formulation appealing to the 'ellipticity on kernel' (Brezzi and Fortin 1991, Chapter 2) expressed in Theorem 4.7, and the estimate

$$\begin{aligned}\sup_{\mathbf{v}_h \in \mathcal{W}_{p,0}^1(\Omega_h)} \frac{(\epsilon \mathbf{v}_h, \mathbf{grad}\, \psi_h)_{\boldsymbol{L}^2(\Omega)}}{\|\mathbf{v}_h\|_{\boldsymbol{H}(\mathbf{curl};\Omega)}} &\geq \frac{(\epsilon\, \mathbf{grad}\, \psi_h, \mathbf{grad}\, \psi_h)_{\boldsymbol{L}^2(\Omega)}}{|\psi_h|_{H^1(\Omega)}} \\ &\geq C\, |\psi_h|_{H^1(\Omega)} \geq C_0\, \|\psi_h\|_{H^1(\Omega)}\end{aligned}$$

for $\psi_h \in \mathcal{W}_{p,0}^0(\Omega_h)$. Standard estimates finish the proof. □

The conditions on the bilinear form $d(\cdot\,,\cdot)$ are very weak. For instance, a lumped $\boldsymbol{L}^2(\Omega)$-inner product according to (3.37) will meet the requirements and lead to a diagonal matrix D. Hence, the matrix $\mathrm{A}_\mu + s\mathrm{G}\mathrm{D}^{-1}\mathrm{G}^T$ in (6.1) remains sparse and can be assembled efficiently. This procedure is called 'discrete regularization', because the manipulations are carried out entirely on the matrix level.

Beside the physically meaningful discrete eigenvalues/eigenfunctions, we get many more from (6.1). The additional non-physical solutions can easily be weeded out by looking at $\varphi_h = \mathrm{D}^{-1}\mathrm{G}^T\vec{\mathbf{e}}_h$. If this is nonzero, the eigenfunction can be dismissed.

Discrete regularization can also be performed for the source problem, for instance in the case of the stabilized problem of Section 5.4. Further, assuming trivial topology, some preprocessing is required, by replacing $\mathbf{j} \leftarrow \mathbf{j} - \mathbf{grad}\, \eta_h$, where $\eta_h \in \mathcal{W}_p^0(\Omega_h)$ satisfies

$$\left(\mathbf{grad}\, \eta_h, \mathbf{grad}\, \eta_h'\right)_{\boldsymbol{L}^2(\Omega)} = \left(\mathbf{j}, \mathbf{grad}\, \eta_h'\right)_{\boldsymbol{L}^2(\Omega)} \quad \forall \eta_h' \in \mathcal{W}_p^0(\Omega_h).$$

Using this modified weakly discretely divergence-free $\mathbf{j}$, the solution for v_h turns out to be zero and the same technique as above can be applied. Then $\mathbf{a}_h$ can be determined by solving a system with the Hermitian matrix

$$\mathrm{A}_\mu + s\mathrm{G}\mathrm{D}^{-1}\mathrm{G}^T - \omega^2 \mathrm{M}_\epsilon.$$

This matrix has a small fixed number of negative eigenvalues, just like the matrix arising from the Helmholtz boundary value problem.

Bibliographical notes
In Bespalov (1988) the discrete regularization idea was introduced for the treatment of electromagnetic eigenproblems. A discussion is given in Adam *et al.* (1997, Section 6). This latter report gives a rather comprehensive account of the numerical treatment of regularized Maxwell eigenvalue problems. Discrete regularization is a fairly natural idea from the perspective of discrete Hodge operators (*cf.* Section 3.3) and is frequently treated in articles on the finite integration technique (Clemens and Weiland 2001, Section 3). Furthermore, an approach resembling discrete regularization is the gist of the paper by Haber, Ascher, Aruliah and Oldenburg (1999).

6.2. grad-div *regularization*

The matrix $\mathrm{GD}^{-1}\mathrm{G}^T$ used above induces an inner product for weakly defined discrete divergences. An analogous regularization can also be carried out on the continuous level. It is motivated by observing that, in the eigenvalue problem (4.1), we necessarily have $\mathbf{e} \in \boldsymbol{Z}_0(\epsilon, \Omega)$, which implies $\operatorname{div}(\epsilon\mathbf{e}) = 0$. We are led to consider the following related variational eigenvalue problem. For $s > 0$ find $\mathbf{e} \in \boldsymbol{X}_0(\epsilon, \Omega)$, $\omega \neq 0$ with

$$\left(\boldsymbol{\mu}^{-1}\,\mathbf{curl}\,\mathbf{e}, \mathbf{curl}\,\mathbf{e}'\right)_{\boldsymbol{L}^2(\Omega)} + s\left(\operatorname{div}(\epsilon\mathbf{e}), \operatorname{div}(\epsilon\mathbf{e}')\right)_{L^2(\Omega)} = \omega^2\left(\epsilon\mathbf{e}, \mathbf{e}'\right)_{\boldsymbol{L}^2(\Omega)} \tag{6.2}$$

for all $\mathbf{e}' \in \boldsymbol{X}_0(\epsilon, \Omega)$. By Corollary 4.3 this is an eigenvalue problem for an operator with a compact resolvent in $\boldsymbol{L}^2(\Omega)$. The true Maxwell eigenvalues will not depend on the regularization parameter s, and the uninteresting solutions can be distinguished by nonzero $\operatorname{div}(\epsilon\cdot)$. The strong form of (6.2) seems to be

$$\mathbf{curl}\,\boldsymbol{\mu}^{-1}\,\mathbf{curl}\,\mathbf{e} - s\epsilon\,\mathbf{grad}\operatorname{div}(\epsilon\mathbf{e}) = \omega^2\epsilon\mathbf{e}. \tag{6.3}$$

This accounts for the parlance 'grad-div regularization'. For the Maxwell source problem a similar strategy can be pursued. For simplicity assume $\operatorname{div}\mathbf{j} = 0$ and trivial topology. Then the grad-div regularized formulation of (5.1) is as follows. Seek $\mathbf{e} \in \boldsymbol{X}_0(\epsilon, \Omega)$ such that

$$\begin{aligned}\left(\boldsymbol{\mu}^{-1}\,\mathbf{curl}\,\mathbf{e}, \mathbf{curl}\,\mathbf{e}'\right)_{\boldsymbol{L}^2(\Omega)} + s\left(\operatorname{div}(\epsilon\mathbf{e}), \operatorname{div}(\epsilon\mathbf{e}')\right)_{\boldsymbol{L}^2(\Omega)} - \omega^2\left(\epsilon\mathbf{e}, \mathbf{e}'\right)_{\boldsymbol{L}^2(\Omega)} \\ = -i\omega\left(\mathbf{j}, \mathbf{e}'\right)_{\boldsymbol{L}^2(\Omega)}\end{aligned} \tag{6.4}$$

for all $\mathbf{e}' \in \boldsymbol{X}_0(\epsilon, \Omega)$. Equivalence of (6.4) and (5.1) is not straightforward, and tied to conditions (*cf.* Costabel and Dauge (1998, Theorem 1.1)).

Lemma 6.2. If ω^2/s is not an eigenvalue of the operator $\operatorname{div}(\epsilon\,\mathbf{grad}\,\cdot)$ on $H_0^1(\Omega)$, then (6.4) and (5.1) have the same solutions.

Proof. It is clear that solutions of (5.1) also solve (6.4). To prove the converse, we only have to show that $\operatorname{div}(\boldsymbol{\epsilon}\mathbf{e}) = 0$ for any solution of (6.4). Testing with gradients of functions in $H_0^1(\Omega)$ shows

$$\big(\operatorname{div}(\boldsymbol{\epsilon}\mathbf{e}), s\operatorname{div}(\boldsymbol{\epsilon}\,\mathbf{grad}\,\psi) + \omega^2\psi\big)_{L^2(\Omega)} = 0 \quad \forall\psi \in H_0^1(\Omega).$$

We have $(s\operatorname{div}(\boldsymbol{\epsilon}\,\mathbf{grad}\,\cdot) + \omega^2\,\mathrm{Id})H_0^1(\Omega) = L^2(\Omega)$, provided that ω, s satisfy the assumptions of the theorem. □

To keep the presentation simple, we set $\boldsymbol{\epsilon} = \mathbf{I}$ for the remainder of this section. We may ask ourselves how (6.2) and (6.4) should be discretized by means of finite elements. Any scheme must be both **curl**- and div-conforming, meaning both tangential and normal continuity. It goes without saying that this amounts to global continuity for any piecewise smooth vector-field. Thus we have to leave the framework of discrete differential forms and approximate $\mathbf{e}$ in the space $\boldsymbol{\mathcal{S}}_{p,0}(\Omega_h) := (\mathcal{W}_p^0(\Omega_h))^3 \cap \boldsymbol{H}_0(\mathbf{curl};\Omega)$ of vector-fields with continuous, piecewise polynomial Cartesian components. Seek $\widetilde{\mathbf{e}} \in \boldsymbol{\mathcal{S}}_{p,0}(\Omega_h)$ such that

$$\begin{aligned}\big(\boldsymbol{\mu}^{-1}\,\mathbf{curl}\,\widetilde{\mathbf{e}}_h, \mathbf{curl}\,\widetilde{\mathbf{e}}'_h\big)_{L^2(\Omega)} + s\,\big(\operatorname{div}(\mathbf{e}), \operatorname{div}(\widetilde{\mathbf{e}}'_h)\big)_{L^2(\Omega)} - \omega^2\,\big(\widetilde{\mathbf{e}}_h, \widetilde{\mathbf{e}}'_h\big)_{L^2(\Omega)} \\ = -i\omega\,\big(\mathbf{j}, \widetilde{\mathbf{e}}'_h\big)_{L^2(\Omega)} \quad (6.5)\end{aligned}$$

for all $\widetilde{\mathbf{e}}'_h \in \boldsymbol{\mathcal{S}}_{p,0}(\Omega_h)$.

The desire to get continuous approximations of $\mathbf{e}$ actually provides a motivation for using the grad-div regularization for the source problem in the case $\operatorname{div}\mathbf{j} = 0$. We saw in Section 4.5 that the space $\boldsymbol{\mathcal{S}}_{p,0}(\Omega_h)$ must be ruled out for the direct discretization of (5.1), because on general meshes the spectrum of the discrete system matrix is polluted by kernel components. This might lead to severely ill-conditioned linear systems, even if ω is far away from any Maxwell eigenvalue. The grad-div regularized formulations can be used to suppress this disastrous pollution effect and thus apparently paves the way for applying Lagrangian finite elements by means of $\boldsymbol{\mathcal{S}}_{p,0}(\Omega_h)$. One can cite several 'practical' reasons why one might prefer $\mathbf{e}_h \in \boldsymbol{\mathcal{S}}_{p,0}(\Omega_h)$ to the approximation by a discrete 1-form.

- The electromagnetic problem should be solved with an existing FEM software package that can only handle vertex-based degrees of freedom.
- One might 'trust' traditional finite elements more than discrete differential forms.
- The result is needed as input data for another computation which might be crippled by discontinuities. Allegedly, this is the case for particle simulations (Maxwell–Vlasov equations).

Is (6.4) actually the continuous problem underlying (6.5)? Evidently, $\boldsymbol{\mathcal{S}}_{p,0}(\Omega_h) \subset \boldsymbol{H}^1(\Omega)$, and so (6.5) can be regarded as a Galerkin discretization of (6.4) posed over the space

$$\boldsymbol{H}^1_\times(\Omega) := \boldsymbol{H}^1(\Omega) \cap \boldsymbol{H}_0(\mathbf{curl};\Omega).$$

This is not a moot point, because any $\widetilde{\mathbf{u}} \in \boldsymbol{H}^1_\times(\Omega)$ can be arbitrarily well approximated by functions in $\boldsymbol{\mathcal{S}}_{p,0}(\Omega_h)$, provided the meshwidth is small enough or the polynomial degree p is large enough (Bonnet-BenDhia *et al.* 1999, Appendix). If the same is true for $\boldsymbol{X}_0(\mathbf{I},\Omega)$, then (6.5) really discretizes (6.4). Be aware that this kind of 'density assumption' is inherent in the notion of Galerkin approximation.

A first observation stirs suspicion. Testing (6.4) with $\mathbf{e}' \in \mathbf{grad}(H^2(\Omega) \cap H^1_0(\Omega)) \subset \boldsymbol{H}^1_\times(\Omega)$, we find

$$\left(\operatorname{div}(\mathbf{e}), s\Delta\psi + \omega^2\psi\right)_{L^2(\Omega)} = 0 \quad \forall \psi \in H^2(\Omega) \cap H^1_0(\Omega).$$

However, it is well known that $(s\Delta + \omega^2 \operatorname{Id})(H^2(\Omega) \cap H^1_0(\Omega))$ is a closed subspace of $L^2(\Omega)$ (Hanna and Smith 1967, Dauge 1988), and a *proper subspace* in the presence of non-convex edges of Ω; then the Dirichlet problem for $s\Delta + \omega^2 \operatorname{Id}$ with right-hand side $\in L^2(\Omega)$ may have solutions that are not in $H^2(\Omega)$. We conclude that $\operatorname{div}\mathbf{e} = 0$ is not guaranteed, if (6.4) is considered on $\boldsymbol{H}^1_\times(\Omega)$. The situation is strikingly disclosed by the following result.

Theorem 6.3. The space $\boldsymbol{H}^1_\times(\Omega)$ is a closed subspace of $\boldsymbol{X}_0(\mathbf{I},\Omega)$, and the inclusion is strict, if Ω has re-entrant edges or corners.

Proof. Let $(\widetilde{\mathbf{u}}_n)_{n\in\mathbb{N}}$ be a Cauchy sequence in $\boldsymbol{X}_0(\mathbf{I},\Omega)$ whose members belong to $\boldsymbol{H}^1_\times(\Omega)$. By Lemma 2.4 we find Cauchy sequences $(\mathbf{v}_n)_{n\in\mathbb{N}}$, $\mathbf{v}_n := \mathsf{R}_0\widetilde{\mathbf{u}}_n$, in $\boldsymbol{H}^1_\times(\Omega)$, and $(\varphi_n)_n$, $\varphi_n := \mathsf{N}_0\widetilde{\mathbf{u}}_n$, in $H(\Delta,\Omega)$, where

$$H(\Delta,\Omega) := \{\phi \in H^1_0(\Omega),\ \Delta\phi \in L^2(\Omega)\},$$

with

$$\widetilde{\mathbf{u}}_n = \mathbf{v}_n + \mathbf{grad}\,\varphi_n \quad \forall n \in \mathbb{N}.$$

As $(\varphi_n)_n$ is also a sequence in $H^2(\Omega) \cap H^1_0(\Omega)$ and this space is closed in $H(\Delta,\Omega)$ (Hanna and Smith 1967), the first assertion follows. The second is immediate from considering gradients of functions in $H(\Delta,\Omega) \setminus (H^2(\Omega) \cap H^1_0(\Omega))$. See Amrouche *et al.* (1998, Section 2.c) for a concrete example. □

This theorem sends the important message that, if Ω has re-entrant edges, we may get different solutions $\mathbf{e}$ and $\widetilde{\mathbf{e}}$ when considering the variational equation (6.4) on $\boldsymbol{X}_0(\mathbf{I},\Omega)$ and $\boldsymbol{H}^1_\times(\Omega)$, respectively. More precisely, $\widetilde{\mathbf{e}}$ is the Galerkin projection of $\mathbf{e}$ onto $\boldsymbol{H}^1_\times(\Omega)$. As regards the finite element approximation in $\boldsymbol{\mathcal{S}}_{p,0}(\Omega_h)$, Theorem 6.3 contends that in $\boldsymbol{\mathcal{S}}_{p,0}(\Omega_h)$ we may not be able to get arbitrarily close to $\mathbf{e}$ in the $\boldsymbol{X}_0(\mathbf{I},\Omega)$-norm, which is

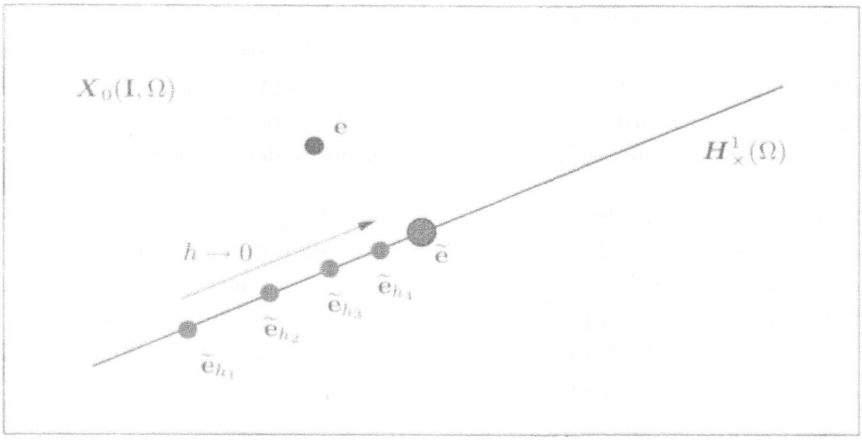

Figure 6.1. 'Convergence' of finite element approximations to the solution of (6.4). The geometry of the plane of the paper is induced by the $\boldsymbol{X}_0(\mathbf{I},\Omega)$-metric

the right energy norm. This behaviour is illustrated in Figure 6.1. The eigenvalue problem (6.2) is afflicted as well. In other words, even if $\mathbb{H}$ is a sequence of meshwidths tending to zero, $\bigcup_{h\in\mathbb{H}} \boldsymbol{\mathcal{S}}_{p,0}(\Omega_h)$ is *not dense* in $\boldsymbol{X}_0(\mathbf{I},\Omega)$, if there are re-entrant edges/corners.

This phenomenon is perilous, because the discrete solutions will 'look' correct and even display nice convergence as the meshes are refined. The unwary will be lulled into a false sense of security and happily accept a possibly wrong result.

Remark 19. Most readers will know the identity

$$-\boldsymbol{\Delta} = \mathbf{curl}\,\mathbf{curl} - \mathbf{grad}\,\mathrm{div},$$

which holds for C^2-vector-fields and, therefore, in the sense of distributions. Crudely speaking, the grad-div regularization aims to exploit this identity and switch to a variational problem in $\boldsymbol{H}^1_\times(\Omega)$. This is prone to failure, because Theorem 6.3 tells us that smooth vector-fields with vanishing tangential components on $\partial\Omega$ are not necessarily dense in $\boldsymbol{X}_0(\mathbf{I},\Omega)$. As a consequence, the variational problem (6.4) on $\boldsymbol{X}_0(\mathbf{I},\Omega)$ cannot be stated as an equivalent partial differential equation in the sense of distributions. Now we understand that the partial differential equation (6.3) actually belongs to the problem on $\boldsymbol{H}^1_\times(\Omega)$ (*cf.* Bonnet-BenDhia *et al.* (1999, Section 4)). $\triangle$

Remark 20. As shown by the investigations in Costabel and Dauge (1998) and Costabel *et al.* (1999), the gap between $\boldsymbol{X}_0(\mathbf{I},\Omega)$ and $\boldsymbol{H}^1_\times(\Omega)$ can be filled with gradients of singular functions associated with re-entrant edges of the domain boundary. Thus, the *singular field method* aims to augment the space $\mathcal{S}_{p,0}(\Omega_h)$ by these singular functions (Bonnet-BenDhia *et al.* 1999). Unfortunately the space spanned by the singular functions has infinite dimension, which thwarts a straightforward application of the idea. On the contrary, it is successful for reduced two-dimensional problems (Hazard and Lohrengel 2000, Assous, Ciarlet Jr and Sonnendrücker 1998). △

Remark 21. Also, the presence of discontinuities in μ and ϵ might rule out the discretization of (6.2), even if finite element spaces are used that belong to $\boldsymbol{H}^1(\Omega_i)$ only for subdomains Ω_i of Ω, where the coefficients are smooth: non-density will strike again. We refer to the discussion in Section 6.2 of Bonnet-BenDhia *et al.* (1999), Costabel *et al.* (1999, Section 2) and Lohrengel and Nicaise (2001). △

Bibliographical notes

Grad-div regularization is an established technique in electromagnetic scattering theory Hazard and Lenoir (1996), Werner (1963) and Leis (1986, Section 8.4). The non-density result of Theorem 6.3 was discovered by M. Costabel (1991). A thorough analysis along with numerical experiments is given in Costabel and Dauge (1999). In this paper it is pointed out that considering (6.2) on $\boldsymbol{H}^1_\times(\Omega)$ amounts to switching from electromagnetism to a setting for linear elasticity. The case of discontinuous coefficients is treated in Lohrengel and Nicaise (2001).

6.3. Weighted regularization

When carrying out grad-div regularization we squandered a lot of freedom by opting for the $\boldsymbol{L}^2(\Omega)$-inner product of the divergence terms right from the beginning. We could have considered a generic regularized problem of the following form. Find $\mathbf{e} \in \boldsymbol{X}_0[Y](\Omega)$ such that

$$\left(\boldsymbol{\mu}^{-1}\,\mathbf{curl}\,\mathbf{e}, \mathbf{curl}\,\mathbf{e}'\right)_{\boldsymbol{L}^2(\Omega)} + \langle \operatorname{div}\mathbf{e}, \operatorname{div}\mathbf{e}'\rangle_Y - \omega^2\left(\mathbf{e},\mathbf{e}'\right)_{\boldsymbol{L}^2(\Omega)} = -i\omega\left(\mathbf{j},\mathbf{e}'\right)_{\boldsymbol{L}^2(\Omega)} \tag{6.6}$$

for all $\mathbf{e}' \in \boldsymbol{X}_0[Y](\Omega)$, where Y is some Hilbert space with inner product $\langle \cdot\,,\cdot\rangle_Y$ and

$$\boldsymbol{X}_0[Y](\Omega) := \{\mathbf{u} \in \boldsymbol{H}_0(\mathbf{curl};\Omega),\ \operatorname{div}\mathbf{u} \in Y\}.$$

In the previous section we saw that $Y = \boldsymbol{L}^2(\Omega)$ fails to ensure density of $\boldsymbol{H}^1_\times(\Omega)$ in $\boldsymbol{X}_0[Y](\Omega)$. On the other hand, $Y = \boldsymbol{H}^{-1}(\Omega)$ recovers $\boldsymbol{X}_0[Y](\Omega) = \boldsymbol{H}_0(\mathbf{curl};\Omega)$, which gives the correct solution, namely that of the non-regularized problem. However, it fails to be compactly embedded into $\boldsymbol{L}^2(\Omega)$ and, therefore, is haunted by kernel pollution when discretized by means of

$\mathcal{S}_{p,0}(\Omega_h)$. Is there a middle ground, that is, a Hilbert space Y,

$$\boldsymbol{L}^2(\Omega) \subset Y \subset \boldsymbol{H}^{-1}(\Omega), \tag{6.7}$$

that meets both

- $\boldsymbol{H}^1_\times(\Omega)$ is dense in $\boldsymbol{X}_0[Y](\Omega)$, and
- $\boldsymbol{X}_0[Y](\Omega)$ is compactly embedded into $\boldsymbol{L}^2(\Omega)$?

To state sufficient and necessary conditions, when this is the case, we rely on the Hilbert space

$$H(\Delta[Y],\Omega) := \{\varphi \in H^1_0(\Omega),\ \Delta\varphi \in Y\},$$

and the associated Riesz operator

$$\mathsf{I} : H(\Delta[Y],\Omega) \mapsto Y, \qquad \langle \psi, \mathsf{I}\varphi\rangle_Y := \langle \varphi, \psi\rangle_{H^1(\Omega)\times H^{-1}(\Omega)} \quad \forall \psi \in Y.$$

As in the case of Lemma 6.2 one proves the equivalence of (6.6) and (5.1) (*cf.* Costabel and Dauge (2002, Theorem 2.1)).

Lemma 6.4. If $\operatorname{div}\mathbf{j} = 0$ and $(s\Delta + \omega^2\mathsf{I})H(\Delta[Y],\Omega)$ is dense in Y, then the solutions of (6.6) and (5.1) coincide.

The main technical tool for further investigations is another regular splitting theorem (Costabel and Dauge 2002, Theorem 2.2).

Lemma 6.5. (Regular splitting for $\boldsymbol{X}_0[Y](\Omega)$) If Y satisfies (6.7) we find continuous mappings $\mathsf{R}_0[Y] : Y \mapsto \boldsymbol{H}^1_\times(\Omega)$, $\mathsf{N}_0[Y] : Y \mapsto H(\Delta[Y],\Omega)$ such that $\mathsf{R}_0[Y] + \mathbf{grad} \circ \mathsf{N}_0[Y] = \mathrm{Id}$.

Proof. We can re-use the operators R_0 and N_0 from Lemma 2.4 after noting that

$$\|\Delta(\mathsf{N}_0\mathbf{u})\|_Y \leq C(\|\operatorname{div}\mathbf{u}\|_Y + \|\operatorname{div}(\mathsf{R}_0\mathbf{u})\|_{\boldsymbol{L}^2(\Omega)}).$$

Thus $\mathsf{R}_0[Y]$ and $\mathsf{N}_0[Y]$ can be chosen as plain restrictions of R_0 and N_0 to Y. □

As remarked in Costabel and Dauge (2002, Section 2.2) this allows two important conclusions.

Corollary 6.6. If the embedding $Y \hookrightarrow \boldsymbol{H}^{-1}(\Omega)$ is compact, the same is true for $\boldsymbol{X}_0[Y](\Omega) \hookrightarrow \boldsymbol{L}^2(\Omega)$.

Corollary 6.7. If $H^2(\Omega) \cap H^1_0(\Omega)$ is a dense subspace of $H(\Delta[Y],\Omega)$, then $\boldsymbol{H}^1_\times(\Omega)$ is dense in $\boldsymbol{X}_0[Y](\Omega)$.

Essentially, issues concerning the choice of Y can be investigated by looking at the Laplacian on polyhedral domains. Its theory is a venerable branch of the theory of elliptic boundary value problems. A key insight gained for second-order elliptic operators on polyhedra (Nazarov and

Plamenevskii 1994, Kondrat'ev 1967) is that it is weighted Sobolev spaces that supply the appropriate functional framework. This is why weighted L^2-spaces are the most promising candidates for Y, suggesting that we define

$$\|u\|_Y := \int_\Omega w(\mathbf{x})^2 |u|^2 \, d\mathbf{x},$$

with a positive weight function ($\gamma_\mathbf{v}, \gamma_e \in \mathbb{R}$)

$$w(\mathbf{x}) := \prod_{\text{corners } \mathbf{v}} \operatorname{dist}(\mathbf{x}; \mathbf{v})^{\gamma_\mathbf{v}} \cdot \prod_{\text{edges } e} \widetilde{\operatorname{dist}}(\mathbf{x}; e)^{\gamma_e},$$

where, with endpoints $\mathbf{v}_1$, $\mathbf{v}_2$ of e,

$$\widetilde{\operatorname{dist}}(\mathbf{x}; e) := \frac{\operatorname{dist}(\mathbf{x}; e)}{\operatorname{dist}(\mathbf{v}_1; e) \operatorname{dist}(\mathbf{v}_2; e)}.$$

Here $\mathbf{v}$, e run through the corners and edges of the boundary of Ω, respectively. Profound results about the singularities of the Laplacian supply a precise recipe for how to choose the weight exponents $\gamma_\mathbf{v}$ and γ_e (Costabel and Dauge 2002, Section 4).

Theorem 6.8. (Choice of weight exponents) If $\frac{1}{2} \leq \gamma_\mathbf{v} < 1$, $0 \leq \gamma_e < 1$ and

$$\gamma_e > 1 - \min_{\mathbf{x} \in e} \frac{\pi}{\omega_e(\mathbf{x})},$$

where $\omega_e(\mathbf{x})$ is the opening angle for the (curved) edge e at point $\mathbf{x} \in e$, then $\boldsymbol{X}_0[Y](\Omega) \hookrightarrow \boldsymbol{L}^2(\Omega)$ compactly and $\boldsymbol{H}^1_\times(\Omega) \subset \boldsymbol{X}_0[Y](\Omega)$ is dense.

Choosing the weight exponents according to the assumptions of the theorem and using a Galerkin discretization of (6.6) based on $\boldsymbol{\mathcal{S}}_{p,0}(\Omega_h)$ gives a scheme for the source problems that is asymptotically optimal on sufficiently fine meshes. This is immediately clear from applying the theory of Galerkin approximations for coercive variational problems. The compact embedding $\boldsymbol{X}_0[Y](\Omega) \hookrightarrow \boldsymbol{L}^2(\Omega)$ makes (6.6) belong to this class. For the same reason the standard theory can also be applied to the regularized eigenvalue problem. Asymptotic optimality can then be concluded with respect to the computed eigenvalues and eigenfunctions.

Remark 22. The idea of weighted regularization is too new to assess its merits. As for the discrete regularization of Section 6.1, it will produce 'nice' linear systems. In addition, we are rewarded with a globally continuous solution for the electric field. On the other hand, when compared with discrete 1-forms, a few drawbacks are obvious.

- Even for constant material parameters, the assembly of the linear system is somewhat complicated and entails using appropriate numerical quadrature.
- Given only raw geometry data in the form of a mesh, it might be costly to compute the weight functions.
- For non-constant coefficient ϵ the evaluation of $\operatorname{div}(\epsilon \mathbf{e}_h)$ requires a suitable approximation.
- If ϵ has discontinuities, demanding that the normal component of $\epsilon \mathbf{e}_h$ is continuous necessitates using a discontinuous trial space instead of $\boldsymbol{\mathcal{S}}_{p,0}(\Omega_h)$. This might rule out the use of standard FEM software, which is a main point in favour of regularized formulations. △

Bibliographical notes

Weighted regularization has been explored in the ground-breaking work by M. Costabel and M. Dauge (2002), which we followed closely in our presentation. There the reader can find all the details, and also refined *a priori* estimates for the discretization errors encountered for the weighted regularized formulation and certain finite element spaces.

7. Conclusion and further issues

This article has aimed to elucidate a few fundamental techniques for the analysis of finite element schemes for electromagnetic problems in the frequency domain. The focus was both on the design of suitable 'physical' finite elements and on the functional analytic techniques necessary to establish results on asymptotic convergence on shape-regular meshes. The discussion has been restricted to a few simple model problems, which, however, display all the subtle difficulties encountered by finite element Galerkin schemes in computational electromagnetism.

Theoretical computational electromagnetism is a huge area and this article can cover only a fraction of it. Key topics that have not been addressed are:

- Numerical methods in the time domain (Taflove 1995, Clemens and Weiland 1999).
- Boundary element methods for scattering problems and eddy current computation and their coupling with finite elements (Nédélec 2001, Colton and Kress 1998, Buffa *et al.* 2002, Hiptmair 2000, Christiansen 2000, Bendali 1984*a*, Bendali 1984*b*, Kuhn and Steinbach 2001).
- Absorbing boundary conditions (Grote and Keller 1998, Teixeira and Chew 1998, Sacks, Kingsland, Lee and Lee 1995).

- Fast iterative solvers and domain decomposition techniques for eddy current and scattering problems (Hiptmair 1999*b*, Hiptmair and Toselli 1999, Gopalakrishnan and Pasciak 2000, Buffa, Mayday and Rapetti 2001, Hoppe 1999).
- the design and analysis of *a posteriori* error estimators (Beck, Hiptmair, Hoppe and Wohlmuth 2000*b*, Beck, Hiptmair and Wohlmuth 2000*a*, Beck, Deuflhard, Hiptmair, Hoppe and Wohlmuth 1998, Monk 1999).

Acknowledgement

I am very grateful to R. Kotiuga, M. Dauge, M. Costabel, D. Boffi, Z. Chen, V. Gradinaru and S. Kurz, who suggested numerous improvements to the first draft of this survey article.

REFERENCES

S. Adam, P. Arbenz and R. Geus (1997), Eigenvalue solvers for electromagnetic fields in cavities. Technical Report 275, Institute of Scientific Computing, ETH Zürich, Zürich, Switzerland.

R. Adams (1975), *Sobolev Spaces*, Academic Press, New York.

M. Ainsworth and J. Coyle (2001*a*), Conditioning of hierarchic *p*-version Nédélec elements on meshes of curvilinear quadrilaterals and hexahedra. Report 26, Department of Mathematics, University of Strathclyde, Glasgow, UK.

M. Ainsworth and J. Coyle (2001*b*), 'Hierarchic *hp*-edge element families for Maxwell's equations on hybrid quadrilateral/triangular meshes', *Comput. Meth. Appl. Mech. Engr.* **190**, 6709–6733.

M. Ainsworth and K. Pinchedez (2001), *hp*-approximation theory for BDM/RT finite elements and applications. Report 15, Department of Mathematics, University of Strathclyde, Glasgow, UK.

A. Alonso and A. Valli (1999), 'An optimal domain decomposition preconditioner for low-frequency time-harmonic Maxwell equations', *Math. Comp.* **68**, 607–631.

C. Amrouche, C. Bernardi, M. Dauge and V. Girault (1998), 'Vector potentials in three–dimensional nonsmooth domains', *Math. Meth. Appl. Sci.* **21**, 823–864.

P. Anselone (1971), *Collectively Compact Operator Approximation Theory and Applications to Integral Equations*, Prentice-Hall, Englewood Cliffs, NJ.

T. Apel (1999), *Anisotropic Finite Elements: Local Estimates and Applications*, Advances in Numerical Mathematics, Teubner, Stuttgart, Germany.

T. Apel and M. Dobrowolski (1992), 'Anisotropic interpolation with applications to the finite element method', *Computing* **47**, 277–293.

T. Apel, S. Nicaise and J. Schöberl (2001), 'Crouzeix–Raviart type finite elements on anisotropic meshes', *Numer. Math.* **89**, 193–223.

F. Assous, P. Ciarlet Jr and E. Sonnendrücker (1998), 'Resolution of Maxwell equations in a domain with reentrant corners', *RAIRO* M2AN **32**, 359–389.

I. Babuška (1971), 'Error bounds for the finite element method', *Numer. Math.* **16**, 322–333.

I. Babuška and J. Osborn (1991), Eigenvalue problems, in *Finite Element Methods, Part 1* (P. G. Ciarlet and J. Lions, eds), Vol. 2 of *Handbook of Numerical Analysis*, Elsevier, Amsterdam, pp. 641–787.

I. Babuška and M. Suri (1994), 'The p and hp versions of the finite element method: Basic principles and properties', *SIAM Review* **36**, 578–632.

I. Babuška, A. Craig, J. Mandel and J. Pitkäranta (1991), 'Efficient preconditioning for the p-version finite element method in two dimensions', *SIAM J. Numer. Anal.* **28**, 624–661.

D. Baldomir (1986), 'Differential forms and electromagnetism in 3-dimensional Euclidean space $\mathbb{R}^3$', *IEE Proc. A* **133**, 139–143.

D. Baldomir and P. Hammond (1996), *Geometry of Electromagnetic Systems*, Clarendon Press, Oxford.

R. Beck, P. Deuflhard, R. Hiptmair, R. Hoppe and B. Wohlmuth (1998), 'Adaptive multilevel methods for edge element discretizations of Maxwell's equations', *Surveys on Mathematics for Industry* **8**, 271–312.

R. Beck, R. Hiptmair and B. Wohlmuth (2000*a*), A hierarchical error estimator for eddy current computation, in *ENUMATH 99: Proceedings of the 3rd European Conference on Numerical Mathematics and Advanced Applications, Jyväskylä, Finland, July 26–30* (P. Neittaanmäki and T. Tiihonen, eds), World Scientific, Singapore, pp. 110–120.

R. Beck, R. Hiptmair, R. Hoppe and B. Wohlmuth (2000*b*), 'Residual based *a-posteriori* error estimators for eddy current computation', *RAIRO* M2AN **34**, 159–182.

A. Bendali (1984*a*), 'Numerical analysis of the exterior boundary value problem for time harmonic Maxwell equations by a boundary finite element method, Part 1: The continuous problem', *Math. Comp.* **43**, 29–46.

A. Bendali (1984*b*), 'Numerical analysis of the exterior boundary value problem for time harmonic Maxwell equations by a boundary finite element method, Part 2: The discrete problem', *Math. Comp.* **43**, 47–68.

A. Bespalov (1988), 'Finite element method for the eigenmode problem of a RF cavity resonator', *Soviet J. Numer. Anal. Math. Model.* **3**, 163–178.

M. Birman and M. Solomyak (1990), 'Construction in a piecewise smooth domain of a function of the class H^2 from the value of the conormal derivative', *J. Math. Sov.* **49**, 1128–1136.

D. Boffi (2000), 'Discrete compactness and Fortin operator for edge elements', *Numer. Math.* **87**, 229–246.

D. Boffi and L. Gastaldi (2001), 'Edge finite elements for the approximation of Maxwell resolvent operator'. Submitted to *RAIRO* M2AN.

D. Boffi, F. Brezzi and L. Gastaldi (2000), 'On the problem of spurious eigenvalues in the approximation of linear elliptic problems in mixed form', *Math. Comp.* **69**, 121–140.

D. Boffi, P. Fernandes, L. Gastaldi and I. Perugia (1999), 'Computational models of electromagnetic resonators: Analysis of edge element approximation', *SIAM J. Numer. Anal.* **36**, 1264–1290.

A. Bonnet-BenDhia, C. Hazard and S. Lohrengel (1999), 'A singular field method for the solution of Maxwell's equations in polyhedral domains', *SIAM J. Appl. Math.* **59**, 2028–2044.

A. Bossavit (1988*a*), Mixed finite elements and the complex of Whitney forms, in *The Mathematics of Finite Elements and Applications VI* (J. Whiteman, ed.), Academic Press, London, pp. 137–144.

A. Bossavit (1988*b*), 'A rationale for edge elements in 3D field computations', *IEEE Trans. Magnetics* **24**, 74–79.

A. Bossavit (1988*c*), 'Whitney forms: A class of finite elements for three-dimensional computations in electromagnetism', *IEE Proc. A* **135**, 493–500.

A. Bossavit (1992), 'A new viewpoint on mixed elements', *Meccanica* **27**, 3–11.

A. Bossavit (1998*a*), *Computational Electromagnetism. Variational Formulation, Complementarity, Edge Elements*, Vol. 2 of *Electromagnetism Series*, Academic Press, San Diego, CA.

A. Bossavit (1998*b*), 'On the geometry of electromagnetism I: Affine space', *J. Japan Soc. Appl. Electromagnetics Mech.* **6**, 17–28.

A. Bossavit (1998*c*), 'On the geometry of electromagnetism II: Geometrical objects', *J. Japan Soc. Appl. Electromagnetics Mech.* **6**, 114–123.

A. Bossavit (1998*d*), 'On the geometry of electromagnetism III: Integration, Stokes', Faraday's law', *J. Japan Soc. Appl. Electromagnetics Mech.* **6**, 233–240.

A. Bossavit (1998*e*), 'On the geometry of electromagnetism IV: "Maxwell's house"', *J. Japan Soc. Appl. Electromagnetics Mech.* **6**, 318–326.

A. Bossavit (2001), Generalized finite differences in computational electromagnetics, in *Geometric Methods for Computational Electromagnetics* (F. Teixeira, ed.), Vol. 32 of *PIER*, PIERS, Boston, MA, pp. 45–64.

A. Bossavit and L. Kettunen (1999*a*), 'Yee-like schemes on a tetrahedral mesh with diagonal lumping', *Internat. J. Numer. Modelling* **12**, 129–142.

A. Bossavit and L. Kettunen (1999*b*), Yee-like schemes on staggered cellular grids: A synthesis between FIT and FEM approaches. Contribution to COMPUMAG '99.

R. Bott and L. Tu (1982), *Differential Forms in Algebraic Topology*, Springer, New York.

D. Braess and C. Schwab (2000), 'Approximation on simplices with respect to weighted Sobolev norms', *J. Approx. Theory* **103**, 329–337.

F. Brezzi and M. Fortin (1991), *Mixed and Hybrid Finite Element Methods*, Springer.

F. Brezzi, J. Douglas and D. Marini (1985), 'Two families of mixed finite elements for 2nd order elliptic problems', *Numer. Math.* **47**, 217–235.

F. Brezzi, J. Rappaz and P. Raviart (1980), 'Finite dimensional approximation of nonlinear problems, Part I: Branches of nonsingular solutions', *Numer. Math.* **36**, 1–25.

A. Buffa, Y. Mayday and F. Rapetti (2001), 'A sliding mesh mortar method for two dimensional eddy currents model for electric engines', *RAIRO* M2AN **35**, 191–228.

A. Buffa, R. Hiptmair, T. von Petersdorff and C. Schwab (2002), 'Boundary element methods for Maxwell equations on Lipschitz domains'. To appear in *Numer. Math.*

W. Burke (1985), *Applied Differential Geometry*, Cambridge University Press, Cambridge.

S. Caorsi, P. Fernandes and M. Raffetto (1999), Approximations of electromagnetic eigenproblems: A general proof of convergence for edge finite elements of any order of both Nédélec's families. Technical Report 16/99, CNR-IMA Genoa, Genoa, Italy.

S. Caorsi, P. Fernandes and M. Raffetto (2000), 'On the convergence of Galerkin finite element approximations of electromagnetic eigenproblems', *SIAM J. Numer. Anal.* **38**, 580–607.

H. Cartan (1967), *Formes Différentielles*, Hermann, Paris.

S. Christiansen (2000), Discrete Fredholm properties and convergence estimates for the EFIE. Technical Report 453, CMAP, Ecole Polytechique, Paris, France.

P. Ciarlet, Jr, and J. Zou (1999), 'Fully discrete finite element approaches for time-dependent Maxwell equations', *Numer. Math.* **82**, 193–219.

P. G. Ciarlet (1978), *The Finite Element Method for Elliptic Problems*, Vol. 4 of *Studies in Mathematics and its Applications*, North-Holland, Amsterdam.

M. Clemens and T. Weiland (1999), 'Transient eddy current calculation with the FI-method', *IEEE Trans. Magnetics* **35**, 1163–1166.

M. Clemens and T. Weiland (2001), Discrete electromagnetism with the finite integration technique, in *Geometric Methods for Computational Electromagnetics* (F. Teixeira, ed.), Vol. 32 of *PIER*, PIERS, Boston, MA, pp. 65–87.

D. Colton and R. Kress (1998), *Inverse Acoustic and Electromagnetic Scattering Theory*, Vol. 93 of *Applied Mathematical Sciences*, 2nd edn, Springer, Heidelberg.

M. Costabel (1990), 'A remark on the regularity of solutions of Maxwell's equations on Lipschitz domains', *Math. Meth. Appl. Sci.* **12**, 365–368.

M. Costabel (1991), 'A coercive bilinear form for Maxwell's equations', *J. Math. Anal. Appl.* **157**, 527–541.

M. Costabel and M. Dauge (1998), Singularities of Maxwell's equations on polyhedral domains, in *Analysis, Numerics and Applications of Differential and Integral Equations* (M. Bach, ed.), Vol. 379 of *Longman Pitman Res. Notes Math. Ser.*, Addison Wesley, Harlow, pp. 69–76.

M. Costabel and M. Dauge (1999), 'Maxwell and Lamé eigenvalues on polyhedra', *Math. Methods Appl. Sci.* **22**, 243–258.

M. Costabel and M. Dauge (2002), 'Weighted regularization of Maxwell equations in polyhedral domains'. To appear in *Numer. Math.*

M. Costabel, M. Dauge and S. Nicaise (1999), 'Singularities of Maxwell interface problems', *RAIRO* M2AN **33**, 627–649.

M. Dauge (1988), *Elliptic Boundary Value Problems on Corner Domains*, Vol. 1341 of *Lecture Notes in Mathematics*, Springer, Berlin.

R. Dautray and J.-L. Lions (1990), *Mathematical Analysis and Numerical Methods for Science and Technology*, Vol. 4, Springer, Berlin.

L. Demkowicz (2001), Edge finite elements of variable order for Maxwell's equations: A discussion, in *Scientific Computing in Electrical Engineering: Proceedings of a workshop held at Rostock, Germany, Aug 20–23 2000* (U. van Rienen, M. Günther and D. Hecht, eds), Vol. 18 of *Lecture Notes in Computer Science and Engineering*, Springer, Berlin, pp. 15–34.

L. Demkowicz and I. Babuška (2001), Optimal p interpolation error estimates for edge finite elements of variable order in 2D. Technical Report 01-11, TICAM, University of Texas, Austin, TX.

L. Demkowicz, P. Monk, C. Schwab and L. Vardapetyan (1999), Maxwell eigenvalues and discrete compactness in two dimensions. Report 99-12, TICAM, University of Texas, Austin, TX.

G. Deschamps (1981), 'Electromagnetics and differential forms', *Proc. IEEE* **69**, 676–695.

A. Dezin (1995), *Multidimensional Analysis and Discrete Models*, CRC Press, Boca Raton, FL, USA.

J. Dodziuk (1976), 'Finite-difference approach to the Hodge theory of harmonic forms', *Amer. J. Math.* **98**, 79–104.

T. Dupont and R. Scott (1980), 'Polynomial approximation of functions in Sobolev spaces', *Math. Comp.* **34**, 441–463.

H. Federer (1969), *Geometric Measure Theory*, Springer, New York.

P. Fernandes and G. Gilardi (1997), 'Magnetostatic and electrostatic problems in inhomogeneous anisotropic media with irregular boundary and mixed boundary conditions', *Math. Models Meth. Appl. Sci.*, M3AS **7**, 957–991.

P. Fernandes and M. Raffetto (2000), Recent developments in the (spurious-free) approximation of electromagnetic eigenproblems by the finite element method. Technical Report 06/00, CNR-IMA Genoa, Genoa, Italy.

R. Fritsch and R. Piccini (1990), *Cellular Structures in Topology*, Vol. 19 of *Cambridge Studies in Advanced Mathematics*, Cambridge University Press.

V. Girault and P. Raviart (1986), *Finite Element Methods for Navier–Stokes Equations*, Springer, Berlin.

J. Gopalakrishnan and J. Pasciak (2000), Overlapping Schwarz preconditioners for indefinite time harmonic Maxwell equations. Technical report, Department of Mathematics, Texas A&M University. Submitted to *Math. Comp.*

V. Gradinaru and R. Hiptmair (1999), 'Whitney elements on pyramids', *Electron. Trans. Numer. Anal.* **8**, 154–168.

H. Grauert and I. Lieb (1977), *Differential- und Integralrechnung III*, 3rd edn, Springer, Berlin.

P. Grisvard (1985), *Elliptic Problems in Nonsmooth Domains*, Pitman, Boston.

P. Gross and P. Kotiuga (2001), Data structures for geometric and topological aspects of finite element algorithms, in *Geometric Methods for Computational Electromagnetics* (F. Teixeira, ed.), Vol. 32 of *PIER*, PIERS, Boston, MA, pp. 151–169.

M. Grote and J. Keller (1998), 'Nonreflecting boundary conditions for Maxwell's equations', *J. Comput. Phys.* **139**, 327–342.

E. Haber, U. Ascher, D. Aruliah and D. Oldenburg (1999), Fast modelling of 3D electromagnetic problems using potentials. Technical report, Department of Computer Science, University of British Columbia, Vancouver, Canada. Submitted to *J. Comput. Phys.*

W. Hackbusch (1993), *Iterative Solution of Large Sparse Systems of Equations*, Vol. 95 of *Applied Mathematical Sciences*, Springer, New York.

M. Hanna and K. Smith (1967), 'Some remarks on the Dirichlet problem in piecewise smooth domains', *Comm. Pure Appl. Math.* **20**, 575–593.

C. Hazard and M. Lenoir (1996), 'On the solution of time-harmonic scattering problems for Maxwell's equations', *SIAM J. Math. Anal.* **27**, 1597–1630.

C. Hazard and S. Lohrengel (2000), A singular field method for Maxwell's equations: Numerical aspects in two dimensions. Submitted.

R. Hiptmair (1999*a*), 'Canonical construction of finite elements', *Math. Comp.* **68**, 1325–1346.

R. Hiptmair (1999*b*), 'Multigrid method for Maxwell's equations', *SIAM J. Numer. Anal.* **36**, 204–225.

R. Hiptmair (2000), Symmetric coupling for eddy current problems. Technical Report 148, Sonderforschungsbereich 382, Universität Tübingen, Tübingen, Germany. To appear in *SIAM J. Numer. Anal.*

R. Hiptmair (2001*a*), 'Discrete Hodge operators'. To appear in *Numer. Math.* Published online May 30, 2001, `http://dx.doi.org/10.1007/s002110100295`.

R. Hiptmair (2001*b*), 'Discrete Hodge operators: An algebraic perspective', in *Geometric Methods for Computational Electromagnetics* (F. Teixeira, ed.), Vol. 32 of *PIER*, EMW Publishing, Cambridge, MA, pp. 247–269.

R. Hiptmair (2001*c*), 'Higher order Whitney forms', in *Geometric Methods for Computational Electromagnetics* (F. Teixeira, ed.), Vol. 32 of *PIER*, EMW Publishing, Cambridge, MA, pp. 271–299.

R. Hiptmair and A. Toselli (1999), Overlapping and multilevel Schwarz methods for vector valued elliptic problems in three dimensions, in *Parallel Solution of Partial Differential Equations* (P. Bjorstad and M. Luskin, eds), Vol 120 of *IMA Volumes in Mathematics and its Applications*, Springer, Berlin, pp. 181–202.

R. Hoppe (1999), 'Mortar edge element methods in $\mathbb{R}^3$', *East–West J. Numer. Anal.* **7**, 159–175.

J. Hyman and M. Shashkov (1999), 'Mimetic discretizations for Maxwell's equations', *J. Comput. Phys.* **151**, 881–909.

J. Hyman and M. Shashkov (2001), Mimetic finite difference methods for Maxwell's equations and the equations of magnetic diffusion, in *Geometric Methods for Computational Electromagnetics* (F. Teixeira, ed.), Vol. 32 of *PIER*, PIERS, Boston, MA, pp. 89–121.

F. Ihlenburg (1998), *Finite Element Analysis of Acoustic Scattering*, Vol. 132 of *Applied Mathematical Sciences*, Springer, New York.

T. Iwaniec (1999), Nonlinear differential forms, in *Lecture Notes of the International Summer School in Jyväskylä, 1998*, Department of Mathematics, University of Jyväskylä, Finland.

F. Jochmann (1997), 'A compactness result for vector fields with divergence and curl in $L^q(\Omega)$ involving mixed boundary conditions', *Appl. Anal.* **66**, 189–203.

B. Khoromskij and J. Melenk (2001), Boundary concentrated finite element methods. Preprint 45, MPI Leipzig, Germany.

F. Kikuch (1987), 'Mixed and penalty formulations for finite element analysis of an eigenvalue problem in electromagnetism', *Comput. Methods Appl. Mech. Engr.* **64**, 509–521.

F. Kikuchi (1989), 'On a discrete compactness property for the Nédélec finite elements', *J. Fac. Sci., Univ. Tokyo, Sect. I A* **36**, 479–490.

A. Kirsch and P. Monk (2000), 'A finite element method for approximating electromagnetic scattering from a conducting object'. To appear in *Numer. Math.*

V. Kondrat'ev (1967), 'Boundary value problems for elliptic equations in domains with conical or angular points', *Trans. Moscow Math. Soc.* **16**, 227–313.

R. Kress (1971), 'Ein kombiniertes Dirichlet–Neumannsches Randwertproblem bei harmonischen Vektorfeldern', *Arch. Rational Mech. Anal.* **42**, 40–49.

R. Kress (1989), *Linear Integral Equations*, Vol. 82 of *Applied Mathematical Sciences*, Springer, Berlin.

M. Kuhn and O. Steinbach (2001), 'FEM-BEM coupling for 3d exterior magnetic field problems'. To appear in *Math. Meth. Appl. Sci.*

S. Lang (1995), *Differential and Riemannian Manifolds*, Vol. 160 of *Graduate Texts in Mathematics*, Springer, New York.

R. Leis (1986), *Initial Boundary Value Problems in Mathematical Physics*, B.G. Teubner, Stuttgart.

S. Lohrengel and S. Nicaise (2001), 'Singularities and density problems for composite materials in electromagnetism'. Submitted to *Comm. Part. Diff. Equ.*

A. Lundell and S. Weingram (1969), *The Topology of CW Complexes*, The University Series in Higher Mathematics, Van Nostrand Reinhold Company, New York.

M. Marrone (2001), Computational aspects of the cell method in electrodynamics, in *Geometric Methods for Computational Electromagnetics* (F. Teixeira, ed.), Vol. 32 of *PIER*, PIERS, Boston, MA, pp. 317–356.

C. Mattiussi (2000), 'The finite volume, finite element and finite difference methods as numerical methods for physical field problems', *Adv. Imaging Electron Physics* **113**, 1–146.

J. Maxwell (1891), *A Treatise on Electricity and Magnetism*, Oxford University Press reprint, 3rd edn (1998).

V. Maz'ya (1985), *Sobolev Spaces*, Springer, Berlin.

C. Mercat (2001), 'Discrete Riemann surfaces and the Ising model', *Commun. Math. Phys.* **218**, 177–216.

P. Monk (1992*a*), 'Analysis of a finite element method for Maxwell's equations', *SIAM J. Numer. Anal.* **29**, 714–729.

P. Monk (1992*b*), 'A finite element method for approximating the time-harmonic Maxwell equations', *Numer. Math.* **63**, 243–261.

P. Monk (1994), 'On the p and hp-extension of Nédélec's conforming elements', *J. Comput. Appl. Math.* **53**, 117–137.

P. Monk (1999), '*A posteriori* error indicators for Maxwell's equations'. To appear in *J. Comput. Appl. Math.*

P. Monk (2001), 'A simple proof of convergence for an edge element discretization of Maxwell's equations'. Submitted to *Math. Comp.*

P. Monk and L. Demkowicz (2001), 'Discrete compactness and the approximation of Maxwell's equations in $\mathbb{R}^3$', *Math. Comp.* **70**, 507–523.

F. Morgan (1995), *Geometric Measure Theory: A Beginner's Guide*, 2nd edn, Academic Press, San Diego, CA.

W. Müller (1978), 'Analytic torsion and R-torsion of Riemannian manifolds', *Adv. Math.* **28**, 233–305.

R. Munoz-Solar (1997), 'Polynomial liftings on a tetrahedron and applications to the *hp*-version of the finite element method in three dimensions', *SIAM J. Numer. Anal.* **34**, 282–314.

S. Nazarov and B. Plamenevskii (1994), *Elliptic Problems in Domains with Piecewise Smooth Boundaries*, Vol. 13 of *Expositions in Mathematics*, Walter de Gruyter, Berlin.

J. Nédélec (1980), 'Mixed finite elements in R^3', *Numer. Math.* **35**, 315–341.

J. Nédélec (1986), 'A new family of mixed finite elements in R^3', *Numer. Math.* **50**, 57–81.

J.-C. Nédélec (2001), *Acoustic and Electromagnetic Equations: Integral Representations for Harmonic Problems*, Vol. 44 of *Applied Mathematical Sciences*, Springer, Berlin.

S. Nicaise (2001), 'Edge elements on anisotropic meshes and approximation of the Maxwell equations', *SIAM J. Numer. Anal.* **39**, 784–816.

R. Nicolaides and D.-Q. Wang (1998), 'Convergence analysis of a covolume scheme for Maxwell's equations in three dimensions', *Math. Comp.* **67**, 947–963.

R. Nicolaides and X. Wu (1997), 'Covolume solutions of three-dimensional div-curl equations', *SIAM J. Numer. Anal.* **34**, 2195–2203.

J. Osborn (1975), 'Spectral approximation for compact operators', *Math. Comp.* **29**, 712–725.

L. Pavarino and O. Widlund (1996), 'A polylogarithmic bound for an iterative substructuring method for spectral elements in three dimensions', *SIAM J. Numer. Anal.* **33**, 1303–1335.

R. Picard (1984), 'An elementary proof for a compact imbedding result in generalized electromagnetic theory', *Math. Z.* **187**, 151–161.

W. Rachowicz and L. Demkowicz (2002), 'An *hp*-adaptive finite element method for electromagnetics, Part II: A 3D implementation', *Internat. J. Numer. Meth. Engr.* **53**, 147–180.

P. A. Raviart and J. M. Thomas (1977), *A Mixed Finite Element Method for Second Order Elliptic Problems*, Vol. 606 of *Springer Lecture Notes in Mathematics*, Springer, New York, pp. 292–315.

Z. Sacks, D. Kingsland, R. Lee and J.-F. Lee (1995), 'A perfectly matched anisotropic absorber for use as an absorbing boundary condition', *IEEE Trans. Antennas Propag.* **43**, 1460–1463.

A. Schatz and J. Wang (1996), 'Some new error estimates for Ritz–Galerkin methods with minimal regularity assumptions', *Math. Comp.* **65**, 19–27.

R. Schuhmann and T. Weiland (2001), Conservation of discrete energy and related laws in the finite integration technique, in *Geometric Methods for Computational Electromagnetics* (F. Teixeira, ed.), Vol. 32 of *PIER*, PIERS, Boston, MA, pp. 301–316.

C. Schwab (1998), *p- and hp-Finite Element Methods: Theory and Applications in Solid and Fluid Mechanics*, Numerical Mathematics and Scientific Computation, Clarendon Press, Oxford.

G. Schwarz (1995), *Hodge Decomposition: A method for Solving Boundary Value Problems*, Vol. 1607 of *Springer Lecture Notes in Mathematics*, Springer, Berlin.

D.-K. Sun, J.-F. Lee and Z. Cendes (2001), 'Construction of nearly orthogonal Nédélec bases for rapid convergence with multilevel preconditioned solvers', *SIAM J. Sci. Comput.* **23**, 1053–1076.

A. Taflove (1995), *The Finite Difference in Time Domain Method*, Artech House, Boston, London.

T. Tarhasaari and L. Kettunen (2001), Topological approach to computational electromagnetism, in *Geometric Methods for Computational Electromagnetics* (F. Teixeira, ed.), Vol. 32 of *PIER*, PIERS, Boston, MA, pp. 189–206.

T. Tarhasaari, L. Kettunen and A. Bossavit (1999), 'Some realizations of a discrete Hodge: A reinterpretation of finite element techniques', *IEEE Trans. Magnetics* **35**, 1494–1497.

F. Teixeira (2001), Geometric aspects of the simplicial discretization of Maxwell's equations, in *Geometric Methods for Computational Electromagnetics* (F. Teixeira, ed.), Vol. 32 of *PIER*, PIERS, Boston, MA, pp. 171–188.

F. Teixeira and C. Chew (1998), 'A general approach to extend Berenger's absorbing boundary conditions to anisotropic and dispersive media', *IEEE Trans. Antennas Propag.* **46**, 1386–1387.

F. Teixeira and W. Chew (1999), 'Lattice electromagnetic theory from a topological viewpoint', *J. Math. Phys.* **40**, 169–187.

E. Tonti (1996), On the geometrical structure of electromagnetism, in *Gravitation, Electromagnetism and Geometrical Structures* (G. Ferrarese, ed.), Pitagora, Bologna, Italy, pp. 281–308.

E. Tonti (2001), Finite formulation of the electromagnetic field, in *Geometric Methods for Computational Electromagnetics* (F. Teixeira, ed.), Vol. 32 of *PIER*, PIERS, Boston, MA, pp. 1–44.

L. Vardapetyan and L. Demkowicz (1999), '*hp*-adaptive finite elements in electromagnetics', *Comput. Meth. Appl. Mech. Engr.* **169**, 331–344.

C. Weber (1980), 'A local compactness theorem for Maxwell's equations', *Math. Meth. Appl. Sci.* **2**, 12–25.

T. Weiland (1996), 'Time domain electromagnetic field computation with finite difference methods', *Internat. J. Numer. Modelling* **9**, 295–319.

P. Werner (1963), 'On the exterior boundary value problem of perfect reflexion for stationary electromagnetic wave fields', *J. Math. Anal. Appl.* **7**, 348–396.

H. Whitney (1957), *Geometric Integration Theory*, Princeton University Press, Princeton.

K. Yee (1966), 'Numerical solution of initial boundary value problems involving Maxwell's equations in isotropic media', *IEEE Trans. Antennas Propag.* **16**, 302–307.

Appendix: Symbols and notation

$A(\mathbb{R}^3)$	Three-dimensional affine space
$\mathbf{e}$	Complex amplitude of electric field
$\mathbf{b}$	Complex amplitude of magnetic induction
$\mathbf{d}$	Complex amplitude of displacement current
$\mathbf{h}$	Complex amplitude of magnetic field
$\boldsymbol{\epsilon}$	Dielectric tensor
$\boldsymbol{\mu}$	Tensorial magnetic permeability
$\boldsymbol{\sigma}$	Conductivity tensor
$\mathcal{M}$	(Piecewise smooth) manifold
$\mathcal{S}_l$	Oriented, piecewise smooth sub-manifolds of $\mathcal{M}$ (Definition 1)
$\mathcal{F}^l(\mathcal{M})$	Space of integral l-forms on manifold $\mathcal{M}$, $l \in \mathbb{N}$ (Definition 1)
$\times$	Cross product of vectors in $\mathbb{R}^3$
$\mathcal{DF}^{l,m}(\mathcal{M})$	Differential forms of class C^m on smooth manifold $\mathcal{M}$ (Definition 2)
$F^l_{\mathcal{M}}$	Injection of differential l-forms into integral l-forms
Υ_l	Association of (piecewise) continuous l-forms and vector proxies in three-dimensional Euclidean space (Table 2.1)
$\mathbf{n}$	(Exterior) unit normal vector-field on some oriented two-dimensional Lipschitz surface
$\boldsymbol{d}$	Exterior derivative for (differential/integral) l-forms
$\boldsymbol{\Phi}^*$	Pullback operator belonging to diffeomorphism $\boldsymbol{\Phi}$ (see (2.12))
$\mathbf{t}_{\mathcal{N}}$	Trace of integral/differential forms onto a sub-manifold (see (2.14))
$\mathfrak{F}^l_{\mathcal{M}}$	Pullbacks for vector proxies of l-forms in three-dimensional Euclidean space (see (2.16)–(2.19))
γ	Standard trace operator, *i.e.*, pointwise restriction
$\gamma_{\mathbf{t}}$	Tangential trace $\mathbf{u} \mapsto \mathbf{u} \times \mathbf{n}_{\vert\Gamma}$
$\gamma_{\mathbf{n}}$	Normal trace $\mathbf{u} \mapsto \mathbf{u} \cdot \mathbf{n}_{\vert\Gamma}$
$\wedge$	Exterior product of differential forms
Ω	(Curvilinear) Lipschitz polyhedron $\subset A(\mathbb{R}^3)$
Γ	Oriented boundary of Ω

$\boldsymbol{H}(\mathbf{curl};\Omega)$	$:= \{\mathbf{v} \in \boldsymbol{L}^2(\Omega),\ \mathbf{curl}\,\mathbf{v} \in \boldsymbol{L}^2(\Omega)\}$
$\boldsymbol{H}(\mathrm{div};\Omega)$	$:= \{\mathbf{v} \in \boldsymbol{L}^2(\Omega),\ \mathrm{div}\,\mathbf{v} \in \boldsymbol{L}^2(\Omega)\}$
$\boldsymbol{H}_0(\mathbf{curl};\Omega)$	$:= \{\mathbf{v} \in \boldsymbol{H}(\mathbf{curl};\Omega),\ \gamma_{\mathbf{t}}\mathbf{v} = 0\}$
$\boldsymbol{H}_0(\mathrm{div};\Omega)$	$:= \{\mathbf{v} \in \boldsymbol{H}(\mathrm{div};\Omega),\ \gamma_{\mathbf{n}}\mathbf{v} = 0\}$
$\boldsymbol{H}(\mathbf{curl}\,0;\Omega)$	$:= \{\mathbf{v} \in \boldsymbol{H}(\mathbf{curl};\Omega),\ \mathbf{curl}\,\mathbf{v} = 0\}$
$\boldsymbol{H}(\mathrm{div}\,0;\Omega)$	$:= \{\mathbf{v} \in \boldsymbol{H}(\mathrm{div};\Omega),\ \mathrm{div}\,\mathbf{v} = 0\}$
$\mathcal{H}^1(\Omega)$	Space of harmonic Neumann vector-fields, contained in $\boldsymbol{H}(\mathbf{curl}\,0;\Omega) \cap \boldsymbol{H}(\mathrm{div}\,0;\Omega)$ (see Lemma 2.2)
$\mathcal{H}^2(\Omega)$	Space of harmonic Dirichlet vector-fields, contained in $\boldsymbol{H}(\mathbf{curl}\,0;\Omega) \cap \boldsymbol{H}(\mathrm{div}\,0;\Omega)$
Ω_h	Triangulation/mesh (see Definition 3) of Ω with meshwidth h according to (3.34)
$\mathcal{S}_l(\Omega_h)$	Set of l-facets of triangulation Ω_h
$\mathcal{C}^l(\Omega_h)$	Vector space of l-cochains on the mesh Ω_h
D^l	Matrix of exterior derivative on l-cochains (see Section 3.1)
I_l	deRham maps $\mathcal{F}^l(\Omega) \mapsto \mathcal{C}^l(\Omega_h)$ (see (3.2))
W^l	Interpolation of l-cochains (Whitney map, see Section 3.2)
$\mathcal{W}^l_p(\Omega_h)$	Space of Whitney l-forms of uniform polynomial degree p on mesh Ω_h (see Definition 5 for $p = 0$, and Section 3.4 for general $p \in \mathbb{N}_0$)
$\mathcal{W}^l_{p,0}(\Omega_h)$	Discrete differential l-forms with vanishing trace on $\partial\Omega$
Π^l	Local interpolation operators for Whitney l-forms (see (3.11))
$\mathcal{P}_p(T)$	Space of multivariate polynomials of total degree $\leq p$ on T
$\widetilde{\mathcal{P}}_p(T)$	Space of multivariate homogeneous polynomials on T of exact total degree p
$\mathcal{DP}^l_p(T)$	Polynomial differential l-forms of polynomial degree $\leq p$ on simplex T (see (3.15))
$\mathcal{X}^l_p(S)$	Local space of p-order Whitney forms associated with facet S
$\mathcal{Z}^l_p(S)$	Polynomial l-forms of degree $\leq p$ and vanishing trace on ∂S (see (3.18))
$\mathsf{E}^l_{p,S}$	Barycentric extension of a polynomial l-form from a facet into a cell (see Section 3.4)
$\mathfrak{K}^l$	Poincaré mapping $\mathcal{DF}^{l,0}(T) \mapsto \mathcal{DF}^{l-1,1}(T)$ (see (3.21))

$\check{\mathcal{W}}^l_p(\Omega_h)$	Second family of Whitney l-forms of polynomial degree p on mesh Ω_h (see (3.28))
$\underline{\mathbf{p}}$	Polynomial degree vector in the variable degree p-version of discrete differential forms (see Section 3.4)
$\Theta^l_{\underline{\mathbf{p}}}(\Omega_h)$	Set of degrees of freedom for $\mathcal{W}^l_{\underline{\mathbf{p}}}(\Omega_h)$ (see Section 3.5)
PL^l_S	Potential liftings of closed polynomial forms on simplex S
PE^l_S	Polynomial extension operators respecting vanishing traces
PQ^l_S	Projections onto closed polynomial forms
$\mathbf{I}$	3×3 identity matrix
$\boldsymbol{X}_0(\boldsymbol{\alpha},\Omega)$	$:= \{\mathbf{u} \in \boldsymbol{H}_0(\mathbf{curl};\Omega),\ \mathrm{div}(\boldsymbol{\alpha}\mathbf{u}) \in L^2(\Omega)\}$ (see Section 4.1)
$\boldsymbol{Z}_0(\boldsymbol{\alpha},\Omega)$	$\boldsymbol{\alpha}$-Orthogonal complement in $\boldsymbol{H}_0(\mathbf{curl};\Omega)$ of kernel of $\mathbf{curl}$ (see (4.3))
$\boldsymbol{Z}_{h,0}(\boldsymbol{\alpha},\Omega_h)$	Discrete counterpart of $\boldsymbol{Z}_0(\boldsymbol{\alpha},\Omega)$
$\mathsf{H}_{\boldsymbol{\epsilon}}$	Hodge mapping $\boldsymbol{H}_0(\mathbf{curl};\Omega) \mapsto \boldsymbol{Z}_0(\boldsymbol{\epsilon},\Omega)$ (see (4.8))
F_h	Fortin projector $\boldsymbol{H}_0(\mathbf{curl};\Omega) \mapsto \boldsymbol{Z}_{h,0}(\boldsymbol{\epsilon},\Omega_h)$ (see (4.10))
$\mathsf{Q}^{\boldsymbol{\alpha}}_h$	$\boldsymbol{\alpha}$-Orthogonal projection onto $\mathbf{curl}\,\mathcal{W}^1_{p,0}(\Omega_h)$
$\mathsf{P}^{\perp}$	Orthogonal projection onto $\boldsymbol{Z}_0(\mathbf{I},\Omega)$
P^0	Orthogonal projection onto $\boldsymbol{H}_0(\mathbf{curl}\,0;\Omega)$
$\boldsymbol{\mathcal{S}}_{p,0}(\Omega_h)$	Space of piecewise polynomial continuous vector-fields on Ω_h
$\boldsymbol{H}^1_{\times}(\Omega)$	Space of vector-fields in $\boldsymbol{H}^1(\Omega)$ with vanishing tangential components on $\partial\Omega$

Acta Numerica (2002), pp. 341–434
DOI: 10.1017/S0962492902000053

Splitting methods

Robert I. McLachlan
IFS, Massey University,
Palmerston North, New Zealand
E-mail: R.McLachlan@massey.ac.nz

G. Reinout W. Quispel
Mathematics Department, La Trobe University,
Bundoora, VIC 3086, Australia
E-mail: R.Quispel@latrobe.edu.au

I thought that instead of the great number of precepts of which logic is composed, I would have enough with the four following ones, provided that I made a firm and unalterable resolution not to violate them even in a single instance. The first rule was never to accept anything as true unless I recognized it to be certainly and evidently such .... The second was to divide each of the difficulties which I encountered into as many parts as possible, and as might be required for an easier solution. (Descartes)

We survey splitting methods for the numerical integration of ordinary differential equations (ODEs). Splitting methods arise when a vector field can be split into a sum of two or more parts that are each simpler to integrate than the original (in a sense to be made precise). One of the main applications of splitting methods is in geometric integration, that is, the integration of vector fields that possess a certain geometric property (*e.g.*, being Hamiltonian, or divergence-free, or possessing a symmetry or first integral) that one wants to preserve. We first survey the classification of geometric properties of dynamical systems, before considering the theory and applications of splitting in each case. Once a splitting is constructed, the pieces are composed to form the integrator; we discuss the theory of such 'composition methods' and summarize the best currently known methods. Finally, we survey applications from celestial mechanics, quantum mechanics, accelerator physics, molecular dynamics, and fluid dynamics, and examples from dynamical systems, biology and reaction–diffusion systems.

CONTENTS

1. Introduction

1.1. What is splitting?

Our topic is a class of methods for the time integration of ODEs and PDEs. With phase space M, differential equation $\dot{x} = X(x)$, $x \in M$, and X a vector field on M, splitting methods involve three equally important steps:

(1) choosing a set of vector fields X_i such that $X = \sum X_i$;

(2) integrating either exactly or approximately each X_i; and

(3) combining these solutions to yield an integrator for X.

For example, writing the flow (*i.e.*, the exact solution) of the ODE $\dot{x} = X$ as $x(t) = \exp(tX)(x(0))$, we might use the composition method

$$\varphi(\tau) = \exp(\tau X_1)\exp(\tau X_2)\ldots\exp(\tau X_n), \tag{1.1}$$

which is first-order accurate, that is,

$$\varphi(\tau) = \exp\Bigl(\tau \sum X_i\Bigr) + \mathcal{O}(\tau^2).$$

Here τ is the time step. In all cases the pieces X_i should be *simpler* than the original vector field X, which can occur in two ways.

(1) The X_i are of a simpler type than X. For example, the Navier–Stokes equations contain advection, diffusion, and pressure (constraint) terms, each with distinct characteristic properties and appropriate numerical methods. In an ODE of the form Hamiltonian plus small dissipation, the Hamiltonian piece has a simpler structure than the combined system.

(2) The X_i are of the same type as X, but are easier to treat numerically. Examples are dimensional splitting for the multidimensional heat equation, Hamiltonian splitting for Hamiltonian ODEs, and the split-step-Fourier method for the Schrödinger equation $i\dot{\psi} = \psi_{xx} + V(x)\psi$ – each piece is linear and Hamiltonian, but the first term can be integrated more quickly (in a Fourier basis, using the FFT) than the combined system.

Splitting methods were originally developed for the traditional numerical motivations of speed, accuracy, and stability. However, it is now clear that they are a very general and flexible way of constructing *geometric integrators* (McLachlan and Quispel 2001*a*, Budd and Iserles 1999, Budd and Piggott 2002, Hairer, Lubich and Wanner 2002) which preserve structural features of the flow of X, conferring qualitative superiority on the integrator, especially when integrating for long times. Examples of such features are symplecticity, volume preservation, integrals, symmetries, and many more. This has led to them being the method of choice, for instance, in celestial mechanics, molecular dynamics, and accelerator physics.

In this review we only consider initial value problems. In geometric integration, it is important to preserve the phase space of the system, so we only consider one-step methods.

1.2. Four examples

We introduce splitting methods with four examples.

Example 1. (Leapfrog) The leapfrog method is the standard example which has motivated much of the work in splitting methods and in geometric integration. Let the phase space be $M = \mathbb{R}^{2n}$ with coordinates (q, p) and consider a Hamiltonian system with energy $H = T + V$ where $T = \frac{1}{2}\|p\|^2$ is the kinetic energy and $V = V(q)$ is the potential energy. Hamilton's equations for H, namely $\dot{x} = X_H = X_T + X_V$, are a sum of two easily

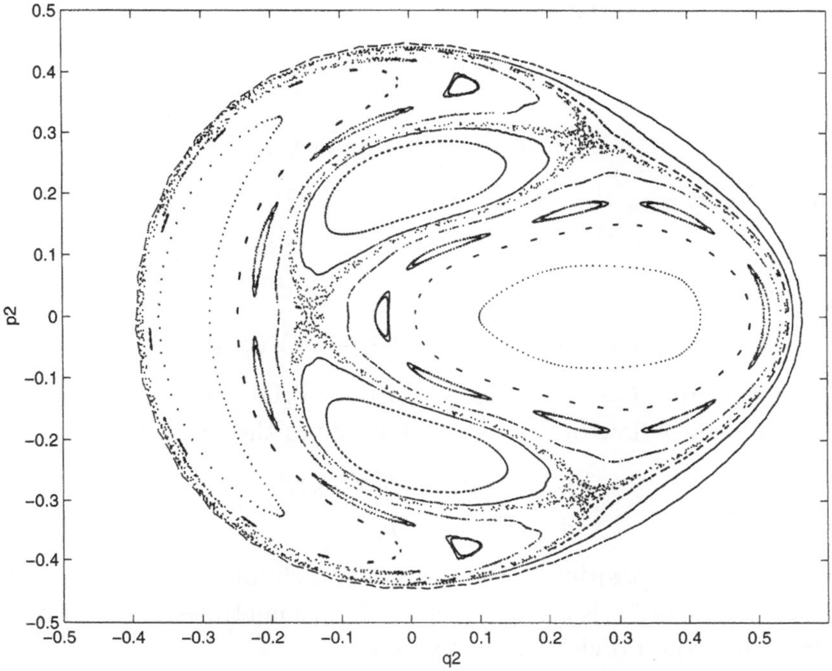

Figure 1.1. The $q_1 = 0$ Poincaré section (q_2, p_2) of the Hénon–Heiles system, calculated by leapfrog, a second-order symplectic splitting method, with time step $\tau = 0.25$

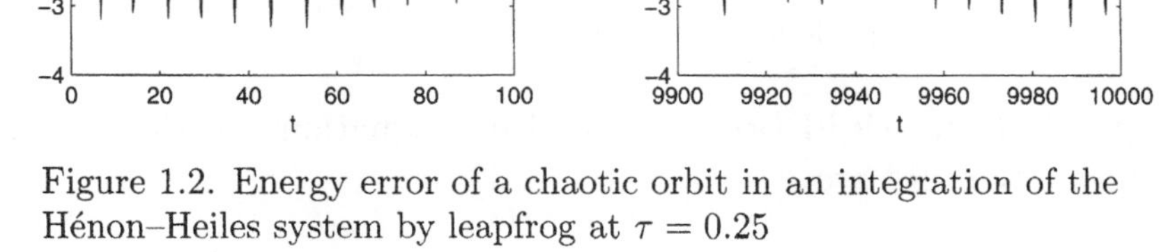

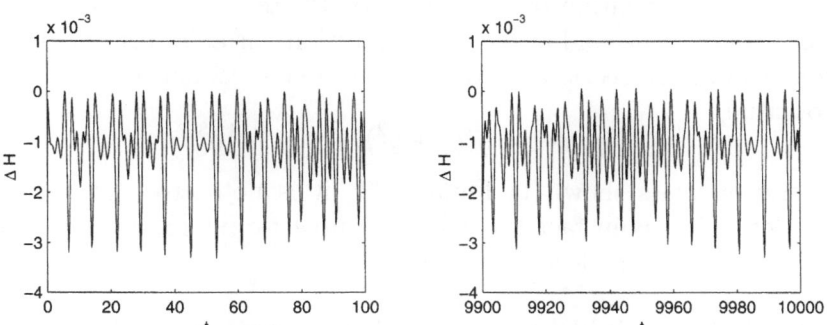

Figure 1.2. Energy error of a chaotic orbit in an integration of the Hénon–Heiles system by leapfrog at $\tau = 0.25$

solvable Hamiltonian equations:

$$X_T\colon\ \dot q = p, \quad \dot p = 0,$$

and

$$X_V\colon\ \dot q = 0, \quad \dot p = -\nabla V(q).$$

The flow of X_T is

$$q(t) = q(0) + tp(0), \quad p(t) = p(0)$$

and the flow of X_V is

$$q(t) = q(0), \quad p(t) = p(0) - t\nabla V(q(0)).$$

Composing the time $t = \tau$ flow of X_V (from initial condition (q_n, p_n)) followed by the time τ flow of X_T, as in (1.1), gives the method

$$\begin{aligned} p_{n+1} &= p_n - \tau\nabla V(q_n), \\ q_{n+1} &= q_n + \tau p_{n+1}. \end{aligned} \tag{1.2}$$

Because it is the composition of the flows of two Hamiltonian systems, it is a symplectic integrator. It is a first-order method, sometimes called symplectic Euler. Including the previous step, namely

$$q_n = q_{n-1} + \tau p_n,$$

and eliminating the p_n gives

$$q_{n+1} - 2q_n + q_{n-1} = -\tau^2\nabla V(q_n), \tag{1.3}$$

which is leapfrog in a more familiar form, the form in which it was first derived as a discretization of $\ddot q = -\nabla V(q)$. (Note that this shows that the method is actually second-order in the position variables q.) In Figure 1.1 we show a Poincaré section for the Hénon–Heiles system which has $n = 2$ and potential

$$V(q_1, q_2) = \tfrac{1}{2}(q_1^2 + q_2^2) + q_1^2 q_2 - \tfrac{1}{3}q_2^3.$$

The energy error is shown in Figure 1.2: it is not zero, but appears to be bounded. This is in fact typical for symplectic integrators on bounded energy surfaces and moderate time steps.

Example 2. (The rigid body) The Euler equations for the motion of a free rigid body in $\mathbb{R}^3$ are

$$\begin{aligned} \dot x_1 &= a_1 x_2 x_3, \\ \dot x_2 &= a_2 x_3 x_1, \\ \dot x_3 &= a_3 x_1 x_2, \end{aligned}$$

where $a_1 = 1/I_2 - 1/I_3$, $a_2 = 1/I_3 - 1/I_1$, $a_3 = 1/I_1 - 1/I_2$, the I_j are the moments of inertia, and x_i is the angular momentum of the body in coordinates fixed in the principal axes of the body (Marsden and Ratiu

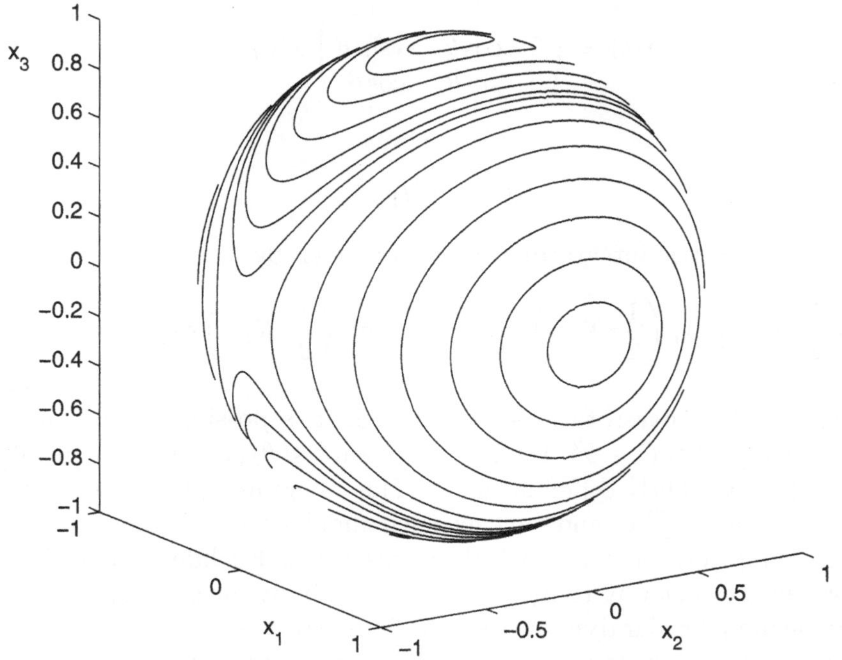

Figure 1.3. The flow of the triaxial rigid body with $I_1 = 1$, $I_2 = 2$, $I_3 = 3$ calculated with the second-order explicit splitting method (1.4) and a time step of 0.05. The maximum relative energy error for all initial conditions is 0.002, or about τ^2

1999). These equations have a lot of geometric structure. They form a Lie–Poisson system of the form

$$\dot{x} = J(x)\nabla H(x),$$

where

$$J(x) = \begin{pmatrix} 0 & x_3 & -x_2 \\ -x_3 & 0 & x_1 \\ x_2 & -x_1 & 0 \end{pmatrix}$$

and $H = \sum_{i=1}^{3} H_i$, $H_i = x_i^2/2I_i$. As in the previous example, splitting the Hamiltonian represents the system as a sum of three Hamiltonian vector fields, each of which is integrable. Furthermore, each preserves the total angular momentum $\sum x_i^2$. For example, the first vector field is $X_{H_1} := J\nabla H_1$:

$$\begin{aligned} \dot{x}_1 &= 0, \\ \dot{x}_2 &= -x_1 x_3/I_1, \\ \dot{x}_3 &= x_1 x_2/I_1, \end{aligned}$$

with solution

$$x(t) = \begin{pmatrix} 1 & 0 & 0 \\ 0 & \cos\theta & -\sin\theta \\ 0 & \sin\theta & \cos\theta \end{pmatrix} x(0),$$

where

$$\theta = tx_1/I_1.$$

An explicit geometric integrator is therefore given by

$$\exp\left(\frac{1}{2}\tau X_{H_1}\right)\exp\left(\frac{1}{2}\tau X_{H_2}\right)\exp(\tau X_{H_3})\exp\left(\frac{1}{2}\tau X_{H_2}\right)\exp\left(\frac{1}{2}\tau X_{H_1}\right), \tag{1.4}$$

a sequence of 5 planar rotations. For speed, it is possible to use the approximations $\cos\theta \approx (1-\theta^2/4)/(1+\theta^2/4)$, $\sin\theta \approx \theta/(1+\theta^2/4)$, equivalent to rotating by a slightly different angle. This integrator preserves the total angular momentum $\sum x_i^2$ and, on each such angular momentum sphere, preserves the symplectic structure which in this case is Euclidean area. This is undoubtedly the best way to integrate the rigid body for most applications, for example in molecular dynamics simulations. (It does not preserve energy, but in applications, systems of rigid bodies are coupled together, and the energy of each body is not individually preserved anyway.) The computed phase portrait for a triaxial rigid body with $I_1 = 1$, $I_2 = 2$, $I_3 = 3$ is shown in Figure 1.3.

Example 3. (The Duffing oscillator) Geometric integrators constructed from splitting are also useful for dissipative systems. We consider the Duffing oscillator (Guckenheimer and Holmes 1983), a forced planar system

$$X\colon \quad \dot q = p, \quad \dot p = q - q^3 + \gamma\cos t - \delta p. \tag{1.5}$$

As shown in Section 3.6, this system is 'conformal Hamiltonian', that is, the linear dissipation $-\delta p$ causes the symplectic structure to contract at a constant rate. Over one period of the forcing, the symplectic structure is scaled by $\mathrm{e}^{-2\pi\delta}$. From (1.5) it can be seen that the trace of the Jacobian of X is $-\delta$, so the sum of the eigenvalues of any fixed point, and more generally the sum of the Lyapunov exponents of any orbit obey $\sigma_1 + \sigma_2 = -\delta$.

It is possible to split X in various ways. To illustrate the standard treatment of nonautonomous terms, let t be a new variable in the extended phase space (q, p, t), and let

$$\begin{aligned} &X = X_1 + X_2, \\ &X_1\colon \ \dot q = p, \quad \dot p = -\delta p, \qquad\qquad\quad \dot t = 1, \\ &X_2\colon \ \dot q = 0, \quad \dot p = q - q^3 + \gamma\cos t, \quad \dot t = 0. \end{aligned}$$

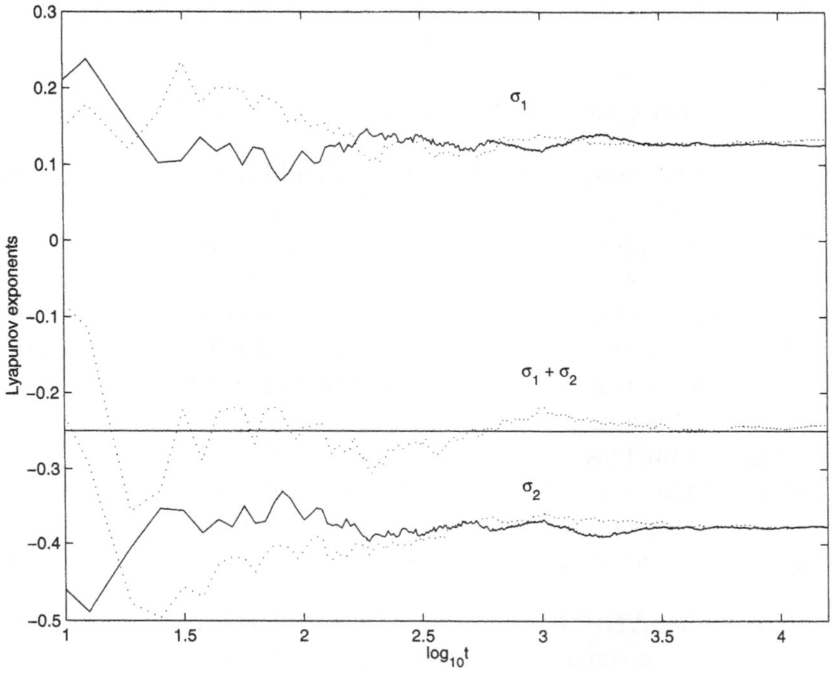

Figure 1.4. Convergence of estimates of the two Lyapunov exponents of the Duffing oscillator for $\delta = -0.25$, $\gamma = 0.3$. The geometric (splitting) method is the solid line, which preserves the sum $\sigma_1 + \sigma_2$; the nongeometric (Taylor series) method is the dotted line

Then the flows of these vector fields are given by

$$\exp(\tau X_1): \quad (q,p,t) \mapsto \left(q + \frac{1-e^{-\delta\tau}}{\delta}p, e^{-\delta\tau}p, t+\tau\right)$$
$$\exp(\tau X_2): \quad (q,p,t) \mapsto (q, p + \tau(q - q^3 + \gamma\cos t), t)$$

and a convenient second-order explicit geometric integrator is given by

$$\exp\left(\frac{1}{2}\tau X_1\right)\exp(\tau X_2)\exp\left(\frac{1}{2}\tau X_1\right).$$

Notice that the time-dependent force is evaluated here half-way through a time step.

In Figure 1.4 we show the results of a calculation of the Lyapunov exponents of a strange attractor of the Duffing oscillator by two methods. The geometric splitting method, above, provides estimates which obey $\sigma_1 + \sigma_2 = -\delta$ for any finite time interval. The non-geometric method (here, a Taylor series method) does not, and although it has similar *local* truncation errors to the geometric integrator, has much larger errors in the Lyapunov exponents. (It is not clear whether the convergence of the exponents as $t \to \infty$ is affected, however.) The exponents are calculated with the discrete method, that is, by calculating the exponents of the integrator itself (Dieci, Russell and van Vleck 1997, McLachlan and Quispel 2001*b*).

We have seen that, in splitting, each piece should have the same (or more) properties as the original system so that they are not destroyed by the integrator. When several properties are present, this is not so easy. The following example preserves phase space volume and has 8 discrete symmetries and 8 discrete reversing symmetries. (Recall that a symmetry of a vector field X is a map $S : M \to M$ that leaves it invariant, *i.e.*, $\mathrm{T}S.X = X \circ S$, where $\mathrm{T}S$ is the tangent map (Jacobian derivative) of the map S, and a reversing symmetry $R : M \to M$ is a map that reverses its direction, *i.e.*, $\mathrm{T}R.X = -X \circ R$. The set of all symmetries and reversing symmetries of a given X forms a group.) One can find a splitting that preserves all 16 (reversing) symmetries, but one of the pieces cannot be integrated in terms of elementary functions; fortunately its flow can be approximated while preserving all the properties.

Example 4. (The ABC flow) The ABC flow has been widely studied as a model volume-preserving three-dimensional flow. It has phase space $\mathbb{T}^3$, the 3-torus:

$$\begin{aligned} \dot{x} &= A \sin z + C \cos y, \\ \dot{y} &= B \sin x + A \cos z, \\ \dot{z} &= C \sin y + B \cos x. \end{aligned} \tag{1.6}$$

We consider the case when two of the parameters are equal, say $B = A$. The system then has a reversing symmetry group with 16 elements, and is divergence-free. The reversing symmetry group is generated by the three elements

$$\begin{aligned} R_1 &: \quad (x, y, z) \mapsto (x, \pi - y, -z), \\ R_2 &: \quad (x, y, z) \mapsto (-x, y, \pi - z), \\ R_3 &: \quad (x, y, z) \mapsto (\tfrac{3\pi}{2} + z, \tfrac{\pi}{2} + y, \tfrac{3\pi}{2} - x). \end{aligned} \tag{1.7}$$

A splitting that preserves these properties is $X = X_1 + X_2$ with

$$\begin{array}{llll} X_1 : & \dot{x} = A \sin z + C \cos y, & \dot{y} = 0, & \dot{z} = C \sin y + A \cos x, \\ X_2 : & \dot{x} = 0, & \dot{y} = A \sin x + A \cos z, & \dot{z} = 0. \end{array} \tag{1.8}$$

Note that X_2 is explicitly integrable but X_1, a 2-dimensional Hamiltonian system, is not. However, the midpoint rule applied to X_1 preserves all the appropriate properties, namely volume and the (reversing) symmetries. Finally, volume and symmetries, being group properties, are preserved by any composition, but reversing symmetries are only preserved by so-called 'symmetric' compositions. A second-order, volume-preserving, reversing

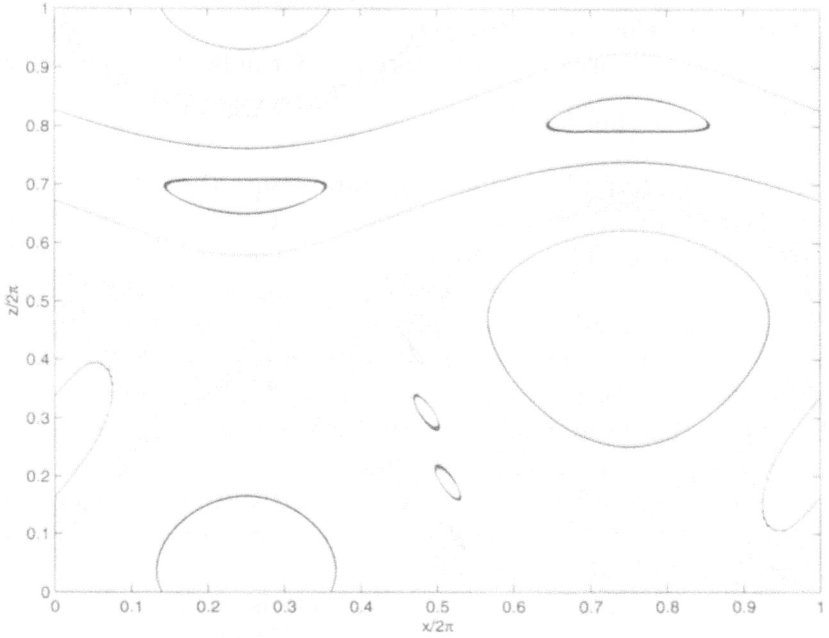

Figure 1.5. The section $y = 0$ of the ABC flow (1.6) with $A = B = 1$, $C = 0.75$, calculated with the symmetry- and volume-preserving splitting method (1.9) at time step $\tau = 0.25$

symmetry group-preserving integrator is therefore

$$\varphi_2(\tau/2)\varphi_1(\tau)\varphi_2(\tau/2), \tag{1.9}$$

where $\varphi_2 = \exp(\tau X_2)$, and φ_1 is the implicit midpoint rule applied to X_1. A sample phase portrait is shown in Figure 1.5. (Applied directly to X, a volume-preserving integrator would not in general preserve the symmetries, and a symmetry-preserving integrator, such as the midpoint rule, would not preserve volume.)

1.3. Historical development

The historical development of splitting methods is difficult to untangle, because it often proceeded independently in different applied fields. Thus we have *dimensional splitting* for parabolic PDEs, *fractional-step* and *operator splitting* methods for the Navier–Stokes equations and reaction–diffusion systems, *split-step* methods in optics and acoustics, *split-Hamiltonian* methods in chemical physics, the *mapping* method in celestial mechanics, and [*Lie–*]*Trotter*[*–Kato*] *formulae* in quantum statistical mechanics (in which leapfrog is called the 'primitive' method). Intermittent cross-fertilization has kept all these fields humming!

Splitting essentially began with the product formula of Trotter (1959),

$$\lim_{n\to\infty} (\mathrm{e}^{-tA/n}\mathrm{e}^{-tB/n})^n = \mathrm{e}^{-t(A+B)}, \tag{1.10}$$

where A and B are self-adjoint linear operators on a Banach space, $A+B$ is essentially self-adjoint, and either $t \in \mathrm{i}\mathbb{R}$ or $t \in \mathbb{R}$, $t \geq 0$ and A and B are bounded above. This includes the cases, for example, where A and B are heat operators, as in dimensional splitting for the heat equation, introduced by Bagrinovskii and Godunov (1957) and by Strang (1963). (Godunov (1999) has explained why he developed this scheme, and why he then dropped it.) For hyperbolic equations key early references are Tappert (1974) and Hardin and Tappert (1973), who introduced the split-step method for the nonlinear Schrödinger equation, the first example of what is now known as a symplectic integrator for a PDE. (In this review we will mostly be concerned with ordinary differential equations, but will briefly survey these related fields in Section 5.)

On the ODE side, the fundamental example is the leapfrog or Verlet method for the Hamiltonian ODE $\ddot{x} = f(x) = -\nabla V(x)$, namely

$$x_{n+1} - 2x_n + x_{n-1} = \tau^2 f(x_n).$$

It is usually credited to Verlet (1967), who used it in a molecular dynamics simulation of 864 particles for 1500 time steps interacting by a Lennard–Jones potential

$$V(x) = \sum_{i>j} (\sigma/\|x_i - x_j\|)^{12} - (\sigma/\|x_i - x_j\|)^6 \tag{1.11}$$

in a three-dimensional periodic box, obtaining excellent agreement with the properties of argon. However, Levesque and Verlet (1993) cite a much earlier use of leapfrog by the French astronomer Jean Baptiste Delambre (De Lambre 1790). We have translated part of this paper in Appendix C, and invite the reader to judge.[1]

Curiously enough, it was first discovered by Delambre's colleague Lagrange at almost exactly the same time that the flow of $\ddot{x} = -\nabla V(x)$ is symplectic (Lagrange 1788). But we had to wait nearly two centuries to put two and two together and appreciate that the leapfrog method is symplectic (suitably interpreted in position-momentum variables). It is hard to say who first realized this, as awareness of it seems to have spread rather slowly from field to field. Devogelaere (1956) notes that the first-order method (1.2) is known to be symplectic, and constructs a general second-order symplectic

[1] The leapfrog method also appears as one of a class of methods introduced by Störmer (1907), most of which are multistep, nonreversible and nonsymplectic; it appears that leapfrog itself was never used by Störmer.

method which reduces to leapfrog on $\ddot{x} = -\nabla V(x,t)$. This is presumably the first written proof that leapfrog is symplectic, although it was never published. In the physics community, a key step was the influential study of the so-called 'standard mappings' of the form (1.2) undertaken by Chirikov (1979), who credits the idea to an unpublished manuscript of Taylor (1968). For example, this inspired Wisdom (1982) to create the first symplectic integrators for celestial mechanics. Ruth (1983) was able to state that the symplecticity of the leapfrog method was already well known. This paper (Ruth 1983) led to a flurry of publications on symplectic integration (see, *e.g.*, the historical introduction to Channell and Scovel (1990)), so that by the early 1990s scientists in many applied fields knew, à la Molière, that they had been doing symplectic integration for years without realizing it.

A second key step was the derivation of the leapfrog method as a composition of flows of elementary Hamiltonians, that is, as a splitting method. As we have noted, splitting methods were already in widespread use by 1990 in numerical PDEs, quantum statistical mechanics, and celestial mechanics. However, their systematic development by numerical analysts was triggered by the work of Neri (1988) who applied the Baker–Campbell–Hausdorff formula to derive composition methods of high order, inspiring the more systematic development by Yoshida (1990). An indication of the growth of the field is given by the fact that Yoshida (1990) has in 10 years received more than 300 citations, mostly for use in diverse applications. In a parallel development, Suzuki (1976) began studying variations of the Trotter formula (1.10), a fourth-order method using derivatives, (4.12) below, was discovered by Takahashi and Imada (1984), and the even-order compositions were discovered by Suzuki (1990).

The third step was the realization that splitting and composition methods could be used to construct integrators for groups other than the symplectic group. This was emphasized by Forest and Ruth (1990), who mentioned the example of $O(3)$ for spin motion in an orbital ring, but was developed for other *infinite-dimensional* groups by Feng (1992), who treated the symplectic, volume-preserving, and contact groups on an equal footing. This point of view was developed further in the context of Cartan's classification of infinite-dimensional transformation groups by the authors (McLachlan and Quispel 2001*b*), so that it can be seen to include methods that preserve integrals, symmetries, orbits of group actions, foliations, in fact a large part of what is now known as 'geometric integration', a term coined by Sanz-Serna (1997). The humble leapfrog has come a long way.

1.4. Survey of this paper

Splitting methods are important in many different areas of mathematics: for instance, Hamiltonian systems, Poisson systems, systems with first integrals such as energy, momentum or angular momentum, systems with continuous or discrete symmetries, and systems with time-reversal symmetries. Similarly, splitting methods find application in many different areas of science: for instance, molecular dynamics, hydrodynamics, quantum mechanics and quantum statistical mechanics, celestial mechanics and accelerator physics. Some readers may only be interested in how splitting methods work in one of these areas of mathematics/science in particular. Those readers are advised to turn directly to the relevant subsection of Section 3 on splitting and Section 5 on applications: see the table of contents. (Some additional applications to Lotka–Volterra equations and similarity reductions of PDEs are discussed in Section 3.14.)

For those readers with a broader interest in splitting methods we now briefly survey the rest of this paper:

As was mentioned, splitting methods are particularly important in constructing geometric integrators for various classes of ODEs. Since a large number of such classes has now been distinguished, it has become important to classify the various ODEs and their corresponding integration methods. This classification proceeds in two stages. At the first stage, ODEs are grouped into 3 classes, depending on whether their flows form a group, a semigroup or a symmetric space (for a definition see below). At the second stage, each of these classes is subdivided further. These classifications (with an emphasis on integrators that form groups, such as the symplectic group and the volume-preserving group) are outlined in Section 2.

Returning to our definition of splitting methods given in the first sentence of the Introduction, we see now that step (1), splitting $X = \sum X_i$, makes sense because so many of the interesting sets of vector fields on M form linear spaces. (It even makes sense for vector fields whose flows lie in a semigroup, which are still closed under positive linear combinations.) We shall see that step (2), integrating the X_i in the appropriate space, is possible because each space of vector fields has a natural decomposition into much simpler vector fields. For some sets of vector fields, splitting is a science: the splitting can be constructed explicitly for all X. For others, it remains an art: one can give some guidelines on how to find a suitable splitting, but no general method is known. We survey these splittings in Section 3.

Step (3), combining the (approximate) flows of the pieces, forms the subject of composition methods and is addressed in Section 4. The group property directly confers a major advantage on geometric integrators. Namely, any composition of flows, even for negative time steps, lies in the group. As we shall see, this is essential for attaining orders higher than 2. In contrast, for general dissipative systems (*e.g.*, those that contract volume (McLachlan and Quispel 2000) or have a Lyapunov function), this is not possible: the dynamics lie only in a semigroup, which is left by negative time steps. This is related to one of the key reasons for the widespread use of splitting methods in applications: they are generally explicit (*i.e.*, faster). Further, while any composition of explicit maps is explicit, and hence suitable for retaining (semi)group properties, explicit flows have the advantage that their inverse is also explicit. No other large class of methods has this property. So, methods based on composition of explicitly integrable flows are uniquely placed to provide geometric integrators. Many geometric integrators exist which are *not* based on splitting (Budd and Iserles 1999, McLachlan and Quispel 2001*a*, Iserles, Munthe-Kaas, Nørsett and Zanna 2000), but we shall not discuss them here.

Some applications of splitting methods (and geometric integration more generally) to physics (molecular dynamics, particle accelerators, quantum (statistical) mechanics), chemistry, biology, celestial mechanics, hydrodynamics and other areas of science are discussed in Section 5.

Finally, some open problems in splitting methods and geometric integration are discussed in Section 6.

We usually work in standard coordinates on $\mathbb{R}^n$. However, from time to time we use coordinate-free notation (Marsden and Ratiu 1999) on an arbitrary manifold M; these parts can be skipped by the reader unfamiliar with differential geometry.

2. Groups of diffeomorphisms

2.1. Classifications of dynamical systems

As discussed in the introduction, splitting methods are particularly useful for the system $\dot{x} = X(x)$ when the flow of X lies in a particular group of diffeomorphisms. Indeed, the classification of such groups was first studied by Lie, who listed the symplectic, volume-preserving, and contact subgroups of the diffeomorphism group of a manifold.[2] However, passing to 'group' is jumping the gun a little, for at least two other algebraic structures come up in dynamical systems. We can make a primary classification of discrete-time dynamical systems into three categories (McLachlan and Quispel 2001*b*):

(1) those which lie in a semigroup (*e.g.*, the set of all maps $\varphi : M \to M$, where M is the phase space);
(2) those which lie in a symmetric space (*e.g.*, sets of diffeomorphisms closed under the composition $\varphi\psi^{-1}\varphi$, such as maps with a given time-reversal symmetry); and
(3) those which lie in a group (*e.g.*, the group of all diffeomorphisms of phase space).

[2] Recall that a diffeomorphism of a manifold M is a map $\varphi : M \to M$ which is differentiable and has a differentiable inverse. Klein (1893), in his Erlangen programme article, also covers infinite-dimensional groups, discussing diffeomorphisms, homeomorphisms, polynomial automorphisms, and contact transformations.

Splitting and composition are relevant, with suitable restrictions, to all three categories. Although we occasionally mention semigroups (*e.g.*, in Section 4.4) and symmetric spaces (*e.g.*, in Section 3.11), we will mostly consider groups. The set of vector fields whose flows lie in a group (respectively, semigroup, symmetric space) form a Lie algebra (respectively, Lie wedge (Hilgert, Hofmann, Heinrich and Lawson 1989), Lie triple system (Munthe-Kaas, Quispel and Zanna 2002)).

2.2. *Examples of diffeomorphism groups*

Geometric integrators can be classified according to their diffeomorphism group.

Definition 1. Let $\mathfrak{X}$ be a Lie algebra of vector fields on a manifold M, *i.e.*, a linear space of vector fields on M closed under the Lie bracket $[X,Y]$ where $[X,Y]f := (XY-YX)f$, whose flows lie in a subgroup $\mathfrak{G}$ of the group of diffeomorphisms of M. A *geometric integrator for* $X \in \mathfrak{X}$ is a 1-parameter family of maps $\varphi(\tau) \in \mathfrak{G}$ satisfying

$$\varphi(\tau) = \exp(\tau X) + \mathcal{O}(\tau^2).$$

We call it a $\mathfrak{G}$*-integrator.* If $X = \sum X_i$ where each X_i is either (i) integrable in terms of elementary functions, or (ii) integrable by quadratures, or (iii) has $\mathfrak{G}$-integrators simpler than those for X, we say X can be *split.*

Diffeomorphism groups can be finite- or infinite-dimensional. For the finite-dimensional case, the flow of the ODE $\dot{x} = X(x,t)$, $X(\cdot,t) \in \mathfrak{X}\ \forall t$, belongs to $\mathfrak{G}$, where in this case $\mathfrak{X}$ is a finite-dimensional Lie algebra. (For example, $\dot{x} = A(t)x$, where $x \in \mathbb{R}^n$ and $A(t) \in \mathfrak{so}(n)$; the flow is orthogonal.) The group orbit through the initial condition x_0 is a homogeneous space; the construction of $\mathfrak{G}$-integrators for ODEs on homogeneous spaces is an important part of geometric integration, for which an extensive and beautiful theory has been developed (Munthe-Kaas and Zanna 1997). When M is 1-dimensional, the only infinite-dimensional group on M is the set of all diffeomorphisms; however, when M is 2-dimensional, several new infinite-dimensional groups appear, such as the area-preserving mappings.

(In fact, there is no general theory of all diffeomorphism groups. One restriction is to study the so-called *Lie pseudogroups*, sets of local diffeomorphisms which are the general solution of a set of local PDEs and which are closed under composition only when the composition is defined. The flows of a Lie algebra of vector fields generally form a pseudogroup, because for a fixed time the flow of a given vector field need not be defined for all $x \in M$. For our applications, the distinction between local and global diffeomorphisms (*i.e.*, between Lie pseudogroups and groups of diffeomorphisms), is not crucial and will not be emphasized.)

Example 5. (Complex maps) Let $M = \mathbb{R}^2$ and write $\varphi = (u, v) \in \mathrm{Diff}(M)$. Then $\mathfrak{G} = \{\varphi : u_x = v_y,\ u_y = -v_x\}$, defined by the Cauchy–Riemann equations, may be identified with the set of complex analytic mappings, an infinite-dimensional group. Any differential equation $\dot{z} = f(z)$, $z \in \mathbb{C}$, f analytic, has a flow in $\mathfrak{G}$; Euler's method in the variable z is a $\mathfrak{G}$-integrator.

Example 6. (Area-preserving maps) $\mathfrak{G} = \{\varphi : u_x v_y - u_y v_x = \det \mathrm{T}\varphi = 1\}$, the symplectic (equivalent in $\mathbb{R}^2$ to area-preserving) mappings, is also infinite-dimensional. (Here $\mathrm{T}\varphi$ is the tangent mapping of φ.) Its Lie algebra is the divergence-free vector fields, which have the form $\dot{x} = H_y(x, y)$, $\dot{y} = -H_x(x, y)$. Symplectic integrators such as the midpoint rule provide $\mathfrak{G}$-integrators.

Diffeomorphism groups can be *primitive* or *nonprimitive*.

Example 7. (A nonprimitive group) $\mathfrak{G} = \{\varphi : v_x = 0\} = \{\varphi : (x, y) \mapsto (u(x, y), v(y))\}$ is infinite-dimensional. $\mathfrak{G}$-integrators have earlier been called 'closed under restriction to closed subsystems' (Bochev and Scovel 1994). All elements of $\mathfrak{G}$ map the set $y = c_1$ (where c_1 is a constant) to the set $y = c_2$ (where $c_2 = v(c_1)$ is another constant). We say that φ leaves the foliation $y =$ const invariant. Groups that leave a foliation invariant are said to be *not primitive*. However, they do arise in geometric integration and we will consider them in Section 2.5.

Definition 2. (Kobayashi 1972) A foliation of M (see Definition 3) is *invariant* under $\mathfrak{G}$ if φ permutes the leaves of the foliation for all $\varphi \in \mathfrak{G}$ (*i.e.*, if $\mathfrak{G}$ maps leaves to leaves). A foliation of M is *fixed* under $\mathfrak{G}$ if φ maps each leaf to itself for all $\varphi \in \mathfrak{G}$. A group $\mathfrak{G}$ is called *primitive* if it leaves no nontrivial foliation invariant.

Example 8. (No nonlinear rotations) Let $M = \mathbb{R}^n$, let G be a Lie subgroup of $\mathrm{GL}(n)$, and let $\mathfrak{G}$ be the group consisting of all diffeomorphisms whose derivative lies in G for all $x \in M$. It can be finite- or infinite-dimensional. If G is the group $\mathrm{Sp}(n)$ of symplectic matrices, $\mathfrak{G}$ is the infinite-dimensional set of symplectic maps; but for $G = \mathrm{SO}(n)$, $\mathfrak{G}$ is finite-dimensional. Indeed, writing $\sum_{i=1}^n f_i(x)\frac{\partial}{\partial x_i}$ for an element of the Lie algebra of $\mathfrak{G}$, we have

$$f_{i,j} + f_{j,i} = 0 \Rightarrow f_{i,jk} = f_{i,kj} = -f_{k,ij} = -f_{k,ji} = f_{j,ki} = f_{j,ik} = -f_{i,jk} = 0,$$

so the general solution is $f(x) = Ax + b$ for $A \in \mathfrak{so}(n)$, $b \in \mathbb{R}^n$.

2.3. The big picture

It is clear that such classifications are absolutely central, not just to geometric integration but also to dynamical systems in general. In each case, one should consider the following.

(1) Classify all instances, *e.g.*, all groups or all symmetric spaces.

(2) Is the structure invariant under diffeomorphisms or homeomorphisms? (For example, many common systems preserve the Euclidean volume. But in another smooth coordinate system, a different volume is preserved, so we should consider systems preserving arbitrary smooth volume forms. Generalizing further leads to continuous-measure or arbitrary measure-preserving systems. The situation is similar for Hamiltonian, reversible, and other systems.)

(3) What is the structure's local normal form? Does it make sense to assume this form? (For example, the local normal form of a symplectic structure is $dq_i \wedge dp_i$; it is certainly worthwhile to consider this case globally, because it is so common in applications and simplifies matters enormously; but doing so throws away all of modern symplectic geometry. Conversely, the local normal form of an integral at a regular point is $I(x_1, \ldots, x_n) = x_1$, but assuming this form would be ridiculous in most settings.)

(4) How can the structure be detected in a given system? (It is of course easy to tell if a system preserves a *given* integral or symplectic form. The harder problem is to tell if a system preserves *any* such structure. There is no known characterization of systems that have an integral or preserve a symplectic form, for example, only necessary conditions, *e.g.*, that all fixed points have a zero eigenvalue in the first case, or $+/-$ eigenvalue pairs in the second case. Conversely, there are algorithms that detect Lie symmetries in many cases.)

(5) How are the groups related to each other, *e.g.*, under intersection? (For example, Hamiltonian systems with symmetry have a richer structure than either class alone.)

(6) How does the structure affect the dynamics? (There is a range of possibilities. In some cases, *e.g.*, Hamiltonian mechanics, it is extremely subtle and so important as to be almost a definition of the field. In others, *e.g.*, systems with an integral, it is obvious.)

(7) What does the neighbourhood of each subgroup look like? (For example, what are the features of Hamiltonian nearly symmetric dynamics, or symmetric nearly Hamiltonian?)

2.4. The Cartan classification: the primitive groups

Cartan developed a structure theory of diffeomorphism groups and gave a classification of the complex primitive infinite-dimensional diffeomorphism groups, finding 6 classes (Cartan 1909).

We give the classification here briefly, and outline how each case arises in geometric integration. In each case it is crucial to consider whether the structure is presented in its local canonical form, the general form being usually much harder to preserve in an integrator.

A primitive infinite-dimensional group of diffeomorphisms $\mathfrak{G}$ on a complex manifold M must be one of the following.

(1) The group of all diffeomorphisms of M. Almost any one-step integrator lies in this group for small enough time step.
(2) The diffeomorphisms preserving a given symplectic 2-form ω. Its Lie algebra consists of the locally Hamiltonian vector fields, X such that $\mathrm{i}_X\omega$ is closed. $\mathfrak{G}$-integrators are called symplectic integrators (Section 3.2. They have only been generally constructed in two cases, when ω is the canonical symplectic 2-form on $\mathbb{R}^n$ and when M is a coadjoint orbit of a Lie algebra (Lie–Poisson integrators (Ge and Marsden 1988), Section 3.3). To classify these groups further depends on classifying the symplectic forms on M, which is an open problem.
(3) The diffeomorphisms preserving a given volume form μ on M. Its Lie algebra consists of the divergence-free vector fields X such that $\operatorname{div}_\mu X = 0$. Volume-preserving integrators have been considered both in the canonical case $M = \mathbb{R}^n$, $\mu = \mathrm{d}x_1 \ldots \mathrm{d}x_n$ (Feng and Wang 1994), and in the general case (Quispel 1995).
(4) The diffeomorphisms preserving a given contact 1-form α up to a scalar function. Contact integrators for the canonical case $\alpha = \mathrm{d}x_0 + \sum x_{2i}\,\mathrm{d}x_{2i+1}$ have been constructed by Feng (1998). A non-canonical example is provided by a Hamiltonian vector field restricted to an energy surface; the theorem of Ge (Ge and Marsden 1988) on energy-symplectic integrators shows that we should not expect to be able to construct $\mathfrak{G}$-integrators in this case.
(5) The diffeomorphisms preserving a given symplectic form ω up to an arbitrary constant multiple. That is, $\varphi^*\omega = c_\varphi\omega$, where the constant c_φ depends on $\varphi \in \mathfrak{G}$. (Here $\varphi^*\omega$ is the pull-back of the 2-form ω by the map φ: see, *e.g.*, Marsden and Ratiu (1999).) We study ODEs and integrators for this *conformal symplectic* group in Section 3.6.
(6) The diffeomorphisms preserving a given volume form μ up to an arbitrary constant multiple. That is, $\varphi^*\mu = c_\varphi\mu$, where the constant c_φ depends on $\varphi \in \mathfrak{G}$. We study this *conformal volume-preserving* case in Section 3.5.

(The case of a real manifold M is subtly different. The above 6 groups are also primitive infinite-dimensional in that case, but when the real manifold also carries a complex structure, there are a further 8 cases (McLachlan and Quispel 2001*b*).)

Note that all these cases are defined by the preservation of a differential form. Of course, the diffeomorphisms that preserve *any* collection of differential forms form a group; the point is that it is usually finite-dimensional.

The theory of dynamical systems has thus far mostly studied general diffeomorphisms, complex diffeomorphisms in complex dynamics, symplectic diffeomorphisms in Hamiltonian dynamics and, to a lesser extent, volume-preserving diffeomorphisms. The conformal and nonprimitive groups have not been studied as much. Clearly, when the flow of a system lies in such a group it has special dynamics; but luckily, the group also provides special tools (*e.g.*, complex analysis!) with which to study that dynamics.

2.5. Nonprimitive groups: systems with integrals, symmetries, and foliations

The largest nonprimitive group is the set of all diffeomorphisms that preserve a given foliation, that is, that map leaves to leaves. It is best to think here of what is called a *simple* foliation, one defined by the level sets of a function. Even regarding constructing integrators, the class of all foliations seems to be too large to admit a useful theory, and we are led (following the example of Lie group integrators (Iserles *et al.* 2000, Munthe-Kaas and Zanna 1997)) to consider foliations defined by the action of a Lie group. We introduce these with an example.

Example 9. (Caesar's laurel wreath) Let $M = \mathbb{R}^2$ and consider the vector field

$$\dot{x} = xy + x(1 - x^2 - y^2), \quad \dot{y} = -x^2 + y(1 - x^2 - y^2). \tag{2.1}$$

In polar coordinates, this becomes

$$\dot{r} = r(1 - r^2), \quad \dot{\theta} = -r\cos\theta,$$

showing that the foliation into circles $r =$ const is invariant under the flow. (In fact, this foliation is *singular*, because the leaf through the origin, a single point, has less than maximal dimension.) A one-step integrator is foliate if the final value of r is independent of the initial value of θ. Of course, this is easy to obtain in polar coordinates, but in fact no standard integrator in Cartesian coordinates is foliate. The leaves of this foliation are the group orbits of the standard action of SO(2) on $\mathbb{R}^2$ (see Figure 2.1).

Another way to describe foliate systems is that they contain a reduced subsystem on the space of leaves. In Example 9, the reduced system is $\dot{r} = r(1 - r^2)$. For a given orbit of the reduced system, one can often find a reconstruction system which describes the motion on the leaves themselves. In Example 9, for a reduced orbit $r(t)$, the reconstruction system is $\dot{\theta} = -r(t)\cos\theta$. For integrators, we do not usually want to construct the reduced system explicitly since the original phase space M is usually linear and easier to work in. We want integrators that preserve the foliation automatically.

For a given foliation, the group of foliate diffeomorphisms has many interesting subgroups, which have not been classified.

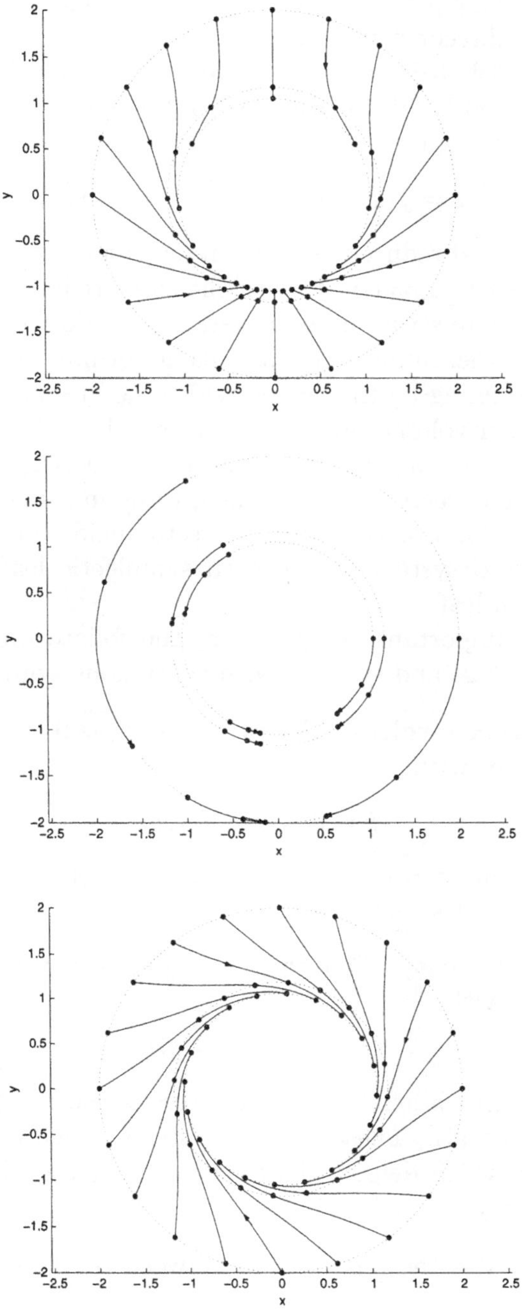

Figure 2.1. Three foliate vector fields that each map circles to circles, that is, they each preserve the same foliation into concentric circles. Top: a general foliate vector field, $\dot{r} = r(1 - r^2)$, $\dot{\theta} = -r\cos\theta$. Middle: a system with an integral, $\dot{r} = 0$, $\dot{\theta} = -r\cos\theta$. Bottom: a system with a continuous symmetry, $\dot{r} = r(1 - r^2)$, $\dot{\theta} = -(1 + r^2/5)$. The dots mark times 0, 0.5, and 1

First, a system may preserve several different foliations. A tree-like structure of reduced and reconstruction systems can be obtained by first reducing by all foliations with 1-dimensional leaves (leaving the largest possible reduced systems), then by all foliations with 2-dimensional leaves, and so on. For example, the system

$$\dot{x} = f(x), \ \dot{y} = g(x, y), \ \dot{z} = h(x, z)$$

has two 2-dimensional reduced systems in (x, y) and (x, z). Each of these is foliate with respect to $x =$ const, with the same reduced system $\dot{x} = f(x)$. Clearly the full foliate structure of a system will affect its dynamics.

Second, we get other infinite-dimensional nonprimitive Lie algebras of vector fields by considering (i) the vector field to lie in some other Lie algebra, as of Hamiltonian or volume-preserving vector fields; (ii) the reduced system to lie in some other Lie algebra; (iii) the reconstruction system, considered as a nonautonomous vector field on a leaf, to lie in some other Lie algebra. For example, the flows of Hamiltonian vector fields on Poisson manifolds have trivial reduced systems, since each symplectic leaf is fixed, but are symplectic on each leaf

The two most important subgroups of the foliate diffeomorphisms are systems with integrals and systems with continuous symmetries.

Example 10. (Just circles) Following Example 9, consider the systems in polar coordinates with

$$\dot{r} = 0, \quad \dot{\theta} = f(r, \theta).$$

The reduced systems $\dot{r} = 0$ are trivial, the function r is a first integral of the system and each leaf is fixed.

Example 11. (The iris) Following Example 9, consider the systems in polar coordinates with

$$\dot{r} = f(r), \quad \dot{\theta} = g(r).$$

The reduced dynamics is arbitrary but the reconstruction dynamics is clearly special (in fact, trivial). These are the systems which are invariant under the group action which defines the foliation, *i.e.*, they have a continuous rotational symmetry.

Figure 2.1 indicates the relationship between the three foliate groups in this case.

2.6. Non-Lie diffeomorphism groups

Finally, all of the above diffeomorphism groups have infinite-dimensional subgroups which are not locally defined (they are not of 'Lie type').

Example 12. (Discrete symmetries) With $\mathfrak{G}$ a diffeomorphism group, G a discrete group acting on M, the G-equivariant maps $\{\varphi \in \mathfrak{G} : \varphi \circ g = g \circ \varphi \; \forall g \in G\}$ form a group.

Example 13. (Weak integrals) The diffeomorphisms with given invariant sets (for example, a given list of fixed points and periodic orbits) form a group for which we would like to construct $\mathfrak{G}$-integrators. Fixed-point-preserving integrators are known (Stuart and Humphries 1996). More generally, suppose for $\dot{x} = X$ there is a function $I : M \to \mathbb{R}^k$ such that $\dot{I} = f(I)g(x)$. Note that I is not an integral, but the levels sets of I which satisfy $f(I) = 0$ are invariant under the flow. Such an I is called a *weak integral* of X. Single weak integrals ($k = 1$) are the most important, for they represent barriers to transport in phase space. A system might have a lot of them, for example $\dot{x}_i = x_i f(x)$, for which all n hyperplanes $x_i = 0$ are invariant. The group of diffeomorphisms with given weak integrals has subgroups obtained by restricting the flow only on the invariant level set to lie in some group of diffeomorphisms of that set. Weak integrals commonly arise as the fixed set of a symmetry. For example, if $x_1 \mapsto -x_1$ is a symmetry then the hyperplane $x_1 = 0$ is invariant.

Example 14. (Polynomial automorphisms) For $M = \mathbb{R}^n$ or $M = \mathbb{C}^n$ the invertible polynomial maps with polynomial inverses form a group (van den Essen 2000). It is infinite-dimensional, but not of Lie type. Its Lie algebra consists of the polynomial vector fields, but clearly the flow of a polynomial vector field is not necessarily a polynomial, that of $\dot{x} = x^2$ for instance. However, as we will see below, for some classes, namely symplectic and volume-preserving polynomial vector fields, we can construct explicit polynomial integrators by splitting. This is desirable for speed, smoothness, and global invertibility.

3. Splitting

3.1. Generating functions

The first step in constructing a splitting, or rather, a general approach to splitting for a given class of systems, is to parametrize all the vector fields in the given linear space of systems. This amounts to finding the general solution of a set of linear PDEs. (In McLachlan and Quispel (2001*b*) we called this solution a *generating function*.) For example, the ODE $\dot{q}_i = f_i(q,p)$, $\dot{p}_i = g_i(q,p)$ is Hamiltonian if $\frac{\partial f_i}{\partial q_j} + \frac{\partial g_i}{\partial p_j} = 0$ for all i and j, a linear PDE which is to be solved for f and g to find the Hamiltonian vector fields. Similarly, the ODE $\dot{x}_i = f_i(x)$ has integral $I(x)$ provided $\sum_i f_i \frac{\partial I}{\partial x_i} = 0$, which is to be solved to find all such f. In the cases of interest these PDEs are very simple and one can always find the general solution *locally*.

Globally it can be more difficult, especially if $M \neq \mathbb{R}^n$. We start with the most common case, the Hamiltonian vector fields.

3.2. Hamiltonian systems

In this case the group in question is the group of symplectic maps, those that preserve a given symplectic 2-form, and its Lie algebra is the set of 'locally Hamiltonian' vector fields, those whose flow is symplectic. In the canonical case a Hamiltonian system on $\mathbb{R}^{2n}$ is defined by the ODE

$$X_H : \quad \begin{aligned} \dot{q} &= \frac{\partial H}{\partial p}, \\ \dot{p} &= -\frac{\partial H}{\partial q}, \end{aligned}$$

where $H(q, p)$ is the energy or Hamiltonian function. That is, there is a bijection between the Hamiltonian vector fields and the scalar functions (H) modulo constants. We say the function H is a generating function for the vector field X_H. Splitting in this case amounts to

$$X_H = X_{\sum H_i} = \sum X_{H_i},$$

so that we must split H into a sum of simpler Hamiltonians. The most important examples of such simple Hamiltonians are $T(p)$ and $V(q)$, as in Example 1, but many others have been proposed and used:

- on $M = \mathbb{R}^{2n}$, $H = x^{\mathrm{T}}Ax$ and X_H linear;
- on $M = T^*Q$, Q Riemannian, the free particle $H = \|p\|^2$, many classically integrable cases of which are known;
- for any free particle on T^*Q, the metric can be (in theory) diagonalized and split into integrable 2D systems;
- integrable two body Hamiltonians $H(q_i, q_j, p_i, p_j)$, such as central force problems, the Kepler problem (Wisdom and Holman 1991), and point vortices;
- on $M = \mathbb{R}^{2n}$, Feng's 'nilpotent of degree 2' Hamiltonians $H(Cx)$, where $CJC^{\mathrm{T}} = 0$, $C \in \mathbb{R}^{n \times 2n}$, which are the most general Hamiltonians whose orbits are straight lines and are computed by Euler's method (Feng and Wang 1998);
- monomials (Channell and Neri 1996);
- various other integrable Hamiltonians arising in accelerator physics and celestial mechanics;
- lattice quantum systems with only near- or nearest-neighbour interactions, which are split into noninteracting parts by, *e.g.*, a checkerboard splitting (De Raedt 1987);

- finite difference spatial discretizations of PDEs, which are treated similarly.

What unifies all these cases? The usual answer is that they are all (Liouville) integrable. That is, there exist n functions $I_1, \dots I_n$, such that $\{H, I_j\} = 0$ for all j, $\{I_i, I_j\} = 0$ for all i, j, and that the $\{\mathrm{d}I_j\}$ are linearly independent for almost all x. However, this does not seem to embody the spirit of the preceding list, the main point of which is that they are all explicitly integrable in terms of elementary functions. Liouville integrability alone does not make it easy to compute the flow of H. Even the free rigid body, which can be integrated in terms of elliptic functions, is so complicated that, as far as we are aware, nobody has bothered to implement it in an integrator. Alternatively, consider the cost of evaluating the solution of an arbitrary planar Hamiltonian system.

With this in mind, we note the following class of systems which unifies and generalizes the class of 'easily' integrable systems. (We use the Poisson bracket $\{F, G\}$, which in the canonical case is defined by $\{F, G\} = \sum \frac{\partial F}{\partial q_i}\frac{\partial G}{\partial p_i} - \frac{\partial F}{\partial p_i}\frac{\partial G}{\partial q_i}$; see also Section 3.3.)

Theorem 1. Let $F_1, \dots, F_k$ be k functions such that X_{F_i} is integrable for all j and $\{F_i, F_j\} = 0$ for all i and j. Then, X_H for $H = H(F_1, \dots, F_k)$ is integrable. Furthermore, if $\frac{\partial H}{\partial F_i}$ and $\exp(X_{F_i})$ can be evaluated in terms of elementary functions, then so can $\exp(X_H)$.

Proof. From the Leibniz rule for Poisson brackets, we have

$$X_H = \sum_{i=1}^{k} \frac{\partial H}{\partial F_i} X_{F_i}. \tag{3.1}$$

Now

$$\begin{aligned}\frac{\mathrm{d}}{\mathrm{d}t}\frac{\partial H}{\partial F_i} &= \left\{\frac{\partial H}{\partial F_i}, H\right\}\\ &= \sum_{j=1}^{k}\sum_{l=1}^{k} \frac{\partial^2 H}{\partial F_i \partial F_j}\frac{\partial H}{\partial F_l}\{F_j, F_l\}\\ &= 0,\end{aligned}$$

so the coefficients of the X_{F_i} in (3.1) are constant along orbits. Furthermore $[X_{F_i}, X_{F_j}] = X_{\{F_j, F_i\}} = 0$ so the flows of the vector fields X_{F_i} commute. The flow of X_H is therefore given by the composition of the time-$\partial H/\partial F_i$ flows of the X_{F_i}. □

This is not quite the same as Liouville integrability for example, when $k < n$ the integrals of F_i are not necessarily shared by H, apart from the F_j. We believe it accounts for most, if not all, Hamiltonian splittings ever proposed, except for one: the Kepler problem seems to be the sole integrable system that is not integrable by elementary functions (it requires the root of a scalar nonlinear equation (Dutka 1997)), which it has been worthwhile implementing in integrators.

Therefore it is important to know sets of suitable commuting functions F_i. In all the cases listed above, these are one of the following:

(1) $F_i = q_i$, $i = 1, \ldots, n$;
(2) $F_i = p_i$, $i = 1, \ldots, n$;
(3) $F_1 = x^{\mathrm{T}} A x$ (*i.e.*, integrate $H(F_1)$);
(4) $F = Cx$ where $CJC^{\mathrm{T}} = 0$, the nilpotent class used by Feng and Wang (1998).

For example, the point vortex Hamiltonian is $\log(F)$ with $F = (q_1 - q_2)^2 + (p_1 - p_2)^2$ quadratic. The nonlinear term in the nonlinear Schrödinger equation is F^2, where $F = |\psi|^2$ is quadratic. The much larger class $H(F_1, \ldots, F_k)$ with $F_i = x^{\mathrm{T}} A_i x$ and $A_i^{\mathrm{T}} J A_j = A_j^{\mathrm{T}} J A_i$ for all i, j has yet to find applications.

The dynamics of Hamiltonian systems is typically very different from that of non-Hamiltonian systems, and many of their typical properties are preserved by symplectic integrators; see, *e.g.*, Reich (1999). We mention just one property here, the preservation of invariant (KAM) tori. As Broer, Huitema and Sevryuk (1996) have remarked, the significance of KAM theory lies not so much in its guarantees that a particular invariant torus is preserved under perturbation, but in its assertion that invariant tori are generic in families of systems of various classes (Hamiltonian, volume-preserving, reversible). Thus, while it is difficult to tell when a given invariant torus of a nonintegrable Hamiltonian system is preserved by an integrator, we do have many more general results, such as the following.

Theorem 2. (Shang 2000) Let there be an analytic, nondegenerate and integrable Hamiltonian system of n degrees of freedom, together with a frequency ω, in the domain of frequencies of the system, which satisfies a Diophantine condition of the form

$$|\langle k, \omega \rangle| \geq \frac{\gamma}{|k|^{\nu}} \quad 0 \neq k = (k_1, \ldots, k_n) \in \mathbb{Z}^n,$$

for some $\gamma > 0$ and $\nu > 0$. Then there exists a Cantor set $I(\omega)$ of $\mathbb{R}$, for any symplectic algorithm applied to the system, and a positive number δ_0, such that, if the step size τ of the algorithm falls into the set $(-\delta_0, \delta_0) \cap I(\omega)$, then the algorithm, if applied to the integrable system, has an invariant torus of frequency $\tau\omega$. The invariant torus of the algorithm approximates the invariant torus of the system in the sense of Hausdorff, with the order equal to the order of accuracy of the algorithm. The Cantor set $I(\omega)$ has density one at the origin in the sense that

$$\lim_{\delta \to 0^+} \frac{m((-\delta, \delta) \cap I(\omega))}{m(-\delta, \delta)} = 1.$$

3.3. Poisson systems

An important generalization of Hamiltonian systems are *Poisson systems* (Weinstein 1983, Marsden and Ratiu 1999) such as the rigid body (Example 2). A Poisson manifold is a manifold equipped with a Poisson bracket, an operation $\{,\} : C^\infty(M) \times C^\infty(M) \to C^\infty(M)$ satisfying (i) bilinearity $\{F, aG + bH\} = a\{F, G\} + b\{F, H\}$; (ii) antisymmetry $\{F, G\} = -\{G, F\}$; (iii) the Jacobi identity $\{F, \{G, H\}\} + \{G, \{H, F\}\} + \{H, \{F, G\}\} = 0$; and (iv) derivation $\{FG, H\} = F\{G, H\} + G\{F, H\}$ Then there exists a 2-vector J, a bilinear antisymmetric map $J : \Lambda^1(M) \times \Lambda^1(M) \to \mathbb{R}$, such that

$$\{F, G\} = J(\mathrm{d}F, \mathrm{d}G).$$

A Poisson vector field is one whose flow is Poisson, that is, it preserves the Poisson bracket,

$$\{F \circ \exp(X), G \circ \exp(X)\} = \{F, G\} \circ \exp(X).$$

The Poisson vector fields form an interesting Lie subalgebra of the algebra of vector fields on M, one that has not been studied a great deal. However, they have an important subalgebra, the Hamiltonian vector fields, defined by

$$X_F := J(\cdot, \mathrm{d}F).$$

The Poisson manifold M is foliated by leaves that carry a symplectic form ω satisfying $\omega(X_F, X_G) = \{F, G\}$. Often, these leaves are the level sets of a function $C : M \to \mathbb{R}^k$, called a *Casimir*, which satisfies $J(\cdot, \mathrm{d}C) \equiv 0$. Poisson vector fields preserve this foliation (*i.e.*, they map leaves to leaves), while Hamiltonian vector fields are tangent to it (*i.e.*, they map each leaf to itself). Thus, Poisson and Hamiltonian vector fields form natural nonprimitive Lie algebras. They arise in mechanics because they are stable under reducing to the leaf space of a system with symmetry, a property which symplectic manifolds lack. They also provide a way of realizing noncanonical symplectic manifolds as a leaf of a linear Poisson manifold, which is useful for computation.

Splitting is crucial to finding integrators in this case, because there are no general Poisson integrators. Theorem 1 also applies in the Poisson case. However, it raises the question of finding commuting sets of integrable Hamiltonians on M, a tall order: one can regard Darboux's theorem as equivalent to constructing these functions locally. Note, however, that if we can find *any* integrable Hamiltonians F_i, not in involution, then we can at least construct explicit Poisson integrators for $H = \sum H_i(F_i)$.

This is true in the important Lie–Poisson case, in which $M = \mathfrak{g}^*$ is the dual of a Lie algebra $\mathfrak{g}$, and $\{F, G\} = \langle x, [\mathrm{d}F, \mathrm{d}G]\rangle$. (Crucially, $J(x)$ is linear in x.) Because $X_{\langle x,c\rangle}$ is linear in x for all $c \in \mathfrak{g}$ and $[X_{\langle x,c\rangle}, X_{\langle x,d\rangle}] = X_{\langle x,[c,d]\rangle}$, Lie–Poisson manifolds carry a finite-dimensional Lie algebra of linear (and hence integrable) vector fields. (These are the analogues of the $2n$-dimensional abelian algebra of constant vector fields on a symplectic vector space.) This gives explicit Poisson integrators for $H = \sum H_i(\langle x, c_i\rangle)$. Further, the splittings constructed in Sections 3.12 and 3.13 give us the following.

Theorem 3. Any polynomial and any trigonometric Lie–Poisson system can be split into a sum of explicitly integrable systems.

The free rigid body (see Example 2) provides an important example of this case, with $F_i = x_i$ and $H_i = x_i^2/2I_i$. Furthermore, we have the following.

Theorem 4. (Reich 1993) Any Euler equation (a Lie–Poisson system with Hamiltonian $\langle x, x\rangle$ for some inner product) can be split into a sum of explicitly integrable systems.

Proof. If the original basis is (x_i), choose a basis (v_i) in which the inner product is diagonal. Then take $F_i = v_i$ and integrate either in the basis x_i or v_i. □

Moreover, let $\mathfrak{g}_1$ be an abelian subalgebra of $\mathfrak{g}$, corresponding to the space of commuting Hamiltonians $\langle x, \mathfrak{g}_1\rangle$. Following Theorem 1, to integrate $H(\langle x, \mathfrak{g}_1\rangle)$ we only need to be able to integrate fixed linear vector fields for arbitrary times. (This is an improvement over McLachlan (1995), which involved integrating linear vector fields containing parameters.)

3.4. Volume-preserving systems

The ODE $\dot{x} = X(x)$ is divergence-free (or source-free) if

$$\nabla \cdot X = \sum_{i=1}^{n} \frac{\partial X_i}{\partial x_i} = 0$$

for all x. The flow of a divergence-free system is volume-preserving.

Some divergence-free systems are very easy to split.

Example 15. (Easy) Consider the ODE $\dot{x}_i = f_i(x)$ where $\frac{\partial f_i}{\partial x_i} = 0$ for all i. This ODE is divergence-free, and each component

$$X_i\colon \quad \dot{x}_i = f_i(x), \ \dot{x}_j = 0, \ j \neq i$$

is integrated exactly by Euler's method. The ABC system (1.6) is an example.

In contrast with Hamiltonian systems, for which there are standard methods, such as the midpoint rule, which are symplectic but for which splittings have to be constructed on a case-by-case basis, no standard methods are known that are volume-preserving (Feng and Wang 1994), but any divergence-free vector field can be split. Informally, this is because the volume-preserving group is a superset of the symplectic group: this makes splitting easier (more pieces to choose from) but general-purpose integration harder (the integrator must cope with more systems). By the same argument, symplectic integrators are not volume-preserving when applied to non-Hamiltonian systems.

Volume-preserving splitting methods were first introduced by Feng (1993). As for Hamiltonian systems, we wish to find a generating function for all divergence-free systems. This can be done as follows.

Theorem 5. Let $M = \mathbb{R}^n$ with the Euclidean volume form. The vector field X is divergence-free if and only if there exists an antisymmetric matrix $S(x)$ such that

$$X = \nabla \cdot S. \tag{3.2}$$

That is, $\dot{x}_i = X_i(x) = \sum_{j=1}^{n} \partial S_{ij}/\partial x_j$. Each such S leads to a splitting of X into a sum of essentially two-dimensional volume-preserving systems.

Proof. First, if S is given, then

$$\nabla \cdot X = \nabla \cdot \nabla \cdot S = \sum_{i,j} \frac{\partial^2 S}{\partial x_i \partial x_j} = 0.$$

The converse is proved in Appendix A, where we construct a specific S for a given X. Finally, any matrix S leads to a splitting of $X := \nabla \cdot S$ into a sum of $n(n-1)/2$ two-dimensional divergence-free ODEs, namely

$$\begin{aligned} \dot{x}_i &= \frac{\partial S_{ij}}{\partial x_j}, \\ \dot{x}_j &= -\frac{\partial S_{ij}}{\partial x_i}, \\ \dot{x}_k &= 0, \quad k \neq i, j, \end{aligned}$$

for $1 \leq i < j \leq n$. (Each of these ODEs is Hamiltonian in the (x_i, x_j) plane and divergence-free in $\mathbb{R}^n$, although they are not usually Hamiltonian in $\mathbb{R}^n$.) □

The splitting of a divergence-free vector field is not at all unique. It corresponds to a general generating function for such systems on arbitrary manifolds given in the next theorem. However, the splitting given in Appendix A has some advantages over earlier methods (McLachlan and Quispel 2001*a*, Feng 1998): it only contains $n-2$ anti-derivatives, contains no multiple integrals, is quite specific, yet allows a lot of freedom with respect to integration constants. This method (and all other known general methods) does have two disadvantages. First, it does not preserve symmetries, not even translational symmetries. (The latter problem can be overcome for trigonometric vector fields: see Section 3.13.) Second, it may not preserve smoothness (see Section 6.3). These disadvantages are reasons to prefer splittings in which each X_i is explicitly integrable.

The advantages of splitting into essentially two-dimensional pieces are that (i) they are all integrable, hence possibly integrable in terms of elementary functions, and (ii) they are area-preserving in their plane, and any symplectic integrator (such as the midpoint rule) can be used to preserve area. Even though such an integrator is not symplectic in the whole space $\mathbb{R}^n$, it *is* volume-preserving.

We finish this subsection by generalizing to systems on an arbitrary volume manifold.

Theorem 6. Let μ be a volume form on a manifold M and let X be a vector field whose flow preserves μ. Then there exist an $(n-2)$-form β and an $(n-1)$-form $\gamma \in H^{(n-1)}(M)$, the equivalence class of closed $(n-1)$-forms modulo exact $(n-1)$-forms, such that

$$\mathrm{i}_X \mu = \mathrm{d}\beta + \gamma.$$

Proof. We have

$$\mathcal{L}_X \mu = \mathrm{d}\mathrm{i}_X \mu + \mathrm{i}_X \,\mathrm{d}\mu = \mathrm{d}\mathrm{i}_X \mu = 0,$$

that is, $\mathrm{i}_X \mu = 0$ is closed, from which the result follows. □

Note that, if M is simply connected, $H^1(M) \cong H^{n-1}(M) = 0$ so $\gamma = 0$. If $M = \mathbb{R}^n$, the $(n-2)$-form β plays the role of the antisymmetric matrix S in Theorem 5.

Example 16. (Volume preserving on a cylinder) Any divergence-free vector field which preserves the standard volume on the cylinder $\mathbb{T}^n \times \mathbb{R}^m$ can be written

$$X = c + \nabla \cdot S,$$

for some antisymmetric matrix S and some constant vector c which has no component along the cylinder.

3.5. Conformal volume-preserving systems

The two most famous conformal volume-preserving dynamical systems are the Lorenz system in $\mathbb{R}^3$ (Example 22) and the Hénon map in $\mathbb{R}^2$. Let μ be a volume form on a manifold M. Let X preserve μ up to a constant and let

Z be any fixed vector field which preserves μ up to a constant 1, that is,

$$\mathcal{L}_X \mu = c\mu, \quad \mathcal{L}_Z \mu = \mu.$$

Then $\mathcal{L}_{X-cZ}\mu = 0$, that is, $X - cZ$ is volume-preserving, the generation of which is given in Theorems 5 and 6. So we can split such systems if we can find an integrable Z. A composition method then leads to the volume expanding or contracting at *exactly* the correct rate.

Note that $\mathcal{L}_Z \mu = \mathrm{d}i_Z \mu + i_Z \, \mathrm{d}\mu = \mathrm{d}i_Z \mu = \mu$, that is, M must be exact if such a Z exists.

On $\mathbb{R}^n$ with the Euclidean volume form, we can take $Z = \sum \frac{1}{n} x_i \frac{\partial}{\partial x_i}$ (certainly integrable), which gives a representation of the constant-divergence vector fields as

$$X = \sum_i \left(\sum_j \frac{\partial S_{ij}}{\partial x_j} - \frac{c}{n} x_i \right) \frac{\partial}{\partial x_i}, \tag{3.3}$$

where $S_{ij}(x) = -S_{ji}(x)$.

The conformal property implies that the sum of the Lyapunov exponents is equal to $\mathrm{div}_\mu X$. A system which contracts some volume element cannot have a completely unstable fixed point, a topological invariant of this class of systems. However, the conformal property is not believed to be a decisive factor in controlling the dynamics in the way that volume preservation itself is. The volume contraction is so strong that all nearby systems may have similar dynamics. Still, the conformal volume-preserving group has infinite codimension in the full diffeomorphism group and staying in it may confer some advantage.

Example 17. (Linear dissipation) On $\mathbb{R}^n$, any volume-preserving (*e.g.*, Hamiltonian) system $\dot{x} = X$ becomes conformal volume-preserving on the addition of linear dissipation Lx. For example, writing the inviscid Euler fluid equations as $\dot{\omega} = N(\omega)$, the Navier–Stokes equations $\dot{\omega} = N(\omega) + \nu \nabla^2 \omega$ are conformal volume-preserving in standard discretizations. Similarly, a Hamiltonian system on T^*Q with the addition of Rayleigh dissipation, namely $\dot{q} = H_p$, $\dot{p} = -H_q - R(q)p$, is conformal volume-preserving if $\mathrm{tr}\, R(q) = \mathrm{const}$.

3.6. Conformal Hamiltonian systems

One finds the vector fields as in the last section. Let (M, ω) be a symplectic manifold. Let X preserve ω up to a constant and let Z be any fixed vector field which preserves ω up to a constant 1, that is,

$$\mathcal{L}_X \omega = c\omega, \quad \mathcal{L}_Z \omega = \omega.$$

Then $\mathcal{L}_{X-cZ} = 0$, that is, $X - cZ$ is Hamiltonian. So we can split such systems if we can find an integrable Z. A composition method then leads to the symplectic form expanding or contracting at *exactly* the correct rate. As before, M must be exact ($\omega = -\,\mathrm{d}\theta$) and not compact.

In the canonical case, $M = \mathbb{R}^{2n}$, $\theta = p\,\mathrm{d}q$, and $\omega = \mathrm{d}q \wedge \mathrm{d}p$, giving the conformal Hamiltonian system

$$\dot{q} = \frac{\partial H}{\partial p}, \quad \dot{p} = -\frac{\partial H}{\partial q} - cp. \tag{3.4}$$

For $H = \frac{1}{2}\|p\|^2 + V(q)$, these are mechanical systems with linear dissipation. (The structure of these systems is studied in McLachlan and Perlmutter (2001).) Since this form of dissipation is special mathematically (forming a group), we argue it must be special physically too.

For the general conformal Hamiltonian system (3.4), the energy obeys $\dot{H} = -cp^{\mathrm{T}}H_p$ which can have any sign. The system can have a 'conformal symplectic attractor', as does the Duffing oscillator, Example 3. For autonomous simple mechanical systems, however, $H = \frac{1}{2}\|p\|^2 + V(q)$, and $\dot{H} = -c\|p\|^2 \leq 0$. The energy becomes a Lyapunov function and all orbits tend to fixed points.

The eigenvalues of the Jacobian of X (and hence the Lyapunov exponents of X) occur in pairs with sum $-c$; the spectrum is as constrained as that of Hamiltonian systems. Consider an invariant set (fixed point, periodic orbit, *etc.*) with stable manifold W^s and unstable manifold W^u. Their dimensions obey

$$\dim W^s \begin{cases} \leq \dim W^u & \text{for } c < 0, \\ \geq \dim W^u & \text{for } c > 0, \\ = \dim W^u & \text{for } c = 0. \end{cases} \tag{3.5}$$

Since these dimensions are invariant under homeomorphisms, the inequality (3.5) is a topological invariant. A system in which one of these three conditions did not hold for all invariant manifolds could not be conformal symplectic.

Conformal Hamiltonian systems also have characteristic properties in the presence of symmetries (see Section 3.10). If there is a momentum map J which evolves under $\dot{J} = 0$ for Hamiltonian systems, it obeys $\dot{J} = -cJ$ for conformal Hamiltonian systems. The two foliations of M into orbits of the symmetry and into level sets of the momentum are still both preserved.

As before, geometric integrators can be constructed by splitting: Z can be integrated exactly, and a symplectic integrator applied to the remainder. Alternatively, since Z is linear, one can split off the entire linear part of X and integrate it exactly (see Example 3). The order can be increased by composition.

Splitting is actually more important in the conformal than in the standard Hamiltonian case, as there are standard methods (the Gaussian Runge–

Kutta methods) that are symplectic, but no standard method is conformal symplectic (McLachlan and Quispel 2001*b*).

Geometric integrators for the conformal cases are particularly useful when the dissipation rate c is small; in particular, when studying the limit $c \to 0$.

3.7. *Contact systems*

Contact geometry arose from the systematic study of first-order PDEs in the nineteenth century. In the words of Felix Klein (1893):

By a contact-transformation is to be understood, analytically speaking, any substitution which expresses the values of the variables x, y, z and their partial derivatives $\frac{dz}{dx} = p$, $\frac{dz}{dy} = q$ in terms of new variables x', y', z', p', q'. It is evident that such substitutions, in general, convert surfaces that are in contact into surfaces that are in contact, and this accounts for the name .... [C]ontact transformations may be defined as *those substitutions of the five variables x, y, z, p, q, by which the relation*

$$\mathrm{d}z - p\,\mathrm{d}x - q\,\mathrm{d}y = 0 \tag{3.6}$$

is converted into itself. In these investigations space is therefore to be regarded as a manifoldness of five dimensions; and this manifoldness is to be treated by taking as fundamental group the totality of the transformations of the variables which leave a certain relation between the differentials unaltered.

(The classification of dynamical systems by diffeomorphism groups is exactly within Klein's Erlangen programme.)

The 1-form (3.6) is an example of a contact form, namely a 1-form α on a manifold M^{2n+1} such that the volume form $\alpha \wedge (\mathrm{d}\alpha)^n \neq 0$. Equivalently, $\ker \alpha$ is a nonintegrable $2n$-dimensional distribution on TM. The diffeomorphism φ is *contact* if it preserves the distribution $\ker \alpha$, that is, if $\varphi^*\alpha = \lambda\alpha$ for some function $\lambda : M \to \mathbb{R}$. Darboux's theorem states that all contact forms are locally equivalent to (3.6), that is, there are coordinates $(x_1, \ldots, x_n, y_1, \ldots, y_n, z) \in M$ in which α takes the canonical form

$$\alpha = \mathrm{d}z - \sum_{i=1}^{n} x_i\,\mathrm{d}y_i.$$

The contact vector fields are generated by the scalar functions $K : M \to \mathbb{R}$. In the canonical case the contact vector field X_K generated by K is

$$\begin{aligned}
\dot{x}_i &= \frac{\partial K}{\partial y_i} + x_i \frac{\partial K}{\partial z}, \\
\dot{y}_i &= -\frac{\partial K}{\partial x_i}, \\
\dot{z} &= K - \sum_{i=1}^{n} x_i \frac{\partial K}{\partial x_i}.
\end{aligned}$$

Then $\mathcal{L}_{X_K}\alpha = \lambda\alpha$ where $\lambda = \partial K/\partial z$. (In contrast to Hamiltonian vector fields, X_K is not invariant under $K \mapsto K + \text{const.}$)

Some of the key properties of contact flows are that:

(1) $\dot{K} = \lambda K$, so the submanifold $K = 0$ is invariant.
(2) If $K \neq 0$ on M, $\mathcal{L}_{X_K} K^{-1}\alpha = 0$ and $\mathcal{L}_{X_K} \mathrm{d}K^{-1}\alpha = 0$; the flow has Hamiltonian-like properties.
(3) If $\lambda \neq 0$ on $K = 0$, then $\mathcal{L}_{\lambda^{-1}X_K} \mathrm{d}\alpha = \mathrm{d}\alpha$; the flow of $\lambda^{-1}X_K$ is conformal symplectic with respect to the symplectic form $\mathrm{d}\alpha|_{K=0}$. Fixed points can only occur on $K = 0$ and if $\lambda \neq 0$ at such a fixed point, it has the same character as a fixed point of a conformal Hamiltonian system.

However, a systematic study of the dynamics of contact flows and maps has yet to be undertaken. (Usually only the case $K \neq 0$, equivalent to Reeb vector fields, is studied.)

Contact integrators were first studied by Feng (1998), who constructed general contact integrators by passing to the 'symplectified' system in $\mathbb{R}^{2n+2}$. Here we are interested in the question: When can contact integrators be constructed by splitting? As for Hamiltonian systems, we have to split the scalar function K. This is much harder than for Hamiltonian systems because there are very few contact vector fields with straight-line flows, *i.e.*, which are integrated by Euler's method. Amongst simple examples, $X_{K(x)}$ and $X_{K(y)}$ are integrated by Euler's method, and $X_{K(y,z)}$ can be integrated by quadratures.

However, we do have that X_K leaves the foliation defined by the level sets of K invariant if and only if $K = K(F(x,y) + z)$. So in these cases the reduced system $\dot{K} = g(K)$ can be integrated by quadratures (although the reconstruction system in (x,y) is not always integrable).

Theorem 7. For the canonical contact structure on $\mathbb{R}^{2n+1}$, the contact vector field X_K is integrable by quadratures for $K = K(a^{\mathrm{T}}x + b^{\mathrm{T}}y + cz)$. Hence any polynomial contact vector field is a sum of integrable contact vector fields.

3.8. Foliations: systems with integrals

The vector field X has integrals $I_1, \ldots, I_k$ if $X(I_j) = 0$. We first need to find a generating function, that is, a representation of all systems with a given set of integrals. This is provided by the following 'skew-gradient' form.

Theorem 8. (McLachlan, Quispel and Robidoux 1999) Given independent $\{I_j\}$ and such a vector field X such that $X(I_j) = 0$, there exists a (nonunique) $(k+1)$-vector S such that

$$X = S(\cdot, \mathrm{d}I_1, \ldots, \mathrm{d}I_k).$$

For example, on $M = \mathbb{R}^n$ and one integral I, there is an antisymmetric matrix $S(x)$ such that $X = S\nabla I$.

On $\mathbb{R}^n$, we can split $S = \sum S_i$ so that each piece has the minimum number of nonzero components. This gives a representation of an arbitrary system with k integrals as a sum of $(k+1)$-dimensional vector fields, each with the same set of k integrals, and hence integrable. However, they may not be integrable in terms of elementary functions. Still, this is a good place to look for splittings.

(Incidentally, this also works if there are no integrals at all! Putting $k = 0$, we find $S = X$, and X splits into its n one-dimensional components, each integrable.)

Example 18. (Energy-preserving methods) A Hamiltonian system has the form $\dot{x} = J\nabla H$, *i.e.*, it is already presented in skew-gradient form. Energy-preserving integrators can therefore be constructed by splitting into the planar systems

$$\begin{aligned}\dot{q}_i &= \frac{\partial H}{\partial p_i},\\ \dot{p}_i &= -\frac{\partial H}{\partial q_i},\\ \dot{q}_j &= \dot{p}_j = 0, \quad \text{for } j \neq i,\end{aligned}$$

for $i = 1, \ldots, n$, each of which is integrable by quadratures (although not necessarily in terms of elementary functions).

Example 19. (Korteweg–de Vries equation) One can construct unconditionally stable finite difference methods for the KdV equation $u_t = 3uu_x + u_{xxx}$ by requiring the space and time discretizations to preserve a positive functional such as $\|u\|^2$. From the above, we see that the space discretization must have the form $\dot{u}_i = \sum_j S_{ij}(u)u_j$ where $S_{ij} = -S_{ji}$. On a grid with uniform spacing h, let Lu be the 4-point central difference approximation of u_{xxx}, so that $L^{\mathrm{T}} = -L$. Let $f(u,v) : \mathbb{R}^2 \to \mathbb{R}$ be any function satisfying $f(u,v) = f(v,u)$ and $f(u,u) = u/h$. Then

$$f(u_i, u_{i+1})u_{i+1} - f(u_{i-1}, u_i)u_{i-1} = 3uu_x + \mathcal{O}(h^2)$$

and the associated matrix with $S_{i,i+1} = -S_{i+1,i} = f(u_i, u_{i+1})$, $S_{i,j} = 0$ for $|j - i| \neq 1$, is antisymmetric. (The discretization can also be chosen to be volume-preserving by taking $f(u,v) = \sqrt{u^2 + v^2}/h$.) The combined system $\dot{u} = (S(u) + L)u$ has integral $\|u\|^2$. This integral can be preserved in the time integration by a quadratic-integral-preserving scheme such as the implicit midpoint rule or by a Lie group integrator (Iserles *et al.* 2000), giving linearly implicit, unconditionally stable schemes.

Splitting methods have to be applied with caution to PDEs (see Section 5.2). The most complete splitting, into planar subsystems, each a rotation, does yield an explicit unconditionally stable method. However, because it does not satisfy the CFL condition that the numerical domain of dependence contains the physical domain of dependence, it cannot be convergent for all τ/h.

When a system has a large number of integrals which it is desirable to preserve, it is better to see if they arise from some structural feature of the equation. When the leaves are the orbits of a group action, each orbit is a *homogeneous space* and it is possible to make big progress using *Lie group integrators* (Iserles *et al.* 2000). Let $\Gamma_g : M \to M$ be the group action for $g \in G$, and write

$$\gamma(v, x) := \frac{\mathrm{d}}{\mathrm{d}t}\Gamma_{\exp(tv)}(x)|_{t=0}$$

so that all DEs tangent to the orbits can be written

$$\dot{x} = X = \gamma(a(x), x)$$

for some function $a : M \to \mathfrak{g}$. This function a generates the given class of DEs, and since it takes values in a linear space $\mathfrak{g}$, it can be split into n components on choosing a basis for $\mathfrak{g}$, say $v_1, \ldots, v_n$ and $a(x) = \sum a_i(x)v_i$. The vector field $\gamma(a_i(x), x)$ is tangent to the one-dimensional group orbit $\exp(tv_j)(x)$; hence, if these are integrable, the ODE $\dot{x} = \gamma(a_i(x), x)$ is integrable by quadratures. Whether they are integrable in terms of elementary functions depends on a and the choice of basis. Example 19 illustrates both cases for the natural action of $\mathrm{SO}(n)$ on $\mathbb{R}^n$; $S \in \mathfrak{so}(n)$ and the orbits $\exp(tv_j)(x)$ are circles in the (u_i, u_{i+1}) plane.

Apart from splitting, there are many other ways of preserving the integrals of such systems, such as Runge–Kutta–Munthe-Kaas methods (Iserles *et al.* 2000, Munthe-Kaas and Owren 1999, Munthe-Kaas and Zanna 1997). Splitting is preferred when the pieces are explicitly integrable or when it is desired to preserve some other property as well as the integrals, such as volume.

3.9. Foliate systems in general

We introduced foliate systems informally in Section 2.5 and gave the example of Caesar's laurel wreath, Example 9. The implicit midpoint rule, which preserves arbitrary quadratic first integrals and arbitrary linear symmetries, does not preserve the quadratic foliation in that example; splitting is necessary. We now introduce foliate systems more formally, leading to two large classes of systems which are amenable to integration by splitting.

Definition 3. (Molino 1988) Let M be a manifold of dimension m. A *singular foliation* F of M is a partition of M into connected immersed submanifolds (the 'leaves'), such that the vector fields on M tangent to the leaves are transitive on each leaf. F is *regular* if each leaf has the same dimension. F has *codimension* q if the maximum dimension of the leaves of F is $m - q$. A diffeomorphism of M is *foliate* with respect to F if it leaves the foliation invariant, *i.e.*, if it maps leaves to leaves. A vector field on M is *foliate* if its flow is foliate. The space of smooth vector fields tangent to the leaves of F is denoted $\mathfrak{X}_{\text{tan}}$. The *space of leaves* (denoted M/F) is obtained by identifying the points in each leaf together with the quotient topology.

Theorem 9. (Molino 1988) $\mathfrak{X}_F$ and $\mathfrak{X}_{\text{tan}}$ form Lie algebras. $\mathfrak{X}_{\text{tan}}$ is an ideal in $\mathfrak{X}_F$. A vector field X is foliate with respect to F if and only if $[X,Y] \in \mathfrak{X}_{\text{tan}}$ for all $Y \in \mathfrak{X}_{\text{tan}}$.

Usually in the study of foliations one begins with an integrable distribution on M, which defines a regular foliation. We do not adopt this point of view because (i) we need global, not local information about the foliation, and (ii) it allows many exotic foliations, *e.g.*, ones with dense leaves, which we are not interested in because they may have no foliate vector fields not tangent to the leaves (the simplest example being the distribution on $\mathbb{T}^2$ defined by a vector field of constant irrational slope).

Theorem 10. (Molino 1988) Let M and N be manifolds of dimension m and n, respectively. Let $I : M \to N$ be a smooth surjection. (If I is not onto, we replace N by $I(M)$.) Then I defines a foliation F whose leaves are given by the connected components of $I^{-1}(y)$ for each $y \in I(M)$. If I is a submersion, that is, if $\mathrm{T}I$ has constant rank n, then F is a regular foliation of codimension n. In this case the space of leaves M/F is diffeomorphic to N.

Such a foliation is called *simple*. Given a vector field, one can search for simple foliations it preserves by looking for functions I such that $\dot{I} = f(I)$.

Example 20. (First integrals) A system with k first integrals $I : M \to \mathbb{R}^k$ is foliate with respect to the level sets of the functions I. Each leaf is in fact fixed by the flow. For this reason we choose the symbol I in Theorem 10 to suggest that simple foliate systems generalize systems with first integrals.

Example 21. (Continuous symmetries) A system with a continuous symmetry is foliate with respect to the orbits of the symmetry. That is, let X admit the Lie group action $\Gamma : G \times M \to M$ as a symmetry, so that its flow φ_t is G-equivariant. Then $\Gamma(g, \varphi_t(x)) = \varphi_t(\Gamma(g,x))$, that is, the foliation with leaves given by the group orbits $\{\Gamma(g,x) : g \in G\}$ is invariant. In this case the reconstruction problem on G is easier to solve than in the general case, because it is G-invariant.

Example 22. (The Lorenz system) The Lorenz system is given by

$$\begin{aligned}\dot{x} &= \sigma y - \sigma x,\\ \dot{y} &= -y - xz - rx,\\ \dot{z} &= xy - bx.\end{aligned}$$

If $b = 2\sigma$, the system is foliate with leaves $x^2 - 2\sigma z = \text{const}$, for

$$\frac{\mathrm{d}}{\mathrm{d}t}(x^2 - 2\sigma z) = -2\sigma(x^2 - 2\sigma z).$$

We split into

$$\begin{aligned}X_1\colon\quad & \dot{x} = \sigma y, & \dot{y} &= -xz - rx, & \dot{z} &= xy - bz,\\ X_2\colon\quad & \dot{x} = -\sigma x, & \dot{y} &= -y, & \dot{z} &= -2\sigma z.\end{aligned}$$

X_1 is tangent to the foliation and may be integrated using the midpoint rule, which preserves the quadratic function $x^2 - 2\sigma z$. X_2 is foliate but linear, and can be solved exactly.

Example 23. (Skew product systems) A special case of the foliations defined by submersions is given by $M = N \times L$, I being projection onto N. Each leaf is then diffeomorphic to L. In coordinates x on N and y on L, any foliate vector field can be written in coordinates as

$$\begin{aligned}\dot{x} &= f(x),\\ \dot{y} &= g(x, y),\end{aligned}$$

and any tangent vector field as

$$\begin{aligned}\dot{x} &= 0,\\ \dot{y} &= g(x, y).\end{aligned}$$

Example 24. (Nonautonomous systems) The extension of a nonautonomous vector field on M to an autonomous vector field on $M \times \mathbb{R}$ preserves the foliation defined by $t = \text{const}$. Most integrators are foliate and, indeed, solve the reduced system $\dot{t} = 1$ exactly.

In a foliate system, one can obtain some information about part of the system (namely, the current leaf) for all time without even knowing the full initial condition. This puts strong dynamical constraints on the whole system.

Example 25. (Three-dimensional foliate systems) Consider a three-dimensional system with a codimension 1 foliation. In local coordinates the system can be written

$$\dot{x} = f(x),\ \dot{y} = g(x, y, z),\ \dot{z} = h(x, y, z).$$

The only possible ω-limit set of the reduced system $\dot{x} = f(x)$ is a point, suggesting that the ω-limit set of the whole system is either a point, a circle (periodic orbit), or a heteroclinic cycle. Similarly, for a three-dimensional system with a codimension 2 foliation, such as

$$\dot{x} = f(x, y), \ \dot{y} = g(x, y), \ \dot{z} = h(x, y, z),$$

the ω-limit set of the reduced system in (x, y) is a point, a circle, or a heteroclinic cycle, suggesting that the ω-limit set of the full system is a point, a circle, a 2-torus, or a heteroclinic cycle. In both of these cases the existence of the foliation suggests that the system cannot be chaotic.

Let G be a Lie group and $\Gamma : G \times M \to M$ be an action of G on M. This group action generates a (possibly singular) foliation whose leaves are the group orbits $\Gamma(G, x)$. The vector field

$$X = X_{\mathrm{tan}} + X_{\mathrm{inv}}$$

is foliate, where X_{tan} is tangent to the leaves, and X_{inv} is G-invariant. We give two important classes of Lie group foliate vector fields.

Example 26. (Natural action) Let $G \subset \mathrm{GL}(n)$ be a matrix group with its natural action on $\mathbb{R}^{n\times k}$,

$$\Gamma(A, L) = AL, \quad A \in G, \ L \in \mathbb{R}^{n\times k}.$$

The ODEs

$$\dot{L} = f(L)L + g(L)$$

are foliate, where $f : \mathbb{R}^{n\times k} \to \mathfrak{g}$ ($f(L)L$ is tangent to the leaves) and $g : \mathbb{R}^{n\times k} \to \mathbb{R}^{n\times k}$ with $g(AL) = g(L)$ for all $A \in G$ (g is invariant). The second term can be written as a function of the invariants of the action, if these are known. For example, for $G = \mathrm{SO}(n)$ we can write $g(L) = h(L^{\mathrm{T}}L)$ and for $G = \mathrm{SL}(n)$ we have $g(L) = h(\det L)$.

Example 27. (Adjoint action) This generalizes the 'isospectral' systems studied in Lie group integrators (Iserles *et al.* 2000, Calvo, Iserles and Zanna 1997) and elsewhere. Let G be a matrix Lie group, let $M = \mathfrak{g}$, the Lie algebra of G, and let G_1 be a subgroup of G which acts on M by adjoint action, that is,

$$\Gamma(U, L) = ULU^{-1}, \quad U \in G_1, \ L \in \mathfrak{g}.$$

The 'isospectral manifolds' of $\mathfrak{g}$ are the sets of matrices similar by an element of $\mathrm{GL}(n)$, while the leaves of the foliation defined by this group action are the sets of matrices in $\mathfrak{g}$ which are similar by an element of G_1, and hence are submanifolds of the isospectral manifolds.

The ODEs

$$\dot{L} = [f(L), L] + g(L),$$
$$f\colon\ \mathfrak{g} \to \mathfrak{g}_1,$$
$$g\colon\ \mathfrak{g} \to \mathfrak{g},\quad g(ULU^{-1}) = Ug(L)U^{-1}\ \forall U \in G_1$$

are foliate. For example,

$$g(L) = p(L)h(\operatorname{tr} L, \operatorname{tr} L^2, \ldots, \operatorname{tr} L^n)$$

is adjoint invariant, where p is an analytic function and $h : \mathbb{R}^n \to \mathbb{R}$.

The decomposition $X = X_{\text{tan}} + X_{\text{inv}}$ gives a way of constructing foliate integrators by splitting: X_{tan} can be integrated by any integrator for vector fields on homogeneous spaces (Munthe-Kaas and Zanna 1997), and X_{inv} by any symmetry-preserving integrator (Section 3.10).

Second, each piece may be decomposed further. This is always possible for X_{tan}, on choosing a basis for the relevant Lie algebra.

3.10. Systems with symmetries

The vector field X has symmetry $S : M \to M$ if

$$(\mathrm{T}S.X)S^{-1} = X. \tag{3.7}$$

The map φ has symmetry S if

$$S\varphi S^{-1} = \varphi. \tag{3.8}$$

If X satisfies (3.7), its flow satisfies (3.8).

There are few general results on preserving nonlinear symmetries (with the exception of Dorodnitsyn (1996)). Here we shall restrict ourselves to linear and affine symmetries.

Theorem 11. (Linear symmetries) Linear (and affine) symmetries are preserved by all Runge–Kutta methods.

This theorem has several corollaries:

(1) if one just wishes to preserve a linear/affine symmetry group, one can use any explicit Runge–Kutta method;
(2) if one wishes to preserve a linear/affine symmetry group *plus* (constant) symplectic structure, one can use any symplectic Runge–Kutta method;
(3) if one wishes to preserve a linear/affine symmetry group plus some other geometry property, one is restricted to using splitting and/or composition.

In a symmetry-preserving composition method, all vector fields that one splits into should be invariant under the entire symmetry group one is interested in (this contrasts with the case of reversing symmetries (McLachlan,

Quispel and Turner 1998)). The possible implementation of point 3 above hence rests on the following result.

Theorem 12. Let $\mathfrak{G}$ be a diffeomorphism group, let $\mathfrak{X}$ be its Lie algebra, let $X \in \mathfrak{X}$, and let $G \subset \mathfrak{G}$ be a finite symmetry group. Let

$$\widetilde{X} = \sum_{S_i \in G} (\mathrm{T}S.X)S_i^{-1}. \tag{3.9}$$

Then $\widetilde{X} \in \mathfrak{X}$, and $\widetilde{X}$ is invariant under G.

We will use Theorem 12 as follows. Assume, for example, that we want to construct a volume-preserving and symmetry-preserving integrator for a given vector field (*i.e.*, $\mathfrak{G}$ is the group of volume-preserving diffeomorphisms.) First assume that we can split off a simplest possible divergence-free vector field X for which we know how to construct a $\mathfrak{G}$-integrator, without worrying about symmetries. Then $\widetilde{X}$ will be divergence-free *and* preserve all the given symmetries. The problem, however, is whether we will be able to construct a volume-preserving symmetry-preserving integrator for $\widetilde{X}$. The two following examples show that, although this will sometimes be the case, in general it will not be.

Example 28. (The AAC flow) The AAC flow is a special case of the ABC flow (1.6) given by $B = A$, that is,

$$\dot{x} = A\sin z + C\cos y, \quad \dot{y} = A\sin x + A\cos z, \quad \dot{z} = C\sin y + A\cos z.$$

Its symmetry group is generated by

$$\begin{aligned} S_1\colon\ & (x,y,z) \mapsto (-x, \pi - y, z - \pi), \\ S_2\colon\ & (x,y,z) \mapsto (\tfrac{3\pi}{2} + z, \tfrac{\pi}{2} - y, x - \tfrac{3\pi}{2}). \end{aligned} \tag{3.10}$$

We start with the simplest divergence-free building block we can think of,

$$X_1\colon\ \dot{x} = A\sin z, \quad \dot{y} = 0, \quad \dot{z} = 0.$$

Applying Theorem 12 to X_1, we see that it is already invariant under S_1, so (ignoring S_1) we get

$$\widetilde{X}_1\colon\ \dot{x} = A\sin z, \quad \dot{y} = 0, \quad \dot{z} = A\cos x.$$

So far so good, because $\widetilde{X}_1$ can be integrated while preserving volume and the symmetries S_1 and S_2, by using the implicit midpoint rule.

To further build up X, the next simple divergence-free building block we start with is

$$X_2\colon\ \dot{x} = C\cos y, \quad \dot{y} = 0, \quad \dot{z} = 0.$$

Applying Theorem 12 again, we get

$$\widetilde{X}_2\colon\ \dot{x} = C\cos y, \quad \dot{y} = 0, \quad \dot{z} = C\sin y.$$

Again we are in luck, because $\widetilde{X}_2$ can also be integrated using the implicit midpoint rule. Finally, starting from

$$X_3\colon\ \dot{x} = 0, \quad \dot{y} = A\sin x, \quad \dot{z} = 0,$$

we obtain

$$\widetilde{X}_3\colon\ \dot{x} = 0, \quad \dot{y} = A\sin x + A\cos z, \quad \dot{z} = 0,$$

which is integrated exactly by Euler's method. Noting that $X = \widetilde{X}_1 + \widetilde{X}_2 + \widetilde{X}_3$, composing these integrators yields a volume-preserving and symmetry-preserving integrator for the AAC flow.

Example 29. (The AAA flow) We now consider the ABC flow (1.6) with $C = B = A$. In addition to the symmetries S_1 and S_2 in (3.10), this flow has the cyclic symmetry

$$S_3\colon\ (x, y, z) \mapsto (y, z, x).$$

We start again from

$$X_1\colon\ \dot{x} = A\sin x, \quad \dot{y} = 0, \quad \dot{z} = 0.$$

But if we now apply Theorem 12, we find that $\widetilde{X}_1 = X$, *i.e.*, this procedure does not allow us to split X into simpler parts. Indeed, as far as we know, it is currently not known whether a volume- and symmetry-preserving integrator for the AAA flow exists.

For continuous symmetry groups G, the ODE preserves the foliation given by the orbits of G (see Example 11). The sum in Theorem 12 becomes an integral. Luckily, in many cases, natural splittings do preserve symmetries. For example, in the Hamiltonian case with $H = \sum H_i$, each H_i should be G-invariant. This occurs with with $\frac{1}{2}p^{\mathrm{T}}M(q)p + V(q)$ splitting when G acts by cotangent lifts, as, *e.g.*, rotational and translational symmetries do. Similarly, if the action of G is linear, then a splitting of X into homogeneous parts preserves G.

3.11. Systems with reversing symmetries

The vector field X has reversing symmetry $R : M \to M$ if (Lamb 1998)

$$(\mathrm{T}R.X) \circ R^{-1} = -X. \tag{3.11}$$

The map φ has reversing symmetry R if

$$R \circ \varphi \circ R^{-1} = \varphi^{-1}. \tag{3.12}$$

(Note that (3.11) is the linearization of (3.12).) If X satisfies (3.11), its flow satisfies (3.12). One reason it is important to preserve reversing symmetries of a system is because (just as in the case of Hamiltonian and divergence-free vector fields) such systems have KAM theorems (Broer *et al.* 1996) guaranteeing the existence of stabilizing tori in phase space.

The combined set of all symmetries and reversing symmetries of a vector fields forms a group, called its reversing symmetry group. While the set of diffeomorphisms preserving a given symmetry group forms a group, discussed in Section 3.10, the set of diffeomorphisms preserving a given reversing symmetry group does not form a group: it is not closed under composition. However, it does form a *symmetric space.*

Definition 4. A symmetric space[3] is a (discrete or continuous, finite-dimensional or infinite-dimensional) subset S of a group G such that

$$\varphi\psi^{-1}\varphi \in S \ \forall \ \varphi, \psi \in S.$$

A Lie triple system is a subspace $\mathfrak{t}$ of a (finite- or infinite-dimensional) Lie algebra $\mathfrak{g}$ such that

$$[X, [Y, Z]] \in \mathfrak{t} \ \forall \ X, Y, Z \in \mathfrak{t}.$$

The linearization of a continuous symmetric space S is a Lie triple system $\mathfrak{t}$, and $\exp(\mathfrak{t}) \subset S$. Clearly one would like to know all the symmetric spaces contained in a given group of diffeomorphisms.

It can be shown that the reversing symmetry group of a given system, if nontrivial, can be generated by the group of symmetries plus a single arbitrarily chosen reversing symmetry (Lamb 1998). This means that, if a system has a number of geometric properties plus a reversing symmetry group, then we can alternatively think of it as possessing the geometric properties plus a symmetry group, plus a single reversing symmetry, as follows.

Theorem 13. Let $\mathfrak{G}$ be a group of diffeomorphisms, let $\mathfrak{S}_G$ be the set of diffeomorphisms preserving the reversing symmetry group G, let $S \subset G$ be the symmetries in G, and let $R \in G$ be any one of the reversing symmetries. Then

$$\mathfrak{G} \cap \mathfrak{S}_G = \mathfrak{G} \cap \mathfrak{S}_S \cap \mathfrak{S}_R.$$

This will enable us to construct $\mathfrak{G} \cap \mathfrak{S}_G$-integrators, *i.e.*, integrators that lie in $\mathfrak{G}$ and preserve the whole reversing symmetry group G.

Theorem 14. (Nonlinear reversing symmetries) Let $\varphi \in \mathfrak{G} \cap \mathfrak{S}_S$, and let the reversing symmetry R be an element of $\mathfrak{G}$. Then the method

$$\chi := \varphi R \varphi^{-1} R^{-1}$$

satisfies

$$\chi \in \mathfrak{G} \cap \mathfrak{S}_G,$$

that is, it is a $\mathfrak{G}$-integrator and preserves the whole reversing symmetry group G.

[3] A more abstract definition is given in Loos (1969).

Proof. Obviously $\chi \in \mathfrak{G}$. Let $\widetilde{R}$ be any reversing symmetry in G. Then

$$\begin{aligned}
\widetilde{R}\chi\widetilde{R}^{-1} &= \widetilde{R}\varphi R\varphi^{-1}R^{-1}\widetilde{R}^{-1} \\
&= \widetilde{R}\varphi RR^{-1}\widetilde{R}^{-1}\varphi^{-1}, \quad \text{since } R^{-1}\widetilde{R}^{-1} \in \mathfrak{G}_G, \\
&= \widetilde{R}\varphi\widetilde{R}^{-1}RR^{-1}\varphi^{-1} \\
&= \widetilde{R}\widetilde{R}^{-1}R\varphi R^{-1}\varphi^{-1}, \quad \text{since } \widetilde{R}^{-1}R \in \mathfrak{G}_G, \\
&= R\varphi R^{-1}\varphi^{-1} \\
&= \chi^{-1}.
\end{aligned}$$

So any $\widetilde{R} \in G$ is a reversing symmetry of χ. Since G is generated by its reversing symmetries (Lamb 1998), this completes the proof. □

This theorem implies that, provided $R \in \mathfrak{G}$, we do not need to worry about preserving reversing symmetries while we are constructing a geometric integrator; they can be incorporated at the final stage. In splitting, it is sufficient to seek a splitting $X = \sum X_i$ with X_i preserving the group properties, that is, $\exp(X_i) \in \mathfrak{G}$ and X_i has symmetry group S, form an integrator from them, and then apply the theorem. Note, however, that if it does happen that X_i has reversing symmetry R and we start with the basic composition (1.1), then we get

$$\chi = \exp(\tau X_1)\ldots\exp(\tau X_n)\exp(\tau X_n)\ldots\exp(\tau X_1),$$

that is, the factors of R all cancel. This is the case in the canonical Example 1: $X = X_T + X_V$, $R : (q, p) \mapsto (q, -p)$, and X_T and X_V are both R-reversible. That is, the leapfrog–Verlet method is reversible.

Note that the above theorem is also true in the case that φ is symplectic and R is antisymplectic.

Theorem 15. (Linear reversing symmetries) Linear (and affine) reversing symmetries are preserved by all self-adjoint Runge–Kutta methods. These also preserve all linear (and affine) symmetries (which are preserved by *all* Runge–Kutta methods).

The fact that all self-adjoint Runge–Kutta methods are implicit implies that explicit splitting methods are preferred if available.

3.12. Splitting for polynomial vector fields

The first splitting methods for Hamiltonian polynomial vector fields were based on splitting into monomials (Channell and Neri 1996). For example, Shi and Yan (1993) partitions the monomials of degree 3, 4, 5, and 6 in 6 variables into 8, 20, 42, and 79 sets, so that the monomials in each set commute. Here we present a new, more efficient method using a different set of basis functions, which has the additional advantage of avoiding the singularities that are associated with monomial splitting. Our starting point is the following result.

Theorem 16. For each m, there exists an N and vectors $k_1, \dots, k_N \in \mathbb{R}^n$ such that the set

$$\{(k_i^{\mathrm{T}} x)^m,\ 1 \le i \le N\} \tag{3.13}$$

forms a basis for the homogeneous polynomials of degree m in $\mathbb{R}^n$. The k_i can be chosen so that some subset of the functions $\{(k_i^{\mathrm{T}} x)^p\}$ forms a basis for the homogeneous polynomials of degree $p < m$.

That is, if P is any polynomial, $P(x) = \sum_m \sum_{i=1}^N a_{im}(k_i^{\mathrm{T}} x)^m$ where the a_{im} (but not the k_i) depend on P; any polynomial in n variables is a sum of polynomials in one variable. The proof is given in Appendix B. Note that N and the k_i can be constructed explicitly.

This result allows us, for example, to construct explicit splittings for Hamiltonian and volume-preserving systems. In the Hamiltonian case, we get

$$H = \sum_{i=1}^N H_i, \quad H_i = \sum_m a_{im}(k_i^{\mathrm{T}} x)^m.$$

Hence

$$X_{H_i} = J\nabla H_i(x) = Jk_i^{\mathrm{T}} \sum_m m a_{im}(k_i^{\mathrm{T}} x)^{m-1}.$$

Since $k_i^{\mathrm{T}} x$ is a first integral for X_{H_i}, the exact flow of X_{H_i} is given by Euler's method. Note that, unlike monomial splitting, this splitting also yields explicit geometric integrators for Poisson systems with constant Poisson tensor J.

In the divergence-free case, we expand each function S_{ij} appearing in the representation (3.2) of divergence-free systems. This gives us the following.

Theorem 17. Let X be a polynomial divergence-free vector field. Then

$$X = \sum_i X_i(k_i^{\mathrm{T}} x),$$

where each X_i is divergence-free and has integral $k_i^{\mathrm{T}} x$. The exact flow of each X_i is given by Euler's method.

Note that, since the pieces here are all volume-preserving, this approach can only be used for the volume-preserving group and its subgroups.

3.13. Splitting for trigonometric vector fields

Every generalized trigonometric polynomial vector field X on $\mathbb{R}^n$ can be written in the form (Quispel and McLaren 2002) $X = \sum X_i$, where

$$X_i(x) = c_i \sin(k_i^{\mathrm{T}} x) + d_i \cos(k_i^{\mathrm{T}} x) \tag{3.14}$$

for certain constant vectors c_i, d_i, $k_i \in \mathbb{R}^n$.

We now consider two cases: the volume-preserving case, and the Hamiltonian case.

(1) *The volume-preserving case*
From the fact that X is divergence-free it follows that

$$\nabla \cdot X = \sum_i k_i^{\mathrm{T}} c_i \cos(k_i^{\mathrm{T}} x) - k_i^{\mathrm{T}} d_i \sin(k_i^{\mathrm{T}} x) = 0, \tag{3.15}$$

and hence, from linear independence,

$$k_i^{\mathrm{T}} c_i = k_i^{\mathrm{T}} d_i = 0 \text{ for all } i. \tag{3.16}$$

Thus each vector field X_i has

$$\frac{\mathrm{d}}{\mathrm{d}t}(k_i^{\mathrm{T}} x) = k_i^{\mathrm{T}} X_i = 0,$$

that is, has integral $k_i^{\mathrm{T}} x$ and is integrated exactly using Euler's method:

$$\exp(tX_i)(x(0)) = x(0) + tX_i(x(0)). \tag{3.17}$$

(2) *The Hamiltonian case*
We get the Hamiltonian case for free. If X is Hamiltonian, all X_i in (3.14) must also be Hamiltonian, and hence their exact flow (3.17) must be symplectic. So if X is Hamiltonian, the method above automatically yields a symplectic integrator!

3.14. Examples

Lotka–Volterra equations

Many well-known families of ODEs may have no special structure in general but contain within them interesting special cases which do have extra structure. We illustrate this for Lotka–Volterra systems, which arise in biology and in economics (Volterra 1931). They have the general form

$$\dot{x}_i = x_i \left(\lambda_i + \sum_{j=1}^{n} a_{ij} x_j \right), \quad i = 1, \ldots n.$$

In the domain $x_i > 0$ we can put $u_i := \log x_i$, to get

$$\dot{u}_i = \lambda_i + \sum_{j=1}^{n} a_{ij} \mathrm{e}^{u_j}$$

or

$$\dot{u} = \lambda + A\mathrm{e}^{u}. \tag{3.18}$$

Each Lotka–Volterra system falls into one or both of the following cases:

(1) $\lambda \in \text{range}(A)$;

(2) $\text{rank}(A) < n$.

In case (1), $\lambda \in \text{range}(A)$, we can rewrite (3.18) in linear-gradient form (McLachlan *et al.* 1999):

$$\dot{u} = A\nabla V(u), \tag{3.19}$$

with $V(u) = \sum_i e^{u_i} + c_i u_i$. Some special cases are:

(i) if A is symmetric positive definite, (3.19) is a gradient system;

(ii) if $A + A^{\mathrm{T}}$ is negative definite (Volterra 1931), (3.19) has V as a Lyapunov function;

(iii) if A is antisymmetric, (3.19) is either a Hamiltonian system (if $\text{rank}(A) = n$) or a Poisson system (if $\text{rank}(A) < n$);

(iv) if $A_{ii} = 0$ for all i (3.19) is divergence-free (Volterra 1931).

In cases (i) and (ii), splitting methods may not be the methods of choice, and we may prefer to use linear-gradient methods (McLachlan *et al.* 1999). Cases (iii) and (iv) are ideal for splitting. We split $X = \sum_{i=1}^{n} X_i$ where

$$X_i = A\nabla(e^{u_i} + c_i u_i).$$

In case (iii) each X_i is Hamiltonian (or Poisson), and in case (iv) each X_i is divergence-free. The X_i will also preserve any Casimirs of A when $\text{rank}(A) < n$. The exact flow of each X_i is given by Euler's method.

In case (2), when $\text{rank}(A) < n$, let w_i^{T}, $i = 1, \ldots, l$, be the left zero eigenvectors of A. It follows from (3.18) that

$$w_i^{\mathrm{T}} \dot{u} = w_i^{\mathrm{T}} \lambda,$$

which can be integrated to

$$w_i^{\mathrm{T}} u(t) = w_i^{\mathrm{T}} u(0) + t w_i^{\mathrm{T}} \lambda.$$

Now suppose $w_i^{\mathrm{T}} \lambda = 0$ for $i = 1, \ldots, k$ and $w_i^{\mathrm{T}} \lambda \neq 0$ for $i = k+1, \ldots, l$. The first k functions $w_i^{\mathrm{T}} u$ are integrals (specifically, Casimirs of A). If $k < l$ then the system has a codimension one foliation with leaves $w_l^{\mathrm{T}} u = \text{const}$ and $l - k - 1$ extra integrals (Casimirs)

$$(w_i^{\mathrm{T}} \lambda w_l^{\mathrm{T}} - w_l^{\mathrm{T}} \lambda w_i^{\mathrm{T}}) u, \quad i = k+1, \ldots, l-1.$$

Because the integrals and foliation are linear, they are preserved by the Runge–Kutta method (McLachlan, Perlmutter and Quispel 2002).

Similarity reductions of PDEs
Conformal volume-preserving ODEs (and their subgroup of conformal Hamiltonian ODEs) commonly arise as similarity reductions of PDEs. This can happen in at least two ways:

(1) travelling wave reductions of PDEs that are linear in the two highest (mixed) derivatives;
(2) spherically symmetric reductions of PDEs involving a Laplacian.

We illustrate each of these with a couple of examples.

Example 30. (Reduction of reaction–diffusion equations) Typically reaction–diffusion equations (Murray 1989) have the form

$$u_t + \nabla^2 u + f(u) = 0, \quad u : \mathbb{R}^{m+1} \to \mathbb{R}^n. \tag{3.20}$$

They admit solutions depending on the travelling wave variable $\xi := a^{\mathrm{T}}x - ct$. Inserting this, we obtain

$$u_\xi = v, \quad v_\xi = |a|^{-2}cv - f(u).$$

This ODE has constant divergence $nc\|a\|^{-2}$. Note that stationary solutions (*i.e.*, $c = 0$) correspond to a divergence-free ODE. A simple splitting is $X = X_1 + X_2$, where

$$\begin{aligned} X_1 &: \; u_\xi = 0, \quad v_\xi = -c\|a\|^{-2}v + f(u), \\ X_2 &: \; u_\xi = v, \quad v_\xi = 0. \end{aligned}$$

Example 31. (Fourth-order PDEs) Equations of the form

$$u_t + u_{xxxx} + \alpha u_{xx} + \beta u + \gamma u^2 + \delta u^3 + \varepsilon (u_x)^2 = 0, \quad u : \mathbb{R}^2 \to \mathbb{R},$$

where $\alpha, \beta, \gamma, \delta, \varepsilon$ are parameters, describe a variety of physical systems. Two special cases are (i) $\beta = 1$, the evolution of a gas flame front (Malomed and Tribelsky 1984), and (ii) $\gamma = \varepsilon = 0$, the Swift–Hohenberg equation (Swift and Hohenberg 1989). The travelling wave reduction $u(x,t) = u(x - ct)$ yields

$$u_\xi = v, \quad v_\xi = w, \quad w_\xi = z, \quad z_\xi = cv - \alpha w - \beta u - \gamma u^2 - \delta u^3 - \varepsilon v^2.$$

This ODE is divergence-free for any choice of the parameters. For $c = 0$ (stationary solutions) it also has the reversing symmetry $(u, v, w, z) \mapsto (u, -v, w, -z)$ (Roberts and Quispel 1992). For $c = \varepsilon = 0$, the system has an additional Hamiltonian structure

$$\begin{pmatrix} u_\xi \\ v_\xi \\ w_\xi \\ z_\xi \end{pmatrix} = \begin{pmatrix} 0 & 0 & 0 & 1 \\ 0 & 0 & -1 & 0 \\ 0 & 1 & 0 & -\alpha \\ -1 & 0 & \alpha & 0 \end{pmatrix} \nabla f,$$

$$\text{where } f(u, v, w, z) = \frac{1}{2}\beta u^2 + \frac{1}{3}\gamma u^3 + \frac{1}{4}\delta u^4 + vz - \frac{1}{2}w^2 + \frac{1}{2}\alpha v^2.$$

For $\gamma = \varepsilon = 0$ there is an additional symmetry $(u, v, w, z) \mapsto -(u, v, w, z)$. A simple splitting which preserves all of these structures is $X = X_1 + X_2$, where

$$X_1: \; u_\xi = v, \; v_\xi = 0, \; w_\xi = z, \; z_\xi = 0,$$

$$X_2: \; u_\xi = 0, \; v_\xi = w, \; w_\xi = 0, \; z_\xi = cv - \alpha w - \beta u - \gamma u^2 - \delta u^3 - \varepsilon v^2.$$

Example 32. (Poisson equations) Poisson equations have the form

$$\nabla^2 u + f(u) = 0.$$

Note that this is identical to the stationary form of the reaction–diffusion equation (3.20). Introducing the radial variable $r := (\sum x_i^2)^{1/2}$ and looking for spherically symmetric solutions, we get

$$u_r = v, \quad v_r = \frac{1-m}{r} v - f(u).$$

This ODE has constant divergence (*i.e.*, independent of u and v) equal to $n(1-m)/r$, *i.e.*, its flow is conformal volume-preserving (with a conformal constant depending on the independent variable r). If $f(u) = \nabla H(u)$, *i.e.*, if the Poisson equation is variational, then the reduced ODE is conformal Hamiltonian.

Example 33. (Stationary NLS equation) The nonlinear Schrödinger equation is given by

$$i\psi_t + \nabla^2 \psi + c|\psi|^2 \psi = 0.$$

Stationary spherically symmetric solutions satisfy

$$\psi_{rr} + \frac{m-1}{r} \psi_r + c|\psi_r|^2 \psi = 0.$$

Defining $q_1 + \mathrm{i} q_2 := \psi$ and $p_1 + \mathrm{i} p_2 := \psi_r$, we obtain

$$q_r = p, \quad p_r = \frac{1-m}{r} p - \frac{1}{2} \nabla((q_1^2 + q_2^2)^2).$$

This ODE is conformal Hamiltonian and can be integrated using the splitting method of Section 3.6.

4. Composition

As only a small fraction of the possibilities to apply the product formula philosophy have been explored, there is much room for further research in this field. I hope that this will encourage the reader to apply the symmetric product formula approach to solve other problems or develop new and more efficient algorithms than the ones proposed in this paper. (De Raedt 1987)

4.1. General theory

The fundamental basis of composition methods is the following. Note that the two methods φ, ψ need not be flows, and can be completely unrelated.

Theorem 18. Let φ be a consistent integrator for X_1, let ψ be a consistent integrator for X_2, and let X_1 be Lipschitz-continuous. Then $\varphi \circ \psi$ is a consistent integrator for $X_1 + X_2$.

The composition of flows (1.1) is only first-order. The order can be increased by including more exponentials in a time step. For a splitting into two parts, $X = A + B$, we have the general nonsymmetric composition

$$\mathrm{e}^{a_m \tau A} \mathrm{e}^{b_m \tau B} \dots \mathrm{e}^{a_1 \tau A} \mathrm{e}^{b_1 \tau B} \mathrm{e}^{a_0 \tau A}. \tag{4.1}$$

By convention, we only count the evaluations of the flow of B, and refer to (4.1) as an m-stage method. The number of stages and the coefficients a_i and b_i are to be chosen to ensure that the method has some order p, that is,

$$\varphi = \exp(\tau(A + B)) + \mathcal{O}(\tau^{p+1}).$$

At least four approaches have been proposed to determine order conditions for the coefficients of methods of high order. The first, very simple, method, works only for a special class of compositions, so it does not always generate the best method of a given order. The other three produce the general order conditions, large systems of polynomials which have to be studied in detail to select methods. They are either reduced and/or solved symbolically if m is small enough, or solved numerically.

(1) The direct method of Suzuki (1990) and Yoshida (1990), which easily produces methods of any even order.

(2) Expansion of (4.1) using the BCH formula (4.7), which gives the order conditions for an m-stage method recursively in terms of those for an $(m-1)$-stage method.

(3) An extension of the theory of rooted trees used in Runge–Kutta theory to composition methods (Murua and Sanz-Serna 1999), which gives the order conditions explicitly.

(4) A method based on time-ordered symmetrized products of noncommuting operators (Tsuboi and Suzuki 1995), which also gives the order conditions explicitly.

We shall present methods (1) and (2) and their extensions.

We start with two facts. First, any map sufficiently close to the identity is close to the flow of some vector field. Specifically, we have the following theorem.

Theorem 19. (Modified equations, backward error; Reich (1999)) Let $\mathfrak{G}$ be a set (*e.g.*, a group) of diffeomorphisms which has a tangent at the identity given by a linear space $\mathfrak{X}$ of vector fields. Let $\varphi(\tau)$ be a curve in $\mathfrak{G}$, analytic in τ, satisfying $\varphi(0) = 1$, the identity. Then there exist vector fields $X_1, X_2, X_3, \dots \in \mathfrak{X}$ such that

$$\varphi(\tau) = \exp\left(\sum_{n=1}^{N-1} \tau^n X_n\right) + \mathcal{O}(\tau^N)$$

for all $N > 1$. (The error can be taken in any coordinate chart.)

Note that $\varphi(\tau)$ is an integrator of X_1 of order p, where $p \geq 1$ is the least integer such that $X_{p+1} \neq 0$.

Second, let $\mathfrak{X}$ be a Lie algebra of vector fields and let $X, Y \in \mathfrak{X}$. Then

$$\mathrm{e}^X \mathrm{e}^Y = \mathrm{e}^{X+Y+o(X,Y)}. \tag{4.2}$$

(The first term in the remainder is $\frac{1}{2}[X,Y]$, but we shall not need this term until the next section.)

Definition 5. The method $\varphi(\tau)$ is *symmetric* or *self-adjoint* if

$$\varphi(\tau)\varphi(-\tau) = 1$$

for all τ.

Then we have the following.

Theorem 20. If $\varphi(\tau)$ is symmetric, then $X_{2i} = 0$ for all i, and $\varphi(\tau)$ necessarily has even order.

It is easy to find symmetric methods, as follows.

Theorem 21. If $\varphi(\tau)$ is any method of order p, then $\varphi(\frac{1}{2}\tau)\varphi^{-1}(-\frac{1}{2}\tau)$ is symmetric and of order at least p (if p is even) or at least $p+1$ (if p is odd).

Applied to the basic composition (1.1), Theorem 21 leads to the symmetrized composition of order 2,

$$\mathrm{e}^{\frac{1}{2}\tau X_1} \dots e^{\frac{1}{2}\tau X_n} \mathrm{e}^{\frac{1}{2}\tau X_n} \dots e^{\frac{1}{2}\tau X_1}, \tag{4.3}$$

which is widely used in many applications: for many purposes it is the most sophisticated method needed. From the flow property $\mathrm{e}^{\tau X}\mathrm{e}^{\sigma X} = \mathrm{e}^{(\tau+\sigma)X}$, the two central stages coalesce, and the last stage coalesces with the first stage of the next time step. When output is not required every time step, the method (4.3) therefore involves evaluating $2n-2$ flows, or $2-2/n$ as much as the first-order method (1.1). This shows the great advantage in searching for splittings with a small number (say 2 or 3) parts.

Applied to the general nonsymmetric composition (4.1), Theorem 21 generates symmetric methods which we denote φ_S.

Theorem 22. Let $\varphi(\tau)$ be a symmetric method of order $2k > 0$. Then the method

$$\varphi(\alpha\tau)^n \varphi(\beta\tau)^m \varphi(\alpha\tau)^n \tag{4.4}$$

is symmetric. It has order $2k+2$ provided

$$2n\alpha + m\beta = 1,$$
$$2n\alpha^{2k+1} + m\beta^{2k+1} = 0.$$

These equations have a unique real solution for all n, m, and k, namely

$$\alpha = \left(2n - (2nm^{2k})^{1/(2k+1)}\right)^{-1}, \quad \beta = (1-2n\alpha)/m. \tag{4.5}$$

Proof. By Theorem 19, we have $\varphi(\tau) = \exp(\tau X_1 + \tau^{2k+1}X_{2k+1} + \mathcal{O}(\tau^{2k+3}))$. So $\varphi(\alpha\tau) = \exp(\alpha\tau X_1 + \alpha^{2k+1}\tau^{2k+1}X_{2k+1} + \mathcal{O}(\tau^{2k+3}))$. Equations (4.5) then follow from (4.2). □

Counting the basic second-order symmetric method as one stage, with $m = n = 1$ this approach uses 3 stages for a 4th-order method, 9 stages for 6th-order, and so on, or $3^{p/2-1} = \mathcal{O}(\sqrt{3}^p)$ stages for order p.

It has been found that it is always best to take $m = 1$, but the best choice of n is not so clear. The shortest methods, with $n = 1$, in fact have notably large error constants. One recent study (McLachlan 1995) found that, for order 4 methods, the work to achieve a given error actually decreased with n right up to $n = 19$. Without going into details, one can say that the order 4 methods with $m = 1$ and $n = 2$ (5 stages) or $n = 3$ (7 stages) are good.

It will be observed from (4.5) that $\beta < 0$ for all n, m, and k. The methods always involve stepping backwards in time. We shall see below that this is unavoidable. However, for geometric integrators in groups, it is of course not a problem. In semigroups, as arise, for instance, in dimensional splitting of diffusion equations, it is a problem. There, splitting was proposed as a cheap way to retain unconditional stability. Methods with backwards time steps can only be conditionally stable; this stumbling block held up the development of high-order compositions for years.

To get methods with fewer stages, Yoshida (1990) and Suzuki (1990) proposed the composition

$$\varphi(a_1\tau)\ldots\varphi(a_m\tau)\ldots\varphi(a_1\tau), \tag{4.6}$$

where $\varphi(\tau)$ is any symmetric method. (We call (4.6) *type SS*, symmetric with symmetric stages.) To analyse such methods, the approximation (4.2) is no longer enough, as we need to keep track of all higher-order terms in the expansion, up to the order of the method. This is provided by the Baker–Campbell–Hausdorff (BCH) formula.

Theorem 23. (BCH) Let $\mathfrak{X}$ be a Lie algebra of vector fields and let $X, Y \in \mathfrak{X}$. Then

$$\mathrm{e}^X\mathrm{e}^Y = \mathrm{e}^Z,$$

where Z is given by the following series, which is asymptotic as X, $Y \to 0$:

$$\begin{aligned} Z &= \sum_{n=1}^{\infty} Z_n = X + Y + \tfrac{1}{2}[X,Y] + \tfrac{1}{12}([X,X,Y] \\ &\quad + [Y,Y,X]) + \tfrac{1}{24}[X,Y,Y,X] + \cdots, \\ Z_1 &= X + Y, \\ (n+1)Z_{n+1} &= \tfrac{1}{2}[X - Y, Z_n] \\ &\quad + \sum_{p=1}^{\lfloor n/2 \rfloor} \frac{B_{2p}}{(2p)!} \sum_{\substack{k_1,\dots,k_{2p} \\ k_i \ge 1,\ \sum k_i = n}} [Z_{k_1}, \dots, Z_{k_{2p}}, X + Y], \quad n \ge 1, \end{aligned} \tag{4.7}$$

where B_j is the jth Bernoulli number and we have defined $[X,Y,Z] := [X,[Y,Z]]$.

That is, the composition of two flows is itself the flow of the vector field Z, which lies in the same Lie algebra as X and Y (*e.g.*, Hamiltonian, divergence-free) and is a linear combination of X, Y and all their iterated Lie brackets. Note that $Z_n \in L_n(X,Y)$, the linear span of all Lie brackets of order n of X and Y. Let

$$c_n := \dim L_n(X,Y).$$

Let us first consider applying the BCH formula to the general composition (4.1). Taking $X = a_i\tau A$, $Y = b_i\tau B$, an element of $L_n(X,Y)$ is $\mathcal{O}(\tau^n)$. Therefore, applying the BCH formula repeatedly to (4.1) gives

$$\varphi = \exp\left(\sum_{n=0}^{\infty} k_n \tau^n L_n(A,B)\right),$$

where the coefficients $k_n(a_1,\dots,b_1,\dots) \in \mathbb{R}^{c_n}$ are c_n polynomials in the variables a_i and b_i. The conditions $k_1 = (1,1)$ (so that the first term is $\tau(A+B)$ as required) and $k_n = 0$ for all $n \le p$ are then sufficient for the method to have order p. In practice, for large p the order conditions are usually calculated symbolically, for example, using the Matlab package DiffMan found at `www.math.ntnu.no/num/diffman`.

For example, one can check directly that

$$e^{\frac{1}{2}\tau A} e^{\tau B} e^{\frac{1}{2}\tau A} = e^{\tau(A+B) + \frac{1}{12}\tau^3[B,B,A] - \frac{1}{24}\tau^3[A,A,B] + \mathcal{O}(\tau^5)}, \tag{4.8}$$

showing that this method is order 2, because the error term in $\tau^2[A,B]$ vanishes. When applied to Hamiltonian systems with $H = \frac{1}{2}p^2 + V(q)$ and choosing $A = X_{\frac{1}{2}p^2}$, $B = X_{V(q)}$ as in Example 1, we get the method

$$Q = q_n + \tfrac{1}{2}\tau p_n,$$
$$p_{n+1} = p_n - \tau \nabla V(Q),$$
$$q_{n+1} = Q + \tfrac{1}{2}\tau p_{n+1},$$

which for the q variables is equivalent to the Delambre–Verlet method (1.3). By extension, we refer to the method (4.8) as leapfrog regardless of A and B.

Similarly, when composing an arbitrary method $\varphi(\tau) = \exp(\tau X_1 + \tau^2 X_2 + \tau^3 X_3 + \cdots)$, we will be working in the Lie algebra $L(X_1, X_2, X_3 \ldots)$, and when composing a symmetric method (type SS, (4.6)) in the Lie algebra $\exp(\tau X_1 + \tau^3 X_3 + \tau^5 X_5 + \cdots)$. Before giving specific integration methods in Section 4.9, we will detour to study these Lie algebras a little.

4.2. Equivalence of methods

It is also possible to consider, following the previous section, three more general compositions, not of flows as in (4.1), but of arbitrary methods. First, we have a nonsymmetric composition of an arbitrary method $\varphi(\tau)$ and its inverse, such as

$$\psi_{\rm NS} := \prod_{i=1}^{m} \varphi^{-1}(-c_i\tau)\varphi(d_i\tau). \qquad (4.9)$$

Expanding $\varphi(\tau) = \exp(X_1 + X_2 + X_3 + \cdots)$, where $X_i = \mathcal{O}(\tau^i)$, we see that $\psi_{\rm NS}$ has an expansion in $L(X_1, X_2, X_3, \ldots)$. For example, there will be 2 order conditions at order 3, corresponding to the coefficients of X_3 and $[X_1, X_2]$. Using Theorem 25 below, one can show that there are c_n order conditions at order $n > 1$, just as for the composition (4.1).

Second, a symmetric composition $\psi_{\rm S}$ with $c_i = d_{n+1-i}$, $i = 1, \ldots, n$. There are still c_n order conditions at order n, but only the odd order conditions need to be enforced, from Theorem 20.

Third, a symmetric composition $\psi_{\rm SS}$ of symmetric methods, (4.6), or (4.9) with $c_i = c_{n+1-i} = d_i = d_{n+1-i}$, $i = 1, \ldots, n$. Expanding $\varphi(\tau) = \exp(X_1 + X_3 + X_5 + \cdots)$, we see that φ(SS has an expansion in $L(X_1, X_3, X_5, \ldots)$. (There is now only one order condition at order 3.)

Recall also the general composition

$$\varphi_{\rm NS} = e^{a_m\tau A} e^{b_m\tau B} \ldots e^{a_1\tau A} e^{b_1\tau B} e^{a_0\tau A}, \qquad (4.10)$$

the symmetric composition

$$\varphi_{\rm S} = e^{a_0\tau A} e^{b_1\tau B} \ldots e^{a_1\tau A} e^{b_1\tau B} e^{a_0\tau A}, \qquad (4.11)$$

(*i.e.*, $a_i = a_{n-i}$, $b_i = b_{n+1-i}$), and $\varphi_{\rm SS}$, the symmetric composition of leapfrog stages.

However, these classes of methods are in fact equivalent.

Theorem 24. (McLachlan 1995, Koseleff 1995) If, for $i = 1, \ldots, m$, we have

$$d_i + c_i = b_i, \quad d_{i+1} + c_i = a_i$$

(setting $d_{m+1} = c_0 = 0$), then the methods ψ_{NS}, ψ_{S}, and ψ_{SS} order $p \geq 1$ if and only if the methods φ_{NS}, φ_{S}, and φ_{SS}, respectively, have order p.

Proof. One direction is trivial: a method of order p of type $\psi_{\mathrm{NS,S,SS}}$, immediately gives a method of order $\geq p$ of type $\varphi_{\mathrm{NS,S,SS}}$, on taking $\psi(\tau) = \mathrm{e}^{\tau B}\mathrm{e}^{\tau A}$. The other direction for the nonsymmetric case NS is proved in McLachlan (1995); we illustrate it here for the conditions at order 2 and 3. Consider finding the order conditions for (4.10) by expressing (4.10) as (4.9) with $\varphi(\tau) = \mathrm{e}^{\tau B}\mathrm{e}^{\tau A}$ (so that the coefficients a_i, b_i, c_i, and d_i are related as in the hypothesis of the theorem). Then $\varphi(\tau) = \exp(\tau X_1 + \tau^2 X_2 + \tau^3 X_3 + \cdots)$ with $X_1 = A + B$, $X_2 = \frac{1}{2}[B, A]$, $X_3 = \frac{1}{12}([B, B, A] + [A, A, B])$ and so on, from the BCH formula (4.7). At order 2, the coefficient of $\frac{1}{2}[B, A]$ in (4.10) is then the same as the coefficient of X_2 in (4.9). At order 3, observe that

$$[X_1, X_2] = \tfrac{1}{2}([B, B, A] - [A, A, B]).$$

Thus, the two order conditions $p_1 = p_2 = 0$ for (4.9) are related to the two order conditions $q_1 = q_2 = 0$ for (4.10) by

$$p_1 = \tfrac{1}{2}(q_1 - q_2), \quad p_2 = \tfrac{1}{12}(q_1 + q_2).$$

So $p_1 = p_2 = 0$ if and only if $q_1 = q_2 = 0$. A similar pattern follows at each order, since from Theorem 25, $\dim L_n(A, B) = \dim L_n(X_1, X_2, X_3, \ldots)$ for $n > 1$. The S and SS cases follow from the symmetries of the coefficients of the methods. □

4.3. Counting order conditions

We have seen that, for methods formed from compositions of flows of the vector fields X_i, there is one order condition for each linearly independent Lie bracket of the X_i. Defining the orders $w(X)$ by $X = \mathcal{O}(\tau^{w(X)})$, we have $w([X, Y]) = w(X) + w(Y)$: the powers of τ add when forming Lie brackets. (The function w is called a *grading* of $L(X_1, X_2, \ldots)$.) Let $L_n(X_1, X_2, \ldots)$ be the linear span of all Lie brackets of the X_i of order n. The dimension of this space (*i.e.*, the number of order conditions of order n), is provided by the following theorem.

Theorem 25. (Munthe-Kaas and Owren 1999, Kang and Kim 1996) Let $p(T) = 1 - \sum_i T^{w(X_i)}$ and let $\log(p(T)) = \sum_{n=0}^{\infty} a_n T^n$. Then

$$\dim L_n(X_1, X_2, \ldots) = \sum_{d|n} \mu(d) a_{n/d},$$

where $\mu(d)$ is the Möbius function $\mu(1) = 1$, $\mu(d) = (-1)^q$ if d is the product of q distinct prime factors, and $\mu(d) = 0$ otherwise. Thus, if $p(T)$ converges to a rational function $q(T)/r(T)$, then

$$\dim L_n(X_1, X_2, \dots) = \frac{1}{n} \sum_{d|n} \mu(d) \left(\left(\sum_j \lambda_j^{-n/d} \right) - \left(\sum_k \rho_k^{-n/d} \right) \right),$$

where the λ_j are the roots of $q(T)$ and the ρ_k are the roots of $r(T)$.

We will consider various cases. First, for compositions of the flows of two vector fields A and B, we take $X_1 = \tau A$ (order 1) and $X_2 = \tau B$ (order 1), to get $p(T) = 1 - 2T$ and $\lambda = \frac{1}{2}$, so

$$c_n := \dim L_n(A, B) = \frac{1}{n} \sum_{d|n} \mu(d) 2^{n/d} = \mathcal{O}\left(\frac{2^n}{n}\right),$$

which is known as the Witt formula, one of many similar formulae counting polynomials, partitions, trees, exterior products, *etc.* The first 10 values of c_n are given in Table 4.1.

For order p type NS methods (4.10), this gives a total of $\sum_{n=1}^{p} c_n = \mathcal{O}(2^p/p)$ order conditions. For type S methods (4.11), the even order conditions are automatically satisfied (Theorem 20), so the number is reduced to $\sum_{k=1}^{p/2} c_{2k-1}$: still $\mathcal{O}(2^p/p)$, but the largest single set of c_p conditions is avoided.

For type SS methods (4.6), we compose flows of $\varphi(\tau) = \exp(X_1 + X_3 + X_5 + \cdots)$, where $w(X_i) = i$. For example, a basis for the 4-dimensional space of brackets of order 7 is $\{X_7, [X_1, X_1, X_5], [X_1, X_1, X_1, X_1, X_3], [X_3, X_1, X_3]\}$. Applying Theorem 25, $p(T) = 1 - T - T^3 - T^5 - \cdots = 1 - T/(1 - T^2) = (1 - T - T^2)/(1 - T^2)$, so $\lambda_1 = (\sqrt{5} + 1)/2$, $\lambda_2 = (\sqrt{5} - 1)/2$, $\rho_1 = 1$, $\rho_2 = -1$. Crucially, $\dim L_n(X_1, X_3, \dots) = \mathcal{O}(\lambda_1^n/n)$, $\lambda_1 \approx 1.618$, and the asymptotic rate of growth has been reduced – compare the first 10 values given in Table 4.1.

Recall that the Yoshida–Suzuki methods (4.4) have $\mathcal{O}(\sqrt{3}^p)$ stages. Thus, type SS methods are sure to beat them for sufficiently high orders p.

However, one can do even better. Consider compositions of symmetric 4th order methods. Then we require $\dim L_n(X_1, X_5, X_7, \dots)$, for which $p(T) = 1 - T^5 - T^7 - \cdots = (1 - T - T^2 + T^3 - T^5)/(1 - T^2)$ and $\lambda_1 \approx 1.4433$. The number of stages for an order p method of this type is $\mathcal{O}(\lambda_1^p/p)$. This is asymptotically smaller than type SS methods, although the break-even value of $p = 12$ is rather large.

Contrast the present situation with that of Runge–Kutta methods. They have one order condition for each elementary differential (f, $f'(f)$, $f''(f, f)$, $f'(f'(f)), \dots$) of order n. The number of these grows extremely quickly: $\mathcal{O}(n^{-3/2}\lambda^n)$ where $\lambda \approx 2.995$ is Otter's tree enumeration constant. However,

Table 4.1. Number of order conditions.
(a) Number of conditions arising at each order: $c_n = \dim L_n(A, B)$, $d_n = \dim L_n(X_1, X_3, X_5, \dots)$, and $e_n = \dim L_n^{\mathrm{RKN}}(A, B)$

Order n	c_n (type NS)	d_n (type SS)	e_n (type RKN)
1	2	1	2
2	1	0	1
3	2	1	2
4	3	1	2
5	6	2	4
6	9	2	5
7	18	4	10
8	30	5	14
9	56	8	25
10	99	11	39

(b) Total number of order conditions for methods of order 4, 6, and 8, with and without using correctors. For example, for SS methods with correctors, the 5 order conditions are the coefficients of X_1, X_3, X_5, X_7, and $[X_3, X_1, X_3]$ in the composition (4.6)

Type of method	Order 4	Order 6	Order 8
NS	8	23	71
S	4	10	28
SRKN	4	8	18
SS	2	4	8
NS with corrector	4	10	31
S with corrector	3	6	15
SRKN with corrector	3	5	10
SS with corrector	2	3	5

the structure of the order conditions is such that they can be satisfied with many fewer stages than this, namely $\mathcal{O}(p)$ for implicit methods of order p and $\mathcal{O}(p^2)$ for explicit methods. Yet, by composing Euler and backward Euler stages, they contain within them the order conditions for type NS, S, and SS methods.

Choosing the actual number of stages to use is still something of an experimental art. Consider m-stage methods of order 6. Type SS methods have 4 order conditions and $(m+1)/2$ parameters, suggesting $m \geq 7$; type S methods have 10 order conditions and $m+1$ parameters, suggesting $m \geq 9$; while type NS methods have 23 order conditions and $2m+1$ parameters, suggesting $m \geq 11$. However, suppose we decide to study 15-stage methods. Then SS methods have 4 free parameters, S methods have 6 free parameters, and NS have 8 free parameters. True, the error in an SS method has fewer independent components than the error in an S method, but this may not be relevant when trying to minimize some norm of the error. Choosing small m will favour type SS, while choosing larger m (which seems to lead to better methods) will favour type S. On the other hand, there is an established resistance from users, who may be used to using leapfrog with $m = 1$, against methods with large m.

4.4. Stability

It might be expected that splitting methods, being nominally explicit, have only modest stability. This is not necessarily the case.

(1) Because they are 1-step methods, they are automatically 0-stable (*i.e.*, stable for $\dot{x} = 0$), unlike multistep methods. Moreover, unlike Runge–Kutta methods, they do not require storage of extra values of X, which is a great memory advantage when solving high-dimensional PDEs.

(2) If $\mathfrak{G}$ is compact, then all $\mathfrak{G}$-integrators are unconditionally stable: you can not go to infinity in a compact space. The most important examples are unitary and orthogonal integrators (see Section 5.3), for which $\mathfrak{G}$ is a finite-dimensional compact Lie group.

(3) If the phase space is compact, all $\mathfrak{G}$-integrators are unconditionally stable, *e.g.*, integrators preserving the integral $\|x\|^2$.

(4) If $\mathfrak{G}$ is noncompact, then most $\mathfrak{G}$ integrators are only conditionally stable. One chooses a simple test problem in $\mathfrak{G}$ with bounded solutions and computes the stability limit of various methods. For Hamiltonian systems, this is usually the harmonic oscillator with $T(p) + V(q)$ splitting. Leapfrog is stable for $\tau < 2$ and higher-order m-stage integrators are stable for $\tau < \tau^*$ where usually $\tau^* \approx \pi$. Special compositions can be found which are more stable. Typically τ^* decreases with increasing order, while τ^*/m increases slightly with m.

(5) If M is a Banach space and the vector fields X_i are linear with

$$\|\exp(tX_i))\| \leq 1, \quad \text{for all } t \geq 0,$$

then composition of flows of such X_i (or their stable approximations) *with positive time steps* is unconditionally stable, for

$$\left\| \prod \exp(\tau_i X_i) \right\| \leq \prod \|\exp(\tau_i X_i)\| \leq 1.$$

The main application is to dimensional splitting in parabolic PDEs such as the heat equation (Strang 1968). However, this limits the order to 2, as Theorem 26 below shows.

(6) Again, if M is a Banach space and φ, ψ are linear operators on M with $\|\varphi\| \leq 1$, $\|\psi\| \leq 1$, then (De Raedt 1987)

$$\begin{aligned} \|\varphi^n - \psi^n\| &= \left\| \sum_{i=0}^{n-1} \varphi^i(\varphi - \psi)\psi^{n-1-i} \right\| \\ &\leq \|\varphi - \psi\| \sum_{i=0}^{n-1} \|\varphi\|^i \|\psi\|^{n-1-i} \\ &\leq n\|\varphi - \psi\|. \end{aligned}$$

That is, errors grow at most linearly in time. This applies for all compositions if $\mathfrak{G}$ is, *e.g.*, the linear action of $\mathrm{U}(n)$ on $\mathbb{C}^n$ (see Section 5.3), or for compositions with positive time steps for, *e.g.*, the heat equation.

(7) Some nonlinear stability can follow merely from the group property, such as preservation of KAM tori in Hamiltonian, volume-preserving, or reversible systems (Broer *et al.* 1996, Shang 2000), preservation of weak integrals which may partition phase space (Example 13), or the nonlinear stability of fixed points of Hamiltonian systems (Skeel and Srinivas 2000).

(8) The modified Hamiltonian of symplectic integrators can confer nonlinear stability. In particular, for splitting methods, the critical points of H do not move or change their value under the perturbation due to the integrator (McLachlan, Perlmutter and Quispel 2001).

(9) Paradoxically, in spite of Theorem 26 below, it can be shown that composition methods can be used to create A_0-stable methods of order 6 and higher (Iserles and Quispel 2002).

(10) Sometimes, a loss of stability is due to a resonance between different linear modes. For example, the time-τ flow of the harmonic oscillator has eigenvalues $\mathrm{e}^{\pm i\tau}$ which meet at $\tau = \pi$. It is possible to design special methods for such resonances, for instance, between fast and slow modes, that do not cause instability (Leimkuhler and Reich 2001).

Theorem 26. (Negative time steps; Sheng (1989)) Let A and B be square matrices. There are no real solutions of the order conditions for the method

$$\sum_{k=1}^{K} \gamma_k \prod_{i=1}^{n} \exp(a_{ik}\tau A)\exp(b_{ik}\tau B)$$

to have order 3 for the ODE $\dot{x} = (A+B)x$ with

$$\gamma_k \geq 0, \quad a_{ik} \geq 0, \quad b_{ik} \geq 0, \quad \text{for all } i \text{ and } k.$$

Although proved for linear systems, the same proof works for for nonlinear ODEs and (by taking $K = 1$) geometric composition methods.

(Note, however, that there can be *complex* solutions with positive real parts. For example, consider the 3-stage 4th-order method (4.4) with $n = m = 1$ and $\alpha = 1/(2 - 2^{1/3}\mathrm{e}^{2\pi \mathrm{i}/3}) \approx 0.3244 + 0.1346\mathrm{i}$, $\beta = 1 - 2\alpha \approx 0.3512 - 0.2692\mathrm{i}$. For linear problems $\dot{x} = Ax + Bx$, $x \in \mathbb{C}^n$, and $\varphi(\tau)$ given by leapfrog, this composition is unconditionally stable provided multiplying by the time steps does not push the eigenvalues of A and B into the right half plane, *e.g.*, if A, B are negative definite. The complex heat equation is easier to integrate than the real heat equation!)

4.5. Correctors

Because of the large number of order conditions, even for type SS methods, various special cases have been considered in order to find better methods. Great progress has been made in a series of studies over the past decade to find better methods for modest orders, say $p = 4$, 6, and 8.

The first special case we consider is the use of a 'corrector' (also known as processing or effective order), introduced by Butcher (1969) for Runge–Kutta methods, by Takahashi and Imada (1984) for compositions of exponentials, and developed greatly for symplectic integrators for solar system dynamics by Wisdom, Holman and Touma (1996). Suppose the method φ can be factored as

$$\varphi = \chi\psi\chi^{-1}.$$

Then, to evaluate n time steps, we have $\varphi^n = \chi\psi^n\chi^{-1}$, so only the cost of ψ is relevant. The maps φ and ψ are conjugate by the map χ, which can be regarded as a change of coordinates. Many dynamical properties of interest (to a theoretical physicist, *all* properties of interest) are invariant under changes of coordinates; in this case we can even omit the χ steps entirely and simply use the method ψ. For example, calculations of Lyapunov exponents, phase space averages, partition functions (Section 5.3), existence and periods of periodic orbits, *etc.*, fall into this class. If the location of individual orbits is important, one still does not need to know χ exactly, but can merely approximate it (López-Marcos, Sanz-Serna and Skeel 1996).

The simplest example of a corrector is the following:

$$e^{\tau A}e^{\tau B} = e^{\frac{1}{2}\tau A}\left(e^{\frac{1}{2}\tau A}e^{\tau B}e^{\frac{1}{2}\tau A}\right)e^{-\frac{1}{2}\tau A},$$

showing that the first-order method (1.1) is conjugate to a second-order symmetric method, namely leapfrog, when X is split into $n = 2$ pieces. Thus, this first-order method has all sorts of serendipitous properties not shared by general first-order methods.

To derive the order conditions in general, we represent the kernel $\psi = e^K$ and corrector $\chi = e^C$ and write

$$\begin{aligned} e^{\tau X} &= e^C e^K e^{-C} \\ &= e^{K+[C,K]+\frac{1}{2!}[C,C,K]+\frac{1}{3!}[C,C,C,K]+\cdots} \end{aligned}$$

which allows one to determine the conditions on C and K first, and then to construct specific C, K that satisfy these conditions. Similarly,

$$\begin{aligned} e^K &= e^{-C}e^{\tau X}e^C \\ &= e^{\tau(X-[C,X]+\frac{1}{2!}[C,C,X]-\frac{1}{3!}[C,C,C,X]-\cdots)}, \end{aligned}$$

which shows that only those terms in the error which are Lie brackets of X (either $X = A+B$ for a method (4.1), or X_1 for a method (4.9)) can possibly be eliminated by a corrector. The second form separates the conditions in C from those in K.

Since all of the terms from $L_{n-1}(A, B)$ are available in C to correct the terms from $L_n(A, B)$ in K, one expects a telescoping sum in counting the number of order conditions on K. This in fact occurs, as the following result shows. The number of order conditions is greatly reduced (although it still has the same asymptotic growth). For symmetric kernels, to avoid introducing even terms into the expansion of K it is necessary to use a corrector which satisfies $\chi(-\tau) = \chi(\tau) + \mathcal{O}(\tau^p)$, which is achieved by iterating $\chi^{(k+1)}(\tau) = \chi^{(k)}(\tau)\chi^{(k)}(-\tau)$.

Theorem 27. (Blanes 2001, Blanes, Casas and Ros 2000*a*) Let $c_n = \dim L_n(A, B)$ and $d_n = \dim L_n(X_1, X_3, X_5, \ldots)$, and define $c_0 = d_0 = 0$. Then the number of order conditions for a type S method of order p, and for a type S or SS method of order $2k$, are as follows.

Type of method	Uncorrected	Corrected
NS	$\sum_{n=1}^{p} c_n$	$c_p + 1$ for $p \geq 2$
S	$\sum_{n=0}^{k-1} c_{2n+1}$	$\sum_{n=0}^{k-1}(c_{2n+1} - c_{2n})$
SS	$\sum_{n=0}^{k-1} d_{2n+1}$	$\sum_{n=0}^{k-1}(d_{2n+1} - d_{2n})$

The total number of order conditions is given in Table 4.1. While the type NS methods look appealing, in that all errors of order $1 < n < p$ can be corrected, it has been found that there are no real solutions of the order conditions with the minimum number of stages.

The error can be substantially reduced by the use of a corrector. However, it can be argued that the part of the error removed by correction was not serious anyway (but at least one should not waste parameters on correctable errors). Therefore, a fair comparison is between optimal correction of a standard method, such as one of those given in the previous section, and optimal correction of an arbitrary method. For example, for type SS methods we have the following result.

Theorem 28. (McLachlan 2002) The Suzuki methods (4.4) of order 4 with $m = 1$ locally minimize the uncorrectable 5th-order error amongst all $(2n+1)$-stage type SS methods.

Some good methods with correctors are given in Section 4.9. The advantages of considering a corrector are even greater in the Runge–Kutta–Nyström and nearly integrable cases, considered below.

4.6. Runge–Kutta–Nyström methods

Let us consider the case of simple mechanical systems, Hamiltonian systems with $H = A + B$ where the kinetic energy $A = A(q,p)$ is quadratic in p and the potential energy $B = B(q)$. (We now work with the Lie algebra of Hamiltonian functions under the Poisson bracket rather than the Lie algebra of Hamiltonian vector fields; the two are isomorphic.) Then each Lie bracket of A and B is homogeneous in p. Let $\deg X$ be the degree of X in p. Then for $\deg X$, $\deg Y$ not both zero, we have

$$\deg[X,Y] = \deg X + \deg Y - 1,$$

while, if $\deg X = \deg Y = 0$ (*i.e.*, if X and Y are both functions of q only), we have

$$[X,Y] = 0$$

and the order conditions corresponding to such a bracket can be dropped. This is a natural generalization of Runge–Kutta–Nyström methods for $\ddot{q} = f(q)$ (in which case $A = \frac{1}{2}p^2$). Let us call the Lie algebra generated by such an A and B, with no further conditions, $L^{\mathrm{RKN}}(A,B)$.

The first bracket that can be dropped is $[B,B,B,A] = 0$. It can be shown that (McLachlan 1995)

$$L^{\mathrm{RKN}}(A,B) \subseteq B + L(A,[B,A],[B,B,A])$$

so that, from Theorem 25,

$$\dim L_n^{\mathrm{RKN}}(A,B) \le \frac{1}{n}\sum_{d|n}\sum_{i=1}^{3}\lambda_i^{-n/d},$$

where λ_i are the 3 roots of $p(T) = 1 - T - T^2 - T^3 = 0$. The only root outside the unit circle is ≈ 1.84. Thus

$$\dim L_n^{\mathrm{RKN}}(A,B) \le \frac{1}{n}1.84^n.$$

So the complexity of these methods is less than type NS or S methods, but still more than type SS. (The RKN case does not simplify type SS methods any further.) The actual dimensions for $n \le 10$ are given in Table 4.1.

Blanes and Moan (2000) have made a detailed study of the order conditions for this case, and find 4th-order methods with error constants 1.32, 0.64, 0.36, and 0.29 for methods with 3, 4, 5, and 6 stages respectively, and 6th-order methods with error constants 1.02, 0.78, and 0.63 with 7, 11, and 14 stages respectively. (See Section 4.9, methods 3(a), (b).)

It should be emphasized that these optimized methods are really very accurate. For example, on the Hénon–Heiles system the best 4th-order 6-stage method beats leapfrog (at constant work) right up to the former's stability limit of $\tau \approx 4.2$. Even for short times these methods are good: integrating Hénon–Heiles for time 1 from initial conditions $(1,1,1,1)$, this 4th-order method has a global error about 0.00175 times that of classical 4th-order Runge–Kutta: that is, it costs $\frac{6}{4}0.00175^{1/4} = 0.31$ as much for a given error. This should be compared to the earliest 4th-order symplectic integrators, (4.4) with $n = m = 1$, which have truncation errors 10 times *larger* than classical Runge–Kutta.

By applying a corrector, one can do better again (Blanes, Casas and Ros 2000*a*).

RKN splitting methods can also be used for non-Hamiltonian systems of the form

$$\begin{aligned}\dot{x} &= f(x) + L(x)y,\\ \dot{y} &= g(x),\end{aligned}$$

where $x \in \mathbb{R}^m$, $y \in \mathbb{R}^n$, and $L(x) \in \mathbb{R}^{m\times n}$. For example, high-order systems $z^{(k)} = f(z, z', \ldots, z^{(k-1)})$ have this form when written as the (divergence-free) first-order system

$$\begin{aligned}\dot{x}_i &= x_{i+1}, \quad i = 1,\ldots,k-2,\\ \dot{x}_{k-1} &= f(x_1,\ldots,x_{k-2}),\end{aligned}$$

where $x_i = z^{(i)}$.

4.7. Nearly integrable systems

Consider the family of systems $\dot{x} = A + \varepsilon B$ involving a small parameter ε, where A and B are integrable. A composition of flows of A and B has an error expansion in $L_n(\tau A, \tau\varepsilon B)$. The error term involving n As and m Bs is $\mathcal{O}(\tau^{n+m}\varepsilon^m)$. In particular, $m \geq 1$ so the error is at most $\mathcal{O}(\varepsilon)$ and vanishes with ε. Splitting is superb for such nearly integrable systems.

But, one can do even better. Typically, $\varepsilon \ll \tau$ and one can preferentially eliminate error terms with small powers of ε. The number of these terms is only polynomial in n, instead of exponential. Thus, we can beat the large cost of high-order methods in this case. For example, there is only 1 error term of each order $\mathcal{O}(\varepsilon\tau^n)$ (namely $\varepsilon\tau^n[A, \ldots, A, B]$), $\lfloor \frac{1}{2}(n-1) \rfloor$ of order $\mathcal{O}(\varepsilon^2\tau^n)$, and $\lfloor \frac{1}{6}(n-1)(n-2) \rfloor$ terms of order $\mathcal{O}(\varepsilon^3\tau^n)$.

Furthermore, we have the following.

Theorem 29. For any kernel of order at least 1, for all n there is a corrector which eliminates the $\mathcal{O}(\varepsilon\tau^p)$ error terms for all $1 < p < n$.

Proof. We have

$$K = \tau A + \varepsilon\tau B + \sum_{n=2}^{\infty} \varepsilon\tau^n k_n [A^{n-1}B] + \mathcal{O}(\varepsilon^2),$$

and take the corrector to be

$$C = \sum_{n=1}^{\infty} \varepsilon\tau^n c_n [A^{n-1}B] + \mathcal{O}(\varepsilon^2).$$

Then

$$[C, K] = -\varepsilon \sum_{n=1}^{\infty} c_n \tau^{n+1} [A^n B] + \mathcal{O}(\varepsilon^2)$$

and $[C, C, K] = \mathcal{O}(\varepsilon^2)$. Therefore

$$\begin{aligned} \mathrm{e}^C \mathrm{e}^K \mathrm{e}^{-C} &= \mathrm{e}^{K + [C,K] + \mathcal{O}(\varepsilon^2)} \\ &= \tau A + \varepsilon\tau B + \varepsilon \sum_{n=2}^{\infty} (k_n - c_{n-1}) \tau^n [A^{n-1}B] + \mathcal{O}(\varepsilon^2), \end{aligned}$$

and all $\mathcal{O}(\varepsilon)$ errors can be corrected by taking $c_n = k_{n+1}$ for $n \geq 1$. □

Thus, any splitting method is 'really' $\mathcal{O}(\varepsilon^2)$ accurate on near-integrable problems. Wisdom and Holman (1991) first derived these correctors for the leapfrog kernel and used them to great effect in their study of the solar system. Blanes, Casas and Ros (2000*a*) have made a systematic study of higher-order methods and construct, for example, a 2-stage method of order $\mathcal{O}(\varepsilon^2\tau^4 + \varepsilon^3\tau^3)$, a 3-stage method of order $\mathcal{O}(\varepsilon^2\tau^6 + \varepsilon^3\tau^4)$, and so on.

4.8. Methods with commutators

Given a splitting $X = \sum X_i$, one can consider constructing a method not just from the flows of X_i but also from the flows of their commutators $[\ldots, X_i, X_j]$, or from approximations of these flows. Using this extra information will always allow one to approximate the flow of X better, but the new information may be costly and the net benefit can only be judged in specific cases.

The most venerable of such methods is due to Takahashi and Imada (1984), for $\dot{x} = A + B$:

$$\mathrm{e}^{C}\mathrm{e}^{\frac{1}{2}\tau A}\mathrm{e}^{\frac{1}{2}\tau B}\mathrm{e}^{-\frac{1}{24}\tau^3[B,B,A]}\mathrm{e}^{\frac{1}{2}\tau B}\mathrm{e}^{\frac{1}{2}\tau A}\mathrm{e}^{-C} = \mathrm{e}^{\tau(A+B)+\mathcal{O}(\tau^5)} \tag{4.12}$$

where the corrector $C = \frac{1}{24}\tau^2[A,B] + \mathcal{O}(\tau^4)$. Without a corrector, one can use

$$\mathrm{e}^{\frac{1}{6}\tau B}\mathrm{e}^{\frac{1}{2}\tau A}\mathrm{e}^{\frac{1}{3}\tau B}\mathrm{e}^{-\frac{1}{72}\tau^3[B,B,A]}\mathrm{e}^{\frac{1}{3}\tau B}\mathrm{e}^{\frac{1}{2}\tau A}\mathrm{e}^{\frac{1}{6}\tau B} = \mathrm{e}^{\tau(A+B)+\mathcal{O}(\tau^5)}. \tag{4.13}$$

Observe that, in the Runge–Kutta–Nyström case with $A = \frac{1}{2}p^{\mathrm{T}}M(q)p$, $B = B(q)$, we have that

$$[B,B,A] = \nabla B^{\mathrm{T}} M \nabla B = f^{\mathrm{T}} M f, \quad f = -\nabla B,$$

is a function of q only, hence integrable by Euler's method. The three central terms in (4.12) and (4.13) then coalesce and only one force evaluation is needed per time step. The flow of $[B,B,A]$ is

$$p(t) = p(0) - t(M'(f,f) + 2f'Mf)(q(0)),$$

which only involves one derivative of the force evaluated in one direction Mf.

This can be very cheap for some problems, for instance, n-body systems with 2-body interactions, for which it costs about the same as one force evaluation, or, if dominated by expensive square root calculations, much less. For, let $V : \mathbb{R}^3 \to \mathbb{R}$ be a potential such that $B = \sum_{j \neq i} V(q_i - q_j)$, $q_i \in \mathbb{R}^3$. Then

$$\sum_j \frac{\partial f_i}{\partial q_j} v_j = \sum_j V''(q_i - q_j)(v_i - v_j),$$

where V'' is the Hessian of V.

Further, for any A and B we have the possibility of splitting $[B,B,A] = \sum C_i$ and using

$$\mathrm{e}^{\tau^3[B,B,A]} = \mathrm{e}^{C_1} \ldots \mathrm{e}^{C_n} + \mathcal{O}(\tau^6).$$

For large systems with local interactions, such as discretizations of PDEs, $[B,B,A]$ is also local and can be split by partitioning the unknowns appropriately.

This method can be extended by (Blanes 2001):

(1) including more stages (as before, this decreases the error constants somewhat);
(2) going to higher order (up to 8th-order methods have been found);
(3) considering near-integrable systems;
(4) including more derivatives (*e.g.*, for RKN systems, $[B, A, B, B, A]$ is a function of q only; its flow involves the third derivative of the force).

4.9. Some good methods

For methods of the types $\varphi_{\mathrm{NS,S,SS}}$, it has been found that if one takes enough stages to provide as many parameters as there are order conditions, the order conditions always have real solutions. However, it is also possible to include more stages and use the free parameters to minimize the error in some sense. This is usually done by choosing a norm on the vector space $L_{p+1}(A, B)$, that is, assigning some weights to the c_{p+1} independent error terms. The inherent arbitrariness in this procedure has not, so far, led to any serious disagreement over which methods have the smallest error. If the norm of the error is e_{p+1}, it is traditional to compare the *effective error* $me_{p+1}^{1/p}$, which, for sufficiently small step sizes, is proportional to the amount of work needed to attain a given error.

For example, consider 4th-order type S methods. There are 4 order conditions, corresponding to A, B, $[A, A, B]$, and $[B, B, A]$. The method (4.11) has $m + 1$ free parameters, so a minimum of three stages is required. There is a unique 3-stage method, namely (4.4) with $n = m = 1$, and $(m - 3)$-parameter families of m-stage methods. Blanes and Moan (2000) find methods with 3, 4, 5, and 6 stages with effective errors of 1.33, 0.71, 0.62, and 0.56, respectively. With 7 stages, order 6 is possible. However, there probably exist 7-stage 4th-order methods with smaller 7th-order errors than the best order 6 method. In truth, there is only a modest range of step sizes, about 0.5 to 1 order of magnitude, in which methods of order p are preferred. Any smaller, and one should switch to order $p + 2$; any larger, and one should switch to order $p - 2$. Within this range, other questions such as the size of each order p error term and of the $p + 2$ errors, and the stability for this τ, come into play.

Some methods with the smallest known effective errors are given below. Methods 1(a), (c) and 3(a), (b) are due to Blanes and Moan (2000), 1(b), 2(a) to Suzuki (1990), 1(d) to McLachlan (1995), 2(b) to Blanes, Casas and Ros (2000*a*), 2(c) to McLachlan (2002), 3(c) to Takahashi and Imada (1984), 3(d) to López-Marcos, Sanz-Serna and Skeel (1997), 3(e), (f) to Blanes, Casas and Ros (2001), and 4(b) to Blanes, Casas and Ros (2000*b*). For simple, easy-to-use methods we particularly recommend 1(a) and 3(a).

1: Methods for arbitrary splittings. These can be applied directly if $X = A+B$ where the flows of A and B are known, or applied to an arbitrary integrator of X by massaging the coefficients as in Theorem 24.

(a) 4th order, type S, $m = 6$ stages, (4.11) with

$$a_1 = 0.0792036964311957, \qquad b_1 = 0.209515106613362,$$
$$a_2 = 0.353172906049774, \qquad b_2 = -0.143851773179818,$$
$$a_3 = -0.0420650803577195, \qquad b_3 = 1/2 - b_1 - b_2,$$
$$a_4 = 1 - 2(a_1 + a_2 + a_3).$$

(b) 4th order, type SS, $m = 5$ stages, (4.4) with $\alpha = 1/(4-4^{1/3})$, $\beta = 1-2\alpha$.

(c) 6th order, type S, $m = 10$ stages, (4.11) with

$$a_1 = 0.0502627644003922, \qquad b_1 = 0.148816447901042,$$
$$a_2 = 0.413514300428344, \qquad b_2 = -0.132385865767784,$$
$$a_3 = 0.0450798897943977, \qquad b_3 = 0.067307604692185,$$
$$a_4 = -0.188054853819569, \qquad b_4 = 0.432666402578175,$$
$$a_5 = 0.541960678450780, \qquad b_5 = 1/2 - \textstyle\sum_{i=1}^4 b_i,$$
$$a_6 = 1 - 2(\textstyle\sum_{i=1}^5 a_i).$$

(d) 6th order, type SS, 9 stages, (4.6) with

$$a_1 = 0.1867, \qquad a_2 = 0.555497023712478399 16,$$
$$a_3 = 0.12946694891347535806, \qquad a_4 = -0.84326562338773460855,$$
$$a_5 = 1 - 2\textstyle\sum_{i=1}^4 a_i.$$

2: Methods with correctors.

(a) 4th order, type SS, $2m+1$ stages, (4.4) with

$$\alpha = 1/(2m - (2m)^{1/3}), \qquad \beta = 1 - 2m\alpha.$$

(b) 4th order, type S, $m = 5$ stages, (4.11) with

$$a_1 = 0, \qquad b_1 = 6/25,$$
$$a_2 = (57 + \sqrt{18069})/300, \qquad b_2 = -1/10,$$
$$a_3 = 1/2 - a_2, \qquad b_3 = 1 - 2(b_1 + b_2).$$

(c) 6th order, type SS, $2m+3$ stages, $a_1 = \cdots = a_m = x$, $a_{m+1} = y$, $a_{m+2} = z = 1 - 2(mx + y)$ where (x, y) is the unique real root of

$$2mx^3 + 2y^3 + z^3 = 2mx^5 + 2y^5 + z^5 = 0.$$

The 9, 11, and 13 stage methods are very good: *e.g.*, for 11 stages,

$$a_{1,2,3,4} = 0.17057688650092221 57, \qquad a_5 = -0.42336614089265804 8.$$

3: Runge–Kutta–Nyström methods. When output is not required, two b_1 stages coalesce in methods 3(a), 3(b), and 3(f), reducing the number of stages by one. Methods with a corrector are also 4th-order for arbitrary (non-RKN) splittings, but are not optimized for that case.

(a) 4th order, $m = 7$ stages, (4.11) with

$$\begin{aligned} a_1 &= 0, & b_1 &= 0.0829844064174052,\\ a_2 &= 0.245298957184271, & b_2 &= 0.396309801498368,\\ a_3 &= 0.604872665711080, & b_3 &= -0.0390563049223486,\\ a_4 &= 1/2 - (a_2 + a_3), & b_4 &= 1 - 2(b_1 + b_2 + b_3). \end{aligned}$$

(b) 6th order, $m = 12$ stages, (4.11) with

$$\begin{aligned} a_1 &= 0, & b_1 &= 0.0414649985182624,\\ a_2 &= 0.123229775946271, & b_2 &= 0.198128671918067,\\ a_3 &= 0.290553797799558, & b_3 &= -0.0400061921041533,\\ a_4 &= -0.127049212625417, & b_4 &= 0.0752539843015807,\\ a_5 &= -0.246331761062075, & b_5 &= -0.0115113874206879,\\ a_6 &= 0.357208872795928, & b_6 &= 1/2 - \textstyle\sum_{i=1}^{5} b_i,\\ a_7 &= 1 - 2\textstyle\sum_{i=2}^{6} a_i. \end{aligned}$$

With a corrector, kernels leading to good methods are

$$e^{a\tau A}e^{b\tau B}e^{(1/2-a)\tau A}e^{(1-2b)\tau B - c\tau^3[B,B,A]}e^{(1/2-a)\tau A}e^{b\tau B}e^{a\tau A},$$

with either

(c) $a = 0$, $b = 0$, $c = \frac{1}{24}$ (4th order),

(d) $a = 0$, $b = \frac{1}{4}$, $c = \frac{1}{96}$ (4th order),

(e) $a = -0.0682610383918630$, $b = 0.2621129352517028$,
$c = 0.0164011128160783$ (6th order),

or

(f) 6th order, (4.11) with $m = 7$ stages and

$$\begin{aligned} a_1 &= 0, & b_1 &= 0.15,\\ a_2 &= 0.316, & b_2 &= 0.3297455985640361,\\ a_3 &= 0.4312992634164797, & b_3 &= -0.04936325705062370 7,\\ a_4 &= 1/2 - (a_2 + a_3), & b_4 &= 1 - 2(b_1 + b_2 + b_3). \end{aligned}$$

4: Methods for nearly integrable systems.

(a) The Takahashi–Imada method 3(c) above is correctable to $\mathcal{O}(\varepsilon^2\tau^4)$ for any splitting with $B = \mathcal{O}(\varepsilon)$ and $A = \mathcal{O}(1)$.

(b) The type S kernel (4.11) with $m = 3$ stages and

$$a_1 = 0.5600879810924619, \qquad b_1 = 1.5171479707207228,$$

is correctable to $\mathcal{O}(\varepsilon^2\tau^6 + \varepsilon^3\tau^4)$ for any splitting.

5. Applications

5.1. Molecular dynamics

In molecular simulation (Allen and Tildesley 1987, Leimkuhler, Reich and Skeel 1996, Kofke 2000, Leimkuhler 2002), one of two methods is used: molecular dynamics (MD) or Monte Carlo. Here we review the former. Two main applications of MD are (i) multibody dynamics, and (ii) macromolecules.

In multibody dynamics one simulates 10 to 100,000 or more atoms (typically 500–1000), viewed as classical point masses. The inter-particle forces have a repulsive core and an attractive tail, and are typically modelled by the Lennard–Jones (1.11) or Coulomb potentials. For the numerical integration of such systems, a fast inexpensive integration method is essential, and generally the Verlet/leapfrog method or one of its variants is used (Example 1).

In a macromolecule such as a protein chain, the potential energies V are functions of the distances between neighbouring pairs of atoms (springs), the angles defined by triples of atoms, and the dihedral angles between the planes defined by successive triples of atoms. One must also take into account long-range forces, as well as 'hydrogen bonds'. The bonds between successive atoms may oscillate hundreds of times faster than the 'dihedral angles'. Since the latter are most important for determining conformational changes, molecular dynamicists generally replace the stiff springs by rigid rods (allowing the use of larger time steps). To solve the resulting constrained Hamiltonian system, the SHAKE algorithm was introduced by Ryckaert, Ciccotti and Berendsen (1977). It was subsequently proved that this algorithm is also symplectic (Leimkuhler and Skeel 1994). If the rods are such as to make each molecule completely rigid, one can split the entire system into rigid body plus potential terms, treating the rigid bodies as in Example 2 (Dullweber, Leimkuhler and McLachlan 1997), giving an explicit symplectic integrator preserving all the constraints.

Recently, variable time step algorithms that preserve time-reversal symmetry, but not symplecticity, have been proposed (Barth, Leimkuhler and Reich 1999).

5.2. Wave equations in fluids, optics, and acoustics

While many of the properties we have considered for ODEs (*e.g.*, symmetries, integrals, Hamiltonian structure) have analogues for PDEs, we are aware of no general classification of PDEs. Therefore one can proceed either by studying particular properties, or by studying particular equations. We have already mentioned PDEs with linear dissipation (Example 17), the KdV equation (Example 19), and symmetry reductions of PDEs (Examples 30, 31, 32 and 33). Here we given some examples of PDEs for which splitting methods are in current use.

Most established splitting methods are used for Lie group integration in the linear case, and symplectic integration in the nonlinear case.

Lie group PDEs

The Lie group usually arises from a quadratic conservation law, for the linear maps that preserve a quadratic form a Lie group. For example, let $u(x,t) \in \mathbb{R}$ and consider the 1-way wave equation $u_t = u_x$. On discretizing $\frac{\partial}{\partial x}$ by an antisymmetric matrix $S \in \mathfrak{so}(n)$ (*e.g.*, by central differences) and $u(x,t)$ by a vector $\mathbf{u}(t) \in \mathbb{R}^n$, we have the system of ODEs $\dot{\mathbf{u}} = S\mathbf{u}$, a Lie group equation on $\mathrm{SO}(n)$. The corresponding conservation law is $\frac{\mathrm{d}}{\mathrm{d}t}\|\mathbf{u}\|^2 = 0$.

However, one has to be careful in splitting such a discretization. The terms that form a finite difference of order r should not be split because each term is $\mathcal{O}(h^{-r})$, where h is the grid size. The local truncation error of a composition method, nominally order p, is then τ^{-1} times the $(p+1)$-order Lie brackets of the pieces, *i.e.*, $\mathcal{O}(\tau^p h^{-r(p+1)})$. To get a method of actual order p for the PDE will then require taking $\tau = \mathcal{O}(h^{r+1})$, which is a very severe restriction on the time step. If one takes $\tau = \mathcal{O}(h)$ in the 1-way wave equation, for example, the method is not even consistent with the PDE – the truncation errors are $\mathcal{O}(1)$. (In fact, from the Lax equivalence theorem, it *has* to be inconsistent, since from preservation of $\|\mathbf{u}\|^2$ it is stable for all τ/h, regardless of the CFL number.)

To avoid this problem, one should split the PDE itself before discretizing.

Many Lie groups can be found lurking in the spatial discretizations of PDEs. Consider any linear ODE $\dot{x} = Fx$, $x \in \mathbb{R}^n$, which has a quadratic integral $x^{\mathrm{T}}Hx$ where H has k positive eigenvalues and $n-k$ negative eigenvalues. There there is a unique antisymmetric matrix S such that $F = SH$. Suppose S is nonsingular. We thus have $F \in \mathfrak{so}(k, n-k)$ (corresponding to preservation of the quadratic form $x^{\mathrm{T}}Hy$), and also $F \in \mathfrak{sp}(n)$, corresponding to preservation of the symplectic form $x^{\mathrm{T}}Sy$. Splitting H provides a symplectic integrator, while splitting S provides an orthogonal integrator; in the case $k = n$, the level set $x^{\mathrm{T}}Hx = \text{const}$ is compact and such an integrator is unconditionally stable. Seeking an *orthosymplectic* splitting, however, is more difficult and depends on the particular S and H. One has to find $F = \sum F_i$ where each F_i is orthosymplectic, *i.e.*, HF_i is antisymmetric and F_iS is symmetric for all i. One case where there are plenty of such matrices is $H = I$ and $S = \begin{pmatrix} 0 & I \\ -I & 0 \end{pmatrix}$: the ODEs $\dot{x} = Fx$ for

$$F = \begin{pmatrix} C & D \\ -D & C \end{pmatrix}, \quad C = -C^{\mathrm{T}},\ D = D^{\mathrm{T}} \tag{5.1}$$

all preserve $x^{\mathrm{T}}x$ and are Hamiltonian with respect to the canonical symplectic structure S.

For example, consider the Maxwell equations. Let $B(x,t) \in \mathbb{R}^3$ be the magnetic field for $x \in \mathbb{R}^3$, $E(x,t) \in \mathbb{R}^3$ the electric field, then (taking units in which $c = 1$) the 3D vacuum Maxwell equations are

$$\begin{pmatrix} B_t \\ E_t \end{pmatrix} = \begin{pmatrix} 0 & -\nabla\times \\ \nabla\times & 0 \end{pmatrix} \begin{pmatrix} B \\ E \end{pmatrix},$$

where the operator on the right is self-adjoint. Letting $D = D^{\mathrm{T}}$ be a symmetric discretization of the curl operator gives a matrix of the form (5.1) with $C = 0$. Therefore, any symmetric splitting of D, such as dimensional splitting, provides symplectic integrators that also preserve the energy $\|E\|^2 + \|B\|^2$. Each piece must now be integrated by a quadratic-preserving symplectic integrator such as the midpoint rule. On the other hand, for Maxwell's equations in an inhomogeneous medium some of the symmetry of the problem is lost; dimensional splitting now preserves energy, but is not symplectic.

Leaving aside splitting methods, the midpoint rule automatically preserves all symmetric and antisymmetric inner products associated with any linear equation and one could argue that is the most geometric integrator around for this class of problems. It does require implementing more sophisticated linear solvers, however.

A very extensively used application is the one-way Helmholtz equation used to model sound propagation in inhomogeneous oceans (Tappert and Brown 1996). The Helmholtz equation

$$\psi_{zz} + \psi_{rr} + n^2(z,r)\psi = 0,$$

where ψ is the amplitude of an acoustic wave, z is depth, r is 'range', and $n(z,r)$ the index of refraction of the medium, is factored to obtain

$$\mathrm{i}\frac{\partial\psi}{\partial r} = H\psi, \tag{5.2}$$

where

$$H = -\sqrt{n^2(z,r) - p^2}$$

and $p = -\mathrm{i}\frac{\partial}{\partial z}$. In this way only out-going waves are retained. Now r is regarded as the time variable and (5.2) is the time-dependent Schrödinger equation with Hamiltonian H. It is a Lie group equation in $\mathrm{U}(n)$, corresponding to preservation of $\int |\psi|^2\, dz$. (Note that the real form of $\mathfrak{u}(n)$ is (5.1).) With a Fourier discretization of p, H cannot be split into a sum of explicitly integrable pieces. The way out is to approximate H by $\widetilde{H}$ so that $\widetilde{H} = \sum \widetilde{H}_i$ and each factor $e^{it\widetilde{H}}$ can be evaluated explicitly. The simplest such approximation is

$$\widetilde{H} = \tfrac{1}{2}p^2 - \tfrac{1}{2}n^2 := H_1 + H_2.$$

The equations corresponding to H_1 are

$$i\dot{\psi} = \tfrac{1}{2}\psi_{xx},$$

which can be solved exactly in a Fourier discretization using the FFT, while the equations corresponding to H_2 are

$$i\dot{\psi} = \tfrac{1}{2}n^2(z,r)\psi,$$

which is an ODE for ψ at each spatial site. (Further splitting removes the r-dependence.) This is the original *split-step Fourier* method of Tappert (1977). Other approximations, valid over a wider range of n and p, can also be used, *e.g.*, $\widetilde{H} = -\sqrt{1 - p^2/n} - n$ (Tappert and Brown 1996). Similar equations and methods arise in optics (Agrawal 1989).

Hamiltonian PDEs

These have the form (Marsden and Ratiu 1999)

$$u_t = \mathcal{D}\frac{\delta \mathcal{H}}{\delta u},$$

where $\mathcal{D}$ is a Poisson operator and $\mathcal{H}$ is the Hamiltonian. (In general, $\mathcal{D}$ can depend on u, as in the Euler fluid equations for example. This case is notoriously difficult to discretize and we do not consider it here.) These can be discretized to obtain systems of Poisson ODEs of the form $u_t = S\nabla H(u)$, where S is an antisymmetric matrix. Finite difference, Fourier, and finite element discretizations on arbitrary grids can be used (McLachlan and Robidoux 2000); the main point is to take care to preserve the symmetry of the differential operators appearing in $\mathcal{D}$ and $\mathcal{H}$. Then, just as in standard symplectic integration, H must be examined to see if it can be split. Usually this is straightforward, and for spatially homogeneous problems even the entire linear part can be integrated exactly as in the split-step Fourier method. See McLachlan (1994) for examples of the nonlinear wave, nonlinear Schrödinger, Boussinesq, KdV and Zakharov equations.

We shall consider just one example, the nonlinear wave equation $q_{tt} = q_{xx} - V'(q)$ with periodic boundary conditions and a Fourier discretization of q_{xx}. In Hamiltonian form, we have $\mathcal{H} = \frac{1}{2}p^2 + \frac{1}{2}q_x^2 + V(q)$ and

$$q_t = p, \quad p_t = q_{xx} - V'(q).$$

The linear part in Fourier space is

$$\dot{\tilde{q}}_n = \tilde{p}_n, \quad \dot{\tilde{p}}_n = -n^2 \tilde{q}_n,$$

which is easily solved exactly, and the nonlinear part in real space is

$$q_t = 0, \quad p_t = -V'(q),$$

also easily solved. Such splittings have the advantage that the highly accurate RKN (Section 4.6), derivative (Section 4.8), and corrector (Section 4.5) methods can be used, while at the same time preserving the near-integrability in case of weak nonlinearities $V'(q)$ (Section 4.7).

Many Hamiltonian PDEs are also *multi*-symplectic and geometric integrators exist which preserve this structure (Marsden and West 2001). For simple cases, such as the nonlinear wave equation discretized with central differences, symplectic integrators are also multisymplectic.

5.3. Quantum mechanics and quantum statistical mechanics

In time-dependent quantum mechanics, one is faced with computing

$$\mathrm{e}^{\mathrm{i}tH}, \tag{5.3}$$

where H is the Hamiltonian operator and t denotes time. In quantum statistical mechanics, by contrast, one must calculate

$$\mathrm{e}^{-\beta H} \tag{5.4}$$

(or rather its trace), where H is again the Hamiltonian, and $\beta = 1/(kT)$, with k being Boltzmann's constant and T being the absolute temperature.

As we shall see, the main difference between (5.3) and (5.4) lies in the factor i which occurs in (5.3), but not in (5.4). This difference has the consequence that operators of the form $\mathrm{e}^{\mathrm{i}tH}$ form a group (the unitary group), while operators of the form $\mathrm{e}^{-\beta H}$ form a symmetric space (the symmetric space of positive definite Hermitian operators). We start with case (5.3) (in spite of the fact that the chronological order is the other way around).

The time-dependent Schrödinger equation (De Raedt 1987)
The time evolution of a non-relativistic quantum mechanical system is governed by the time-dependent Schrödinger equation

$$\frac{\partial \psi(r,t)}{\partial t} = -\mathrm{i}H\psi(r,t), \tag{5.5}$$

where H is the Hamiltonian of the system, $\psi(r,t)$ is the normalized, complex-valued wave function, $\psi(r,0)$ is the initial state at time $t=0$, and the units are such that $\hbar = 1$.

For simplicity we will restrict our discussion to the case of a particle moving on a 1-dimensional interval $0 \le x \le a$, and take

$$H = -\frac{\mathrm{d}^2}{\mathrm{d}x^2} + V(x),$$

where $V(x)$ represents the (real) potential energy at position x. It is clear that, with this choice, (5.5) is a linear hyperbolic partial differential equation.

Being linear, (5.5) is of course always integrable. So a splitting into integrable pieces is not the question here. Rather, the pieces should be much faster to solve than the full system. Numerous discretizations and splittings of H into easily integrated pieces are possible, the two most popular being a Fourier discretization (so that $\exp\big(\mathrm{i}t\frac{\mathrm{d}^2}{\mathrm{d}x^2}\big)$ can be evaluated using the FFT) combined with splitting, and finite differences combined with a unitary integrator such as the midpoint rule.

Note that $\mathrm{e}^{\mathrm{i}tH}$ is not only unitary, it is also symplectic; the canonical coordinates are $Re(\psi)$ and $Im(\psi)$ and the system evolves in $\mathrm{U}(n) \subset \mathrm{Sp}(2n)$. So it is also possible to use q–p splitting on the system $\dot{q} = Hp$, $\dot{p} = -Hq$. The integrator is no longer unitary, merely symplectic, and hence no longer unconditionally stable; but this does allow one to handle arbitrary Hamiltonians, which H-splitting does not (*e.g.*, if H contains $V(x, \frac{\mathrm{d}}{\mathrm{d}x})$). (Note that the Schrödinger equation, the 1-way Helmholtz equation, and the vacuum Maxwell equations (Section 5.2) all have essentially the same structure, and the same Lie group $\mathrm{U}(n)$.)

Quantum statistical mechanics (De Raedt and Lagendijk 1985)
The central object in quantum statistical mechanics is the partition function

$$Z := \mathrm{Tr}\,\mathrm{e}^{-\beta H},$$

from which thermodynamic functions such as energy and specific heat can be derived. Here 'Tr' means one has to calculate the trace of the operator $\mathrm{e}^{-\beta H}$. (One can think of H as a matrix. For spin systems and other discrete systems, this is immediate; continuous systems can be approximated by discrete systems by, *e.g.*, finite differences.)

In practical applications, one must generally resort to using an approximation to $\mathrm{e}^{-\beta H}$. Since H is an element of the Lie triple system of Hermitian operators, it follows that $\mathrm{e}^{-\beta H}$ is an element of the symmetric space of *positive definite* Hermitian operators. Suitable approximations to $\mathrm{e}^{-\beta H}$ are therefore obtained using *symmetric* compositions (as in Theorem 20) (De Raedt and De Raedt 1983) such as (4.3), (4.4), (4.11), and (4.12). For discrete spin systems, H is split by a (*e.g.*, odd/even) partitioning of the lattice sites into uncoupled subsets. In each case, one approximates $\mathrm{e}^{-\beta H}$ by composing m steps of the method with 'time' step $\tau = \beta/m$. Note that methods based on correctors, Section 4.5, are preferred, as the trace in Z eliminates any correction term. Even though, after splitting $H = \sum H_i$, each $e^{-\beta H_i}$ can in principle be evaluated in closed form, even this is often too expensive, so Monte Carlo is applied to the entire symmetric composition (so that only matrix-vector products $e^{-\beta H_i}v$ have to be evaluated). Numerous variations of this idea have been successfully implemented by Suzuki (1990) and others.

Note that the operators itH also lie in a Lie triple system, of imaginary skew-Hermitian matrices, corresponding to the reversing symmetry $\psi \mapsto \overline{\psi}$ of (5.5). Therefore it is desirable but not essential to use symmetric compositions to approximate e^{itH} as well, to stay in the symmetric space of symmetric unitary matrices.

5.4. Celestial mechanics

Symplectic integrators based on splitting have been used to great effect in celestial mechanics by Wisdom and others. In Sussman and Wisdom (1992), 100 million year integrations of the whole solar system were performed, yielding a positive Lyapunov exponent, which suggests that the solar system is chaotic. (This chaos in the solar system is reviewed in Lecar, Franklin, Holman and Murray (2001).) The Hamiltonian for the n-body problem is

$$H = \sum_{i=0}^{n-1} \frac{p_i^2}{m_i} - \sum_{i<j} \frac{Gm_i m_j}{r_{ij}},$$

where $r_{ij} := \|q_i - q_j\|$. The first fundamental idea (Wisdom and Holman 1991) is not to split $H = T(p) + V(q)$, which could easily be done, but to split

$$H = H_{\text{Kepler}} + \varepsilon H_{\text{interaction}}, \tag{5.6}$$

where H_{Kepler} represents the $n-1$ independent Keplerian motions of the planets/satellites with respect to the central body, $H_{\text{interaction}}$ represents the perturbation of the outer bodies on one another, and ε is the ratio of the mass of the largest outer body to the mass of the central body. For the solar system, $\varepsilon \approx 10^{-3}$. That is, the splitting preserves the near-integrable character of the system, see Section 4.7. Standard leapfrog, as used in Sussman and Wisdom (1992), leads to $\mathcal{O}(\varepsilon\tau^2)$ errors, while corrected leapfrog leads to $\mathcal{O}(\varepsilon^2\tau^2)$ errors with no additional work. In practice, even with a fairly large time step of 7.2 days (Wisdom, Holman and Touma 1996), leapfrog had a bounded relative energy error of 2×10^{-9} and corrected leapfrog a linear relative energy error (presumably due to round-off) of 2×10^{-11} per 100 million years.

To end this subsection, we will give the derivation of H_{Kepler} and $H_{\text{interaction}}$

in (5.6). One first transforms to Jacobi coordinates:

$$\begin{aligned} \tilde{q}_0 &:= \frac{\sum_{j=0}^{n-1} m_j q_j}{\sum_{j=0}^{n-1} m_j}, \\ \tilde{q}_i &:= q_i - \frac{\sum_{j=0}^{i-1} m_j q_j}{\sum_{j=0}^{i-1} m_j}. \end{aligned} \tag{5.7}$$

In these coordinates,

$$H = \frac{\tilde{p}_0^2}{2\sum_{j=0}^{n-1} m_j} + H_{\text{Kepler}} + \varepsilon H_{\text{interaction}},$$

where

$$\begin{aligned} H_{\text{Kepler}} &= \sum_{i=1}^{n-1} \frac{\tilde{p}_i^2}{2\tilde{m}_i} - \frac{G m_i m_o}{\|\tilde{q}_i\|}, \\ \varepsilon H_{\text{interaction}} &= \sum_{i=1}^{n-1} G m_i m_0 \left(\frac{1}{\|\tilde{q}_i\|} - \frac{1}{\|q_i - q_0\|} \right) - \sum_{0<i<j\le n-1} \frac{G m_i m_j}{\|q_i - q_j\|}, \end{aligned}$$

with

$$\tilde{m}_i := m_i \frac{\sum_{j=0}^{i-1} m_j}{\sum_{j=0}^{i} m_j}.$$

The first term in H represents the free motion of the centre of mass. It commutes with H_{Kepler} and $H_{\text{interaction}}$ and can hence be ignored. Inverting (5.7), $H_{\text{interaction}}$ can be expressed in terms of the Jacobi coordinates only, and it can be shown that the first term in $H_{\text{interaction}}$ is of the same order as the second, and hence much smaller than H_{Kepler}.

Thus, it is worth going to some trouble to preserve the near-integrability of the system. Other developments include multiple time-stepping (Hardy, Okunbor and Skeel 1999), to take advantage of the range of frequencies of the planets, and the 'smooth switch' of Kværnø and Leimkuhler (2000). In the latter, one considers a general n-body problem and wants to integrate any close encounters of two bodies exactly. Depending on which bodies are close to each other, one may prefer a splitting $H = A_1 + B_1$ in region R_1 of phase space, and a splitting $H = A_2 + B_2$ in region R_2. Changing abruptly from one to another destroys all geometric properties, but by introducing a buffer zone and sufficiently smooth (*e.g.*, piecewise polynomial) interpolation between the two splittings in this zone, the geometric properties can be retained.

5.5. Advection–reaction–diffusion equations

We consider systems of partial differential evolution equations of the form

$$u_t = X := X_1 + X_2 + X_3 + X_4 + X_5, \tag{5.8}$$

where $u : D \times \mathbb{R} \to \mathbb{R}^k$ are the field variables, D is the spatial domain, and $X_1 = \nabla \cdot f(u)$, where $f : \mathbb{R}^k \to \mathbb{R}^k$ is a flux (so that $u_t = \nabla \cdot f(u)$ is a hyperbolic system of conservation laws); $X_2 = g(u)$, where $g : \mathbb{R}^k \to \mathbb{R}^k$ are source or reaction terms (in reacting chemical systems, the ODEs $u_t = g(u)$ are typically extremely stiff); $X_3 = \nabla \cdot (A(x)\nabla u)$, where $A : D \to \mathbb{R}^{k\times k}$ is a matrix of diffusion constants; $X_4 = h(x,t)$, where $h : D \times \mathbb{R} \to \mathbb{R}^k$ are external forces; and X_5 is a Lagrange multiplier for any constraints that may be present, as in the incompressible Navier–Stokes equations. Many 'operator-splitting' schemes have been proposed for equations with various combinations of these terms present, splitting into various sums of the X_i. In addition, dimension- or Strang-splitting (Strang 1968) is widely used, in which, for instance, the diffusion terms are diagonalized and split as $X_3 = \sum A_{ii}(x)u_{x_i x_i}$. If φ_i is an unconditionally stable method for the one-dimensional heat equation $u_t = A_{ii}u_{x_i x_i}$, then either leapfrog or

$$\sum_\sigma \prod_i \varphi_{\sigma_i},$$

where σ runs over all permutations of the spatial dimensions, is unconditionally stable. Since integration methods for these large systems are usually only second order in time, the restriction to second order (see Section 4.4) has not been regarded as onerous.

We shall not survey this huge field here; see representative applications in reaction–diffusion systems (Karlsen and Lie 1999), fluid-particle systems (Glowinski, Pan, Hesla and Joseph 1999), chemotaxis (Tyson, Stern and LeVeque 2000), magnetohydrodynamics (Ryu, Jones and Frank 1995), image processing (Weickert, Romeny and Viergever 1998), combustion chemistry (Yang and Pope 1998), meteorology (Leonard, Lock and MacVean 1996), and porous media (Barry, Bajracharya and Miller 1996). Rather, we are interested in potential similarities between this form of splitting and that which we have reviewed for ODEs.

A recent comparison of many integration methods for the Navier–Stokes equations is interesting here (Turek 1996). (Note that the Navier–Stokes equations contain all terms above except X_2.) Let $\varphi(\tau, z, w)$ be the method

$$\begin{aligned} \frac{u_{n+1} - u_n}{\tau} &= z(X_1 + X_3)(u_n) + (1 - z)(X_1 + X_3)(u_{n+1}) + \\ &\qquad wX_4(t_n) + (1 - w)X_4(t_n + \tau) + X_5(u_{n+1}), \\ \nabla \cdot u_{n+1} &= 0. \end{aligned}$$

Then the method

$$\varphi(\theta\tau, 2\theta, 1)\varphi((1-2\theta)\tau, 1-2\theta, 0)\varphi(\theta\tau, 2\theta, 1),$$

$\theta = 1 - \frac{1}{\sqrt{2}}$, is second-order and stiffly A-stable. Turek (1996) finds this method superior to the widely used backward Euler and Crank–Nicolson methods, and to more explicit multi-step treatments of the nonlinear advection term, without requiring more storage or more work (when applied with 3 times the time step of these competitors). The parallels with ODE splitting and composition methods are striking and it seems that the two fields could usefully learn from each other.

When the diffusion terms in (5.8) are absent, we have a first-order hyperbolic system of conservation laws with source terms. Shocks can form and dimension splitting runs into an apparently fundamental obstacle; its order is at most 1 (Crandall and Majda 1986), and in fact in most versions (*e.g.*, in Tang and Teng (1995)) is only proved to be $\frac{1}{2}$. Practical methods for multidimensional conservation laws (LeVeque 1998), while using a form of splitting, are in fact significantly more complicated than simple leapfrog.

5.6. Accelerator physics (Forest 1998, Dragt and Abell 1996, Dragt 2002)

There are many fields of charged particle dynamics where a single particle description is useful: storage rings, linear accelerators, and electron microscopes, to name a few. Here we will concentrate on particle storage rings.

In large storage rings, particles typically make 10^8 or more revolutions. This is an instance of the first of the following two remarkable facts of ring dynamics (Forest 1998):

(1) the motion is nearly stable: particles seem to spend a long time in the ring;
(2) the motion is symplectic in hadron (*e.g.*, proton) machines and nearly symplectic in electron machines (but also slightly stochastic).

To model this situation, one then has a non-autonomous (periodic) dynamical system of three degrees of freedom (*i.e.*, 6 coupled equations). In comparison with the solar system, for instance, one might think this system should be much simpler. The problem, however, is that a single ring can contain of the order of 5000 magnets for bending, focusing, and for correcting the orbit. Hence, though the dimensionality of the system is low, the equations of motion themselves are actually exceedingly complicated. One therefore certainly aims to use *explicit* methods.

Traditional treatments assume that it is possible to write down a global (Hamiltonian) differential equation for the entire ring; more recently descriptions have been developed that take discrete symplectic maps as the (complementary) starting point. In the latter approach, the return map for a particle revolution around the entire ring is often only available as a Taylor series expansion up to a certain order. A problem then is that such a Taylor series is in general not symplectic. This conundrum, dubbed the *symplectic completion problem*, has been investigated in detail by Dragt and collaborators. Starting from the Hamiltonian description, another method is to split the Hamiltonian into monomials that can each be explicitly integrated (Channell and Neri 1996).

5.7. *Other applications*

There are many other applications, including the following.

- Constructing lattice maps. Maps evaluated in standard floating point arithmetic are not usually invertible, which can be a small but noticeable source of numerical dissipation. One way around this is to replace phase space $\mathbb{R}^n$ by a lattice, say $\varepsilon\mathbb{Z}^n$, and require the map to be a bijection of the lattice (Levesque and Verlet 1993). (This has been used in the simulation of classical mechanical systems by quantum computers (Georgeot and Shepelyansky 2001).) The only known practical method for constructing such bijections is to compose shears of the form $(x, y) \mapsto (x, y + \varepsilon\lfloor f(x)/\varepsilon \rfloor)$. So it is easy to construct lattice maps for any vector field X that is a sum of vector fields integrated by Euler's method.
- Splitting is an excellent way to construct integrators for stochastic ODEs (Misawa 2001), especially when they are geometric.
- Nonautonomous systems. The usual way to treat these is to split the corresponding autonomous systems in the extended phase space, as in Example 3. However, Blanes and Moan (2001) have proposed an interesting alternative based on the Magnus expansion. To integrate the ODE $\dot{x} = X(x,t)$ on $[t_0, t_0 + \tau]$, first calculate the autonomous vector fields

$$X_0 := \int_{t_0}^{t_0+\tau} X(x,t)\ \mathrm{d}t, \quad X_1 := \tfrac{1}{\tau} \int_{t_0}^{t_0+\tau} (t - \tfrac{1}{2}\tau) t X(x,t)\ \mathrm{d}t.$$

  Then a second-order approximation of the flow of X is given by $\exp(\tau X_0)$, and a fourth-order approximation is given by

$$\exp\big(\tfrac{1}{2}X_0 - 2X_1\big) \exp\big(\tfrac{1}{2}X_0 + 2X_1\big).$$

  Each of these vector fields is then split and an integrator constructed by composition. This can be cost-effective because more information about the t-dependence of X is used.

6. Open problems

We present some open problems varying in difficulty from fairly straightforward to perhaps impossible. In the latter case, one could aim to prove that indeed the problem is impossible to solve, as in the 'no-go' results that (under rather general conditions) (i) there are no symplectic energy-preserving integrators for nonintegrable systems (Ge and Marsden 1988), and (ii) there are no general analytic volume-preserving methods (Feng and Wang 1994).

6.1. Weak integrals

A nontrivial n-dimensional vector field can have at most $n-1$ independent first integrals. (If it has n, there would be no motion.) In contrast, it can have an arbitrarily large number of weak integrals. The integration method given in Section 3.8 works for both integrals and weak integrals (Example 13). The maximum total number of integrals and weak integrals that can be accommodated, however, is $n-1$. This leads to the following problem.

Problem 1. How does one preserve more than $n-1$ integrals and weak integrals?

6.2. Hamiltonian splitting

McLachlan and Scovel (1996) posed a number of problems in symplectic integration. A number of these are still open; among them is the following.

Problem 2. Which Hamiltonians can be written as $H = \sum_{i=1}^{n} H_i$ where one of the following holds: (i) each H_i is completely integrable; (ii) each H_i is integrable in terms of elementary functions?

Problem 3. What is the structure of the Lie algebra generated by the Hamiltonians p^2 and $V(q)$ under Poisson brackets? What is the dimension of each graded subspace and the asymptotic behaviour of this dimension as $n \to \infty$?

6.3. Volume preservation

The two known general volume-preserving methods (that is, the splitting method presented in Section 3.4 and the correction method of Shang (1994) and Quispel (1995)) give, for a C^r vector field, at best a C^{r-1} integration method (instead of a C^r one). This could lead to problems, for instance, in the preservation of KAM-tori, in certain cases. We therefore pose the following problem.

Problem 4. For which C^r divergence-free vector fields can C^r volume-preserving integrators be constructed?

6.4. Preserving two geometric properties simultaneously

Many open problems concern the simultaneous preservation of two or more geometric properties, for instance, symmetry and any other property: that is, for groups of diffeomorphisms $\mathfrak{G}$ and $\mathfrak{H}$, to construct $\mathfrak{G} \cap \mathfrak{H}$-integrators. Some problems of this type now follow.

Here we will restrict to Lie or discrete, linear or affine symmetries, and to the divergence-free, Hamiltonian and polynomial cases. We first discuss the divergence-free case. To our knowledge, for the AAA flow (Example 29), it is not known how to preserve volume and the whole reversing symmetry group. This leads us to state the following problem.

Problem 5. Which (affine) symmetries can be preserved simultaneously with volume?

As far as we know, no efficient integrator has been constructed that even preserves volume and any translation symmetry.

In the Hamiltonian case, symplectic Runge–Kutta methods also preserve all affine symmetries. They have the drawback, however, of being implicit. We therefore pose the following problem.

Problem 6. Which (affine) symmetries can be preserved by explicit symplectic integrators?

For polynomial vector fields, the following problem has not been solved in full generality.

Problem 7. For which (affine) symmetries and polynomial vector fields can explicit symmetry-preserving integrators be constructed?

Of course, one can further restrict this problem to the divergence-free or Hamiltonian case, for instance.

Problem 8. Construct geometric integrators for systems preserving volume and an integral.

This is solved in McLachlan and Quispel (2001*b*) in the special case in which the topology of the level sets of the integral is completely understood, by splitting into integrable 3-dimensional pieces. Some systems can be written $\dot{x} = S(x)\nabla I(x)$, where $S^{\mathrm{T}} = -S$, such that after splitting S leading to two-dimensional pieces, each (x_i, x_j) system is area-preserving. Then we need to solve the following.

Problem 9. Develop general area- and integral-preserving integrators for two-dimensional systems.

6.5. Splitting and composition

Problem 10. For systems that evolve in a semigroup, such as the heat equation, develop effective methods of order higher than 2.

Problem 11. In a composition method of order p, explore the relationship between the leading and subsequent error terms. Reconcile the conflicting demands of small principal errors, few stages, stability, and robustness at large step sizes. Obtain a theoretical bound for the work-error 'envelope' below which no method can operate.

Problem 12. Find optimal corrected methods for near-integrable systems $A + \varepsilon B$ of each principal error $\sum \varepsilon^n \tau^{p_n}$.

Acknowledgements

We are grateful to our families for accepting our frequent absences from home. We would also like to thank our long-suffering editor Arieh Iserles for his forbearance and Per Christian Moan and Mark Sofroniou for their useful comments. This work was partly supported by the Australian Research Council. Sections 3.5, 3.6, and 3.9 are joint work with Matthew Perlmutter, the results of Section 3.7 are joint work with Seung-Hee Joo, and the first part of Section 3.14 is joint work with Per Christian Moan.

REFERENCES

G. P. Agrawal (1989), *Nonlinear Fiber Optics*, Academic Press, Boston.

M. P. Allen and D. J. Tildesley (1987), *Computer Simulation of Liquids*, Oxford University Press, Oxford.

K. A. Bagrinovskii and S. K. Godunov (1957), Difference schemes for multidimensional problems, *Dokl. Acad. Nauk SSSR* **115**, 431–433.

D. A. Barry, K. Bajracharya and C. T. Miller (1996), Alternative split-operator approach for solving chemical reaction groundwater transport models, *Adv. Water Res.* **19**, 261–275.

E. Barth, B. Leimkuhler and S. Reich (1999), A time-reversible variable-stepsize integrator for constrained dynamics, *SIAM J. Sci. Comput.* **21**, 1027–1044.

S. Blanes (2001), High order numerical integrators for differential equations using composition and processing of low order methods, *Appl. Numer. Math.* **37**, 289–306.

S. Blanes and P. C. Moan (2000), Practical symplectic partitioned Runge–Kutta and Runge–Kutta–Nyström methods. Preprint.

S. Blanes and P. C. Moan (2001), Splitting methods for non-autonomous Hamiltonian equations, *J. Comput. Phys.* **170**, 205–230.

S. Blanes, F. Casas and J. Ros (2000*a*), Symplectic integrators with processing: A general study, *SIAM J. Sci. Comput.* **21**, 711–727.

S. Blanes, F. Casas and J. Ros (2000*b*), Processing symplectic methods for near-integrable Hamiltonian systems, *Celest. Mech. Dyn. Astr.* **77**, 17–35.

S. Blanes, F. Casas and J. Ros (2001), High-order Runge–Kutta–Nyström geometric methods with processing, *Appl. Numer. Math.* **39**, 245–259.

P. Bochev and C. Scovel (1994), On quadratic invariants and symplectic structure, *BIT* **34**, 337–345.

H. W. Broer, G. B. Huitema and M. B. Sevryuk (1996), *Quasi-Periodic Motions in Families of Dynamical Systems*, Vol. 1645 of *Lecture Notes in Mathematics*, Springer, Berlin.

C. J. Budd and A. Iserles (1999), Geometric integration: Numerical solution of differential equations on manifolds, *R. Soc. London Philos. Trans. A* **357**, 943–1133.

C. J. Budd and M. D. Piggott (2002), Geometric integration and its applications, lecture notes available at `http://www.bath.ac.uk/~mascjb/home.html`.

J. Butcher (1969), The effective order of Runge–Kutta methods, in *Conf. Numer. Sol. Diff. Eqs.*, Vol. 109 of *Lecture Notes in Mathematics*, Springer, Berlin, pp. 133–139.

M. P. Calvo, A. Iserles and A. Zanna (1997), Numerical solution of isospectral flows, *Math. Comput.* **66**, 1461–1486.

E. Cartan (1909), Les groupes de transformation continus, infinis, simples, *Ann. Sci. Ecole Norm. Sup.* **26**, 93–161; *Œuvres Complètes d'Elie Cartan*, Springer, Berlin, 1984.

P. J. Channell and F. R. Neri (1996), A brief introduction to symplectic integrators, in *Integration Algorithms and Classical Mechanics* (J. E. Marsden, G. W. Patrick and W. F. Shadwick, eds), AMS, pp. 45–58.

P. J. Channell and J. C. Scovel (1990), Symplectic integration of Hamiltonian systems, *Nonlinearity* **3**, 231–259.

S. A. Chin and D. W. Kidwell (2000), Higher-order force gradient symplectic algorithms, *Phys. Rev. E* **62**, 8746–8752, Part B.

B. V. Chirikov (1979), A universal instability of many-dimensional oscillators systems, *Phys. Rep.* **52**, 263–379.

M. Crandall and A. Majda (1986), The method of fractional steps for conservation laws, *Numer. Math.* **34**, 285–314.

J. B. De Lambre (1790–91), De l'usage du calcul différentiel dans la construction des tables astronomiques, *Mémoires de L'Académie Royale des Sciences de Turin* **5**, 143–180.

H. De Raedt (1987), Product formula algorithms for solving the time dependent Schrödinger equation, *Comput. Phys. Rep.* **7**, 1–72.

H. De Raedt and B. De Raedt (1983), Applications of the generalized Trotter formula, *Phys. Rev. A* **28**, 3575–3580.

H. De Raedt and A. Lagendijk (1985), Monte Carlo simulations of quantum statistical lattice models, *Phys. Rep.* **127**, 233–307.

R. Descartes (1637), *Discours de la Méthode*, Leiden. English edition *Discourse on Method*, translated by L. J. Lafleur: Bobbs-Merrill (1964), p. 15.

R. Devogelaere (1956), Methods of integration which preserve the contact transformation property of the Hamiltonian equations, Report 4, Center of Numerical Analysis, University of Notre Dame.

L. Dieci, R. D. Russell and E. S. van Vleck (1997), On the computation of Lyapunov exponents for continuous dynamical systems, *SIAM J. Numer. Anal.* **34**, 402–423.

V. Dorodnitsyn (1996), Continuous symmetries of finite-difference evolution equations and grids, in *Symmetries and Integrability of Difference Equations* (Estérel, PQ, 1994), AMS, Providence, pp. 103–112.

A. J. Dragt (2002) *Lie Methods for Nonlinear Dynamics with Applications to Accelerator Physics.* Technical Report, Physics Department, University of Maryland.

A. J. Dragt and D. T. Abell (1996), Symplectic maps and computation of orbits in particle accelerators, in *Integration Algorithms and Classical Mechanics* (J. E. Marsden, G. W. Patrick and W. F. Shadwick, eds), AMS, Providence, pp. 59–85.

A. Dullweber, B. Leimkuhler and R. I. McLachlan (1997), Split-Hamiltonian methods for rigid body molecular dynamics, *J. Chem. Phys.* **107**, 5840–5851.

J. Dutka (1997), A note on 'Kepler's equation', *Arch. Hist. Exact Sci.* **51**, 59–65.

K. Feng (1992), Formal power series and numerical algorithms for dynamical systems, in *Proc. Conf. Scientific Computation Hangzhou, 1991* (T. Chan and Z.-C. Shi, eds), World Scientific, Singapore, pp. 28–35.

K. Feng (1993), Symplectic, contact and volume-preserving algorithms, in *Proc 1st China–Japan Conf. Numer. Math.* (Z. C. Shi, T. Ushijima, eds), World Scientific, Singapore.

K. Feng (1998), Contact algorithms for contact dynamical systems, *J. Comput. Math.* **16**, 1–14.

K. Feng and D.–L. Wang (1994), Dynamical systems and geometric construction of algorithms, in Vol. 163 of *Contemporary Mathematics*, AMS, Providence, pp. 1–32.

K. Feng and D.-L. Wang (1998), Variations on a theme by Euler, *J. Comput. Math.* **16**, 97–106.

E. Forest (1998), *Beam Dynamics*, Harwood Academic Publishers.

E. Forest and R. Ruth (1990), Fourth-order symplectic integration, *Physica D* **43**, 105–117.

T. Fukushima (1999), Fast procedure solving universal Kepler's equation, *Cel. Mech. Dyn. Astron.* **75**, 201–226.

Z. Ge and J. Marsden (1988), Lie–Poisson Hamilton–Jacobi theory and Lie–Poisson integrators, *Phys. Lett. A* **133**, 135–139.

B. Georgeot and D. L. Shepelyansky (2001), Stable quantum computation of unstable classical chaos, *Phys. Rev. Lett.* **86**, 5393–5396.

R. Glowinski, T. W. Pan, T. I. Hesla and D. D. Joseph (1999), A distributed Lagrange multiplier fictitious domain method for particulate flows, *Int. J. Multiphase Flow* **25**, 755–794.

S. K. Godunov (1999), Reminiscences about difference schemes, *J. Comput. Phys.* **163**, 6–25.

J. Guckenheimer and P. Holmes (1983), *Nonlinear Oscillations, Dynamical Systems, and Bifurcations of Vector Fields*, Vol. 42 of *Applied Mathematical Sciences*, Springer, New York.

E. Hairer, Ch. Lubich and G. Wanner (2002), *Geometric Numerical Integration: Structure-Preserving Algorithms for Ordinary Differential Equations*, Vol. 31 of *Springer Series in Computational Mathematics*, Springer.

R. H. Hardin and F. D. Tappert (1973), Applications of the split-step Fourier method to the numerical solution of nonlinear and variable coefficient wave equations, *SIAM Review* **15**, 423.

D. J. Hardy, D. I. Okunbor and R. D. Skeel (1999), Symplectic variable step size integration for N-body problems, *Appl. Numer. Math.* **29**, 19–30.

J. Hilgert, K. Hofmann, K. Heinrich and J. D. Lawson, *Lie Groups, Convex Cones, and Semigroups*, Oxford University Press, New York.

A. Iserles (1984), Composite methods for numerical solution of stiff systems of ODE's, *SIAM J. Numer. Anal.* **21**, 340–351.

A. Iserles and G. R. W. Quispel (2002), Stable compositions for parabolic PDEs and stiff ODEs. In preparation.

A. Iserles, H. Z. Munthe-Kaas, S. P. Nørsett and A. Zanna (2000), Lie group methods, *Acta Numerica*, Vol. 9, Cambridge University Press, pp. 215–365.

S.-J. Kang and M.-H. Kim (1996), Free Lie algebras, generalized Witt formula, and the denominator identity, *J. Algebra* **183**, 560–594.

K. H. Karlsen and K.-A. Lie (1999), An unconditionally stable splitting scheme for a class of nonlinear parabolic equations, *IMA J. Numer. Anal.* **19**, 609–635.

F. Klein (1893), A comparative review of recent researches in geometry, *Bull. New York Math. Soc.* **2**, 215–249. Translated by M. W. Haskell.

S. Kobayashi (1972), *Transformation Groups in Differential Geometry*, Springer, Berlin.

D. A. Kofke (2000), Molecular Simulation, available from:
`http://www.cheme.buffalo.edu/courses/ce530/Lectures/`

P.-V. Koseleff (1995), About approximations of exponentials, in *Applied Algebra, Algebraic Algorithms and Error-Correcting Codes* (Paris, 1995), Vol. 948 of *Lecture Notes in Computer Science*, Springer, Berlin, pp. 323–333.

A. Kværnø and B. Leimkuhler (2000), A time-reversible, regularized, switching integrator for the N-body problem, *SIAM J. Sci. Comput.* **22**, 1016–1035.

J. L. Lagrange (1788), *Mécanique Analytique*, Chez la Veuve Disaint, Paris.

J. S. W. Lamb, ed. (1998), Time-reversal symmetry in dynamical systems, *Physica D* **112**, 1–328.

M. Lecar, F. A. Franklin, M. J. Holman and M. W. Murray (2001), Chaos in the solar system, *Ann. Rev. Astron. Astrophys.* **39**, 581–631.

B. Leimkuhler (2002), Molecular Dynamics, available from:
`http://www.math.ukans.edu/~leimkuhl/molecular.html`

B. Leimkuhler and S. Reich (2001), A reversible averaging integrator for multiple time-scale dynamics, *J. Comput. Phys.* **171**, 95–114.

B. J. Leimkuhler and R. D. Skeel (1994), Symplectic numerical integrators in constrained Hamiltonian systems, *J. Comput. Phys.* **112**, 117–125.

B. J. Leimkuhler, S. Reich and R. D. Skeel (1996), Integration methods for molecular dynamics, in *Mathematical Approaches to Biomolecular Structure and Dynamics* (Minneapolis, MN, 1994), Vol. 82 of *IMA Volumes in Mathematics and its Applications*, Springer, New York, pp. 161–185.

B. P. Leonard, A. P. Lock and M. K. MacVean (1996), Conservative explicit unrestricted-time-step multidimensional constancy-preserving advection schemes, *Monthly Weather Rev.* **124**, 2588–2606.

R. J. LeVeque (1998), Balancing source terms and flux gradients in high-resolution Godunov methods: The quasi-steady wave-propagation algorithm, *J. Comput. Phys.* **146**, 346–365.

D. Levesque and L. Verlet (1993), Molecular dynamics and time reversibility, *J. Statist. Phys.* **72**, 519–537.

O. Loos (1969), *Symmetric Spaces: I General Theory; II Compact Spaces and Classification*, W. A. Benjamin, Inc., New York/Amsterdam.

M. A. López-Marcos, J. M. Sanz-Serna and R. D. Skeel (1996), Cheap enhancement of symplectic integrators, in *Numerical Analysis 1995* (Dundee, 1995), Vol. 344 of *Pitman Research Notes in Mathematics*, Longman, Harlow, pp. 107–122.

M. A. López-Marcos, J. M. Sanz-Serna and R. D. Skeel (1997), Explicit symplectic integrators using Hessian-vector products, *SIAM J. Sci. Comput.* **18**, 223–238.

B. A. Malomed and M. I. Tribelsky (1984), Bifurcations in distributed kinetic systems with aperiodic instability, *Physica D* **14**, 67–87.

J. E. Marsden and T. S. Ratiu (1999), *Introduction to Mechanics and Symmetry*, 2nd edn, Springer, New York.

J. E. Marsden and M. West (2001), Discrete mechanics and variational integrators, *Acta Numerica*, Vol. 10, Cambridge University Press, pp. 357–514.

R. I. McLachlan (1993), Explicit Lie–Poisson integration and the Euler equations, *Phys. Rev. Lett.* **71**, 3043–3046.

R. I. McLachlan (1994), Symplectic integration of Hamiltonian wave equations, *Numer. Math.* **66**, 465–492.

R. I. McLachlan (1995), On the numerical integration of ordinary differential equations by symmetric composition methods, *SIAM J. Sci. Comput.* **16**, 151–168.

R. I. McLachlan (2002), Families of high-order composition methods. To appear.

R. I. McLachlan and M. Perlmutter (2001), Conformal Hamiltonian systems, *J. Geom. Phys.* **39**, 276–300.

R. I. McLachlan and G. R. W. Quispel (2000), Numerical integrators that contract volume, *Appl. Numer. Math.* **34**, 253–260.

R. I. McLachlan and G. R. W. Quispel (2001*a*), Six lectures on geometric integration, in *Foundations of Computational Mathematics* (R. DeVore, A. Iserles and E. Süli, eds), Cambridge University Press, Cambridge, pp. 155–210.

R. I. McLachlan and G. R. W. Quispel (2001*b*), What kinds of dynamics are there? Lie pseudogroups, dynamical systems, and geometric integration, *Nonlinearity* **14**, 1689–1706.

R. I. McLachlan and N. Robidoux (2000), Antisymmetry, pseudospectral methods, and conservative PDEs, in *Proc. Int. Conf. EQUADIFF 99* (B. Fiedler, K. Gröger and J. Sprekels, eds), World Scientific, pp. 994–999.

R. I. McLachlan and C. Scovel (1996), Open problems in symplectic integration, in *Integration Algorithms and Classical Mechanics* (J. E. Marsden, G. W. Patrick and W. F. Shadwick, eds), AMS, pp. 151–180.

R. I. McLachlan, G. R. W. Quispel and G. S. Turner (1998), Numerical integrators that preserve symmetries and reversing symmetries, *SIAM J. Numer. Anal.* **35**, 586–599.

R. I. McLachlan, G. R. W. Quispel and N. Robidoux (1999), Geometric integration using discrete gradients, *Phil. Trans. Roy. Soc. A* **357**, 1021–1046.

R. I. McLachlan, M. Perlmutter and G. R. W. Quispel (2001), On the nonlinear stability of symplectic integrators. Preprint.

R. I. McLachlan, M. Perlmutter and G. R. W. Quispel (2002), Lie group foliate integrators. In preparation.

T. Misawa (2001), A Lie algebraic approach to numerical integration of stochastic differential equations, *SIAM J. Sci. Comput.* **23**, 866–890.

P. Molino (1988), *Riemannian Foliations*, Birkhäuser, Boston.

H. Munthe-Kaas and B. Owren (1999), Computations in a free Lie algebra, *R. Soc. London Philos. Trans. A* **357**, 957–981.

H. Munthe-Kaas and A. Zanna (1997), Numerical integration of differential equations on homogeneous manifolds, *Foundations of Computational Mathematics* (Rio de Janeiro, 1997; F. Cucker and M. Shub, eds), Springer, Berlin. pp. 305–315.

H. Z. Munthe-Kaas, G. R. W. Quispel and A. Zanna (2002), Applications of symmetric spaces and Lie triple systems in numerical analysis. Preprint.

A. Murua and J. M Sanz-Serna (1999), Order conditions for numerical integrators obtained by composing simpler integrators, *R. Soc. London Philos. Trans. A* **357**, 1079–1100.

J. D. Murray (1989), *Mathematical Biology*, Springer, Berlin.

F. Neri (1988), Lie algebras and canonical integration. Preprint, Department of Physics, University of Maryland.

G. R. W. Quispel (1995), Volume-preserving integrators, *Phys. Lett. A* **206**, 226–230.

G. R. W. Quispel and D. I. McLaren (2002), Explicit volume-preserving and symplectic integrators for trigonometric polynomial flows. Preprint.

S. Reich (1993), Numerical integration of the generalized Euler equations. Technical report 93–20, Department of Computer Science, University of British Columbia.

S. Reich (1999), Backward error analysis for numerical integrators, *SIAM J. Numer. Anal.* **36**, 1549–1570.

J. A. G. Roberts and G. R. W. Quispel (1992), Chaos and time-reversal symmetry. Order and chaos in reversible dynamical systems, *Phys. Rep.* **216**, 63–177.

R. D. Ruth (1983), A canonical integration technique, *IEEE Trans. Nucl. Sci.* **NS–30**, 2669–2671.

J. P. Ryckaert, G. Ciccotti and H. J. C. Berendsen (1977), Numerical integration of the Cartesian equations of motion of a system with constraints: Molecular dynamics of n-alkanes, *J. Comput. Phys.* **23**, 327–341.

D. S. Ryu, T. W. Jones and A. Frank (1995), Numerical magnetohydrodynamics in astrophysics: Algorithm and tests for multidimensional flow, *Astrophys. J.* **452**, 785–796.

J. M. Sanz-Serna (1997), Geometric integration, in *The State of the Art in Numerical Analysis* (York, 1996), Vol. 63 of *Inst. Math. Appl. Conf. Ser. New Ser.*, Oxford University Press, New York, pp. 121–143.

Z. J. Shang (1994), Construction of volume-preserving difference schemes for source-free systems via generating functions, *J. Comput. Math.* **12**, 265–272.

Z. J. Shang (2000), Resonant and diophantine step sizes in computing invariant tori of Hamiltonian systems, *Nonlinearity* **13**, 299–308.

Q. Sheng (1989), Solving linear partial differential equations by exponential splitting, *IMA J. Numer. Anal.* **9**, 199–212.

J. Shi and Y. T. Yan (1993), Explicitly integrable polynomial Hamiltonians and evaluation of Lie transformations, *Phys. Rev. E* **48**, 3943–3951.

R. D. Skeel and K. Srinivas (2000), Nonlinear stability analysis of area-preserving integrators, *SIAM J. Numer. Anal.* **38**, 129–148 (electronic).

C. Störmer (1907), Sur les trajectoires des corpuscules électrisés, *Arch. Sci. Phys. Nat.* **24**, pp. 5–18, 113–158, 221–247.

G. Strang (1963), Accurate partial difference methods, I: Linear Cauchy problems, *Arch. Rat. Mech.* **12**, 392–402.

G. Strang (1968), On the construction and comparison of difference schemes, *SIAM J. Numer. Anal.* **5**, 507–517.

A. M. Stuart and A. R. Humphries (1996), *Dynamical Systems and Numerical Analysis*, Cambridge University Press, Cambridge.

G. J. Sussman and J. Wisdom (1992), Chaotic evolution of the solar system, *Science* **257**, 56–62.

M. Suzuki (1976), Generalized Trotter's formula and systematic approximants of exponential operators and inner derivations with applications to many-body problems, *Comm. Math. Phys.* **51**, 183–190.

M. Suzuki (1990), Fractal decomposition of exponential operators with applications to many-body theories and Monte Carlo simulations, *Phys. Lett. A* **146**, 319–323.

J. B. Swift and P. C. Hohenberg (1989), Rayleigh–Bénard convection with time-dependent boundary conditions, *Phys. Rev. A* (3) **39**, 4132–4136.

M. Takahashi and M. Imada (1984), Monte Carlo calculations of quantum systems, II: Higher order correction, *J. Phys. Soc. Japan* **53**, 3765–3769.

T. Tang and Z. H. Teng (1995), Error bounds for fractional step methods for conservation laws with source terms, *SIAM J. Numer. Anal.* **32**, 110–127.

F. D. Tappert (1974), Numerical solutions of the Korteweg–de Vries equation and its generalizations by the split-step Fourier method, in *Nonlinear Wave Motion* (A. C. Newell, ed.), Vol. 15 of *Lectures in Applied Mathematics*, AMS, pp. 215–216.

F. D. Tappert (1977), The parabolic approximation method, in *Wave Propagation and Underwater Acoustics* (J. B. Keller and J. S. Papadakis, eds), Vol. 70 of *Lecture Notes in Physics*, Springer, New York, pp. 224–287.

F. D. Tappert and M. G. Brown (1996), Asymptotic phase errors in parabolic approximations to the one-way Helmholtz equation, *J. Acoust. Soc. Amer.* **99**, 1405–1413.

J. B. Taylor (1968), unpublished.

H. F. Trotter (1959), On the product of semi-groups of operators, *Proc. Amer. Math. Soc.* **10**, 545–551.

Z. Tsuboi and M. Suzuki (1995), Determining equations for higher-order decompositions of exponential operators, *Int. J. Mod. Phys. B* **25**, 3241–3268.

S. Turek (1996), A comparative study of time-stepping techniques for the incompressible Navier–Stokes equations: From fully implicit non-linear schemes to semi-implicit projection methods, *Int. J. Numer. Meth. Fluids* **22**, 987–1011.

R. Tyson, L. G. Stern and R. J. LeVeque (2000), Fractional step methods applied to a chemotaxis model, *J. Math. Biol.* **41**, 455–475.

A. van den Essen (2000), *Polynomial Automorphisms and the Jacobian Conjecture*, Vol. 190 of *Progress in Mathematics*, Birkhäuser, Basel.

L. Verlet (1967), Computer 'experiments' on classical fluids, I: Thermodynamical properties of Lennard–Jones molecules, *Phys. Rev.* **159**, 98–103.

V. Volterra (1931), *Leçons sur la Théorie Mathématique de la Lutte pour la Vie*, reprinted by Éditions Jacques Gabay, Sceaux (1990).

J. Weickert, B. M. T. Romeny and M. A. Viergever (1998), Efficient and reliable schemes for nonlinear diffusion filtering, *IEEE Trans. Image Proc.* **7**, 398–410.

A. Weinstein (1983), The local structure of Poisson manifolds, *J. Diff. Geom.* **18**, 523–557.

J. Wisdom (1982), The origin of the Kirkwood gaps: A mapping for asteroidal motion near the 3/1 commensurability, *Astron. J.* **87**, 577–593.

J. Wisdom and M. Holman (1991), Symplectic maps for the N-body problem, *Astron. J.* **102**, 1528–1538.

J. Wisdom, M. Holman and J. Touma (1996), Symplectic correctors, in *Integration Algorithms and Classical Mechanics* (J. E. Marsden, G. W. Patrick and W. F. Shadwick, eds), AMS, Providence, pp. 217–244.

B. Yang and S. B. Pope (1998), An investigation of the accuracy of manifold methods and splitting schemes in the computational implementation of combustion chemistry, *Combust. Flame* **112**, 16–32.

H. Yoshida (1990), Construction of higher order symplectic integrators, *Phys. Lett. A* **150**, 262–268.

Appendix A: Proof of Theorem 5

We need to show that a given divergence-free X can be written as $X = \nabla \cdot S$. To do this we shall first construct a splitting into two-dimensional volume-preserving systems, and then construct the associated matrix S.

Start with the system of ODEs

$$\dot{x}_i = f_i(x), \quad i = 1, \ldots, n.$$

This can be rewritten equivalently as

$$\begin{aligned} \dot{x}_i &= f_i(x), \quad i = 1, \ldots, n-1, \\ \dot{x}_n &= \left(f_n(x) + \sum_{j=1}^{n-2} \int \frac{\partial f_j}{\partial x_j} \, \mathrm{d}x_n \right) - \sum_{j=1}^{n-2} \int \frac{\partial f_j}{\partial x_j} \, \mathrm{d}x_n. \end{aligned} \tag{A.1}$$

We now split f as the sum of $n-1$ two-dimensional divergence-free vector fields. The first $n-2$ are

$$\begin{aligned} \dot{x}_i &= 0, \quad i \neq j, n, \\ \dot{x}_j &= f_j(x), \\ \dot{x}_n &= -\int \frac{\partial f_j}{\partial x_j} \, \mathrm{d}x_n, \end{aligned} \tag{A.2}$$

for $j = 1, \dots, n-2$, and the final one is

$$\begin{aligned} \dot{x}_i &= 0, \quad i = 1, \dots, n-2, \\ \dot{x}_{n-1} &= f_{n-1}(x), \\ \dot{x}_n &= f_n(x) + \sum_{j=1}^{n-2} \int \frac{\partial f_j}{\partial x_j} \, \mathrm{d}x_n. \end{aligned} \tag{A.3}$$

It is trivial to check that the vector fields (A.2) are divergence-free. To verify that (A.3) is also divergence-free, take its divergence and use $\nabla \cdot f = 0$. One can take any arbitrary integration constraints in (A.2) as long as the same ones are used in the corresponding anti-derivatives in (A.3).

Now observe that each subsystem is Hamiltonian in (x_i, x_n) variables, the Hamiltonians being the entries in $S(x)$; we have $X = \nabla S$ for

$$S = \begin{pmatrix} 0 & \dots & 0 & H_1 \\ \vdots & & \vdots & \vdots \\ 0 & \dots & 0 & H_{n-1} \\ -H_1 & \dots & -H_{n-1} & 0 \end{pmatrix},$$

where $H_j = \int f_j(x) \, \mathrm{d}x_n$ for $j = 1, \dots, n-2$, while H_{n-1} is determined by

$$\begin{aligned} \frac{\partial H_{n-1}}{\partial x_n} &= f_{n-1}(x), \\ \frac{\partial H_{n-1}}{\partial x_{n-1}} &= -f_n(x) - \sum_{j=1}^{n-2} \int \frac{\partial f_j}{\partial x_j} \, \mathrm{d}x_n, \end{aligned} \tag{A.4}$$

or

$$H_{n-1} = \int_C \left[f_{n-1}(x) \, \mathrm{d}\ell_{x_n} - \left(f_n(x) + \sum_{j=1}^{n-2} \int \frac{\partial f_j}{\partial x_j} \, \mathrm{d}x_n \right) \mathrm{d}C \right], \tag{A.5}$$

in which C denotes an arbitrary curve in the (x_{n-1}, x_n) plane going from $(0,0)$ to (x_{n-1}, x_n).

Appendix B: Splitting polynomials

Proof of Theorem 16. The usual basis for polynomials is the monomials $x_1^{i_1} x_2^{i_2} \dots x_n^{i_n}$, which we write using multi-indices as $\mathbf{x}^{\mathbf{i}}$. Here we want to derive a basis in terms of the perfect powers $(a_1 x_1 + \dots + a_n x_n)^m = (\mathbf{a}^{\mathrm{T}} \mathbf{x})^m$. For example, we can write $x_1 x_2 = \frac{1}{2}(x_1 + x_2)^2 - \frac{1}{2} x_1^2 - \frac{1}{2} x_2^2$. We shall show that there exists a vector $\mathbf{a} \in \mathbb{R}^n$ such that the required basis (3.13) exists with $k_{ji} = a_i^j$, $k = 0, \dots, d_m$, where $d_m + 1 = \frac{(n+m-1)!}{m!(n-1)!}$ is the dimension of the space of homogeneous polynomials of degree m. This is true if each monomial can be uniquely expressed in the basis. This gives a system of $d_m + 1$ equations, namely

$$(\mathbf{a}^{\mathrm{T}}\mathbf{x})^j = \sum_{\mathbf{i}} c_{\mathbf{i}} \mathbf{a}^{j\mathbf{i}} \mathbf{x}^{\mathbf{i}}, \quad j = 0, \ldots, d_m, \tag{B.1}$$

where the $c_{\mathbf{i}}$ are the multinomial coefficients defined by the expansion of $(x_1 + \cdots + x_n)^m$. Regarding these as linear equations in the $d_m + 1$ unknowns $c_{\mathbf{i}}\mathbf{x}^{\mathbf{i}}$, they have a unique solution if and only if the determinant of the coefficient matrix is nonzero. The coefficient matrix $\mathbf{a}^{j\mathbf{i}}$ is Vandermonde, with determinant

$$D = \prod_{\mathbf{k} \neq \mathbf{l}} (\mathbf{a}^{\mathbf{k}} - \mathbf{a}^{\mathbf{l}}).$$

Therefore, $D \neq 0$ if and only if $\mathbf{a}^{\mathbf{k}} \neq \mathbf{a}^{\mathbf{l}}$ for all $\mathbf{k} \neq \mathbf{l}$. Taking logs, $D \neq 0$ if

$$(\mathbf{k} - \mathbf{l})^{\mathrm{T}} \log \mathbf{a} \neq 0$$

for all $\mathbf{k} \neq \mathbf{l}$ with $k_i \geq 0$, $l_i \geq 0$, and $\sum k_i = \sum l_i = m$. The precise set of valid **a**s depends on m, but certainly $D \neq 0$ if *no* integer combination $\sum c_i \log a_i$ with $\sum c_i = 0$ is zero; or, setting without loss of generality $a_1 = 1$, if $\log a_2, \ldots, \log a_n$ are linearly independent over $\mathbb{Z}$. For example, choosing $a_2, \ldots, a_n$ to be the first $n-1$ prime numbers is sufficient, by unique factorization. That is, $D \neq 0$ if none of the a_i is a nonzero rational power of the others.

Having chosen such an **a**, solving the linear equations (B.1) lets one express any homogeneous degree m polynomial in the basis $(\mathbf{a}^j)^{\mathrm{T}}\mathbf{x}$.

By construction, the initial portions of this basis, *i.e.*, $(\mathbf{a}^j)^{\mathrm{T}}\mathbf{x}$ for $j = 0, \ldots, d_p$, form a basis for the monomials of any degree $p < m$. □

We remark that the basis constructed here may not be the best one to use in practice, which remains an interesting problem.

Appendix C: 'On the usage of differential calculus in the construction of astronomical tables', by M. Delambre

Jean Baptiste Delambre (DL), *1749–1822, did not even begin studying astronomy until his early 30s. In 1771 he tutored the son of M. d'Assy, Receiver General of Finances, and in 1788 d'Assy built an observatory for DL. Here in 1792 he published* Tables du Soleil, de Jupiter, de Saturne, d'Uranus et des satellites de Jupiter. *He won a prize for his work in determining the orbit of the recently discovered Uranus. We translate here a little of an article he wrote at about this time (De Lambre 1790). DL is also noted for measuring (in order to establish the metre unit of length) a baseline from Dunkirk to Barcelona, with trigonometric tables in degrees and new, metric survey instruments in gradians, during a revolution; and for completing an enormous technical history of all astronomy.*

In the construction of astronomical tables one is ordinarily content to determine a certain number of terms exactly, at larger or smaller intervals, depending on whether the progression of the differences is large or small.

If the first differences are relatively equal, one fills in the gaps by simple proportional parts, but this case is rather rare when one aspires to high precision.

If uniformity is only found in the second differences, there are formulae & easy tables to correct the errors in the simple proportional parts; but this method, which entails a sometimes tiring precision in the fundamental calculations, itself becomes sometimes insufficient; & the methods proposed to make up for this have seemed to me long & painful. This is what has led me to search for more certain and faster methods.

I. The idea that presents itself first is to differentiate the formula with which one calculates the table. In this way one obtains differentials of as many orders as needed. One can calculate them in advance & form subsidiary tables of them that considerably diminish the work. This method has always worked for me & often much beyond my expectations; I am going to apply it to the construction of the most used tables in the practice of astronomy.

II. There are no more useful ones than those of logarithms. The ones we have appear exact & sufficient; but one can be curious to see what would be the most certain & the most easy methods to reconstruct them, perfect them, or extend them.

Let N be any number, M the base of the common tables, that is, the number

$$0.4342944819032518276511289189166050822943970058 04.$$

One knows that the differential of the logarithm of N, or $d\log N = M[\frac{dN}{N} - \frac{1}{2}(\frac{dN}{N})^2 + \frac{1}{3}(\frac{dN}{N})^3 + \&c.]$. If one supposes dN infinitely small,[4] the expression reduces to $d\log N = M\frac{dN}{N}$.

III. The second difference will have the expression $dd\log N = -M(\frac{dN}{N})^2$. Similarly for the third difference, we have $ddd\log N = 2M(\frac{dN}{N})^3$, & for the fourth $dddd\log N = -3M(\frac{dN}{N})^4$. If N is very large with respect to

[4] DL interprets dN as either the differential or forward difference of N, as the occasion demands.

dN, the error of these formulas will hardly be felt. Suppose therefore that $dN = 1$, as is necessary to construct a table, and we will have[5] $dd \log N = -\frac{M}{N^2}$.[6]

...

X. Let us now propose to construct tables of the equation of the centre and of the ray vector for the planets.[7]

We call z the mean anomaly, u the true anomaly, q the equation of the centre, a & b the two semi-axes of the ellipse, e the eccentricity and s the sinus of $1''$.[8] One generally has $q = z - u$, hence $dq = dz - du = 1^\circ - du$ when one constructs a table.

Hence also $ddq = -ddu = -d(\frac{abdz}{rr}) = +\frac{2abdzdr}{r^3}$; moreover $r = \frac{bb}{a - e\cos u}$ & $dr = -\frac{aesdz \sin u}{b}$; then $ddq = -\frac{2a^2esd^2z \sin u(a-e\cos u)^3}{b^6}$, or for simplicity putting $a = 1$, & giving to the other letters the values corresponding to this assumption, & you will have $ddq = -\frac{2esd^2z \sin u(1-e\cos u)^3}{b^6}$. This expression can easily be made into a table taking u as argument; it suffices to calculate in steps of 5° except when $\sin u$ is very large.[9] One fills in the gaps by proportional parts.

The infinitesimal formula $du = \frac{bdz}{rr} = \frac{dz}{b^3}(1 - e\cos u)^2$ is not at all exact enough in practice,[10] even putting $(r + \frac{1}{2}dr)^2$ & $(u + \frac{1}{2}du)$ in lieu of r^2 & u.[11] It is easy to demonstrate that the true expression is $\sin du = \frac{bdz}{r(r+dr)}$, and this formula will be more accurate to the extent that one can consider the little elliptic arc between the two ray vectors r & $(r + dr)$ to be a straight

[5] The first appearance of the leapfrog method? DL calculates values of $u(t)$ by applying the leapfrog method to $\ddot{u} = f(t)$, where $f(t)$ is easier to calculate than $u(t)$.

[6] DL gives various calculations of $\log N$ by this method. In IV, he adds more terms in the Taylor series of $dd \log N$. In V, he expands $d \log N$ in inverse odd powers of $2N + 1$. In VI, he illustrates how the first term in his series suffices when N is large enough. In VII–IX he extends the method to trigonometric functions, using trigonometric identities to simplify and improve the methods.

[7] The ODEs that DL studies are special in that they can be solved implicitly, which allows one to check and correct the error. This is still done today when we solve, say, $f(u, \alpha) = 0$ by continuation in the parameter α. For the history of solving Kepler's and related equations, see Dutka (1997) and Fukushima (1999) and references therein.

[8] Angles are measured in seconds, and $d \sin u = s \cos u \, du$.

[9] That is, DL wants to solve the ODE $\dot{q} = f(q)$. He does this by applying the leapfrog method (III) to $\ddot{q} = f'(q)f(q)$, which is essentially a symplectic lift of the original ODE, although not the usual one $\dot{q} = f(q)$, $\dot{p} = -f'(q)p$. He is tantalizingly close to applying leapfrog to a genuine mechanical system $\ddot{q} = -\nabla V(q)$.

[10] That is, Euler's method is not accurate enough.

[11] An early appearance of the implicit midpoint rule? DL now goes on to approximate this equation further for his example, arriving at essentially a second-order Runge–Kutta method.

line. One can always, even for Mercury, put du in lieu of $\sin du$, & one has

$$du = \frac{dz}{b^3}\left[1 - e\cos u - e\cos(u+du) + e^2\cos u\cos(u+du)\right],$$

hence

$$dq = dz - \tfrac{dz}{b^3} + \tfrac{edz}{b^3}\cos u + \tfrac{edz}{b^3}\cos(u+du) - \tfrac{e^2dz}{b^3}\cos u\cos(u+du).$$

The two first terms are constant, the next are easy to calculate because the progress of differences always gives the value of du up to a few seconds, which is sufficient for the terms which depend on $(u+du)$. Only for the first of the first differences is it necessary to use a little trial and error.[12]

...

[T]his[13] is indeed the equation that one finds in the new table of Monsieur De la Lande. From this one can judge the precision of this method that without trial and error corrects an error of $3^\circ\, 8'\, 31''$.

This solution of Kepler's problem appears to me the shortest of all the ones I know. I invite those who doubt this to calculate the same example by the methods of Cassini, Simpson & La Caille. The last says in his astronomy lessons that for no planet of the solar system can the adjustment of the method go to three iterations. Apparently he has done all his trials with the mean anomaly in the first quadrant, & then he may be right, but this is not nearly true in the second quadrant & in our example six iterations would be necessary.[14,15]

...

One could apply the same methods to the construction of several other tables, such as those of the 90th & its height, those of refractions &c., but what we have said is more than sufficient. I therefore suppress what I have done for those tables, & end here this memoir, which without any doubt is too long.

12 DL gives various methods for checking and correcting the error in this case.

13 The value of the true anomaly calculated by the Runge–Kutta method.

14 We see that little has changed in the world of numerical analysis in 200 years.

15 In XI, DL extends the method to calculating $\log r$. In XII, he applies his methods to Jupiter, Saturn, and Herschel (Uranus). In XIII, he relates the true anomaly, true longitude, and aphelion. In XIV, for the parabolic orbits of comets, he studies $du = \cos^4 \frac{1}{2}u$, for which one M. Cagnoli had foolishly proposed to use $\cos^4 \frac{1}{2}(u+du)$. DL considers instead $\cos^2(\frac{1}{2}u)\cos^2\frac{1}{2}(u+du)$, together with an Euler estimate of du, but concludes that 'the work will still be considerable; it is greatly shortened by using second differences... $ddu = -2\cos^7\frac{1}{2}u\sin\frac{1}{2}u$.' XV–XVIII consider various other trigonometric approximations. The article then concludes.

Acta Numerica (2002), pp. 435–477
DOI: 10.1017/S0962492902000065

Topological techniques for efficient rigorous computation in dynamics

Konstantin Mischaikow*
Center for Dynamical Systems and Nonlinear Studies,
Georgia Institute of Technology,
Atlanta, GA 30332, USA
E-mail: mischaik@math.gatech.edu

We describe topological methods for the efficient, rigorous computation of dynamical systems. In particular, we indicate how Conley's Fundamental Decomposition Theorem is naturally related to combinatorial approximations of dynamical systems. Furthermore, we show that computations of Morse decompositions and isolating blocks can be performed efficiently. We conclude with examples indicating how these ideas can be applied to finite- and infinite-dimensional discrete and continuous dynamical systems.

CONTENTS

1. Introduction

This paper is an expository article on using topological methods for the *efficient*, *rigorous* computation of dynamical systems. Of course, since its inception the computer has been used for the purpose of simulating nonlinear models. However, in recent years there has been a rapid development in numerical methods specifically designed to study these models from a dynamical systems point of view, that is, with a particular emphasis on the structures which capture the long-term or asymptotic states of the system. At the risk of greatly simplifying these results, this work has followed two themes: *indirect methods* and *direct methods*.

* Research supported in part by NSF grant DMS-9805584 and DMS-0107395.

The indirect methods are most closely associated with simulations, and as such are extremely important because they tend to be the cheapest computationally. The emphasis is on developing numerical schemes whose solutions exhibit the same dynamics as the original system: for example, if one is given a Hamiltonian system, then it is reasonable to want a numerical method that preserves the integrals of the original system. A comprehensive introduction to these questions can be found in Stuart and Humphries (1996).

The direct methods focus on the development of numerical techniques that find particular dynamical structures, for instance fixed points, periodic orbits, heteroclinic orbits, invariant manifolds, *etc.*, and are often associated with continuation methods (see Cliffe, Spence and Tavener (2000), Doedel, Champneys, Fairgrieve, Kuznetsov, Sandstede and Wang (1999) and Dieci and Eirola (2001) and references therein). To paraphrase Poincaré, these techniques provide us with a window into the rich structures that nonlinear systems exhibit.

There is no question that these methods are essential. However, they cannot capture the full dynamics. As pointed out in Stuart and Humphries (1996, p. xiii) a fundamental question for the indirect method, requiring a positive answer, is:

> Assume that the differential equation has a particular invariant set. Does the numerical method have a corresponding invariant set which converges to the true invariant set as $\Delta t \to 0$?

For non-hyperbolic systems the answer is almost surely 'no'. The set of parameters at which global bifurcations occur can be dense, which suggests that any numerical error will lead to significant distinctions in the structure of the invariant sets (Palis and Takens 1993).

Similarly, for the direct methods to compute these invariant objects effectively, they typically need to be hyperbolic, which limits the applicability for general systems. Further, one of the more interesting features of nonlinear systems is the existence of complicated or chaotic dynamics, the structure of which has not yet been captured by the above-mentioned methods.

A final observation is that, while these dynamical objects are ubiquitous, the number of specific nonlinear systems for which mathematicians have been able to prove the existence of these structures, especially as it relates to chaotic dynamics, is quite small. Therefore, it appears that there is room for a *complementary* computational method which can provide accurate information about global structures of nonlinear dynamical systems and is rigorous in the sense of a mathematical proof, albeit involving the assistance of a computer.

At the same time, such a method is of little value unless it can be shown to be computationally efficient. Hence, throughout this paper the discussion of the topological techniques will be presented in the context of the computational effort required.

For the sake of simplicity we will restrict most of our discussion to the dynamics of continuous maps. However, we will begin on a very general level, specializing as we proceed. Let X be a metric space and let $f : X \to X$ be a continuous function. Since we are not assuming that f is a homeomorphism we will make extensive use of the following definition. A *full trajectory* of f through $x \in X$ is a function $\sigma_x : \mathbb{Z} \to X$ satisfying:

(1) $\sigma_x(0) = x$, and
(2) $\sigma_x(n+1) = f(\sigma_x(n))$ for all $n \in \mathbb{Z}$.

In other words, $\sigma_x(k) = f^k(x)$, where f^k denotes the kth iterate of f. Recall that $S \subset X$ is *invariant* under f if, for every $x \in S$, there exists a full trajectory $\sigma_x : \mathbb{Z} \to S$.

Ideally we want to understand the existence and topological structure of invariant sets. The same above-mentioned conundrum shows that, as stated, this goal is completely unrealistic. Therefore, we step back and ask:

Q1. *What should one compute?*

The theme of Section 2 is that ϵ-chain-recurrent sets and isolating blocks are very computable objects.

The ϵ-chain-recurrent set consists, essentially, of those points which, having allowed for a finite number of errors of size ϵ, return to themselves under the dynamics. Conley's Fundamental Decomposition Theorem, Theorem 2.1, states that, away from the chain-recurrent set, the dynamics is gradient-like. Thus, finding the chain-recurrent set provides, on a very crude level, global information about the dynamics. This is described in detail in Section 2.1.

A compact set $N \subset X$ is an *isolating neighbourhood* for f if

$$\operatorname{Inv}(N, f) := \{x \in X \mid \exists \sigma_x : \mathbb{Z} \to N\} \subset \operatorname{int}(N),$$

where $\operatorname{int}(N)$ denotes the interior of N. $\operatorname{Inv}(N, f)$ is the *maximal invariant set* in N. An invariant set S is *isolated* if there exists an isolating neighbourhood N such that $S = \operatorname{Inv}(N, f)$.

Observe that if N is an isolating neighbourhood then $\operatorname{Inv}(N, f) \cap \partial N = \emptyset$, where ∂N denotes the topological boundary of N. An *isolating block* is a special form of an isolating neighbourhood that satisfies the additional condition

$$f^{-1}(N) \cap N \cap f(N) \subset \operatorname{int}(N).$$

If N is an isolating block and $x \in \partial N$, then either $f(x) \cap N = \emptyset$ or $f^{-1}(x) \cap N = \emptyset$, that is, points on the boundary either leave N immediately or all its pre-images lie outside N. Isolating blocks and their relation to chain-recurrent sets will be discussed in Section 2.2.

Isolating neighbourhoods serve two purposes. The first is to localize the dynamics being considered. As will be made clear later, once an isolating neighbourhood N has been chosen, the focus is on the structure of the dynamics of $\operatorname{Inv}(N, f)$. The second point follows from the observation that continuity and compactness implies that, if N is an isolating neighbourhood for f, then N is an isolating neighbourhood for perturbations of f. This suggests that isolating neighbourhoods are robust objects and therefore computable.

Though the entire discussion could be carried out in greater generality, from now on $X \subset \mathbb{R}^n$ is compact and $f : \mathbb{R}^n \to \mathbb{R}^n$. Also, assume that X is an isolating neighbourhood for f. Then we can consider $f : X \to \mathbb{R}^n$.

Of course, the method of computation of isolating neighbourhoods is essential. In Section 2.3, motivated by keeping track of numerically induced errors, we consider *topological multivalued maps.* To be more precise, consider the continuous function f which generates the dynamics we are interested in analysing. Given $x \in X$ and using the computer to evaluate f, we cannot expect to obtain the value $f(x)$. Because of the associated numerical errors, at best we obtain a numerical value $f_n(x)$ and an error bound $\epsilon > 0$, from which we can conclude that $f(x) \in B_\epsilon(f_n(x))$, *i.e.*, that the true image lies in an ϵ-ball centred at the numerically generated image. In essence, to each point $x \in f^{-1}(X)$ we can associate a subset $F(x) \subset X$ with the property that $f(x) \in F(x) \subset X$. To emphasize the fact that the images of F are subsets of a topological space, we refer to F as a topological multivalued map and use the notation $F : X \rightrightarrows X$.

The next step is to obtain a combinatorial description of F. This is done by discretizing the space X via a finite grid $\mathcal{G}$, to obtain a purely combinatorial multivalued map $\mathcal{F} : \mathcal{G} \rightrightarrows \mathcal{G}$. Some of the material in these sections may appear redundant, as we generalize concepts from dynamical systems for continuous maps such as f to a topological multivalued map F, and then to a combinatorial multivalued map $\mathcal{F}$. However, we want to stress that the topological multivalued map provides a straightforward relationship between the dynamical structures of interest: those corresponding to f, and the computable structures of the combinatorial multivalued map $\mathcal{F}$.

Finally, in Section 2.4 we discuss the graph algorithms that are used to perform the computations. As will be made precise, the key operations are essentially linear in the number of grid elements used to do the approximations.

The discussion in all these subsections is done on the level of maps. From the perspective of modelling, the dynamics of flows generated by differential equations is at least as important. In Section 2.5 we discuss three different approaches: time τ maps, Poincaré maps, and flow-transverse polygonal approximations. The first two are fairly standard and therefore will be only briefly mentioned. The third approach is still speculative. It is presented here because the same algorithms that apply to the map case work in this setting. What is lacking are computational geometry algorithms for obtaining simplicial decompositions of the phase space that lead to good approximations of the flow. Not surprisingly, this appears to be a difficult problem when the dimension of the phase space is greater than or equal to three.

Having argued that ϵ-chain-recurrent sets and isolating blocks are computable, the obvious question is:

Q2. *How can one interpret the results of the computations?*

Algebraic topology and, in particular, the Conley index are used at this stage. The Conley index is the topological generalization of Morse theory to maps and flows which do not necessarily arise from variational problems. The Conley index is very briefly described in Section 3; more complete discussions can be found elsewhere. What we wish to emphasize now is that the Conley index has three important characteristics.

(1) The Conley index is an index of isolating neighbourhoods. Furthermore, if N and N' are isolating neighbourhoods for the map f and $\mathrm{Inv}(N, f) = \mathrm{Inv}(N', f)$, then the Conley index of N is the same as the Conley index of N'. Observe that this implies that we can also consider the Conley index as an index of isolated invariant sets. Both interpretations are convenient, the first for applications and the latter for presenting definitions.

(2) **Ważewski Property:** If the Conley index of N is not trivial, then $\mathrm{Inv}(N, f) \neq \emptyset$.

(3) **Continuation:** If N is an isolating neighbourhood for a continuous family of maps f_λ, for instance,

$$\mathrm{Inv}(N, f_\lambda) \subset \mathrm{int}\, N \quad \text{for } \lambda \in [0, 1],$$

then the Conley index of N under f_0 is the same as the Conley index of N under f_1.

The first point suggests that to detect specific invariant sets does not require one to compute a unique isolating block. The second is the simplest example of how the Conley index is used to make assertions about the dynamics of f. The third point leads to the conclusion that the Conley index of the dynamical system generated by a sufficiently good numerical approximation will be the same as that of the original system. In a weak, but universal, sense this is the positive answer to the question posed for indirect methods.

As was mentioned above, little detail about the index theory is presented in this paper, though references are provided. The focus is on the computability. As will be emphasized, having computed chain-recurrent sets and isolating blocks, the input for the Conley index computations is available at essentially no cost. However, to determine the Conley index, homology groups and homology maps need to be computed. The cost of the homology groups seems to be reasonable, though for higher-dimensional complexes the known complexity bounds seem rather pessimistic. Currently the cost of computing homology maps is expensive. However, algorithms in this subject are fairly new and there appears to be considerable room for improvement.

We finish the paper with four examples that demonstrate that this approach can be applied in nontrivial settings. The first example is the Hénon map. While the results are not surprising, that is, they agree with results obtained by simulations, they are mathematically rigorous and provide information about the structure of the chaotic dynamics that is observed. In particular, this approach provides the greatest rigorous lower bound on the entropy of the global attractor at the classical parameter values. The next example involves the Lorenz equations, and is included to demonstrate that, using Poincaré sections, these methods can also be applied to differential equations.

A criticism of these topological methods is that the cost grows very rapidly with dimension. This is a serious issue, which cannot be dismissed lightly. For this reason we have included two recent results that involve infinite-dimensional systems. In both examples, even though the ambient phase space is infinite-dimensional, the dynamic objects of interest are low-dimensional, for instance, fixed points, periodic orbits, horseshoes, *etc.* From the purely topological point of view one might hope that it is the dimension of the invariant set that determines the complexity of the computations. These rather special examples provide evidence that this might actually be the case.

2. Chain recurrence and isolating blocks

In this section we discuss the computation of isolating blocks and ϵ-chain-recurrent sets. The presentation begins with classical definitions and results from dynamical systems and ends with graph algorithms. The focus is on the computability of these objects and on the fact that there is a mathematically valid interpretation of the graph-theoretic structures as objects in continuous dynamical systems.

2.1. Chain recurrence

The following is a standard definition from dynamical systems (Easton 1998, Robinson 1995).

Definition 1. Let S be a subset of X. An *ϵ-chain* from x to y in S for the map f is a finite sequence of points $\{z_0, z_1, \ldots, z_n\} \subset S$, such that $x = z_0$, $y = z_n$ and $\|f(z_i) - z_{i+1}\| < \epsilon$.

Though it may seem strange at the moment, we prefer to think of an ϵ-chain of f from x to y in S as a map $\mu_x : \{0, \ldots, n\} \to S$ satisfying:

(1) $\mu_x(0) = x$ and $\mu_x(n) = y$,
(2) $\|\mu_x(i+1) - f(\mu_x(i))\| < \epsilon$ for $i = 1, \ldots, n-1$.

If $\mu_x : \{0, \ldots, n\} \to S$ is an ϵ-chain from x to itself, then $\mu_x(0) = \mu_x(n) = x$. In this case, we can extend $\mu_x : \mathbb{Z} \to S$ periodically. This leads to the following concept.

Definition 2. The *ϵ-chain-recurrent set* of S under f is defined by

$$\mathcal{R}(S, f, \epsilon) = \{x \in X \mid \text{there exists a periodic } \epsilon\text{-chain } \mu_x : \mathbb{Z} \to S\}.$$

There is a natural partition of $\mathcal{R}(S, f, \epsilon)$ into equivalence classes defined as follows.

Definition 3. Let $x, y \in S$. Set $x \sim_\epsilon y$ if and only if there exists a periodic ϵ-chain $\mu_x : \mathbb{Z} \to S$ such that $\mu_x(j) = y$ for some $j \in \mathbb{Z}$.

Let $\mathcal{P}$ be an indexing set for the resulting collection of equivalence classes, that is, let

$$\mathcal{R}(S, f, \epsilon) = \bigcup_{p \in \mathcal{P}} \mathcal{R}_p(S, f, \epsilon),$$

where $x, y \in \mathcal{R}_p(S, f, \epsilon)$ if and only if $x \sim_\epsilon y$.

Recall that a *strict partial order* $\succ$ is a nonreflexive, nonsymmetric, transitive relation. A strict partial order $\succ$ is an *admissible* order on $\mathcal{P}(S, f, \epsilon)$ if, for $p \neq q$, the existence of an ϵ-chain from an element of $\mathcal{R}_p(S, f, \epsilon)$ to an element of $\mathcal{R}_q(S, f, \epsilon)$ implies that $p \succ q$. Observe that it need not be the case that any two equivalence classes are necessarily related by $\succ$.

We are particularly interested in the ϵ-chain-recurrent sets in the case where S is a compact invariant set. In this case, simple arguments based on the continuity of f show that not only is $\mathcal{R}(S, f, \epsilon)$ open, but each equivalence class $\mathcal{R}_p(S, f, \epsilon)$ is also open. Since $\{\mathcal{R}_p(S, f, \epsilon) \mid p \in \mathcal{P}\}$ covers S, this implies that, for any given $\epsilon > 0$, there is at most a finite number of elements in $\mathcal{P}$.

Definition 4. The *chain-recurrent set* of S under f is given by

$$\mathcal{R}(S, f) := \bigcap_{\epsilon > 0} \mathcal{R}(S, f, \epsilon).$$

Again, one can define equivalence classes $\mathcal{R}_p(S, f)$, $p \in \mathcal{P}$ of $\mathcal{R}(S, f)$ by $x \sim y$ if and only if $x \sim_\epsilon y$ for all $\epsilon > 0$.

There are cases where $S = \mathcal{R}(S, f)$. For example, if S were a strange attractor, then it would be an invariant set, but every point in S is chain-recurrent to every other point.

Theorem 2.1. (Fundamental Decomposition Theorem) Let S be a compact invariant set for f. Then there exists a continuous Lyapunov function $L : S \to [0,1]$ satisfying:

(1) $L(f(x)) < L(x)$ for all $x \in S \setminus \mathcal{R}(S, f)$,

(2) for each $p \in \mathcal{P}$ there exists $c_p \in [0,1]$ such that $\mathcal{R}_p(S, f) \subset L^{-1}(c_p)$.

Furthermore, $\mathcal{R}(S, f)$ is a compact invariant set and

$$\mathcal{R}(\mathcal{R}(S, f), f) = \mathcal{R}(S, f). \tag{2.1}$$

To understand the importance of this theorem, observe that (2.1) implies that the chain-recurrent set is the minimal recurrent set. In particular, computing the chain-recurrent set while restricting ourselves to $\mathcal{R}(S, f)$ does not result in a smaller set. On the other hand, the existence of the Lyapunov function L indicates that we have captured all the recurrent dynamics. This theorem suggests that, to understand the global dynamics of f, it is sufficient to understand the dynamics in the equivalence classes $\mathcal{R}_p(S, f)$ of $\mathcal{R}(S, f)$ and the structure of the set of connecting orbits between these equivalence classes.

The Fundamental Decomposition Theorem provides a starting point for the investigation of *any* continuous dynamical system. However, within the context of this paper the question we now need to address is whether, from a practical point of view, the chain-recurrent set is computable. With this in mind, consider the following example.

Example 1. Let $f : \mathbb{R} \to \mathbb{R}$ be given by $f(x) = x + x^2 \sin(\frac{1}{x})$ Observe that $S = [-\frac{1}{\pi}, \frac{1}{\pi}]$ is a compact invariant set, and

$$\mathcal{R}(S, \varphi) = \left\{ \pm \frac{1}{n\pi} \mid n \in \mathbb{N} \right\} \cup \{0\}.$$

In particular, in this example $\mathcal{R}(S, \varphi)$ consists of an infinite number of connected components. This suggests that its explicit computation will be difficult.

To go a step further, consider a simple perturbation of this equation. In particular, let f^λ be given by

$$f^\lambda = x + x^2 \sin\left(\frac{1}{x}\right) + \lambda\left(x^2 - \frac{1}{\pi^2}\right).$$

Again, $S = [-\frac{1}{\pi}, \frac{1}{\pi}]$ is a compact invariant set for f^λ; however, for $\lambda \neq 0$ the structure of $\mathcal{R}(S, \varphi^\lambda)$ changes dramatically in the sense that the number of equivalence classes in $\mathcal{R}(S, f)$ is finite. The point to be emphasized is that small perturbations in the dynamical system can lead to significant qualitative changes in the recurrent set.

This example is meant to suggest that while the chain-recurrent set is of fundamental importance to our understanding of the structure of invariant sets it should not be the focus of our computational efforts.

Therefore, we scale back our ambition concerning the detail of dynamical information that we seek. The simplest possibility is to preclude the study of an infinite collection of 'recurrent' sets. This leads to the following concept.

Definition 5. Let S be a compact invariant set for the map f. A *Morse decomposition* of S consists of a finite collection of mutually disjoint compact invariant sets,

$$\mathcal{M}(S) := \{M(p) \mid p \in \mathcal{P}\},$$

for which there exists a Lyapunov function $L : S \to [0,1]$ satisfying:

(1) $L(f(x)) < L(x)$ for all $x \in S \setminus \bigcup_{p\in\mathcal{P}} M(p)$,

(2) for each $p \in \mathcal{P}$ there exists $c_p \in [0,1]$ such that $M(p) \subset L^{-1}(c_p)$.

The individual invariant sets $M(p)$ are called *Morse sets.*

Though the motivation for this definition may appear somewhat artificial, we claim that Morse decompositions are, from the computational point of view, the *natural* global structures to study. To begin justifying this statement we will now work our way back to the ϵ-chain-recurrent set, beginning with some definitions.

Recall that the *ω-limit set* of x under f is

$$\omega(x) := \bigcap_{n>0} \mathrm{cl}\left(\bigcup_{i=n}^{\infty} f^i(x)\right),$$

where cl denotes the topological closure. We can define an *α-limit set* for x using a full trajectory, that is,

$$\alpha(\sigma_x) := \bigcap_{n<0} \mathrm{cl}\left(\bigcup_{i=n}^{-\infty} \sigma_x(i)\right).$$

Returning to the setting of a Morse decomposition, the existence of a Lyapunov function leads to the existence of partial orders on the indexing set $\mathcal{P}$. More precisely, since S is invariant, $x \in S \setminus \cup_{p\in\mathcal{P}} M(p)$ implies that there exists a full trajectory $\sigma_x : \mathbb{Z} \to S$ and Morse sets $M(p)$ and $M(q)$ such that:

$$\omega(x) \subset M(q) \qquad \text{and} \qquad \alpha(\sigma_x) \subset M(p). \tag{2.2}$$

The following definition will be used to describe the set of points with this property.

Definition 6. The set of *connecting orbits* from $M(p)$ to $M(q)$ is

$$C(p,q) := \{x \in S \mid \exists \sigma_x : \mathbb{Z} \to S, \alpha(\sigma_x) \subset M(p),\ \omega(x) \subset M(q)\}.$$

An *admissible order* is a strict partial order $\succ$ satisfying

$$C(p,q) \neq \emptyset \quad \Rightarrow \quad p \succ q.$$

This definition suggests that, given a Morse decomposition, the fundamental questions are: What is the dynamics within the Morse sets? and What is the structure of the set of connecting orbits between the Morse sets?

By choosing the same notation for the indexing sets and the partial orders we have strongly suggested that there is a relation between the equivalence classes of the ϵ-chain-recurrent set and Morse sets. The following theorem makes this precise.

Theorem 2.2. Let S be an invariant set for f. Let $\{\mathcal{R}_p(S,f,\epsilon) \mid p \in \mathcal{P}\}$ be the set of equivalence classes of the ϵ-chain-recurrent set of S. Define

$$M(p) := \operatorname{Inv}(\mathcal{R}_p(S,f,\epsilon), f).$$

Then $\mathcal{M}(S) := \{M(p) \mid p \in \mathcal{P}\}$ is a Morse decomposition of S. Furthermore, if $\succ$ is an admissible order for the equivalence classes of $\mathcal{R}(S,f,\epsilon)$, then $\succ$ is an admissible order for $\mathcal{M}(S)$.

Theorem 2.2 shows that an ϵ-chain-recurrent set produces a Morse decomposition. However, it is not the case that every Morse decomposition arises via an ϵ-chain-recurrent set. To explain this we shall explore a little further the structure of Morse decompositions.

Consider a Morse decomposition $\mathcal{M}(S) = \{M(p) \mid p \in (\mathcal{P},\succ)\}$ where $\succ$ is an admissible order. A subset $I \subset \mathcal{P}$ is called an *interval* if $p,q \in I$ and $p \succ r \succ q$ implies that $r \in I$. An *attracting* interval I satisfies the additional condition that $p \in I$, and $p \succ q$ implies that $q \in I$. The set of intervals on $(\mathcal{P},\succ)$ will be denoted by $\mathcal{I}(\mathcal{P},\succ)$ and the set of attracting intervals by $\mathcal{A}(\mathcal{P},\succ)$.

Proposition 2.3. Let $I \in \mathcal{I}(\mathcal{P},\succ)$ and define

$$M(I) = \left(\bigcup_{p\in I} M(p)\right) \cup \left(\bigcup_{p,q\in I} C(M(p),M(q))\right).$$

Then, $M(I)$ is an isolated invariant set.

The proof is fairly straightforward and follows from the compactness of S. The following proposition shows that a given Morse decomposition can give rise to a coarser Morse decomposition.

Proposition 2.4. $\mathcal{M}(S) = \{M(p) \mid p \in \mathcal{P}\backslash I\} \cup \{M(I)\}$ defines a Morse decomposition of S. Furthermore, an admissible partial order $\succ'$ is given by

$$\begin{array}{lll} p \succ' q & \Leftrightarrow & p \succ q \text{ if } p, q \in \mathcal{P}\backslash I, \\ p \succ' I & \text{if} & \text{there exists } q \in I \text{ such that } p \succ q, \\ I \succ' p & \text{if} & \text{there exists } q \in I \text{ such that } q \succ p. \end{array}$$

Bibliographical notes

The results described in this section are by now classical. Conley's proof of the decomposition theorem in the setting of flows can be found in Conley (1978). This is also an excellent reference for Morse decompositions. For another proof and further references see Robinson (1995). The presentation in Easton (1998) (see also Norton (1989)) is closest in spirit to that of this paper.

2.2. Isolating blocks

As was mentioned in the Introduction, isolating blocks provide us with a means of localizing the dynamics of interest. The most important property of an isolating block is that it is robust with respect to perturbation.

Proposition 2.5. Let $f, g : X \to X$ be continuous functions. Assume N is an isolating block for f. Then there exists $\epsilon > 0$ such that, if $\|f - g\| < \epsilon$, then N is an isolating block for g.

The proof follows directly from the continuity of f and g and the compactness of N. From the perspective of numerics, this suggests that isolating blocks are stable with respect to numerical errors. The problem is that, to do rigorous computations, we would like to have a sufficient ϵ *a priori*, so that we know what error tolerance can be permitted. We will return to this issue in the next section; for the moment we will present a very simple example to indicate that verification of the existence of an isolating block is possible even in the presence of sizeable errors.

Example 2. Consider the linear map $f : \mathbb{R}^2 \to \mathbb{R}^2$ given by

$$f(x) = \begin{bmatrix} 2 & 0 \\ 0 & \frac{1}{3} \end{bmatrix} x, \quad x \in \mathbb{R}^2. \tag{2.3}$$

Obviously, the origin $(0, 0)$ is a fixed point and the square $N = [-1, 1]^2$ is an isolating block with $(0, 0) = \operatorname{Inv}(N, f)$. Let $g : \mathbb{R}^2 \to \mathbb{R}^2$ satisfy $\|g(x) - f(x)\| \leq \frac{1}{2}$ for all $x \in N$. Then N is an isolating block for g.

While this trivial example shows that isolating blocks persist in the presence of fairly large perturbations, it does not address the question of how to find them. Again, we will address this issue from an algorithmic point of view later; however, the following proposition suggests that, viewed in the context of global dynamics, they are natural objects.

Proposition 2.6. Let S be an invariant set under f. Let $\mathcal{R}_p(S,f,\epsilon)$ be an equivalence class of the ϵ-chain-recurrent set. Then $\mathrm{cl}(\mathcal{R}_p(S,f,\epsilon))$ is an isolating block.

Bibliographical notes
There is a variety of references on isolating blocks, beginning with Conley and Easton (1971) for flows, and Easton (1975). Explicit relations between ϵ-chain recurrence and isolation can be found in Norton (1989) and Easton (1989, 1998).

2.3. Multivalued maps

As was made clear in the previous section, the notion of an ϵ-chain is fundamental in dynamical systems. Unfortunately, the classical definition sheds little light on how to systematically compute chain-recurrent sets. With this in mind, let us change our perspective slightly. Consider a point $x \in X$ and a continuous map f. Let $\sigma_x : \{0,\ldots,n\} \to X$ be an ϵ-chain starting at x. By definition, this implies that $\sigma_x(1) \in B_\epsilon(f(x))$ and, more generally, that $\sigma_x(n+1) \in B_\epsilon(f(\sigma_x(n)))$. This suggests that multivalued or set-valued maps provide a convenient language in which to describe the set of possible elements in an ϵ-chain.

Definition 7. A *multivalued map* $F : X \rightrightarrows X$ is a function whose values are subsets of X, *i.e.*, for every $x \in X$, $F(x) \subset X$. Because X is a topological space, we shall on occasion refer to F as a *topological* multivalued map. A *continuous selector* of F is a continuous function $g : X \to X$ for which $g(x) \in F(x)$ for all $x \in X$.

We will treat F as a dynamical system. A *trajectory of F through x* is a function $\mu_x : I \to X$ defined on an interval $I \subset \mathbb{Z}$ and satisfying the following conditions:

(1) $\mu_x(0) = x$,
(2) $\mu_x(n+1) \in F(\mu_x(n))$.

If $I = \mathbb{Z}$, then μ_x is a *full trajectory*. $S \subset X$ is an *invariant set* of F if for every $x \in S$ there exists a full trajectory $\mu_x : \mathbb{Z} \to S$. A compact set $N \subset X$ is an *isolating block* for F if

$$F^{*-1}(N) \cap N \cap F(N) \subset \mathrm{int}(N),$$

where $F^{*-1}(N) := \{x \in X \mid F(x) \cap N \neq \emptyset\}$. At the risk of being redundant, observe that, if $x \in \partial N$, then either

$$F(x) \cap N = \emptyset \qquad \text{or} \qquad F^{*-1}(x) \cap N = \emptyset. \tag{2.4}$$

An invariant set S is *isolated* if there exists an isolating block N such that

$$S = \text{Inv}(N, F) := \{x \mid \text{there exists } \mu_x : \mathbb{Z} \to N\}.$$

Example 3. In the context of this paper, the most natural way to generate a multivalued map is to begin with a continuous map $f : X \to X$ and define $F : X \rightrightarrows X$ by

$$x \mapsto B_\epsilon(f(x)).$$

The value of ϵ can be chosen to be an upper bound on the round-off errors that arise in the evaluation of the map f. Observe that $\mathcal{R}(X, f, \epsilon)$ is the same as the set of $x \in X$ for which there exists a periodic trajectory $\mu_x : \mathbb{Z} \to X$ of F. Thus, the set of periodic trajectories of F determines a Morse decomposition of f. In fact, if one assumes that ϵ is determined by the round-off error, then the resulting Morse decomposition is the finest that can be obtained given the specific level of numerical accuracy.

It is also easy to check that, if N is an isolating block for F, then by (2.4) N is an isolating block for f. However, the following observation is more important. The same argument also shows that N is an isolating block for any continuous selector g of F. In particular, if we assume that F was obtained numerically, then the values of the map f are not known precisely. Therefore, it is essential that the information that can be extracted from F be valid for all possible continuous selectors.

While this example is supposed to be suggestive of the value of using multivalued maps, it should be clear that, as presented, it is still not a computationally effective tool. In particular, the values of F differ at each point $x \in X$ and, hence, F is not a combinatorial object. To overcome this problem we discretize the phase space.

Definition 8. Let $X \subset \mathbb{R}^n$ and let $\mathcal{G} := \{G_i \subset \mathbb{R} \mid i \in \mathcal{I}\}$. $\mathcal{G}$ is a *grid covering* X if the following conditions are satisfied.

(1) $X \subset \bigcup_{i \in \mathcal{I}} G_i$.
(2) For every compact set $K \subset X$, there are only finitely many elements of $\mathcal{G}$ which intersect K nontrivially.
(3) For every $G_i \in \mathcal{G}$,
$$\emptyset \neq G_i = \text{cl}(\text{int}(G_i)).$$
(4) If $G_i \neq G_j$, then $\text{int}(G_i) \cap \text{int}(G_j) = \emptyset$.

$\mathcal{G}$ is a *convex* grid if each grid element G_i is a convex set.

Example 4. The classical example of a convex grid comes from a simplicial complex $\mathcal{K}$ in $\mathbb{R}^n$. $\mathcal{K}$ is a *full simplicial complex* if every simplex in $\mathcal{K}$ is the face of an n-dimensional simplex. Let

$$\mathcal{K}^{(l)} := \{K \in \mathcal{K} \mid \dim K = l\}.$$

Set $\mathcal{S} := \mathcal{K}^{(n)}$. Then, $\mathcal{S}$ is a *simplicial grid.*

We can generalize this as follows. Define a *polygon* P to be a connected set obtained from the union of a set of elements of $\mathcal{K}^{(n)}$ that share $n-1$-dimensional faces. The elements of a *polygonal grid* are polygons.

Example 5. From the perspective of data structures, a particularly nice grid is the cubical grid. Consider, for example, $X = \prod_{i=1}^{n}[0,1]$. We can define a sequence of grids by

$$\mathcal{G}^{(s)} := \left\{ \prod_{i=1}^{n} \left[\frac{i_k}{2^s}, \frac{i_k+1}{2^s} \right] \;\middle|\; i_k \in \{0,\ldots,2^s-1\} \right\}.$$

Definition 9. A *combinatorial multivalued map* is a multivalued map $\mathcal{F} : \mathcal{G} \rightrightarrows \mathcal{G}$ defined on a finite grid $\mathcal{G}$.

Again, we wish to view $\mathcal{F}$ as a dynamical system, which leads to the following definitions.

Definition 10. A *full trajectory* through $G \in \mathcal{G}$ of a combinatorial multivalued map $\mathcal{F} : \mathcal{G} \rightrightarrows \mathcal{G}$ is a function $\gamma_G : \mathbb{Z} \to \mathcal{G}$ satisfying:

(1) $\gamma_G(0) = G$,

(2) for every $n \in \mathbb{Z}$, $\gamma_G(n+1) \in \mathcal{F}(\gamma_G(n))$.

Let $\mathcal{U} \subset \mathcal{G}$. The *maximal invariant set* in $\mathcal{U}$ is given by

$$\operatorname{Inv}(\mathcal{U}, \mathcal{F}) := \{G \in \mathcal{U} \mid \text{there exists } \gamma_G : \mathbb{Z} \to \mathcal{U}\}.$$

There are two ways to view the grid $\mathcal{G}$. The first is as a collection of subsets of $\mathbb{R}^n$, that is, $\mathcal{G} = \{G_i \mid G_i \subset X\}$. In this case each grid element G_i carries topological information. The second is simply as a finite list of elements $\mathcal{G} = \{G_i\}$. This is the data that the computer can manipulate. We will adopt both perspectives. The latter is used to develop algorithms, while the former is used to give mathematical interpretation to the computations. In practice, we begin with the set X and define a grid $\mathcal{G} = \{G_i \mid G_i \subset X\}$. However, it is the list $\mathcal{G} = \{G_i\}$ that is entered into the computer (the computer knows no topology) and the output typically contains a sublist of $\mathcal{G}$. To use this output to draw conclusions about the dynamics will require re-introducing the topological structure of the grid elements. To emphasize this we will talk about the *support* of an element of the list, which in an abuse of notation is denoted by $|G_i| = G_i \subset X$. Similarly, given a set of elements of the list, $\mathcal{U} \subset \mathcal{G}$, the support of $\mathcal{U}$ is

$$|\mathcal{U}| := \bigcup_{G \in \mathcal{U}} |G| \subset X.$$

A combinatorial multivalued map $\mathcal{F} : \mathcal{G} \rightrightarrows \mathcal{G}$ on a grid $\mathcal{G}$ (here $\mathcal{G}$ is viewed as a list) induces a topological multivalued map $F : X \rightrightarrows X$ by

$$F(x) := \bigcup_{x \in G} |\mathcal{F}(G)|. \tag{2.5}$$

Using this relationship we can impose geometric and topological constraints on $\mathcal{F}$. For example, the combinatorial multivalued map $\mathcal{F}$ is *convex* if $F(x)$ is convex for every $x \in X$.

Beginning with an arbitrary topological multivalued map, it is not evident that there is a simple procedure to produce a combinatorial multivalued map. Fortunately, we do not need to work on this level of generality. We are only interested in combinatorial representations of continuous functions. Furthermore, we are interested in combinatorial representations that will lead to isolating blocks, which is why the following condition is imposed.

Definition 11. Let $\mathcal{G}$ be a grid covering X. A combinatorial multivalued map $\mathcal{F} : \mathcal{G} \rightrightarrows \mathcal{G}$ is an *outer approximation* of the continuous map $f : X \to X$ if, for every $G \in \mathcal{G}$,

$$\bigcup_{x \in G} f(x) \subset \operatorname{int}(|\mathcal{F}(G)|).$$

Observe that, if $\mathcal{F}$ is generated via rigorous numerical computations, then this condition is almost automatically satisfied.

We also need to be able to recognize on the combinatorial level when we have obtained an isolating neighbourhood, or better, an isolating block. To do this we need to make sense of neighbourhoods on the grid level. Let $\mathcal{U} \subset \mathcal{G}$,

$$o(\mathcal{U}) := \{G \in \mathcal{G} \mid G \cap |\mathcal{U}| \neq \emptyset\}.$$

Observe that $|o(\mathcal{U})|$ is the smallest neighbourhood of $|\mathcal{U}|$ that can be represented using elements of the grid $\mathcal{G}$. Let

$$d(\mathcal{U}) := o(\mathcal{U}) \setminus \mathcal{U}.$$

Remark 1. Recall that in the Introduction we made the assumption that $X \subset \mathbb{R}^n$ was an isolating neighbourhood under f. Since we have discretized the phase space we need to strengthen this assumption. From now on we assume the following:

(A) X is a neighbourhood of $|\operatorname{Inv}(\mathcal{G}, \mathcal{F})|$ relative to $\mathbb{R}^n$.

Theorem 2.7. Let $\mathcal{U} \subset \mathcal{G}$. If $\operatorname{Inv}(o(\mathcal{U}), \mathcal{F}) \subset \mathcal{U}$, then $|\mathcal{U}|$ is an isolating neighbourhood for f.

In Section 2.1 we asserted that Morse decompositions are the natural global objects to study. To see how they arise in the context of combinatorial multivalued maps we only need to recast the definitions. Set

$$\mathcal{R}(\mathcal{G},\mathcal{F}) := \{G \in \mathcal{G} \mid \text{there exists a periodic trajectory } \gamma_G : \mathbb{Z} \to \mathcal{G}\}.$$

We can decompose $\mathcal{R}(\mathcal{G},\mathcal{F})$ into equivalence classes $\mathcal{R}_p(\mathcal{G},\mathcal{F})$, $p \in \mathcal{P}$, by setting $G \sim G'$ if there exists a periodic trajectory $\gamma_G : \mathbb{Z} \to \mathcal{G}$ such that $\gamma_G(j) = G'$ for some $j \in \mathbb{Z}$. Similarly, we can define an admissible order $\succ$ on $\mathcal{P}$.

Theorem 2.8. Let $\mathcal{F} : \mathcal{G} \rightrightarrows \mathcal{G}$ be a combinatorial multivalued map which is an outer approximation of $f : X \to \mathbb{R}^n$. Let $M(p) := \operatorname{Inv}(\mathcal{R}_p(\mathcal{G},\mathcal{F}), f)$, $p \in \mathcal{P}$, and let $\succ$ be an admissible order for $\mathcal{P}$. Then, $\{M(p) \mid p \in \mathcal{P}\}$ is a Morse decomposition for f with admissible order $\succ$.

The final point that needs to be discussed is how well continuous maps can be approximated by combinatorial multivalued maps. This leads to the following definition.

Definition 12. Let $\mathcal{G}^{(n)}$ be a convex grid covering X; let $\mathcal{F}^{(n)} : \mathcal{G}^{(n)} \rightrightarrows \mathcal{G}^{(n)}$ be a combinatorial multivalued map. Set $F^{(n)} : X \rightrightarrows X$ to be the topological multivalued map satisfying (2.5). A sequence of outer approximations $\mathcal{F}^{(n)}$ *converges* to f if, for each n, $\mathcal{F}^{(n)}$ is an outer approximation of f and for every $\epsilon > 0$ there exists an integer N such that $n \geq N$ and $y \in F^{(n)}(x)$ implies that $\|f(x) - y\| < \epsilon$.

Theorem 2.9. If $f : X \to X$ is Lipschitz-continuous, then there exists a sequence of convex grids $\mathcal{G}^{(n)}$ and convex combinatorial multivalued maps $\mathcal{F}^{(n)} : \mathcal{G}^{(n)} \rightrightarrows \mathcal{G}^{(n)}$ converging to f.

Bibliographical notes

The use of multivalued maps as a means of translating between the combinatorial structures used to compute and the continuous dynamics of interest was first used by Mrozek and the author to provide a computer-assisted proof of chaotic dynamics in the Lorenz equation (Mischaikow and Mrozek 1995, 1998). For a systematic discussion of the use of combinatorial multivalued maps to understand continuous functions see Mrozek (1996). A proof of Theorem 2.7 can be found in Szymczak (1997). Assumption (A) can be removed but then additional (verifiable) conditions need to be checked to obtain the conclusions of Theorems 2.7 and 2.8. See Szymczak (1997) and Boczko, Kalies and Mischaikow (2002).

2.4. Graph algorithms

We now turn to the problem of determining the global structure of the dynamics. The general strategy is as follows. We begin with a continuous map $f : X \to X$. We choose a grid $\mathcal{G}$ that covers X and compute a combinatorial multivalued map $\mathcal{F} : \mathcal{G} \rightrightarrows \mathcal{G}$ that is an outer approximation of f. The goal is to determine either the global structure of the dynamics, that is, to compute a Morse decomposition, or to find an isolating block. To do this efficiently we make use of classical graph algorithms. To explain these we begin by introducing some standard definitions.

Definition 13. A *directed graph* is a pair $(\mathcal{V}, \mathcal{E})$, where the collection of *vertices* $\mathcal{V}$ consists of a finite set and the *edges* $\mathcal{E}$ are ordered pairs of vertices from $\mathcal{V}$. The number of elements in $\mathcal{V}$ and $\mathcal{E}$ are denoted by $|\mathcal{V}|$ and $|\mathcal{E}|$, respectively. Furthermore,

$$\mathcal{E}^T := \{(u,v) \mid (v,u) \in \mathcal{E}\}.$$

Observe that the combinatorial multivalued map $\mathcal{F} : \mathcal{G} \rightrightarrows \mathcal{G}$ defines a directed graph $(\mathcal{V}, \mathcal{E})$ as follows. Set $\mathcal{V} = \mathcal{G}$ and let

$$(G_i, G_j) \in \mathcal{E} \quad \Leftrightarrow \quad G_j \in \mathcal{F}(G_i).$$

Throughout this paper we shall always assume that the directed graph $(\mathcal{V}, \mathcal{E})$ was generated by a combinatorial multivalued map $\mathcal{F} : \mathcal{G} \rightrightarrows \mathcal{G}$ that is an outer approximation of $f : X \to X$.

Definition 14. A *path* in a directed graph $(\mathcal{V}, \mathcal{E})$ is a function $\gamma_G : \{0, \ldots, n\} \to \mathcal{V}$ with the property that $(\gamma_G(i), \gamma_G(i+1)) \in \mathcal{E}$ for all $i = 0, \ldots, n-1$; γ_{G_i} is a path from G_i to G_j if $\gamma_{G_i}(0) = G_i$ and $\gamma_{G_i}(n) = G_j$. It is a *cycle* if $\gamma_{G_i}(0) = \gamma_{G_i}(n)$.

Observe that a path in the directed graph $(\mathcal{V}, \mathcal{E})$ is equivalent to a trajectory of $\mathcal{F}$.

Because we are interested in using these directed graphs to perform the computations, it is important to consider how this information is represented. An obvious method is through an adjacency matrix $E = [e_{i,j}]$, where $e_{i,j} = 1$ if and only if $(G_j, G_i) \in \mathcal{E}$. This, of course, entails a cost of $|\mathcal{V}|^2$. However, since our directed graphs are generated by combinatorial multivalued maps which are meant to be reasonable outer approximations of a continuous function, one expects that there are only a few edges associated with any given vertex. This in turn implies that the matrix E is sparse. Therefore, for applications it is probably preferable to represent the graph as an adjacency list.

Definition 15. An *adjacency list representation* of a directed graph $(\mathcal{V}, \mathcal{E})$ is an array of $|\mathcal{V}|$ lists, one list for each vertex. For each vertex G, the list consists of all vertices such that $(G, G') \in \mathcal{E}$.

Observe that the size of an adjacency list representation is given by $|\mathcal{E}|$.

In what follows we assume that an adjacency list representation is used. This implies that we have an ordering of the vertices $\mathcal{V} = \{G_i \mid i = 1, \ldots, |\mathcal{V}|\}$

and then, for each vertex G_i, an ordering of the edges in the list associated with G_i.

Global dynamics

As was demonstrated in the previous subsection, a Morse decomposition for f can be obtained by determining the components of the recurrent set of $\mathcal{F}$. A standard definition from graph theory that nearly corresponds to components of recurrent sets is as follows.

Definition 16. A *strongly connected component* (SCC) of $(\mathcal{V}, \mathcal{E})$ is a maximal set of vertices $\mathcal{U} \subset \mathcal{V}$ such that, for every pair $G_i, G_j \in \mathcal{U}$, there exist paths from G_i to G_j and from G_j to G_i.

We can restate Proposition 2.6 as follows.

Theorem 2.10. Every SCC is an isolating block for f.

The problem with this definition is that, if $G \in \mathcal{V}$ and does not belong to any cycle, then vacuously $\{G\}$ defines a strongly connected component. For this reason we introduce a slightly stronger condition.

Definition 17. A *strongly connected path component* (SCPC) of $(\mathcal{V}, \mathcal{E})$ is a strongly connected component that contains at least one edge.

Example 6. Consider the directed graph $(\mathcal{V}, \mathcal{E})$ of Figure 2.1. The set of strongly connected components consists of

$$\{\{a\}, \{c, d, g\}, \{h\}, \{e\}, \{f\}, \{b\}\},$$

while the set of strongly connected path components consists of

$$\{\{c, d, g\}, \{f\}\}.$$

Observe that, if $(\mathcal{V}, \mathcal{E})$ was generated by a combinatorial multivalued map $\mathcal{F} : \mathcal{G} \rightrightarrows \mathcal{G}$, then the components of the chain-recurrent set would correspond to the strongly connected path components. Furthermore, the maximal invariant set consists of $\{c, d, g, h, f\}$.

The following theorem says that Morse decompositions are the optimal global objects to compute.

Theorem 2.11. Let $(\mathcal{V}, \mathcal{G})$ be a directed graph. There is a linear-time algorithm (more precisely, the computation time is proportional to the number of vertices and edges, *i.e.*, $|\mathcal{V}| + |\mathcal{E}|$) that identifies the strongly connected path components

$$\{\mathcal{R}_p \subset \mathcal{V} \mid p = 1, \dots n\},$$

and produces a function

$$L : \mathcal{V} \to \mathbb{Z}$$

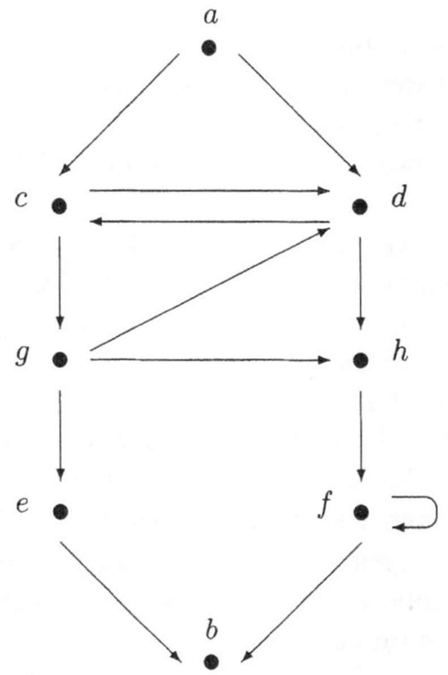

Figure 2.1. Directed graph $(\mathcal{V}, \mathcal{E})$

with the property that L is constant on each strongly connected path component and, if $G \in \mathcal{V} \setminus \cup_{p=1}^{n} \mathcal{R}_p$ and $(G, G') \in \mathcal{E}$, then

$$L(G) > L(G').$$

We shall not give a complete proof here, but, instead, briefly describe the depth-first search (DFS) algorithm which is the key step.

Depth-first search is a fairly simple recursive algorithm. To implement it we need a counter ι which takes integer values. We will also assign to each vertex G three values. The first value is $\xi(G) \in \{0, 1, 2\}$. If $\xi(G) = 0$, this indicates that the vertex G has not yet been viewed by the algorithm. The second value is an integer $\alpha(G)$ that marks the initial step at which the vertex is viewed. The third value is, also, an integer $\omega(G)$ which marks the final step at which the vertex is viewed. Finally, we can keep an ordered list of unviewed vertices, that is, the set of vertices for which $\xi = 0$. At the beginning this list is all of $\mathcal{V}$; however, as each vertex is viewed it is removed from this list.

Recall that we are assuming an adjacency list representation of $(\mathcal{V}, \mathcal{E})$ and, hence, have an ordering on the vertices and on the edges from each vertex. Initially, $\iota = 0$, and for each $G_i \in \mathcal{V}$,

$$\xi(G_i) = 0, \alpha(G_i) = 0, \omega(G_i) = 0.$$

The first rule is that ι is increased by one if and only if $\xi(G)$ is changed for some $G \in \mathcal{V}$.

The algorithm begins on the first vertex G_1. $\xi(G_1) = 0$, which means that this is the first time that the vertex has been viewed. Reset $\xi(G_1) = 1$, augment the counter to $\iota = 1$, and set $\alpha(G_1) = \iota$. Let (G_1, G_i) be the first edge in the list associated to G_1. View G_i. (Observe that if $i \neq 1$ then $\xi(G_i) = 0$.)

We now describe a general recursive step. Assume that $\iota = k$, and we view a vertex G_i such that $\xi(G_i) = 0$. Then, reset $\xi(G_i) = 1$, augment the counter to $\iota = k+1$, and set $\alpha(G_i) = \iota$. Let (G_i, G_j) be the first edge in the list associated to G_i and view G_j.

An alternative is that $\iota = k$, and we view a vertex G_i such that $\xi(G_i) \in \{1, 2\}$. Observe that to be in this situation we must have arrived at G_i by considering an edge (G_l, G_i). In this case we return to the list of edges from G_l. We did not change the value of any $\xi(G)$, therefore $\iota = k$. As was remarked before, the edges from G_l are in an ordered list, so we consider the edge (G_l, G_j) which follows the edge (G_l, G_i) in that list. View G_j. (Notice that if we are performing this step then $\xi(G_l) = 1$.)

It is possible that the edge (G_l, G_i) is the last in the list. In this case, set $\xi(G_l) = 2$, augment the counter to $\iota = k+1$, and set $\omega(G_l) = \iota$. There are now two cases to consider. The first is that we first viewed G_l via an edge (G_m, G_l). In this case view G_m. Otherwise view the first vertex on the list of unviewed vertices.

We will add one additional action to this algorithm. Observe that one of the ways we move through the vertices is by viewing edges (G_i, G_j) where $\xi(G_j) = 0$. If this is the case we label this edge as a *forward edge*. Let $\mathcal{E}^f \subset \mathcal{E}$ denote the set of forward edges obtained by running DFS. It is easy to check that the directed graph $(\mathcal{V}, \mathcal{E}^f)$ is a forest, that is, a disjoint collection of trees. Observe that we began constructing a tree within this forest each time we chose a vertex from the list of unviewed vertices. We declare this vertex to be the root of the tree. Thus, having run DFS, to each $G \in \mathcal{V}$ we can assign a unique vertex $\rho(G)$ that denotes the root of the tree to which it belongs. Notice that $\rho(G)$ has the lowest α value and largest ω value of any vertex G in the tree.

If we apply DFS to a directed graph $(\mathcal{V}, \mathcal{E})$ the outputs of interest are the values $\{\alpha(G_i), \omega(G_i) \mid G_i \in \mathcal{V}\}$, all of which are distinct, and the trees of $(\mathcal{V}, \mathcal{E}^f)$.

The following algorithm, which we shall refer to as SCC, satisfies Theorem 2.11.

(1) Apply DFS to the directed graph $(\mathcal{V}, \mathcal{E})$ to compute $\{\omega(G) \mid G \in \mathcal{V}\}$.
(2) Compute $(\mathcal{V}, \mathcal{E}^T)$.
(3) Re-order the vertices of $\mathcal{V}$ by decreasing values of $\omega(G)$. Using this order, apply DFS to the directed graph $(\mathcal{V}, \mathcal{E}^T)$. This computes a new set of values $\{\omega(G) \mid G \in \mathcal{V}\}$ and a new set of trees.

(4) Each tree of $(\mathcal{V}, (\mathcal{E}^T)^f)$ is a strongly connected component of $(\mathcal{V}, \mathcal{E})$. Define

$$L(G) = |\mathcal{V}| - \omega(\rho(G)).$$

Since this algorithm produces the trees associated with the strongly connected components, we can use this information to produce a new directed graph $(\mathcal{V}^{SCC}, \mathcal{E}^{SCC})$ called the *component graph*, which is defined as follows. Let $\{\mathcal{C}_1, \ldots, \mathcal{C}_k\}$ be the set of strongly connected components of $(\mathcal{V}, \mathcal{G})$. Set $\mathcal{V}^{SCC} = \{C_1, \ldots, C_k\}$, that is, one vertex for each component. $(C_i, C_j) \in \mathcal{E}^{SCC}$ if there exists $G \in \mathcal{C}_i$, $G' \in \mathcal{C}_j$, and $(G, G') \in \mathcal{E}$.

This construction of the component graph demands a brief digression back to the realm of continuous dynamics. Conley (1978) defined a *chain-recurrent flow* (*map*) to be a flow (map) whose chain-recurrent set was the entire phase space. At the other extreme, he declared a flow (map) to be *strongly gradient-like* if the chain-recurrent set is totally disconnected and consists only of equilibria. Using this terminology he then recast his Fundamental Decomposition Theorem into the following statement:

> Every flow on a compact space is uniquely represented as the extension of a chain recurrent flow by a strongly gradient-like flow; that is the flow admits a unique subflow which is chain recurrent and such that the quotient flow is strongly gradient-like.

(The same is true for continuous maps.) The quotient flow is obtained by collapsing each connected component of the chain-recurrent set to a distinct point. Observe that the procedure for constructing the component graph is the discrete analogue of the construction of the strongly gradient-like dynamical system. The point which is meant to be emphasized is that there is a natural correspondence between Conley's approach to decomposing continuous dynamical systems and the basic decompositions of directed graphs.

It was observed earlier that paths in $(\mathcal{V}, \mathcal{E})$ correspond to trajectories of $\mathcal{F}$. Of course, from the perspective of dynamics we are interested in full trajectories of $\mathcal{F}$. The following proposition characterizes those vertices of $\mathcal{V}$ for which one can define a path $\gamma_G : \mathbb{Z} \to \mathcal{V}$ which corresponds to a full trajectory.

Proposition 2.12. Let $(\mathcal{V}, \mathcal{E})$ be a directed graph with strongly connected path components $\{\mathcal{R}_p \mid p \in \mathcal{P}\}$. Let $G \in \mathcal{V}$. There exists a path $\gamma_G : \mathbb{Z} \to \mathcal{V}$ if and only if $G \in \cup_{p \in \mathcal{P}} \mathcal{R}_p$, or there exists a path $\gamma_{G_0} : \{0, \ldots, n\} \to \mathcal{V}$ from G_0 to G_1 such that $G_i \in \cup_{p \in \mathcal{P}} \mathcal{R}_p$ and $\gamma_{G_0}(j) = G$ for some $j \in \{0, \ldots, n\}$.

Isolation within recurrent sets

Of considerable interest in nonlinear systems is the existence of complicated or chaotic dynamics. Of course, this behaviour is recurrent and will therefore be found within the strongly connected path components. As was indicated in the Introduction we use the Conley index theory to extract lower bounds on the structure of the dynamics of f. The details of this will be discussed in the next section. For the moment, we remind the reader that effective computation of the index depends upon isolating blocks. In particular, the more 'structure' that the blocks possess, the more information about the dynamics can be extracted. Therefore, in this subsection we will discuss two different approaches for finding isolating blocks within strongly connected path components. The first begins with the full SCPC and reduces the number of grid elements in the complex. The second approach adopts the opposite strategy. One begins with a minimal number of grid elements which may possess a specific dynamics, and then adds grid elements until isolation is achieved.

Essential to both approaches is the ability to compute invariant sets for combinatorial multivalued maps. More precisely, given $\mathcal{F} : \mathcal{G} \rightrightarrows \mathcal{G}$ and $\mathcal{U} \subset \mathcal{G}$, we need to be able to determine $\operatorname{Inv}(\mathcal{U}, \mathcal{F})$. As before, we let $(\mathcal{V}, \mathcal{E})$ be the directed graph associated to $\mathcal{F}$. Let $(\mathcal{U}, \mathcal{E}')$ be the subgraph of $(\mathcal{V}, \mathcal{E})$ obtained by restricting to those vertices associated with $\mathcal{U}$.

Proposition 2.13. $\operatorname{Inv}(\mathcal{U}, \mathcal{F})$ can be computed in linear time: more precisely, in time proportional to the number of vertices and edges in $(\mathcal{U}, \mathcal{E}')$.

As before, we will not provide a formal proof of this proposition, but, instead, briefly describe an algorithm that will perform the task. Furthermore, the algorithm we describe is presented to emphasize a graph-theoretic counterpart to the Fundamental Decomposition Theorem: a vertex is in the maximal invariant set if and only if it belongs to a SCPC or it lies on a path from one SCPC to another SCPC. The first step is to run SCC$(\mathcal{U}, \mathcal{E}')$ and create $(\mathcal{U}^{SCC}, \mathcal{E}'^{SCC})$. Within $\mathcal{U}^{SCC}$, identify those which correspond to the strongly connected path components of $(\mathcal{U}, \mathcal{E}')$ (these are the vertices which have an edge to themselves). Now apply the following modified version of DFS. To each vertex $G \in \mathcal{U}^{SCC}$ we assign an additional value $\zeta(G) \in \{0, 1\}$. Initially, $\zeta(G) = 0$, unless G corresponds to a strongly connected path component, in which case $\zeta(G) = 1$. This modified DFS is only initiated on vertices that correspond to strongly connected path components. There are two occasions when $\zeta(G)$ may be changed. The first is that we are at a vertex G_i and considering an edge (G_i, G_j) where $\zeta(G_j) = 1$. Then, set $\zeta(G_i) = 1$. The second is that $\zeta(G_l) = 1$, and we first viewed G_l via an edge (G_m, G_l). Then, set $\zeta(G_m) = 1$.

At the completion of this modified DFS, $\operatorname{Inv}(\mathcal{U}, \mathcal{F})$ is given by the set of vertices for which $\zeta = 1$.

Throughout the remainder of this subsection, we assume that $(\mathcal{V}, \mathcal{E})$ is a directed graph that has a unique strongly connected component.

The first approach is based on the following theorem.

Theorem 2.14. Let $\mathcal{U} \subset \mathcal{V}$. Let $\mathcal{U}_0 := \mathcal{U}$. For $i = 1, \ldots, m$, let $\mathcal{U}_i, \mathcal{W}_i \subset \mathcal{V}$, which satisfies the following conditions:

(1) $\operatorname{Inv}(\mathcal{U}_{i-1}, \mathcal{F}) = \mathcal{U}_i \cup \mathcal{W}_i$,
(2) $o(\mathcal{U}_i) \cap \mathcal{W}_i = \emptyset$,
(3) $o(\mathcal{U}_m) \subset \mathcal{U}$.

Then, $\mathcal{U}_m = \operatorname{Inv}(o(\mathcal{U}_m) \cup \mathcal{W}_m) \cap o(\mathcal{U}_m)$.

Using Theorem 2.14 there is a variety of specific algorithms that can be created (see Szymczak (1997, Section 4)). The essential point is to generate the sets $\mathcal{U}_i$ and $\mathcal{W}_i$ inductively in such a way that conditions (1) and (2) are satisfied. The algorithm then halts when condition (3) is satisfied. Observe that in this method we begin with a 'large' set of vertices, *i.e.*, $\mathcal{U}$, and at each step we reduce the set of vertices being considered. To emphasize this point, assume that $|\mathcal{V}|$ is a connected set and that we choose $\mathcal{U} = \mathcal{V}$. Then $\mathcal{U} = \operatorname{Inv}(U_0, \mathcal{F}) = \mathcal{U}_1 \cup \mathcal{W}_1$. Therefore, because of condition (2), $\mathcal{W}_1 = \emptyset$ and $\mathcal{U}_1 = U_0$. In this case, no progress has been made.

In principle, these algorithms can be applied with no knowledge of paths within $(\mathcal{V}, \mathcal{E})$. The next strategy takes the opposite approach. Consider Figure 2.2. This is meant to represent that within $(\mathcal{V}, \mathcal{E})$ we have identified a cycle of length one and a cycle of length two. One might suspect that $|G_3|$ contains a fixed point of f and $|G_1| \cup |G_2|$ contains a period two orbit of f. (In the next section we will provide theorems that allow one to check such claims rigorously.) However, since $(\mathcal{V}, \mathcal{E})$ has a unique strongly connected

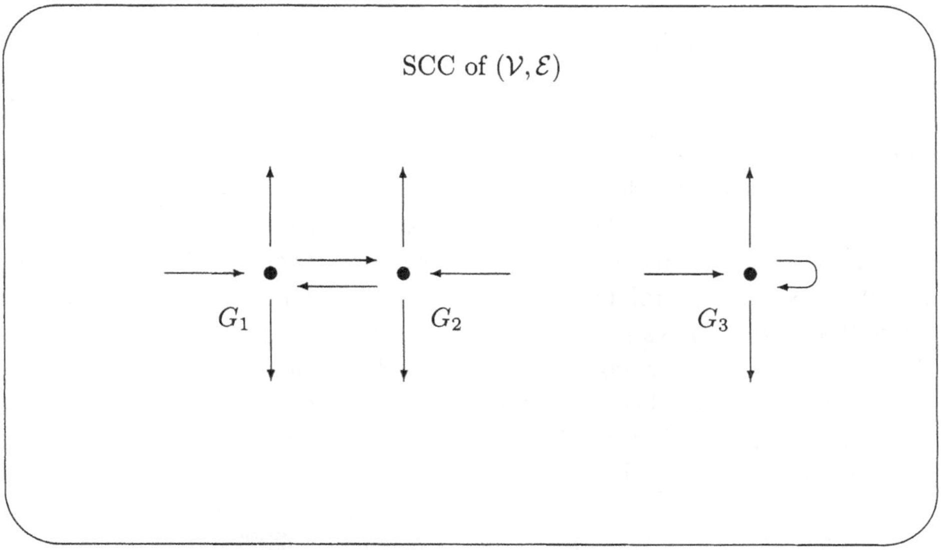

Figure 2.2. A strongly connected component which contains a period one path and a period two path

component, we know that there are paths from G_3 to G_1 and from G_1 to G_3. This in turn suggests the existence of horseshoe or symbolic dynamics based on trajectories of f that pass through $|G_1| \cup |G_2|$ and $|G_3|$. To prove this requires an appropriate isolating neighbourhood and is the purpose of the next algorithm.

The first step of the process is to generate an initial guess for the isolating block that will capture the desired dynamics. Let $\mathcal{B}_i \subset \mathcal{V}$, $i = 0, 1$. Let $(\mathcal{B}_i, \mathcal{E}_i)$ be the corresponding subgraphs of $(\mathcal{V}, \mathcal{E})$. Assume that $(\mathcal{B}_i, \mathcal{E}_i)$ have unique strongly connected components. Let $\gamma_0 : \{0, \ldots, n_0\} \to \mathcal{V}$ be a minimal path from any element of $\mathcal{B}_0$ to any element of $\mathcal{B}_1$ and let $\gamma_1 : \{0, \ldots, n_1\} \to \mathcal{V}$ be a minimal path from any element of $\mathcal{B}_1$ to any element of $\mathcal{B}_0$. Let $\mathcal{W}_i = \gamma_i(\{0, \ldots, n_i\})$ and let

$$\mathcal{U} := \mathcal{B}_0 \cup \mathcal{B}_1 \cup \mathcal{W}_0 \cup \mathcal{W}_1.$$

Proposition 2.15. Under the assumption that the number of elements of $\mathcal{B}_i$ is small compared to the number of elements in $\mathcal{V}$, Dijkstra's algorithm (Corman, Leiserson, Rivest and Stein 2001) will determine $\mathcal{W}_0$ and $\mathcal{W}_1$ with a running time $O(|\mathcal{V}| \ln |\mathcal{V}| + |\mathcal{E}|)$.

Observe that the resulting directed subgraph $(\mathcal{U}, \mathcal{E}')$ has a unique strongly connected component and therefore $\operatorname{Inv}(\mathcal{U}, \mathcal{F}) = \mathcal{U}$. There is no reason *a priori* to believe that $|\mathcal{U}|$ is an isolating block for f. Thus, the next step of the process is to isolate the dynamics we have captured. Let $\mathcal{U}_0 := \mathcal{U}$. Given $\mathcal{U}_i$, define $\mathcal{X}_i = o(\mathcal{U}_i)$. Set $\mathcal{U}_{i+1} = \operatorname{Inv}(X_i, \mathcal{F})$. If

$$o(\mathcal{U}_{i+1}) \subset \mathcal{X}_i$$

then $\mathcal{U}_{i+1}$ is an isolating block for f and we stop. If not we repeat the process.

Bibliographical notes

Definitions and explicit descriptions of the graph algorithms can be found in Corman *et al.* (2001). The use of these algorithms to find chain-recurrent sets, isolating neighbourhoods and index pairs (see Section 3) was first implemented by M. Eidenschink (1995) in the context of simplicial approximations of two-dimensional flows (see Section 2.5).

We did not provide any complexity bounds for either algorithm of this subsection. In part this is because there is no guarantee that these algorithms will produce an isolating block. In the first case, it is possible to reduce $\mathcal{U}$ to the empty set. In the second case, it is possible that $\mathcal{U}$ will grow to be the entire SCPC. In practice, however, these methods seem to work well (Szymczak 1997, Allili, Day, Junge and Mischaikow 2002).

The usefulness of these ideas are only as good as the software that can support them. Fortunately, M. Dellnitz, A. Hohmann and O. Junge have *independently* developed an excellent general purpose package GAIO based on cubical grids (Dellnitz and Hohmann 1997, Dellnitz and Junge 2001, Dellnitz, Froyland and Junge 2000). Some of the above-mentioned graph algorithms are already incorporated in the code. Furthermore, the grid is constructed in an adaptive method which has significant implications for the cost of computing the combinatorial multivalued map. In particular, one can expect that for many problems the cost is determined by the dimension of the invariant set rather than the dimension of the phase space.

2.5. Flows

We now turn to a discussion of computations in the context of flows. Let $\dot{x} = g(x)$ be an ordinary differential equation defined on $\mathbb{R}^n$ that generates a flow $\varphi : \mathbb{R} \times \mathbb{R}^n \to \mathbb{R}^n$. Currently, the only systematic approach to problems of this nature is to change it to a problem involving continuous maps. A completely general procedure is to fix a time $\tau > 0$ and define $f : \mathbb{R}^n \to \mathbb{R}^n$ by $f(x) := \varphi(\tau, x)$. The following result indicates that, on a theoretical level, the invariant sets of f which we can hope to recover using these topological techniques are equivalent to those of φ.

Proposition 2.16. Consider a flow $\varphi : \mathbb{R} \times \mathbb{R}^n \to \mathbb{R}^n$. Let $\varphi_\tau : \mathbb{R}^n \to \mathbb{R}^n$ be defined by $\varphi_\tau(x) = \varphi(\tau, x)$. Then, the following are equivalent:

(1) S is an isolated invariant set for φ,

(2) S is an isolated invariant set for φ_t for all $t > 0$,

(3) S is an isolated invariant set for φ_t for any $t > 0$.

Unfortunately, while theoretically simple, its implementation is less obvious. In particular, some choice of τ needs to be made and the map f needs to be computed, most probably by numerically integrating the differential equation. Therefore, if τ is large, then the numerical errors will force the combinatorial multivalued map to have exponentially large values. This in turn will make it difficult to find isolating neighbourhoods. On the other hand, if τ is small then f will be a near identity map and an extremely fine grid size will be required. This suggests that one needs to look for an 'optimal' choice of τ. Unfortunately, such a value is probably dependent on the location in phase space. It should be mentioned that, in spite of these negative comments, this method can be employed successfully (Pilarczyk 1999), though at a significant computational price.

A more successful variation of this strategy involves the use of local Poincaré sections for the flow. In this case we study the dynamics of the Poincaré map $f : \Xi \to \Xi$, where Ξ consists of $(n-1)$-dimensional hypersurfaces that are transverse to the flow. Again, f is determined by integrating the differential equation and therefore one is faced with exponential growth in error.

However, in this setting it is easier to implement multistep methods and multiple hypersurfaces to control the error. Furthermore, and this leads to significant computational savings, the dimension of the grid is reduced by one. This approach has been successfully applied to a variety of low-dimensional problems. Since, aside from the numerical issue of computing f, the technical aspects are similar to the map case, we will not discuss this approach further.

In Example 4 a polygonal grid $\mathcal{T}$ was defined in terms of a full simplicial complex $\mathcal{K}$ in $\mathbb{R}^n$. In particular, a polygon P is a connected set that can written as

$$P = \bigcup_{i=1}^{j} K_i,$$

where $K_i \in \mathcal{K}^{(n)}$. In an abuse of notation we will write $K_i \in P$. Observe that the boundary of P, denoted by ∂P, can be written as the union of elements of $\mathcal{K}^{(n-1)}$, that is,

$$\partial P = \bigcup_{i=1}^{m} L_i,$$

where $L_i \in \mathcal{K}^{(n-1)}$. We will continue the abuse of notation by writing $L_i \in \partial P$.

Given $L \in \mathcal{K}^{(n-1)}$, let

$$\nu(L)$$

denote one of the two unit normal vectors to L. To determine a unique choice of $\nu(L)$, let $K \in \mathcal{K}^{(n)}$ such that L is a face of K. Then,

$$\nu_K(L)$$

is defined to be the inward unit normal of L with respect to K.

Definition 18. $L \in \mathcal{K}^{(n-1)}$ is a *flow-transverse face* with respect to the vector field g, if

$$\nu(L) \cdot g(x) \neq 0$$

for every $x \in L$. A polygon P is *flow-transverse* if every $L \in \partial P$ is flow-transverse.

Definition 19. Let $\mathcal{T} = \{P_1, \ldots, P_N\}$ be a polygonal decomposition of X. $\mathcal{T}$ is a *flow-transverse polygonal decomposition* of X if every polygon P_i is flow-transverse.

As we shall explain, given a simplicial complex, constructing a flow-transverse polygonal decomposition can be done in linear time in the number of n-dimensional simplices. For this discussion we shall assume that X

is the polygonal region defined by the full simplicial complex $\mathcal{K}$, and for the sake of simplicity that X is flow-transverse (see Boczko *et al.* (2002) where this latter assumption is dropped). Let $K_i, K_j \in \mathcal{K}^{(n)}$ such that $K_i \cap K_j = L \in \mathcal{K}^{(n-1)}$. Set $K_i \sim K_j$ if $\nu(L) \cdot f(x) = 0$ for some $x \in L$. Extend this relation by transitivity. Then $\sim$ is an equivalence relation on $\mathcal{K}$. Define $\mathcal{T} := \{P_1, \dots, P_N\}$ to be the polygons defined by the equivalence classes. This grid is a flow-transverse polygonal decomposition of X.

The next step is to define a combinatorial multivalued map $\mathcal{F} : \mathcal{T} \rightrightarrows \mathcal{T}$ that captures the dynamics of the flow. Observe that if P contains recurrent dynamics under the flow φ, then we want $P \in \mathcal{F}(P)$. The simplest way to ensure this is as follows:

$$P \in \mathcal{F}(P) \quad \Leftrightarrow \quad \|g(x)\| < A \text{ for some } x \in P, \tag{2.6}$$

where the constant A is determined by the diameter of P and $\|Dg(x)\|$. Otherwise, set

$$P_i \in \mathcal{F}(P_j) \quad \Leftrightarrow \quad \nu_{P_i}(L) \cdot g(x) > 0, \text{ for some } x \in L \in \partial P_i \cap \partial P_j \cap \mathcal{K}^{(n-1)}. \tag{2.7}$$

Observe that this implies that a polygon can be in the image of another polygon only if they share an $(n-1)$-dimensional simplex. This implies that the directed graph $(\mathcal{V}, \mathcal{E})$ generated by the combinatorial multivalued map $\mathcal{F} : \mathcal{T} \rightrightarrows \mathcal{T}$ is, in general, sparse. At this point we are, of course, free to apply the graph algorithms of the previous subsection. As the following result indicates, this allows us to compute a Morse decomposition[1] efficiently.

Theorem 2.17. Let $M(p) := \operatorname{Inv}(\mathcal{R}_p(\mathcal{T}, \mathcal{F}))$, $p \in \mathcal{P}$, and let $\succ$ be an admissible order for $\mathcal{P}$. Then, $\{M(p) \mid p \in \mathcal{P}\}$ is a Morse decomposition for $\operatorname{Inv}(X, \varphi)$ with admissible order $\succ$.

The other important structure that we need to be able to compute is an isolating block which we shall define in the context of polygonal approximations.

Definition 20. A polygon P is an *isolating block* if, for every point $x \in \partial P$ and any $\epsilon > 0$,

$$\varphi((-\epsilon, \epsilon), x) \not\subset P.$$

The fact that φ is a smooth flow implies the following result.

Proposition 2.18. If P is a convex flow-transverse polygon, then it is automatically an isolating block.

[1] The definition of a Morse decomposition for a flow is the obvious analogue of Definition 5, or see Conley (1978), Robinson (1995) and Arnold, Jones, Mischaikow and Raugel (1995).

Theorem 2.19. Let $\mathcal{U}$ be a strongly connected component of $(\mathcal{V}, \mathcal{E})$. Then $|\mathcal{U}|$ is a isolating block for φ.

Thus, we regain two out of the three important computational results. What remains is the question of obtaining arbitrarily good approximations of the dynamics. This depends heavily on the initial simplicial complex and remains a difficult open question in computational geometry.

Bibliographical notes

References to rigorous computational results for ordinary differential equations using Poincaré maps can be found in Mischaikow and Mrozek (2001). The results on flow-transverse polygonal approximations will appear in the forthcoming paper by Boczko *et al.* (2002). Though not presented in this generality, in his thesis, Eidenschink (1995) provided code that provides good simplicial approximations for two-dimensional systems.

3. Conley index

The numerical computations and algorithms discussed in the previous section do not in themselves provide rigorous information concerning the existence and structure of invariant sets. To do this we will invoke the Conley index theory.

3.1. Topological preliminaries

Before providing a definition of the index we recall a few elementary notations from topology.

Definition 21. A *pointed space* (X, x_0) is a topological space X with a distinguished point $x_0 \in X$. A *continuous map between pointed spaces* $f : (X, x_0) \to (Y, y_0)$ is a continuous map $f : X \to Y$ with the additional condition that $f(x_0) = y_0$.

Given a pair (N_1, N_0) of spaces with $N_0 \subset N_1$, set

$$N_1/N_0 := (N_1 \setminus N_0) \cup [N_0],$$

where $[N_0]$ denotes the equivalence class of points in N_0 in the equivalence relation: $x \sim y$ if and only if $x = y$ or $x, y \in N_0$. Hereafter, we will usually use N_1/N_0 to denote the pointed space $(N_1/N_0, [N_0])$. The topology on $(N_1/N_0, [N_0])$ is defined as follows: a set $U \subset N_1/N_0$ is open if U is open in N_1 and $[N_0] \notin U$, or the set $(U \cap (N_1 \setminus N_0)) \cup N_0$ is open in N_1. If $N_0 = \emptyset$, then

$$(N_1/N_0, [N_0]) := (N_1 \cup \{*\}, \{*\}),$$

where $\{*\}$ denotes the equivalence class consisting of the empty set.

Definition 22. Let (X, x_0) and (Y, y_0) be pointed topological spaces and let $f, g : (X, x_0) \to (Y, y_0)$ be continuous functions. f is *homotopic* to g, denoted by

$$f \sim g,$$

if there exists a continuous function $F : X \times [0,1] \to Y$ such that

$$\begin{aligned} F(x,0) &= f(x), \\ F(x,1) &= g(x), \\ F(x_0,s) &= y_0, \quad 0 \le s \le 1. \end{aligned}$$

Obviously $\sim$ is an equivalence relation. The equivalence class of f in this relation is called the *homotopy class* of f and denoted by $[f]$.

Definition 23. Two pointed topological spaces (X, x_0) and (Y, y_0) are *homotopic*

$$(X, x_0) \sim (Y, y_0)$$

if there exists $f : (X, x_0) \to (Y, y_0)$ and $g : (Y, y_0) \to (X, x_0)$ such that

$$f \circ g \sim \mathrm{id}_Y \quad \text{and} \quad g \circ f \sim \mathrm{id}_X.$$

Observe that homotopy defines an equivalence class on the set of topological spaces. Homotopy classes of topological spaces are extremely difficult to work with directly. One of the most useful tools in this area is algebraic topology, in particular, homology theory. As was indicated in the Introduction we use homology to provide a rigorous interpretation of the objects that have been computed numerically. For the purposes of this presentation it is sufficient to know that, given a pair of spaces $N_0 \subset N_1$, which are described in terms of grid elements or a simplicial complex, it is possible to compute the *relative homology groups*

$$H_*(N_1, N_0) = \{H_k(N_1, N_0) \mid k = 0, 1, 2, \ldots\}.$$

For each $k \in \mathbb{N}$, $H_k(N_1, N_0)$ is a finitely generated abelian group and, for the types of calculations we are interested in performing, $H_k(N_1, N_0) \cong 0$ for sufficiently large k. Furthermore, if $f : (N_1, N_0) \to (N_1', N_0')$ is a continuous map of pairs, that is, $f : N_1 \to N_1'$ is continuous and $f(N_0) \subset N_0'$, then there is a collection of induced group homomorphisms, the *homology maps*,

$$f_* : H_*(N_1, N_0) \to H_*(N_1', N_0'),$$

where $f_* = \{f_k : H_k(N_1, N_0) \to H_k(N_1', N_0') \mid k = 0, 1, 2, \ldots\}$.

As will become clear shortly, the sets N_i for which we wish to compute homologies arise as large sets of grid elements. To further complicate matters, we do not know the map f, but rather a multivalued combinatorial map $\mathcal{F}$ which is an outer approximation of f. Because of the size of these objects, the homology computations must be done by computer. It is only recently that reasonably efficient algorithms for these types of computations have been developed and, in the case of computing homology maps, there is considerable room for improvement. However, the important point is that there are algorithms and code available to perform these computations (see Kaczyński, Mischaikow and Mrozek (2001) and references therein).

The following set of definitions is required to define the Conley index for maps.

Definition 24. Let $f : X \to X$ and $g : Y \to Y$ be continuous maps between topological spaces (group homomorphism between abelian groups). They are *shift-equivalent* if there exist continuous maps (group homomorphism) $r : X \to Y$, $s : Y \to X$ and a natural number m such that

$$r \circ f = g \circ r, \quad s \circ g = f \circ s, \quad r \circ s = g^m, \quad s \circ r = f^m.$$

The homotopy classes of f and g are shift-equivalent if there exist continuous maps $s : X \to Y$, $r : Y \to X$ and a natural number m such that

$$r \circ f \sim g \circ r, \quad s \circ g \sim f \circ s, \quad r \circ s \sim g^m, \quad s \circ r \sim f^m.$$

Bibliographical notes

The terminology present in this subsection can be found in any algebraic topology book (Spanier 1982). An undergraduate-level introduction to computational homology is Kaczyński, Mischaikow and Mrozek (200x).

3.2. Conley index for flows

As a means of introducing the Conley index we begin with a discussion in the context of flows, since this is the simpler case. Furthermore, since it is for motivational purposes, we shall present the definitions in a very restrictive setting. For a more general presentation the reader should consult Conley (1978), Smoller (1980), Salamon (1985), Arnold *et al.* (1995) and Mischaikow and Mrozek (2001).

As before, $\varphi : \mathbb{R} \times \mathbb{R}^n \to \mathbb{R}^n$ denotes the flow generated by the differential equation $\dot{x} = g(x)$. Let $\mathcal{T}$ be a flow-transverse polygonal decomposition of $X \subset \mathbb{R}^n$.

Let $P \in \mathcal{T}$. If $L \in \partial P$, then there exists a unique n-dimensional simplex $K \in P$ such that L is face of K. L is an *exit face* of P if

$$\nu_K(L) \cdot f(x) < 0 \quad \text{for all } x \in L.$$

Definition 25. A pair $N = (N_1, N_0)$ of compact sets with $N_0 \subset N_1$ is an *index pair* if N_1 is a flow-transverse polygonal isolating block and

$$N_0 = \{L \in \partial N_1 \mid L \text{ is an exit face of } N_1\}.$$

Observe that by Proposition 2.18 polygonal isolating blocks, and hence polygonal index pairs, can be computed efficiently.

Let $N = (N_1, N_0)$ be an index pair. Let

$$S := \mathrm{Inv}\,(\mathrm{cl}\,(\mathrm{int}\, N_1 \setminus N_0), \varphi) = \mathrm{Inv}\,(N_1, \varphi) \subset \mathrm{int}\,(N_1 \setminus N_0). \tag{3.1}$$

Definition 26. The *(homotopy) Conley index* of S is

$$h(S) \sim (N_1/N_0, [N_0]).$$

Remark 2. Here we have presented the Conley index as an index of the isolated invariant set S. It can be proved that $h(S)$ is independent of the index pair used to compute it. In other words, if $N' = (N_1', N_0')$ is another index pair with the property that

$$S = \mathrm{Inv}\,(\mathrm{cl}\,(\mathrm{int}\, N_1' \setminus N_0'), \varphi),$$

then

$$(N_1/N_0, [N_0]) \sim (N_1'/N_0', [N_0']).$$

Observe that, as presented in Definition 26, the index is the homotopy equivalence class of a pointed topological space. This is an extremely difficult object to compute. For this reason we work with a coarser invariant, the *homological Conley index*

$$CH_*(S) :\cong H_*(N_1/N_0, [N_0]) \cong H_*(N_1, N_0).$$

The isomorphism on the right is due to the fact that N_1 and N_0 are simplicial complexes.

3.3. Index pairs and index filtrations

Having defined the Conley index for flows, we return to the setting of continuous maps. Let $N = (N_1, N_0)$ be a pair of compact sets with $N_0 \subset N_1$. Define $f_N : (N_1/N_0, [N_0]) \to (N_1/N_0, [N_0])$ by

$$f_N(x) = \begin{cases} f(x), & \text{if } x, f(x) \in N_1 \setminus N_0, \\ [N_0], & \text{otherwise.} \end{cases}$$

Definition 27. The pair of compact sets $N = (N_1, N_0)$ is an *index pair* if $\mathrm{cl}\,(N_1 \setminus N_0)$ is an isolating neighbourhood and the *index map* f_N is continuous.

The Conley index for maps will be defined in terms of the index map f_N. Therefore, it is essential that we be able to compute index pairs. As the theorems of this section indicate, the same algorithms that determined isolating blocks and Morse decompositions provide us with index pairs.

Theorem 3.1. Let $\mathcal{F}$ be an outer approximation of f. Let $\mathcal{U} = \mathrm{Inv}\,(o(\mathcal{U}) \cup \mathcal{V}, \mathcal{F}) \cap o(\mathcal{U})$ where $\mathcal{V} \cap o(\mathcal{U}) = \emptyset$. Then the pair

$$N = (N_1, N_0) := (|(d(\mathcal{U}) \cap \mathcal{F}(\mathcal{U})) \cup \mathcal{U}|, |d(\mathcal{U}) \cap \mathcal{F}(\mathcal{U})|)$$

is an index pair for f.

If we let $\mathcal{V} = \emptyset$, then we obtain the following corollary. Observe that the assumption is precisely the halting condition for our algorithm to find isolating blocks within recurrent sets.

Corollary 3.2. Let $\mathcal{U} = \mathrm{Inv}\,(o(\mathcal{U}), \mathcal{F})$. Then the pair

$$N = (N_1, N_0) := (|(d(\mathcal{U}) \cap \mathcal{F}(\mathcal{U})) \cup \mathcal{U}|, |d(\mathcal{U}) \cap \mathcal{F}(\mathcal{U})|)$$

is an index pair for f.

As will be indicated in the next subsection, we can use these theorems to gain additional information about the structure of the dynamics within recurrent sets.

The construction which follows will allow us to understand the structure of connecting orbits between Morse sets.

Definition 28. Let $\mathcal{M}(S) := \{M(p) \mid p \in (\mathcal{P}, \succ)\}$ be a Morse decomposition of S with admissible order $\succ$. An *index filtration* of $\mathcal{M}(S)$ is a collection of compact sets $\mathcal{N} := \{N(I) \mid I \in \mathcal{A}(P, \succ)\}$ satisfying the following properties:

(1) for each $I \in \mathcal{A}(\mathcal{P}, \succ)$, $(N(I), N(\emptyset))$ is an index pair for $M(I)$,
(2) for any $I, J \in \mathcal{A}(\mathcal{P}, \succ)$,

$$N(I \cap J) = N(I) \cap N(J) \qquad \text{and} \qquad N(I) \cup N(J) = N(I \cup J).$$

Theorem 3.3. Let $\mathcal{F} : \mathcal{G} \rightrightarrows \mathcal{G}$ be an outer approximation of f. Let $\mathcal{S}$ be an isolated invariant set for $\mathcal{F}$ and let $\{\mathcal{R}_p \mid p \in (\mathcal{P}, >)\}$ be the set of SCPC for the directed graph associated to $\mathcal{F}$ restricted to the isolating neighbourhood of $\mathcal{S}$. Let $\mathcal{M}(S) := \{M(p) \mid p \in (\mathcal{P}, \succ)\}$ be the associated Morse decomposition for f, that is, $S = \mathrm{Inv}\,(|\mathcal{S}|, f)$ and $M(p) := \mathrm{Inv}\,(|\mathcal{R}_p|, f)$. Then there exists a collection of subsets of $\mathcal{G}$ such that $\{|\mathcal{N}(I)| \mid I \in \mathcal{A}(P, \succ)\}$ is an index filtration of S under f.

A modification of the depth-first search algorithm that was used to compute the set of SCPC and the discrete Lyapunov function also provides for an index filtration. In particular, let $(\mathcal{N}(\mathcal{P}), \mathcal{N}(\emptyset))$ be an index pair for $\mathcal{S}$. For each strongly connected path component $\mathcal{R}_p$ of $\mathcal{F}$ restricted to $\mathcal{N}(\mathcal{P})$, define $\mathcal{E}_p$ to be the set of grid elements G for which there exists a path $\gamma_G : \{0, 1, \ldots, n\} \to \mathcal{N}(\mathcal{P})$ such that $\gamma_G(n) \in \mathcal{R}_p$. One obtains an index filtration by defining

$$\mathcal{N}(I) := \mathcal{N}(\mathcal{P}) \setminus \bigcup_{p \notin I} \mathcal{E}_p.$$

Bibliographical notes

The proof of Theorem 3.1 is due to Szymczak (1997). Index filtrations were first developed by R. Franzosa (1986) in the context of flows. The particular construction described here has been used on several occasions (Franzosa and Mischaikow 1988, Eidenschink 1995, Richeson 1997).

3.4. Conley index for maps

Let

$$S := \mathrm{Inv}\,(\mathrm{cl}\,(\mathrm{int}\,N_1 \setminus N_0), f) \subset \mathrm{int}\,(N_1 \setminus N_0). \tag{3.2}$$

(compare with (3.1)). One is tempted at this point to define the Conley index of S in terms of the homotopy type of the pointed space $(N_1/N_0, [N_0])$. Unfortunately, it is easy to create examples where (N_1, N_0) and (N_1', N_0') are index pairs for S, that is,

$$\mathrm{Inv}\,(\mathrm{cl}\,(\mathrm{int}\,N_1 \setminus N_0), f) = S = \mathrm{Inv}\,(\mathrm{cl}\,(\mathrm{int}\,N_1' \setminus N_0'), f),$$

but

$$(N_1/N_0, [N_0]) \not\sim (N_1'/N_0', [N_0']).$$

The appropriate definition is as follows.

Definition 29. Let $N = (N_1, N_0)$ be an *index pair* for S. The (*homotopy*) *Conley index* for S is the homotopy shift equivalence class of the index map $f_N : (N_1/N_0, [N_0]) \to (N_1/N_0, [N_0])$.

For computational reasons we use the (*homology*) *Conley index* which is the shift equivalence class for group homomorphisms of

$$f_{N*} : H_*(N_1/N_0, [N_0]) \to H_*(N_1/N_0, [N_0]),$$

which in our context is equivalent to the shift equivalence class of

$$f_{N*} : H_*(N_1, N_0) \to H_*(N_1, N_0),$$

since N_1 and N_0 are constructed from grid elements.

Bibliographical notes

Defining the Conley index in terms of shift equivalence is due to Franks and Richeson (2000).

It should be noted that, because of the nature of our approximations, we cannot compute f_{N*} directly. Therefore, it is important to remark that, if $\mathcal{F}$ is a convex-valued outer approximation of f, then it is possible to compute f_{N*} from the data of $\mathcal{F}$. Details can be found in Kaczyński *et al.* (2001) and Kaczyński *et al.* (200x).

3.5. The structure of invariant sets

As was indicated in the Introduction, there is a variety of other references that provide information about the Conley index. What we provide here is a very curt description of how knowledge of the index provides information about the existence and structure of dynamics.

The first theorem is the most fundamental: it implies that there exists a nontrivial invariant set.

Theorem 3.4. (Ważewski Property) Let $N = (N_1, N_0)$ be an index pair for f. If f_{N*} is not shift-equivalent to the trivial map, then

$$S := \operatorname{Inv}(\operatorname{cl}(\operatorname{int} N_1 \setminus N_0), f) \neq \emptyset.$$

There are additional theorems based on finer topological invariants that allow one to deduce the existence of fixed points and periodic orbits (McCord, Mischaikow and Mrozek 1995, McCord 1988, Mrozek 1989) or even that the invariant set has positive topological entropy (Baker 1998).

The next result provides a method by which one can move from local results to global results.

Theorem 3.5. (Summation property) Let $N = (N_1, N_0)$ and $N' = (N_1', N_0')$ be index pairs where $N_1 \cap N_1' = \emptyset$. Assume $N \cup N'$ is also an index pair for f. Then $f_{N\cup N'*}$ is shift-equivalent to $f_{N*} \oplus f_{N'*}$.

The simplest application is to assume that one has a Morse decomposition that consists of two Morse sets, that is, $\mathcal{M}(S) = \{M(0), M(1) \mid 1 \succ 0\}$. If the Conley index of S is not the direct sum of the Conley index of $M(0)$ and the Conley index of $M(1)$, then there must exist a connecting orbit from $M(1)$ to $M(0)$. This type of calculation can be extended to general Morse decompositions via the *connection matrix* (Franzosa 1989, Franzosa 1988, Arnold *et al.* 1995, Richeson 1997, Robbin and Salamon 1992, Mischaikow and Mrozek 2001) which is a generalization of the Morse inequalities. These types of arguments can also be used to describe the structure of the set of connecting orbits (Arnold *et al.* 1995, Mischaikow and Mrozek 2001, McCord and Mischaikow 1996, Mischaikow 1995, Mischaikow and Morita 1994) and global bifurcations (McCord and Mischaikow 1992, Franzosa and Mischaikow 1998, Mischaikow 1989).

In an alternate direction, by comparing the structure of $f_{N\cup N'*}$ with that of $f_{N*} \oplus f_{N'*}$ one can derive conditions under which one must have chaotic symbolic dynamics where the symbols correspond to the regions $\operatorname{cl}(\operatorname{int} N_1 \setminus N_0)$ and $\operatorname{cl}(\operatorname{int} N_1' \setminus N_0')$ (Mischaikow and Mrozek 1995, Szymczak 1996, Szymczak 1995, Carbinatto, Kwapisz and Mischaikow 2000, Carbinatto and Mischaikow 1999).

4. Examples

In this section we present four examples that show how the ideas of the previous sections can be applied. Space does not permit a full description of the work; therefore we only present the most pertinent aspects of each result.

4.1. Hénon map

Recall that the Hénon map is given by the formula

$$f(x,y) = (1 - ax^2 + y, bx).$$

Szymczak (1997) obtained a computer-assisted proof of the following theorem.

Theorem 4.1. For $a = 1.4$ and $b = 0.3$ the Hénon map admits periodic orbits of all minimal periods except for 3 and 5.

We take this opportunity to present an outline of the proof that indicates how the constructions and results of the previous sections are applied.

The first step is to choose a region X of phase space for which a multivalued map is computed. Let

$$X = \left(\left[-\frac{85}{69}, \frac{85}{69}\right] \times \left[-\frac{86}{230}, \frac{86}{230}\right]\right) \setminus \left(\left[-\frac{85}{69}, \frac{10}{69}\right) \times \left[-\frac{86}{230}, 0\right)\right).$$

The second step is to construct a grid $\mathcal{G}$ that covers X. This was done using rectangles of size $\frac{1}{69} \times \frac{1}{230}$. Simple estimates allow one to construct a multivalued map $\mathcal{F} : \mathcal{G} \rightrightarrows \mathcal{G}$ that is an outer approximation of f.

The third step is to find an interesting isolating block that is a subset of $\mathcal{G}$. This was done using the reduction technique described in Theorem 2.14. An isolating block with five components was chosen, from which an index pair was constructed using Theorem 3.1. From this a Conley index map f_N was computed and the Conley index was used to guarantee the existence of periodic orbits with all minimal periods except 3 and 5.

The final step is to exclude the periodic orbits of period 3 and 5. This was done by examining all cubes G with the property that $G \in \mathcal{F}^3(G)$ or $G \in \mathcal{F}^5(G)$, and then showing that, within the region of phase space containing these cubes, the only recurrent dynamics consisted of a fixed point.

4.2. Lorenz equations

As was indicated earlier, one way to study complicated dynamics for ordinary differential equations is to study the dynamics of a Poincaré map $f : \Xi \to \Xi$ by computing a multivalued outer approximation $\mathcal{F}$ and following similar steps as described for the Hénon map computation. Of course the computational expense is considerably more than in the previous example, since one needs to rigorously bound the trajectories of a differential equation. This approach was used by M. Mrozek, A. Szymczak and the author (Mischaikow, Mrozek and Szymczak 2001) to study the Lorenz equations with the following results.

Let $\Sigma(k)$ be the set of bi-infinite sequences on k symbols. For a $k \times k$ matrix $A = (A_{ij})$ over $\mathbb{Z}_2$ let $\Sigma(A)$ be the set of bi-infinite sequences $\alpha : \mathbb{Z} \to \{1, 2, \ldots, k\}$ on k symbols that satisfy the following restriction:

$$A_{\alpha(i)\alpha(i+1)} = 1 \quad \text{for all } i \in \mathbb{Z}.$$

Clearly, $\Sigma(A) \subset \Sigma(k)$. Let $\sigma : \Sigma(A) \to \Sigma(A)$ be the shift map given by

$$\sigma(\alpha)(i) = \alpha(i+1).$$

Let $n_{(10,28,8/3)} = 6$, $n_{(10,60,8/3)} = n_{(10,54,45)} = 4$,

$$A_{(10,28,8/3)} = \begin{bmatrix} 0 & 1 & 1 & 0 & 0 & 0 \\ 0 & 0 & 0 & 1 & 1 & 0 \\ 0 & 0 & 0 & 0 & 0 & 1 \\ 1 & 0 & 0 & 0 & 0 & 0 \\ 0 & 1 & 1 & 0 & 0 & 0 \\ 0 & 0 & 0 & 1 & 1 & 0 \end{bmatrix},$$

$$A_{(10,60,8/3)} = \begin{bmatrix} 0 & 0 & 1 & 0 \\ 0 & 0 & 0 & 1 \\ 1 & 1 & 0 & 0 \\ 0 & 0 & 1 & 0 \end{bmatrix}, \qquad A_{(10,54,45)} = \begin{bmatrix} 1 & 1 & 0 & 0 \\ 0 & 0 & 1 & 1 \\ 1 & 1 & 0 & 0 \\ 0 & 0 & 1 & 1 \end{bmatrix}.$$

Theorem 4.2. Consider the Lorenz equations

$$\begin{aligned} \dot{x} &= s(y - x), \\ \dot{y} &= Rx - y - xz, \\ \dot{z} &= xy - qz, \end{aligned} \tag{4.1}$$

and the plane $P := \{(x, y, z) \mid z = R - 1\}$. For all parameter values in a sufficiently small neighbourhood of (σ, R, b) there exists a Poincaré section $N \subset P$ such that the associated Poincaré map f is Lipschitz and well defined. Furthermore, there is a continuous map $\rho : \mathrm{Inv}(N, f) \to \Sigma(n_{(\sigma,R,b)})$ such that

$$\rho \circ f = \sigma \circ \rho$$

and $\Sigma(A_{(\sigma,R,b)}) \subset \rho(\mathrm{Inv}(N, f))$. Moreover, for each periodic sequence $\alpha \in \Sigma(A)$ there exists an $x \in \mathrm{Inv}(N, f)$ on a periodic trajectory of the same minimal period such that $\rho(x) = \alpha$.

4.3. A partial differential equation

As was indicated in the Introduction, these ideas can be carried out for infinite-dimensional systems. A long-range goal is to be able to perform efficient rigorous computations to prove the existence of chaotic dynamics in partial differential equations. There are three obvious difficulties that need to be overcome to achieve our goal.

D1 Because of the finite nature of a computer it is impossible to compute directly on an infinite-dimensional system. Therefore, it is necessary to use an appropriate finite-dimensional reduction.

D2 Given a finite-dimensional system, *i.e.*, an ordinary differential equation, we need to be able to find isolating blocks and compute Conley indices.

D3 The index results of the finite-dimensional system need to be lifted to the full infinite-dimensional system.

However, as was discussed in the subsection of flows, we still do not have efficient methods for producing flow-transverse polygonal grids that are good approximations of the dynamics. Therefore, in general, we are not yet able to overcome difficulty D2. On the other hand, recent work by P. Zgliczyński and the author (Zgliczyński and Mischaikow 2001) shows that efficient, accurate and rigorous computation of fixed points is possible.

They considered the Kuramoto–Sivashinsky equation

$$u_t = -\nu u_{xxxx} - u_{xx} + 2uu_x \qquad (t,x) \in [0,\infty) \times (-\pi,\pi)$$

subject to periodic and odd boundary conditions

$$u(t,-\pi) = u(t,\pi) \quad \text{and} \quad u(t,-x) = -u(t,x). \tag{4.2}$$

The finite-dimensional reduction is done using a standard Galerkin projection based on a Fourier series expansion. In particular, one obtains the following infinite set of ordinary differential equations:

$$\dot{a}_k = k^2(1-\nu k^2)a_k - k\sum_{n=1}^{k-1} a_n a_{k-n} + 2k\sum_{n=1}^{\infty} a_n a_{n+k} \quad k = 1,2,3,\ldots, \tag{4.3}$$

which is then truncated to the finite-dimensional system

$$\dot{a}_k = k^2(1-\nu k^2)a_k - k\sum_{n=1}^{k-1} a_n a_{k-n} + 2k\sum_{n=1}^{m-k} a_n a_{n+k} \quad k = 1,\ldots,m. \tag{4.4}$$

Let $W \subset \mathbb{R}^m$ be a compact region. Within W, fixed points for (4.4) are identified. One can linearize around these points and use the eigenvectors to identify m-dimensional cubes with the property that the vector field generated by (4.4) is normal to the boundary faces.

Of course, (4.4) is just an approximation of the true vector field given by (4.3). To control the errors, it is assumed that

$$|a_k| \le \pm\frac{C_s}{k^s} \quad \text{for all } k > m,$$

where $s \geq 4$. Observe that this is a regularity condition. However, solutions to the Kuramoto–Sivashinsky equation that exist for all time are analytic, and therefore the higher Fourier coefficients of any solution that lies in a bounded invariant set will satisfy these constraints for sufficiently large C_s. It should also be remarked that this high degree of regularity is not particular to Kuramoto–Sivashinsky (see Hale and Raugel (2001) and references therein).

These constraints imply that, in the function space, we are studying the dynamics on the compact region

$$W \times \prod_{k=m+1}^{\infty} \left[-\frac{C_s}{k^s}, \frac{C_s}{k^s}\right].$$

Restricted to this region, one can bound the difference between the vector fields generated by (4.4) and (4.3), and therefore one can check if the transversality conditions on the faces of the isolating blocks in $\mathbb{R}^n$ are valid, not only for the vector field (4.4) but also for (4.3).

Finally, since Kuramoto–Sivashinsky is a strongly dissipative system it is easy to believe that, for sufficiently large k, the vector field is essentially of the form $\dot{a}_k = \lambda_k a_k$, where $\lambda < 0$. In other words, for sufficiently large k, the vector field is transverse to the faces defined by the product $\prod_{k=m+1}^{\infty}\left[-\frac{C_s}{k^s}, \frac{C_s}{k^s}\right]$ and furthermore, point inward. Since the Conley index is determined by the boundary pieces where the vector field points outwards, the index is determined by the low modes. In this way one can compute the index by a finite-dimensional approximation.

The final step is to show that this information can be lifted to the full system. The reader is referred to Zgliczyński and Mischaikow (2001) for the details.

4.4. An infinite-dimensional map

This final example is included to indicate that in the case of maps one can rigorously determine a wide range of dynamical objects even in the infinite-dimensional setting. Recently, M. Allili, S. Day, O. Junge and the author (Allili *et al.* 2002) have studied the Kot–Schaffer growth-dispersal model for plants. This is a map $\Phi : L^2([-\pi,\pi]) \to L^2([-\pi,\pi])$ of the form

$$\Phi(a)(y) := \frac{1}{2\pi}\int_{-\pi}^{\pi} b(x,y)\,\mu\,a(x)\left(1 - \frac{a(x)}{c(x)}\right)\mathrm{d}x, \qquad (4.5)$$

where $\mu > 0$ and $b(x,y) = b(|x-y|)$. Observe that the regularity of this map is determined by the regularity of the dispersal kernel b and the spatial heterogeneity of c in the nonlinear term.

The same difficulties D1–D3 need to be confronted with this problem. Again, using Fourier series one can rewrite the problem as an infinite system of maps:

$$a'_k = f_k(a) := \mu b_k \left[a_k - \sum_{\substack{j+l+n=k \\ j,l,k\geq 0}} c_j a_l a_n \right], \quad k = 0, 1, 2, \ldots. \tag{4.6}$$

Simple projection results in the finite-dimensional system $f^{(m)} : \mathbb{R}^m \to \mathbb{R}^m$

$$a'_k = f_k^{(m)}(a_0, \ldots, a_{m-1}) := \mu b_k \left[a_k - \sum_{\substack{j+l+n=k \\ 0\leq j,l,n\leq m-1}} c_j a_l a_n \right], \tag{4.7}$$

where $k = 0, 1, \ldots, m-1$, upon which the numerical computations are based.

Let

$$W := \prod_{k=0}^{m-1} [a_k^-, a_k^+] \subset \mathbb{R}^m,$$

and consider $f^{(m)} : W \to \mathbb{R}^m$. Assuming that the Fourier coefficients of the dispersal kernel b decay at an exponential rate, and taking advantage of the resulting regularity constraint on solutions which lie in bounded invariant sets, one can restrict attention to a compact set

$$Z := W \times \prod_{k=m}^{\infty} \left[-\frac{C_s}{s^k}, \frac{C_s}{s^k} \right]. \tag{4.8}$$

A grid $\mathcal{G}$ covering W is constructed and a multivalued map $\mathcal{F} : \mathcal{G} \rightrightarrows \mathcal{G}$ that is an outer approximation of f is constructed. It should be emphasized that the error bounds used to determine $\mathcal{F}$ include the errors obtained by truncation to the finite-dimensional system for elements in Z. This implies that, with the proper choice of m and C_s, the index computations for the finite-dimensional approximation are valid in the infinite-dimensional setting.

Using the techniques described in the latter part of Section 2.4, one can find isolating blocks for fixed points, periodic orbits, connecting orbits, and horseshoes. At this point the cost of computing the Conley index becomes an issue. Computation of homology grows in expense with the dimension of the complex. On the other hand, to obtain precise estimates on the solutions requires the use of more modes.

The strategy adopted in Allili *et al.* (2002) is as follows. A coarse (but mathematically rigorous) computation is done to determine an isolating block using a minimal number of modes. At this point the index is computed. One then increases the number of modes and refines the isolating block. As was described in Section 2, these computations are essentially linear in the number of boxes, and therefore the computational complexity is determined by the dimension of the invariant set. In the case of heteroclinic orbits or horseshoes, these are at most 1-dimensional objects. Since the blocks isolate the same invariant set, the Conley index remains unchanged. In this way the number of modes used in the approximation can be increased until the significant errors are those introduced by the floating point errors. The reader is referred to Allili *et al.* (2002) for details.

Acknowledgements
The author would like to thank Thomas Wanner, Anthony Baker, and the editors for carefully reading the orginal manuscript and providing a multitude of corrections and useful suggestions.

REFERENCES

M. Allili, S. Day, O. Junge and K. Mischaikow (2002), A rigorous numerical method for the global analysis of an infinite dimensional discrete dynamical system. In preparation.

L. Arnold, C. Jones, K. Mischaikow and G. Raugel (1995), *Dynamical Systems* (R. Johnson, ed.), Vol. 1609 of *Lecture Notes in Mathematics*, Springer.

A. Baker (1998), Topological entropy and the first homological Conley index map. Preprint.

E. Boczko, W. Kalies and K. Mischaikow (2002), Polygonal approximations of flows. In preparation.

M. Carbinatto and K. Mischaikow (1999), 'Horseshoes and the Conley index spectrum II: The theorem is sharp', *Dis. Cont. Dyn. Sys.* **5**, 599–616.

M. Carbinatto, J. Kwapisz and K. Mischaikow (2000), 'Horseshoes and the Conley index spectrum', *Erg. Theory Dyn. Sys.* **20**, 365–377.

K. A. Cliffe, A. Spence and S. J. Tavener (2000), 'The numerical analysis of bifurcation problems with application to fluid mechanics', in *Acta Numerica*, Vol. 9, Cambridge University Press, pp. 39–131.

C. Conley (1978), *Isolated Invariant Sets and the Morse Index*. Vol. 38 of *CBMS Lecture Notes*, AMS, Providence, RI.

C. Conley and R. Easton (1971), 'Isolated invariant sets and isolating blocks', *Trans. Amer. Math. Soc.* **158**, 35–61.

T. Corman, C. Leiserson, R. Rivest, C. Stein (2001), *Introduction to Algorithms*, MIT Press.

M. Dellnitz and A. Hohmann (1997), 'A subdivision algorithm for the computation of unstable manifolds and global attractors', *Numer. Math.* **75**, 293–317.

M. Dellnitz and O. Junge (2001), Set oriented numerical methods for dynamical systems, in *Handbook of Dynamical Systems* III: *Towards Applications* (B. Fiedler, G. Iooss and N. Kopell, eds), World Scientific. To appear.

M. Dellnitz, G. Froyland and O. Junge (2000), The algorithms behind GAIO: Set oriented numerical methods for dynamical systems. Preprint.

L. Dieci and T. Eirola (2001), *Numerical Dynamical Systems.* Notes.

E. J. Doedel, A. R. Champneys, T. F. Fairgrieve, Yu. A. Kuznetsov, B. Sandstede and X. J. Wang (1999), AUTO: *Continuation and Bifurcation Software with Ordinary Differential Equations, User's Guide*, Concordia University.

R. Easton (1975), 'Isolating blocks and symbolic dynamics', *J. Diff. Equations* **17**, 96–118.

R. Easton (1989), 'Isolating blocks and epsilon chains for maps', *Physica D* **39**, 95–110.

R. Easton (1998), *Geometric Methods for Discrete Dynamical Systems*, Oxford University Press.

M. Eidenschink (1995), Exploring global dynamics: A numerical algorithm based on the Conley index theory, PhD Thesis, Georgia Institute of Technology.

A. Floer (1987), 'A refinement of the Conley index and an application to the stability of hyperbolic invariant sets', *Erg. Theory Dyn. Sys.* **7**, 93–103.

J. Franks and D. Richeson (2000), Shift equivalence and the Conley index. *Trans. Amer. Math. Soc.* **352**, 3305–3322.

R. Franzosa (1986), 'Index filtrations and the homology index braid for partially ordered Morse decompositions', *Trans. Amer. Math. Soc.* **298**, 193–213.

R. Franzosa (1988), 'The continuation theory for Morse decompositions and connection matrices', *Trans. Amer. Math. Soc.* **310**, 781–803.

R. Franzosa (1989), 'The connection matrix theory for Morse decompositions', *Trans. Amer. Math. Soc.* **311**, 561–592.

R. Franzosa and K. Mischaikow (1988), 'The connection matrix theory for semiflows on (not necessarily locally compact) metric spaces', *J. Diff. Equations* **71**, 270–287.

R. Franzosa and K. Mischaikow (1998), 'Algebraic transition matrices in the Conley index theory', *Trans. Amer. Math. Soc.* **350**, 889–912.

J. Hale and G. Raugel (2001), Regularity, Determining Modes and Galerkin Methods, Preprint.

T. Kaczyński and M. Mrozek (1995), 'Conley index for discrete multivalued dynamical systems', *Topology Appl.* **65**, 83–96.

T. Kaczyński and M. Mrozek (1997), 'Stable index pairs for discrete dynamical systems', *Canadian Math. Bull.* **40**, 448–455.

T. Kaczyński, K. Mischaikow and M. Mrozek (2001), Computing homology. Preprint.

T. Kaczyński, K. Mischaikow and M. Mrozek (200x), *Computational Homology*, Springer. In preparation.

C. McCord (1988), 'Mappings and homological properties in the homology Conley index', *Erg. Theory Dyn. Sys.* **8***, 175–198.

C. McCord and K. Mischaikow (1992), 'Connected simple systems, transition matrices, and heteroclinic bifurcations', *Trans. Amer. Math. Soc.* **333**, 379–422.

C. McCord and K. Mischaikow (1996), 'On the global dynamics of attractors for scalar delay equations', *J. Amer. Math. Soc.* **9**, 1095–1133.

C. McCord, K. Mischaikow and M. Mrozek (1995), Zeta functions, periodic trajectories, and the Conley index, *J. Diff. Equations* **121**, 258–292.

K. Mischaikow (1989), 'Transition systems', *Proc. Roy. Soc. Edin.* **112A**, 155–175.

K. Mischaikow (1995), 'Global asymptotic dynamics of gradient-like bistable equations', *SIAM Math. Anal.* **26**, 1199–1224.

K. Mischaikow and Y. Morita (1994), 'Dynamics on the global attractor of a gradient flow arising in the Ginzburg–Landau equation', *Jap. J. Ind. Appl. Math.* **11**, 185–202.

K. Mischaikow and M. Mrozek (1995), 'Isolating neighborhoods and chaos', *Jap. J. Ind. Appl. Math.* **12**, 205–236.

K. Mischaikow and M. Mrozek (1995), 'Chaos in Lorenz equations: A computer assisted proof', *Bull. Amer. Math. Soc. (N.S.)* **33**, 66–72.

K. Mischaikow and M. Mrozek (1998), 'Chaos in Lorenz equations: A computer assisted proof, Part II: Details', *Math. Comput.* **67**, 1023–1046.

K. Mischaikow and M. Mrozek (2001), Conley index, in *Handbook of Dynamical Systems*, Vol. 2 (B. Fiedler, ed.) Elsevier.

K. Mischaikow, M. Mrozek and A. Szymczak (2001), 'Chaos in Lorenz equations: A computer assisted proof, Part III: The classic case'. *J. Diff. Equations* **169** 17–56.

M. Mrozek (1989), 'Index pairs and the fixed point index for semidynamical systems with discrete time', *Fund. Mathematicae* **133**, 177–192.

M. Mrozek (1990), 'Leray functor and the cohomological Conley index for discrete dynamical systems', *Trans. Amer. Math. Soc.* **318**, 149–178.

M. Mrozek (1996), 'Topological invariants, multivalued maps and computer assisted proofs in dynamics', *Comput. Math.* **32**, 83–104.

M. Mrozek (1999), 'An algorithmic approach to the Conley index theory', *J. Dyn. Diff. Equations* **11** 711-734.

M. Mrozek, M. Żelawski (1997), 'Heteroclinic connections in the Kuramoto–Sivashinsky equation', *Reliable Computing* **3**, 277–285.

D. Norton (1989), A metric approach to the Conley decomposition theorem, PhD thesis, University of Minnesota.

J. Palis and F. Takens (1993), *Hyperbolicity and Sensitive Chaotic Dynamics at Homoclinic Bifurcations*, Cambridge University Press.

P. Pilarczyk (1999), Computer assisted method for proving existence of periodic orbits. Preprint.

M. Poźniak (1994), 'Lusternik–Schnirelmann category of an isolated invariant set', *Univ. Jag. Acta Math.* **XXXI**, 129–139.

J. Reineck (1988), 'Connecting orbits in one-parameter families of flows', *Erg. Theory Dyn. Sys.* **8***, 359–374.

D. Richeson (1997), Connection matrix pairs for the discrete Conley index. Preprint.

J. W. Robbin and D. Salamon (1988), 'Dynamical systems, shape theory and the Conley index', *Erg. Theory Dyn. Sys.* **8***, 375–393.

J. W. Robbin and D. Salamon (1992), 'Lyapunov maps, simplicial complexes and the Stone functor', *Erg. Theory Dyn. Sys.* **12**, 153–183.

C. Robinson (1995), *Dynamical Systems: Stability, Symbolic Dynamics, and Chaos*, CRC Press.

C. Robinson (1977), 'Stability theorems and hyperbolicity in dynamical systems', *Rocky Mount. J. Math.* **7**, 425–437.

D. Salamon (1985), 'Connected simple systems and the Conley index of isolated invariant sets'. *Trans. Amer. Math. Soc.* **291**, 1–41.

J. Smoller (1980), *Shock Waves and Reaction Diffusion Equations*, Springer, New York.

E. H. Spanier (1982), *Algebraic Topology*, Springer. (McGraw-Hill 1966.)

A. M. Stuart and A. R. Humphries (1996), *Dynamical Systems and Numerical Analysis*, Vol. 2 of *Cambridge Monographs on Applied and Computational Mathematics*, Cambridge University Press.

A. Szymczak (1995), 'The Conley index for discrete dynamical systems', *Topology and its Applications* **66**, 215–240.

A. Szymczak (1995), 'The Conley index for decompositions of isolated invariant sets', *Fund. Math.* **148**, 71–90.

A. Szymczak (1996), 'The Conley index and symbolic dynamics', *Topology* **35**, 287–299.

A. Szymczak (1997), 'A combinatorial procedure for finding isolating neighborhoods and index pairs', *Proc. Roy. Soc. Edin.* **127A**.

P. Zgliczyński and K. Mischaikow (2001), 'Rigorous numerics for partial differential equations: The Kuramoto–Sivashinsky equation'. *J. FoCM* **1**, 255–288.

Acta Numerica (2002), pp. 479–517
DOI: 10.1017/S0962492902000077

The immersed boundary method

Charles S. Peskin
Courant Institute of Mathematical Sciences,
New York University, 251 Mercer Street,
New York, NY10012-1185, USA
E-mail: peskin@cims.nyu.edu

To Dora and Volodya

This paper is concerned with the mathematical structure of the immersed boundary (IB) method, which is intended for the computer simulation of fluid–structure interaction, especially in biological fluid dynamics. The IB formulation of such problems, derived here from the principle of least action, involves both Eulerian and Lagrangian variables, linked by the Dirac delta function. Spatial discretization of the IB equations is based on a fixed Cartesian mesh for the Eulerian variables, and a moving curvilinear mesh for the Lagrangian variables. The two types of variables are linked by interaction equations that involve a smoothed approximation to the Dirac delta function. Eulerian/Lagrangian identities govern the transfer of data from one mesh to the other. Temporal discretization is by a second-order Runge–Kutta method. Current and future research directions are pointed out, and applications of the IB method are briefly discussed.

CONTENTS

1. Introduction

The immersed boundary (IB) method was introduced to study flow patterns around heart valves and has evolved into a generally useful method for problems of fluid–structure interaction. The IB method is both a mathematical formulation and a numerical scheme. The mathematical formulation employs a mixture of Eulerian and Lagrangian variables. These are related by interaction equations in which the Dirac delta function plays a prominent role. In the numerical scheme motivated by the IB formulation, the Eulerian variables are defined on a fixed Cartesian mesh, and the Lagrangian variables are defined on a curvilinear mesh that moves freely through the fixed Cartesian mesh without being constrained to adapt to it in any way at all. The interaction equations of the numerical scheme involve a smoothed approximation to the Dirac delta function, constructed according to certain principles that we shall discuss.

This paper is concerned primarily with the mathematical structure of the IB method. This includes both the IB form of the the equations of motion and also the IB numerical scheme. Details of implementation are omitted, and applications are discussed only in summary fashion, although references to these topics are provided for the interested reader.

2. Equations of motion

Our purpose in this section is to derive the IB formulation of the equations of motion of an incompressible elastic material. The approach that we take is similar to that of Ebin and Saxton (1986, 1987). The starting point of this derivation is the principle of least action. Although we begin in Lagrangian variables, the key step is to introduce Eulerian variables along the way, and to do so in a manner that brings in the Dirac delta function. Roughly speaking, our goal here is to make the equations of elasticity look as much as possible like the equations of fluid dynamics. Once we have done that, fluid–structure interaction will be easier to handle.

Consider, then, an elastic incompressible material filling three-dimensional space. Let (q, r, s) be curvilinear coordinates attached to the material, so that fixed values of (q, r, s) label a material point. Let $\mathbf{X}(q, r, s, t)$ be the position at time t in Cartesian coordinates of the material point whose label is (q, r, s). Let $M(q, r, s)$ be the mass density of the material in the sense that $\int_Q M(q, r, s)\, \mathrm{d}q\, \mathrm{d}r\, \mathrm{d}s$ is the mass of the part of the material defined by $(q, r, s) \in Q$. Note that M is independent of time, since mass is conserved.

Note that $\mathbf{X}(\ ,\ ,\ , t)$ describes the configuration in space of the whole material at the particular time t. We assume that this determines the elastic (potential) energy of the material according to an energy functional $E[\mathbf{X}]$ such that $E[\mathbf{X}(\ ,\ ,\ , t)]$ is the elastic energy stored in the material at time t.

A prominent role in the following will be played by the Fréchet derivative of E, which is implicitly defined as follows.

Consider a perturbation $\wp\mathbf{X}(\,,\,,t)$ of the configuration $\mathbf{X}(\,,\,,t)$. (We denote perturbations by the symbol $\wp$ instead of the traditional calculus-of-variations symbol δ. That is because we need δ for the Dirac delta function that will soon make its appearance.) Up to terms of first order, the resulting perturbation in elastic energy will be a *linear* functional of the perturbation in configuration of the material. Such a functional can always be put in the form

$$\wp E\left[\mathbf{X}(\,,\,,t)\right] = \int (-\mathbf{F}(q,r,s,t)) \cdot \wp\mathbf{X}(q,r,s,t)\, \mathrm{d}q\, \mathrm{d}r\, \mathrm{d}s \tag{2.1}$$

The function $-\mathbf{F}(\,,\,,t)$ which appears in this equation is the Fréchet derivative of E evaluated at the configuration $\mathbf{X}(\,,\,,t)$. The physical interpretation of the foregoing is that $\mathbf{F}$ is the force density (with respect to q,r,s) generated by the elasticity of the material. This is essentially the principle of virtual work. As shorthand for (2.1) we shall write

$$\mathbf{F} = -\frac{\wp E}{\wp\mathbf{X}}. \tag{2.2}$$

We digress here to give an example of an elastic energy functional E and the elastic force $\mathbf{F}$ that it generates. Consider a system of elastic fibres, with the fibre direction $\boldsymbol{\tau}$ varying smoothly as a function of position. Let q,r,s be material (Lagrangian) curvilinear coordinates chosen in such a manner that $q,r =$ constant along each fibre. We assume that the elastic energy is of the form

$$E = \int \mathcal{E}\left(\left|\frac{\partial\mathbf{X}}{\partial s}\right|\right) \mathrm{d}q\, \mathrm{d}r\, \mathrm{d}s. \tag{2.3}$$

The meaning of this is that the elastic energy depends only on the strain in the fibre direction, and not at all on the cross-fibre strain. In other words, we have an extreme case of an anisotropic material. Also, since there is no restriction on $|\partial\mathbf{X}/\partial s|$, which determines the local fibre strain, and since the local energy density $\mathcal{E}$ is an arbitrary function of this quantity, we are dealing here with a case of nonlinear elasticity. To evaluate $\mathbf{F}$, we apply the perturbation operator $\wp$ to E and then use integration by parts with respect to s to put the result in the following form:

$$\wp E = -\int \frac{\partial}{\partial s}\left(\mathcal{E}'\left(\left|\frac{\partial\mathbf{X}}{\partial s}\right|\right)\frac{\partial\mathbf{X}/\partial s}{|\partial\mathbf{X}/\partial s|}\right) \cdot \wp\mathbf{X}\, \mathrm{d}q\, \mathrm{d}r\, \mathrm{d}s, \tag{2.4}$$

where $\mathcal{E}'$ is the derivative of $\mathcal{E}$. By definition, then,

$$\mathbf{F} = \frac{\partial}{\partial s}\left(\mathcal{E}'\left(\left|\frac{\partial\mathbf{X}}{\partial s}\right|\right)\frac{\partial\mathbf{X}/\partial s}{|\partial\mathbf{X}/\partial s|}\right). \tag{2.5}$$

Let

$$T = \mathcal{E}'\left(\left|\frac{\partial \mathbf{X}}{\partial s}\right|\right) = \text{fibre tension,} \tag{2.6}$$

and

$$\boldsymbol{\tau} = \frac{\partial \mathbf{X}/\partial s}{|\partial \mathbf{X}/\partial s|} = \text{unit tangent to the fibres.} \tag{2.7}$$

Then

$$\mathbf{F} = \frac{\partial}{\partial s}(T\boldsymbol{\tau}). \tag{2.8}$$

This formula can also be derived without reference to the elastic energy, by starting from the assumption that $\pm T\boldsymbol{\tau}\,\mathrm{d}q\,\mathrm{d}r$ is the force transmitted by the bundle of fibres $\mathrm{d}q\,\mathrm{d}r$, and by considering force balance on an arbitrary segment of such a bundle given by $s \in (s_1, s_2)$.

Expanding the derivative in (2.8), we reach the important conclusion that the elastic force density generated by a system of elastic fibres is locally parallel to the osculating plane of the fibres, that is, the plane spanned by $\boldsymbol{\tau}$ and $\partial\boldsymbol{\tau}/\partial s$. There is no elastic force density in the binormal direction.

Returning to our main task, we consider the constraint of incompressibility. Let

$$J(q,r,s,t) = \det\left(\frac{\partial \mathbf{X}}{\partial q}, \frac{\partial \mathbf{X}}{\partial r}, \frac{\partial \mathbf{X}}{\partial s}\right). \tag{2.9}$$

The volume occupied at time t by the part of the material defined by $(q,r,s) \in Q$ is given by $\int_Q J(q,r,s,t)\,\mathrm{d}q\,\mathrm{d}r\,\mathrm{d}s$. Since the material is incompressible, this should be independent of time for every choice of Q, which is only possible if

$$\frac{\partial J}{\partial t} = 0. \tag{2.10}$$

From now on, we shall write $J(q,r,s)$ instead of $J(q,r,s,t)$.

The principle of least action states that our system will evolve over the time interval $(0,T)$ in such a manner as to minimize the action S defined by

$$S = \int_0^T L(t)\,\mathrm{d}t, \tag{2.11}$$

where L is the Lagrangian (defined below). The minimization is to be done subject to the constraint of incompressibility (equation (2.10), above), and also subject to given initial and final configurations:

$$\mathbf{X}(q,r,s,0) = \mathbf{X}_0(q,r,s), \tag{2.12}$$

$$\mathbf{X}(q,r,s,T) = \mathbf{X}_T(q,r,s). \tag{2.13}$$

In general, the Lagrangian L is the difference between the kinetic and potential energies. In our case, this reads

$$L(t) = \frac{1}{2} \int M(q, r, s) \left| \frac{\partial \mathbf{X}}{\partial t}(q, r, s, t) \right|^2 \mathrm{d}q \, \mathrm{d}r \, \mathrm{d}s - E[\mathbf{X}(\ ,\ ,\ , t)]. \tag{2.14}$$

Thus, for arbitrary $\wp\mathbf{X}$ consistent with the constraints, we require

$$0 = -\wp S = \int_0^T \int \left(M \frac{\partial^2 \mathbf{X}}{\partial t^2} - \mathbf{F} \right) \cdot \wp\mathbf{X} \, \mathrm{d}q \, \mathrm{d}r \, \mathrm{d}s \, \mathrm{d}t. \tag{2.15}$$

To arrive at this result, we have used integration by parts with respect to t in the first term. In the second term we have made use of the definition of $-\mathbf{F}$ as the Fréchet derivative of the elastic energy E.

At this point, if $\wp\mathbf{X}$ were arbitrary we would be done, and the conclusion would be nothing more than Newton's law that force equals mass times acceleration. But $\wp\mathbf{X}$ is not arbitrary: it must be consistent with the constraints, in particular with the constraint of incompressibility (equations (2.9)–(2.10)). This constraint seems difficult to deal with in its present form.

Perhaps we can overcome this difficulty by introducing Eulerian variables. As is well known, the constraint of incompressibility takes on a very simple form in terms of the Eulerian velocity field $\mathbf{u}(\mathbf{x}, t)$, namely $\nabla \cdot \mathbf{u} = 0$. The following definitions of Eulerian variables are mostly standard. The new feature, however, is to introduce a pseudo-velocity $\mathbf{v}$ that corresponds to the perturbation $\wp\mathbf{X}$ in the same way as the physical velocity field $\mathbf{u}$ corresponds to the rate of change of the material's configuration $\partial\mathbf{X}/\partial t$.

More precisely, let $\mathbf{u}$ and $\mathbf{v}$ be implicitly defined as follows:

$$\mathbf{u}\left(\mathbf{X}(q, r, s, t), t\right) = \frac{\partial \mathbf{X}}{\partial t}(q, r, s, t), \tag{2.16}$$

$$\mathbf{v}\left(\mathbf{X}(q, r, s, t), t\right) = \wp\mathbf{X}(q, r, s, t). \tag{2.17}$$

Thus $\mathbf{u}(\mathbf{x}, t)$ is the velocity of whatever material point happens to be at position $\mathbf{x}$ at time t, and $\mathbf{v}(\mathbf{x}, t)$ is the perturbation experienced by whatever material point happens to be at position $\mathbf{x}$ at time t (according to the unperturbed motion). We shall also make use of the familiar material derivative defined as follows:

$$\frac{D\mathbf{u}}{Dt} = \frac{\partial \mathbf{u}}{\partial t} + \mathbf{u} \cdot \nabla \mathbf{u}, \tag{2.18}$$

for which we have the identity

$$\frac{D\mathbf{u}}{Dt}\left(\mathbf{X}(q, r, s, t), t\right) = \frac{\partial^2 \mathbf{X}}{\partial t^2}(q, r, s, t). \tag{2.19}$$

Thus $\frac{D\mathbf{u}}{Dt}(\mathbf{x}, t)$ is the acceleration of whatever material point happens to be at position x at time t.

In the following, we shall need the form taken by the constraints in the new variables. There is no difficulty about the initial and final value constraints, (2.12)–(2.13). Since the perturbation must vanish at the initial and final times to be consistent with these constraints, we simply have $\mathbf{v}(\mathbf{x},0) = \mathbf{v}(\mathbf{x},T) = 0$. The incompressibility constraint requires further discussion, however. The perturbations that we consider are required to be volume-preserving (for every piece of the material, not just for the material as a whole). This means that, if $\mathbf{X} + \wp\mathbf{X}$ is substituted for $\mathbf{X}$ in the formula for J, equation (2.9), then the value of $J(q,r,s)$ should be unchanged, at any particular q,r,s, up to terms of first order in $\wp\mathbf{X}$.

To see what this implies about $\mathbf{v}$, it is helpful to use matrix notation. Thus, let

$$\mathbf{a} = (q,r,s), \tag{2.20}$$

and let $\partial\mathbf{X}/\partial\mathbf{a}$ be the 3×3 matrix whose entries are the components of $\mathbf{X}$ differentiated with respect to q, r, or s. In this notation

$$J = \det\left(\frac{\partial\mathbf{X}}{\partial\mathbf{a}}\right). \tag{2.21}$$

Now, we need to apply the perturbation operator $\wp$ to both sides of this equation, but to do so we need the following identity involving the perturbation of the determinant of an arbitrary nonsingular square matrix A:

$$\wp\log\left(\det(A)\right) = \operatorname{trace}\left((\wp A)A^{-1}\right). \tag{2.22}$$

Although this identity is well known, we give a brief derivation of it for completeness. The starting point is the familiar formula for the inverse of a matrix in terms of determinants, written in the following possibly unfamiliar way:

$$\left(A^{-1}\right)_{ji} = \frac{1}{\det(\mathrm{A})}\frac{\partial\det(A)}{\partial A_{ij}}, \tag{2.23}$$

where we have written the signed cofactor of A_{ij} as $\partial\det(A)/\partial A_{ij}$. That we may do so follows from the expansion of the determinant by minors of row i (or column j). Making use of this formula (in the next-to-last step, below), we find

$$\begin{aligned}\wp\log\left(\det(A)\right) &= \sum_{ij}(\wp A)_{ij}\frac{\partial\log\left(\det(A)\right)}{\partial A_{ij}}\\ &= \sum_{ij}(\wp A)_{ij}\frac{1}{\det(A)}\frac{\partial\det(A)}{\partial A_{ij}}\\ &= \sum_{ij}(\wp A)_{ij}(A^{-1})_{ji}\\ &= \operatorname{trace}\left((\wp A)A^{-1}\right).\end{aligned} \tag{2.24}$$

This establishes the identity in (2.22).

Now consider the equation that implicitly defines $\mathbf{v}$, (2.17). Differentiating on both sides with respect to q, r, or s and using the chain rule, we get (in matrix notation)

$$\frac{\partial \wp \mathbf{X}}{\partial \mathbf{a}} = \frac{\partial \mathbf{v}}{\partial \mathbf{x}} \frac{\partial \mathbf{X}}{\partial \mathbf{a}}. \tag{2.25}$$

Now interchange the order of the operators $\wp$ and $\partial/\partial\mathbf{a}$, and then multiply both sides of this equation on the right by the inverse of $\partial\mathbf{X}/\partial\mathbf{a}$. The result is

$$\left(\wp \frac{\partial \mathbf{X}}{\partial \mathbf{a}}\right) \left(\frac{\partial \mathbf{X}}{\partial \mathbf{a}}\right)^{-1} = \frac{\partial \mathbf{v}}{\partial \mathbf{x}}. \tag{2.26}$$

Taking the trace of both sides, and making use of the identity derived above (equation (2.22)) as well as the definition of J (equation (2.21)), we see that

$$\wp \log\left(J(q, r, s)\right) = \nabla \cdot \mathbf{v}(\mathbf{X}(q, r, s, t), t). \tag{2.27}$$

Thus, the constraint of incompressibility ($\wp J = 0$) is equivalent to

$$\nabla \cdot \mathbf{v} = 0. \tag{2.28}$$

An argument that is essentially identical to the foregoing yields the familiar expression of the incompressibility constraint in terms of the Eulerian velocity field:

$$\nabla \cdot \mathbf{u} = 0. \tag{2.29}$$

This completes the discussion of the Eulerian form of the constraint of incompressibility.

We are now ready to introduce Eulerian variables into our formula for the variation of the action $\wp S$. We do so by using the defining property of the Dirac delta function, that one can evaluate a function at a point by multiplying it by an appropriately shifted delta function and integrating over all of space. We use this twice, once for the perturbation in configuration of the elastic body, and once for the dot product of that perturbation with the acceleration of the elastic body:

$$\wp \mathbf{X}(q, r, s, t) = \int \mathbf{v}(\mathbf{x}, t)\, \delta\left(\mathbf{x} - \mathbf{X}(q, r, s, t)\right) d\mathbf{x}, \tag{2.30}$$

$$\frac{\partial^2 \mathbf{X}}{\partial t^2}(q, r, s, t) \cdot \wp \mathbf{X}(q, r, s, t) = \int \frac{D\mathbf{u}}{Dt} \cdot \mathbf{v}(\mathbf{x}, t)\, \delta\left(\mathbf{x} - \mathbf{X}(q, r, s, t)\right) d\mathbf{x}. \tag{2.31}$$

In these equations (and throughout this paper), $\delta(\mathbf{x})$ denotes the three-dimensional delta function $\delta(x_1)\delta(x_2)\delta(x_3)$, where x_1, x_2, x_3 are the Cartesian components of the vector $\mathbf{x}$.

Substituting (2.30)–(2.31) into (2.15), we get

$$0 = \int_0^T \iint \left(M(q,r,s) \frac{D\mathbf{u}}{Dt}(\mathbf{x},t) - \mathbf{F}(q,r,s,t) \right) \cdot \mathbf{v}(\mathbf{x},t) \\ \delta\left(\mathbf{x} - \mathbf{X}(q,r,s,t)\right) \mathrm{d}\mathbf{x}\, \mathrm{d}q\, \mathrm{d}r\, \mathrm{d}s\, \mathrm{d}t. \quad (2.32)$$

Note that this expression contains a mixture of Lagrangian and Eulerian variables. The Lagrangian variables that remain in it are the mass density $M(q,r,s)$ and the elastic force density $\mathbf{F}(q,r,s,t)$. We can eliminate these (and get rid of the integral over q,r,s) by making the following definitions:

$$\rho(\mathbf{x},t) = \int M(q,r,s)\, \delta\left(\mathbf{x} - \mathbf{X}(q,r,s,t)\right) \mathrm{d}q\, \mathrm{d}r\, \mathrm{d}s, \quad (2.33)$$

$$\mathbf{f}(\mathbf{x},t) = \int \mathbf{F}(q,r,s,t)\, \delta\left(\mathbf{x} - \mathbf{X}(q,r,s,t)\right) \mathrm{d}q\, \mathrm{d}r\, \mathrm{d}s. \quad (2.34)$$

To see the meaning of the Eulerian variables $\rho(\mathbf{x},t)$ and $\mathbf{f}(\mathbf{x},t)$ defined by these two equations, integrate both sides of each equation over an arbitrary region Ω. Since $\int_\Omega \delta\left(\mathbf{x} - \mathbf{X}(q,r,s,t)\right) \mathrm{d}\mathbf{x}$ is equal to 1 or 0 depending on whether or not $\mathbf{X}(q,r,s,t) \in \Omega$, the results are

$$\int_\Omega \rho(\mathbf{x},t)\, \mathrm{d}\mathbf{x} = \int_{\mathbf{X}(q,r,s,t)\in\Omega} M(q,r,s)\, \mathrm{d}q\, \mathrm{d}r\, \mathrm{d}s, \quad (2.35)$$

$$\int_\Omega \mathbf{f}(\mathbf{x},t)\, \mathrm{d}\mathbf{x} = \int_{\mathbf{X}(q,r,s,t)\in\Omega} \mathbf{F}(q,r,s,t)\, \mathrm{d}q\, \mathrm{d}r\, \mathrm{d}s. \quad (2.36)$$

These equations show that ρ is the Eulerian mass density and that $\mathbf{f}$ is the Eulerian density of the elastic force.

Making use of the definitions of ρ and $\mathbf{f}$, we see that (2.32) can be rewritten

$$0 = -\wp S = \int_0^T \int \left(\rho(\mathbf{x},t) \frac{D\mathbf{u}}{Dt}(\mathbf{x},t) - \mathbf{f}(\mathbf{x},t) \right) \cdot \mathbf{v}(\mathbf{x},t)\, \mathrm{d}\mathbf{x}\, \mathrm{d}t. \quad (2.37)$$

This completes the transition from Lagrangian to Eulerian variables.

Equation (2.37) holds for arbitrary $\mathbf{v}(\mathbf{x},t)$ subject to the constraints. As shown above, the constraints on $\mathbf{v}(\mathbf{x},t)$ are that $\mathbf{v}(\mathbf{x},0) = \mathbf{v}(\mathbf{x},T) = 0$ and that $\nabla \cdot \mathbf{v} = 0$. Otherwise, $\mathbf{v}$ is arbitrary.

We now appeal to the Hodge decomposition: an arbitrary vector field may be written as the sum of a gradient and a divergence-free vector field. In particular, it is always possible to write

$$\rho \frac{D\mathbf{u}}{Dt} - \mathbf{f} = -\nabla p + \mathbf{w}, \quad (2.38)$$

where

$$\nabla \cdot \mathbf{w} = 0. \quad (2.39)$$

We shall show that $\mathbf{w} = 0$. To do so, we make use of the freedom in the choice of $\mathbf{v}$ by setting

$$\mathbf{v}(\mathbf{x}, t) = \phi(t)\mathbf{w}(\mathbf{x}, t). \tag{2.40}$$

Clearly, this satisfies all of the constraints on $\mathbf{v}$ provided that $\phi(0) = \phi(T) = 0$. This leaves a lot of freedom in the choice of ϕ, and we use this freedom to make $\phi(t) > 0$ for all $t \in (0, T)$.

With these choices, (2.37) becomes

$$0 = \int_0^T \phi(t) \int (-\nabla p(\mathbf{x}, t) + \mathbf{w}(\mathbf{x}, t)) \cdot \mathbf{w}(\mathbf{x}, t)\, \mathrm{d}\mathbf{x}\, \mathrm{d}t. \tag{2.41}$$

Now the term involving ∇p drops out, as can easily be shown using integration by parts, since $\nabla \cdot \mathbf{w} = 0$. This leaves us with

$$0 = \int_0^T \phi(t) \int |\mathbf{w}(\mathbf{x}, t)|^2 \, \mathrm{d}\mathbf{x}\, \mathrm{d}t. \tag{2.42}$$

Recalling that ϕ is positive on $(0, T)$, we conclude that $\mathbf{w} = 0$, as claimed.

In conclusion, we have the IB form of the equations of motion of an incompressible elastic material. We shall collect these equations here, taking the liberty of adding a viscous term which was omitted in the derivation (since the principle of least action applies to conservative systems) but which is important in applications. We assume a uniform viscosity of the kind that appears in a Newtonian fluid, although this could of course be generalized. Thus, the viscous term that we add is of the form $\mu\Delta\mathbf{u}$, where μ is the (dynamic) viscosity, and Δ is the Laplace operator (which is applied separately to each Cartesian component of the velocity field $\mathbf{u}$).

The equations of motion with the viscous term thrown in but otherwise as derived above read as follows:

$$\rho\left(\frac{\partial \mathbf{u}}{\partial t} + \mathbf{u} \cdot \nabla \mathbf{u}\right) + \nabla p = \mu \Delta \mathbf{u} + \mathbf{f}, \tag{2.43}$$

$$\nabla \cdot \mathbf{u} = 0, \tag{2.44}$$

$$\rho(\mathbf{x}, t) = \int M(q, r, s)\, \delta\left(\mathbf{x} - \mathbf{X}(q, r, s, t)\right) \mathrm{d}q\, \mathrm{d}r\, \mathrm{d}s, \tag{2.45}$$

$$\mathbf{f}(\mathbf{x}, t) = \int \mathbf{F}(q, r, s, t)\, \delta\left(\mathbf{x} - \mathbf{X}(q, r, s, t)\right) \mathrm{d}q\, \mathrm{d}r\, \mathrm{d}s, \tag{2.46}$$

$$\begin{aligned} \frac{\partial \mathbf{X}}{\partial t}(q, r, s, t) &= \mathbf{u}\left(\mathbf{X}(q, r, s, t), t\right) \\ &= \int \mathbf{u}(\mathbf{x}, t)\, \delta\left(\mathbf{x} - \mathbf{X}(q, r, s, t)\right) \mathrm{d}\mathbf{x}, \end{aligned} \tag{2.47}$$

$$\mathbf{F} = -\frac{\wp E}{\wp \mathbf{X}}. \tag{2.48}$$

In these equations, $M(q, r, s)$ is a given function, the Lagrangian mass density of the material, and $E[\mathbf{X}]$ is a given functional, the elastic potential energy of the material in configuration $\mathbf{X}$. The notation $\wp E/\wp \mathbf{X}$ is shorthand for the Fréchet derivative of E.

Note that (2.43)–(2.44) are completely in Eulerian form. In fact, they are the equations of a viscous incompressible fluid of non-uniform mass density $\rho(\mathbf{x}, t)$ subject to an applied body force (*i.e.*, a force per unit volume) $\mathbf{f}(\mathbf{x}, t)$. Equation (2.48) is completely in Lagrangian form. It expresses the elasticity of the material.

Equations (2.45)–(2.47) are interaction equations. They convert from Lagrangian to Eulerian variables (equations (2.45)–(2.46)) and *vice versa* (equation (2.47)). The first line of (2.47) is just the definition of the Eulerian velocity field, and the second line uses nothing more than the defining property of the Dirac delta function. All of the interaction equations involve integral operators with the same kernel $\delta\left(\mathbf{x} - \mathbf{X}(q, r, s, t)\right)$ but there is this subtle difference between (2.45)–(2.46) and (2.47). The interaction equations that define ρ and $\mathbf{f}$ are relationships between corresponding densities: the numerical values of ρ and M are *not* the same at corresponding points, and similarly the numerical values of $\mathbf{f}$ and $\mathbf{F}$ are *not* the same at corresponding points. But the numerical value of $\partial \mathbf{X}/\partial t$ and $\mathbf{u}(\mathbf{x}, t)$ *are* the same at corresponding points, as stated explicitly on the first line of (2.47). This difference comes about because of the mapping $(q, r, s) \to \mathbf{X}(q, r, s, t)$, which appears within the argument of the Dirac delta function. This complicates the integrals over q, r, s in (2.45)–(2.46); there is no such complication in the integral over $\mathbf{x}$ in (2.47). The distinction that we have just described is especially important in a special case that we shall discuss later, in which the elastic material is confined to a surface immersed in a three-dimensional fluid. Then ρ and $\mathbf{f}$ become singular, while $\partial \mathbf{X}/\partial t$ remains finite.

A feature of the foregoing equations of motion that the reader may find puzzling is that they do not include the equation

$$\frac{\partial \rho}{\partial t} + \mathbf{u} \cdot \nabla \rho = 0, \tag{2.49}$$

which asserts that the mass density at any given material point is independent of time, *i.e.*, that $D\rho/Dt = 0$. In fact (2.49) is a consequence of (2.44), (2.45) and (2.47), as we now show.

Applying $\partial/\partial t$ to both sides of (2.45), we get

$$\begin{aligned}\frac{\partial \rho}{\partial t} &= -\int M(q, r, s)\frac{\partial \mathbf{X}}{\partial t}(q, r, s, t) \cdot \nabla\delta(\mathbf{x} - \mathbf{X}(q, r, s, t))\, \mathrm{d}q\, \mathrm{d}r\, \mathrm{d}s \\ &= -\int M(q, r, s)\mathbf{u}(\mathbf{X}(q, r, s, t), t) \cdot \nabla\delta(\mathbf{x} - \mathbf{X}(q, r, s, t))\, \mathrm{d}q\, \mathrm{d}r\, \mathrm{d}s.\end{aligned} \tag{2.50}$$

On the other hand,

$$\mathbf{u}\cdot\nabla\rho(\mathbf{x},t)=\int M(q,r,s)\mathbf{u}(\mathbf{x},t)\cdot\nabla\delta\left(\mathbf{x}-\mathbf{X}(q,r,s,t)\right)\mathrm{d}q\,\mathrm{d}r\,\mathrm{d}s. \tag{2.51}$$

To complete the proof, we need the equation

$$\mathbf{u}(\mathbf{x})\cdot\nabla\delta(\mathbf{x}-\mathbf{X})=\mathbf{u}(\mathbf{X})\cdot\nabla\delta(\mathbf{x}-\mathbf{X}). \tag{2.52}$$

This is actually not true in general, but it is true when $\nabla\cdot\mathbf{u}=0$. In that case we have, for any suitable test function ϕ,

$$\nabla\cdot(\mathbf{u}\phi)=\mathbf{u}\cdot\nabla\phi, \tag{2.53}$$

$$\nabla\cdot(\mathbf{u}\phi)(\mathbf{X})=\mathbf{u}(\mathbf{X})\cdot\nabla\phi(\mathbf{X}), \tag{2.54}$$

$$\int\nabla\cdot(\mathbf{u}\phi)(\mathbf{x})\delta(\mathbf{x}-\mathbf{X})\,\mathrm{d}\mathbf{x}=\mathbf{u}(\mathbf{X})\cdot\int\nabla\phi(\mathbf{x})\delta(\mathbf{x}-\mathbf{X})\,\mathrm{d}\mathbf{x}, \tag{2.55}$$

$$-\int\phi(\mathbf{x})\mathbf{u}(\mathbf{x})\cdot\nabla\delta(\mathbf{x}-\mathbf{X})\,\mathrm{d}\mathbf{x}=-\mathbf{u}(\mathbf{X})\cdot\int\phi(\mathbf{x})\nabla\delta(\mathbf{x}-\mathbf{X})\,\mathrm{d}\mathbf{x}. \tag{2.56}$$

Since ϕ is arbitrary, this implies (2.52). Then, with the help of (2.52), it is easy to combine (2.50)–(2.51) to yield (2.49).

3. Fluid–structure interaction

In the previous section, we envisioned a viscoelastic incompressible material filling all of space. Let us now consider the situation in which only part of space is filled by the viscoelastic material and the rest by a viscous incompressible fluid. (As before, we assume that the viscosity μ is constant everywhere.) This actually requires no change at all in the equations of motion as we have formulated them. The only change is in the elastic energy functional $E[\mathbf{X}(\,,\,,\,,t)]$, which now includes no contribution from those points (q,r,s) that are in the fluid regions. As a result of this, the Lagrangian elastic force density $\mathbf{F}=-\wp E/\wp\mathbf{X}$ is zero at such fluid points (q,r,s), and the Eulerian elastic force density $\mathbf{f}(\mathbf{x},t)$ is zero at all space points $\mathbf{x}$ that happen to lie in fluid regions at time t.

Note that the Lagrangian coordinates (q,r,s) are still used in the fluid regions, since they carry the mass density $M(q,r,s)$, from which $\rho(\mathbf{x},t)$ is computed. This formulation allows for the possibility of a *stratified* fluid, in which $\rho(\mathbf{x},t)$ is non-uniform, even in the fluid regions.

If, on the other hand, the fluid has a uniform mass density ρ_0, then we can dispense with the Lagrangian variables in the fluid regions and let the Lagrangian coordinates (q,r,s) correspond to elastic material points only. In that case, in the spirit of Archimedes, we replace the Lagrangian mass density $M(q,r,s)$ by the *excess* Lagrangian mass density $\tilde{M}(q,r,s)$ such that

$$\rho(\mathbf{x},t) = \rho_0 + \int \tilde{M}(q,r,s)\,\delta\left(\mathbf{x} - \mathbf{X}(q,r,s,t)\right) \mathrm{d}q\,\mathrm{d}r\,\mathrm{d}s. \tag{3.1}$$

Note that $\tilde{M}$ may be negative; that would represent elastic material that is less dense than the ambient fluid. The integral of $\tilde{M}$ over any part of the elastic material is the difference between the mass of that elastic material and the mass of the fluid displaced by it.

Now we come to the important special case that gives the immersed *boundary* method its name. Suppose that all or part of the elastic material is confined to certain surfaces that are immersed in fluid. Examples include heart valve leaflets, parachute canopies, thin airfoils (including bird, insect, or bat wings), sails, kites, flags, and weather vanes. For any such surface, only two Lagrangian parameters, say (r,s) are needed. Then the interaction equations become

$$\rho(\mathbf{x},t) = \rho_0 + \int M(r,s)\,\delta\left(\mathbf{x} - \mathbf{X}(r,s,t)\right) \mathrm{d}r\,\mathrm{d}s, \tag{3.2}$$

$$\mathbf{f}(\mathbf{x},t) = \int \mathbf{F}(r,s,t)\,\delta\left(\mathbf{x} - \mathbf{X}(r,s,t)\right) \mathrm{d}r\,\mathrm{d}s, \tag{3.3}$$

$$\frac{\partial \mathbf{X}}{\partial t} = \mathbf{u}\left(\mathbf{X}(r,s,t),t\right) = \int \mathbf{u}(\mathbf{x},t)\,\delta\left(\mathbf{x} - \mathbf{X}(r,s,t)\right) \mathrm{d}\mathbf{x}. \tag{3.4}$$

Note that there is no distinction here between $M(r,s)$ and $\tilde{M}(r,s)$, since the volume displaced by an elastic material *surface* is zero, by definition, so the excess mass of the immersed boundary is simply its mass. In (3.2) and (3.3), the delta function $\delta(\mathbf{x}) = \delta(x_1)\delta(x_2)\delta(x_3)$ is still three-dimensional, but there are only two integrals $\mathrm{d}r\,\mathrm{d}s$. As a result of this discrepancy, $\rho(\mathbf{x},t)$ and $\mathbf{f}(\mathbf{x},t)$ are each singular like a one-dimensional delta function, the singularity being supported on the immersed elastic boundary. Although $\rho(\mathbf{x},t)$ and $\mathbf{f}(\mathbf{x},t)$ are infinite on the immersed boundary, their integrals over any finite volume are finite. Specifically, the integral of $\rho(\mathbf{x},t)$ over such a volume is the sum of the mass of the fluid contained within that volume and the mass of that part of the elastic boundary that is contained within that volume. Similarly, the integral of $\mathbf{f}(\mathbf{x},t)$ over such a volume is the total force applied to the fluid by the part of the immersed boundary contained within the volume in question.

Unlike (3.2) and (3.3), the integral in (3.4) is three-dimensional and gives a finite result, the velocity of the immersed elastic boundary. Here again we see an important distinction between the interaction equations that convert from Lagrangian to Eulerian variables and the interaction equation that converts in the other direction.

An important remark is that the equations of motion that we have just derived for a viscous incompressible fluid containing an immersed elastic boundary are mathematically equivalent to the conventional equations that one would write down involving the jump in the fluid stress across that boundary. This can be shown by integrating the Navier–Stokes equation across the boundary and noting the contribution from the delta function layers given by (3.2) and (3.3). For proofs of this kind, albeit in the case of a massless boundary, see Peskin and Printz (1993)and Lai and Li (2001).

In summary, the equations of motion derived in the previous section are easily adapted to problems of fluid–structure interaction, including problems involving stratified fluids (with or without structures) and also problems involving immersed elastic boundaries as well as immersed elastic solids.

4. Spatial discretization

This section begins the description of the *numerical* IB method. We shall present this method in the context of the equations of motion of an incompressible viscoelastic material, as derived above: see (2.43)–(2.48). The modifications of these equations needed for fluid–structure interaction have been described in the previous section and require corresponding minor changes in the numerical IB method. These will not be written out, since they are straightforward.

The plan that we shall follow in presenting the numerical IB method is to discuss spatial discretization first. Then (in the next section) we shall derive some identities satisfied by the spatially discretized scheme. Some of these motivate requirements that should be imposed on the approximate form of the Dirac delta function that is used in the interaction equations. The construction of that approximate delta function is discussed next, and finally we consider the temporal discretization of the scheme.

The spatial discretization that we shall describe employs two independent grids, one for the Eulerian and the other for the Lagrangian variables. The Eulerian grid, denoted g_h, is the set of points of the form $\mathbf{x} = \mathbf{j}h$, where $\mathbf{j} = (j_1, j_2, j_3)$ is a 3-vector with integer components. Similarly, the Lagrangian grid, denoted G_h is the set of (q, r, s) of the form $(k_q\Delta q, k_r\Delta r, k_s\Delta s)$, where $(k_q, k_r, k_s) = \mathbf{k}$ has integer components. To avoid leaks, we impose the restriction that

$$|\mathbf{X}(q + \Delta q, r, s, t) - \mathbf{X}(q, r, s, t)| < \frac{h}{2} \tag{4.1}$$

for all (q, r, s, t), and similarly for Δr and Δs.

In the continuous formulation, the elastic energy functional $E[\mathbf{X}]$ is typically an integral over a local energy density $\mathcal{E}$. Corresponding to this, we have in the discrete case

$$E_h = \sum_{\mathbf{k}'} \mathcal{E}_{\mathbf{k}'}(\cdots \mathbf{X}_{\mathbf{k}} \cdots)\, \Delta q\, \Delta r\, \Delta s. \tag{4.2}$$

Perturbing this, we find

$$\wp E_h = \sum_{\mathbf{k}} \sum_{\mathbf{k}'} \frac{\partial \mathcal{E}_{\mathbf{k}'}}{\partial \mathbf{X}_{\mathbf{k}}} \cdot \wp \mathbf{X}_{\mathbf{k}} \, \Delta q \, \Delta r \, \Delta s, \tag{4.3}$$

where $\partial / \partial \mathbf{X}_{\mathbf{k}}$ denotes the gradient with respect to $\mathbf{X}_{\mathbf{k}}$. This is of the form

$$\wp E_h = -\sum_{\mathbf{k}} \mathbf{F}_{\mathbf{k}} \cdot \wp \mathbf{X}_{\mathbf{k}} \, \Delta q \, \Delta r \, \Delta s, \tag{4.4}$$

provided that we set

$$\mathbf{F}_{\mathbf{k}} = -\sum_{\mathbf{k}'} \frac{\partial \mathcal{E}_{\mathbf{k}'}}{\partial \mathbf{X}_{\mathbf{k}}}. \tag{4.5}$$

Note that this is equivalent to

$$\mathbf{F}_{\mathbf{k}} \, \Delta q \, \Delta r \, \Delta s = -\frac{\partial E_h}{\partial \mathbf{X}_{\mathbf{k}}}. \tag{4.6}$$

Thus $\mathbf{F}_{\mathbf{k}}$ is the discrete Lagrangian force density associated with the material point $\mathbf{k} = (k_q \Delta q, k_r \Delta r, k_s \Delta s)$, and $\mathbf{F}_{\mathbf{k}} \, \Delta q \, \Delta r \, \Delta s$ is the actual *force* associated with that point. In a similar way, if $M_{\mathbf{k}}$ is the given Lagrangian mass density at the point $\mathbf{k}$, then $M_{\mathbf{k}} \, \Delta q \, \Delta r \, \Delta s$ is the actual *mass* associated with that point in the discrete formulation.

Let us now consider the spatial finite difference operators that are used on the Eulerian grid g_h. For the most part, these are built from the central difference operators $D^0_{h,\beta}$, $\beta = 1, 2, 3$, defined as follows:

$$\left(D^0_{h,\beta}\right)(\mathbf{x}) = \frac{\phi(\mathbf{x} + h\mathbf{e}_\beta) - \phi(\mathbf{x} - h\mathbf{e}_\beta)}{2h}, \tag{4.7}$$

where $\mathbf{e}_1, \mathbf{e}_2, \mathbf{e}_3$ are the standard basis vectors. We shall also use the notation that $\mathbf{D}^0_h$ is the vector difference operator whose components are $D^0_{h,\beta}$. Thus, $\mathbf{D}^0_h \phi$ is the central difference approximation to the gradient of ϕ, and $\mathbf{D}^0_h \cdot \mathbf{u}$ is the central difference approximation to the divergence of $\mathbf{u}$.

For the Laplacian that appears in the viscous terms, however, we do not use $\mathbf{D}^0_h \cdot \mathbf{D}^0_h$, since this would entail a staggered stencil of total width equal to 4 meshwidths. Instead, we use the 'tight' Laplacian L_h defined as follows:

$$(L_h \phi)(\mathbf{x}) = \sum_{\beta=1}^{3} \frac{\phi(\mathbf{x} + h\mathbf{e}_\beta) + \phi(\mathbf{x} - h\mathbf{e}_\beta) - 2\phi(\mathbf{x})}{h^2}. \tag{4.8}$$

Finally, we consider the operator $\mathbf{u} \cdot \nabla$ which appears in the nonlinear terms of the Navier–Stokes equations. Since $\nabla \cdot \mathbf{u} = 0$, we have the identity

$$\mathbf{u} \cdot \nabla \phi = \nabla \cdot (\mathbf{u} \phi). \tag{4.9}$$

We take advantage of this by introducing the skew-symmetric difference operator $S_h(\mathbf{u})$ defined as follows:

$$S_h(\mathbf{u})\phi = \frac{1}{2}\mathbf{u} \cdot \mathbf{D}_h^0\phi + \frac{1}{2}\mathbf{D}_h^0 \cdot (\mathbf{u}\phi). \tag{4.10}$$

Note that the two quantities that are averaged on the right-hand side are discretizations of the equal expressions $\mathbf{u} \cdot \nabla\phi$ and $\nabla \cdot (\mathbf{u}\phi)$, respectively. Nevertheless, the two discretizations are *not* in general equal to each other, not even if $\mathbf{D}_h^0 \cdot \mathbf{u} = 0$. That is because the product rule that holds for derivatives does not apply to the difference operator $\mathbf{D}_h^0$.

The skew symmetry of $S_h(\mathbf{u})$ is shown as follows. Let $(\ ,\ )_h$ be the discrete inner product defined by

$$(\phi, \psi)_h = \sum_{\mathbf{x}\in g_h} \phi(\mathbf{x})\psi(\mathbf{x})h^3 \tag{4.11}$$

for scalars, and

$$(\mathbf{u}, \mathbf{v})_h = \sum_{\mathbf{x}\in g_h} \mathbf{u}(\mathbf{x}) \cdot \mathbf{v}(\mathbf{x})h^3 \tag{4.12}$$

for vectors. It is easy to show by 'summation by parts' (which is merely re-indexing) that

$$\left(\phi, \mathbf{D}_h^0 \cdot \mathbf{u}\right)_h = -\left(\mathbf{D}_h^0\phi, \mathbf{u}\right)_h. \tag{4.13}$$

In other words, $\mathbf{D}_h^0$ is skew-symmetric with respect to this inner product. With the help of this identity, we see that

$$\begin{aligned}(\psi, S_h(\mathbf{u})\phi)_h &= \frac{1}{2}\left(\psi, \mathbf{u} \cdot \mathbf{D}_h^0\phi\right)_h + \frac{1}{2}\left(\psi, \mathbf{D}_h^0 \cdot (\mathbf{u}\phi)\right)_h \\ &= \frac{1}{2}\left(\mathbf{u}\psi, \mathbf{D}_h^0\phi\right)_h - \frac{1}{2}\left(\mathbf{D}_h^0\psi, \mathbf{u}\phi\right)_h \\ &= -\frac{1}{2}\left(\mathbf{D}_h^0 \cdot (\mathbf{u}\psi), \phi\right)_h - \frac{1}{2}\left(\mathbf{u} \cdot \mathbf{D}_h^0\psi, \phi\right)_h \\ &= -(S_h(\mathbf{u})\psi, \phi)_h, \end{aligned} \tag{4.14}$$

so $S_h(\mathbf{u})$ is indeed skew-symmetric, as claimed. Note that the skew symmetry of $S_h(\mathbf{u})$ holds for all $\mathbf{u}$; it does not require $\mathbf{D}_h^0 \cdot \mathbf{u} = 0$. This completes the discussion of the difference operators that are needed on the Eulerian grid g_h.

Finally, we consider the interaction equations, which involve the Dirac delta function via the kernel $\delta\left(\mathbf{x} - \mathbf{X}(q, r, s, t)\right)$. The integrals in these equations are, of course, replaced by sums over the appropriate grid (g_h or G_h), but what about the delta function itself? What is clearly needed is a function $\delta_h(\mathbf{x})$ that is nonsingular for each h but approaches $\delta(\mathbf{x})$ as $h \to 0$. There are, of course, many ways to construct such δ_h, and we shall describe the method of construction that we recommend in a subsequent section. For now, however, we proceed on the assumption that the function $\delta_h(\mathbf{x})$ is known.

With the apparatus introduced above, we may spatially discretize (2.43)–(2.48) as follows:

$$\rho\left(\frac{\mathrm{d}\mathbf{u}}{\mathrm{d}t} + S_h(\mathbf{u})\mathbf{u}\right) + \mathbf{D}_h^0 p = \mu L_h \mathbf{u} + \mathbf{f}, \tag{4.15}$$

$$\mathbf{D}_h^0 \cdot \mathbf{u} = 0, \tag{4.16}$$

$$\rho(\mathbf{x}, t) = \sum_{(q,r,s)\in G_h} M(q,r,s)\,\delta_h\left(\mathbf{x} - \mathbf{X}(q,r,s,t)\right)\,\Delta q\,\Delta r\,\Delta s, \tag{4.17}$$

$$\mathbf{f}(\mathbf{x}, t) = \sum_{(q,r,s)\in G_h} \mathbf{F}(q,r,s,t)\,\delta_h\left(\mathbf{x} - \mathbf{X}(q,r,s,t)\right)\,\Delta q\,\Delta r\,\Delta s, \tag{4.18}$$

$$\frac{\mathrm{d}\mathbf{X}}{\mathrm{d}t}(q,r,s,t) = \sum_{\mathbf{x}\in g_h} \mathbf{u}(\mathbf{x},t)\,\delta_h\left(\mathbf{x} - \mathbf{X}(q,r,s,t)\right)\,h^3, \tag{4.19}$$

$$\mathbf{F}(q,r,s,t)\,\Delta q\,\Delta r\,\Delta s = -\frac{\partial}{\partial \mathbf{X}(q,r,s)} E_h\big(\cdots \mathbf{X}(q',r',s',t)\cdots\big). \tag{4.20}$$

These comprise a system of ordinary differential equations, with $\mathbf{x}$ restricted to g_h, and with (q,r,s) or (q',r',s') restricted to G_h.

5. Eulerian/Lagrangian identities

First, we show that total momentum and total power each come out the same whether they are evaluated in Eulerian or in Lagrangian form. These two identities require no assumptions at all on the form of δ_h. They are important because they reveal the dual nature of the operations defined by (4.17) and (4.19), and similarly between the operations defined by (4.18) and (4.19). The operations that take Lagrangian input and generate Eulerian output (equations (4.17) and (4.18)) define relationships between corresponding densities, and we shall refer to them as 'spreading' operations. Thus (4.17) spreads the Lagrangian mass out onto the Eulerian grid, and the operation defined by (4.18) does the same for the Lagrangian force. The operation defined by (4.19), on the other hand, is best described as 'interpolation', since it averages the numerical values on nearby Eulerian grid points to produce a corresponding numerical value at a Lagrangian material point. The mathematical content of the two identities that we shall now derive is that spreading is the *adjoint* of interpolation.

Starting from the Eulerian formula for the momentum, we find

$$\begin{aligned}
&\sum_{\mathbf{x}\in g_h} \rho(\mathbf{x},t)\mathbf{u}(\mathbf{x},t)h^3 \\
&\quad = \sum_{\mathbf{x}\in g_h} \sum_{(q,r,s)\in G_h} M(q,r,s)\,\delta_h\left(\mathbf{x}-\mathbf{X}(q,r,s,t)\right)\,\Delta q\,\Delta r\,\Delta s\,\mathbf{u}(\mathbf{x},t)\,h^3 \\
&\quad = \sum_{(q,r,s)\in G_h} M(q,r,s)\frac{\mathrm{d}\mathbf{X}}{\mathrm{d}t}(q,r,s,t)\,\Delta q\,\Delta r\,\Delta s, \qquad (5.1)
\end{aligned}$$

which is the corresponding Lagrangian formula for the momentum. In the first step we used (4.17) to eliminate ρ and in the second step we used (4.19) to eliminate $\mathbf{u}$. Note that the argument only works because the same kernel $\delta_h\left(\mathbf{x}-\mathbf{X}(q,r,s,t)\right)$ is used the spreading step and in the interpolation step.

In exactly the same way, starting from the Eulerian formula for the power that is generated by the elastic force, we find

$$\begin{aligned}
&\sum_{\mathbf{x}\in g_h} \mathbf{f}(\mathbf{x},t)\cdot\mathbf{u}(\mathbf{x},t)\,h^3 \\
&\quad = \sum_{\mathbf{x}\in g_h} \sum_{(q,r,s)\in G_h} \mathbf{F}(q,r,s,t)\,\delta_h\left(\mathbf{x}-\mathbf{X}(q,r,s,t)\right)\,\Delta q\,\Delta r\,\Delta s\cdot\mathbf{u}(\mathbf{x},t)h^3 \\
&\quad = \sum_{(q,r,s)\in G_h} \mathbf{F}(q,r,s,t)\cdot\frac{\mathrm{d}\mathbf{X}}{\mathrm{d}t}(q,r,s,t)\,\Delta q\,\Delta r\,\Delta s, \qquad (5.2)
\end{aligned}$$

which is the corresponding Lagrangian expression. Here we used (4.18) and (4.19) in the same manner as above. Again, the argument only works because the same δ_h appears in both interaction equations.

Next, we show that mass, force, and torque all come out the same whether they are evaluated in terms of Eulerian or Lagrangian variables. These identities differ from those derived above in that they rely on certain properties of the function δ_h. We state these properties now, so that we may use them as needed in the derivations that follow. They will play a central role in the construction of δ_h as described in the following section. The two key properties that we need to impose on δ_h to obtain the mass, force, and torque identities are:

$$\sum_{\mathbf{x}\in g_h} \delta_h(\mathbf{x}-\mathbf{X})h^3 = 1, \quad \text{all } \mathbf{X}, \qquad (5.3)$$

$$\sum_{\mathbf{x}\in g_h} (\mathbf{x}-\mathbf{X})\,\delta_h(\mathbf{x}-\mathbf{X})h^3 = 0, \quad \text{all } \mathbf{X}. \qquad (5.4)$$

It is important that these properties hold for all real shifts $\mathbf{X}$, not merely for $\mathbf{X}\in g_h$. Note that these properties are the discrete analogues of $\int \delta(\mathbf{x}-\mathbf{X})\,\mathrm{d}\mathbf{x} = 1$ and $\int(\mathbf{x}-\mathbf{X})\,\delta(\mathbf{x}-\mathbf{X})\,\mathrm{d}\mathbf{x} = 0$. Finally, we remark that the two properties stated above can be combined to yield

$$\sum_{\mathbf{x}\in g_h} \mathbf{x}\,\delta_h(\mathbf{x}-\mathbf{X})h^3 = \mathbf{X}, \quad \text{all } \mathbf{X}. \tag{5.5}$$

The mass and force identities need only (5.3) and are derived as follows:

$$\begin{aligned} &\sum_{\mathbf{x}\in g_h} \rho(\mathbf{x},t)h^3 \\ &\quad = \sum_{\mathbf{x}\in g_h} \sum_{(q,r,s)\in G_h} M(q,r,s)\,\delta_h\left(\mathbf{x}-\mathbf{X}(q,r,s,t)\right)\,\Delta q\,\Delta r\,\Delta s\,h^3 \\ &\quad = \sum_{(q,r,s)\in G_h} M(q,r,s)\,\Delta q\,\Delta r\,\Delta s. \end{aligned} \tag{5.6}$$

By the way, this establishes conservation of mass, since the last line is independent of time. Exactly the same manipulations yield the force identity:

$$\begin{aligned} &\sum_{\mathbf{x}\in g_h} \mathbf{f}(\mathbf{x},t)h^3 \\ &\quad = \sum_{\mathbf{x}\in g_h} \sum_{(q,r,s)\in G_h} \mathbf{F}(q,r,s,t)\,\delta_h\left(\mathbf{x}-\mathbf{X}(q,r,s,t)\right)\,\Delta q\,\Delta r\,\Delta s\,h^3 \\ &\quad = \sum_{(q,r,s)\in G_h} \mathbf{F}(q,r,s,t)\,\Delta q\,\Delta r\,\Delta s. \end{aligned} \tag{5.7}$$

The torque identity is similar but relies on (5.5):

$$\begin{aligned} &\sum_{\mathbf{x}\in g_h} \mathbf{x}\times\mathbf{f}(\mathbf{x},t)h^3 \\ &\quad = \sum_{\mathbf{x}\in g_h} \sum_{(q,r,s)\in G_h} \mathbf{x}\times\mathbf{F}(q,r,s,t)\,\delta_h\left(\mathbf{x}-\mathbf{X}(q,r,s,t)\right)\,\Delta q\,\Delta r\,\Delta s\,h^3 \\ &\quad = \sum_{(q,r,s)\in G_h} \mathbf{X}(q,r,s,t)\times\mathbf{F}(q,r,s,t)\,\Delta q\,\Delta r\,\Delta s. \end{aligned} \tag{5.8}$$

In summary, we have established the equivalence of the Lagrangian and Eulerian expressions for mass, momentum, force, torque and power. (There do not seem to be similar identities for angular momentum or kinetic energy, though the reader is welcome to look for them.) The mass identity also establishes *conservation* of mass, since the Lagrangian expression for the mass is time-independent. The significance of the force, torque, and power identities is that momentum, angular momentum, and energy are not spuriously created or destroyed by the interaction equations. This is not the same thing as saying that these quantities are conserved, since that will depend on other aspects of the numerical scheme.

We conclude this section by showing in an important special case that momentum and energy are in fact conserved by the spatially discretized IB method. (In the case of energy, the rate of production of heat by viscosity has to be taken into account.) The special case that we consider is the one in which the whole system has a uniform mass density ρ. In other words, the immersed elastic material is neutrally buoyant in the ambient fluid, or it occupies so little volume that its mass can be neglected. In that case, we can dispense with (4.17) and instead treat ρ as a given constant.

Conservation of momentum requires the additional hypothesis that the elastic energy functional E be translation invariant, that is,

$$E[\mathbf{X}(\ ,\ ,\ ,t)+\mathbf{Z}]=E[\mathbf{X}(\ ,\ ,\ ,t)], \tag{5.9}$$

where $\mathbf{Z}$ is an arbitrary fixed shift, independent of the Lagrangian parameters (q,r,s). It is easy to show that this is equivalent to the condition that the total force generated by the elastic energy be identically zero. Thus, we assume here that

$$\sum_{(q,r,s)\in G_h}\mathbf{F}(q,r,s,t)\,\Delta q\,\Delta r\,\Delta s=0. \tag{5.10}$$

According to our force identity, (5.7), this property is inherited by the Eulerian force density so that

$$\sum_{\mathbf{x}\in g_h}\mathbf{f}(\mathbf{x},t)h^3=0. \tag{5.11}$$

Now, since ρ is constant, we can find the rate of change of momentum simply by summing (4.15) over the entire Eulerian grid. It is easy to see that this yields zero for all three of the terms $\rho S_h(\mathbf{u})\mathbf{u}$, $\mathbf{D}_h^0 p$, and $\mu L_h\mathbf{u}$ (recall that μ is constant throughout this paper). The reason is that summing over the grid is equivalent to taking the inner product with a constant. The operators involved in these three terms are all either antisymmetric or symmetric. Thus, they can be made to apply to the constant, where they all yield zero. The conclusion is that

$$\frac{\mathrm{d}}{\mathrm{d}t}\sum_{\mathbf{x}\in g_h}\rho\mathbf{u}(\mathbf{x},t)h^3=0, \tag{5.12}$$

which is conservation of momentum.

Conservation of energy does not require a translation-invariant elastic energy function, so we no longer make use of (5.9)–(5.11). Taking the inner product of both sides of (4.15) with $\mathbf{u}$, we find

$$\frac{\mathrm{d}}{\mathrm{d}t}\frac{1}{2}\rho(\mathbf{u},\mathbf{u})_h=-\rho(\mathbf{u},S_h(\mathbf{u})\mathbf{u})_h-(\mathbf{u},\mathbf{D}_h^0 p)_h+\mu(\mathbf{u},L_h\mathbf{u})_h+(\mathbf{u},\mathbf{f})_h. \tag{5.13}$$

Now S_h and $\mathbf{D}_h^0$ are both skew-symmetric, so the term involving S_h is immediately zero, and the term involving $\mathbf{D}_h^0$ can be rewritten $\left(\mathbf{D}_h^0\cdot\mathbf{u},p\right)$, which is zero because the discrete divergence of $\mathbf{u}$ is zero (equation (4.16)). Also, we can use the power identity (equation (5.2)) to rewrite the last term in Lagrangian variables. The result is

$$\frac{\mathrm{d}}{\mathrm{d}t}\frac{1}{2}\rho(\mathbf{u},\mathbf{u})_h = +\mu\left(\mathbf{u},L_h\mathbf{u}\right)_h + \sum_{(q,r,s)\in G_h}\mathbf{F}(q,r,s,t)\cdot\frac{\mathrm{d}\mathbf{X}}{\mathrm{d}t}(q,r,s,t)\,\Delta q\,\Delta r\,\Delta s. \tag{5.14}$$

The last step is to make use of (4.20) together with the chain rule to show that the last term is simply $-\mathrm{d}E_h/\mathrm{d}t$. Thus,

$$\frac{\mathrm{d}}{\mathrm{d}t}\left(\frac{1}{2}\rho\left(\mathbf{u},\mathbf{u}\right)_h + E_h(\cdots\mathbf{X}(q',r',s',t)\cdots)\right) = \mu(\mathbf{u},L_h\mathbf{u})_h. \tag{5.15}$$

Because L_h is a symmetric nonpositive operator the null space of which contains only constants, the term on the right-hand side is negative (unless $\mathbf{u}$ is constant, in which case that term is zero). Its magnitude is the rate of heat production by viscosity. This completes the proof of conservation of energy for the spatially discretized IB method with uniform mass density.

Although it would be nice to generalize the momentum and energy conservation results to the case of non-uniform mass density, this does not seem to be possible within the current framework. The equation for $\mathrm{d}\rho/\mathrm{d}t$, which is then needed, is a complicating factor. We can get such an equation by differentiating (4.17), but it involves the gradient of δ_h and does not seem to fit in well with the other equations. Nevertheless, we have at least shown that the force, torque, and power (which are, of course, the rates of change of momentum, angular momentum, and energy) are correctly converted back and forth between Lagrangian and Eulerian form by our scheme. These results hold even in the case of non-uniform mass density.

6. Construction of δ_h

In this section we describe the construction of the function δ_h that is used in the interaction equations (equations (4.17)–(4.19)). We first give the form of δ_h that is most commonly used, and then briefly discuss some possible variants. The postulates that we use to determine the function $\delta_h(\mathbf{x})$ are as follows. (The motivation for these postulates will be given after they have all been stated.)

First, we assume that the three-dimensional δ_h is given by a product of one-variable functions that scale with the meshwidth h in the following manner:

$$\delta_h(\mathbf{x}) = \frac{1}{h^3}\phi\left(\frac{x_1}{h}\right)\phi\left(\frac{x_2}{h}\right)\phi\left(\frac{x_3}{h}\right), \tag{6.1}$$

where x_1,x_2,x_3 are the Cartesian components of $\mathbf{x}$.

Having said this, we can state the rest of our postulates in terms of the function $\phi(r)$, where r can denote x_1/h, x_2/h, or x_3/h. These postulates are:

$$\phi(r) \text{ is continuous for all real } r, \tag{6.2}$$

$$\phi(r) = 0 \text{ for } |r| \geq 2, \tag{6.3}$$

$$\sum_{j \text{ even}} \phi(r-j) = \sum_{j \text{ odd}} \phi(r-j) = \frac{1}{2} \text{ for all real } r, \tag{6.4}$$

$$\sum_j (r-j)\phi(r-j) = 0 \text{ for all real } r, \tag{6.5}$$

$$\sum_j (\phi(r-j))^2 = C \text{ for all real } r, \tag{6.6}$$

where the constant C is independent of r. Its numerical value will be determined later. Even though the sums are over integer values of j, it is important that the above equations hold for all real shifts r, not just for integer values of r.

For future reference, we note that (6.4) and (6.5) can be combined to yield

$$\sum_j j\phi(r-j) = r \text{ for all real } r. \tag{6.7}$$

Before solving the above equations for $\phi(r)$, let us consider the motivation for each of these postulates in turn. The product form given by (6.1) is not essential, but it makes a tremendous simplification, since it reduces all subsequent considerations to the one-dimensional case. Also, the scaling introduced in (6.1) has the simplifying effect of removing the parameter h from all subsequent formulae. Moreover, this scaling is the natural one, so that $\delta_h \to \delta$ as $h \to 0$.

Generally speaking, the purpose of the other postulates is to hide the presence of the Eulerian computational lattice as much as possible. The continuity of ϕ, for example, is imposed to avoid jumps in velocity or in applied force as the Lagrangian marker points cross over the Eulerian grid planes.

The postulate of bounded support (equation (6.3)), however, is made for a different reason: computational efficiency. There are functions with unbounded support that one might well like to use for $\phi(r)$, such as $\exp(-r^2/2)$ or $\sin(r)/r$, but these would entail enormous computational cost, since *each* Lagrangian marker point would then interact directly with *all* grid points of the Eulerian computational lattice. The support of $\phi(r)$ as specified in (6.3) is the smallest possible while maintaining consistency with the other postulates.

An immediate consequence of (6.4) is that

$$\sum_j \phi(r-j) = 1 \text{ for all real } r, \tag{6.8}$$

and this, in turn, implies the identity (5.3), which was used in the derivation of the mass, force, and torque identities of the previous section. What requires further explanation, however, is why we impose the stronger conditions that the sum over the even grid points should separately give 1/2, and similarly the sum over odd grid points. This is a technical issue relating to the use of the central difference operator $\mathbf{D}^0$ in our numerical scheme. The null space of the gradient operator based on $\mathbf{D}^0$ is eight-dimensional. It contains not only the constants but any function that is constant on each of the 8 'chains' of points $\{j_1 \text{ even}, j_2 \text{ even}, j_3 \text{ even}\}$, $\{j_1 \text{ even}, j_2 \text{ even}, j_3 \text{ odd}\}$, *etc.* The separate conditions given in (6.4) ensure that all eight chains get the same amount of force from each Lagrangian marker point, and also that each Lagrangian marker point assigns equal weight to all eight chains when computing its interpolated velocity. This avoids oscillations from one grid point to the next that would otherwise occur, especially when localized forces are applied.

Equation (6.5) is postulated because it implies the three-dimensional identity (5.4), which was used to derive the torque identity in the previous section. Notice, too, that (6.4) and (6.5) together imply that, when δ_h is used for interpolation, linear functions are interpolated exactly, and smooth functions are therefore interpolated with second-order accuracy. Having said this, however, we must also point out that, in those cases in which the IB method is used with true boundaries (*i.e.*, immersed elastic structures of zero thickness), the velocity field that must be interpolated at the boundary is *not*, in fact, smooth. It is continuous but suffers a jump in its normal derivative. In this situation, the interpolation formula involving δ_h is only first-order accurate. The 'immersed interface method' (LeVeque and Li 1994, LeVeque and Li 1997, Li and Lai 2001) overcomes this difficulty.

The final postulate, (6.6), is different in character from the others and requires more explanation. It comes from a condition that we would like to impose but cannot (for a reason that will be explained below), namely

$$\sum_j \phi(r_1-j)\phi(r_2-j) = \Phi(r_1-r_2) \text{ for all real } r_1, r_2, \tag{6.9}$$

where Φ is some other function related to ϕ. If a condition of this form could be imposed (regardless of the function Φ), it would imply (proof left as an exercise for the reader) the exact translation invariance of the IB method as applied to a translation-invariant linear system like the Stokes equations, that is, the IB results would remain exactly the same despite shifts in position of all immersed elastic structures by fixed (possibly fractional) amounts relative to the Eulerian grid. This would be the ultimate in 'hiding the grid'. Even though we are solving the nonlinear Navier–Stokes equations, exact translation invariance in the case of the Stokes equations would suggest a good approximation to translation invariance in the case of the Navier–Stokes equations.

It is easy to prove, however, that (6.9) is incompatible with any condition of bounded support, like (6.3). The idea of the proof is to choose $r_1 - r_2$ such that the width of the overlap of the support of the functions of j given by $\phi(r_1 - j)$ and $\phi(r_2 - j)$ is some fraction θ that is strictly between 0 and 1. Then, by varying r_1 and r_2 in such a manner that their difference remains constant, one will find that sometimes the sum on the left-hand side of (6.9) contains exactly one nonzero term, so it cannot be zero, and sometimes that sum contains no nonzero terms, so it has to be zero. This contradicts the assumption that this sum depends only on $r_1 - r_2$.

Here is a more detailed version of the above argument. Suppose, for example, that $\phi(r) \neq 0$ for $a < r < b$, and that $\phi(r) = 0$ otherwise. Then $\phi(r_1 - j)\phi(r_2 - j)$ is nonzero if and only if

$$a < (r_1 - j) < b \quad \text{and} \quad a < (r_2 - j) < b, \tag{6.10}$$

which is equivalent to

$$r_1 - b < j < r_1 - a \quad \text{and} \quad r_2 - b < j < r_2 - a. \tag{6.11}$$

If $r_2 > r_1$, this is equivalent to

$$r_2 - b < j < r_1 - a. \tag{6.12}$$

Now fix the difference between r_2 and r_1 as follows:

$$r_2 = r_1 + (b - a) - \theta, \tag{6.13}$$

where $0 < \theta < 1$ and $\theta < (b - a)$. Under these conditions, $\phi(r_1 - j)\phi(r_2 - j)$ is nonzero if and only if

$$r_1 - a - \theta < j < r_1 - a. \tag{6.14}$$

Note that the interval which must contain the integer j is of length $\theta < 1$, so there is at most one integer j that satisfies this condition. By varying the real number r_1 while keeping $r_2 - r_1$ constant, we can clearly find cases where $\sum_j \phi(r_1 - j)\phi(r_2 - j)$ has exactly one nonzero term, and is therefore nonzero, and other cases in which that sum has no such term, and is therefore zero. This contradicts (6.9), which asserts that this sum depends only on the difference $r_2 - r_1$, which has indeed been held constant as we varied r_1.

Since (6.9) cannot be imposed, we content ourselves with the weaker condition that is derived from (6.9) by setting $r_1 = r_2$, namely

$$\sum_j (\phi(r-j))^2 = \Phi(0) = C, \tag{6.15}$$

which is (6.6). Although weaker than (6.9), our postulate (6.6) does give some information about the sum that appears on the left-hand side of (6.9), since it implies the inequality

$$\left| \sum_j \phi(r_1 - j)\phi(r_2 - j) \right| \leq C \text{ for all real } r_1, r_2 \tag{6.16}$$

with equality (see (6.9)) when $r_1 = r_2$. This is just the Schwarz inequality, since the left-hand side is the magnitude of an inner product, and the right-hand side $C = C^{1/2}C^{1/2}$ is the product of the norms of the vectors in question: see (6.9).

The significance of this inequality is as follows. Expressions like the sum that appears on the left-hand side arise when one considers the interaction between two Lagrangian marker points via the Eulerian grid. The inequality (6.16) guarantees that such coupling is strongest when the two Lagrangian points coincide, and moreover (according to (6.9)) that the self-coupling in that case is independent of the position of the Lagrangian marker with respect to the grid.

This completes the discussion of the motivation for the postulates given by (6.1)–(6.6). We now show how these postulates determine the form of ϕ and hence δ_h. The key step is to restrict r to the interval $[0, 1]$ and then to write out the equations for $\phi(r)$ explicitly as follows:

$$\phi(r-2) + \phi(r) = \frac{1}{2}, \tag{6.17}$$

$$\phi(r-1) + \phi(r+1) = \frac{1}{2}, \tag{6.18}$$

$$2\phi(r-2) + \phi(r-1) - \phi(r+1) = r, \tag{6.19}$$

$$(\phi(r-2))^2 + (\phi(r-1))^2 + (\phi(r))^2 + (\phi(r+1))^2 = C. \tag{6.20}$$

Note that (6.19) is the transcription of (6.7).

These are 4 equations in the 4 unknowns $\phi(r-2)$, $\phi(r-1)$, $\phi(r)$, and $\phi(r+1)$. (A complication is that we still do not know C, but that will be remedied shortly.) One way to solve this system is to use the first three (linear) equations to express all of the other unknowns in terms of $\phi(r)$. Then (6.20) becomes a quadratic equation for $\phi(r)$. The choice of root is uniquely determined by the continuity of ϕ and the condition that $\phi(\pm 2) = 0$. Here are the details.

The three linear equations yield:

$$\phi(r-2) = \frac{1}{2} - \phi(r), \tag{6.21}$$

$$\phi(r-1) = \frac{r}{2} - \frac{1}{4} + \phi(r), \tag{6.22}$$

$$\phi(r+1) = -\frac{r}{2} + \frac{3}{4} - \phi(r). \tag{6.23}$$

At this point we can determine C. Consider the special case $r = 0$. Then $\phi(r-2) = \phi(-2) = 0$ (see (6.2) and (6.3)). It follows immediately from the above results that $\phi(0) = \frac{1}{2}$ and $\phi(\pm 1) = \frac{1}{4}$. Substituting these results into (6.20), we get

$$C = 0 + \left(\frac{1}{4}\right)^2 + \left(\frac{1}{2}\right)^2 + \left(\frac{1}{4}\right)^2 = \frac{3}{8}. \tag{6.24}$$

Substituting all of these results into (6.20), we get the following quadratic equation for $\phi(r)$:

$$4\left(\phi(r)\right)^2 - (3-2r)\phi(r) + \frac{1}{2}(1-r)^2 = 0. \tag{6.25}$$

We may choose a root based on the requirement found above that $\phi(0) = \frac{1}{2}$. With this choice, the unique solution is

$$\phi(r) = \frac{3 - 2r + \sqrt{1 + 4r - 4r^2}}{8}, \quad 0 \le r \le 1. \tag{6.26}$$

It is important to note the restriction that this formula is only valid for $r \in [0,1]$. To determine ϕ on the intervals $[-2,-1]$, $[-1,0]$, and $[1,2]$, use (6.26) together with (6.21)–(6.23), always bearing in mind the restriction that all of these equations are only valid for $r \in [0,1]$. Simple changes of variable are needed to get explicit formulae for $\phi(r)$ on the other intervals. Finally, of course, we have by hypothesis that $\phi(r) = 0$ for $|r| \ge 2$. Combining all these results, we get

$$\begin{aligned}
\phi(r) &= 0, \;\; r \le -2 \\
&= \frac{1}{8}\left(5 + 2r - \sqrt{-7 - 12r - 4r^2}\right), \quad -2 \le r \le -1 \\
&= \frac{1}{8}\left(3 + 2r + \sqrt{1 - 4r - 4r^2}\right), \quad -1 \le r \le 0 \\
&= \frac{1}{8}\left(3 - 2r + \sqrt{1 + 4r - 4r^2}\right), \quad 0 \le r \le 1 \\
&= \frac{1}{8}\left(5 - 2r - \sqrt{-7 + 12r - 4r^2}\right), \quad 1 \le r \le 2 \\
&= 0, \quad 2 \le r
\end{aligned} \tag{6.27}$$

Comparison of the corresponding formulae for negative and positive r reveal that $\phi(-r) = \phi(r)$, a property we did not postulate, although it is certainly expected. One can check that the function $\phi(r)$ defined by the above is not only continuous (as postulated) but actually has a continuous first derivative (a property we did not postulate but got for free) at the points $r = -2, -1, 0, 1, 2$, and indeed everywhere.

It may be worth mentioning that the function $\phi(r)$ that we have just constructed is *extremely* well approximated by the simple formula

$$\tilde{\phi}(r) = \begin{cases} \frac{1}{4}\left(1 + \cos\left(\frac{\pi r}{2}\right)\right), & |r| \leq 2, \\ 0, & \text{otherwise}, \end{cases} \tag{6.28}$$

which satisfies all of the above postulates except for the first-moment condition, (6.5). This function was discovered earlier and is still sometimes used in IB computations.

We conclude this section with some remarks on possible modifications of the above construction. First, it may be desirable to narrow the support of δ_h, both for reasons of computational cost, and to improve the resolution of the boundary. One way to do this is to dispense with the separate 'even/odd' conditions in (6.4) and just use (6.8) instead. This makes it possible to reduce the width of the support of $\phi(r)$ from 4 to 3. In a three-dimensional computation, this is a considerable saving, since it reduces the volume of the support of $\delta_h(\mathbf{x})$ from 64 to 27 cubic meshwidths. The construction of ϕ is similar to that given above, and the reader is invited to have the fun of working out the details. This 3-point delta function is only recommended, however, for use in conjunction with a numerical scheme that does not suffer from the problems described above concerning the eight-dimensional null space of the central difference gradient operator. An example of the use of the 3-point delta function can be found in the work of A. Roma (Roma, Peskin and Berger 1999), who used it in conjunction with the MAC scheme (Harlow and Welch 1965).

Another way to try to get better boundary resolution is to impose more moment conditions. This requires *broadening* the support of δ_h which sounds like going in the wrong direction. To see why it might work, though, consider the second-moment condition:

$$\sum_j (r - j)^2 \phi(r - j) = 0 \text{ for all real } r. \tag{6.29}$$

This essentially states that the mean square thickness of the boundary is zero! Note that the imposition of this condition requires $\phi(r)$ to change sign. Imposing higher moments has been investigated by J. Stockie (1997), who has written a Maple program that automates the process, and independently by D. M. McQueen and the present author (unpublished). It turns out that the moments have to be introduced two at a time, and the introduction of each new pair of moments increases the width of the support of $\phi(r)$ by 2. Thus, starting from the 4-point delta function introduced above, one can impose second and third moment conditions and thereby arrive at a 6-point delta function, *etc.* The 6-point delta function has negative tails, and resembles the 'Mexican hat' functions that are used to describe receptive fields in natural and machine vision. Preliminary numerical experiments indicate some improvement in boundary resolution despite the increased width of the support of the delta function. Note that it is also possible to start from the 3-point delta function mentioned above and impose higher moment conditions, two at a time, to arrive at a 5-point delta function, a 7-point delta function, *etc.*

Another direction would be to aim at improved translation invariance. Here is a possible strategy, which has not yet been tried. Recall the condition given by (6.9) which would guarantee translation invariance (in the Stokes case), but which could not be imposed because it is incompatible with bounded support. Our strategy in the foregoing was simply to impose the special case of (6.9) that arises by setting $r_1 = r_2$. This resulted in our postulate given by (6.6). Now suppose that, instead of merely doing this, we also differentiate (6.9) one or more times before setting $r_1 = r_2$. It turns out that differentiating once gives nothing new, just a condition that can be derived from (6.6). But if we differentiate (6.9) twice, once with respect to r_1 and once with respect to r_2, and then set $r_1 = r_2 = r$, we get the new condition,

$$\sum_j (\phi'(r))^2 = -\Phi''(0) = C_1 \text{ for all real } r, \tag{6.30}$$

where $'$ denotes differentiation, and where C_1, like C, is a constant that has to be determined. It remains to be seen how to mix this condition in with the moment conditions, but that seems like a worthwhile subject of future investigation.

7. Temporal discretization

The temporal discretization that we currently use (Lai and Peskin 2000, McQueen and Peskin 2001) is a second-order accurate Runge–Kutta method, based primarily on the midpoint rule. For a system of ordinary differential equations of the form

$$\frac{\mathrm{d}y}{\mathrm{d}t} = f(y), \tag{7.1}$$

such a scheme looks like this:

$$\frac{y^{n+\frac{1}{2}} - y^n}{\Delta t/2} = f(y^n), \tag{7.2}$$

$$\frac{y^{n+1} - y^n}{\Delta t} = f(y^{n+\frac{1}{2}}), \tag{7.3}$$

where the superscript is the time-step index. The salient feature of this scheme is that each time-step involves a 'preliminary substep' to the half-time level followed by a 'final substep' from time level n to $n+1$, in which the results of the preliminary substep are used. The preliminary substep involves a first-order accurate scheme (forward Euler in the above example), and the final substep is done by a second-order accurate scheme (here, the midpoint rule). One would think that the accuracy would be limited by the least accurate substep, but the magic of Runge–Kutta is that the overall scheme is second-order accurate.

Our time-stepping scheme in the IB method follows the general idea of the above, but with some modifications. Instead of using pure forward Euler for the preliminary substep, we use a scheme that is a mixture of forward and backward Euler. In the final substep of each time-step, we use not only the midpoint rule but also (in one place) the trapezoidal rule. These changes are made to improve numerical stability by using implicit methods wherever that can be done at reasonable cost. But we retain the important feature that the preliminary substep is done by a first-order accurate method, and the final substep is done by a second-order accurate method that makes use of the results of the preliminary substep.

The preliminary substep, from time level n to $n+\frac{1}{2}$, proceeds as follows. First, find the positions of the Lagrangian markers at time level $n+\frac{1}{2}$:

$$\mathbf{X}^{n+\frac{1}{2}}(q,r,s) = \mathbf{X}^n(q,r,s) + \frac{\Delta t}{2} \sum_{\mathbf{x}\in g_h} \mathbf{u}^n(\mathbf{x})\, \delta_h\left(\mathbf{x} - \mathbf{X}^n(q,r,s)\right)\, h^3. \tag{7.4}$$

Next, use the configuration of the Lagrangian markers at the half-time level to calculate the elastic force by taking the gradient of the elastic energy function E_h evaluated at that configuration:

$$\mathbf{F}^{n+\frac{1}{2}}(q,r,s,)\, \Delta q\, \Delta r\, \Delta s = -\frac{\partial}{\partial \mathbf{X}(q,r,s)} E_h\big(\cdots \mathbf{X}^{n+\frac{1}{2}}(q',r',s')\cdots\big). \tag{7.5}$$

Now spread the Lagrangian force and mass densities onto the Eulerian grid:

$$\mathbf{f}^{n+\frac{1}{2}}(\mathbf{x}) = \sum_{(q,r,s)\in G_h} \mathbf{F}^{n+\frac{1}{2}}(q,r,s)\, \delta_h\big(\mathbf{x} - \mathbf{X}^{n+\frac{1}{2}}(q,r,s)\big)\, \Delta q\, \Delta r\, \Delta s, \tag{7.6}$$

$$\rho^{n+\frac{1}{2}}(\mathbf{x}) = \sum_{(q,r,s)\in G_h} M(q,r,s)\, \delta_h\big(\mathbf{x} - \mathbf{X}^{n+\frac{1}{2}}(q,r,s)\big)\, \Delta q\, \Delta r\, \Delta s. \tag{7.7}$$

With these Eulerian quantities defined, we are ready to integrate the Navier–Stokes equations on the Eulerian grid g_h from time level n to time level $n+\frac{1}{2}$. This is done by the following implicit scheme, which is forward Euler in the nonlinear terms, but backward Euler otherwise:

$$\rho^{n+\frac{1}{2}}\left(\frac{\mathbf{u}^{n+\frac{1}{2}}-\mathbf{u}^n}{\Delta t/2}+S_h(\mathbf{u}^n)\mathbf{u}^n\right)+\mathbf{D}_h^0\tilde{p}^{n+\frac{1}{2}}=\mu L_h\mathbf{u}^{n+\frac{1}{2}}+\mathbf{f}^{n+\frac{1}{2}}, \quad (7.8)$$

$$\mathbf{D}_h^0\cdot\mathbf{u}^{n+\frac{1}{2}}=0. \quad (7.9)$$

Note that (7.8)–(7.9) form a system of equations in the unknowns $\mathbf{u}^{n+\frac{1}{2}}(\mathbf{x})$ and $\tilde{p}^{n+\frac{1}{2}}(\mathbf{x})$, for $\mathbf{x}\in g_h$. (The only reason for the 'tilde' over the p is to distinguish it from another $p^{n+\frac{1}{2}}$ that will appear later.) The method(s) used to solve this system will be discussed below, after we have finished the statement of the time-stepping scheme.

The preliminary substep to the half-time level is now complete. The final substep uses the results of the preliminary substep and proceeds as follows. Note how everything in it is centred in time, thus achieving second-order accuracy.

First, we move the Lagrangian configuration from $\mathbf{X}^n$ to $\mathbf{X}^{n+1}$ by interpolating the velocity $\mathbf{u}^{n+\frac{1}{2}}$ to the configuration $\mathbf{X}^{n+\frac{1}{2}}$:

$$\mathbf{X}^{n+1}(q,r,s)=\mathbf{X}^n(q,r,s)+\Delta t\sum_{\mathbf{x}\in g_h}\mathbf{u}^{n+\frac{1}{2}}(\mathbf{x})\,\delta_h\big(\mathbf{x}-\mathbf{X}^{n+\frac{1}{2}}(q,r,s)\big)\,h^3. \quad (7.10)$$

Now we can proceed directly to the Navier–Stokes equations, since we are going to use the same $\rho^{n+\frac{1}{2}}$ and $\mathbf{f}^{n+\frac{1}{2}}$ as we did in the preliminary substep. Our task here is to integrate the Navier–Stokes equations from time level n to time level$n+1$, and we do so by a scheme that is very similar to (7.8)–(7.9):

$$\rho^{n+\frac{1}{2}}\left(\frac{\mathbf{u}^{n+1}-\mathbf{u}^n}{\Delta t}+S_h(\mathbf{u}^{n+\frac{1}{2}})\mathbf{u}^{n+\frac{1}{2}}\right)+\mathbf{D}_h^0p^{n+\frac{1}{2}}=\mu L_h\frac{\mathbf{u}^n+\mathbf{u}^{n+1}}{2}+\mathbf{f}^{n+\frac{1}{2}}, \quad (7.11)$$

$$\mathbf{D}_h^0\cdot\mathbf{u}^{n+1}=0. \quad (7.12)$$

Again, this is a system of equations in the unknowns $\mathbf{u}^{n+1}(\mathbf{x})$ and $p^{n+\frac{1}{2}}(\mathbf{x})$, for $\mathbf{x}\in g_h$. The method(s) of solution will be discussed below. Note that the nonlinear terms are here evaluated at the half-time level, thus using the result of the preliminary substep. In the viscous terms, however, we do not do this but instead make the scheme more implicit (and presumably more stable) by using the trapezoidal rule. The pressures $\tilde{p}^{n+\frac{1}{2}}$ and $p^{n+\frac{1}{2}}$ are just two different approximations to the pressure at the half-time level.

Although it might appear in the foregoing that the (7.12) is not centred in time between time levels n and $n+1$, this is illusory. The equation holds for all n and so we have $\mathbf{D}_h^0 \cdot \mathbf{u}^n = 0$ as well as $\mathbf{D}_h^0 \cdot \mathbf{u}^{n+1} = 0$. Similarly, one might wonder looking at the above scheme whether $\mathbf{u}^{n+1}$ is actually used for anything or just computed to be thrown away, since it is not used in the computation of the final configuration $\mathbf{X}^{n+1}$. The answer is that it is very much used, in both the preliminary and final substeps of the *next* time-step, where it plays the role of $\mathbf{u}^n$.

To complete the discussion of the time-stepping scheme, we have to say how to solve the two very similar systems of equations (7.8)–(7.9) and (7.11)–(7.12). These are linear systems of equations, since the nonlinear terms are known and can be evaluated before solving the system in each case. Similarly, the coefficient $\rho^{n+\frac{1}{2}}(\mathbf{x})$ is known (see (7.7)), so the terms in which ρ appears do not involve products of unknowns. The method of choice for solving these Stokes systems depends, however, on whether the mass density ρ is constant or not.

In many applications of the IB method, the immersed elastic material is neutrally buoyant in the ambient fluid, so ρ is a constant, independent of space and time, and (7.7) is not used. In this case, the linear systems of equations that we have to solve on the Eulerian grid g_h are difference equations with constant coefficients. As such, they are easily solved by the discrete Fourier transform, implemented by the Fast Fourier Transform algorithm. (To facilitate this, we typically formulate our problem on a periodic domain, *i.e.*, a cube with opposite faces identified.) For details, see Peskin and McQueen (1996).

When ρ is not constant, these linear systems of equations contain a nonconstant (albeit known) coefficient, namely $\rho^{n+\frac{1}{2}}(\mathbf{x})$, and the Fourier transform is no longer useful, since different modes get coupled together. In this case, some iterative method has to be used. For IB computations of this type, see Fogelson and J. Zhu (1996), L. Zhu (2001) and L. Zhu and Peskin (2002), in which the multigrid method is used, but actually to solve the equations of a somewhat different time-stepping scheme. The specific time-stepping scheme stated here has not yet been implemented in the case of nonconstant mass density.

8. Research directions

The purpose of this section is to outline some of the ways in which the IB method (as stated above) might be improved. For most of the research directions suggested here, some work has already been done, as will be pointed out with references.

The time-stepping scheme presented in the previous section is mainly explicit, although it does contain a partly implicit Navier–Stokes solver. In particular, the time-stepping scheme is explicit in the computation of the elastic force. In most applications, this results in a severe restriction on the time-step duration, and it would be a huge improvement if this could be overcome. For research on implicit and semi-implicit versions of the IB or related methods, see Peskin (1992), Tu and Peskin (1992), Mayo and Peskin (1993), Fauci and Fogelson (1993) and LeVeque and Li (1997).

The IB method described in this paper employs a uniform Eulerian grid. It would obviously be an advantage to be able to concentrate the computational effort where it is most needed. This is especially true for problems involving high Reynolds number flow, where high resolution is needed primarily in thin boundary layers or in other regions of limited extent where vorticity is concentrated. Several approaches to this problem have been and are being tried. The IB method has been combined with vortex methods (McCracken and Peskin 1980, Peskin 1981), with impulse methods (Cortez 1996, Cortez and Varela 1997, Cortez 2000), and with adaptive mesh refinement (Roma *et al.* 1999).

The IB method described in this paper is second-order accurate in space and time for problems with smooth solutions, such as the vibrations of an incompressible elastic material filling all of space, but problems with sharp interfaces do not have smooth solutions, and for these the IB method seems to be limited to first-order spatial accuracy. Another way to say this is that the IB method smears out sharp interfaces to a thickness which is of the order of the meshwidth. As mentioned above, the 'immersed interface method' (LeVeque and Li 1994, LeVeque and Li 1997, Li and Lai 2001) avoids this smearing and maintains second-order accuracy by modifying the difference equations near the interface. Another route to high accuracy is the 'blob-projection method' (Cortez and Minion 2000). It remains to be seen whether there is some simple way to modify the IB method to achieve this same goal.

One of the ways that the numerical error of the IB method shows up is a lack of volume conservation. Even though the velocity field on the Eulerian computational lattice is discretely divergence-free, this does not guarantee that the interpolated velocity field, in which the Lagrangian markers move, is continuously divergence-free. This error, of course, goes to zero as $h \to 0$, but it would be nice to reduce it at any particular h. Peskin and Printz (1993) (see also Rosar and Peskin (2001)) have shown that this can be done by using divergence and gradient operators that are 'tuned' to the interpolation scheme. Perhaps these principles can be applied to the Navier–Stokes solver as a whole, not just to the discrete divergence and gradient operators.

Although implementation issues have not been discussed in this paper, it is clearly important to parallelize the IB method. Shared-memory parallelization has been done (McQueen and Peskin 1997), and a distributed implementation in Titanium (Yelick *et al.* 1998) is under development (Yau 2002).

With regard to the mathematical formulation, it is not clear how to handle variable viscosity. The issue is not merely how to write the Navier–Stokes equations with a variable viscosity $\mu(\mathbf{x}, t)$, which is well known, but rather how to determine $\mu(\mathbf{x}, t)$ from the Lagrangian configuration of the material $\mathbf{X}(q, r, s, t)$. Also, in the case of an anisotropic elastic material, the viscoelasticity may be anisotropic, too. How should the formulation given here be modified to include such effects?

Can a turbulence model be effectively combined with the IB method (Schmid 200x)? This would make it possible to study the interaction of turbulent flows with immersed elastic bodies or boundaries.

Finally, there is not yet any convergence proof for the IB method (but see the stability analysis of Stockie and Wetton (1995, 1999)). Perhaps the reader will be inspired to discover such a proof.

9. Applications

The IB method was created in order to study the fluid dynamics of heart valves, and it has been applied to both natural (Peskin 1972, McQueen, Peskin and Yellin 1982, Peskin 1982, Meisner *et al.* 1985) and prosthetic (McQueen and Peskin 1983, 1985, 1991) cardiac valves. An outgrowth of this work has been a three-dimensional computer model of the whole heart (Peskin and McQueen 1996, McQueen and Peskin 1997, 2000, 2001, Kovacs, McQueen and Peskin 2001, McQueen, Peskin and Zhu 2001). Some other applications related to cardiovascular physiology include platelet aggregation during blood clotting (Fogelson 1984, Fauci and Fogelson 1993, Fogelson 1992, Wang and Fogelson 1999, Haoyu 2000), the deformation of red blood cells in a shear flow (Eggleton and Popel 1998), flow in elastic blood vessels (Vesier and Yoganathan 1992), flow in collapsible thin-walled vessels (Rosar and Peskin 2001), and flow in arterioles subject to the influence of a vasodilator transported by the flow itself (Arthurs, Moore, Peskin, Pitman and Layton 1998).

The swimming of eels, sperm, and bacteria have been studied by the IB method (Fauci and Peskin 1988, Fauci and McDonald 1994, Dillon, Fauci and Gaver 1995), as have the crawling locomotion of amoeba (Bottino 1998, Bottino and Fauci 1998), the waving motions of cilia beating in fluid (Dillon and Fauci 2000*a*), and the rotary motions and interactions of bacterial flagella immersed in fluid (Lim and Peskin 200x). Computer simulations of biofilms have been done in which the swimming of multiple microorganisms is handled by the IB method (Dillon, Fauci, Fogelson and Gaver 1996, Dillon and Fauci 2000*b*). Subtle fluid-dynamic interactions between nearby swimmers have been elucidated (Fauci 1990, Fauci and McDonald 1994). Embryological motions, too, have been studied by the IB method (Dillon and Othmer 1999).

Flows with suspended particles have been simulated by the IB method (Fogelson and Peskin 1988). Some noteworthy features of the IB method as applied to this problem are that the computational effort grows only linearly with the number of suspended particles, that the particles can be flexible and of arbitrary shape (*e.g.*, wood pulp fibres (Stockie and Green 1998)), and that the methodology is not restricted to Stokes flow but can handle nonzero Reynolds number. A variant (Sulsky and Brackbill 1991) of the IB method has been developed specifically for this problem. It has many interesting features including the direct transfer of stress (rather than force) from the immersed elastic body to the fluid, and it facilitates the use of conventional elasticity theory in describing the material properties of the suspended particles.

There are creatures that eat suspended (food) particles but they have to catch them first. The fluid dynamics of this has been studied by the IB method (Grunbaum, Eyre and Fogelson 1998).

The basilar membrane of the inner ear is an elastic shell immersed in fluid that vibrates in response to the incident sound. Its mechanical properties have the effect of sorting the sound energy into its different frequency components. This process has been simulated by the IB method, first in a simplified model (Beyer 1992), but more recently in the context of a three-dimensional model that includes the spiral geometry of the cochlea (Givelberg 1997).

Flow past a cylinder has been studied by the IB method (Lai and Peskin 2000). This traditional (fixed-boundary) fluid dynamics problem is mentioned for two reasons. The first reason is to emphasize that the IB method can do such problems: the technique is simply to put Lagrangian markers on the boundary and to tie each of these markers by a stiff spring to its target position in space. Small deviations of the markers from their target positions then generate the force needed to keep the boundary (approximately) fixed in place. (This idea was actually used before the introduction of the IB method – see Sirovich (1967) – and has recently been further developed (Goldstein, Handler and Sirovich 1993).) Note that this technique is as easily applied to a boundary of arbitrarily complicated geometry as it is to a cylinder. Also, there is no added difficulty if the boundary moves in a prescribed manner – just move the target points.

The second reason for mentioning flow past a cylinder is that it is a benchmark problem that provides validation for the IB method. The formation of the von Karmen vortex street in the wake of the cylinder is characterized by a Strouhal number (dimensionless frequency of vortex shedding) which varies with the Reynolds number (dimensionless reciprocal viscosity) in a manner that is well characterized experimentally. The IB method results reported in Lai and Peskin (2000) are in close agreement with experiment over a range of Reynolds numbers (100–200) in which the Strouhal number varies significantly as the Reynolds number is changed.

Valveless pumping has been simulated by the IB method (Jung and Peskin 2001). This is a peculiar phenomenon in which a loop of fluid-filled tubing comprised of a flexible part and a stiff part is driven by periodic compression near one end of the flexible part. Despite the absence of valves, a robust mean flow (superimposed on an oscillating flow at the driving frequency) can be driven around the tube in this manner. Although this qualitative phenomenon was known before the IB simulations were done, those simulations led to an unexpected prediction: that the direction of the mean flow would be dependent on the frequency (and for some frequencies also on the amplitude) of the imposed periodic compression. This prediction has since been experimentally confirmed (Zhang 200x).

A laboratory experiment (Zhang, Childress, Libchaber and Shelley 2000) involving a flexible filament that flaps in a flowing soap film (like a flag in the wind) has been simulated by the IB method (Zhu 2001, Zhu and Peskin 2002, McQueen, Peskin and Zhu 2001). From a methodological standpoint, this computation is particularly significant because it takes into account the mass of the immersed elastic boundary. Indeed, an important discovery, made first in the simulation and only later confirmed experimentally, is that filament mass is essential for sustained flapping. The computation reproduces the observed frequency of flapping (about 50 Hz), and also the observed bistability of the filament: For the same parameters, different initial conditions can lead either to a flapping state or to a straight, stationary state of the filament.

Several aerodynamic applications of the IB method are well underway. They include sails, parachutes, flags, and insect flight. Airborne seeds (like the maple seed that spins as it falls) are natural candidates for IB simulation, but this has not yet been tried.

10. Conclusions

The immersed boundary (IB) method is for problems in which immersed incompressible viscoelastic bodies or immersed elastic boundaries interact with viscous incompressible fluid. The immersed viscoelastic bodies do not need to be neutrally buoyant in the ambient fluid, and the immersed elastic boundaries do not have to be massless. Indeed, the fluid itself does not need to have a uniform mass density.

The philosophy of the IB method is to blur the distinction between fluid dynamics and elasticity. Both the Eulerian variables that are conventionally used in fluid dynamics and also the Lagrangian variables that are conventionally used in elasticity theory are employed. Upon discretization, these two kinds of variables are defined on a fixed Cartesian grid and a moving curvilinear grid, respectively. These grids are linked by a smoothed approximation to the Dirac delta function.

The simplicity of the Cartesian grid for the Eulerian variables facilitates the numerical integration of the Navier–Stokes equations, and the generality of the curvilinear grid for the Lagrangian variables makes it easy to model anisotropic (*e.g.*, fibre-reinforced) elastic material. It is an important feature of the IB method that these two grids are not constrained to relate to each other in any way at all, except that the moving curvilinear grid must be sufficiently dense with respect to the fixed Cartesian grid. A broad range of applications, mostly but not exclusively in biofluid dynamics, attest to the versatility of the immersed boundary method.

Acknowledgement

I had the good fortune to meet Alexandre Chorin at the time when I was just beginning to think about doing something in biological fluid dynamics. We were introduced by my cousins, Deborah and Vladimir Lipski, over a glass of tea. From Chorin, I learned fluid mechanics, and also his projection method (Chorin 1968, Chorin 1969) for incompressible flow, upon which foundation the immersed boundary method was built. Chorin made it possible for me to work at the Courant Institute, even though I was at the time a graduate student in Physiology at the Albert Einstein College of Medicine. He got other people at Courant interested in my work, especially Peter Lax and Cathleen Morawetz, who have been inspiring influences ever since, and Olof Widlund, who took me under his wing and saw to my (informal) education in numerical analysis. The skew-symmetric differencing of the nonlinear terms described in this paper was a suggestion of Widlund's. He made the suggestion years ago, but I only got around to trying it recently. It is the key to stable, accurate computation at the Reynolds number of the human heart.

The work described in this paper has been conducted over many years, during which time it was supported by several grants from the National Institutes of Health and the National Science Foundation. Current support is from the National Science Foundation under research grant DMS-9980069.

REFERENCES

K. M. Arthurs, L. C. Moore, C. S. Peskin, E. B. Pitman and H. E. Layton (1998), Modeling arteriolar flow and mass transport using the immersed boundary method. *J. Comput. Phys.* **147** 402–440.

R. P. Beyer (1992), A computational model of the cochlea using the immersed boundary method. *J. Comput. Phys.* **98** 145–162.

D. C. Bottino (1998), Modeling viscoelastic networks and cell deformation in the context of the immersed boundary method. *J. Comput. Phys.* **147** 86–113.

D. Bottino and L. J. Fauci (1998), A computational model of ameboid deformation and locomotion. *Eur. Biophys. J.* **27** 532–539.

A. J. Chorin (1968) Numerical solution of the Navier–Stokes equations. *Math. Comput.* **22** 745–762.

A. J. Chorin (1969) On the convergence of discrete approximations to the Navier–Stokes equations. *Math. Comput.* **23** 341–353.

R. Cortez (1996), An impulse-based approximation of fluid motion due to boundary forces. *J. Comput. Phys.* **123** 341–353.

R. Cortez (2000), A vortex/impulse method for immersed boundary motion in high Reynolds number flows. *J. Comput. Phys.* **160** 385–400.

R. Cortez and M. Minion (2000), The blob projection method for immersed boundary problems. *J. Comput. Phys.* **161** 428–453.

R. Cortez and D. A. Varela (1997), The dynamics of an elastic membrane using the impulse method. *J. Comput. Phys.* **138** 224–247.

R. Dillon and L. J. Fauci (2000*a*), An integrative model of internal axoneme mechanics and external fluid dynamics in ciliary beating. *J. Theor. Biol.* **207** 415–430.

R. Dillon and L. J. Fauci (2000*b*), A microscale model of bacterial and biofilm dynamics in porous media. *Biotechnol. Bioeng.* **68** 536–547.

R. Dillon and H. Othmer (1999), A mathematical model for the outgrowth and spatial patterning of the vertebrate limb bud. *J. Theor. Biol.* **197** 295–330.

R. Dillon, L. J. Fauci and D. Gaver (1995), A microscale model of bacterial swimming, chemotaxis and substrate transport. *J. Theor. Biol.* **177** 325–340.

R. Dillon, L. J. Fauci, A. L. Fogelson and D. Gaver (1996), Modeling biofilm processes using the immersed boundary method. *J. Comput. Phys.* **129** 57–73.

D. G. Ebin and R. A. Saxton (1986), The initial-value problem for elastodynamics of incompressible bodies. *Arch. Rat. Mech. Anal.* **94** 15–38.

D. G. Ebin and R. A. Saxton (1987), The equations of incompressible elasticity. In *Nonstrictly Hyperbolic Conservation Laws: Proc. AMS Special Session*, Vol. 60 of *Contemporary Mathematics*, AMS, pp. 25–34.

C. D. Eggleton and A. S. Popel (1998), Large deformation of red blood cell ghosts in a simple shear flow. *Phys. Fluids* **10** 1834–1845.

L. J. Fauci (1990), Interaction of oscillating filaments: A computational study. *J. Comput. Phys.* **86** 294–313.

L. J. Fauci and A. L. Fogelson (1993), Truncated Newton methods and the modeling of complex immersed elastic structures. *Comm. Pur. Appl. Math.* **46** 787–818.

L. J. Fauci and A. McDonald (1994), Sperm motility in the presence of boundaries. *B. Math. Biol.* **57** 679–699.

L. Fauci and C. S. Peskin (1988), A computational model of aquatic animal locomotion. *J. Comput. Phys.* **77** 85–108.

A. L. Fogelson (1984), A mathematical model and numerical method for studying platelet adhesion and aggregation during blood clotting. *J. Comput. Phys.* **56** 111–134.

A. L. Fogelson (1992) Continuum models of platelet aggregation: Formulation and mechanical properties. *SIAM J. Applied Math.* **52** 1089–1110.

A. L. Fogelson and C. S. Peskin (1988), A fast numerical method for solving the three-dimensional Stokes' equations in the presence of suspended particles. *J. Comput. Phys.* **79** 50–69.

A. L. Fogelson and J. Zhu (1996), Implementation of a variable-density immersed boundary method, unpublished. Available from: `http:/www.math.utah.edu/~fogelson`

E. Givelberg (1997), Modeling elastic shells immersed in fluid. PhD Thesis, Mathematics, New York University. Available from: `http://www.umi.com/hp/Products/DisExpress.html`, order no. 9808292.

D. Goldstein, R. Handler and L. Sirovich (1993) Modeling a no-slip flow boundary with an external force-field. *J. Comput. Phys.* **105** 354–366.

D. Grunbaum, D. Eyre, and A. Fogelson (1998), Functional geometry of ciliated tentacular arrays in active suspension feeders. *J. Exp. Biol.* **201** 2575–2589.

Y. Haoyu (2000) Three dimensional computational modeling and simulation of platelet aggregation on parallel computers. PhD thesis, University of Utah.

F. H. Harlow and E. Welch (1965), Numerical calculation of time-dependent viscous incompressible flow of fluids with free surface. *Phys. Fluids* **8** 2182–2189.

E. Jung and C. S. Peskin (2001), Two-dimensional simulations of valveless pumping using the immersed boundary method. *SIAM J. Sci. Comput.* **23** 19–45.

S. J. Kovacs, D. M. McQueen and C. S. Peskin (2001), Modeling cardiac fluid dynamics and diastolic function. *Phil. Trans. R. Soc. Lond. A* **359** 1299–1314.

M.-C. Lai and Z. L. Li (2001), A remark on jump conditions for the three-dimensional Navier–Stokes equations involving an immersed moving membrane. *Applied Math. Letters* **14** 149–154.

M.-C. Lai and C. S. Peskin (2000), An immersed boundary method with formal second order accuracy and reduced numerical viscosity. *J. Comput. Phys.* **160** 705–719.

R. J. LeVeque and Z. L. Li (1994), The immersed interface method for elliptic equations with discontinuous coefficients and singular sources. *SIAM J. Numer. Anal.* **31** 1019–1044.

R. J. LeVeque and Z. Li (1997), Immersed interface methods for Stokes flow with elastic boundaries or surface tension. *SIAM J. Sci. Comput.* **18** 709–735.

Z. L. Li and M.-C. Lai (2001), Immersed interface method for the Navier–Stokes equations with singular forces. *J. Comput. Phys.* **171** 822–842.

S. K. Lim and C. S. Peskin (200x), Unpublished.

M. F. McCracken and C. S. Peskin (1980), A vortex method for blood flow through heart valves. *J. Comput. Phys.* **35** 183–205.

D. M. McQueen and C. S. Peskin (1983), Computer-assisted design of pivoting-disc prosthetic mitral valves. *J. Thorac. Cardiovasc. Surg.* **86** 126–135.

D. M. McQueen and C. S. Peskin (1985), Computer-assisted design of butterfly bileaflet valves for the mitral position. *Scand. J. Thorac. Cardiovasc. Surg.* **19** 139–148.

D. M. McQueen and C. S. Peskin (1991), Curved Butterfly Bileaflet Prosthetic Cardiac Valve. US Patent Number 5,026,391.

D. M. McQueen and C. S. Peskin (1997), Shared-memory parallel vector implementation of the immersed boundary method for the computation of blood flow in the beating mammalian heart. *J. Supercomputing* **11** 213–236.

D. M. McQueen and C. S. Peskin (2000), A three-dimensional computer model of the human heart for studying cardiac fluid dynamics. *Computer Graphics* **34** 56–60.

D. M. McQueen and C. S. Peskin (2001), Heart simulation by an immersed boundary method with formal second-order accuracy and reduced numerical viscosity. In *Mechanics for a New Millennium, Proceedings of the International Conference on Theoretical and Applied Mechanics* (ICTAM) *2000* (H. Aref and J. W. Phillips, eds.), Kluwer Academic Publishers.

D. M. McQueen, C. S Peskin, and E. L. Yellin (1982), Fluid dynamics of the mitral valve: Physiological aspects of a mathematical model. *Amer. J. Physiol.* **242** H1095–H1110.

D. M. McQueen, C. S. Peskin and L. Zhu (2001), The immersed boundary method for incompressible fluid–structure interaction. In *Proceedings of the First M.I.T. Conference on Computational Fluid and Solid Mechanics, June 12–15, 2001* (K. J. Bathe, ed.), Elsevier Science Ltd, Oxford, UK, Vol. 1, pp. 26–30.

A. A. Mayo and C. S. Peskin (1993), An implicit numerical method for fluid dynamics problems with immersed elastic boundaries. In *Fluid Dynamics in Biology: Proc. AMS-IMS-SIAM Joint Summer Research Conf. Biofluiddynamics*, Vol. 141 of *Contemporary Mathematics*, AMS, pp. 261–277.

J. S. Meisner, D. M. McQueen, Y. Ishida, H. O. Vetter, U. Bortolotti, J. A. Strom, R. W. M. Frater, C. S. Peskin, and E. L. Yellin (1985), Effects of timing of atrial systole on LV filling and mitral valve closure: Computer and dog studies. *Amer. J. Physiol.* **249** H604–H619.

C. S. Peskin (1972), Flow patterns around heart valves: A digital computer method for solving the equations of motion. PhD Thesis, Albert Einstein College of Medicine, July, 211pp. Available from:
`http://www.umi.com/hp/Products/DisExpress.html`, order no. 7230378.

C. S. Peskin (1981), Vortex dynamics of the aortic sinus. In *Mathematical Aspects of Physiology* (Hoppensteadt FC, ed.), Vol. 19 of *Lectures in Applied Mathematics*, AMS, Providence, RI, pp. 93–104.

C. S. Peskin (1982), The fluid dynamics of heart valves: Experimental, theoretical, and computational methods. *Ann. Rev. Fluid Mech.* **14** 235–259.

C. S. Peskin (1992), Two examples of mathematics and computing in the biological sciences: Blood flow in the heart and molecular dynamics. In *Mathematics into the Twenty-first Century* (F. E. Browder, ed.), AMS, pp. 395–415.

C. S. Peskin and D. M. McQueen (1996), Fluid dynamics of the heart and its valves. In *Case Studies in Mathematical Modeling: Ecology, Physiology, and Cell Biology* (H. G. Othmer, F. R. Adler, M. A. Lewis and J. C. Dallon, eds.), Prentice-Hall, Englewood Cliffs, NJ, pp. 309–337.

C. S. Peskin and B. F. Printz (1993), Improved volume conservation in the computation of flows with immersed elastic boundaries. *J. Comput. Phys.* **105** 33–46.

A. M. Roma, C. S. Peskin and M. J. Berger (1999), An adaptive version of the immersed boundary method. *J. Comput. Phys.* **153** 509–534.

M. E. Rosar and C. S. Peskin (2001), Fluid flow in collapsible elastic tubes: A three-dimensional numerical model. *New York J. Math.* **7** 281–302.

P. Schmid (200x), Unpublished.

L. Sirovich (1967) Initial and boundary value problems in dissipative gas dynamics. *Phys. Fluids* **10** 24–34.

J. M. Stockie (1997), Analysis and computation of immersed boundaries, with application to pulp fibres. PhD thesis, Institute of Applied Mathematics, University of British Columbia, Vancouver, BC, Canada. Available from: `http://www.iam.ubc.ca/theses/stockie/stockie.html`

J. M. Stockie and S. I. Green (1998), Simulating the motion of flexible pulp fibres using the immersed boundary method. *J. Comput. Phys.* **147** 147–165.

J. M. Stockie and B. T. R. Wetton (1995) Stability analysis for the immersed fiber problem. *SIAM J. Applied Math.* **55** 1577–1591.

J. M. Stockie and B. T. R. Wetton (1999) Analysis of stiffness in the immersed boundary method and implications for time-stepping schemes. *J. Comput. Phys.* **154** 41–64.

D. Sulsky and J. U. Brackbill (1991), A numerical method for suspension flow. *J. Comput. Phys.* **96** 339–368.

C. Tu and C. S. Peskin (1992), Stability and instability in the computation of flows with moving immersed boundaries: A comparison of three methods. *SIAM J. Sci. Statist. Comput.* **13** 1361–1376.

C. C. Vesier and A. P. Yoganathan (1992), A computer method for simulation of cardiovascular flow fields: Validation of approach. *J. Comput. Phys.* **99** 271–287.

N. T. Wang and A. L. Fogelson (1999) Computational methods for continuum models of platelet aggregation. *J. Comput. Phys.* **151** 649–675.

S. M. Yau (2002), A generic immersed boundary method package in titanium. Master's Report, Computer Science Division, University of California, Berkeley.

K. A. Yelick, L. Semenzato, G. Pike, C. Miyamoto, B. Liblit, A. Krishnamurthy, P. N. Hilfinger, S. L. Graham, D. Gay, P. Colella and A. Aiken (1998), Titanium: A high-performance Java dialect. *Concurrency: Practice and Experience* **10**.

J. Zhang (200x), Unpublished.

J. Zhang, S. Childress, A. Libchaber and M. Shelley (2000), Flexible filaments in a flowing soap film as a model for one-dimensional flags in a two-dimensional wind. *Nature* **408** 835.

L. Zhu (2001), Simulation of a flapping filament in a flowing soap film by the immersed boundary method. PhD Thesis, New York University.

L. Zhu and C. S. Peskin (2002), Simulation of a flapping filament in a flowing soap film by the immersed boundary method. Submitted to *J. Comput. Phys.*

Acta Numerica (2002), pp. 519–584
DOI: 10.1017/S0962492902000089

Numerical methods for large eigenvalue problems

Danny C. Sorensen
Department of Computational and Applied Mathematics,
Rice University,
6100 Main St., MS134,
Houston, TX 77005-1892, USA
E-mail: sorensen@rice.edu

Over the past decade considerable progress has been made towards the numerical solution of large-scale eigenvalue problems, particularly for nonsymmetric matrices. Krylov methods and variants of subspace iteration have been improved to the point that problems of the order of several million variables can be solved. The methods and software that have led to these advances are surveyed.

CONTENTS

1. Introduction

The algebraic eigenvalue problem

$$\mathbf{Ax} = \mathbf{x}\lambda$$

is fundamental to scientific computing. Large-scale problems are of increasing importance, and recent advances in the area of nonsymmetric problems have enormously expanded capabilities in areas such as linear stability and bifurcation analysis. Considerable progress has been made over the past decade towards the numerical solution of large-scale nonsymmetric problems. However, there is still a great deal to be done. This is a very challenging area of research that is still very active.

This survey is an attempt to introduce some of these advances. It emphasizes two main approaches: Krylov subspace projection and a variant of subspace iteration. Within these two classes, the implicitly restarted Arnoldi method (Sorensen 1992) and the Jacobi–Davidson method (Sleijpen and van der Vorst 1995) are featured. There are several important competing methods but these are discussed in far less detail. Availability of reliable software for large symmetric and nonsymmetric problems has enabled many significant advances in applications. Problems of the order of several million variables are now being solved on massively parallel machines. Problems of order ten thousand can now be solved on a laptop computer. Software and performance issues are therefore a third component of this survey.

Large eigenvalue problems arise in a variety of settings. Two important areas are vibrational analysis of structures and linear stability analysis of fluid flow. The former analysis usually leads to symmetric eigenproblems where the goal typically is to determine the lowest modes. The latter analysis leads to nonsymmetric eigenproblems and the interest is in determining if the eigenvalues lie in a particular half of the complex plane. In both of these settings the discrete problem can become extremely large, but only a few eigenvalues are needed to answer the question of interest.

A typical source of large-scale problems is the discretization of a partial differential equation, for example,

$$\begin{aligned} \mathcal{L}u &= u\lambda \ \text{ for } \ u \in \Omega, \\ u &= 0 \ \text{ for } \ u \in \partial\Omega, \end{aligned} \tag{1.1}$$

where $\mathcal{L}$ is some linear differential operator. Often, $\mathcal{L}$ is a linearization of a nonlinear operator about a particular solution to the nonlinear equation, such as a steady state. A number of techniques may be used to discretize $\mathcal{L}$. The finite element method provides an elegant discretization, and an oversimplified sketch of this discretization follows. If $\mathcal{W}$ is a linear space (or vector space) of functions in which the solution to (1.1) may be found, and $\mathcal{W}_n \subset \mathcal{W}$ is an n-dimensional subspace with basis functions $\{\phi_j\}$, then an approximate solution u_n can be expanded in the form

$$u_n = \sum_{j=1}^{n} \phi_j \xi_j.$$

A variational or Galerkin principle is used, depending on whether $\mathcal{L}$ is self-adjoint, to obtain

$$\left\langle \phi_i, \mathcal{L}\left(\sum_{j=1}^{n} \phi_j \xi_j \right) \right\rangle = \left\langle \phi_i, \sum_{j=1}^{n} \phi_j \xi_j \right\rangle \lambda,$$

where $\langle\cdot,\cdot\rangle$ is an inner product on $\mathcal{W}_n$. This leads to the following systems of equations:

$$\sum_{j=1}^{n}\langle\phi_i,\mathcal{L}\phi_j\rangle\xi_j=\sum_{j=1}^{n}\langle\phi_i,\phi_j\rangle\xi_j\lambda, \tag{1.2}$$

for $1\le i\le n$. We may rewrite (1.2) and obtain the matrix equation

$$\mathbf{Ax}=\mathbf{Bx}\lambda,$$

where

$$\begin{aligned}\mathbf{A}_{i,j}&=\langle\phi_i,\mathcal{L}\phi_j\rangle,\\ \mathbf{B}_{i,j}&=\langle\phi_i,\phi_j\rangle,\\ \mathbf{x}^T&=[\xi_1,\ldots,\xi_n]^T,\end{aligned}$$

for $1\le i,j\le n$.

There are several attractive features of a FEM discretization. The boundary conditions are naturally and systematically imposed in a consistent way in the discrete problem. Other important physical properties can also be incorporated into the finite element spaces. Rayleigh quotients with respect to $(\mathbf{A},\mathbf{B})$ give Rayleigh quotients for $\mathcal{L}$:

$$\frac{\mathbf{v}^*\mathbf{Av}}{\mathbf{v}^*\mathbf{Bv}}=\frac{\langle\phi,\mathcal{L}\phi\rangle}{\langle\phi,\phi\rangle},$$

where $\phi\in\mathcal{W}_n$ is the function defined by the components of $\mathbf{v}$ as expansion coefficients. Since $\phi\in\mathcal{W}_n\subset\mathcal{W}$, in the self-adjoint case the smallest generalized eigenvalue of $(\mathbf{A},\mathbf{B})$ is an upper bound for the smallest eigenvalue of the continuous operator $\mathcal{L}$. Typically the basis functions are chosen so that $\mathbf{A}$ and $\mathbf{B}$ are sparse matrices, that is, only a few of the entries in a typical row are nonzero.

In particular, methods for solving the eigenproblem that avoid matrix factorizations and similarity transformations are of interest. The methods discussed here only require matrix-vector products, or perhaps a single sparse direct matrix factorization. Typically, only a few eigenpairs are sought and these methods only require storage proportional to $n\cdot k$, where k is the number of eigenpairs desired. Advantages of such methods are obvious and we list a few:

- sparsity of the matrices is exploited,
- matrices need not be stored – we only need a subroutine for computing the necessary matrix-vector product,
- parallelism is easy.

2. Notation and background

Before discussing methods, we give a brief review to fix notation and introduce basic ideas. We shall consider $n \times n$ square matrices $\mathbf{A}$ with complex entries. The notation $\mathbf{v}^*, \mathbf{A}^*$ will denote the complex conjugate-transpose of a vector (if complex), or the transpose (if real), and likewise for matrices. We shall use $\|\mathbf{v}\|$ to denote the Euclidean norm of a vector $\mathbf{v}$ and $\|\mathbf{A}\|$ to denote the induced matrix two-norm. The real and complex number fields will be denoted by $\mathbb{R}$ and $\mathbb{C}$ respectively. The set of numbers $\sigma(\mathbf{A}) := \{\lambda \in \mathbb{C} : \text{rank}(\lambda\mathbf{I} - \mathbf{A}) < n)\}$ is called the *spectrum* of $\mathbf{A}$. The elements of $\sigma(\mathbf{A})$ are the *eigenvalues* of $\mathbf{A}$ and are the n roots of the *characteristic polynomial* $p_A(\lambda) := \det(\lambda\mathbf{I} - \mathbf{A})$. To each distinct eigenvalue $\lambda \in \sigma(\mathbf{A})$ corresponds at least one nonzero right eigenvector $\mathbf{x}$ such that $\mathbf{A}\mathbf{x} = \mathbf{x}\lambda$. A nonzero vector $\mathbf{y}$ such that $\mathbf{y}^*\mathbf{A} = \lambda\mathbf{y}^*$ is called a *left eigenvector.* The *algebraic* multiplicity $n_a(\lambda)$ is the multiplicity of λ as a root of p_A, and the dimension $n_g(\lambda)$ of $\text{Null}(\lambda\mathbf{I} - \mathbf{A})$ is the *geometric* multiplicity of λ. A matrix is *defective* if $n_g(\lambda) < n_a(\lambda)$, for some λ, and otherwise $\mathbf{A}$ is *nondefective.* The eigenvalue λ is *simple* if $n_a(\lambda) = 1$, and $\mathbf{A}$ is *derogatory* if $n_g(\lambda) > 1$ for some λ.

A subspace $\mathcal{S}$ of $\mathbb{C}^{n\times n}$ is an *invariant subspace* of $\mathbf{A}$ if $\mathbf{A}\mathcal{S} \subset \mathcal{S}$. It is straightforward to show that if $\mathbf{A} \in \mathbb{C}^{n\times n}$, $\mathbf{V} \in \mathbb{C}^{n\times k}$ and $\mathbf{H} \in \mathbb{C}^{k\times k}$ satisfy

$$\mathbf{A}\mathbf{V} = \mathbf{V}\mathbf{H}, \tag{2.1}$$

then $\mathcal{S} := \text{Range}(\mathbf{V})$ is an invariant subspace of $\mathbf{A}$. Moreover, if $\mathbf{V}$ has full column rank k, then the columns of $\mathbf{V}$ form a basis for this subspace and $\sigma(\mathbf{H}) \subset \sigma(\mathbf{A})$. If $k = n$ then $\sigma(\mathbf{H}) = \sigma(\mathbf{A})$, and $\mathbf{A}$ is said to be *similar* to $\mathbf{H}$. The matrix $\mathbf{A}$ is *diagonalizable* if it is similar to a diagonal matrix. We use the notation $\mathcal{S} = \mathcal{S}_1 \oplus \mathcal{S}_2$ to denote that $\mathcal{S}$ is a *direct sum* of subspaces $\mathcal{S}_1$ and $\mathcal{S}_2$ ($\mathcal{S} = \mathcal{S}_1 + \mathcal{S}_2$ and $\mathcal{S}_1 \cap \mathcal{S}_2 = \{0\}$).

The *Schur decomposition* is fundamental to this discussion and is relevant to some very successful numerical algorithms.

Theorem 2.1. Every square matrix $\mathbf{A}$ possesses a Schur decomposition

$$\mathbf{A}\mathbf{Q} = \mathbf{Q}\mathbf{R}, \tag{2.2}$$

where $\mathbf{Q}$ is unitary ($\mathbf{Q}^*\mathbf{Q} = \mathbf{I}$) and $\mathbf{R}$ is upper triangular. The diagonal elements of $\mathbf{R}$ are the eigenvalues of $\mathbf{A}$.

Schur decompositions are not unique: the eigenvalues of $\mathbf{A}$ may appear on the diagonal of $\mathbf{R}$ in any specified order. From the Schur decomposition, it is easily seen that:

- the matrix $\mathbf{A}$ is normal ($\mathbf{A}\mathbf{A}^* = \mathbf{A}^*\mathbf{A}$) if and only if $\mathbf{A} = \mathbf{Q}\mathbf{\Lambda}\mathbf{Q}^*$ with $\mathbf{Q}$ unitary, and $\mathbf{\Lambda}$ diagonal,
- the matrix $\mathbf{A}$ is Hermitian ($\mathbf{A} = \mathbf{A}^*$) if and only if $\mathbf{A} = \mathbf{Q}\mathbf{\Lambda}\mathbf{Q}^*$ with $\mathbf{Q}$ unitary, and $\mathbf{\Lambda}$ is diagonal with real diagonal elements.

In either case the eigenvectors of $\mathbf{A}$ are the orthonormal columns of $\mathbf{Q}$ and the eigenvalues are the diagonal elements of $\mathbf{\Lambda}$.

If $\mathbf{V}_k$ represents the leading k columns of $\mathbf{Q}$, and $\mathbf{R}_k$ the leading principal $k \times k$ submatrix of $\mathbf{R}$, then

$$\mathbf{AV}_k = \mathbf{V}_k\mathbf{R}_k.$$

This is called a *partial Schur decomposition* of $\mathbf{A}$, and there is always a partial Schur decomposition of $\mathbf{A}$ with the diagonal elements of $\mathbf{R}_k$ consisting of any specified subset of k eigenvalues of $\mathbf{A}$. Moreover, Range$\{\mathbf{V}_k\}$ is an invariant subspace of $\mathbf{A}$ corresponding to these eigenvalues.

3. Single-vector iterations

Single-vector iterations are the simplest and most storage-efficient ways to compute a single eigenvalue and its corresponding eigenvector. The classic power method is the simplest of these and underlies the behaviour of virtually all methods for large-scale problems. This stems from the fact that one is generally restricted to repeated application of a fixed operator to produce a sequence of vectors. The power method is shown in Algorithm 1.

```
Given a nonzero v;
for k = 1, 2, 3, ..., until convergence
    w = Av
    j = i_max(w)
    λ = w(j)
    v ← w/λ
end
```

Algorithm 1. The power method

This method is suggested by the observation

$$\mathbf{A}^k\mathbf{v}_1 = \sum_{j=1}^{n} \mathbf{q}_j\lambda_j^k\gamma_j,$$

where $\mathbf{Aq}_j = \mathbf{q}_j\lambda_j$ and $\mathbf{v}_1 = \sum_{j=1}^{n} \mathbf{q}_j\gamma_j$, and this leads to a straightforward convergence analysis when $\mathbf{A}$ is diagonalizable.

If the eigenvalues of $\mathbf{A}$ are indexed such that $|\lambda_1| > |\lambda_2| \geq |\lambda_3| \geq \cdots \geq |\lambda_n|$, then we have

$$\frac{1}{\lambda_1^k}\mathbf{A}^k\mathbf{v}_1 = \mathbf{q}_1\gamma_1 + \left(\frac{\lambda_2}{\lambda_1}\right)^k \mathbf{z}_k, \tag{3.1}$$

where $\mathbf{z}_k := \sum_{j=2}^{n} \mathbf{q}_j (\frac{\lambda_j}{\lambda_2})^k \gamma_j$. The ordering of λ_j implies that $\|\mathbf{z}_k\|$ is uniformly bounded. Of course, λ_1 is not available, but it is easily seen that, after k iterations, the contents of $\mathbf{v}$ are

$$\begin{aligned}
\mathbf{v} &= \frac{\mathbf{A}^k \mathbf{v}_1}{\mathbf{e}_{j_o}^T \mathbf{A}^k \mathbf{v}_1} \\
&= \frac{\lambda_1^{-k} \mathbf{A}^k \mathbf{v}_1}{\lambda_1^{-k} \mathbf{e}_{j_o}^T \mathbf{A}^k \mathbf{v}_1} \\
&= \frac{\mathbf{q}_1 \gamma_1 + (\frac{\lambda_2}{\lambda_1})^k \mathbf{z}_k}{\mathbf{e}_{j_o}^T (\mathbf{q}_1 \gamma_1 + (\frac{\lambda_2}{\lambda_1})^k \mathbf{z}_k)} \\
&= \mathbf{q}_1 + \mathcal{O}\left(\left| \frac{\lambda_2}{\lambda_1} \right|^k \right) \to \mathbf{q}_1, \qquad \text{as } k \to \infty,
\end{aligned}$$

where $j_o = \text{i\_max}(\mathbf{q}_1)$ and we assume $\mathbf{q}_1(j_o) = 1$. The function $\text{i\_max}(\mathbf{w})$ selects the index of the first element of largest magnitude. For sufficiently large k, the selection of $j = \text{i\_max}(\mathbf{w})$ in Algorithm 1 returns $j = j_o$ (except in one annoying case that is of no real consequence).

This simple analysis must be modified when $\mathbf{A}$ is defective. In this case the behaviour of powers of Jordan blocks of spectral radius less than one replace the powers of ratios of eigenvalues.

Scaling by the component of largest magnitude facilitates the convergence analysis. We could just as easily scale to the unit ball in any of the standard vector norms. The directions of the vectors $\mathbf{v}$ are the same regardless. Often the eigenvalue estimate is taken to be the Rayleigh quotient $\lambda = \mathbf{v}^* \mathbf{A} \mathbf{v}$, where $\mathbf{v} = \mathbf{w}/\|\mathbf{w}\|$, and this is certainly recommended when $\mathbf{A}$ is Hermitian, since the eigenvalue estimates converge about twice as fast with this estimate.

The two major drawbacks to the power method are the rate of convergence, which is proportional to $|\frac{\lambda_2}{\lambda_1}|$ and can be arbitrarily slow, and that only one eigenvector can be computed.

The problem of slow convergence and convergence to interior eigenvalues may, of course, be remedied by replacing $\mathbf{A}$ by $(\mathbf{A} - \sigma \mathbf{I})^{-1}$, where σ is near an eigenvalue of interest. Later, more will be said about such spectral transformations. To address the problem of obtaining several eigenvectors, deflation schemes have been devised to find a subsequent eigenvector once the first one has converged (Saad 1992). Wielandt deflation is one of these. However, this scheme is not suitable for the nonsymmetric problem.

It is clear that various linear combinations of power iterates might be devised to approximate additional eigenvectors. For example, $\hat{\mathbf{v}}_j = (\mathbf{v}_j - \mathbf{v}_{j-1}\lambda_1)/\lambda_2$ will converge to a multiple of $\mathbf{q}_2$. However, there is a systematic way to consider all such possibilities at once and pick the optimal one automatically.

4. Krylov subspace projection methods

A systematic way to approach this question is to consider all possible linear combinations of the leading k vectors in the power sequence and ask how the best possible approximate eigeninformation might be extracted. The successive vectors produced by a power iteration may contain considerable information along eigenvector directions corresponding to eigenvalues near the one with largest magnitude. A single-vector power iteration simply ignores this information. Subspace projection provides a way to extract this additional information. Rather than discard the vectors produced during the power iteration, additional eigen-information is obtained by looking at various linear combinations of the power sequence. This immediately leads to consideration of the *Krylov subspace*

$$\mathcal{K}_k(\mathbf{A}, \mathbf{v}) := \text{Span}\{\mathbf{v}, \mathbf{A}\mathbf{v}, \mathbf{A}^2\mathbf{v}, \ldots, \mathbf{A}^{k-1}\mathbf{v}\},$$

and to seek the best approximate eigenvector that can be constructed from this subspace.

Approximate eigenpairs are constructed by imposing a Galerkin condition. Given any k-dimensional subspace $\mathcal{S}$ of $\mathbb{C}^n$, we define a vector $\mathbf{x} \in \mathcal{S}$ to be a *Ritz vector*, with corresponding *Ritz value* θ, if the Galerkin condition

$$\langle \mathbf{w}, \mathbf{A}\mathbf{x} - \mathbf{x}\theta \rangle = 0, \quad \text{for all} \quad \mathbf{w} \in \mathcal{S}, \tag{4.1}$$

is satisfied, with $\langle \cdot, \cdot \rangle$ denoting some inner product on $\mathbb{C}^n$. In this setting, we are interested in $\mathcal{S} = \mathcal{K}_k(\mathbf{A}, \mathbf{v}_1)$. More general subspaces will be considered later.

The definition of $\mathcal{K}_k := \mathcal{K}_k(\mathbf{A}, \mathbf{v})$ implies that every $\mathbf{w} \in \mathcal{K}_k$ is of the form $\mathbf{w} = \phi(\mathbf{A})\mathbf{v}_1$ for some polynomial ϕ of degree less than k and also that $\mathcal{K}_{j-1} \subset \mathcal{K}_j$ for $j = 2, 3, \ldots, k$. If a sequence of orthogonal bases $\mathbf{V}_j = [\mathbf{v}_1, \mathbf{v}_2, \ldots, \mathbf{v}_j]$ has been constructed with $\mathcal{K}_j = \text{Range}(\mathbf{V}_j)$ and $\mathbf{V}_j^*\mathbf{V}_j = \mathbf{I}_j$, then it is fairly straightforward to see that $\mathbf{v}_j = \phi_{j-1}(\mathbf{A})\mathbf{v}_1$ where ϕ_{j-1} is a polynomial of degree $j-1$. To extend the basis for $\mathcal{K}_k$ to one for $\mathcal{K}_{k+1}$, a new vector must be constructed with a component in the direction of $\mathbf{A}^k\mathbf{v}_1$ and then orthogonalized with respect to the previous basis vectors. Since $\mathbf{v}_k$ is the only basis vector available with a component in the direction of $\mathbf{A}^{k-1}\mathbf{v}_1$, the new basis vector $\mathbf{v}_{k+1}$ is obtained by

$$\mathbf{f}_k = \mathbf{A}\mathbf{v}_k - \mathbf{V}_k\mathbf{h}_k, \tag{4.2}$$

$$\mathbf{v}_{k+1} = \mathbf{f}_k / \|\mathbf{f}_k\|, \tag{4.3}$$

where the vector $\mathbf{h}_k$ is constructed to achieve $\mathbf{V}_k^*\mathbf{f}_k = 0$. Of course, the orthogonality of the columns of $\mathbf{V}_k$ gives the formula $\mathbf{h}_k = \mathbf{V}_k^*\mathbf{A}\mathbf{v}_k$.

This construction provides a crucial fact concerning $\mathbf{f}_k$:

$$\|\mathbf{f}_k\| = \min_{\mathbf{h}} \|\mathbf{A}\mathbf{v}_k - \mathbf{V}_k\mathbf{h}\| = \min \|\phi(\mathbf{A})\mathbf{v}_1\|, \tag{4.4}$$

where the second minimization is over all polynomials ϕ of degree k with the same leading coefficient as ϕ_{k-1} (*i.e.*, $\lim_{\tau\to\infty} \frac{\tau\phi_{k-1}(\tau)}{\phi(\tau)} = 1$, where $\mathbf{v}_k = \phi_{k-1}(\mathbf{A})\mathbf{v}_1$).

This construction fails when $\mathbf{f}_k = 0$, but in this case

$$\mathbf{A}\mathbf{V}_k = \mathbf{V}_k\mathbf{H}_k,$$

where $\mathbf{H}_k = \mathbf{V}_k^*\mathbf{A}\mathbf{V}_k = [\mathbf{h}_1, \mathbf{h}_2, \ldots, \mathbf{h}_k]$ (with a slight abuse of notation). Hence, this 'good breakdown' happens precisely when $\mathcal{K}_k$ is an invariant subspace of $\mathbf{A}$. The precise conditions that cause $\mathbf{f}_k = 0$ are introduced later in connection with restarting.

4.1. The Arnoldi factorization

The construction leading to the formulas in (4.2) results in the fundamental Arnoldi method for constructing an orthonormal basis for $\mathcal{K}_k$. It expresses a relation between the matrix $\mathbf{A}$, the basis matrix $\mathbf{V}_k$ and the residual vector $\mathbf{f}_k$ of the form

$$\mathbf{A}\mathbf{V}_k = \mathbf{V}_k\mathbf{H}_k + \mathbf{f}_k\mathbf{e}_k^*,$$

where $\mathbf{V}_k \in \mathbb{C}^{n\times k}$ has orthonormal columns, $\mathbf{V}_k^*\mathbf{f}_k = 0$ and $\mathbf{H}_k = \mathbf{V}_k^*\mathbf{A}\mathbf{V}_k$ is a $k \times k$ upper Hessenberg matrix with nonnegative subdiagonal elements. This will be called a *k-step Arnoldi factorization* of $\mathbf{A}$. When $\mathbf{A}$ is Hermitian this implies $\mathbf{H}_k$ is real, symmetric and tridiagonal and then the relation is called a *k-step Lanczos factorization* of $\mathbf{A}$. The columns of $\mathbf{V}_k$ are referred to as the *Arnoldi vectors* or *Lanczos vectors*, respectively.

Ritz pairs satisfying the Galerkin condition (4.1) are derived from the eigenpairs of the small projected matrix $\mathbf{H}_k$. If $\mathbf{H}_k\mathbf{y} = \mathbf{y}\theta$, then the vector $\mathbf{x} = \mathbf{V}_k\mathbf{y}$ satisfies

$$\|\mathbf{A}\mathbf{x} - \mathbf{x}\theta\| = \|(\mathbf{A}\mathbf{V}_k - \mathbf{V}_k\mathbf{H}_k)\mathbf{y}\| = |\beta_k\mathbf{e}_k^*\mathbf{y}|,$$

where $\beta_k := \|\mathbf{f}_k\|$. Observe that if $(\mathbf{x}, \theta)$ is a Ritz pair then

$$\theta = \mathbf{y}^*\mathbf{H}_k\mathbf{y} = (\mathbf{V}_k\mathbf{y})^*\mathbf{A}(\mathbf{V}_k\mathbf{y}) = \mathbf{x}^*\mathbf{A}\mathbf{x}$$

is a Rayleigh quotient (assuming $\|\mathbf{y}\| = 1$), and the associated Rayleigh quotient residual $\mathbf{r}(\mathbf{x}) := \mathbf{A}\mathbf{x} - \mathbf{x}\theta$ satisfies

$$\|\mathbf{r}(\mathbf{x})\| = |\beta_k\mathbf{e}_k^*\mathbf{y}|.$$

When $\mathbf{A}$ is Hermitian, this relation may be used to provide computable rigorous bounds on the accuracy of the eigenvalues of $\mathbf{H}_k$ as approximations to eigenvalues of $\mathbf{A}$ (Parlett 1980). Of course, when $\mathbf{A}$ is non-Hermitian, a small residual does not necessarily imply an accurate approximate eigenpair. Nonnormality effects may corrupt the accuracy. In any case, in exact arithmetic, when $\mathbf{f}_k = 0$ these Ritz pairs become exact eigenpairs of $\mathbf{A}$.

The explicit steps needed to form a k-step Arnoldi factorization are given in Algorithm 2. The factorization is represented visually in Figure 1.

$\mathbf{v}_1 = \mathbf{v}/\|\mathbf{v}\|;$
$\mathbf{w} = \mathbf{A}\mathbf{v}_1;\ \ \alpha_1 = \mathbf{v}_1^*\mathbf{w};$
$\mathbf{f}_1 \leftarrow \mathbf{w} - \mathbf{v}_1\alpha_1;$
$\mathbf{V}_1 \leftarrow [\mathbf{v}_1];\ \mathbf{H}_1 \leftarrow [\alpha_1];$
for $j = 1, 2, 3, \ldots, k-1,$
  $\beta_j = \|\mathbf{f}_j\|;\ \mathbf{v}_{j+1} \leftarrow \mathbf{f}_j/\beta_j;$
  $\mathbf{V}_{j+1} \leftarrow [\mathbf{V}_j, \mathbf{v}_{j+1}];$
  $\hat{\mathbf{H}}_j \leftarrow \begin{bmatrix} \mathbf{H}_j \\ \beta_j \mathbf{e}_j^* \end{bmatrix};$
  $\mathbf{w} \leftarrow \mathbf{A}\mathbf{v}_{j+1};$
  $\mathbf{h} \leftarrow \mathbf{V}_{j+1}^*\mathbf{w};$
  $\mathbf{f}_{j+1} \leftarrow \mathbf{w} - \mathbf{V}_{j+1}\mathbf{h};$
  $\mathbf{H}_{j+1} \leftarrow [\hat{\mathbf{H}}_j, \mathbf{h}];$
end

Algorithm 2. k-step Arnoldi factorization

Figure 1. Arnoldi visualization

The formulas given here are based on the *classical Gram–Schmidt* (CGS) orthogonalization process. Often, the orthogonalization is expressed in terms of the *modified Gram–Schmidt* (MGS) process. When the Arnoldi factorization is used to approximate the solution of a linear system, MGS is usually adequate. However, for eigenvalue calculations, the orthogonal basis is very important numerically. In finite precision, MGS does not provide an orthogonal basis and the orthogonality deteriorates in proportion to the condition number of the matrix $[\mathbf{v}, \mathbf{A}\mathbf{v}, \ldots, \mathbf{A}^{k-1}\mathbf{v}]$. In the restarting schemes we shall devise, it is a goal to reach a state of dependence in order to obtain $\mathbf{f}_k = 0$, and MGS is inappropriate for this situation. A second drawback for MGS is that it must be expressed in terms of Level 1 BLAS (Lawson, Hanson, Kincaid and Krogh 1979).

When expressed in terms of CGS, the dense matrix-vector products $\mathbf{V}_{j+1}^*\mathbf{w}$ and $\mathbf{V}_{j+1}\mathbf{h}$ may be coded in terms of the Level 2 BLAS operation _GEMV (Dongarra, DuCroz, Hammarling and Hanson 1988). This provides a significant performance advantage on virtually every platform from workstation to supercomputer.

Unfortunately, the CGS process is notoriously unstable and will fail miserably in this setting without modification. However, it can be rescued via a technique proposed by Daniel, Gragg, Kaufman and Stewart (DGKS) in 1976. This provides an excellent way to construct a vector $\mathbf{f}_{j+1}$ that is numerically orthogonal to $\mathbf{V}_{j+1}$. It amounts to computing a correction

$$\mathbf{c} = \mathbf{V}_{j+1}^*\mathbf{f}_{j+1}; \quad \mathbf{f}_{j+1} \leftarrow \mathbf{f}_{j+1} - \mathbf{V}_{j+1}\mathbf{c}; \quad \mathbf{h} \leftarrow \mathbf{h} + \mathbf{c};$$

just after the construction of $\mathbf{f}_{j+1}$ if necessary. One may perform a simple test to avoid this DGKS correction if it is not needed. The correction only needs to be computed if $\|\mathbf{h}\| < \eta(\|\mathbf{h}\|^2 + \|\mathbf{f}_{j+1}\|^2)^{1/2}$, where $0 < \eta < 1$ is a specified parameter. The test ensures that the new vector $\mathbf{A}\mathbf{v}$ makes an angle greater than $\cos^{-1}\eta$ with the existing Krylov subspace. This mechanism maintains orthogonality to full working precision at very reasonable cost. The special situation imposed by the restarting scheme we are about to discuss makes this modification essential for obtaining accurate eigenvalues and numerically orthogonal Schur vectors (eigenvectors in the Hermitian case). This scheme is visualized in Figure 2, where it is shown that the initial projection $\mathbf{V}\mathbf{h}$ of $\mathbf{w} = \mathbf{A}\mathbf{v}$ is the exact projection of a perturbed vector. The correction vector $\mathbf{c}$ then corrects the non-orthogonal vector $\mathbf{f} = \mathbf{w} - \mathbf{V}\mathbf{h}$ to a new one $\mathbf{f}_+ \leftarrow \mathbf{f} - \mathbf{V}\mathbf{c}$ that is orthogonal to Range($\mathbf{V}$).

It has been well documented that failure to maintain orthogonality leads to numerical difficulties. In the Hermitian case, Paige (1971) showed that the loss of orthogonality occurs precisely when an eigenvalue of $\mathbf{H}_j$ is close to an eigenvalue of $\mathbf{A}$. In fact, the Lanczos vectors lose orthogonality in the direction of the associated approximate eigenvector. Failure to maintain orthogonality results in spurious copies of the approximate eigenvalue

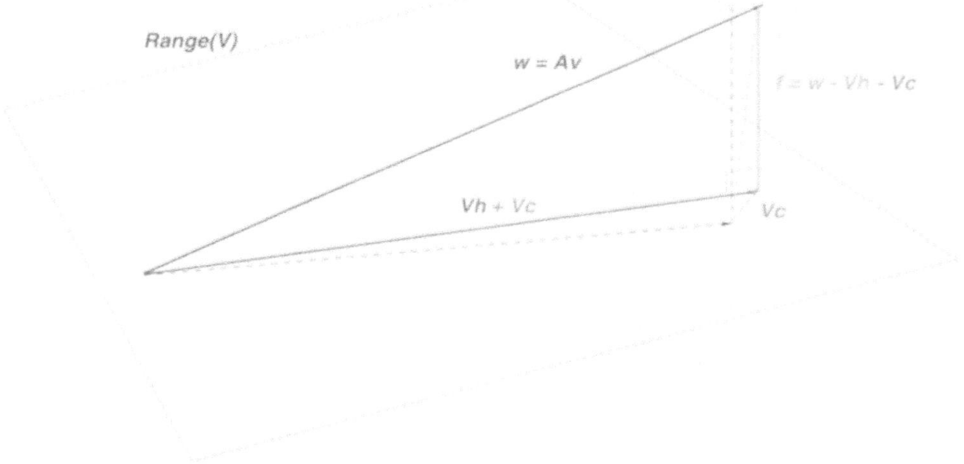

Figure 2. DGKS correction

produced by the Lanczos method (Algorithm 4). Implementations based on selective and partial orthogonalization (Grimes, Lewis and Simon 1994, Parlett and Scott 1979, Simon 1984) monitor the loss of orthogonality and perform additional orthogonalization steps only when necessary. The methods developed in Cullum and Willoughby (1981, 1985) and in Parlett and Reid (1981) use the three-term recurrence with no re-orthogonalization steps. Once a level of accuracy has been achieved, the spurious copies of computed eigenvalues are located and deleted. Then the Lanczos basis vectors are regenerated from the three-term recurrence and Ritz vectors are recursively constructed in place. This is a very competitive strategy when the matrix-vector product $\mathbf{w} \leftarrow \mathbf{A}\mathbf{v}$ is relatively inexpensive.

4.2. Restarting the Arnoldi process

A clear difficulty with the Lanczos/Arnoldi process is that the number of steps required to calculate eigenvalues of interest within a specified accuracy cannot be predetermined. This depends completely on the starting vector $\mathbf{v}_1$, and generally eigen-information of interest does not appear until k gets very large. In Figure 3 the distribution in the complex plane of the Ritz values (shown in grey dots) is compared with the spectrum (shown as +s). The original matrix is a normally distributed random matrix of order 200 and the Ritz values are from a ($k = 50$)-step Arnoldi factorization. Note that hardly any Ritz values appear in the interior and also that very few eigenvalues of $\mathbf{A}$ are well approximated. Eigenvalues at the extremes of the spectrum of $\mathbf{A}$ are clearly better approximated than the others.

Figure 3. Typical distribution of Ritz values

For large problems, it is clearly intractable to compute and store a numerically orthogonal basis set $\mathbf{V}_k$ for large k. Storage requirements are $\mathcal{O}(n \cdot k)$ and arithmetic costs are $\mathcal{O}(n \cdot k^2)$ flops to compute the basis vectors plus $\mathcal{O}(k^3)$ flops to compute the eigensystem of $\mathbf{H}_k$.

To control this cost, restarting schemes have been developed that iteratively replace the starting vector $\mathbf{v}_1$ with an 'improved' starting vector $\mathbf{v}_1^+$, and then compute a new Arnoldi factorization of fixed length k. Beyond the obvious motivation to control computational cost and storage overheads, there is a clear interest in forcing $\mathbf{f}_k = 0$. However, this is useful only if the spectrum $\sigma(\mathbf{H}_k)$ has the desired properties. The structure of $\mathbf{f}_k$ guides the strategy. The goal is to iteratively force $\mathbf{v}_1$ to be a linear combination of eigenvectors of interest.

Since $\mathbf{v}_1$ determines the subspace $\mathcal{K}_k$, this vector must be constructed to select the eigenvalues of interest. The following lemmas serve as a guide.

Lemma 4.1. If $\mathbf{v} = \sum_{j=1}^{k} \mathbf{q}_j \gamma_j$ where $\mathbf{A}\mathbf{q}_j = \mathbf{q}_j \lambda_j$, and

$$\mathbf{AV} = \mathbf{VH} + \mathbf{f}\mathbf{e}_k^T$$

is a k-step Arnoldi factorization with unreduced $\mathbf{H}$, then $\mathbf{f} = 0$ and $\sigma(\mathbf{H}) = \{\lambda_1, \lambda_2, \ldots, \lambda_k\}$.

This lemma follows easily from the observation that $\phi(\mathbf{A})\mathbf{v}_1 = 0$ with $\phi(\tau) = \prod_{i=1}^{k}(\tau - \lambda_j)$ together with the minimization property (4.4), which implies $\mathbf{f}_k = 0$. (An upper Hessenberg matrix $\mathbf{H}$ is *unreduced* if no element of the first subdiagonal is zero.) A more precise statement is as follows.

Lemma 4.2. $\mathbf{f}_k = 0$ if and only if $\mathbf{v}_1 = \mathbf{Q}_k\mathbf{y}$, where $\mathbf{AQ}_k = \mathbf{Q}_k\mathbf{R}_k$ is a partial Schur decomposition of $\mathbf{A}$ with $\mathbf{R}_k$ non-derogatory. Moreover, the Ritz values of $\mathbf{A}$ with respect to $\mathcal{K}_k$ are eigenvalues of $\mathbf{A}$, and are given by the diagonal elements of $\mathbf{R}_k$.

Thus, a more general and superior numerical strategy is to force the starting vector to be a linear combination of Schur vectors that span the desired invariant subspace.

Restarting was initially proposed by Karush (1951) soon after the Lanczos algorithm appeared (Lanczos 1950). Subsequently, there were developments by Paige (1971) Cullum and Donath (1974) and Golub and Underwood (1977) Then, Saad (1984) developed a polynomial restarting scheme for eigenvalue computation based on the acceleration scheme of Manteuffel (1978) for the iterative solution of linear systems.

4.3. Polynomial restarting

Polynomial restarting strategies replace $\mathbf{v}_1$ by

$$\mathbf{v}_1 \leftarrow \psi(\mathbf{A})\mathbf{v}_1,$$

where ψ is a polynomial constructed to damp unwanted components from the starting vector. If $\mathbf{v}_1 = \sum_{j=1}^{n} \mathbf{q}_j\gamma_j$ where $\mathbf{Aq}_j = \mathbf{q}_j\lambda_j$, then

$$\mathbf{v}_1^+ = \psi(\mathbf{A})\mathbf{v}_1 = \sum_{j=1}^{n} \mathbf{q}_j\gamma_j\psi(\lambda_j).$$

The idea is to force the starting vector to be ever closer to an invariant subspace, by constructing ψ so that $\psi(\lambda)$ is as small as possible on a region containing the unwanted eigenvalues. This is motivated by Lemmas 4.1 and 4.2. Because of this effect of filtering out (damping) the unwanted components, we often refer to these polynomials as *filter polynomials*, and we refer to their roots as *filter shifts*. The reason for this terminology will become clear when we introduce implicit restarting.

An iteration is defined by repeatedly restarting until the updated Arnoldi factorization eventually contains the desired eigenspace. For more information on the selection of effective restarting vectors, see Saad (1992). One of the more successful approaches is to use Chebyshev polynomials in order to damp unwanted eigenvector components in the available subspace.

Explicit restarting techniques are easily parallelized, in contrast to the overheads involved in implicit restarting (Section 4.4). The reason is that a major part of the work is in matrix-vector products. When we have to solve the eigenproblem on a massively parallel computer for a matrix that allows inexpensive matrix-vector products, this may be an attractive property.

Two possibilities for constructing ψ suggest themselves immediately. One is to construct the polynomial to be 'small' in magnitude on the unwanted set of eigenvalues and large on the wanted set. This criterion can be met by constructing a polynomial that best approximates 0 on a specified set that encloses the unwanted set and excludes the wanted set of eigenvalues. The other possibility is to use the best available approximation to the wanted eigenvectors. These are the Ritz vectors, and so it makes sense to select the current Ritz vectors corresponding to Ritz values that best approximate the wanted eigenvalues, and form

$$\mathbf{v}_+ = \sum_{j=1}^{k} \hat{\mathbf{q}}_j \gamma_j. \tag{4.5}$$

Since each Ritz vector is of the form $\hat{\mathbf{q}}_j = \phi_j(\mathbf{A})\mathbf{v}$, where ϕ_j is a polynomial of degree $j - 1 < m$, this mechanism is also a polynomial restart. In Saad (1992), heuristics are given for choosing the weights γ_j.

A third way is to specify the polynomial ψ by its roots. A fairly obvious choice is to find the eigenvalues θ_j of the projected matrix $\mathbf{H}$ and sort these into two sets according to a given criterion: the wanted set $\mathcal{W} = \{\theta_j : j = 1, 2, \ldots, k\}$ and the unwanted set $\mathcal{U} = \{\theta_j : j = k+1, k+2, \ldots, k+p\}$. Then we specify the polynomial ψ as the polynomial with these unwanted Ritz values as its roots. This choice of roots, called *exact shifts*, was suggested in Sorensen (1992).

Morgan (1996) found a remarkable property of this strategy. If exact shifts are used to define $\psi(\tau) = \prod_{j=k+1}^{k+p}(\tau - \theta_j)$, then the Krylov space generated by $\mathbf{v}_1^+ = \psi(\mathbf{A})\mathbf{v}_1$ satisfies

$$\mathcal{K}_m(\mathbf{A}, \mathbf{v}_1^+) = \mathrm{Span}\{\hat{\mathbf{q}}_1, \hat{\mathbf{q}}_2, \ldots, \hat{\mathbf{q}}_k, \mathbf{A}\hat{\mathbf{q}}_j, \mathbf{A}^2\hat{\mathbf{q}}_j, \ldots, \mathbf{A}^p\hat{\mathbf{q}}_j\},$$

for any $j = 1, 2, \ldots, k$. Thus polynomial restarting with exact shifts will generate a new subspace that contains all of the possible choices in (4.5).

This property follows from the fact that $\mathcal{K}_m(\mathbf{A}, \mathbf{v}_1^+) = \psi(\mathbf{A})\mathcal{K}_m(\mathbf{A}, \mathbf{v}_1)$, together with the fact that a Ritz vector $\hat{\mathbf{q}}_j$ has the form

$$\hat{\mathbf{q}}_j = \prod_{\substack{i=1\\ i\neq j}}^{k}(\mathbf{A} - \theta_i\mathbf{I})\psi(\mathbf{A})\mathbf{v}_1,$$

and thus

$$\mathbf{A}^\ell\hat{\mathbf{q}}_j = \mathbf{A}^\ell \prod_{\substack{i=1\\ i\neq j}}^{k}(\mathbf{A} - \theta_i\mathbf{I})\mathbf{v}_1^+ \in \mathcal{K}_m(\mathbf{A}, \mathbf{v}_1^+), \quad \text{for } \ell = 1, 2, \ldots, p.$$

Hence

$$\mathrm{Span}\{\hat{\mathbf{q}}_1, \hat{\mathbf{q}}_2, \ldots, \hat{\mathbf{q}}_k, \mathbf{A}\hat{\mathbf{q}}_j, \mathbf{A}^2\hat{\mathbf{q}}_j, \ldots, \mathbf{A}^p\hat{\mathbf{q}}_j\} \subset \mathcal{K}_m(\mathbf{A}, \mathbf{v}_1^+).$$

A minimal polynomial argument may then be used to establish the linear independence of $\{\hat{\mathbf{q}}_1, \hat{\mathbf{q}}_2, \ldots, \hat{\mathbf{q}}_k, \mathbf{A}\hat{\mathbf{q}}_j, \mathbf{A}^2\hat{\mathbf{q}}_j, \ldots, \mathbf{A}^p\hat{\mathbf{q}}_j\}$, and thus a dimension argument establishes the desired equality. When wanted Ritz values are not distinct, generalized eigenvectors enter into this discussion.

Exact shifts have proved to perform remarkably well in practice and have been adopted as the shift selection of choice when no other information is available. However, there are many other possibilities. For example, if we knew of a region containing the wanted eigenvalues, we might be able to construct filter shifts designed to assure that the filter polynomial would ultimately be very small (in absolute value) over that region. If the containment region were a line segment or an ellipse, we could construct the Chebyshev points related to that region. Another distribution of filter shifts that can be designed for very general containment regions are the Leja points. These have been studied extensively in the literature and have been applied very successfully in the context of an implicitly restarted Lanczos method (IRLM) by Baglama, Calvetti and Reichel (1996). These points figure prominently in the convergence analysis we give in Section 5.

4.4. Implicit restarting

A straightforward way to implement polynomial restarting is to explicitly construct the starting vector $\mathbf{v}_1^+ = \psi(\mathbf{A})\mathbf{v}_1$ by applying $\psi(\mathbf{A})$ through a sequence of matrix-vector products. However, there is an alternative implementation that provides a more efficient and numerically stable formulation. This approach, called *implicit restarting*, uses a sequence of implicitly shifted QR steps to an m-step Arnoldi or Lanczos factorization to obtain a truncated form of the implicitly shifted QR-iteration. Numerical difficulties and storage problems normally associated with Arnoldi and Lanczos processes are avoided. The algorithm is capable of computing a small number k of eigenvalues with user-specified features such as largest real part or largest magnitude using $2nk + \mathcal{O}(k^2)$ storage. The computed Schur basis vectors for the desired k-dimensional eigenspace are numerically orthogonal to working precision.

Implicit restarting enables the extraction of desired eigenvalues and vectors from high-dimensional Krylov subspaces while avoiding the standard storage and numerical difficulties. Desired eigen-information is continually compressed into a fixed-size k-dimensional subspace through an implicitly shifted QR mechanism. An Arnoldi factorization of length $m = k + p$,

$$\mathbf{A}\mathbf{V}_m = \mathbf{V}_m\mathbf{H}_m + \mathbf{f}_m\mathbf{e}_m^*, \tag{4.6}$$

is compressed to a factorization of length k that retains the eigen-information of interest.

QR steps are used to apply p linear polynomial factors $\mathbf{A} - \mu_j\mathbf{I}$ implicitly to the starting vector $\mathbf{v}_1$. The first stage of this shift process results in

$$\mathbf{A}\mathbf{V}_m^+ = \mathbf{V}_m^+\mathbf{H}_m^+ + \mathbf{f}_m\mathbf{e}_m^*\mathbf{Q}, \tag{4.7}$$

where $\mathbf{V}_m^+ = \mathbf{V}_m\mathbf{Q}$, $\mathbf{H}_m^+ = \mathbf{Q}^*\mathbf{H}_m\mathbf{Q}$, and $\mathbf{Q} = \mathbf{Q}_1\mathbf{Q}_2\cdots\mathbf{Q}_p$. Each $\mathbf{Q}_j$ is the orthogonal matrix associated with implicit application of the shift $\mu_j = \theta_{k+j}$. Since each of the matrices $\mathbf{Q}_j$ is Hessenberg, it turns out that the first $k-1$ entries of the vector $\mathbf{e}_m^*\mathbf{Q}$ are zero (*i.e.*, $\mathbf{e}_m^*\mathbf{Q} = [\sigma\mathbf{e}_k^T, \hat{\mathbf{q}}^*]$). Hence, the leading k columns in equation (4.7) remain in an Arnoldi relation and provide an updated k-step Arnoldi factorization

$$\mathbf{A}\mathbf{V}_k^+ = \mathbf{V}_k^+\mathbf{H}_k^+ + \mathbf{f}_k^+\mathbf{e}_k^*, \tag{4.8}$$

with an updated residual of the form $\mathbf{f}_k^+ = \mathbf{V}_m^+\mathbf{e}_{k+1}\hat{\beta}_k + \mathbf{f}_m\sigma$. Using this as a starting point, it is possible to apply p additional steps of the Arnoldi process to return to the original m-step form.

Virtually any explicit polynomial restarting scheme can be applied with implicit restarting, but considerable success has been obtained with exact shifts. Exact shifts result in $\mathbf{H}_k^+$ having the k wanted Ritz values as its spectrum. As convergence takes place, the subdiagonals of $\mathbf{H}_k$ tend to zero and the most desired eigenvalue approximations appear as eigenvalues of the leading $k \times k$ block of $\mathbf{R}$ as a partial Schur decomposition of $\mathbf{A}$. The basis vectors $\mathbf{V}_k$ tend to numerically orthogonal Schur vectors.

The basic IRAM iteration is shown in Algorithm 3. The expansion and contraction process of the IRAM iteration is visualized in Figure 4.

4.5. Convergence of IRAM

There is a fairly straightforward intuitive explanation of how this repeated updating of the starting vector $\mathbf{v}_1$ through implicit restarting might lead to convergence. If $\mathbf{v}_1$ is expressed as a linear combination of eigenvectors $\{\mathbf{q}_j\}$ of $\mathbf{A}$, then

$$\mathbf{v}_1 = \sum_{j=1}^{n} \mathbf{q}_j\gamma_j \Rightarrow \psi(\mathbf{A})\mathbf{v}_1 = \sum_{j=1}^{n} \mathbf{q}_j\psi(\lambda_j)\gamma_j.$$

Applying the same polynomial (*i.e.*, using the same shifts) repeatedly for ℓ iterations will result in the jth original expansion coefficient being attenuated by a factor

$$\left(\frac{\psi(\lambda_j)}{\psi(\lambda_1)}\right)^{\ell},$$

where the eigenvalues have been ordered according to decreasing values of $|\psi(\lambda_j)|$. The leading k eigenvalues become dominant in this expansion and the remaining eigenvalues become less and less significant as the iteration proceeds. Hence, the starting vector $\mathbf{v}_1$ is forced into an invariant subspace as desired. The adaptive choice of ψ provided with the exact shift mechanism further enhances the isolation of the wanted components in this expansion. Hence, the wanted eigenvalues are approximated ever better as the iteration

Compute an $m = k + p$ step Arnoldi factorization
$\quad \mathbf{A}\mathbf{V}_m = \mathbf{V}_m\mathbf{H}_m + \mathbf{f}_m\mathbf{e}_m^*.$
repeat until convergence,
$\quad$ Compute $\sigma(\mathbf{H}_m)$ and select p
$\quad$ shifts $\mu_1, \mu_2, \ldots, \mu_p$;
$\quad \mathbf{Q} = \mathbf{I}_m$;
$\quad$ **for** $j = 1, 2, \ldots, p$,
$\quad\quad$ Factor $[\mathbf{Q}_j, \mathbf{R}_j] = \mathrm{qr}(\mathbf{H}_m - \mu_j\mathbf{I})$;
$\quad\quad \mathbf{H}_m \leftarrow \mathbf{Q}_j^*\mathbf{H}_m\mathbf{Q}_j$;
$\quad\quad \mathbf{Q} \leftarrow \mathbf{Q}\mathbf{Q}_j$;
$\quad$ **end**
$\quad \hat{\beta}_k = \mathbf{H}_m(k+1, k); \;\; \sigma_k = \mathbf{Q}(m, k)$;
$\quad \mathbf{f}_k \leftarrow \mathbf{v}_{k+1}\hat{\beta}_k + \mathbf{f}_m\sigma_k$;
$\quad \mathbf{V}_k \leftarrow \mathbf{V}_m\mathbf{Q}(:, 1{:}\, k); \;\; \mathbf{H}_k \leftarrow \mathbf{H}_m(1{:}\, k, 1{:}\, k)$;
$\quad$ Beginning with the k-step Arnoldi factorization
$\quad\quad \mathbf{A}\mathbf{V}_k = \mathbf{V}_k\mathbf{H}_k + \mathbf{f}_k\mathbf{e}_k^*,$
$\quad$ apply p additional steps of the Arnoldi process
$\quad$ to obtain a new m-step Arnoldi factorization
$\quad\quad \mathbf{A}\mathbf{V}_m = \mathbf{V}_m\mathbf{H}_m + \mathbf{f}_m\mathbf{e}_m^*.$
end

Algorithm 3. Implicitly restarted Arnoldi method (IRAM)

Figure 4. Visualization of IRAM

Figure 5. Final filter polynomial from IRAM

proceeds. Unfortunately, making this heuristic argument precise has turned out to be quite difficult. Some fairly sophisticated analysis is required to understand convergence of these methods. In Section 5 we sketch such an analysis for polynomial restarting.

Another way to look at this procedure is to consider the aggregate polynomial

$$\mathbf{v}_{\text{final}} = \Phi(\mathbf{A})\mathbf{v}_1,$$

where $\Phi(\tau)$ is the product of all the polynomials that were applied during the course of the computation. In Figure 5, the plot shows the surface of $\log|\Phi(\tau)|$ for τ in a region of the complex plane containing the eigenvalues of $\mathbf{A}$ (shown by +s). The circled eigenvalues are the five eigenvalues of largest real part that were computed. The filter polynomial Φ was automatically constructed, through the choice of filter shifts, to be small on the unwanted portion of the spectrum and to enhance the wanted portion (the five eigenvalues of largest real part).

We can also learn a great deal by considering a plot of the totality of all the filter shifts in relation to the final converged eigenvalues. This is shown in Figure 6 (a different example than shown in Figure 5). The plot shows all

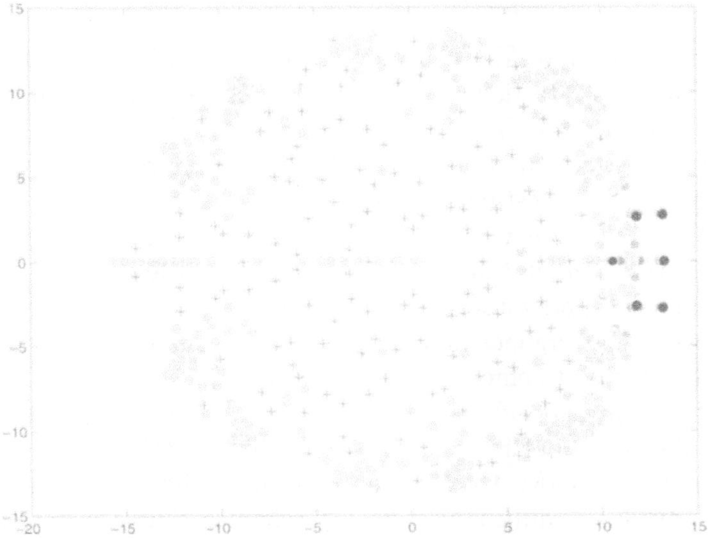

Figure 6. Distribution of filter shifts in IRAM

of the filter shifts (the light dots) applied, and the converged eigenvalues as the five darkest points to the right. The actual eigenvalues of $\mathbf{A}$ are shown as dark +s.

It is worth noting that if $m = n$ then $\mathbf{f}_m = 0$, and this iteration is precisely the same as the implicitly shifted QR iteration. Even for $m < n$, the first k columns of $\mathbf{V}_m$ and the Hessenberg submatrix $\mathbf{H}_m(1\colon k, 1\colon k)$ are mathematically equivalent to the matrices that would appear in the full implicitly shifted QR iteration using the same shifts μ_j. In this sense, the implicitly restarted Arnoldi method may be viewed as a truncation of the implicitly shifted QR iteration. The fundamental difference is that the standard implicitly shifted QR iteration selects shifts to drive subdiagonal elements of $\mathbf{H}_n$ to zero from the bottom up, while the shift selection in the implicitly restarted Arnoldi method is made to drive subdiagonal elements of $\mathbf{H}_m$ to zero from the top down.

This implicit scheme costs p rather than the $k+p$ matrix-vector products the explicit scheme would require. Thus the exact shift strategy can be viewed both as a means for damping unwanted components from the starting vector and also for directly forcing the starting vector to be a linear combination of wanted eigenvectors. See Sorensen (1992) for information on the convergence of IRAM and Baglama *et al.* (1996) and Stathopoulos, Saad and Wu (1998) for other possible shift strategies for Hermitian $\mathbf{A}$. The reader is referred to Lehoucq and Scott (1996) and Morgan (1996) for studies comparing implicit restarting with other schemes.

4.6. Deflation schemes for IRAM

The performance of IRAM can be improved considerably with the introduction of appropriate deflation schemes to isolate approximate invariant subspaces associated with converged Ritz values. These deflation strategies can make it possible to compute multiple or clustered eigenvalues with a single-vector implicit restart method.

Since IRAM may be viewed as a truncation of the standard implicitly shifted QR-iteration, it inherits a number of desirable properties. These include some well-understood deflation rules that are extremely important with respect to convergence and stability. These deflation rules are essential for the QR-iteration to efficiently compute multiple or clustered eigenvalues. These rules simply specify a numerically stable criterion for setting small subdiagonal elements of $\mathbf{H}$ to zero. While these existing QR deflation rules are applicable to IRAM, they are not the most effective schemes possible. Here, we introduce additional deflation schemes that are better suited to implicit restarting.

In the large-scale setting, it is highly desirable to provide schemes that can deflate with user-specified relative error tolerances ϵ_D that are perhaps considerably greater than working precision ϵ_M. Without this capability, excessive and unnecessary computational effort is often required to detect and deflate converged approximate eigenvalues. The ability to deflate at relaxed tolerances provides an effective way to compute multiple or clustered eigenvalues with a single-vector implicitly restarted Arnoldi method.

We shall introduce two forms of deflation. The first, a *locking* operation, decouples converged approximate eigenvalues and associated invariant subspaces from the active part of the IRAM iteration. The second, a *purging* operation, removes unwanted but converged eigenpairs. Locking has the effect of isolating an approximate eigenspace once it has converged to a certain level of accuracy and then forcing subsequent Arnoldi vectors to be orthogonal to the converged subspace. With this capability, additional instances of a multiple eigenvalue can be computed to the same specified accuracy without the expense of converging them to unnecessarily high accuracy. Purging allows the deletion of converged but unwanted Ritz values and vectors from the Krylov space when they are not purged naturally by the restarting scheme. With the aid of these deflation schemes, convergence of the IRAM iteration can be greatly improved. Computational effort is also reduced. These notions and appropriate methods were developed in Lehoucq (1995) and Lehoucq and Sorensen (1996). Here, we present a slightly improved variant of those deflation schemes.

Small subdiagonal elements of $\mathbf{H}$ may occur during implicit restarting. However, it is usually the case that there are converged Ritz values appearing in the spectrum of $\mathbf{H}$ long before small subdiagonal elements appear.

This convergence is usually detected through observation of a small last component in an eigenvector $\mathbf{y}$ of $\mathbf{H}$.

It turns out that in the case of a small last component of $\mathbf{y}$, there is an orthogonal similarity transformation of $\mathbf{H}$ that will give an equivalent Arnoldi factorization with a slightly perturbed $\mathbf{H}$ that does indeed have a zero subdiagonal, and this is the basis of our deflation schemes.

Orthogonal deflating transformations

Our deflation schemes rely on the construction of a special orthogonal transformation. As in Lehoucq (1995) and Lehoucq and Sorensen (1996), the deflation is related to an eigenvector $\mathbf{y}$ associated with the Ritz value to be deflated. In the following discussion, it is *very important* to note that the eigenvector $\mathbf{y}$ in either the locking or purging *need not be accurate.* All that is required for successful deflation schemes is that $\|\mathbf{H}\mathbf{y} - \mathbf{y}\theta\| \leq \epsilon_M \|\mathbf{H}\|$ in the case of locking, and that $\|\mathbf{y}^*\mathbf{H} - \theta\mathbf{y}^*\| \leq \epsilon_M \|\mathbf{H}\|$ in the case of purging, to obtain backward stable deflation rules.

The construction is based on a sequence of Givens transformations. If $\mathbf{y}$ is a given vector of unit length, compute a sequence of plane rotations $\mathbf{G}_{1,j}$, $j = 2, \ldots, n$ such that

$$\mathbf{y}_j^* = \mathbf{y}_{j-1}^* \mathbf{G}_{1,j} = (\tau_j, 0, \ldots, 0, \eta_{j+1}, \ldots, \eta_n),$$

beginning with $\mathbf{y}_1 := \mathbf{y}$ and ending with $\mathbf{y}_n = \mathbf{e}_1$. Thus, the orthogonal matrix $\mathbf{Q} = \mathbf{G}_{1,2}\mathbf{G}_{1,3} \cdots \mathbf{G}_{1,n}$ satisfies $\mathbf{y}^*\mathbf{Q} = \mathbf{e}_1^*$. Recalling that each $\mathbf{G}_{1,j}$ is the identity matrix $\mathbf{I}_n$ with the $(1,1), (j,j)$ entries replaced with γ_j, $\bar{\gamma}_j$ and the $(1,j), (j,1)$ entries replaced with $-\bar{\sigma}_j$, σ_j, where $|\gamma_j|^2 + |\sigma_j|^2 = 1$, it is easily seen that

$$\mathbf{Q}\mathbf{e}_1 = \mathbf{y} \quad \text{and} \quad \mathbf{e}_k^*\mathbf{Q} = (\eta, \tau\mathbf{e}_{k-1}^*), \tag{4.9}$$

with $|\eta|^2 + |\tau|^2 = 1$.

Locking or purging a single eigenvalue

Now, we shall use the orthogonal transformations developed above to construct stable and efficient transformations needed to implement locking and purging. The simplest case to consider is the treatment of a single eigenvalue. When working in complex arithmetic, this will suffice. Handling complex conjugate eigenvalues of a real nonsymmetric matrix in real arithmetic is a bit more complicated but essentially follows the same theme to deflate two vectors at once.

Locking $\boldsymbol{\theta}$. The first instance to discuss is the locking of a single converged Ritz value. Assume that

$$\mathbf{H}\mathbf{y} = \mathbf{y}\theta, \quad \|\mathbf{y}\| = 1,$$

with $\mathbf{e}_k^*\mathbf{y} = \eta$, where $|\eta| \leq \epsilon_D \|\mathbf{H}\|$. Here, it is understood that $\epsilon_M \leq \epsilon_D < 1$ is a specified relative accuracy tolerance between ϵ_M and 1.

If θ is 'wanted' then it is desirable to lock θ. However, in order to accomplish this it will be necessary to arrange a transformation of the current Arnoldi factorization to one with a small subdiagonal to isolate θ. This may be accomplished by constructing a $k \times k$ orthogonal matrix $\mathbf{Q} = \mathbf{Q}(\mathbf{y})$ as above with properties (4.9).

Consider the matrix $\mathbf{H}_+ = \mathbf{Q}^*\mathbf{H}\mathbf{Q}$. The first column of this matrix is

$$\mathbf{H}_+\mathbf{e}_1 = \mathbf{Q}^*\mathbf{H}\mathbf{Q}\mathbf{e}_1 = \mathbf{Q}^*\mathbf{H}\mathbf{y} = \mathbf{Q}^*\mathbf{y}\theta = \mathbf{e}_1\theta.$$

Thus $\mathbf{H}_+$ is of the form

$$\mathbf{H}_+ = \begin{bmatrix} \theta & \mathbf{h}^* \\ 0 & \widehat{\mathbf{H}} \end{bmatrix}.$$

We may return the matrix to Hessenberg form using orthogonal similarity transformations without destroying the desirable structure of the last row of $\mathbf{Q}$. One way to accomplish this is to apply a succession of orthogonal transformations of the form

$$\widehat{\mathbf{U}}_j = \begin{bmatrix} 1 & 0 & 0 \\ 0 & \mathbf{U}_j & 0 \\ 0 & 0 & \mathbf{I}_{k-j} \end{bmatrix},$$

so that $\mathbf{H}_+ \leftarrow \widehat{\mathbf{U}}_j^*\mathbf{H}\widehat{\mathbf{U}}_j$ is constructed to introduce zeros in positions $2, \ldots, j-1$ of row $j+1$ for $j = k-1, k-2, \ldots, 3$. This is a standard Householder reduction to Hessenberg form working from the bottom up. Of course, the orthogonal matrix $\mathbf{Q}$ must be updated in the same way to give $\mathbf{Q} \leftarrow \mathbf{Q}\widehat{\mathbf{U}}_j$, $j = k-1, k-2, \ldots, 3$. On completion, the kth row of $\mathbf{Q}$ remains undisturbed from the original construction.

The end result of these transformations is

$$\mathbf{A}[\mathbf{v}_1, \mathbf{V}_2] = [\mathbf{v}_1, \mathbf{V}_2] \begin{bmatrix} \theta & \mathbf{h}^* \\ 0 & \mathbf{H}_2 \end{bmatrix} + [\mathbf{f}\eta, \mathbf{f}\tau\mathbf{e}_{k-1}^*],$$

where $[\mathbf{v}_1, \mathbf{V}_2] = \mathbf{V}\mathbf{Q}$. Moreover, this relation may be rearranged to give

$$[\mathbf{A} - \mathbf{f}\eta\mathbf{v}_1^*][\mathbf{v}_1, \mathbf{V}_2] = [\mathbf{v}_1, \mathbf{V}_2] \begin{bmatrix} \theta & \mathbf{h}^* \\ 0 & \mathbf{H}_2 \end{bmatrix} + \mathbf{f}\tau\mathbf{e}_k^*,$$

to see that we have an exactly deflated Arnoldi factorization of a nearby matrix $\widehat{\mathbf{A}} = \mathbf{A} - \mathbf{f}\eta\mathbf{v}_1^*$ (remember, η is small).

Now, subsequent implicit restarting steps take place only in the last $k-1$ columns, as if we had

$$\mathbf{A}\mathbf{V}_2 = \mathbf{V}_2\mathbf{H}_2 + \mathbf{f}\tau\mathbf{e}_{k-1}^*,$$

with all the subsequent orthogonal matrices and column deletions associated with implicit restarting applied to $\mathbf{h}^*$, and never disturbing the relation $\mathbf{A}\mathbf{v}_1 = \mathbf{v}_1\theta + \mathbf{f}\eta$. Therefore, if $\widehat{\mathbf{Q}}$ represents a $(k-1) \times (k-1)$ orthogonal matrix associated with an implicit restart, then

$$\mathbf{A}\mathbf{V}_2\widehat{\mathbf{Q}} = (\mathbf{v}_1, \mathbf{V}_2\widehat{\mathbf{Q}}) \begin{pmatrix} \mathbf{h}^*\widehat{\mathbf{Q}} \\ \widehat{\mathbf{Q}}^*\mathbf{H}_2\widehat{\mathbf{Q}} \end{pmatrix} + \mathbf{f}\tau\mathbf{e}_{k-1}^*\widehat{\mathbf{Q}}.$$

In subsequent Arnoldi steps, $\mathbf{v}_1$ participates in the orthogonalization so that the selective orthogonalization recommended by Parlett and Scott (Parlett and Scott 1979, Parlett 1980) is accomplished automatically.

Purging $\boldsymbol{\theta}$. If θ is 'unwanted' then we may wish to remove θ from the spectrum of the projected matrix $\mathbf{H}$. This purging process is required since the implicit restart strategy using exact shifts will sometimes fail to purge a converged unwanted Ritz value (Lehoucq and Sorensen 1996).

The purging process is quite analogous to the locking process just described. However, in this case it is advantageous to use a left eigenvector to obtain the deflation. Let $\mathbf{y}$ be a left eigenvector of $\mathbf{H}$ corresponding to θ, that is,

$$\mathbf{y}^*\mathbf{H} = \theta\mathbf{y}^*.$$

Just as before, we construct a $(k \times k)$ orthogonal matrix $\mathbf{Q}$ such that

$$\mathbf{y}^*\mathbf{Q} = \mathbf{e}_1^*, \quad \text{and} \quad \mathbf{e}_k^*\mathbf{Q} = (\eta, 0, \ldots, 0, \tau),$$

where $\eta = \mathbf{e}_k^*\mathbf{y}$ and $|\tau|^2 + |\eta|^2 = 1$.

Again, consider the matrix $\mathbf{H}_+ = \mathbf{Q}^*\mathbf{H}\mathbf{Q}$. The first row of this matrix is

$$\mathbf{e}_1^*\mathbf{H}_+ = \mathbf{e}_1^*\mathbf{Q}^*\mathbf{H}\mathbf{Q}\mathbf{e}_1 = \mathbf{y}^*\mathbf{H}\mathbf{Q} = \theta\mathbf{y}^*\mathbf{Q} = \theta\mathbf{e}_1^*.$$

Thus $\mathbf{H}_+$ is of the form

$$\mathbf{H}_+ = \begin{bmatrix} \theta & 0 \\ \mathbf{h} & \widehat{\mathbf{H}} \end{bmatrix},$$

and thus

$$\mathbf{A}[\mathbf{v}_k, \widehat{\mathbf{V}}] = [\mathbf{v}_k, \widehat{\mathbf{V}}] \begin{bmatrix} \theta & 0 \\ \mathbf{h} & \widehat{\mathbf{H}} \end{bmatrix} + \mathbf{f}(\eta, \tau\mathbf{e}_{k-1}^*),$$

where $[\mathbf{v}_k, \widehat{\mathbf{V}}] = \mathbf{V}\mathbf{Q}$. Now, simply delete the first column on both sides to get

$$\mathbf{A}\widehat{\mathbf{V}} = \widehat{\mathbf{V}}\widehat{\mathbf{H}} + \mathbf{f}\tau\mathbf{e}_{k-1}^*.$$

We may return this to an Arnoldi factorization as before by constructing an orthogonal $\widehat{\mathbf{Q}}$ with $\mathbf{e}_{k-1}^*\widehat{\mathbf{Q}} = \mathbf{e}_{k-1}^*$ such that $\widehat{\mathbf{Q}}^*\widehat{\mathbf{H}}\widehat{\mathbf{Q}}$ is upper Hessenberg.

In fact, we can use the structure of $\mathbf{Q}$ to show that $\widehat{\mathbf{H}}$, surprisingly, must be upper Hessenberg automatically. However, there are subtleties in achieving an implementation that attains this numerically.

Recently, Stewart has introduced an implicit restarting method that may well resolve the issue of locking and purging. This is presented in Stewart (2001).

4.7. The Lanczos method

As previously mentioned, when $\mathbf{A}$ is Hermitian ($\mathbf{A} = \mathbf{A}^*$) then the projected matrix $\mathbf{H}$ is tridiagonal and the Arnoldi process reduces to the Lanczos method. Historically, the Lanczos process preceded the Arnoldi process.

In the Hermitian case, if we denote the subdiagonal elements of $\mathbf{H}$ by $\beta_1, \beta_2, \ldots, \beta_{n-1}$ and the diagonal elements by $\alpha_1, \alpha_2, \ldots, \alpha_n$, then the relation

$$\mathbf{A}\mathbf{V}_k = \mathbf{V}_k\mathbf{H}_k + \mathbf{f}_k\mathbf{e}_k^*$$

gives

$$\begin{aligned}\mathbf{f}_k &= \mathbf{v}_{k+1}\beta_k \\ &= \mathbf{A}\mathbf{v}_k - \mathbf{v}_k\alpha_k - \mathbf{v}_{k-1}\beta_{k-1}.\end{aligned}$$

This famous three-term recurrence has been studied extensively since its inception. The numerical difficulties are legendary, with the two main issues being the numerical orthogonality of the sequence of basis vectors and the almost certain occurrence of 'spurious' copies of converged eigenvalues reappearing as eigenvalues of the projected matrix $\mathbf{H}_k$.

The most favourable form of the recurrence, in the absence of any additional attempt at achieving orthogonality, is displayed in Algorithm 4. This organization amounts to the last two steps of a modified Gram–Schmidt variant of the Arnoldi process. Mathematically, all of the other coefficients that would ordinarily appear in the Arnoldi process are zero in the Hermitian case and this condition is forced to obtain the Lanczos process.

Once the tridiagonal matrix $\mathbf{H}_m$ has been constructed, analysed and found to possess k converged eigenvalues $\{\theta_1, \theta_2, \ldots, \theta_k\}$, with corresponding eigenvectors $\mathbf{Y} = [\mathbf{y}_1, \mathbf{y}_2, \ldots, \mathbf{y}_k]$, we may construct the eigenvectors with the recursion given in Algorithm 5.

This mechanism is quite attractive when the matrix-vector product $\mathbf{w} \leftarrow \mathbf{A}\mathbf{v}$ is relatively inexpensive. However, there are considerable numerical difficulties to overcome. Cullum and Willoughby developed schemes for analysing the projected matrix $\mathbf{H}_m$ and modifying it to get rid of the spurious eigenvalue cases. Briefly, this analysis consists of deleting the first row and column of $\mathbf{H}$ and then comparing the Ritz values of the new $\widehat{\mathbf{H}}$ with those of the original $\mathbf{H}$. Those that are the same are the spurious ones. The heuristic idea is that convergence of the 'good' Ritz values is triggered by significant components in the starting vector. A converged Ritz vector is composed from basis vectors in the Krylov subspace, and these basis vectors only contain (at least in exact arithmetic) components of the converged Ritz vector if the starting vector has a nonzero component in that direction. Since the starting vector is one of the orthogonal basis vectors $\mathbf{V}$ for the Krylov subspace, deleting it from the basis should tend

$\mathbf{v}_1 = \mathbf{v}/\|\mathbf{v}\|;$
$\mathbf{w} = \mathbf{A}\mathbf{v}_1;$
$\alpha_1 = \mathbf{v}_1^*\mathbf{w};$
$\mathbf{f}_1 \leftarrow \mathbf{w} - \mathbf{v}_1\alpha_1;$
for $j = 1, 2, 3, \dots, m-1,$
    $\beta_j = \|\mathbf{f}_j\|;$
    $\mathbf{v}_{j+1} \leftarrow \mathbf{f}_j/\beta_j;$
    $\mathbf{w} \leftarrow \mathbf{A}\mathbf{v}_{j+1} - \mathbf{v}_j\beta_j;$
    $\alpha_{j+1} = \mathbf{v}_{j+1}^*\mathbf{w};$
    $\mathbf{f}_{j+1} \leftarrow \mathbf{w} - \mathbf{v}_{j+1}\alpha_{j+1};$
end

Algorithm 4. The Lanczos process

$\mathbf{X} = \mathbf{v}_1\mathbf{Y}(1,:);$
$\mathbf{w} = \mathbf{A}\mathbf{v}_1;$
$\mathbf{f}_1 \leftarrow \mathbf{w} - \mathbf{v}_1\alpha_1;$
for $j = 1, 2, 3, \dots, m-1,$
    $\beta_j = \|\mathbf{f}_j\|;$
    $\mathbf{v}_{j+1} \leftarrow \mathbf{f}_j/\beta_j;$
    $\mathbf{X} \leftarrow \mathbf{X} + \mathbf{v}_{j+1}\mathbf{Y}(j+1,:);$
    $\mathbf{f}_{j+1} \leftarrow \mathbf{A}\mathbf{v}_{j+1} - \mathbf{v}_{j+1}\alpha_{j+1} - \mathbf{v}_j\beta_j;$
end

Algorithm 5. Eigenvector recovery in the Lanczos process

to remove important components that triggered convergence. Deleting the first row and column of $\mathbf{H} = \mathbf{V}^*\mathbf{A}\mathbf{V}$ gives an orthogonal projection $\widehat{\mathbf{H}}$ of $\mathbf{A}$ onto a subspace that is orthogonal to the starting vector. Consequently, if a Ritz value persists as an eigenvalue of $\widehat{\mathbf{H}}$, it must be spurious and therefore is the result of rounding errors.

Parlett and Reid (1981) suggested another mechanism to detect convergence of Ritz values, by constructing intervals that must contain an eigenvalue. The advantage of this approach is that it also identifies true eigenvalues that are discovered as a result of rounding errors (for instance when the starting vector was unintentionally orthogonal to the corresponding eigenvector).

Even with this convergence test, there is no assurance of numerical orthogonality of the converged eigenvectors. Parlett and Scott advocate a selected orthogonalization procedure (Parlett and Scott 1979) that orthogonalizes against converged Ritz vectors as they appear. An excellent account of the complete process is given in Parlett (1980). Grimes *et al.* (1994) advocate always using shift-invert, so that the Lanczos sequence is relatively short, and the separation of the transformed eigenvalues aids in the orthogonality, so that a selective orthogonalization scheme is quite successful.

4.8. Harmonic Ritz values and vectors

As we have seen, Ritz values usually approximate the extremal values of the spectrum well, but give poor approximations to interior eigenvalues. One attempt to better approximate interior eigenvalues has been the introduction of *harmonic Ritz values.* These were formally introduced in Paige, Parlett and van der Vorst (1995) for symmetric matrices, but have previously been used for analysis and computation in Morgan (1991) and Freund (1992). In particular, harmonic Ritz values have been proposed for restarting strategies when interior eigenvalues are sought.

There are various ways to introduce this notion. A fairly intuitive idea is to consider Rayleigh quotients of $\mathbf{A}^{-1}$ of the form

$$\theta = \frac{\mathbf{w}^*\mathbf{A}^{-1}\mathbf{w}}{\mathbf{w}^*\mathbf{w}} \quad \text{with} \quad \mathbf{w} \in \mathcal{S},$$

where $\mathcal{S}$ is a well-chosen subspace. A convenient choice is $\mathcal{S} = \mathbf{A}\mathcal{K}_k(\mathbf{A}, \mathbf{v})$. If $\mathbf{w} \in \mathcal{S}$ then $\mathbf{w} = \mathbf{AVy}$ for some $\mathbf{y}$, and

$$\begin{aligned}
\theta &= \frac{\mathbf{w}^*\mathbf{A}^{-1}\mathbf{w}}{\mathbf{w}^*\mathbf{w}} \\
&= \frac{\mathbf{y}^*\mathbf{V}^*\mathbf{A}^*\mathbf{A}^{-1}\mathbf{AVy}}{\mathbf{y}^*\mathbf{V}^*\mathbf{A}^*\mathbf{AVy}} \\
&= \frac{\mathbf{y}^*\mathbf{V}^*\mathbf{A}^*\mathbf{Vy}}{\mathbf{y}^*\mathbf{V}^*\mathbf{A}^*\mathbf{AVy}} \\
&= \frac{\mathbf{y}^*\mathbf{H}^*\mathbf{y}}{\mathbf{y}^*[\mathbf{VH} + \mathbf{fe}_k^*]^*[\mathbf{VH} + \mathbf{fe}_k^*]\mathbf{y}} \\
&= \frac{\mathbf{y}^*\mathbf{H}^*\mathbf{y}}{\mathbf{y}^*(\mathbf{H}^*\mathbf{H} + \beta^2\mathbf{e}_k\mathbf{e}_k^*)\mathbf{y}},
\end{aligned}$$

where $\beta = \|\mathbf{f}\|$. Thus θ is a generalized Rayleigh quotient for the matrix pencil $(\mathbf{H}^*, \mathbf{H}^*\mathbf{H} + \beta^2\mathbf{e}_k\mathbf{e}_k^*)$. The harmonic Ritz values are defined as the generalized eigenvalues associated with this pencil. Since θ is related to eigenvalues of $\mathbf{A}^{-1}$, it is natural to define the harmonic Ritz values μ to be the reciprocals of the critical points θ of this Rayleigh quotient. Thus the harmonic Ritz values are the eigenvalues

$$(\mathbf{H}^*\mathbf{H} + \beta^2\mathbf{e}_k\mathbf{e}_k^*)\mathbf{y} = \mathbf{H}^*\mathbf{y}\mu,$$

and the corresponding harmonic Ritz vectors are

$$\mathbf{x} = \mathbf{AVy} = \mathbf{VHy} + \mathbf{f}(\mathbf{e}_k^*\mathbf{y}).$$

When $\mathbf{H}$ is nonsingular, this simplifies to

$$(\mathbf{H} + \mathbf{g}\mathbf{e}_k^*)\mathbf{y} = \mathbf{y}\mu \quad \text{with} \quad \mathbf{g} = \beta^2\mathbf{H}^{-*}\mathbf{e}_k.$$

When $\mathbf{A}$ is Hermitian and indefinite, with λ_- the largest negative eigenvalue and λ_+ the smallest positive eigenvalue of $\mathbf{A}$, then

$$\mu_- \leq \lambda_- < 0 \quad \text{and} \quad 0 < \lambda_+ \leq \mu_+,$$

with μ_- the largest negative and μ_+ the smallest positive harmonic Ritz values. In other words, the largest interval containing 0 but no eigenvalues of $\mathbf{A}$ is also devoid of any harmonic Ritz values.

The harmonic Ritz values have some interesting properties. For symmetric matrices, the Ritz values converge monotonically to exterior eigenvalues of $\mathbf{A}$. In contrast, the harmonic Ritz values converge monotonically, albeit often at a slow rate, to the interior eigenvalues of $\mathbf{A}$ closest to the origin. This property is intuitively attractive, but has not really resulted in an effective way to compute interior eigenvalues from a Krylov subspace generated by $\mathbf{A}$. We need somehow to generate a subspace that really does contain vectors that can approximate eigenvectors associated with interior eigenvalues. Morgan (1991) has suggested harmonic Ritz vectors for restarts, and the idea has also been incorporated in the Jacobi–Davidson algorithm (Sleijpen and van der Vorst 1996). The latter method does introduce vectors into the subspace that approximate inverse iteration directions, and hence the harmonic Ritz vectors have a better chance of being effective in that setting.

5. Convergence of polynomial restart methods

For nonsymmetric problems, convergence of Krylov projection methods has been studied extensively, but the general case has been elusive. Saad (1980) developed a bound for matrices with simple eigenvalues for the gap between a single eigenvector and the Krylov subspace (gap will be defined below). This result was generalized in Jia (1995) to include defective matrices, but the bounds explicitly involve the Jordan canonical form and derivatives of approximating polynomials. Simoncini (1996) analyses convergence of a block Arnoldi method for defective matrices using pseudospectra. Lehoucq (2001) relates IRAM to subspace iteration to analyse convergence to an invariant subspace. Calvetti, Reichel and Sorensen (1994) introduce concepts from potential theory to analyse IRLM convergence to a single eigenvector for Hermitian matrices.

In the nonsymmetric case, the possibility of nonnormality complicates the analysis considerably. The possibility of derogatory matrices (an eigenvalue with geometric multiplicity greater than one) may even render certain invariant subspaces unreachable. These concepts are introduced in Beattie, Embree and Rossi (2001). They employ various ideas from functional analysis, pseudospectra and potential theory. Their analysis focuses on convergence in gap (a generalized notion of angle) of a (restarted) Krylov space to a desired invariant subspace of $\mathbf{A}$, and they are able to treat convergence in full generality.

In this section, we give a modified version of their results. Our purpose here is to lay out the main ideas as simply as possible while retaining the general theme and content of those excellent results. In doing so, we sacrifice some rigour and our convergence estimates are not as refined. We strongly recommend that the interested reader consult that reference for a thorough and insightful treatment of the convergence issues.

Before launching into this discussion, some motivation is in order. If the matrix $\mathbf{A}$ is normal, then its eigensystem is perfectly conditioned (insensitive to perturbations). In this case, it makes perfect sense to phrase convergence analysis in terms of eigenvalues. However, even in this case, when there are multiple eigenvalues, it makes more sense numerically to phrase such results in terms of convergence to invariant subspaces. In the nonsymmetric case, there is a possibility of a nontrivial Jordan form. If, for example, $\mathbf{A}$ has a Jordan form with a block of order $\ell > 1$, then certain eigenvalues of $\mathbf{A} + \mathbf{E}$ are likely to be perturbed by as much as $\|\mathbf{E}\|^{(1/\ell)}$ from the eigenvalues of $\mathbf{A}$. The best we can hope for in a numerical algorithm is to compute the exact eigensystem of a slightly perturbed matrix of this form with $\|\mathbf{E}\| = \|\mathbf{A}\|\mathcal{O}(\epsilon_M)$ (machine precision). Convergence results based on damping out specific eigenvalues (the unwanted set) in the presence of such perturbations are numerically meaningless. We must, instead, phrase such results in terms of convergence of invariant subspaces and also provide a mechanism to encompass such perturbations. This perturbation theory for nonnormal matrices is perhaps best described in Trefethen (1992, 1999). However, there are several important related papers. For really fascinating computational studies on this topic we heartily recommend the software `MATLAB Pseudospectra GUI, 2000-2001` by T. G. Wright, available at `www.comlab.ox.uk/pseudospectra/pasgui`. Pseudospectra will play a fundamental role in the following discussion.

The following is an attempt to provide a completely general convergence analysis based on the theory presented in Beattie *et al.* (2001). The development here is, admittedly, not entirely rigorous. The intent is to sketch the main ideas in a comprehensive way that can be readily understood.

5.1. Some preliminaries

We are naturally concerned with the Krylov subspace generated by a given starting vector. Note that, for any starting vector $\mathbf{v}_1$, there is a least positive integer k such that

$$\mathcal{K}_k(\mathbf{A}, \mathbf{v}_1) = \mathcal{K}(\mathbf{A}, \mathbf{v}_1) = \mathrm{Span}\{\mathbf{v}_1, \mathbf{A}\mathbf{v}_1, \mathbf{A}^2\mathbf{v}_1, \ldots\}.$$

Moreover, k is the degree of the minimal polynomial of $\mathbf{A}$ with respect to $\mathbf{v}_1$. This is the monic polynomial ϕ of least degree such that $\phi(\mathbf{A})\mathbf{v}_1 = 0$. From this property, it is straightforward to see that

$$\mathbf{A}\mathbf{K} = \mathbf{K}\mathbf{A}_k, \quad \text{with} \quad \mathbf{K} = [\mathbf{v}_1, \mathbf{A}\mathbf{v}_1, \ldots, \mathbf{A}^{k-1}\mathbf{v}_1],$$

where $\mathbf{A}_k = \mathbf{J} + \mathbf{g}\mathbf{e}_k^*$, where $\mathbf{J}$ is a Jordan matrix of order k with ones on the first subdiagonal and zeros elsewhere, and where $\mathbf{g}^T = (\gamma_0, \gamma_1, \ldots, \gamma_{k-1})$ with $\phi(\tau) = \tau^k - \gamma_{k-1}\tau^{k-1} - \cdots - \gamma_1\tau - \gamma_0$. This implies that $\mathcal{K}_k(\mathbf{A}, \mathbf{v}_1)$ is an invariant subspace with respect to $\mathbf{A}$ and that $\mathcal{K}_j(\mathbf{A}, \mathbf{v}_1) \subset \mathcal{K}_k(\mathbf{A}, \mathbf{v}_1)$ for all positive integers j.

Since $\mathbf{A}_k$ is non-derogatory (every eigenvalue of $\mathbf{A}_k$ has geometric multiplicity 1), this observation shows that it is impossible to capture more than a single Jordan block associated with a given eigenvalue. Indeed, if $\mathbf{A}$ is derogatory, then it is technically impossible to compute the entire invariant subspace corresponding to an eigenvalue of geometric multiplicity greater than one, from such a Krylov space. We would necessarily need to employ deflation and restart techniques in this case. In practice, round-off error usually blurs this situation.

To develop an understanding of convergence, a minimal amount of machinery needs to be established. Let λ_j, $1 \leq j \leq N$ be the distinct eigenvalues of $\mathbf{A}$ and let n_j be the algebraic multiplicity of λ_j. From a Schur decomposition $\mathbf{A} = \mathbf{Q}\mathbf{R}\mathbf{Q}^*$ (recall that the eigenvalues λ_j may appear in any specified order on the diagonal of $\mathbf{R}$), we can construct a spectral decomposition

$$\mathbf{A} = \mathbf{X}\widehat{\mathbf{R}}\mathbf{Y}^*, \quad \text{with} \quad \mathbf{Y}^*\mathbf{X} = \mathbf{X}\mathbf{Y}^* = \mathbf{I},$$

where $\widehat{\mathbf{R}}$ is block diagonal with upper triangular blocks $\mathbf{R}_j = \lambda_j\mathbf{I}_{n_j} + \mathbf{U}_j$, and

$$\mathbf{X} = [\mathbf{X}_1, \mathbf{X}_2, \ldots, \mathbf{X}_N], \quad \mathbf{Y} = [\mathbf{Y}_1, \mathbf{Y}_2, \ldots, \mathbf{Y}_N].$$

This construction is detailed in Golub and Van Loan (1996).

The property $\mathbf{Y}^*\mathbf{X} = \mathbf{X}\mathbf{Y}^* = \mathbf{I}$ implies that each $\mathbf{P}_j := \mathbf{X}_j\mathbf{Y}_j^*$ is a projector with the following properties:

$$\begin{aligned}
\mathbf{A}\mathbf{P}_j &= \mathbf{P}_j\mathbf{A} = \mathbf{X}_j\mathbf{R}_j\mathbf{Y}_j^*, \\
\mathbf{A}\mathcal{S}_j &\subset \mathcal{S}_j := \mathrm{Range}(\mathbf{P}_j), \\
\mathbf{I} &= \mathbf{P}_1 + \mathbf{P}_2 + \cdots + \mathbf{P}_N, \\
\mathbb{C}_n &= \oplus_{j=1}^N \mathcal{S}_j.
\end{aligned}$$

With polynomial restart techniques, we attempt to modify the starting vector $\mathbf{v}_1$ in a systematic way to force the invariant subspace $\mathcal{K} := \mathcal{K}_k(\mathbf{A}, \mathbf{v}_1)$ ever closer to a desired invariant subspace $\mathcal{X}_g$ corresponding to wanted eigenvalues λ_j, $1 \leq j \leq L$. In keeping with the notation of Beattie *et al.* (2001) we shall denote this selected set as 'good' eigenvalues and the remaining ones will be called 'bad' eigenvalues. It is important to note that there is no assumption about algebraic or geometric multiplicity of these eigenvalues.

Naturally, we are interested in some measure of nearness of $\mathcal{K}$ to $\mathcal{X}_g$. One such device is the *gap* between subspaces. The quantity

$$\delta(\mathcal{W}, \mathcal{V}) := \sup_{\mathbf{w} \in \mathcal{W}} \inf_{\mathbf{v} \in \mathcal{V}} \frac{\|\mathbf{w} - \mathbf{v}\|}{\|\mathbf{w}\|}$$

is called the *containment gap* between subspaces $\mathcal{W}$ and $\mathcal{V}$. We can show that $\delta(\mathcal{W}, \mathcal{V}) = \sin(\theta)$, where θ is the largest canonical angle between a closest subspace $\hat{\mathcal{V}}$ of $\mathcal{V}$ having the same dimension as $\mathcal{W}$. If $\mathcal{W}$ and $\mathcal{V}$ are both one-dimensional, then θ is the angle between unit vectors $\mathbf{v} \in \mathcal{V}$ and $\mathbf{w} \in \mathcal{W}$. We shall use this notion to describe the relation between a Krylov subspace and a desired invariant subspace $\mathcal{X}_g$ corresponding to good (or wanted) eigenvalues of $\mathbf{A}$.

The following lemma will provide a decomposition of $\mathcal{K}(\mathbf{A}, \mathbf{v}_1)$ associated with a given starting vector $\mathbf{v}_1$. This lemma provides a fundamental step towards understanding convergence in gap between $\mathcal{X}_g$ and $\mathcal{K}(\mathbf{A}, \mathbf{v}_1)$.

Lemma 5.1.

$$\mathcal{K} = \oplus_{j=1}^{N} \mathcal{K}_{k_j}(\mathbf{A}, \mathbf{P}_j \mathbf{v}_1),$$

where $k_j \leq n_j$ is the degree of the minimal polynomial of $\mathbf{A}$ with respect to $\mathbf{P}_j \mathbf{v}_1$.

Proof. Since $\mathbf{P}_j \mathbf{v}_1$ is in the invariant subspace $\mathcal{S}_j$, we have $\mathcal{K}_\ell(\mathbf{A}, \mathbf{P}_j \mathbf{v}_1) \subset \mathcal{S}_j$ for all ℓ. Given any $\mathbf{x} = \psi(\mathbf{A})\mathbf{v}_1 \in \mathcal{K}$, we have

$$\mathbf{x} = \psi(\mathbf{A}) \left(\sum_{j=1}^{N} \mathbf{P}_j \mathbf{v}_1 \right) = \sum_{j=1}^{N} \psi(\mathbf{A}) \mathbf{P}_j \mathbf{v}_1 \in \oplus_{j=1}^{N} \mathcal{K}_{k_j}(\mathbf{A}, \mathbf{P}_j \mathbf{v}_1)$$

(this is a direct sum since $\mathcal{S}_i \cap \mathcal{S}_j = \{\mathbf{0}\}$, $i \neq j$). To demonstrate the opposite containment, let $\mathbf{x} \in \oplus_{j=1}^{N} \mathcal{K}_{k_j}(\mathbf{A}, \mathbf{P}_j \mathbf{v}_1)$. Then $\mathbf{x} = \sum_{j=1}^{N} \psi_j(\mathbf{A}) \mathbf{P}_j \mathbf{v}_1$ with $\deg(\psi_j) < k_j$. Let ϕ be the unique polynomial of degree $\hat{k} < k_1 + k_2 + \cdots + k_N$ that interpolates the specified Hermite data $(\lambda_j, \psi_j^{(\ell)}(\lambda_j))$ for $0 \leq \ell \leq k_j - 1$ and for $1 \leq j \leq N$. At each j, after expanding ϕ in a Taylor series about λ_j, we see that

$$\phi(\mathbf{A})\mathbf{P}_j\mathbf{v}_1 = \mathbf{X}_j\phi(\mathbf{R}_j)\mathbf{Y}_j^*\mathbf{v}_1 = \mathbf{X}_j \sum_{\ell=0}^{k_j-1} \frac{\phi_j^{(\ell)}(\lambda_j)}{\ell!}(\mathbf{R}_j - \lambda_j\mathbf{I}_{n_j})^\ell\mathbf{Y}_j^*\mathbf{v}_1,$$

since $\mathbf{X}_j(\mathbf{R}_j - \lambda_j\mathbf{I}_{n_j})^\ell\mathbf{Y}_j^*\mathbf{v}_1 = (\mathbf{A} - \lambda_j\mathbf{I})^\ell\mathbf{P}_j\mathbf{v}_1 = 0$, for $\ell \geq k_j$.

Thus, $\psi_j(\mathbf{A})\mathbf{P}_j\mathbf{v}_1 = \phi(\mathbf{A})\mathbf{P}_j\mathbf{v}_1$, since the Hermite interpolation conditions imply that the leading k_j terms of the Taylor expansion of ψ_j and ϕ about λ_j will agree. Hence

$$\mathbf{x} = \sum_{j=1}^{N} \psi_j(\mathbf{A})\mathbf{P}_j\mathbf{v}_1 = \sum_{j=1}^{N} \phi(\mathbf{A})\mathbf{P}_j\mathbf{v}_1 = \phi(\mathbf{A})\left(\sum_{j=1}^{N} \mathbf{P}_j\mathbf{v}_1\right) \in \mathcal{K}. \qquad \square$$

We define $\mathbf{P}_g := \sum_{j=1}^{L} \mathbf{P}_j$ and $\mathbf{P}_b := \sum_{j=L+1}^{N} \mathbf{P}_j$, and we use the notation Ω_g and Ω_b to denote two open sets containing the good and bad eigenvalues respectively. We assume the closures of these regions are two disjoint sets with the appropriate connectedness and regularity of boundaries to make all of the contour integrals appearing below well defined. (Note: in those integrals, the factor $1/2\pi i$ has been absorbed into the $\mathrm{d}\zeta$ term.)

With polynomial restart techniques, we attempt to modify the starting vector $\mathbf{v}_1$ in a systematic way to force the invariant subspace $\mathcal{K}_k(\mathbf{A}, \mathbf{v}_1)$ ever closer to a desired invariant subspace $\mathcal{X}_g$ corresponding to the desired eigenvalues λ_j, $1 \leq j \leq L$. From the spectral decomposition, we know that

$$\mathcal{X}_g = \oplus_{j=1}^{L}\mathcal{S}_j = \mathrm{Range}(\mathbf{P}_g),$$

where $\mathcal{S}_j = \mathrm{Range}(\mathbf{P}_j)$. We shall define the complementary space $\mathcal{X}_b := \mathrm{Range}(\mathbf{P}_b)$. From Lemma 5.1 we have

$$\mathcal{K}(\mathbf{A}, \mathbf{v}_1) = \mathcal{U}_g \oplus \mathcal{U}_b,$$

where $\mathcal{U}_g := \oplus_{j=1}^{L}\mathcal{K}_{k_j}(\mathbf{A}, \mathbf{P}_j\mathbf{v}_1)$ and $\mathcal{U}_b := \oplus_{j=L+1}^{N}\mathcal{K}_{k_j}(\mathbf{A}, \mathbf{P}_j\mathbf{v}_1)$.

The questions we hope to answer are:

What is the gap $\delta(\mathcal{X}_g, \mathcal{K}(\mathbf{A}, \mathbf{v}_1))$?
What is the gap $\delta(\mathcal{X}_g, \mathcal{K}(\mathbf{A}, \hat{\mathbf{v}}_1)$ with $\hat{\mathbf{v}}_1 = \Phi(\mathbf{A})\mathbf{v}_1$?

The following discussion attempts to answer these questions.

Definition. Given a starting vector $\mathbf{v}_1$ and a selection of 'good' eigenvalues $\{\lambda_j : 1 \leq j \leq L\}$ with corresponding invariant subspace $\mathcal{X}_g$, we define the maximal reachable set $\mathcal{U}_{\max}$ to be

$$\mathcal{U}_{\max} := \mathcal{K}(\mathbf{A}, \mathbf{v}_1) \cap \mathcal{X}_g.$$

It is easily seen that $\mathcal{U}_{\max}$ is invariant with respect to $\mathbf{A}$ and the following lemma will characterize this invariant subspace precisely.

Lemma 5.2. Given a starting vector $\mathbf{v}_1$ and a selection of 'good' eigenvalues $\{\lambda_j : 1 \le j \le L\}$, the maximal reachable set $\mathcal{U}_{\max}$ is

$$\mathcal{U}_{\max} = \oplus_{j=1}^{L} \mathcal{K}_{k_j}(\mathbf{A}, \mathbf{P}_j\mathbf{v}_1),$$

and therefore

$$\mathcal{U}_{\max} = \mathcal{U}_g.$$

Proof. The proof is immediate from the characterization of $\mathcal{K}(\mathbf{A}, \mathbf{v}_1)$ and the fact that $\mathcal{K}_\ell(\mathbf{A}, \mathbf{P}_j\mathbf{v}_1) \subset \mathcal{S}_j$ for all ℓ. □

Unfortunately, there are situations where it is impossible to produce a Krylov space that contains a good approximating subspace to all of $\mathcal{X}_g$. Note that $\mathcal{U}_g \subset \mathcal{X}_g$, and that the only possibility for this containment to be proper is if $k_j < n_j$ for some $1 \le j \le L$. That is to say, at least one good eigenvalue must be derogatory. The following lemma establishes that it is impossible to converge to all of $\mathcal{X}_g$ whenever there is a derogatory eigenvalue amongst the good eigenvalues.

Lemma 5.3. Suppose $\mathcal{U}_g \subset \mathcal{X}_g$ is a proper subset of $\mathcal{X}_g$. Then

$$\delta(\mathcal{X}_g, \mathcal{K}(\mathbf{A}, \mathbf{v}_1)) \ge \frac{1}{\|\mathbf{P}_g\|}.$$

Proof. Since $\mathcal{U}_g$ is a proper subset of $\mathcal{X}_g$, there is a $\mathbf{z} \in \mathcal{X}_g$ such that $\|\mathbf{z}\| = 1$ and $\mathbf{z} \in \mathcal{U}_g^{\perp}$. Thus, for any $\mathbf{v}_g \in \mathcal{U}_g$ we must have

$$\|\mathbf{v}_g - \mathbf{z}\|^2 = \|\mathbf{v}_g\|^2 + \|\mathbf{z}\|^2 \ge \|\mathbf{z}\|^2 = 1.$$

Now, since any $\mathbf{v} \in \mathcal{K} := \mathcal{K}(\mathbf{A}, \mathbf{v}_1)$ can be written uniquely as $\mathbf{v} = \mathbf{v}_g + \mathbf{v}_b$ with $\mathbf{v}_g \in \mathcal{U}_g$ and $\mathbf{v}_b \in \mathcal{U}_b$, we have

$$\begin{aligned}
\delta(\mathcal{X}_g, \mathcal{K}(\mathbf{A}, \mathbf{v}_1)) &= \max_{\mathbf{u}\in\mathcal{X}_g} \min_{\mathbf{v}\in\mathcal{K}} \frac{\|\mathbf{v} - \mathbf{u}\|}{\|\mathbf{u}\|} \ge \min_{\mathbf{v}\in\mathcal{K}} \frac{\|\mathbf{v} - \mathbf{z}\|}{\|\mathbf{z}\|} \\
&\ge \min_{\substack{\mathbf{v}_g\in\mathcal{U}_g \\ \mathbf{v}_b\in\mathcal{U}_b}} \frac{\|\mathbf{v}_g + \mathbf{v}_b - \mathbf{z}\|}{\|\mathbf{z}\|} \ge \min_{\substack{\mathbf{v}_g\in\mathcal{U}_g \\ \mathbf{v}_b\in\mathcal{U}_b}} \frac{\|(\mathbf{v}_g - \mathbf{z}) + \mathbf{v}_b\|}{\|\mathbf{v}_g - \mathbf{z}\|} \\
&\ge \min_{\substack{\mathbf{y}\in\mathcal{X}_g \\ \mathbf{v}_b\in\mathcal{X}_b}} \frac{\|\mathbf{v}_b - \mathbf{y}\|}{\|\mathbf{y}\|} = \min_{\substack{\mathbf{y}\in\mathcal{X}_g \\ \mathbf{v}_b\in\mathcal{X}_b}} \frac{\|\mathbf{v}_b - \mathbf{y}\|}{\|\mathbf{P}_g(\mathbf{v}_b - \mathbf{y})\|} \\
&\ge \min_{\mathbf{x}} \frac{\|\mathbf{x}\|}{\|\mathbf{P}_g\mathbf{x}\|} = \frac{1}{\|\mathbf{P}_g\|}.
\end{aligned}$$

□

In this lemma, we could just as well have replaced $\mathcal{X}_g$ with any subspace $\mathcal{U}$ of $\mathcal{X}_g$ that properly contains $\mathcal{U}_g$. This justifies calling $\mathcal{U}_g$ the maximal reachable subspace. Moreover, since $\mathcal{K}(\mathbf{A}, \Phi(\mathbf{A})\mathbf{v}_1)$ is a subspace of $\mathcal{K}(\mathbf{A}, \mathbf{v}_1)$, the result also applies to all possible subspaces obtained by polynomial restarting.

The best we can hope for is to produce a Krylov space that contains an approximation to $\mathcal{U}_g$. Of course, when the dimension is sufficiently large, $\mathcal{U}_g$ will be captured exactly, since $\mathcal{U}_g \subset \mathcal{K}(\mathbf{A}, \mathbf{v}_1)$. We are interested in cases where the dimension of the Krylov space is reasonably small.

We begin with a discussion of the distance of a Krylov space of dimension ℓ from $\mathcal{U}_g$, and then introduce the consequences for restarting.

Lemma 5.4. Let $\ell \geq m = \dim\{\mathcal{U}_g\}$. Then

$$\delta(\mathcal{U}_g, \mathcal{K}_\ell(\mathbf{A}, \mathbf{v}_1)) \leq \max_{\psi} \min_{\phi} \frac{\|\phi(\mathbf{A})\mathbf{P}_b\mathbf{v}_1\|}{\|\psi(\mathbf{A})\mathbf{P}_g\mathbf{v}_1\|}$$

such that $\phi(\mathbf{A})\mathbf{P}_g\mathbf{v}_1 = \psi(\mathbf{A})\mathbf{P}_g\mathbf{v}_1$ and $\deg(\phi) < \ell$, $\deg(\psi) < m$.

Proof. Since $\mathcal{U}_g = \oplus_{j=1}^{L} \mathcal{K}_{k_j}(\mathbf{A}, \mathbf{P}_j\mathbf{v}_1)$, $\mathbf{x} \in \mathcal{U}_g$ implies

$$\mathbf{x} = \sum_{j=1}^{L} \psi_{k_j}(\mathbf{A})\mathbf{P}_j\mathbf{v}_1 = \psi(\mathbf{A})\mathbf{P}_g\mathbf{v}_1,$$

where ψ is the unique polynomial of degree less than m that interpolates the Hermite data defining ψ_{k_j}, $1 \leq j \leq L$. Also, $\mathbf{v} \in \mathcal{K}_\ell(\mathbf{A}, \mathbf{v}_1)$ implies $\mathbf{v} = \phi(\mathbf{A})\mathbf{v}_1$ with $\deg(\phi) < \ell$. Thus

$$\begin{aligned}
\delta(\mathcal{U}_g, \mathcal{K}_\ell(\mathbf{A}, \mathbf{v}_1)) &= \max_{\psi} \min_{\phi} \frac{\|\phi(\mathbf{A})\mathbf{v}_1 - \psi(\mathbf{A})\mathbf{P}_g\mathbf{v}_1\|}{\|\psi(\mathbf{A})\mathbf{P}_g\mathbf{v}_1\|} \\
&= \max_{\psi} \min_{\phi} \frac{\|[\phi(\mathbf{A}) - \psi(\mathbf{A})]\mathbf{P}_g\mathbf{v}_1 + \phi(\mathbf{A})\mathbf{P}_b\mathbf{v}_1\|}{\|\psi(\mathbf{A})\mathbf{P}_g\mathbf{v}_1\|} \\
&\leq \max_{\psi} \min_{\phi} \frac{\|\phi(\mathbf{A})\mathbf{P}_b\mathbf{v}_1\|}{\|\psi(\mathbf{A})\mathbf{P}_g\mathbf{v}_1\|},
\end{aligned}$$

where the final inequality is obtained by restricting ϕ to satisfy the Hermite interpolation data defining ψ on λ_j for $1 \leq j \leq L$. □

We wish to refine this estimate into a more quantitative one. Recall $m = \dim\{\mathcal{U}_g\} = \sum_{j=1}^{L} k_j$ and $\ell \geq m$. Define $\alpha(\tau)$ to be the minimal polynomial of $\mathbf{A}$ with respect to $\mathbf{P}_g\mathbf{v}_1$. It is straightforward to show $\alpha(\tau) = \prod_{j=1}^{L}(\tau - \lambda_j)^{k_j}$. Moreover, any polynomial ϕ of degree $\ell - 1 \geq m - 1$ satisfying the constraint of Lemma 5.4 must be of the form

$$\phi(\tau) = \psi(\tau) + \hat{\phi}(\tau)\alpha(\tau).$$

Intuitively, this means that the matrix $\phi(\mathbf{A}) - \psi(\mathbf{A})$ must annihilate $\mathcal{U}_g$. We have the following result.

Corollary 5.5.

$$\delta(\mathcal{U}_g, \mathcal{K}_\ell(\mathbf{A}, \mathbf{v}_1)) \leq \max_{\|\psi(\mathbf{A})\mathbf{P}_g\mathbf{v}_1\|=1} \min_{\hat{\phi}} \|[\psi(\mathbf{A}) + \hat{\phi}(\mathbf{A})\alpha(\mathbf{A})]\mathbf{P}_b\mathbf{v}_1\|.$$

Thus, our gap estimate amounts to a question of how well a polynomial $\hat{\phi}$ of degree $\ell - m$ can approximate the rational function $\frac{\psi(\tau)}{\alpha(\tau)}$ over certain regions of the complex plane and, in particular, over the region Ω_b.

We can easily verify

$$\|[\psi(\mathbf{A}) + \hat{\phi}(\mathbf{A})\alpha(\mathbf{A})]\mathbf{P}_b\mathbf{v}_1\| = \left\| \oint_{\partial\Omega_b} [\psi(\zeta) + \hat{\phi}(\zeta)\alpha(\zeta)](\zeta\mathbf{I} - \mathbf{A})^{-1}\mathbf{P}_b\mathbf{v}_1 \, d\zeta \right\|,$$

where (as specified above) Ω_b includes the bad eigenvalues and excludes the good ones (with sufficient regularity conditions on connectedness and smoothness of the boundary).

The previous discussion gives a qualitative idea of how the bounds will be obtained, but does not really lead to a concrete bound since ψ may be arbitrarily large.

5.2. Bounding $\|\psi(\mathbf{A})\mathbf{P}_g\mathbf{v}_1\|$ from below

We first consider the case that there is just one wanted eigenvalue λ_1 and that $\alpha(\tau) = (\tau - \lambda_1)^{k_1}$ is the minimal polynomial of $\mathbf{A}$ with respect to $\mathbf{P}_g\mathbf{v}_1$. We may conclude that $\|(\mathbf{A} - \lambda_1\mathbf{I})^j\mathbf{P}_g\mathbf{v}_1\| > 0$ for $0 \le j < k_1$. Given ψ of degree less than k_1, set

$$\psi(\tau) = \hat{\psi}(\tau)(\tau - \lambda_1)^j, \ (j < k_1),$$

where we assume that $\hat{\psi}(\lambda_1) = 1$ since the numerator and denominator in Lemma 5.4 may be simultaneously scaled by the same nonzero constant. Now, let $\mathbf{\Lambda}_\epsilon = \{\zeta \in \mathbb{C} : \|(\zeta\mathbf{I} - \mathbf{A})^{-1}\| \ge \frac{1}{\epsilon}\}$. The set $\mathbf{\Lambda}_\epsilon$ is called the ϵ-pseudospectrum of $\mathbf{A}$ (Trefethen 1992, 1999). This is one of several equivalent descriptions. The boundaries of this family of sets are level curves of the function $\|(\zeta\mathbf{I} - \mathbf{A})^{-1}\|$ and these are called lemniscates. Let ϵ be sufficiently small that there is a connected component of $\mathbf{\Lambda}_\epsilon$ denoted by Ω_g that contains λ_1 and no other eigenvalue of $\mathbf{A}$. Then $\|(\zeta\mathbf{I} - \mathbf{A})^{-1}\| = 1/\epsilon$ on $\partial\Omega_g$ and since $\hat{\psi}(\lambda_1) = 1$, we may take ϵ sufficiently small to ensure $|\hat{\psi}(\zeta)| > 1/2$ on $\partial\Omega_g$. Then

$$\begin{aligned}
\|(\mathbf{A} - \lambda_1\mathbf{I})^j\mathbf{P}_g\mathbf{v}_1\| &= \left\| \left(\oint_{\partial\Omega_g} \frac{1}{\hat{\psi}(\zeta)}(\zeta\mathbf{I} - \mathbf{A})^{-1} \, d\zeta \right) \hat{\psi}(\mathbf{A})\mathbf{P}_g(\mathbf{A} - \lambda_1\mathbf{I})^j\mathbf{P}_g\mathbf{v}_1 \right\| \\
&\le \left\| \oint_{\partial\Omega_g} \frac{1}{\hat{\psi}(\zeta)}(\zeta\mathbf{I} - \mathbf{A})^{-1} \, d\zeta \right\| \|\hat{\psi}(\mathbf{A})(\mathbf{A} - \lambda_1\mathbf{I})^j\mathbf{P}_g\mathbf{v}_1\|.
\end{aligned}$$

Thus,

$$\begin{aligned}
\frac{\|(\mathbf{A} - \lambda_1\mathbf{I})^j\mathbf{P}_g\mathbf{v}_1\|}{\|\psi(\mathbf{A})\mathbf{P}_g\mathbf{v}_1\|} &\le \left(\max_{\xi\in\partial\Omega_g} \frac{1}{|\hat{\psi}(\xi)|} \right) \oint_{\partial\Omega_g} \|(\zeta\mathbf{I} - \mathbf{A})^{-1}\| |d\zeta| \\
&\le \frac{\mathcal{L}_g}{\pi\epsilon} =: C_1,
\end{aligned}$$

where $\mathcal{L}_g$ is the length of the boundary of Ω_g.

With a little more work, this argument may be extended to L eigenvalues with $\psi(\tau) = \hat{\psi}(\tau)\hat{\alpha}(\tau)$ where $\hat{\alpha}(\tau) = \prod_{j=1}^{L}(\tau - \lambda_j)^{\ell_j}$ with $\ell_j < k_j$. In this case Ω_g is the union of the disjoint ϵ-lemniscates enclosing the good eigenvalues, and $\psi(\zeta) = \hat{\psi}(\zeta)\hat{\alpha}(\zeta)$, where $\hat{\psi}$ has been normalized to have absolute value greater than or equal to one at all of the good eigenvalues. As before, we assume that ϵ is sufficiently small to ensure that $|\hat{\psi}(\zeta)| > 1/2$ on $\partial\Omega_g$. Then, the bound becomes

$$\frac{\|\hat{\alpha}(\mathbf{A})\mathbf{P}_g\mathbf{v}_1\|}{\|\psi(\mathbf{A})\mathbf{P}_g\mathbf{v}_1\|} \le \frac{\mathcal{L}_g}{\pi\epsilon} =: C_1.$$

5.3. Bounding $\|\phi(\mathbf{A})\mathbf{P}_b\mathbf{v}_1\|$ from above

We impose the restriction $\phi(\tau) = \psi(\tau) + \hat{\phi}(\tau)\alpha(\tau)$ and consider

$$\begin{aligned}\phi(\mathbf{A})\mathbf{P}_b\mathbf{v}_1 &= [\psi(\mathbf{A}) + \hat{\phi}(\mathbf{A})\alpha(\mathbf{A})]\mathbf{P}_b\mathbf{v}_1 \\ &= \left(\oint_{\partial\Omega_b}\left[\frac{\psi(\zeta)}{\alpha(\zeta)} + \hat{\phi}(\zeta)\right](\zeta\mathbf{I} - \mathbf{A})^{-1}\,\mathrm{d}\zeta\right)\alpha(\mathbf{A})\mathbf{P}_b\mathbf{v}_1,\end{aligned}$$

which is valid since $\frac{\psi(\zeta)}{\alpha(\zeta)}$ is analytic on Ω_b. Hence,

$$\begin{aligned}\|\phi(\mathbf{A})\mathbf{P}_b\mathbf{v}_1\| &\le \left\|\oint_{\partial\Omega_b}\left[\frac{\psi(\zeta)}{\alpha(\zeta)} + \hat{\phi}(\zeta)\right](\zeta\mathbf{I} - \mathbf{A})^{-1}\,\mathrm{d}\zeta\right\|\|\alpha(\mathbf{A})\mathbf{P}_b\mathbf{v}_1\| \\ &\le \max_{\zeta\in\partial\Omega_b}\left|\frac{\psi(\zeta)}{\alpha(\zeta)} + \hat{\phi}(\zeta)\right|\oint_{\partial\Omega_b}\|(\zeta\mathbf{I} - \mathbf{A})^{-1}\|\,|\mathrm{d}\zeta|\|\alpha(\mathbf{A})\mathbf{P}_b\mathbf{v}_1\|.\end{aligned}$$

5.4. Gap estimates for polynomial restarting

We now consider the possibilities for achieving convergence in gap through polynomial restarting. This will be analysed by revising the previous estimates when we replace $\mathbf{v}_1$ with $\hat{\mathbf{v}}_1 = \Phi(\mathbf{A})\mathbf{v}_1$, where Φ is the aggregate restart polynomial. We shall assume that all of the roots of Φ are in $\mathbb{C}\ \Omega_b$. In this case we have

$$\begin{aligned}\delta(\mathcal{U}_g, \mathcal{K}_\ell(\mathbf{A}, \hat{\mathbf{v}}_1)) &= \max_{\psi}\min_{\phi}\frac{\|\phi(\mathbf{A})\Phi(\mathbf{A})\mathbf{v}_1 - \psi(\mathbf{A})\mathbf{P}_g\mathbf{v}_1\|}{\|\psi(\mathbf{A})\mathbf{P}_g\mathbf{v}_1\|} \\ &= \max_{\psi}\min_{\phi}\frac{\|[\phi(\mathbf{A})\Phi(\mathbf{A}) - \psi(\mathbf{A})]\mathbf{P}_g\mathbf{v}_1 + \phi(\mathbf{A})\Phi(\mathbf{A})\mathbf{P}_b\mathbf{v}_1\|}{\|\psi(\mathbf{A})\mathbf{P}_g\mathbf{v}_1\|} \\ &\le \max_{\psi}\min_{\phi}\frac{\|\phi(\mathbf{A})\Phi(\mathbf{A})\mathbf{P}_b\mathbf{v}_1\|}{\|\psi(\mathbf{A})\mathbf{P}_g\mathbf{v}_1\|},\end{aligned}$$

if $\phi \cdot \Phi$ is restricted to satisfy the Hermite interpolation data defining ψ on λ_j for $1 \le j \le L$.

Motivated by the arguments above, we put

$$\phi(\tau)\Phi(\tau) = \psi(\tau) + \Psi(\tau)\alpha(\tau).$$

This is accomplished by requiring ϕ to be specified so that $\phi \cdot \Phi$ does indeed satisfy the Hermite interpolation data defining ψ on λ_j for $1 \le j \le L$. (This is possible since Φ has no zeros in Ω_g.)

Once we have ϕ defined, observe that $\Phi(\tau_j) = 0$ will imply that

$$\Psi(\tau_j) = -\frac{\psi(\tau_j)}{\alpha(\tau_j)}.$$

Hence, Ψ interpolates $-\frac{\psi}{\alpha}$ at each root τ_j of Φ. (This is also true at the roots of ϕ, but we have no control over the placement of those.) Note: this interpolation property is automatic and nothing need be done to enforce it.

Again, converting to integral form gives

$$\begin{aligned}\phi(\mathbf{A})\Phi(\mathbf{A})\mathbf{P}_b\mathbf{v}_1 &= [\psi(\mathbf{A}) + \Psi(\mathbf{A})\alpha(\mathbf{A})]\mathbf{P}_b\mathbf{v}_1 \\ &= \left(\oint_{\partial\Omega_b} \left[\frac{\psi(\zeta)}{\alpha(\zeta)} + \Psi(\zeta)\right](\zeta\mathbf{I} - \mathbf{A})^{-1}\,\mathrm{d}\zeta\right)\alpha(\mathbf{A})\mathbf{P}_b\mathbf{v}_1,\end{aligned}$$

and this allows us to obtain the estimate

$$\begin{aligned}&\|\phi(\mathbf{A})\Phi(\mathbf{A})\mathbf{P}_b\mathbf{v}_1\| \\ &\quad\le \left\|\oint_{\partial\Omega_b}\left[\frac{\psi(\zeta)}{\alpha(\zeta)} + \Psi(\zeta)\right](\zeta\mathbf{I} - \mathbf{A})^{-1}\,\mathrm{d}\zeta\right\| \|\alpha(\mathbf{A})\mathbf{P}_b\mathbf{v}_1\| \\ &\quad\le \max_{\zeta\in\partial\Omega_b}\left|\frac{\psi(\zeta)}{\alpha(\zeta)} + \Psi(\zeta)\right| \oint_{\partial\Omega_b} \|(\zeta\mathbf{I} - \mathbf{A})^{-1}\|\,|\,\mathrm{d}\zeta|\,\|\alpha(\mathbf{A})\mathbf{P}_b\mathbf{v}_1\|.\end{aligned}$$

If we assume $\Omega_g \cup \Omega_b$ consists of the ϵ-pseudospectrum of $\mathbf{A}$ with ϵ sufficiently small that the closures of these sets do not intersect, then $\|(\zeta\mathbf{I} - \mathbf{A})^{-1}\| = \frac{1}{\epsilon}$ for $\zeta \in \partial\Omega_b$, and we obtain

$$\|\phi(\mathbf{A})\Phi(\mathbf{A})\mathbf{P}_b\mathbf{v}_1\| \le \max_{\zeta\in\partial\Omega_b}\left|\frac{\psi(\zeta)}{\alpha(\zeta)} + \Psi(\zeta)\right|\frac{\mathcal{L}_b}{2\pi\epsilon}\|\alpha(\mathbf{A})\mathbf{P}_b\mathbf{v}_1\|,$$

where $\mathcal{L}_b$ is the length of the boundary of Ω_b.

Since we are free to choose the roots of Φ, we should be able to make this estimate arbitrarily and uniformly small. The key to this will be the selection of points that have desirable asymptotic approximation properties with respect to interpolation of a given rational function on Ω_b at an increasing number of points. Leja points (and also Fejér or Fekete points) are known to have such properties but they are expensive to compute. A more attractive option is the use of so-called fast Leja points, introduced in Baglama, Calvetti and Reichel (1998). Fast Leja points give almost the same interpolation behaviour as Leja points but they are efficiently computed (as shown in Baglama *et al.* (1998)). The construction amounts to a recursively defined distribution of the points on $\partial\Omega_b$ in a way that is nearly optimal with respect to asymptotic interpolation properties. There are several additional properties that make these points very attractive computationally (see Baglama *et al.* (1998)).

There is no asymptotic rate of convergence available for fast Leja points, but there is one for Leja points that does ensure a linear rate of convergence for our application. Let us suppose for the moment that ψ is a fixed polynomial of degree $\ell - 1$ and α is as specified above. The following result may be found in Gaier (1987) and in related papers (Reichel 1990, Fischer and Reichel 1989).

Theorem 5.6. Assume that $\partial\Omega_b$ is a Jordan curve. Let $\mathcal{G}(\omega)$ be the conformal mapping from the exterior of the unit disk to the exterior of Ω_b, such that $\mathcal{G}(\infty) = \infty$ and $\mathcal{G}'(\infty) > 0$ (guaranteed to exist by the Riemann mapping theorem). We want to approximate $f(\zeta) = \frac{\psi(\zeta)}{\alpha(\zeta)}$ on Ω_b. Since α has all zeros outside Ω_b, there is a circle with radius $\rho > 1$ and centre at the origin, such that its image $\mathcal{C}$ under $\mathcal{G}$ goes through a zero of α and there is no zero of α in the interior of the curve $\mathcal{C}$. Let q_M be the polynomial of degree $< M$ that interpolates f at M Leja (Fejér or Fekete) points on the boundary of Ω_b. Then

$$\limsup_{M\to\infty} \max_{\zeta\in\Omega_b} |f(\zeta) - q_M(\zeta)|^{\frac{1}{M}} = \frac{1}{\rho}.$$

Thus we expect a linear rate of convergence with ratio ρ^{-1}. If we apply p shifts at a time the convergence factor should be ρ^{-p}. The convergence is, of course, faster for larger ρ: *i.e.*, as the distance of the zeros of α from Ω_b increases, so does ρ, and the convergence is correspondingly faster. Recalling that the zeros of α are the desired eigenvalues, this confirms and makes precise the intuitive notion that convergence should be faster when the wanted eigenvalues are well separated from the rest of the spectrum.

The final convergence result will be of the form

$$\begin{aligned}\delta(\mathcal{U}_g, \mathcal{K}_\ell(\mathbf{A}, \hat{\mathbf{v}}_1)) &\le \max_{\psi} \min_{\phi} \frac{\|\phi(\mathbf{A})\Phi(\mathbf{A})\mathbf{P}_b\mathbf{v}_1\|}{\|\psi(\mathbf{A})\mathbf{P}_g\mathbf{v}_1\|} \\ &\le \left(\frac{1}{\rho}\right)^M C_0C_1C_2 \max_{\hat{\alpha}} \frac{\|\alpha(\mathbf{A})\mathbf{P}_b\mathbf{v}_1\|}{\|\hat{\alpha}(\mathbf{A})\mathbf{P}_g\mathbf{v}_1\|},\end{aligned}$$

where $C_1 = \frac{\mathcal{L}_g}{\pi\epsilon}$, $C_2 = \frac{\mathcal{L}_b}{2\pi\epsilon}$, and where $\hat{\alpha}(\tau) = \prod_{j=1}^{L}(\tau - \lambda_j)^{\ell_j}$ with $\ell_j < k_j$. The positive constant C_0 is associated with converting the *lim-sup* statement to a convergence rate. The integer $M = \deg(\Phi) = \nu p$ if there have been ν restarts of degree p.

These terms have very natural interpretations. In particular, ρ is determined by the distance of the good eigenvalues λ_j, $1 \le j \le L$ from the set Ω_b enclosing the bad eigenvalues. The constants C_1 and C_2 are related to the nearness to nonnormality through the behaviour of the ϵ-pseudospectra. Finally, the ratio $\frac{\|\alpha(\mathbf{A})\mathbf{P}_b\mathbf{v}_1\|}{\|\hat{\alpha}(\mathbf{A})\mathbf{P}_g\mathbf{v}_1\|}$ reflects the influence of bias in the starting vector towards $\mathcal{U}_g$. A pleasing consequence of this term is that, whenever $\mathbf{v}_1 \in \mathcal{U}_g$, then there is termination as soon as $\ell \ge m$ in exact arithmetic.

Certain lemniscates of the ϵ-pseudospectra will form the boundary of Ω_b, and hence (unless they just touch) this boundary will be a union of Jordan curves. In certain cases, we can obtain concrete estimates by replacing Ω_b with another set that encloses all of the bad eigenvalues, and with a positive distance from Ω_g. If this new set can be constructed so that the integrals can be calculated or estimated, then actual convergence rates follow. In practice, it is unusual to have advanced knowledge of such a set. In the symmetric case, such sets are ϵ-balls centred at the eigenvalues, and this leads to containment intervals on the real line. There is a method for constructing Leja points for an adaptively defined containment interval. This has been quite successful, as demonstrated in Baglama *et al.* (1996). Exact shifts tend to discover such regions adaptively. As we have seen in prior examples, they distribute themselves near the boundary of the adaptively discovered containment region. This is one heuristic reason why exact shifts seem to be successful in many cases. See Beattie *et al.* (2001) for a convincing computational example of this.

6. Subspace iteration methods

There is another generalization of the power method that is perhaps more straightforward than Krylov subspace projection. This is the generalized power method or subspace iteration. It treats a block of vectors simultaneously in a direct analogy to the power method. In Algorithm 6 a shift-invert variant of this method is described.

```
Factor VR = W (with W ∈ ℂ^{n×k} arbitrary);
Set H = 0;
while (||AV − VH|| > tol||H||),
    μ = Select_shift(H);
    Solve (A − μI)W = V;
    Factor [V+, R] = qr(W);
    H = V+* V R^{-1} + μI;
    V ← V+;
end
```

Algorithm 6. Generalized shifted inverse power method

Noting that $\mathbf{H} = \mathbf{V}^*\mathbf{AV}$ in Algorithm 6, it is evident that the Ritz pairs $(\mathbf{x}, \theta)$ may be obtained from the eigensystem of $\mathbf{H}$ just as in the Krylov setting. However, in this case the subspace $\mathcal{S} = \text{Range}(\mathbf{V})$ will be dominated by eigenvector directions corresponding to the eigenvalues nearest to the selected shifts μ_j. The stopping rule can be modified so that additional matrix-vector products to obtain $\mathbf{AV}$ are not explicitly required. Also, in practice, it will most likely be more reliable to compute the final value of $\mathbf{H}$ (after convergence) by computing $\mathbf{V}^*(\mathbf{AV})$ directly.

Typically a single shift μ is selected and a single sparse direct factorization of $\mathbf{A} - \mu\mathbf{I}$ is computed initially and re-used to solve the systems

$$(\mathbf{A} - \mu\mathbf{I})\mathbf{W} = \mathbf{V}$$

repeatedly as needed. In this case, it is easily seen that the result on convergence is a partial Schur decomposition,

$$\mathbf{AV} = \mathbf{V}(\mathbf{R}^{-1} + \mu\mathbf{I}).$$

When $k = n$ this iteration becomes the well-known and very important shifted QR iteration. To see this, suppose an initial orthogonal similarity transformation of $\mathbf{A}$ to upper Hessenberg form has been made so that

$$\mathbf{AV} = \mathbf{VH} \text{ with } \mathbf{V}^*\mathbf{V} = \mathbf{I}, \ \mathbf{H} \text{ upper Hessenberg.}$$

If $\mathbf{H} = \mathbf{QR}$ is the QR factorization of $\mathbf{H}$, then $\mathbf{W} = (\mathbf{VQ})\mathbf{R}$ is the QR factorization of $\mathbf{W} = \mathbf{AV}$. Moreover,

$$\begin{aligned}(\mathbf{A} - \mu\mathbf{I})(\mathbf{VQ}) &= (\mathbf{VQ})(\mathbf{RQ}),\\ \mathbf{AV}_+ &= \mathbf{V}_+\mathbf{H}_+ \text{ with } \mathbf{V}_+^*\mathbf{V}_+ = \mathbf{I}, \ \mathbf{H}_+ = \mathbf{RQ} + \mu\mathbf{I}.\end{aligned}$$

Of course, the amazing thing is that $\mathbf{H}_+$ remains upper Hessenberg if $\mathbf{H}$ is originally upper Hessenberg. Moreover, the QR factorization of $\mathbf{H}$ by Givens' method and the associated updating $\mathbf{V}_+ = \mathbf{VQ}$ requires $O(n^2)$ flops rather than $O(n^3)$ for a dense QR factorization.

The important observation to make with this iteration is that the construction of the subspace is divorced from the construction of Ritz vectors. Therefore, the system $(\mathbf{A} - \mu\mathbf{I})\mathbf{W} = \mathbf{V}$ could just as well be solved (approximately) with an iterative method. The projected matrix $\mathbf{H}$ would then be obtained directly by forming $\mathbf{H} \leftarrow \mathbf{V}_+^*\mathbf{AV}_+$. The Krylov structure would be lost with this approach, but there are trade-offs.

A downside to abandoning the Krylov structure is a loss of efficiency in obtaining Ritz approximations and associated error estimates directly from $\mathbf{H}$. Also, certain very powerful polynomial approximation properties are lost. However, there are some significant advantages.

- There is the possibility of admitting approximate solutions to the block linear system $(\mathbf{A} - \mu\mathbf{I})\mathbf{W} = \mathbf{V}$. Other than the effect on convergence, there is no set accuracy requirement for these solves. This is in contrast to the Krylov setting, where important theoretical properties are lost if these solves are not accurate enough.

- If a sequence of closely related problems is being solved, as in a parameter study, the entire subspace basis from the previous problem can be used as the initial basis for the next problem. In the (single-vector) Krylov setting we must be content with a linear combination of the previous basis vectors (or Ritz vectors) to form a single starting vector for the next problem.
- If an iterative method is used to solve $(\mathbf{A} - \mu\mathbf{I})\mathbf{w} = \mathbf{v}$ approximately, then several matrix-vector products are performed for each access to the matrix $\mathbf{A}$. This can be quite important on high-performance architectures where it is desirable to perform as many floating point operations as possible per each memory access.
- It is possible to be very general in constructing vectors to adjoin to the subspace. Schemes may be devised that do not attempt to solve the shift-invert equations directly, but instead attempt to construct defect corrections as vectors to adjoin to the subspace. Davidson's method (Davidson 1975) and its variants are based on this idea.

6.1. Davidson's method

Davidson's method has been a mainstay in computational chemistry, where it is generally preferred over the Lanczos method. Typically, *ab initio* calculations in chemistry result in large symmetric positive definite matrices, which are strongly diagonally dominant.

Davidson's method attempts to exploit that structure. Given a subspace $\mathcal{S}_k = \text{Range}(\mathbf{V}_k)$ of dimension k with orthogonal basis matrix $\mathbf{V}_k$ and a selected Ritz value $\theta \in \sigma(\mathbf{V}_k^*\mathbf{A}\mathbf{V}_k)$ with corresponding Ritz vector $\hat{\mathbf{x}}$, the strategy is to expand the space with a residual defect correction designed to improve the selected Ritz value. In the following discussion one should regard $\hat{\mathbf{x}}$ as the current approximation to an eigenvector $\mathbf{x}$ and θ as the current approximation to the corresponding eigenvalue θ.

If λ is the closest eigenvalue to θ and $\mathbf{x}$ is a corresponding eigenvector, putting $\mathbf{x} = \hat{\mathbf{x}} + \mathbf{z}$ and $\lambda = \theta + \delta$ and expanding gives the standard second-order perturbation equation

$$\begin{aligned}
(\mathbf{A} - \theta\mathbf{I})\mathbf{z} &= -(\mathbf{A} - \theta\mathbf{I})\hat{\mathbf{x}} + \hat{\mathbf{x}}\delta + \mathbf{z}\delta && (6.1)\\
&= -\mathbf{r} + \hat{\mathbf{x}}\delta + \mathcal{O}(\|\mathbf{z}\delta\|) && (6.2)\\
&\approx -\mathbf{r} + \hat{\mathbf{x}}\delta. && (6.3)
\end{aligned}$$

Typically, this second-order residual correction equation is completed by forcing a condition such as $\mathbf{z}^*\hat{\mathbf{x}} = 0$. Davidson (1975) chooses to approximate $\mathbf{D}_A - \theta\mathbf{I} \approx \mathbf{A} - \theta\mathbf{I}$ on the left side and ignore both the first- and second-order terms on the right side, using the equation

$$(\mathbf{D}_A - \theta\mathbf{I})\mathbf{z} = -\mathbf{r} \quad \text{where} \quad \mathbf{r} = (\mathbf{A} - \theta\mathbf{I})\hat{\mathbf{x}}$$

to obtain an approximate residual correction step $\mathbf{z}$. A novelty of the Davidson approach was to orthogonalize $\mathbf{z}$ against the existing basis set to obtain a new basis vector $\mathbf{v}_{k+1}$ in the direction $(\mathbf{I}-\mathbf{V}_k\mathbf{V}_k^*)\mathbf{z}$, and expand the space to $\mathcal{S}_{k+1} = \text{Range}(\mathbf{V}_{k+1})$ where $\mathbf{V}_{k+1} = [\mathbf{V}_k, \mathbf{v}_{k+1}]$. A new Ritz value and vector are obtained from the updated space, and then this process is repeated until a storage limit is reached and the method is restarted.

This method has been quite successful in finding dominant eigenvalues of strongly diagonally dominant matrices. Evidently, from the second-order expansion and as suggested in Davidson (1993), this scheme is related to a Newton–Raphson iteration, and this has been used as a heuristic to explain its fast convergence.

Numerical analysts have attempted to explain the success of this approach by viewing $(\mathbf{D}_A - \theta\mathbf{I})^{-1}$ as a preconditioner, or as an approximation to $(\mathbf{A} - \theta\mathbf{I})^{-1}$. With this interpretation, improvements to Davidson's method have been attempted through the introduction of more sophisticated preconditioners (Crouzeix, Philippe and Sadkane 1994, Morgan 1991, Morgan and Scott 1993). However, a perplexing aspect of this interpretation has been that the ultimate preconditioner, namely $(\mathbf{A}-\theta\mathbf{I})^{-1}$, would just result in $\mathbf{z} = \hat{\mathbf{x}}$ and would not expand the search space.

6.2. The Jacobi–Davidson method

Progress towards improving on Davidson was finally made after recognizing that the correction should be restricted to the orthogonal complement of the existing space. This notion follows almost directly from reconsidering the second-order correction equation (6.1) and completing the equations by forcing the correction $\mathbf{z}$ to be orthogonal to $\hat{\mathbf{x}}$. This may be accomplished by forming a bordered set of equations or by explicit projection. Multiplying on the left of (6.1) by the projection $(\mathbf{I} - \hat{\mathbf{x}}\hat{\mathbf{x}}^*)$ and requiring $\hat{\mathbf{x}}^*\mathbf{z} = 0$ gives

$$
\begin{aligned}
(\mathbf{I} - \hat{\mathbf{x}}\hat{\mathbf{x}}^*)(\mathbf{A} - \theta\mathbf{I})(\mathbf{I} - \hat{\mathbf{x}}\hat{\mathbf{x}}^*)\mathbf{z} &= (\mathbf{I} - \hat{\mathbf{x}}\hat{\mathbf{x}}^*)(-\mathbf{r} + \hat{\mathbf{x}}\delta + \mathcal{O}(\|\mathbf{z}\delta\|)) && (6.4)\\
&= -\mathbf{r} + \mathcal{O}(\|\mathbf{z}\delta\|) && (6.5)\\
&\approx -\mathbf{r}. && (6.6)
\end{aligned}
$$

The second equality follows from the observation that $\theta = \hat{\mathbf{x}}^*\mathbf{A}\hat{\mathbf{x}}$ is a Rayleigh quotient, and thus $\mathbf{r} = \mathbf{A}\hat{\mathbf{x}} - \mathbf{x}\theta = (\mathbf{I} - \hat{\mathbf{x}}\hat{\mathbf{x}}^*)\mathbf{A}\mathbf{x}$.

This formulation actually results in a second-order correction $\mathbf{z} = (\mathbf{I} - \hat{\mathbf{x}}\hat{\mathbf{x}}^*)\mathbf{z}$ that is orthogonal to $\hat{\mathbf{x}}$. The coefficient matrix $(\mathbf{I} - \hat{\mathbf{x}}\hat{\mathbf{x}}^*)(\mathbf{A} - \theta\mathbf{I})(\mathbf{I} - \hat{\mathbf{x}}\hat{\mathbf{x}}^*)$ is indeed singular, but the linear system is consistent and offers no fundamental difficulty to an iterative method. Moreover, if θ approximates a simple eigenvalue that is moderately separated from the rest of the spectrum of $\mathbf{A}$, this system is likely to be better conditioned than one involving $\mathbf{A} - \theta\mathbf{I}$ as a coefficient matrix, since the nearly singular subspace has been projected out.

Now, the Davidson idea can be fully realized. The projected correction equation (6.4) is solved iteratively. Typically, this is done with a preconditioned iterative method for linear systems. Then the correction is used as with the original Davidson idea to expand the search space. Note that one step of the preconditioned GMRES method using $\mathbf{D}_A - \theta\mathbf{I}$ as a preconditioner would result in Davidson's method.

This approach due to Sleijpen and van der Vorst (1995) is called the Jacobi–Davidson (JD) method. The method applied to a symmetric $\mathbf{A} = \mathbf{A}^*$ is outlined in Algorithm 7. The hat notation $\hat{\mathbf{x}}$ is dropped in that description and $\mathbf{x}$ is the current approximate eigenvector. The update to obtain $\mathbf{H}_k$ from $\mathbf{H}_{k-1}$ is just slightly more complicated for nonsymmetric $\mathbf{A}$.

Set $\mathbf{x} = \mathbf{v}_1 = \mathbf{v}/\|\mathbf{v}\|_2$ for some initial guess $\mathbf{v}$;
$\mathbf{w} = \mathbf{A}\mathbf{v}_1$, $\theta = \mathbf{H}(1,1) = [\mathbf{v}_1^*\mathbf{w}]$, $\mathbf{r} = \mathbf{w} - \theta\mathbf{x}$;
  while $\|\mathbf{r}\|_2 > \varepsilon$
    Solve (approximately) for $\mathbf{z} \perp \mathbf{x}$:
      $(\mathbf{I} - \mathbf{x}\mathbf{x}^*)(\mathbf{A} - \theta\mathbf{I})(\mathbf{I} - \mathbf{x}\mathbf{x}^*)\mathbf{z} = -\mathbf{r}$;
    $\mathbf{c} = \mathbf{V}_{k-1}^*\mathbf{z}$; $\mathbf{z} = \mathbf{z} - \mathbf{V}_{k-1}\mathbf{c}$;
    $\mathbf{v}_k = \mathbf{z}/\|\mathbf{z}\|_2$; $\mathbf{V}_k = [\mathbf{V}_{k-1}, \mathbf{v}_k]$;
    $\mathbf{w} = \mathbf{A}\mathbf{v}_k$;

$$\begin{bmatrix} \mathbf{h} \\ \alpha \end{bmatrix} = \mathbf{V}_k^*\mathbf{w} \; ; \; \mathbf{H}_k = \begin{bmatrix} \mathbf{H}_{k-1} & \mathbf{h} \\ \mathbf{h}^* & \alpha \end{bmatrix};$$

    Compute $\mathbf{H}_k\mathbf{y} = \mathbf{y}\theta$;
    (θ the largest eigenvalue of $\mathbf{H}_k$, $\|\mathbf{y}\|_2 = 1$)
    $\mathbf{x} \leftarrow \mathbf{V}_k\mathbf{y}$;
    $\mathbf{r} = \mathbf{A}\mathbf{x} - \mathbf{x}\theta$;
  end

Algorithm 7. The Jacobi–Davidson method for $\lambda_{\max}(\mathbf{A})$ with $\mathbf{A} = \mathbf{A}^*$

There are several ways to approximately solve the correction equation. Returning to the second-order expansion (6.1),

$$(\mathbf{A} - \theta\mathbf{I})\mathbf{z} = -\mathbf{r} + \hat{\mathbf{x}}\delta,$$

to get $\mathbf{z}$ orthogonal to $\hat{\mathbf{x}}$, choose

$$\delta = \frac{\hat{\mathbf{x}}^*(\mathbf{A} - \theta\mathbf{I})^{-1}\mathbf{r}}{\hat{\mathbf{x}}^*(\mathbf{A} - \theta\mathbf{I})^{-1}\hat{\mathbf{x}}}.$$

If $(\mathbf{A} - \theta\mathbf{I})$ is replaced with a preconditioner $\mathbf{K}$, then we set

$$\mathbf{z} = -\mathbf{K}^{-1}\mathbf{r} + \mathbf{K}^{-1}\hat{\mathbf{x}} \quad \text{with} \quad \delta = \frac{\hat{\mathbf{x}}^*\mathbf{K}^{-1}\mathbf{r}}{\hat{\mathbf{x}}^*\mathbf{K}^{-1}\hat{\mathbf{x}}}.$$

If the basis vectors $\mathbf{V}_k$ are not retained and $\mathbf{z}$ is not orthogonalized against them, this becomes the method proposed by Olsen, Jørgensen and Simons (1990).

If this correction equation is to be solved approximately with a preconditioned iterative method, care must be taken to obtain efficiency. Left-preconditioning can be applied efficiently, and it is common to have a left preconditioner $\mathbf{K}_o$ for $\mathbf{A}$ available (*e.g.*, to solve linear systems required to track the dynamics). We can then take $\mathbf{K} := \mathbf{K}_o - \mu\mathbf{I}$ as a preconditioner for $\mathbf{A} - \mu\mathbf{I}$, where μ is a value reasonably close to the desired eigenvalue. Of course, it is possible to update $\mu = \theta_k$ at each JD iteration, but the cost of construction and factorization of a new preconditioner for each value of θ_k may overcome the gains from accelerated convergence. To be effective, the preconditioner should be restricted to a subspace that is orthogonal to $\hat{\mathbf{x}}$. Thus, it is desirable to work with the restricted preconditioner

$$\widetilde{\mathbf{K}} := (\mathbf{I} - \hat{\mathbf{x}}\hat{\mathbf{x}}^*)\mathbf{K}(\mathbf{I} - \hat{\mathbf{x}}\hat{\mathbf{x}}^*).$$

This is likely to be a good preonditioner for the restricted operator $\widetilde{\mathbf{A}} := (\mathbf{I} - \hat{\mathbf{x}}\hat{\mathbf{x}}^*)(\mathbf{A} - \theta\mathbf{I})(\mathbf{I} - \hat{\mathbf{x}}\hat{\mathbf{x}}^*)$. If we use a Krylov subspace iteration method for solving

$$\widetilde{\mathbf{A}}\mathbf{z} = -\mathbf{r}$$

that is initialized with $\mathbf{z}_0 = 0$, then all vectors occurring in the iterative solution process will be automatically orthogonal to $\hat{\mathbf{x}}$.

Typically, the iterative solver will require repeated evaluation of expressions like

$$\mathbf{w} = \widetilde{\mathbf{K}}^{-1}\widetilde{\mathbf{A}}\mathbf{v}$$

for vectors $\mathbf{v}$ generated during the iteration.

Since $\hat{\mathbf{x}}^*\mathbf{v} = 0$, we first compute $\mathbf{y} = (\mathbf{A} - \theta\mathbf{I})\mathbf{v}$. Then we have to solve $\mathbf{w} \perp \hat{\mathbf{x}}$ from $\widetilde{\mathbf{K}}\mathbf{w} = (\mathbf{I} - \hat{\mathbf{x}}\hat{\mathbf{x}}^*)\mathbf{y}$. This amounts to solving

$$(\mathbf{I} - \hat{\mathbf{x}}\hat{\mathbf{x}}^*)\mathbf{K}(\mathbf{I} - \hat{\mathbf{x}}\hat{\mathbf{x}}^*)\mathbf{w} = (\mathbf{I} - \hat{\mathbf{x}}\hat{\mathbf{x}}^*)\mathbf{y}.$$

Observe that this equation will be satisfied by $\mathbf{w}$ if we are able to solve

$$(\mathbf{I} - \hat{\mathbf{x}}\hat{\mathbf{x}}^*)\mathbf{K}\mathbf{w} = (\mathbf{I} - \hat{\mathbf{x}}\hat{\mathbf{x}}^*)\mathbf{y} \quad \text{with} \quad \mathbf{w}^*\hat{\mathbf{x}} = 0.$$

This is easily accomplished by solving the bordered equation

$$\begin{bmatrix} \mathbf{K} & \hat{\mathbf{x}} \\ \hat{\mathbf{x}}^* & 0 \end{bmatrix} \begin{bmatrix} \mathbf{w} \\ \delta \end{bmatrix} = \begin{bmatrix} \mathbf{y} \\ 0 \end{bmatrix}.$$

From this equation, it follows that

$$\text{(i)}\ \ \mathbf{w}^*\hat{\mathbf{x}} = 0 \qquad \text{and} \qquad \text{(ii)}\ \ \mathbf{K}\mathbf{w} = \mathbf{y} - \hat{\mathbf{x}}\delta.$$

Hence,

$$(\mathbf{I} - \hat{\mathbf{x}}\hat{\mathbf{x}}^*)\mathbf{K}\mathbf{w} = (\mathbf{I} - \hat{\mathbf{x}}\hat{\mathbf{x}}^*)(\mathbf{y} - \hat{\mathbf{x}}\delta) = (\mathbf{I} - \hat{\mathbf{x}}\hat{\mathbf{x}}^*)\mathbf{y},$$

with $\mathbf{w}^*\hat{\mathbf{x}} = 0$ as required.

This block system need not be formed explicitly. Block elimination will give the solution through the following steps:

> Given $\mathbf{y} = (\mathbf{A} - \mu\mathbf{I})\mathbf{v}$;
> Solve $\mathbf{K}[\mathbf{t}_y\ \mathbf{t}_x] = [\mathbf{y}\ \hat{\mathbf{x}}]$;
> Set $\delta = \frac{\hat{\mathbf{x}}^*\mathbf{t}_y}{\hat{\mathbf{x}}^*\mathbf{t}_x}$;
> Set $\mathbf{w} = \mathbf{t}_y - \mathbf{t}_x\delta$.

Since we are interested in solving this for several $\mathbf{v}_j$ during the course of the iterative method for solving the correction equation $\widetilde{\mathbf{A}}\mathbf{z} = -\mathbf{r}$, this can be re-organized for efficiency by computing $\mathbf{t}_x$ once and for all and then re-using it for each of the $\mathbf{v}_j$.

> Solve $\mathbf{K}\mathbf{t}_x = \hat{\mathbf{x}}$;
> **for** j = 1,2, ..., **until** convergence,
> Produce $\mathbf{v}_j$ from the iterative method;
> $\mathbf{y} \leftarrow (\mathbf{A} - \mu\mathbf{I})\mathbf{v}_j$;
> Solve $\mathbf{K}\mathbf{t}_y = \mathbf{y}$;
> Set $\delta = \frac{\hat{\mathbf{x}}^*\mathbf{t}_y}{\hat{\mathbf{x}}^*\mathbf{t}_x}$;
> Set $\mathbf{w} = \mathbf{t}_y - \mathbf{t}_x\delta$.

With this scheme we only need to solve a linear system involving $\mathbf{K}$ once per step of the Krylov iteration for solving $\widetilde{\mathbf{A}}\mathbf{z} = -\mathbf{r}$. Additional details on this implementation are specified in Sleijpen and van der Vorst (1995).

With respect to parallel computation, the Jacobi–Davidson method has the same computational structure as a Krylov method. Successful parallel implementation largely depends on how well an effective preconditioner can be parallelized. An additional complication is that, even if a good preconditioner $\mathbf{K}$ exists for $\mathbf{A}$, there is no assurance that $\mathbf{K} - \theta\mathbf{I}$ will be a good one for $\mathbf{A} - \theta\mathbf{I}$. Moreover, since this operator is usually indefinite, there is often difficulty with incomplete factorization preconditioners.

6.3. JDQR: restart and deflation

The Jacobi–Davidson method can be extended to find more than one eigenpair by using *deflation* techniques. As eigenvectors converge, the iteration is continued in a subspace forced to be orthogonal to the converged eigenvectors.

Such an extension is developed in Fokkema, Sleijpen and van der Vorst (1996) to obtain an algorithm called JDQR, for computing several eigenpairs at once. The algorithm is based on the computation of a partial Schur form of $\mathbf{A}$,

$$\mathbf{A}\mathbf{Q}_k = \mathbf{Q}_k\mathbf{R}_k,$$

where $\mathbf{Q}_k$ is an $(n \times k)$ orthonormal matrix, and $\mathbf{R}_k$ is a $(k \times k)$ upper triangular matrix, with $k \ll n$. As noted previously, if $(\mathbf{y}, \lambda)$ is an eigenpair of $\mathbf{R}_k$, then $(\mathbf{Q}_k\mathbf{y}, \lambda)$ is an eigenpair of $\mathbf{A}$.

To develop this algorithm, we need to derive conditions required of a new column $\mathbf{q}$ in order to expand an existing decomposition with $\mathbf{q}$ to obtain an updated partial Schur decomposition:

$$\mathbf{A}\,[\mathbf{Q}_k, \mathbf{q}] = [\mathbf{Q}_k, \mathbf{q}] \begin{bmatrix} \mathbf{R}_k & \mathbf{s} \\ 0 & \lambda \end{bmatrix}$$

with $\mathbf{Q}^*\mathbf{q} = 0$.

Equating the last column on both sides gives

$$\mathbf{A}\mathbf{q} = \mathbf{Q}_k\mathbf{s} + \mathbf{q}\lambda.$$

Multiplying this equation on the left by $\mathbf{I} - \mathbf{Q}_k\mathbf{Q}_k^*$ and enforcing the requirement $\mathbf{Q}_k^*\mathbf{q} = 0$ gives

$$(\mathbf{I} - \mathbf{Q}_k\mathbf{Q}_k^*)\mathbf{A}\mathbf{q} = (\mathbf{I} - \mathbf{Q}_k\mathbf{Q}_k^*)(\mathbf{Q}_k\mathbf{s} + \mathbf{q}\lambda) = \mathbf{q}\lambda.$$

Finally, we arrive at

$$(\mathbf{I} - \mathbf{Q}_k\mathbf{Q}_k^*)\mathbf{A}(\mathbf{I} - \mathbf{Q}_k\mathbf{Q}_k^*)\mathbf{q} = \mathbf{q}\lambda,$$

which shows that the new pair $(\mathbf{q}, \lambda)$ must be an eigenpair of

$$\widetilde{\mathbf{A}} = (\mathbf{I} - \mathbf{Q}_k\mathbf{Q}_k^*)\mathbf{A}(\mathbf{I} - \mathbf{Q}_k\mathbf{Q}_k^*).$$

Now, we are prepared to apply the JD method so that the partial Schur decomposition may be updated.

The JDQR iteration. Assume we have $\mathbf{A}\mathbf{Q}_k = \mathbf{Q}_k\mathbf{R}_k$. Apply the JD iteration to $\widetilde{\mathbf{A}}$ and construct an orthonormal subspace basis $\mathbf{V}_\ell := [\mathbf{v}_1, \ldots, \mathbf{v}_\ell]$. Then a projected $\ell \times \ell$ matrix $\mathbf{M} = \mathbf{V}_\ell^*\widetilde{\mathbf{A}}\mathbf{V}_\ell$ is formed. We then compute the complete Schur form $\mathbf{M}\mathbf{U} = \mathbf{U}\mathbf{S}$, with $\mathbf{U}^*\mathbf{U} = \mathbf{I}$, and $\mathbf{S}$ upper triangular. This can be done with the standard QR algorithm (Golub and Van Loan 1996).

Next, $\mathbf{S}$ is reordered (using Givens' similarity transformations) to remain upper triangular but with $|\mathbf{S}_{i,i} - \tau|$ now nondecreasing with i. The first few diagonal elements of $\mathbf{S}$ then represent the eigen-approximations closest to τ, and the first few of the correspondingly reordered columns of $\mathbf{V}_k$ represent the subspace of best eigenvector approximations. If memory is limited then this subset can be used for restart. The other columns are simply discarded. The remaining subspace is expanded according to the Jacobi–Davidson method. This is repeated until sufficiently accurate Ritz values and vectors have been obtained.

After convergence of this procedure, we have $(\mathbf{q}, \lambda)$ with $\widetilde{\mathbf{A}}\mathbf{q} = \mathbf{q}\lambda$. Since $\mathbf{Q}_k^*\widetilde{\mathbf{A}} = 0$, we have $\mathbf{Q}_k^*\mathbf{q} = 0$ automatically. Now, set $\mathbf{s} = \mathbf{Q}_k^*\mathbf{A}\mathbf{q}$ to obtain

$$\mathbf{A}\mathbf{q} = \mathbf{Q}_k\mathbf{s} + \mathbf{q}\lambda,$$

and update

$$\mathbf{Q}_{k+1} := [\mathbf{Q}_k, \mathbf{q}] \text{ and } \mathbf{R}_{k+1} := \begin{bmatrix} \mathbf{R}_k & \mathbf{s} \\ 0 & \lambda \end{bmatrix}$$

to obtain a new partial Schur decomposition of dimension $k+1$.

This process is repeated until the desired number of eigenvalues has been obtained.

7. The generalized eigenproblem

In many applications, the generalized eigenproblem $\mathbf{A}\mathbf{x} = \mathbf{B}\mathbf{x}\lambda$ arises naturally. A typical setting is a finite element discretization of a continuous problem where the matrix $\mathbf{B}$ arises from inner products of basis functions. In this case, $\mathbf{B}$ is symmetric and positive (semi-) definite, and for some algorithms this property is a necessary condition. Generally, algorithms are based on transforming the generalized problem to a standard problem. The details of how this is done are clearly important to efficiency and robustness. However, the fundamentals and performance of the algorithms for the standard problem carry over directly to the generalized case.

7.1. Krylov methods with spectral transformations

Perhaps the most successful general scheme for converting the generalized problem to a standard problem that is amenable to a Krylov or a subspace iteration method is to use the *spectral transformation* suggested by Ericsson and Ruhe (1980):

$$(\mathbf{A} - \sigma\mathbf{B})^{-1}\mathbf{B}x = \mathbf{x}\nu. \tag{7.1}$$

An eigenvector $\mathbf{x}$ of this transformed problem is also an eigenvector of the original problem $\mathbf{A}\mathbf{x} = \mathbf{B}\mathbf{x}\lambda$, with the corresponding eigenvalue given by $\lambda = \sigma + \frac{1}{\nu}$. With this transformation there is generally rapid convergence to eigenvalues near the shift σ because they are transformed to extremal well-separated eigenvalues. Perhaps an even more influential aspect of this transformation is that eigenvalues far from σ are damped (mapped near zero). It is often the case in applications that the discrete operator has eigenvalues that are large in magnitude but nonphysical and uninteresting with respect to the computation. The spectral transformation automatically overcomes the effect of these. A typical strategy is to choose σ to be a point in the complex plane that is near eigenvalues of interest and then compute the eigenvalues ν of largest magnitude in equation (7.1). It is not necessary to have σ really close to an eigenvalue. This transformation together with the implicit restarting technique is usually adequate for computing a significant number of eigenvalues near σ.

It is important to note that, even when $\mathbf{B} = \mathbf{I}$, we must generally use the shift-invert spectral transformation to find interior eigenvalues. The extreme eigenvalues of the transformed operator $\mathbf{A}_\sigma$ are generally large and well separated from the rest of the spectrum. The eigenvalues ν of largest magnitude will transform back to eigenvalues λ of the original $\mathbf{A}$ that are in a disk about the point σ. This is illustrated in Figure 7, where the $+$ symbols are the eigenvalues of $\mathbf{A}$, and the circled ones are the computed eigenvalues in the disk (dashed circle) centred at the point σ.

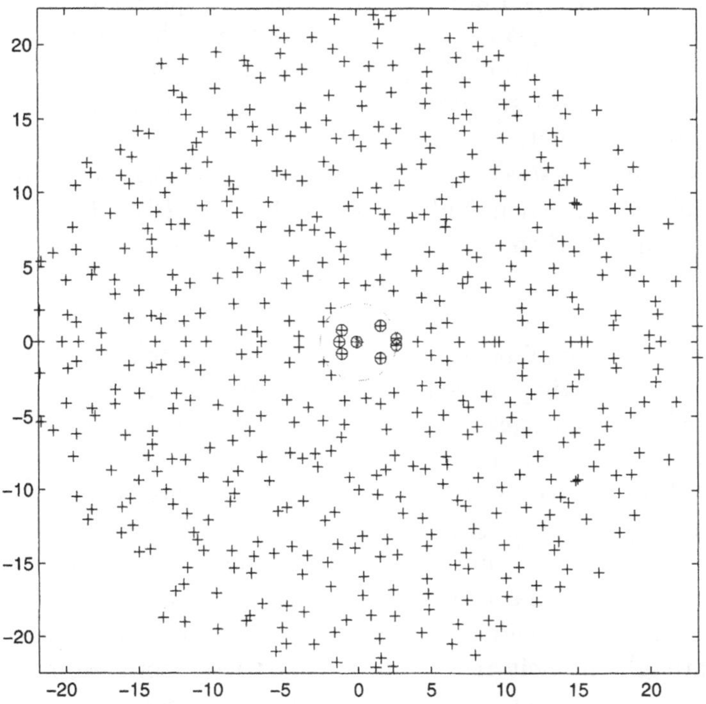

Figure 7. Eigenvalues from shift-invert

The Arnoldi process may be applied to the matrix $\mathbf{A}_\sigma := (\mathbf{A} - \sigma\mathbf{B})^{-1}\mathbf{B}$. Whenever a matrix-vector product $\mathbf{w} \leftarrow \mathbf{A}_\sigma\mathbf{v}$ is required, the following steps are performed:

$\mathbf{z} = \mathbf{B}\mathbf{v}$;
solve $(\mathbf{A} - \sigma\mathbf{B})\mathbf{w} = \mathbf{z}$ for $\mathbf{w}$.

The matrix $\mathbf{A} - \sigma\mathbf{B}$ is factored initially with a sparse direct *LU*-decomposition or in a symmetric indefinite factorization, and this single factorization is used repeatedly to apply the matrix operator $\mathbf{A}_\sigma$ as required.

When $\mathbf{A}$ and $\mathbf{B}$ are both symmetric and $\mathbf{B}$ is positive (semi-) definite, this approach needs to be modified slightly to preserve symmetry. In this case we can use a weighted $\mathbf{B}$ (semi-) inner product in the Lanczos/Arnoldi process (Ericsson and Ruhe 1980, Grimes *et al.* 1994, Meerbergen and Spence 1997). This amounts to replacing the computation of $\mathbf{h} \leftarrow \mathbf{V}_{j+1}^*\mathbf{w}$; and $\beta_j = \|\mathbf{f}_j\|$ with $\mathbf{h} \leftarrow \mathbf{V}_{j+1}^*\mathbf{B}\mathbf{w}$; and

$$\beta_j = \sqrt{\mathbf{f}_j^*\mathbf{B}\mathbf{f}_j},$$

respectively, in the Arnoldi process shown in Algorithm 2.

When $\mathbf{A}$ is symmetric and $\mathbf{B}$ is symmetric positive (semi-) definite, the matrix operator $\mathbf{A}_\sigma$ is self-adjoint with respect to this (semi-) inner product, that is, $\langle \mathbf{A}_\sigma\mathbf{x}, \mathbf{y}\rangle = \langle \mathbf{x}, \mathbf{A}_\sigma\mathbf{y}\rangle$ for all vectors $\mathbf{x}, \mathbf{y}$, where $\langle \mathbf{w}, \mathbf{v}\rangle := \sqrt{\mathbf{w}^*\mathbf{B}\mathbf{v}}$. This implies that the projected Hessenberg matrix $\mathbf{H}$ is actually symmetric and tridiagonal and the standard three-term Lanczos recurrence is recovered with this inner product.

There is a subtle aspect to this approach when $\mathbf{B}$ is singular. The most pathological case is when $\mathrm{Null}(\mathbf{A}) \cap \mathrm{Null}(\mathbf{B}) \neq \{0\}$. If $x \in \mathrm{Null}(\mathbf{A}) \cap \mathrm{Null}(\mathbf{B})$ is nonzero, then

$$\mathbf{A}\mathbf{x} = \mathbf{B}\mathbf{x}\lambda$$

for every complex number λ. This case is not treated here. A challenging but far less devastating situation occurs when this intersection is just the zero vector. In this case, $\mathrm{Null}(\mathbf{A}_\sigma) = \mathrm{Null}(\mathbf{B})$ for any σ that is not a generalized eigenvalue of the pair $(\mathbf{A}, \mathbf{B})$. Unfortunately, any nonzero vector $x \in \mathrm{Null}(\mathbf{B})$ corresponds to an *infinite eigenvalue*, since any such $\mathbf{x}$ will be an eigenvector of $\mathbf{A}_\sigma$ corresponding to the eigenvalue $\nu = 0$, and the formula $\lambda = \sigma + \frac{1}{\nu}$ indicates that $\mathbf{x}$ must correspond to an infinite eigenvalue of the original problem. Using the $\mathbf{B}$ inner product in the shift-invert Arnoldi process and requesting the eigenvalues ν of largest magnitude for $\mathbf{A}_\sigma$ through implicit restarting avoids these troublesome eigenvalues. In theory (*i.e.*, in exact arithmetic), if the starting vector $\mathbf{v}_1 = \mathbf{A}_\sigma\mathbf{v}$ is in $\mathrm{Range}(\mathbf{A}_\sigma)$ then the method cannot converge to a zero eigenvalue of $\mathbf{A}_\sigma$. However, eigenvectors are only computed approximately and these may have components in directions corresponding to infinite eigenvalues. Such components can be *purged* from a computed eigenvector $\mathbf{x}$ by replacing it with $\mathbf{x} \leftarrow \mathbf{A}_\sigma\mathbf{x}$ and renormalizing. Therefore, the recommendation is to begin the Arnoldi process with a starting vector that has been multiplied by $\mathbf{A}_\sigma$ and, after convergence, to perform a purging step on the converged approximate eigenvectors.

A clever way to perform this operation has been suggested by Ericsson and Ruhe (1980). If $\mathbf{x} = \mathbf{V}\mathbf{y}$ with $\mathbf{H}\mathbf{y} = \mathbf{y}\theta$, then

$$\mathbf{A}_\sigma \mathbf{x} = \mathbf{V}\mathbf{H}\mathbf{y} + \mathbf{f}\mathbf{e}_k^T\mathbf{y} = \mathbf{x}\theta + \mathbf{f}\mathbf{e}_k^T\mathbf{y}.$$

Replacing the $\mathbf{x}$ with the improved eigenvector approximation $\mathbf{x} \leftarrow (\mathbf{x}\theta + \mathbf{f}\mathbf{e}_k^T\mathbf{y})$ and renormalizing has the effect of purging the undesirable components without requiring any additional matrix-vector products with $\mathbf{A}_\sigma$. The residual error of the computed Ritz vector with respect to the original problem is

$$\|\mathbf{A}\mathbf{x} - \mathbf{B}\mathbf{x}\lambda\| = \|\mathbf{B}\mathbf{f}\| \frac{|\mathbf{e}_k^T\mathbf{y}|}{|\theta|^2}, \tag{7.2}$$

where $\lambda = \sigma + 1/\theta$. Since $|\theta|$ is usually quite large under the spectral transformation, this new residual is generally considerably smaller than the original.

7.2. Additional methods and accelerations

When a sparse direct factorization is possible, the shift-invert spectral transformation combined with implicitly restarted Arnoldi is probably the method of choice. However, this may not be practical in many applications. In a parallel computing environment, success of this approach also depends critically on how well the solution process for the shift-invert equations can be parallelized. Finally, if applying $\mathbf{A}_\sigma$ is very cheap then one may wish to avoid the expense of implicit restarting and complete orthogonalization.

One approach that avoids the need to keep a complete set of basis vectors is the *Bi-Lanczos method.* This biorthogonal dual-basis approach is based on a three-term recurrence that results in a nonsymmetric tridiagonal projected matrix instead of an upper Hessenberg projection. This Bi-Lanczos method is related to the QMR and Bi-CG methods for linear systems. Both methods lead to the same projected tridiagonal matrix. Freund and Nachtigal (1991) and Cullum and Willoughby (1986) have published codes for the computation of eigenvalues using this approach. However, the accuracy of these methods is a point of concern, since the projections are oblique rather than orthogonal as they are in the Arnoldi process. Also, there is no particular advantage in having the projected matrix in nonsymmetric tridiagonal form, since the only algorithms that can take advantage of the structure are generally based on hyperbolic rotations, and are of questionable numerical stability.

An alternative spectral transformation that can be effective in linear stability analysis in CFD problems is the generalized Cayley transformation

$$\mathbf{C} := (\mathbf{A} - \sigma\mathbf{B})^{-1}(\mathbf{A} - \lambda\mathbf{B}).$$

An important aspect of this transformation is the additional control on the image of the left half plane (say) under the transformation. A detailed study may be found in Garratt (1991) and Meerbergen and Spence (1997). Lehoucq and Salinger (2001) make particularly effective use of this transformation in a stability analysis of a simulation of a CVD reactor with over four million variables. More recently, 16 million variable problems of this type have been solved (Burroughs, Romero, Lehoucq and Salinger 2001).

The use of inexact forms of the Cayley transform is studied in Meerbergen (1996), where the required inverse operation is approximated by a few steps of an iterative method, for Arnoldi's method. The wanted eigensolutions are solutions of $\mathbf{Cx} = 0$. The essential part $\mathbf{A} - \lambda\mathbf{B}$ is computed exactly and the inexact inversion of $\mathbf{A} - \sigma\mathbf{B}$ may be viewed as a kind of preconditioning. Indeed, this technique has a close relationship to polynomial preconditioning. The inexact Cayley transform is well suited to parallel computing since the dominant computational elements are matrix-vector products instead of direct linear solves.

Ruhe (1994*b*) introduced a remarkable generalization of the Krylov space that admits the application of several different shift-invert transforms within the same iteration. This is called *rational Krylov subspace* (RKS) iteration and the transformations are of the form

$$(\delta_j \mathbf{A} - \gamma_j \mathbf{B})^{-1}(\sigma_j \mathbf{A} - \rho_j \mathbf{B}),$$

in which the coefficients may be different for each iteration step j. With respect to the subspace with these operators, the given problem is projected onto a small generalized system

$$(\zeta \mathbf{K}_{j,j} - \eta \mathbf{L}_{j,j})\mathbf{y} = 0,$$

where $\mathbf{K}_{j,j}$ and $\mathbf{L}_{j,j}$ are upper Hessenberg matrices of dimension j. This small system may be solved by the QZ algorithm in order to obtain approximate values for an eigenpair. The parameters in RKS can be chosen to obtain faster convergence to interior eigenvalues. When combined with a certain deflation scheme, a considerable number of eigenvalues can be computed without being forced to construct a large basis set. Eigenvectors can be written to auxiliary storage as needed. For a comparison of RKS and Arnoldi, see Ruhe (1994*a*, 1994*b*). Again, successful parallelization for this approach depends on how well linear systems with the matrix $\delta_j \mathbf{A} - \gamma_j \mathbf{B}$ can be solved to sufficiently high accuracy.

Clearly, the most straightforward alternative to solving the shift-invert equations directly is to use a preconditioned iterative method to solve them approximately. However, there are several difficulties. The shifted matrix is often ill-conditioned because σ will be chosen near an interesting eigenvalue. Moreover, this shifted matrix will usually be indefinite (or have indefinite symmetric part). These are the conditions that are most difficult for iterative solution of linear systems. A further difficulty is that each linear system must be solved to a greater accuracy than the desired accuracy of the eigenvalue calculation. Otherwise, each step of the Lanczos/Arnoldi process will essentially involve a different matrix operator. The approach can be quite successful, however, if done with care. A good example of this may be found in Lehoucq and Salinger (2001).

Subspace iteration methods are more amenable to admitting inaccurate approximate solutions to the shift-invert equations. This has already been discussed in Section 6.2. The Jacobi–Davidson approach can be adapted nicely to the generalized problem and is particularly well suited to the introduction of inaccurate approximate solutions.

For a good overview of subspace iteration methods, see Saad (1992). There are several other methods that allow the possibility of inexact solves and preconditioning in eigenvalue computations. Two of these are the LOBPCG method developed in Knyazev (2001) and the TRQ method developed in Sorensen and Yang (1998).

Knyazev (2001) presents numerical evidence to suggest that LOBPCG performs for symmetric positive definite eigenproblems as the preconditioned conjugate gradient method performs for symmetric positive definite linear systems. In Sorensen and Yang (1998), the TRQ method is derived as a truncation of the RQ iteration. This is just like the QR method with the exception that the shifted matrices are factored into an orthogonal $\mathbf{Q}$ times an upper triangular $\mathbf{R}$. Quadratic convergence takes place in the leading column of $\mathbf{Q}$ and preconditioned inexact solves are possible to complete the update equations. This scheme is very closely related to the JDQR method.

7.3. The Jacobi–Davidson QZ algorithm

The Jacobi–Davidson method can be modified for the generalized eigenproblem without having to transform the given problem to a standard eigenproblem. In this formulation, called JDQZ (Fokkema *et al.* 1996), explicit inversion of matrices is not required. The method is developed with orthogonal projections and the theme is once again to compute a partial (generalized) Schur decomposition. A subspace is generated onto which the given eigenproblem is projected. The much smaller projected eigenproblem is solved by standard direct methods, and this leads to approximations for the eigensolutions of the given large problem. Then, a correction equation for a selected eigenpair is set up. The solution of the correction equation defines an orthogonal correction for the current eigenvector approximation. The correction, or an approximation for it, is used for the expansion of the subspace and the procedure is repeated.

The subspace projection leads to a formulation that may be viewed as an inexact truncated form of the QZ factorization. The algorithm is designed to compute a few eigenvalues of $\mathbf{Ax} = \mathbf{Bx}\lambda$ close to a given target $\tau \in \mathbb{C}$. Given a low-dimensional subspace $\mathcal{S} = \text{Range}(\mathbf{V}_k)$, eigenvector approximations called Petrov–Ritz values are extracted from a small projected problem obtained through a Petrov–Galerkin projection.

Here $\mathbf{V}_k$ is an $(n \times k)$ matrix with orthonormal columns $\mathbf{v}_j$. A Petrov–Galerkin condition is to define a Petrov–Ritz pair $(\mathbf{x}, \theta)$, where $\mathbf{x} \in \mathcal{S}$. We require

$$\langle \mathbf{w}, \mathbf{A}\mathbf{x} - \mathbf{B}\mathbf{x}\theta \rangle = 0, \quad \text{for all} \;\; \mathbf{w} \in \text{Range}(\mathbf{W}_k),$$

where $\mathbf{W}_k$ is another $(n \times k)$ matrix with orthonormal columns $\mathbf{w}_j$. This gives a small projected problem of order k:

$$\mathbf{W}_k^*\mathbf{A}\mathbf{V}_k\mathbf{y} - \mathbf{W}_k^*\mathbf{B}\mathbf{V}_k\mathbf{y}\theta = 0. \tag{7.3}$$

For each eigenpair $(\mathbf{y}, \theta)$, we obtain a Petrov–Ritz vector $\mathbf{x} = \mathbf{V}_k\mathbf{y}$ and Petrov–Ritz value θ as approximate eigenpairs for the original problem.

Using essentially the same approach described for the standard problem (see Section 6.3), the basis sets $\mathbf{V}_k$ and $\mathbf{W}_k$ are each increased one dimension by including directions obtained from a residual correction equation. The method is briefly described here. For full details one should consult Fokkema *et al.* (1996).

First the QZ algorithm (Golub and Van Loan 1996) is used to reduce (7.3) to a generalized Schur form. This provides orthogonal $(k \times k)$ matrices $\mathbf{U}_R$ and $\mathbf{U}_L$, and upper triangular $(k \times k)$ matrices $\mathbf{S}_A$ and $\mathbf{S}_B$, such that

$$\mathbf{U}_L^* \left(\mathbf{W}_k^*\mathbf{A}\mathbf{V}_k\right) \mathbf{U}_R = \mathbf{S}_A, \tag{7.4}$$

$$\mathbf{U}_L^* \left(\mathbf{W}_k^*\mathbf{B}\mathbf{V}_k\right) \mathbf{U}_R = \mathbf{S}_B. \tag{7.5}$$

The decomposition is ordered (by similarity transformations) so that the leading diagonal elements of $\mathbf{S}_A$ and $\mathbf{S}_B$ represent the eigenvalue approximation closest to the target value τ. The approximation for the eigenpair is then taken as

$$(\tilde{\mathbf{q}}, \theta) := (\mathbf{V}_k\mathbf{U}_R(:,1), \mathbf{S}_B(1,1)/\mathbf{S}_A(1,1)), \tag{7.6}$$

assuming that $\mathbf{S}_A(1,1) \neq 0$. This gives a *residual vector*:

$$\mathbf{r} := \mathbf{A}\tilde{\mathbf{q}} - \mathbf{B}\tilde{\mathbf{q}}\theta.$$

To obtain a correction equation analogous to (6.4), we define an auxiliary vector $\tilde{\mathbf{w}}\gamma = \mathbf{A}\tilde{\mathbf{q}} - \mathbf{B}\tilde{\mathbf{q}}\tau$, where γ is such that $\|\tilde{\mathbf{w}}\|_2 = 1$. Then a correction equation is defined to provide a correction $\mathbf{z} \perp \tilde{\mathbf{q}}$:

$$(\mathbf{I} - \tilde{\mathbf{w}}\tilde{\mathbf{w}}^*)(\mathbf{A} - \theta\mathbf{B})(\mathbf{I} - \tilde{\mathbf{q}}\tilde{\mathbf{q}}^*)\mathbf{z} = -\mathbf{r}. \tag{7.7}$$

In practice, only a few steps of a preconditioned iterative method are done to get an approximate solution to (7.7).

The approximation for $\mathbf{z}$ is then further orthogonalized with respect to $\mathbf{V}_k$ to get the new basis vector $\mathbf{v}_{k+1}$ in the direction of $(\mathbf{I} - \mathbf{V}_k\mathbf{V}_k^*)\mathbf{z}$. The expansion vector $\mathbf{w}_{k+1}$ is taken in the direction $(\mathbf{I} - \mathbf{W}_k\mathbf{W}_k^*)(\mathbf{A}\mathbf{z} - \mathbf{B}\mathbf{z}\tau)$. This gives a brief description of the *harmonic Petrov value approach* proposed in Fokkema *et al.* (1996).

7.4. JDQZ: restart and deflation

As in JDQR, deflation and restarting must be employed to find several eigenvalues and vectors simultaneously. As eigenvectors converge, the iteration is continued in a subspace orthogonal to the converged vectors. The algorithm is based on the computation of a partial generalized Schur form for the matrix pair $(\mathbf{A}, \mathbf{B})$:

$$\mathbf{A}\mathbf{Q}_k = \mathbf{Z}_k\mathbf{S}_k \quad \text{and} \quad \mathbf{B}\mathbf{Q}_k = \mathbf{Z}_k\mathbf{T}_k,$$

in which $\mathbf{Q}_k$ and $\mathbf{Z}_k$ are $(n \times k)$ orthonormal matrices and $\mathbf{S}_k$, $\mathbf{T}_k$ are $(k \times k)$ upper triangular matrices, with $k \ll n$. The scheme is more complicated but essentially follows the ideas described previously for JDQR. The full details may be found in Fokkema *et al.* (1996).

8. Eigenvalue software

Several software packages were developed during the 1980s for large-scale symmetric problems. Perhaps the most influential of these was Grimes *et al.* (1994). This block Lanczos code has been a mainstay of structural analysis calculations in industrial applications. It has been updated many times and is still the most heavily used code in this field. Considerable progress has been made over the past decade on the production of high-quality mathematical software for large nonsymmetric eigenvalue problems. Many packages are freely available online, and may be found via netlib.

A few of these are:

Lanczos (`http://www.netlib.org/`)

Authors: Jane Cullum and Ralph A. Willoughby
Description: Lanczos Algorithms for computing a few eigenvalues and eigenvectors of a large (sparse) symmetric matrix, real symmetric and Hermitian matrices; singular values and vectors of real, rectangular matrices (Fortran)
Reference: Cullum and Willoughby (1985)

SRRIT (`http://www.netlib.org/`)

Authors: Z. Bai and G. W. Stewart
Description: Subspace iteration to calculate the dominant invariant subspace of a nonsymmetric matrix (Fortran)
Reference: Bai and Stewart (1997)

ARNCHEB (`http://www.cerfacs.fr/~chatelin/`)

Authors: T. Braconnier and F. Chatelin
Description: Arnoldi–Chebyshev restarted method for computing a few eigenvalues and vectors of large, unsymmetric sparse matrices (Fortran)
Reference: Users' Guide (`http://www.cerfacs.fr/~chatelin/`)

LOBPCG (`http://www-math.cudenver.edu/~aknyazev/software/CG`)

Author: A. Knyazev
Description: Locally optimal block preconditioned conjugate gradient method for a few eigenvalues and vectors of large symmetric (or Hermitian) matrices (Matlab)
Reference: Knyazev (2001)

Laso (`http://www.netlib.org/`)

Author: D. Scott
Description: Lanczos method for a few eigenvalues and eigenvectors of a large (sparse) symmetric matrix (Fortran)
Reference: Parlett and Scott (1979)

SVDpack (`http://www.netlib.org/`)

Authors: M. W. Berry and M. Liang
Description: Computes a partial SVD of large sparse non-Hermitian complex matrices using the Lanczos algorithm for $\mathbf{A}^*\mathbf{A}$ with selective reorthogonalization (Fortran)
Reference: Berry (1992)

IRBL (`http://www.cs.bsu.edu/~jbaglama/#Software`)

Authors: J. Baglama, D. Calvetti and L. Reichel
Description: Block implicitly restarted Lanczos with Leja points as shifts.

JDQR, JDQZ (`http://www.math.uu.nl/people/sleijpen/JD_software`)

Author: G. L. G. Sleijpen
Description: JDQR and JDQZ implementations of Jacobi–Davidson method for a partial Schur decomposition corresponding to a selected subset of eigenvalues (eigenvectors also computed on request). Symmetric, nonsymmetric, generalized problems solved (Matlab)
Reference: Sleijpen and van der Vorst (1995), Fokkema *et al.* (1996)

ARPACK (`http://www.caam.rice.edu/software/ARPACK/`)

Authors: R. B. Lehoucq, D. C. Sorensen, and C. Yang
Description: Implicitly restarted Arnoldi method for computing a partial Schur decomposition corresponding to a selected subset of eigenvalues (eigenvectors also computed on request). Symmetric, nonsymmetric, generalized and SVD problems solved (Fortran)
Reference: Lehoucq, Sorensen and Yang (1998)

We should also mention the codes available in the Harwell Subroutine Library (HSL). These are freely available to UK academics, but not in general. In particular, the code EB13 based on Scott (200x) is available for nonsymmetric problems.

8.1. Software design

Today's software designers are faced with many new options in languages, design options, and computational platforms. However, certain principles can lead to robust software that is both portable and efficient over a wide variety of computing platforms.

When designing general-purpose software for use in the public domain, it is important to adopt a development strategy that will meet the goals of robustness, efficiency, and portability. Two very important principles are modularity and independence from specific vendor-supplied communication and performance libraries.

In this final section, we discuss some design and performance features of the eigenvalue software ARPACK. This is a collection of Fortran77 subroutines based on the IRAM described in Algorithm 3. This software can solve large-scale non-Hermitian or Hermitian (standard and generalized) eigenvalue problems. It has been used on a wide range of applications. P_ARPACK is a parallel extension to the ARPACK library and is designed for distributed memory message passing systems. The message passing layers currently supported are BLACS and MPI (MPI Forum 1994, Dongarra and Whaley 1995). Performance and portability are attained simultaneously because of the modular construction of the dense linear algebra operations. These are based on the Level 2 and Level 3 BLAS (Dongarra *et al.* 1988, Dongarra, DuCroz, Duff and Hammarling 1990) for matrix-vector and matrix-matrix operations and on LAPACK (Anderson *et al.* 1992) for higher-level dense linear algebra routines.

The important features of ARPACK and P_ARPACK are as follows.

- A reverse communication interface.
- Computes k eigenvalues that satisfy a user-specified criterion such as largest real part, largest absolute value, *etc.*

- A fixed predetermined storage requirement of $n \cdot \mathcal{O}(k) + \mathcal{O}(k^2)$ bytes.
- Driver routines are included as templates for implementing various spectral transformations to enhance convergence and to solve the generalized eigenvalue problem, or the SVD problem.
- Special consideration is given to the generalized problem $\mathbf{Ax} = \mathbf{Bx}\lambda$ for singular or ill-conditioned symmetric positive semi-definite $\mathbf{B}$.
- A Schur basis of dimension k that is numerically orthogonal to working precision is always computed. These are also eigenvectors in the Hermitian case. In the non-Hermitian case eigenvectors are available on request. Eigenvalues are computed to a user-specified accuracy.

Reverse communication

Reverse communication is an artifact of certain restrictions in the Fortran language; with reverse communication, control is returned to the calling program when interaction with the matrix is required. (For the C++ programmer, reverse communication is the Fortran substitute for defining functions specific to the class of matrices.) This is a convenient interface for experienced users. However, it seems to be a difficult concept to grasp for inexperienced users. Even though it is extremely useful for interfacing with large application codes, the software maintenance problems imposed on the developers are very demanding.

This interface avoids having to express a matrix-vector product through a subroutine with a fixed calling sequence. This means that the user is free to choose any convenient data structure for the matrix representation. Also, it is up to the user to partition the matrix-vector product in the most favourable way for parallel efficiency. Moreover, if the matrix is not available explicitly, the user is free to express the action of the matrix on a vector through a subroutine call or a code segment. It is not necessary to conform to a fixed format for a subroutine interface, and hence there is no need to communicate data through the use of `COMMON`.

A typical use of this interface is illustrated as follows:

```
10  continue
    call snaupd (ido, bmat, n, which,...,workd,..., info)
    if (ido .eq. newprod) then
       call matvec ('A', n, workd(ipntr(1)), workd(ipntr(2)))
    else
       return
    endif
    go to 10
```

This shows a code segment of the routine the user must write to set up the reverse communication call to the top level ARPACK routine `snaupd` to solve a nonsymmetric eigenvalue problem. The action requested of the calling program is specified by the reverse communication parameter `ido`. In this case the requested action is multiply the vector held in the array `workd` beginning at location `ipntr(1)` and and then to insert into the array `workd` beginning at location `ipntr(2)`. Here a call is made to a subroutine `matvec`. However, it is only necessary to supply the action of the matrix on the specified vector and put the result in the designated location. Because of this, reverse communication is very flexible and even provides a convenient way to use ARPACK interfaced with code written in another language such as C or C++.

8.2. Parallel aspects

The parallelization paradigm found to be most effective for ARPACK on distributed memory machines was to provide the user with a *single program multiple data* (SPMD) template. This means there are many copies of the same program running on multiple processors executing the same instruction streams on different data. The parallelization scheme described here is well suited to all of the methods discussed earlier, since they all share the basic needs of orthogonalizing a new vector with respect to a current basis for a subspace. They also share the need to apply a linear operator to a vector.

The reverse communication interface provides a means for a very simple SPMD parallelization strategy. Reverse communication allows the P_ARPACK codes to be parallelized internally without imposing a fixed parallel decomposition on the matrix or the user-supplied matrix-vector product. Memory and communication management for the matrix-vector product $\mathbf{w} \leftarrow \mathbf{Av}$ can be optimized independently of P_ARPACK. This feature enables the use of various matrix storage formats as well as calculation of the matrix elements as needed.

The calling sequence to ARPACK remains unchanged except for the addition of an MPI communicator (MPI Forum 1994, Dongarra and Whaley 1995). Inclusion of the communicator is necessary for global communication as well as managing I/O.

The numerically stable generation of the Arnoldi factorization

$$\mathbf{AV}_k = \mathbf{V}_k\mathbf{H}_k + \mathbf{f}_k\mathbf{e}_k^T$$

coupled with an implicit restarting mechanism is the basis of the ARPACK codes. The simple parallelization scheme used for P_ARPACK is as follows:

- $\mathbf{H}_k$ replicated on every processor
- $\mathbf{V}_k$ is distributed across a 1D processor grid (blocked by rows)
- $\mathbf{f}_k$ and workspace distributed accordingly.

The SPMD code looks essentially like the serial code except that the local block of the set of Arnoldi vectors, $\mathbf{V}_{\text{loc}}$, is passed in place of $\mathbf{V}$, and n_{loc}, the dimension of the local block, is passed instead of n.

With this approach there are only two communication points within the construction of the Arnoldi factorization inside P_ARPACK: computation of the 2-norm of the distributed vector $\mathbf{f}_k$ and the orthogonalization of $\mathbf{f}_k$ to $\mathbf{V}_k$ using classical Gram–Schmidt with DGKS correction (Daniel, Gragg, Kaufman and Stewart 1976). Additional communication will typically occur in the user-supplied matrix-vector product operation as well. Ideally, this product will only require nearest neighbour communication among the processes. Typically, the blocking of $\mathbf{V}$ coincides with the parallel decomposition of the matrix $\mathbf{A}$. The user is free to select an appropriate blocking of $\mathbf{V}$ to achieve optimal balance between the parallel performance of P_ARPACK and the user-supplied matrix-vector product.

The SPMD parallel code looks very similar to that of the serial code. Assuming a parallel version of the subroutine `matvec`, an example of the application of the distributed interface is illustrated as follows:

```
10   continue
     call psnaupd (comm, ido, bmat, nloc, which, ...,
   *                        Vloc , ... lworkl, info)
     if (ido .eq. newprod) then
        call matvec ('A', nloc, workd(ipntr(1)), workd(ipntr(2)))
     else
        return
     endif
     go to 10
```

Here, `nloc` is the number of rows in the block `Vloc` of $\mathbf{V}$ that has been assigned to this node process.

The blocking of $\mathbf{V}$ is generally determined by the parallel decomposition of the matrix A. For parallel efficiency, this blocking must respect the configuration of the distributed memory and interconnection network. Logically, the $\mathbf{V}$ matrix is partitioned by blocks

$$\mathbf{V}^T = (\mathbf{V}^{(1)^T}, \mathbf{V}^{(2)^T}, \ldots, \mathbf{V}^{(\text{nproc})^T}),$$

with one block per processor and with $\mathbf{H}$ replicated on each processor. The explicit steps of the CGS process taking place on the jth processor are shown in Algorithm 8.

Note that the function `gnorm` at step (1) is meant to represent the global reduction operation of computing the norm of the distributed vector $\mathbf{f}_k$ from

$$
\begin{array}{ll}
(1) & \beta_k \leftarrow \text{gnorm}(\|\mathbf{f}_k^{(*)}\|);\ \ \mathbf{v}_{k+1}^{(j)} \leftarrow \mathbf{f}_k^{(j)} \cdot \frac{1}{\beta_k}; \\
(2) & \mathbf{w}^{(j)} \leftarrow (\mathbf{A}_{\text{loc}})v_{k+1}^{(j)}; \\
(3) & \begin{pmatrix} \mathbf{h} \\ \alpha \end{pmatrix}^{(j)} \leftarrow \begin{pmatrix} \mathbf{V}_k^{(j)T} \\ \mathbf{v}_{k+1}^{(j)T} \end{pmatrix} \mathbf{w}^{(j)};\ \begin{pmatrix} \mathbf{h} \\ \alpha \end{pmatrix} \leftarrow \text{gsum}\left[\begin{pmatrix} \mathbf{h} \\ \alpha \end{pmatrix}^{(*)}\right] \\
(4) & \mathbf{f}_{k+1}^{(j)} \leftarrow \mathbf{w}^{(j)} - (\mathbf{V}_k, \mathbf{v}_{k+1})^{(j)} \begin{pmatrix} \mathbf{h} \\ \alpha \end{pmatrix}; \\
(5) & \mathbf{H}_{k+1} \leftarrow \begin{pmatrix} \mathbf{H}_k & \mathbf{h} \\ \beta_k & \mathbf{e}_k^T \end{pmatrix}; \\
(6) & \mathbf{V}_{k+1}^{(j)} \leftarrow (\mathbf{V}_k, \mathbf{v}_{k+1})^{(j)};
\end{array}
$$

Algorithm 8. The explicit steps of the process responsible for the j block

the norms of the local segments $\mathbf{f}_k^{(j)}$, and the function **gsum** at step (3) is meant to represent the global sum of the local vectors $\mathbf{h}^{(j)}$ so that the quantity $\mathbf{h} = \sum_{j=1}^{\text{nproc}} \mathbf{h}^{(j)}$ is available to each process on completion. These are the only two communication points within this algorithm. The remainder is perfectly parallel. Additional communication will typically occur at step (2). Here the operation $(\mathbf{A}_{\text{loc}})\mathbf{v}$ is meant to indicate that the user-supplied matrix-vector product is able to compute the local segment of the matrix-vector product $\mathbf{Av}$ that is consistent with the partition of $\mathbf{V}$. Ideally, this would only involve nearest neighbour communication among the processes.

Since $\mathbf{H}$ is replicated on each processor, the implicit restart mechanism described in Section 4.4 remains untouched. The only difference is that the local block $\mathbf{V}^{(j)}$ is in place of the full matrix $\mathbf{V}$. Operations associated with implicit restarting are perfectly parallel with this strategy.

All operations on the matrix $\mathbf{H}$ are replicated on each processor. Thus there are no communication overheads. However, the replication of $\mathbf{H}$ and the shift selection and application to $\mathbf{H}$ on each processor amount to a serial bottleneck limiting the scalability of this scheme when k grows with n. Nevertheless, if k is fixed as n increases then this scheme scales linearly with n, as we shall demonstrate with some computational results. In the actual implementation, separate storage is not required for the Q_i. Instead, it is represented as a product of 2×2 Givens or 3×3 Householder transformations that are applied directly to update Q. On completion of this accumulation of Q, the operation $\mathbf{V}_m^{(j)} \leftarrow \mathbf{V}_m^{(j)}\mathbf{Q}$ occurs independently on each processor j using the Level 3 BLAS operation _GEMM.

An important aspect to this approach is that changes to the serial version of ARPACK were minimal. Only eight routines were affected in a minimal way. These routines either required a change in norm calculation to accommodate distributed vectors (step (1)), modification of the distributed dense matrix-vector product (step (4)), or inclusion of the context or communicator for I/O (debugging/tracing).

8.3. Communication and synchronization

On many shared memory MIMD architectures, a level of parallelization can be accomplished through compiler options alone, without requiring any modifications to the source code. For example, on the SGI Power Challenge architecture, the MIPSpro F77 compiler uses a Power Fortran Accelerator (PFA) preprocessor to uncover the parallelism in the source code automatically. PFA is an optimizing Fortran preprocessor that discovers parallelism in Fortran code and converts those programs to parallel code. A brief discussion of implementation details for ARPACK using PFA preprocessing may be found in Debicki, Jedrzejewski, Mielewski, Przybyszewski and Mrozowski (1995). The effectiveness of this preprocessing step is still dependent on how suitable the source code is for parallelization. Since most of the vector and matrix operations for ARPACK are accomplished via BLAS and LAPACK routines, access to efficient parallel versions of these libraries alone will provide a reasonable level of parallelization.

For distributed memory implementations, message passing between processes must be explicitly addressed within the source code, and numerical computations must take into account the distribution of data. In addition, for the parallel code to be portable, the communication interface used for message passing must be supported on a wide range of parallel machines and platforms. For P_ARPACK, this portability is achieved via the *basic linear algebra communication subprograms* (BLACS) (Dongarra and Whaley 1995) developed for the ScaLAPACK project and *message passing interface* (MPI) (MPI Forum 1994).

8.4. Parallel performance

P_ARPACK has been run on a wide variety of parallel processors. The simple SPMD strategy has proved to be very effective. Near-linear scalability has been demonstrated on massively parallel machines for the internal dense linear algebra operations required to implement the IRAM. However, such scalability relies entirely on the parallel efficiency of the user-supplied matrix-vector product or linear solves when shift-invert is used. A synopsis of such performance results is available in Maschhoff and Sorensen (1996).

Perhaps more important is the ability to solve real problems. A very impressive computation has been done by Lehoucq and Salinger (2001) on a linear stability analysis of a CVD reactor. The problem involved four million variables resulting from a 3D finite element model. They used P_ARPACK on the Sandia-Intel 1024 processor Teraflop machine. A Cayley transformation $(\mathbf{A} - \sigma_1\mathbf{B})\mathbf{w} = (\mathbf{A} + \sigma_2\mathbf{B})\mathbf{v}$ was used to accelerate convergence and to better isolate the rightmost eigenvalues. The AZTEC package for iterative solution of linear systems was used to implement this. They selected an ILUT preconditioner with GMRES. In this calculation, P_ARPACK only contributed to about 5% of the total computation time. This is typical of many applications. The application of the linear operator (in this case the Cayley-transformed matrix) usually dominates the computation. The internal operations required for IRAM are generally inconsequential when compared to the application of the linear operator.

The Lehoucq and Salinger paper reports some very impressive results on bifurcation as well as stability analysis. They also give a very interesting study of the two-step CGS orthogonalization scheme in the context of the GMRES calculations required to solve the linear systems for the Cayley transformation. This is pertinent to all of the methods discussed here and is of particular interest in the implementation of the Arnoldi factorization that underlies GMRES and also ARPACK. Two-step CGS orthogonalization is classical Gram–Schmidt followed by one step of the DGKS correction described previously. This is done at every orthogonalization step. Considerable experience with this option for CGS has demonstrated completely reliable orthogonalization properties of many orthogonalization steps. It completely resolves the numerical problems with CGS.

Lehoucq and Salinger compare the performance of CGS to that of modified Gram–Schmidt. A comparison of computational times is shown in Figure 8.

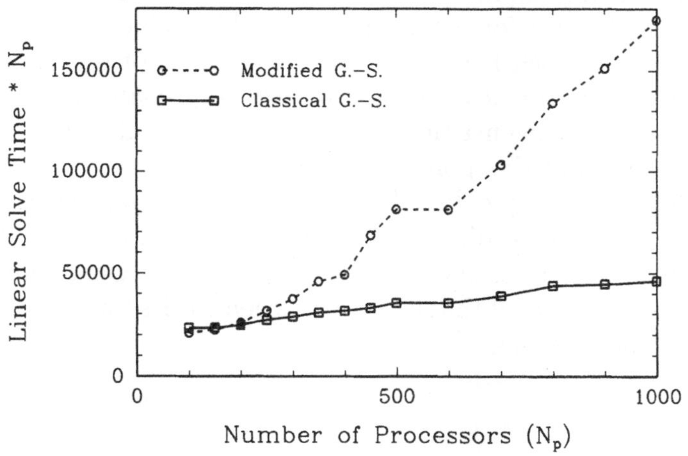

Figure 8. DGKS correction

This comparison shows that two-step CGS scales almost linearly, while MGS has very poor scalability properties. This is due to the many additional communication points needed for vector-vector operations (Level 1 BLAS) in comparison to the matrix-vector (Level 2 BLAS) formulation available with CGS. In these calculations, problem size is increased in proportion to the number of processors. Perfect scaling would give a flat horizontal graph indicating a constant computational time.

It should be noted that (unrestarted) GMRES will give the same numerical result for the linear system with either orthogonalization scheme. However, the Intel machine (called ASCI Red) has very fast communication and hence these results would be even more dramatic on most other massively parallel platforms.

8.5. Summary

The implementation of P_ARPACK is portable across a wide range of distributed memory platforms. The portability of P_ARPACK is achieved by use of the BLACS and MPI. With this strategy, it takes very little effort to port P_ARPACK to a wide variety of parallel platforms. It has been installed and successfully tested on many massively parallel systems.

9. Conclusions and acknowledgements

This introduction to the current state of the art in methods and software for large-scale eigenvalue problems has necessarily been limited. There are many excellent researchers working in the area. This discussion has focused on IRLM and JDQR methods. We have tried to include brief descriptions of most of the techniques that have been developed recently, but there are certainly unintentional omissions. The author apologizes for these.

The recent advances for nonsymmetric problems have been considerable. However, there is much left to be done. The areas of preconditioning and other forms of convergence acceleration are very challenging. The ability to compute interior eigenvalues reliably, without shift and invert spectral transformations is, at this point, out of reach.

The author owes many debts of gratitude to other researchers in this area. Several have contributed directly to this work. Of particular note are Chris Beattie, Mark Embree, Lothar Reichel, Rich Lehoucq, Chao Yang and Kristi Maschhoff. A final note of thanks goes to Arieh Iserles for his encouragement and unbelievable patience.

REFERENCES

E. Anderson, Z. Bai, C. Bischof, J. Demmel, J. Dongarra, J. DuCroz, A. Greenbaum, S. Hammarling, A. McKenney, S. Ostrouchov and D. Sorensen (1992), *LAPACK User's Guide*, SIAM, Philadelphia, PA.

J. Baglama, D. Calvetti and L. Reichel (1996), 'Iterative methods for the computation of a few eigenvalues of a large symmetric matrix', *BIT* **36**, 400–440.

J. Baglama, D. Calvetti and L. Reichel (1998), 'Fast Leja points', *ETNA* **7**, 124–140.

Z. Bai and G. W. Stewart (1997), 'SRRIT: A FORTRAN subroutine to calculate the dominant invariant subspace of a nonsymmetric matrix', *ACM Trans. Math. Software* **23**, 494.

C. Beattie, M. Embree and J. Rossi (2001), Convergence of restarted Krylov subspaces to invariant subspaces, Numerical Analysis Technical Report 01/21, OUCL, Oxford, UK.

M. Berry (1992), 'Large scale singular value computations', *Supercomput. Appl.* **6**, 13–49.

E. A. Burroughs, L. A. Romero, R. B. Lehoucq and A. J. Salinger (2001), Large scale eigenvalue calculations for computing the stability of buoyancy driven flows, Technical Report 2001-0113J, Sandia National Laboratories. Submitted to *J. Comput. Phys.*

D. Calvetti, L. Reichel and D. Sorensen (1994), 'An implicitly restarted Lanczos method for large symmetric eigenvalue problems', *ETNA* **2**, 1–21.

M. Crouzeix, B. Philippe and M. Sadkane (1994), 'The Davidson method', *SIAM J. Sci. Comput.* **15**, 62–76.

J. Cullum and W. E. Donath (1974), A block Lanczos algorithm for computing the q algebraically largest eigenvalues and a corresponding eigenspace for large, sparse symmetric matrices, in *Proc. 1974 IEEE Conference on Decision and Control*, New York, pp. 505–509.

J. Cullum and R. A. Willoughby (1981), 'Computing eigenvalues of very large symmetric matrices: An implementation of a Lanczos algorithm with no reorthogonalization', *J. Comput. Phys.* **434**, 329–358.

J. Cullum and R. A. Willoughby (1985), *Lanczos Algorithms for Large Symmetric Eigenvalue Computations*, Vol. 1: Theory, Birkhäuser, Boston, MA.

J. Cullum and R. A. Willoughby (1986), A practical procedure for computing eigenvalues of large sparse nonsymmetric matrices, in *Large Scale Eigenvalue Problems* (J. Cullum and R. A. Willoughby, eds), North-Holland, Amsterdam, pp. 193–240.

J. Daniel, W. B. Gragg, L. Kaufman and G. W. Stewart (1976), 'Reorthogonalization and stable algorithms for updating the Gram–Schmidt QR factorization', *Math. Comput.* **30**, 772–795.

E. R. Davidson (1975), 'The iterative calculation of a few of the lowest eigenvalues and corresponding eigenvectors of large real symmetric matrices', *J. Comput. Phys.* **17**, 87–94.

E. R. Davidson (1993), 'Monster matrices: Their eigenvalues and eigenvectors', *Comput. Phys.* **7**, 519–522.

M. P. Debicki, P. Jedrzejewski, J. Mielewski, P. Przybyszewski and M. Mrozowski (1995), Application of the Arnoldi method to the solution of electromagnetic eigenproblems on the multiprocessor power challenge architecture, Preprint 19/95, Department of Electronics, Technical University of Gdansk, Gdansk, Poland.

J. Dongarra and R. C. Whaley (1995), A User's Guide to the BLACS v1.0, Technical Report UT CS-95-281, LAPACK Working Note #94, University of Tennessee.

J. J. Dongarra, J. DuCroz, I. Duff and S. Hammarling (1990), 'A set of Level 3 Basic Linear Algebra Subprograms: Model implementation and test programs', *ACM Trans. Math. Software* **16**, 18–28.

J. J. Dongarra, J. DuCroz, S. Hammarling and R. Hanson (1988), 'An extended set of Fortran Basic Linear Algebra Subprograms', *ACM Trans. Math. Software* **14**, 1–17.

T. Ericsson and A. Ruhe (1980), 'The spectral transformation Lanczos method for the numerical solution of large sparse generalized symmetric eigenvalue problems', *Math. Comput.* **35**, 1251–1268.

B. Fischer and L. Reichel (1989), 'Newton interpolation in Chebyshev and Fejér points', *Math. Comput.* **53**, 265–278.

D. R. Fokkema, G. L. G. Sleijpen and H. A. van der Vorst (1996), Jacobi–Davidson style QR and QZ algorithms for the partial reduction of matrix pencils, Technical Report 941, Mathematical Institute, Utrecht University.

R. W. Freund (1992), 'Conjugate gradient-type methods for a linear systems with complex symmetric coefficient matrices', *SIAM J. Sci. Comput.* **13**, 425–448.

R. W. Freund and N. M. Nachtigal (1991), 'QMR: A quasi-minimal residual method for non-Hermitian linear systems', *Numer. Math.* **60**, 315–339.

D. Gaier (1987), *Lectures on Complex Approximation*, Birkhäuser.

T. J. Garratt (1991), The numerical detection of Hopf bifurcations in large systems arising in fluid mechanics, PhD thesis, University of Bath, School of Mathematical Sciences, Bath, UK.

G. H. Golub and R. Underwood (1977), The block Lanczos method for computing eigenvalues, in *Mathematical Software III* (J. Rice, ed.), Academic Press, New York, pp. 361–377.

G. H. Golub and C. F. Van Loan (1996), *Matrix Computations*, The Johns Hopkins University Press, Baltimore.

R. G. Grimes, J. G. Lewis and H. D. Simon (1994), 'A shifted block Lanczos algorithm for solving sparse symmetric generalized eigenproblems', *SIAM J. Matrix Anal. Appl.* **15**, 228–272.

Z. Jia (1995), 'The convergence of generalized Lanczos methods for large unsymmetric eigenproblems', *SIAM J. Matrix Anal. Appl.* **16**, 843–862.

W. Karush (1951), 'An iterative method for finding characteristics vectors of a symmetric matrix', *Pacific J. Math.* **1**, 233–248.

A. Knyazev (2001), 'Toward the optimal preconditioned eigensolver: Locally optimal block preconditioned conjugate gradient method', *SIAM J. Sci. Comput.* **23**, 517–541.

C. Lanczos (1950), 'An iteration method for the solution of the eigenvalue problem of linear differential and integral operators', *J. Res. Nat. Bur. Standards* **45**, 255–282. Research Paper 2133.

C. Lawson, R. Hanson, D. Kincaid and F. Krogh (1979), 'Basic Linear Algebra Subprograms for Fortran usage.', *ACM Trans. Math. Software* **5**, 308–329.

R. B. Lehoucq (1995), Analysis and Implementation of an Implicitly Restarted Iteration, PhD thesis, Rice University, Houston, TX. Also available as Technical report TR95-13, Department of Computational and Applied Mathematics.

R. B. Lehoucq (2001), 'Implicitly restarted Arnoldi methods and subspace iteration', *SIAM J. Matrix Anal. Appl.* **23**, 551–562.

R. B. Lehoucq and A. G. Salinger (2001), 'Large-scale eigenvalue calculations for stability analysis of steady flows on massively parallel computers', *Internat. J. Numer. Methods Fluids* **36**, 309–327.

R. B. Lehoucq and J. A. Scott (1996), An evaluation of software for computing eigenvalues of sparse nonsymmetric matrices, Preprint MCS-P547-1195, Argonne National Laboratory, Argonne, IL.

R. B. Lehoucq and D. C. Sorensen (1996), 'Deflation techniques for an implicitly restarted Arnoldi iteration', *SIAM J. Matrix Anal. Appl.* **17**, 789–821.

R. B. Lehoucq, D. C. Sorensen and C. Yang (1998), *ARPACK Users Guide: Solution of Large Scale Eigenvalue Problems with Implicitly Restarted Arnoldi methods*, SIAM Publications, Philadelphia, PA.

T. A. Manteuffel (1978), 'Adaptive procedure for estimating parameters for the nonsymmetric Tchebychev iteration', *Numer. Math.* **31**, 183–208.

K. J. Maschhoff and D. C. Sorensen (1996), P_ARPACK: An efficient portable large scale eigenvalue package for distributed memory parallel architectures, in *Applied Parallel Computing in Industrial Problems and Optimization*, Springer, Berlin, pp. 478–486.

K. Meerbergen (1996), Robust methods for the calculation of rightmost eigenvalues of nonsymmetric eigenvalue problems, PhD thesis, Katholieke Universiteit Leuven, Belgium.

K. Meerbergen and A. Spence (1997), 'Implicitly restarted Arnoldi with purification for the shift-invert transformation', *Math. Comput.* **218**, 667–689.

R. B. Morgan (1991), 'Computing interior eigenvalues of large matrices', *Lin. Alg. Appl.* **154–156**, 289–309.

R. B. Morgan (1996), 'On restarting the Arnoldi method for large nonsymmetric eigenvalue problems', *Math. Comput.* **65**, 1213–1230.

R. B. Morgan and D. S. Scott (1993), 'Preconditioning the Lanczos algorithm for sparse symmetric eigenvalue problems', *SIAM J. Sci. Comput.* **14**, 585–593.

MPI Forum (1994), 'MPI: A Message-Passing Interface standard', *Internat. J. Supercomput. Appl. High Performance Comput.* Special issue on MPI. Electronic form: `ftp://www.netlib.org/mpi/mpi-report.ps`.

J. Olsen, P. Jørgensen and J. Simons (1990), 'Passing the one-billion limit in full configuration-interaction (FCI) calculations', *Chem. Phys. Lett.* **169**, 463–472.

C. C. Paige (1971), The computation of eigenvalues and eigenvectors of very large sparse matrices, PhD thesis, University of London.

C. C. Paige, B. N. Parlett and H. A. van der Vorst (1995), 'Approximate solutions and eigenvalue bounds from Krylov subspaces', *Numer. Lin. Alg. Appl.* **2**, 115–134.

B. N. Parlett (1980), *The Symmetric Eigenvalue Problem*, Prentice-Hall, Englewood Cliffs, NJ.

B. N. Parlett and J. K. Reid (1981), 'Tracking the progress of the Lanczos algorithm for large symmetric eigenproblems', *IMA J. Numer. Anal.* **1**, 135–155.

B. N. Parlett and D. Scott (1979), 'The Lanczos algorithm with selective orthogonalization', *Math. Comput.* **33**, 217–238.

L. Reichel (1990), 'Newton interpolation at Leja points', *BIT* **30**, 332–346.

A. Ruhe (1994*a*), 'The rational Krylov algorithm for nonsymmetric eigenvalue problems, III: Complex shifts for real matrices', *BIT* **34**, 165–176.

A. Ruhe (1994*b*), 'Rational Krylov algorithms for nonsymmetric eigenvalue problems, II: Matrix pairs', *Lin. Alg. Appl.* **197–198**, 283–295.

Y. Saad (1980), 'Variations on Arnoldi's method for computing eigenelements of large unsymmetric matrices', *Lin. Alg. Appl.* **34**, 269–295.

Y. Saad (1984), 'Chebyshev acceleration techniques for solving nonsymmetric eigenvalue problems', *Math. Comput.* **42**, 567–588.

Y. Saad (1992), *Numerical Methods for Large Eigenvalue Problems*, Manchester University Press, Manchester, UK.

Y. Saad (1994), 'ILUT: A dual threshold incomplete LU factorization', *Numer. Lin. Alg. Appl.* **1**, 387–402.

J. A. Scott (200x), 'An Arnoldi code for computing selected eigenvalues of sparse real unsymmetric matrices', *ACM Trans. Math. Software.*

H. Simon (1984), 'Analysis of the symmetric Lanczos algorithm with reorthogonalization methods', *Lin. Alg. Appl.* **61**, 101–131.

V. Simoncini (1996), 'Ritz and pseudo-Ritz values using matrix polynomials', *Lin. Alg. Appl.* **241–243**, 787–801.

G. L. G. Sleijpen and H. A. van der Vorst (1995), 'An overview of approaches for the stable computation of hybrid BiCG methods', *Appl. Numer. Math.* **19**, 235–254.

G. L. G. Sleijpen and H. A. van der Vorst (1996), 'A Jacobi–Davidson iteration method for linear eigenvalue problems', *SIAM J. Matrix Anal. Appl.* **17**, 401–425.

D. C. Sorensen (1992), 'Implicit application of polynomial filters in a k-step Arnoldi method', *SIAM J. Matrix Anal. Appl.* **13**, 357–385.

D. C. Sorensen and C. Yang (1998), 'A truncated RQ-iteration for large scale eigenvalue calculations', *SIAM J. Matrix Anal. Appl.* **19**, 1045–1073.

A. Stathopoulos, Y. Saad and K. Wu (1998), 'Dynamic thick restarting of the Davidson, and the implicitly restarted Arnoldi methods', *SIAM J. Sci. Comput.* **19**, 227–245.

G. W. Stewart (2001), 'A Krylov–Schur algorithm for large eigenproblems', *SIAM J. Matrix Anal. Appl.* **23**, 601–614.

W. J. Stewart and A. Jennings (1981), 'Algorithm 570: LOPSI, A Fortran subroutine for approximations to right or left eigenvectors corresponding to the dominant set of eigenvalues of a real symmetric matrix', *ACM Trans. Math. Software* **7**, 230–232.

L. N. Trefethen (1992), Pseudospectra of matrices, in *Numerical Analysis 1991* (D. F. Griffiths and G. A. Watson, eds), Longman, pp. 234–266.

L. N. Trefethen (1999), Computation of pseudospectra, in *Acta Numerica*, Vol. 9, Cambridge University Press, pp. 247–296.